Oxidative Stress and Antioxidant Protection

The Science of Free Radical Biology and Disease

Oxidative Stress and Antioxidant Protection

The Science of Free Radical Biology and Disease

EDITED BY

Donald Armstrong
Robert D. Stratton

WILEY Blackwell

Published by John Wiley & Sons, Inc., Hoboken, New Jersey
Published simultaneously in Canada

For general information on our other products and services or for technical support, please contact our Customer Care Department within the United States at (800) 762-2974, outside the United States at (317) 572-3993 or fax (317) 572-4002.

Wiley also publishes its books in a variety of electronic formats. Some content that appears in print may not be available in electronic formats. For more information about Wiley products, visit our web site at www.wiley.com.

Library of Congress Cataloging-in-Publication Data applied for.

9781118832486 (hardback)

Cover image: Getty/Luk Cox

Typeset in 10/12pt Meridien LT Std by SPi Global, Chennai, India
Printed and bound in Malaysia by Vivar Printing Sdn Bhd

1   2016

Contents

List of contributors

Aneela Afzal
Advanced Imaging Research Center
Oregon Health and Science University
Portland, OR, USA

Mohammad Afzal
Department of Biological Sciences
Faculty of Science
Kuwait University
Safat, Kuwait

Ashok Agarwal
American Center for Reproductive Medicine
Cleveland Clinic
Cleveland, OH 44195, USA

María José Alcaraz
Department of Pharmacology and IDM
University of Valencia
Valencia, Spain

Ryyan Alobaidi
Pathology Department
King Saud University
Riyadh, KSA

Juan A. Ardura
Bone and Mineral Metabolism Laboratory
Instituto de Investigación Sanitaria (IIS) –
Fundación Jiménez Díaz and UAM
Madrid, Spain

Donald Armstrong
Department of Biotechnical and
Clinical Laboratory Sciences
State University of New York at Buffalo
Buffalo, NY, USA
Departmte of Ophthalmology
University of Florida College of Medicine
Gainesville, FL, USA

Ines Batinic-Haberle
Department of Radiation Oncology
Duke University School of Medicine
Durham, NC, USA

Maurizio Battino
Department of Dentistry and
Specialized Clinical Sciences
Biochemistry Section
Università Politecnica delle Marche
Ancona, Italy

Bogdan Calenic
Department of Biochemistry
Faculty of Dental Medicine
University of Medicine and
Pharmacy 'CAROL DAVILA'
Bucharest, Romania

Jing Chen
Department of Pathology, Immunology, and
Laboratory Medicine
University of Florida College of Medicine
Gainesville, FL, USA

Xiwei Chen
Department of Biostatistics
School of Public Health and Health
Professions State University of New York
at Buffalo

Renan C. Chisté
Departamento de Ciências Químicas
Faculdade de Farmácia
Universidade do Porto
REQUIMTE 4050-313
Porto, Portugal

Patrick Colahan
Department of Large Animal Clinical Sciences
College of Veterinary Medicine
University of Florida
Gainesville, FL, USA

D. Scott Covington
Healogics, Inc.
Jacksonville, FL, USA

Anna Dietrich-Muszalska
Department of Biological Psychiatry
The Chair of Experimental and
Clinical Physiology
Medical University of Lodz
Lodz, Poland

Herminia Domínguez
Departamento de Enxeñería Química
Facultad de Ciencias
Universidade de Vigo (Campus Ourense)
Ourense, Spain

Hassan A. N. El-Fawal
The Pharmaceutical Research Institute
and Neurotoxicology Laboratory
Albany College of Pharmacy
and Health Sciences
Albany, NY, USA

Pedro Esbrit
Bone and Mineral Metabolism Laboratory
Instituto de Investigación Sanitaria (IIS) –
Fundación Jiménez Díaz and UAM
Madrid, Spain

Elena Falqué
Departamento de Química Analítica
Facultad de Ciencias
Universidade de Vigo (Campus Ourense)
Ourense, Spain

Eduarda Fernandes
Departamento de Ciências Químicas
Faculdade de Farmácia
Universidade do Porto
REQUIMTE 4050-313
Porto, Portugal

Marisa Freitas
Departamento de Ciências Químicas,
Faculdade de Farmácia
Universidade do Porto
REQUIMTE 4050-313
Porto, Portugal

Natan Gadoth
Department of Neurology
Maynei Hayeshua Medical Center and
The Sackler Faculty of Medicine
Tel Aviv University
Tel Aviv, Israel

Reza Ghiasvand
Department of Community Nutrition
School of Nutrition and Food Sciences
Isfahan University of Medical Sciences
Isfahan, Iran

Steve Ghivizzani
Department of Orthopedics and
Rehabilitation
University of Florida College of Medicine
Gainesville, FL, USA

Maria Greabu
Department of Biochemistry
Faculty of Dental Medicine
University of Medicine and
Pharmacy 'CAROL DAVILA'
Bucharest, Romania

Mitra Hariri
Department of Community Nutrition
School of Nutrition and Food Sciences
Isfahan University of Medical Sciences
Isfahan, Iran

M. Elizabeth Hartnett
Department of Ophthalmology
Moran Eye Center
University of Utah
Salt Lake City, UT, USA

Hanbo Hu
Pulmonary Division
Department of Medicine University of Florida
Gainesville, FL, USA
Malcom Randall Veteran Affairs Medical
Center
Gainesville, FL, USA

Priyanka M. Jadhav
Department of Pathology, Immunology and
Laboratory Medicine
UF Diabetes Institute
College of Medicine
University of Florida College of Medicine
Gainesville, FL, USA

Surinder K. Jindal
Jindal Clinics
Chandigarh, India

Colleen G. Le Prell
Callier Center for Communication Disorders
School of Behavioral and Brain Sciences
University of Texas at Dallas
Dallas, TX, USA

Judith Lightsey
Department of Radiation Oncology
University of Florida College of Medicine
Gainesville, FL, USA

Bin Liu
Department of Pharmacodynamics
College of Pharmacy
University of Florida
Gainesville, FL, USA

Chao Liu
Department of Pathology,
Immunology, and Laboratory Medicine
University of Florida
Gainesville, FL, USA

Clayton E. Mathews
Department of Pathology,
Immunology, and Laboratory Medicine
University of Florida
Gainesville, FL, USA

Josef M. Miller
Department of Otolaryngology
Kresge Hearing Research Institute
University of Michigan
Ann Arbor, MI, USA

Christina L. Mitchell
Department of Dermatology
University of Florida College of Medicine
Gainesville, FL, USA

Arshag D. Mooradian
Department of Medicine
University of Florida College of Medicine
Jacksonville, FL, USA

Shaker A. Mousa
The Pharmaceutical Research Institute
Albany College of Pharmacy and
Health Sciences
Rensselaer, NY, USA

Carlos Palacio
Department of Medicine
University of Florida College of Medicine
Jacksonville, FL, USA

Aaron Panicker
Department of Urology
Wayne State University
Detroit, MI, USA

Sergio Portal-Núñez
Bone and Mineral Metabolism Laboratory
Instituto de Investigación Sanitaria (IIS) –
Fundación Jiménez Díaz and UAM
Madrid, Spain

Robert Rembisz
The Pharmaceutical Research Institute and
Neurotoxicology Laboratory
Albany College of Pharmacy and Health
Sciences
Albany, NY, USA

Heinrich Sauer
Department of Physiology
Faculty of Medicine
Justus-Liebig University
Giessen, Germany

Fatemeh Sharifpanah
Department of Physiology
Faculty of Medicine
Justus-Liebig University
Giessen, Germany

Bechan Sharma
Department of Biochemistry
Faculty of Science
University of Allahabad
Allahabad, India

Janet H. Silverstein
Department of Pediatrics
University of Florida College of Medicine
Gainesville, FL, USA

Shweta Singh
Department of Biochemistry
Faculty of Science
University of Allahabad
Allahabad, India

Robert D. Stratton
Department of Ophthamology
University of Florida College of Medicine
Department of Small Animal Clinical Sciences
College of Veterinary Medicine
University of Florida
Gainesville, FL, US

Hanna Tadros
Center for Reproductive Medicine
Glickman Urological and Kidney Institute
Cleveland Clinic
Cleveland, OH, USA

Rajiv Tikamdas
Department of Pharmacodynamics
College of Pharmacy
University of Florida
Gainesville, FL, USA

Artak Tovmasyan
Department of Radiation Oncology
Duke University School of Medicine
Durham, NC, USA

Hirokazu Tsukahara
Department of Pediatrics
Okayama University Graduate School of
Medicine, Dentistry and Pharmaceutical
Sciences, Okayama, Japan

Eva Tvrdá
Department of Animal Physiology
Slovak University of Agriculture
Nitra, Slovakia

Albert Vexler
Department of Biostatistics
School of Public Health and
Health Professions
State University of New York at Buffalo
Buffalo, NY, USA

Rachael Watson
Department of Orthopedics and
Rehabilitation
University of Florida
College of Medicine
Gainesville, FL, USA

James R. Wilcox
Department of Clinical Research & Physician
Education
Serena Group
Cambridge, MA, USA

William E. Winter
Department of Pathology, Immunology and
Laboratory Medicine
University of Florida College of Medicine
Gainesville, FL, USA

Justin Wray
Department of Radiation Oncology
University of Florida College of Medicine
Gainesville, FL, USA

Masato Yashiro
Department of Pediatrics
Okayama University Graduate School of
Medicine, Dentistry and Pharmaceutical
Sciences, Okayama, Japan

Ping Zhang
Department of Pharmacodynamics
College of Pharmacy
University of Florida
Gainesville, FL, USA

Special recognition

In memory of Christine M. Armstrong the editors and authors of this textbook unanimously acknowledge her for her dedication to support the concept of oxidative stress and antioxidant therapy.

She was instrumental in typing the first International Symposium in 1982 and over many years helped prepare other manuscripts from her knowledge of English grammar and sentence structure into the series on Methods and Protocols and the series on Oxidative Stress in Basic Research and Clinical Practice. She worked laboriously in transcribing numerous symposia.

Anyone who met "Chris" knew her persona. She showed a gracious and genuine interest in the diverse efforts of many colleagues who wrote in the Protocols and Clinical Practice series. She was always an advocate for natural foods and antioxidant supplements. Chris died on May 27, 2014, from pancreatic cancer and will be sorely missed by everyone who were fortunate to know her on a personal and spiritual level. We dedicate this textbook as a token of our esteem.

For his help in research for this textbook, the editors thank Dennis E. Armstrong.

Foreword

Don Armstrong and I grew up together in Oregon and Colorado, both at the lab bench and during shared family gatherings over several decades. Half a century ago, we coauthored two papers on sulfatase deficiencies responsible for two human neurological diseases. It is both a special privilege and an honor to add a personal observation to my long-esteemed colleague's latest textbook.

In 1963, neither he nor I could have anticipated the remarkable scope and depth of his many scientific interests, research publications, careers as a teacher, and related editorial accomplishments. His efforts over that time span have helped clarify the intricate biochemical and subcellular mechanisms of oxidative damage, in both human disease and animal models. These pivotal mechanisms have a practical application – helping us all defer that condition overly simplified in the old phrase, "normal aging."

I hope that as you read the chapters of this extraordinary text written by a cadre of international experts, the diversity of topics reflects the inspired efforts of an exceptionally foresighted person who envisioned long ago how crucial this whole field would become in biology and medicine.

James H. Austin, MD
Professor Emeritus
University of Colorado Health and Science Center
Columbia, Missouri

Preface

This textbook starts with the principles of the oxidative stress process and a historical perspective of oxidative stress pioneers to explain the basic principles related to biochemistry and molecular biology showing pathways and biomarkers. The section also gives an explanation of differential diagnosis and brief descriptions of the use of diagnostic imaging.

The second section covers clinical correlations on acute and chronic disease and discusses our novel approach that bridges the gap between these two concepts. It shows how oxidative damage and inflammation trigger the disease process and how compromised immunity then leads to the proliferation of disease, with a bench-to-bedside approach to clinical application. These discussions focus on immune responses in early and chronic diseases. There is a glossary and explanation on medical terminology and a set of questions for the student. The effects of aging, senescence, and life span are presented. A thematic summary box is included for each chapter so that students know what is required of them.

Chapters on Clinical Correlations cover the most common medical diseases and cover developmental change. The goal is to stimulate new directions for student education in their professional career. The clinical topics are taken from a series of books on specific diseases whose emphasis was on oxidative stress and antioxidants entitled, *Oxidative Stress in Applied Basic Research and Clinical Practice*. This book represents a compilation of scientific and clinical data summarized from books in the series that emphasize result-oriented findings in (1) neurology, psychiatry, and behavioral data including the eye and ear, (2) pediatrics, (3) disorders of skin and musculoskeletal system, (4) gender-related issues, (5) chronic diseases, (6) pain and inflammation management, (7) wellness issues, and (8) biostatistics.

The book can be used as a college text as well as a graduate level course of pre-professional level, providing an overview covering oxidative damage and antioxidant imbalance. This section links science and medicine at the pre-professional and graduate levels and is appropriate for medical and veterinary residents, providing an overview of disease mechanisms related to oxidative damage and antioxidant imbalance.

Our authors have extensive experience in the laboratory or in medical therapy. Each of the chapters provides an understanding of appropriate *mechanistic* information and what biomarker(s) and physiological functions may be relevant to the over-riding concept of redox issues leading to critical mass of stress oxidants in early disease to chronic conditions including potential alternative therapy. Areas of research that are not yet fully developed are indicated, helping students in choosing a direction for their careers.

The textbook is the first of its kind and should have wide appeal to students majoring in life science, biology, and allied health professions, as well as to physician assistants and nurse practitioners, with a secondary appeal to those in pharmacy, food and nutrition, exercise, sports medicine, psychology, and public health. This textbook will also be of use to programs in herbalist and holistic schools and could likewise be used for residency training and Continuing Medical Education courses.

Our goal was to present the most authoritative source of current and emerging knowledge on oxidative stress-related disorders. To this end, we are confident we have achieved our aim as editors. We are pleased with the collegial support given to us by these authors. We sincerely hope students who study the information presented herein, will gain from this perspective and it will be helpful in their future professional careers.

Donald Armstrong and Robert D. Stratton

SECTION I
Introduction

CHAPTER 1

Introduction to free radicals, inflammation, and recycling

Donald Armstrong[1,2]

[1] Department of Biotechnical and Clinical Laboratory Sciences, State University of New York at Buffalo, Buffalo, NY, USA
[2] Department of Ophthalmology, University of Florida School of Medicine, Gainesville, FL, USA

This introductory chapter will give you information to fill in the gaps and understand the complexities reported in Chapters 3–31.

THEMATIC SUMMARY BOX

At the end of this chapter, students should be able to:

- Show free radical formation

- Show how endogenous and exogenous free radicals stimulate and potentially initiate disease

- Define inflammation and the immune response

- Differentiate between acute and chronic inflammation

- Describe pathways leading to apoptosis, necrosis, cell death, and disease

- Define pathogenesis

- Show how antioxidants scavenge free radicals and participate in recycling pathways

- Describe biomarker measurements using their abbreviations

Historical perspective

In 1993, an International Symposium on *Free Radicals in Diagnostic Medicine: a systems approach to laboratory technology, clinical correlations, and antioxidant therapy* was organized, and in 1994 it was published as volume 366 in *Advances in Experimental Medicine and Biology*.[1] This was the first attempt to coordinate the various laboratory findings from research publications that were divided into subsections on pathophysiology, analysis, organ-specific disorders, systemic involvement, and therapeutic intervention. In 2007, another International Symposium on Free Radicals in Biosystems was conducted.[2] These two conferences set the standard for the present textbook, which is an extension of those meetings in the application and understanding of free radical (FR) methods and protocols, and is once

---

Oxidative Stress and Antioxidant Protection: The Science of Free Radical Biology and Disease, First Edition.
Edited by Donald Armstrong and Robert D. Stratton.
© 2016 John Wiley & Sons, Inc. Published 2016 by John Wiley & Sons, Inc.

again timely 20 years later. The author of this chapter has multiple books in the field.

In 1990, a new series of books covering laboratory techniques in *Methods in Molecular Biology* (MiMB) was initiated and later became Advanced Protocols 1–3 including separate volumes on lipidomics and nanotechnology that contain specific sections on oxidative stress and antioxidants. This was followed in 2000 by a new series devoted to clinical studies (*Oxidative Stress in Applied Basic Research and Clinical Practice*), which increased our oxidative stress library database to over 1300 research collaborators from 34 countries and 20 states in the United States.

These books have been a major effort to cover subject areas in advanced detail, which are germane to background information that supports perturbations of lipids and proteins in our concept of oxidative stress. In addition, platform technology on metabolomics and transcriptomics using high-pressure liquid chromatography (HPLC) and mass spectroscopy (MS) can be read in the literature by the student separately and integrated with the state-of-the-art coverage. The result of these activities is the basis for the present textbook, which illustrates the concept of *educational research* in stress-induced disease reactions.

Oxidative stress concept

The concept of oxidative stress in disease means an environmental stimulus is able to create an FR by random chance. The number of "hits" per day is estimated at 100,000 coming from the mitochondria as molecular oxygen moves through the electron respiratory chain and from environmental radiation. FRs are characterized by loss of an electron-making species highly reactive to other biochemicals resulting in cellular damage (Figure 1.1). We think of FRs as tipping the balance toward disease, so that as FRs increase along the up slope as a function of progressive disease, antioxidants (AOXs) decrease along the down slope as they are being consumed due to oxidative stress, so that homeostasis tips toward disease. Supplementation with AOX nutraceuticals leads to protection and eventual homeostasis when the two processes become equal, and no disease is clinically evident by conventional physiological testing in the patient populations.

The most damaging FRs are the hydroxyl ($HO^\bullet$) and hydrogen radicals produced during ionizing radiation or environmental toxicology reactions. Superoxide FRs

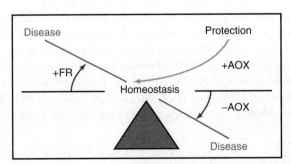

Figure 1.1 Homeostasis is a balance between levels of free radicals (FR) and antioxidants (AOX).

($O_2^{\bullet-}$) are produced in mitochondrial electron transport reactions. Superoxide dismutase rapidly forms hydrogen peroxide (H_2O_2) from $O_2^{\bullet-}$. Reduced free iron Fe^{2+} and copper Cu^{1+} participate in Fenton reactions with H_2O_2 in the millisecond range to produce hydroxyl radicals and hydroxide anions, oxidizing the metals to Fe^{3+} and Cu^{2+}. Oxidized free iron or copper can then oxidize H_2O_2 to form hydroperoxyl ($HOO^{\bullet}$) radicals and protons (H^+), reducing the metals to Fe^{2+} and Cu^{1+}, so that new peroxidation reactions can occur in a cyclical manner. Thiol ($SH^{\bullet}$) and lipidperoxide radicals ($L^{\bullet}/LO^{\bullet}/LOO^{\bullet}$) are other radicals that cause FR damage to cellular components.

Hydrogen peroxide also interacts at Fe^{2+} metal-binding sites on heme-containing enzymes to generate the reactive hydroxyl radical. The hydroxyl radical damage to proteins has been shown to produce covalently bound protein aggregates, and when disulfide bridges are causing the aggregation, cross-linked protein adducts are formed. These reactive oxygen radicals modify amino acids at metal-binding sites and facilitate proteolytic attack. Other FR damage includes DNA strand scission and lipid hydroperoxidation (LHP).

Polyunsaturated fatty acids (PUFA) contain multiple carbon–carbon double bonds. These fatty acids provide mobility and fluidity to the plasma membrane, properties which are known to be essential for the proper function of biological membranes. The process of *lipid peroxidation (LHP)* is a step-wise process with the removal of an electron at the initiation step and at subsequent propagation reactions. Iron salts and other iron complexes help initiate the process by forming alkoxy or peroxy radicals upon reaction with oxygen species.

In general, there are three damaging consequences of LHP within the plasma membrane *in vivo*. The first consequence is a decrease in membrane fluidity. Saturated fatty acids are structurally more rigid than the flexible PUFA. The second consequence of LHP is an increase in the "leakiness" of the membrane pores to substances, which normally do not pass through the membrane. Finally, membrane-bound proteins are damaged by propagation reactions of LHP. Hydroxyl radicals remove a hydrogen atom from methylene groups in a PUFA resulting in a PUFA lipid radical. The lipid radical then reacts with "normal" oxygen ("triplet" O_2) to form a lipid peroxyl radical, which then reacts with another PUFA to form another lipid radical. This propagates the production of reactive oxygen species (ROS) in what is called the peroxidative chain reaction (Figure 1.2).

Reactive nitrogen species (RNS) can also participate in nitrosative oxidative damage. Nitric oxide synthase is the enzyme that drives this reaction. FRs can attack proteins, lipids, carbohydrates, and nucleic acids causing extracellular matrix, cellular, and subcellular damage. Nitrous oxide + superoxide $\rightarrow$ peroxynitrite ($ONOO^-$), which yields the $NO_2^{\bullet}$ reactive oxidant.

Oxidative stress is highest in the plasma membrane, mitochondria, nucleus, golgi, and lysosomes. AOXs represent about 50% of total Internet citations and cover anti-cancer, anti-inflammatory, and anti-proliferation. FRs can also act in signaling and function as messenger agents.

Inflammation is most often the initial step in a disease process followed by an immune response, but the immune response may occur at nearly the same time. With time, various molecular dysfunctions develop into a chronic disease such as documented in diabetes, cardiovascular disease, organ failure, and cancer.

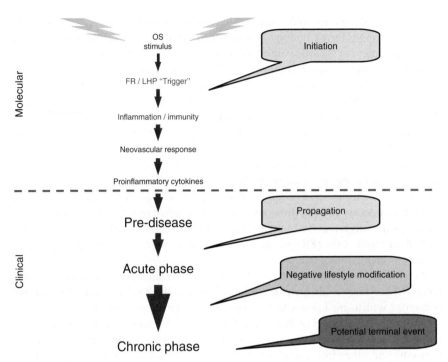

Figure 1.2 Oxidative stress starts a cascade that can lead to chronic disease if not modified by corrective actions.

The first step is called *initiation,* followed by *propagation, termination,* and/or *protection.* The AOX pathways are generated by endogenous (internally synthesized) or exogenous sources that come from dietary preferences and lifestyle controlling modifications.

The pathway originates from stimuli that trigger radical formation. These molecular events lead to inflammation and immune responses, provoke neovascularization, and upregulate proinflammatory cytokines. Protective scavenger AOXs are divided into small water-soluble molecules (vitamin C, glutathione (GSH)), lipid-soluble molecules (vitamin E, lipoic acid, carotenoids, and coenzyme Q_{10} (CoQ_{10})), and larger enzyme molecules that need to be synthesized internally (superoxide dismutase, catalase, and GSH peroxidase). These detoxify aqueous and lipid-soluble peroxides. Preventive AOXs bind to essential proteins (albumin, metallothionein, transferrin, ceruloplasmin, myoglobin, and ferritin).

These factors can lead to an "imbalance" – a lack of AOXs because of their underproduction, misdistribution, or environmental stressor depletion. In the clinical realm, these can lead to pre, acute, and chronic disease and may advance to a potentially terminal event.

Biomarkers of oxidative stress can be analyzed in cells; tissues; blood; urine; CSF; synovial fluid; saliva; tears; and many other substances such as botanical nutraceuticals, marine algae, and food samples. Key markers used in most scientific publications are shown by the following oxidative stress metrics (Table 1.1).

Table 1.1 Biomarkers for oxidative stress.

Organelle	Biomarker activity
Nucleus	8-Hydroxy deoxyguanosine
Mitochondria	Catalase, Cu/Zn-SOD, Mn-SOD
Endoplasmic reticulum and golgi	PEG-SOD, F2-isoprostanes, HNE
Plasma	TBARS, CUPRAC, 8-iso-PGF(2α), LHP
Total cellular constituents	Cytokines, chaperones, telomeres

Cytokines are immunomodulatory agents that act as intracellular chemical mediators. They activate antigens and carry signals to adjacent cells of the immune system, thus magnifying the response to disease. Chaperones are functionally related groups of proteins synthesized in the endoplasmic reticulum. They are cellular machines that assist in protein folding and protect against degradation. Under physiological stress, cells respond to an increase by less than 5°C of temperature to produce heat shock proteins, which participate in anoxia, inflammation, and oxidant injury. They can be analyzed by electron microscopy.

Oxidative stress plays a *major* role in many human diseases and may well become the salient feature in most diseases. To date, involvement of oxidative stress has been confirmed in over *100* disorders. Oxidative stress has been previously linked to a plethora of changes induced during aging as well as in specific diseases such as obesity, diabetes, cancer, cardiovascular disease, stroke, neurodegenerative disease, trauma, hypoxia, psychological behavior, pain, chronic fatigue, fibromyalgia, pulmonary disease, hepatic disease, renal disease, gastrointestinal disease, macular degeneration of the eye, disorders of noise-induced hearing loss, fertility, menopause, osteoporosis, endocrine disorders, skin disease, musculoskeletal disorders, bone marrow abnormalities, oral health, nutrition, environmental health, and complications following extended space travel. Genetics may also play a role in the overall pathology of oxidative stress. Many of these topics have extensive publications in peer-reviewed scientific and clinical journals, corroborating the role of oxidative stress. Therefore, oxidative stress should be considered a *primary* cause of most diseases, or at least the result of several compounded processes that require more data and are currently under investigation. The student should consult the Internet, PubMed, Citation Index, or ISI Web of Science to study the many oxidative stress-related activities present in tissues and organs measured with new appropriate biomarkers.

Free radicals

Oxidation reactions cause the formation of a variety of FRs, which are unstable substances, that can initiate chain reactions in *microseconds*, leading to disease and programmed apoptotic cell death. Cells may recover or may undergo apoptotic autophagy or uncontrolled necrosis. Necrosis is when the tissue cannot regulate the influx of fluids and the loss of electrolytes, most notably in mitochondria and is associated with extensive damage resulting in an inflammatory response.

Apoptosis is by definition a phenomenon of programmed cell death under normal homeostasis control, and consequently no inflammatory response is observed. The cell shows membrane blebbing, shrinkage in size, nuclear condensation, DNA chromatin fragmentation, aggregation of chromatin, nuclear condensation, and partition of cytoplasm into membrane-bound vesicles, which contain ribosomes and nuclear material. These are phagocytized by macrophages, but there is no inflammatory response. Chronic apoptosis may cause widespread atrophy.

Free radicals may occur from a specific stimulus such as ultraviolet radiation, multiple environmental factors, poor nutrition, or sedentary lifestyle. Oxidative stress can stimulate neutrophils in the blood to ingest pathogens, but these are replaced on a daily basis by younger cells. FRs can attack proteins, lipids, carbohydrates, and nucleic acids causing cellular and subcellular damage to cells by ROS or RNS (Figure 1.3).

Note that the amount of oxidized protein is proposed as the tipping point in the FR and AOX balance scheme. Oxidized proteins activate the caspase enzyme cascade in the proteasome, form intracellular aggregates, and are predominately nonrepairable because cross-links limit repair mechanisms so that recycling of amino acids for continuous protein synthesis is diminished.

The master AOX for recycling and FR inactivation is reduced GSH, but lutein and phenols with hydroxyl groups can readily take up unpaired electrons together with ascorbate; α-tocopherol/tocotrienol; and enzymes such as superoxide dismutase, catalase, GSH peroxidase, GSH reductase, and CoQ_{10}.

Inflammatory pathways

Inflammation facilitates healing from noxious or foreign stimuli. The initiating event is tissue damage. It involves the formation of nuclear factor kappa B (NFκB) and systemic cytokine by-products such as TNF-α and prostaglandin E2-α. Biomarkers are thiobarbituric acid reactive substances (TBARS), derivatives of reactive oxygen metabolites (dROMS), oxygen radical absorbance capacity (ORAC), hydroxyl radical antioxidant capacity (HORAC), arachidonic acid, thromboxane, lipopolysaccharides, and trolox equivalent antioxidant capacity (TEAC)). Many biomarkers degrade over

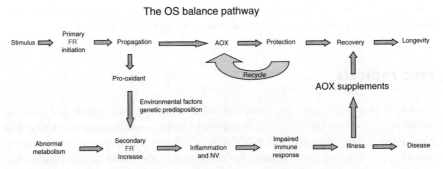

Figure 1.3 Free radicals (FR) are key to initiation and propagation of the paths that lead to disease. Antioxidants (AOX) are key to protection.

time, so it is advised to use fresh or freshly frozen samples. For example, in type 2 diabetic patients, oxidative stress is closely associated with chronic inflammation by upregulating key vascular ROS- and RNS-producing enzymes and the corresponding endogenous AOXs. Heme oxygenases (HO-1) utilize NADPH and oxygen to rupture the heme moiety, causing modulation of cellular bioenergetics and leading to apoptosis and inflammation. These are my interpretations of oxidative stress and AOX events in disease to gain a mechanistic summary and thus an approach to therapy.

Mitochondria

Molecular oxygen diffuses into the mitochondrial inner membrane where ROS sources are actively produced through the electron transport chain and nitric oxide synthase reactions. Cytochromes are present in mitochondria and transfer electrons along the respiratory chain that involves oxidation and reduction of iron. CoQ_{10} is a naturally occurring AOX and a prominent component of mitochondrial electron transport chain. CoQ_{10} is recognized as an obligatory cofactor for the function of thermogenesis uncoupling proteins and a modulator of the mitochondrial transition pore. It was also observed that CoQ_{10} is part of an endogenous AOX defense that increases SOD2 and GSH peroxidase. In disease-prone cells, mitochondrial superoxide is exported to adjacent cells, triggering lipid peroxidation, propagation reactions, and inflammation.

Educational redox

The following list of information on products for redox therapy as an experimental therapy is based on results-oriented data. The following list has contact information from prominent companies specializing in pro-oxidant and AOX agents in cells. These also cover alternative and holistic medicine.

1 Life Extension Foundation, 5th edition, 2013 (www.lifeextension.com/track)
2 Integrative Therapeutics, Inc (www.integrativepro.com)
3 (www.naturalmedicines.therapeuticresearch.com)
4 Advanced Bionutritionals (www.advancedbionutritionals.com)
5 Oxford Biomedical Research (http://www.oxfordbiomed.com)
6 OXIS Research International (www.oxisresearch.com)
7 Cayman Chemical Co. (www.caymanchem.com)
8 ALPCO Diagnostics (www.alpko.com)
9 The Japan Institute for the Control of Aging, Nikken SEIL Corp. (www.biotech@jaica.com), Genox is the USA distributor.
10 ALEXIS Biochemicals (www.alexis-biochemicals.com)
11 INOVA Diagnostics (www.inovadx.com)
12 Molecular Probes by Life Technologies (www.lifetechnologies.com)
13 PROBIOX SA, Belgium (www.probiox.com)

14 The National Center for Comparative Alternative Medicine (nccam.nih.gov) is an investigator-initiated project on advanced research covering complementary and alternative medicine.

Great variability in the activities of AOXs is present in over-the-counter nutraceuticals. Disclaimers for food products that are not under FDA regulation are treated as a food, not as a pharmaceutical. The caveat that must be put on the label is "These statements have not been evaluated by the FDA and the product is not intended to diagnose, treat, cure, or prevent any disease." Therefore, be careful with sources of AOX treatments.

Stay current with new research initiatives. The student is also encouraged to keep a record of these applications. In searching the Internet, the student should key in on *MiMB* and *Oxidative Stress in Applied Basic Research and Clinical Practice*, published by Humana Press, a brand of Springer and part of Springer Science + Business Media.

References

1 Armstrong, D. (1994) Free radicals in diagnostic medicine: a systems approach to laboratory technology. In: *Advances in Experimental Medicine and Biology*. Vol. 366. Plenum Press.

2 Afzal, M. & Armstrong, D. (eds) (2007) Molecular biotechnology: an international symposium on free radicals in biosystems. In: *Molecular Biotechnology*. Vol. 37. Humana Press.

3 Armstrong D. 1998. *Free radical and antioxidant protocols, Methods in Molecular Biology*, 108. Humana Press.

4 Armstrong D. 2002. *Oxidative stress biomarkers and antioxidant protocols, Methods in Molecular Biology* 186. Humana Press.

5 Armstrong D. 2002. *Oxidants and antioxidants: ultrastructure and molecular biology: ultrastructure and molecular biology protocols. Methods in Molecular Biology* 196. Humana Press.

6 Armstrong, D. (ed) (2008) Advanced protocols in oxidative stress I. In: *Methods in Molecular Biology*. Vol. 477. Humana Press, a part of Springer Science + Business Media, LLC, NY.

7 Armstrong, D. (ed) (2010) Advanced protocols in oxidative stress II. In: *Methods in Molecular Biology*. Vol. 594. Humana Press, a part of Springer Science + Business Media, LLC, NY.

8 Armstrong D. 2014. *Advanced protocols in oxidative stress III Methods in Molecular Biology* 1208. Humana Press, a part of Springer Science + Business Media, LLC, NY.

9 Armstrong, D. (2009) Lipidomics: methods and protocols. In: *Methods in Molecular Biology*. Vol. 579–580. Humana Press, a part of Springer Science + Business Media, LLC, NY.

10 Armstrong, D. & Bharali, D.J. (2013) Oxidative stress and nanotechnology. In: *Methods and Protocols*. Vol. 1028. Humana Press, a part of Springer Science + Business Media, LLC, NY.

CHAPTER 2

Diagnostic imaging and differential diagnosis

Robert D. Stratton[1,2]

[1] Department of Ophthalmology, College of Medicine, University of Florida, Gainesville, FL, USA
[2] Department of Small Animal Sciences, College of Veterinary Medicine, University of Florida, Gainesville, FL, USA

THEMATIC SUMMARY BOX

At the end of this chapter, students should be able to:

- Describe the main diagnostic methods
- Describe the main methods of measuring bioelectrical signals
- Describe the main methods of imaging using electromagnetic energy
- Describe the main methods of imaging using radioisotopes

Diagnostic method in clinical practice

History

Clinicians are trained to employ the differential diagnostic method using a wide range of diagnostic methods and tests to eliminate diagnostic possibilities and narrow the diagnostic list. The history and physical exam are the most important first steps in the differential diagnosis. The history is a conversation with the patient to ascertain the *chief complaint*, which may be more generally described as the *presenting problem* so as to include unresponsive or uncooperative patients. A *history of the present illness* is then taken to find the symptom locations, the character of pain the problem may be causing, including location of any radiating discomfort, the time of onset of the symptom and duration, the severity of discomfort, and aggravating or relieving factors. Other complaints are investigated in a similar manner. To elicit other complaints, a *past medical and surgical history* is taken, including major past and present illnesses, hospitalizations, treatments, procedures and surgeries, allergic reactions, and present medications. A *review of systems* is next taken, which is a systematic review of the major organ systems, including the head, eyes, ears, nose, throat, teeth, tongue, cardiovascular system, respiratory system, gastrointestinal system, hematological system, neurological system, musculoskeletal system, genitourinary system,

---

Oxidative Stress and Antioxidant Protection: The Science of Free Radical Biology and Disease, First Edition.
Edited by Donald Armstrong and Robert D. Stratton.
© 2016 John Wiley & Sons, Inc. Published 2016 by John Wiley & Sons, Inc.

endocrine system, skin, psychiatric problems, and orientation. A *family history* elicits diseases that may be passed on to progeny, and a *social history* elicits workplace and home dynamics.

Physical examination

Equipped with data from the history, a focused physical exam is then performed. A standard physical examination includes *vital signs, alertness assessment, general body assessment*, and a systematic organ systems examination guided by the history. Many of the organ system examinations are performed in much greater detail by medical and surgical specialists, but all clinicians evaluate head, eyes, ears, nose, and throat (HEENT), respiratory system, cardiovascular system, abdomen, neurological system, genitourinary system, integument, and musculoskeletal system.

Specialized examinations are performed when indicated. *Neurological* examination may include coma assessment, detailed motor and sensory testing of somatic and cranial nerves including reflexes, gait evaluation, and cognitive evaluation. *Ophthalmological* examination may include detailed vision testing, visual field examination (perimetry and color perimetry), contrast sensitivity, extraocular muscle evaluation, slit lamp examination of the anterior and posterior segments of the eye including optic nerve head evaluation, and indirect ophthalmoscopy with scleral indentation to examine the peripheral retina and ciliary body. *Otologic* examination may include otoscopy with the manipulation of the tympanic membrane, evaluation of hearing acuity and tuning fork, and evaluation of the vestibulo-ocular reflex. Laryngoscopy, with the familiar ENT head mirror being replaced with direct light sources, is performed to evaluate the larynx and vocal cords. Auscultation, percussion, and palpation are performed to evaluate the chest and abdomen. Musculoskeletal examination can be one of the most complex of the general physical exam but, focused by the history, evaluates the part in question.

Differential diagnosis

Once the history and physical exams are complete, the clinician must then consider all the possible compatible diagnoses drawn from the clinician's fund of knowledge. Diagnostic testing is then considered if the diagnosis is in doubt. In diseases of either systemic or organ-specific damage owing to oxidative stress, diagnostic testing usually depends on biomarker measurements and imaging techniques. A biomarker is almost any measurement reflecting an interaction between a biological system and its environment, which may be chemical, physical, or biological, and the measured response may be a functional, physiological, biochemical, or molecular interaction.[1] These measurements are used to detect a current state.

Biomarkers in biological investigation

Biochemical biomarkers

Clinical chemistry laboratories provide measurements of chemicals, proteins, DNA, RNA, and many organic macromolecules in any biological fluid, including plasma, serum, and cerebrospinal fluid. Measurements include enzymes, antibodies, hormones, cytokines, antigens, cytosol proteins, and levels of pharmaceuticals. Trace amounts of DNA or RNA can be detected using *polymerase chain reaction* (PCR) technology, making early diagnosis of some infections and cancers possible. An increasing array of PCR techniques are used in medical, forensic, and genetic investigation.[2]

Electromagnetic biomarkers

Biomarkers from the whole range of electromagnetic radiation (photons) are used for imaging and detection, from a Hertz of 0 (standing potentials), to a Hertz of up to 10^{20} (γ radiation).

Electrooculography (EOG) measures the electrical potential across the eye from the cornea to the choroid, and is mostly a measurement of the electrical potential generated by the monolayer of retinal pigment epithelial (RPE) cells, called the standing potential because of its nontransient character (near 0 Hz). This potential can be directly measured accurately with invasive placement of measuring electrodes, but in order to find this value noninvasively, measurements of electrical potential are made external to the eye and compared to a reference electrode. The potentials measured with the eye gazing far right is compared to the potential with the eye gazing far left, and a calculation of the vector of electrical potential of the RPE is made. Although the resulting calculated standing potential is not very accurate, a change in this standing potential in response to dark and light adaptation accurately reflects the function of the RPE in normal and dysfunctional states.

Electroretinography (ERG) measures the transient electrical potential generated by the neurosensory retina in response to a brief whole field simulation with light (Figure 2.1). This can be done in a dark adapted state to assess the scotopic retinal function (mainly the rod-based neural network) or in the light adapted state to assess the photopic retinal function (mainly the cone-based neural network). An electrode placed on the anesthetized cornea is compared to a reference electrode (usually on the forehead). The initial negative response, called the A wave, is generated by the photoreceptor cell that has the first response to the light stimulus. The following positive B wave is a combined response from the photoreceptor, bipolar, and Mueller cells. The A wave in ERG done during dark adaptation is mostly from the rod photoreceptor cells. The A wave done during light adaptation is mostly from the cone photoreceptor cells. A more definitive way to differentiate between slow recovering rods and more quickly recovering cones is to use a flickering light stimulus of 30 Hz so that the rods cannot recover before the next flash, whereas normal cones recover in time to give a measurable electrical impulse. Differentiation among various retinal degenerations is greatly aided by results of EOG and ERG testings. Progression and prevention of retinal degeneration can be measured by serial testing. *Multifocal electroretinography* measures the same transient electrical signal from the retina, but a carefully planned and patterned stimulus rather than a whole field light stimulus is used and the ERG pattern from focal retinal areas are computed from the corneal electrode recorded signals. *Dark adaptometry* measures subjective light sensitivity thresholds at a focal spot in the retina as the eye adapts from bright light to dim light levels. Several subjective color vision tests are available

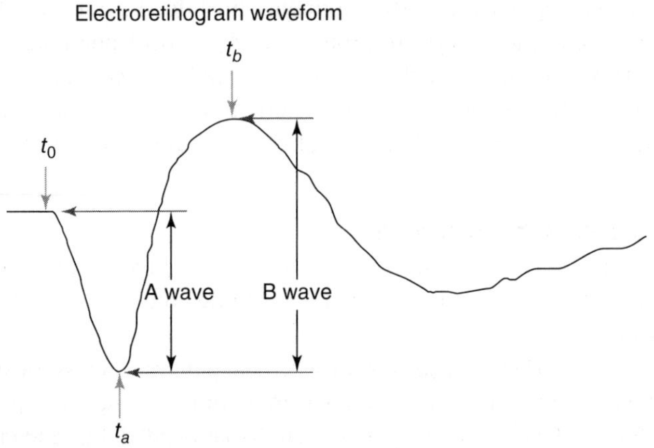

Electroretinogram waveform

Figure 2.1 Electroretinogram waveform. t_0 is the time of light stimulus, t_a is the time to the peak of the negative deflection and t_b is the time to the peak of the positive deflection. A wave is the amplitude of the negative deflection, mainly from the photoreceptor cell depolarization; B wave is the amplitude of the positive deflection from the neurosensory retinal response to the photoreceptor cell depolarization. Times vary with species, but are generally in the 10–300 ms range. Amplitudes also vary widely, but are in the 20–500 mV range.

including Ishihara pseudochromatic plates, Farnsworth-Munsell D-15 panel, and a more detailed F-M 100 hue test.

The *electrocardiogram* is a recording of the electrical signals of myocardial contraction from electrodes on the skin, familiar to most as an ECG. The more invasive processes of introducing electrodes behind the heart through a *transesophageal ECG* yield additional information. An ECG of an electrode attached to the needle used to drain fluid from the pericardial sac can indicate contact with the myocardium by an abrupt change in the ECG and guide the placement of the aspirating needle.

Electrocorticography (ECOG) is the direct recording of cerebral cortical electrical potentials during craniotomy using electrodes. *Single neuron recording* is achieved using a wide variety of microelectrodes and microelectrode arrays. The *electroencephalogram* (EEG) records mass activity of the brain and is, therefore, very useful in detecting seizures where there is mass recruitment of neuronal activity or in confirming brain death where there is no activity. A strong stimulus can be used to evoke an EEG response, called *sensory evoked potential* (SEP). *Visual evoked potential* (VEP) or visual evoked response (VER) is an EEG recording of the scalp overlying the visual cortex. Flash whole field stimuli evoke strong signals with inconsistent timing. Pattern stimulus evoked responses are weaker but are much more consistent. Delays in the evoked signal show pathology in the neural pathways. Optic nerve lesions most often cause unilateral delays. The auditory system is assessed by *auditory brainstem response* (ABR) with a click auditory stimulus and scalp EEG electrodes. The initial response (Wave I) is predominantly the auditory nerve, and later responses (i.e., Wave V) are predominantly from the brainstem. *Magnetoencephalography* (MEG) also records mass electrical signals from the cerebral cortex by detecting the magnetic fields generated by the mass neuronal activity. The changes in signals over time give a way to measure functional brain activity.

Electromyography (EMG) is the clinical recording of the mass signal of muscle from skin electrodes overlying a muscle or transcutaneous needle electrode in a muscle. *Nerve conduction velocity* (NCV) is the measurement of the speed and strength of a peripheral nerve from an electrical stimulus. EMG and NCV together help to differentiate between myopathic and neuropathic diseases and also measure responses to drugs or treatment.

Spectrographic biomarkers

Chromatography is the separation of mixtures based on some differing characteristics of the populations in the mixture, typically molecular size, charge, and shape. *Gas chromatography, gel column chromatography, affinity chromatography,* and *high performance liquid chromatography* separate by elution. *Gel electrophoresis* separation of DNA is a common technique in modern laboratories. Many specialized forms of chromatography have been devised and are used extensively in research laboratories.

Mass spectrometry measures the mass-to-charge ratio of ions produced from a sample by using the deflection of accelerated charged particle by an electromagnetic field. The measurement is the relative abundance of ions of a particular m/z ratio so that the data are presented as a graph or image (see Figure 2.2). Different ionization and fragmentation techniques have been developed to help analyze samples. Separation procedures with gas chromatography or with tandem mass spectrometers and quadrupole filtering allow for the identification of complex samples. Rasterizing tissue samples with laser allows for imaging by mass spectrometry.[3]

Biological imaging

When the amount of biomarker information increases to the point of being hard to conceptualize, the data are presented as an image. Imaging can be a way of presenting

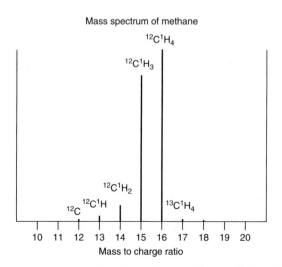

Figure 2.2 Simple mass spectrogram of methane. The less abundant ^{13}C is hidden in the ^{12}C peaks except for $^{13}C^1H_4$ visible right of $^{12}C^1H_4$ peak.

a large amount of data so that trends and relationships can be seen. This is used not only with electromagnetic data but also in other large data sets as in genomics, proteomics, and brain mapping.

Acoustic imaging

Ultrasonography or *echography* is a technique that uses acoustic reflections to make measurements and construct images. There is a trade-off between higher frequency giving better resolution and lower frequency giving better penetration. Sound frequencies of 2–20 MHz are typically used. *A-scan* is recording of the echo of a series of pings along a single vector for the purpose of measuring distance or sound attenuation. *B-scan* is recording of a series of pings along a series of adjacent vectors for the purpose of constructing a 2D image. Multiple B-scan images can be computationally constructed into a 3D image, sometimes called a C-scan. *Doppler scan* identifies moving media and is useful in visualizing blood flow velocity and direction. Doppler information is often displayed as a color palette overlying a B-scan image. Ultrasound has become a widely used technology due to the rapid and dynamic assessment, noninvasiveness, and low cost. Computer enhancement of images and Doppler data have made evaluation of blood vessels with ultrasound a good alternative to angiography, and *intravascular ultrasound* (IVUS) has enhanced angiography.

Light imaging

Photography has been expanded to a vast array of technologies. *Fundus photography* and *retinal fluorescein angiography* capture images of the retina, optic nerve head, and blood vessels in the posterior segment of the eye. *Photomicroscopy* can be enhanced in many different ways. *Stains* can enhance the structural details of tissue on histologic sections, with hematoxylin staining the nucleus and eosin staining the cytoplasm in the most common H&E stain. Special stains can identify various cellular and extracellular components in tissue sections and cell preparations. Antibody-linked *immunofluorescent staining* has become a very powerful investigative tool with advances in new techniques for antibody production.

   Phase contrast microscopy can provide details of living cells by the enhancement of the changes in phase as light passes through transparent cellular structures. *Optical coherence tomography (OCT)* is an imaging technique that uses interference patterns from reflected light to construct an image, somewhat analogous to ultrasonography, but with a much higher resolution. OCT is most easily used in investigating surfaces or the optically clear structures of the eye, but investigation into using lower frequency light to penetrate tissue and higher frequency light to give higher resolution will continue. Intravascular OCT has increased the optical resolution of vascular imaging. *Diffuse optical imaging* is an infrared light technique using computer reconstruction of images similar to the familiar computed tomography (CT) scan imaging technique. This is able to detect a range of metabolic biomarkers including oxyhemoglobin, deoxyhemoglobin, and cytochrome C oxidase redox status.[4] *Transmission electron microscopy* (TEM) uses magnetic fields to accelerate and focus electrons to produce a 2D absorbance image on a fluorescent screen recorded by a charge-coupled device. Resolution of TEM depends on the wavelength of the electrons, giving resolutions that approach 1 nm and below. Tissue samples are stained with heavy metals to increase absorption contrast. Osmium, uranium, and gold are often used. Specific antibody stains including colloidal gold–antibody complexes are used to stain specific antigens. *Scanning electron microscopy* (SEM) produces an image of electrons from a conductive surface, usually sputter coated with a thin metal layer, stimulated by a raster scanning narrow electron beam. Scanning a specimen with a physical probe rather than a beam of light or electrons is called *scanning probe microscopy* (SPM). Many variations of SPM have been developed that give very high resolution imaging of atomic structures. Scanning tunneling microscopy and atomic force microscopy are two of the more familiar techniques used in oxidative stress research.

Endoscopy has transformed diagnosis and surgery, with an endoscopic instrument available for almost every organ system, even the eye. Diagnostic ultrasound and OCT catheters are becoming available, along with numerous tissue sampling techniques and treatment techniques. Moore's law is at work, with steady improvements in imaging systems and miniaturization.

X-ray projection (plain) radiology was the only imaging system for internal organs for many years, and is still widely used in clinical medicine, primarily in instances when there is good contrast, such as bone lesions, lung tissue lesions, and fluid levels. Contrast material is use to increase the range of usefulness of plain films. Barium is used to visualize the gastrointestinal tract, and iodinated compounds are used to visualize arteries (angiography), veins (venography), lymph vessels and nodes (lymphography), spinal cord (myelography), brain (ventriculography), bile and pancreatic ducts (cholangiography), and joints (arthrography). *X-ray computed tomography (CT)* allows for more detailed imaging by using computer analysis of X-ray images along multiple vectors. CT scan images may be reconstructed through computer processing to give 2D and 3D images. Iodinated contrast material is used intravenously with CT in a technique called *computed tomography angiography (CTA)*. X-ray diffraction techniques (*crystallography*) allow determination of molecular structures of a wide variety of complex molecules, including DNA. Dual-energy X-ray absorptiometry (*DEXA scan*) measures a bone density assessment to evaluate osteopenia and osteoporosis.

Nuclear magnetic resonance spectroscopy (NMRS) aids in identification of organic molecules, with techniques developed for protein and DNA analyses. *Magnetic resonance imaging (MRI)* uses computer analysis of large data sets of NMRS relaxation energy measurements along multiple vectors to construct images using a process similar to the one used in CT scanning. Techniques to improve image contrast have been found using differences in relaxation times of various tissue types. Contrast material is also used to enhance images in MRI. Chelated gadolinium, iron oxide nanoparticles, chemical exchange saturation transfer (CEST) compounds, and mangafodipir (containing paramagnetic Mn^{2+}) can be used. To obtain images of vessels, a technique called magnetic resonance angiography (*MRA*) is used with a contrast agent to make blood vessels stand out.

Electron paramagnetic resonance spectroscopy (electron spin resonance) detects unpaired electrons in a way much similar to NMRS. A free radical has an unpaired electron, so EPRS is used extensively in free radical chemistry research. In a strong magnetic field, a paramagnetic unpaired electron will align its magnetic spin moment in either a positive parallel or negative (anti-) parallel direction, creating an energy state difference between each alignment, the strength of which is proportional to the strength of the magnetic field. Absorption or emission of photons is measured to detect energy transformations between the two energy states.

Carbon-13 (^{13}C) isotopes are commercially available in a wide variety of compounds for a nonradioactive method of labeling carbon atoms in reactions. Because ^{12}C is more abundant (99%) and ^{13}C is less abundant (1%), mass spectrometry can be used to identify where a labeled carbon (^{13}C) is located in a reaction product. Unlike the abundant ^{12}C that has a spin quantum number of 0 and cannot be detected with NMRS, ^{13}C is paramagnetic with a net spin of 1/2. Both NMRS and NMI (2D and 3D) can be done.

Radioactive isotope imaging (scintigraphy)

Radioactive isotopes are used in laboratory investigation in a wide range of applications to label molecules in cellular structures and biochemical reactions. This allows for imaging of specific metabolic activities rather than just tissue anatomy. *Nuclear medicine* uses radioactive isotopes to detect, measure, or image by measuring α, β, or γ emissions from the radioactive decays. *Radioactive thyroid uptake scans* image the amount of iodine the thyroid gland concentrates by measuring γ radiation emission after intravenous radioactive iodine injection. Uptake of iodine in thyroid nodules, which are detected by ultrasound, can be quantified, helping to differentiate between benign and cancerous nodules. The technetium isotope ^{99m}Tc is a metastable isotope that emits γ radiation with a half-life of 6 h. ^{99m}Tc is used to label a variety of compounds for *scintigraphy* of various organs. An increase in the uptake of phosphate labeled with ^{99m}Tc indicates an area of increased osteoblast metabolism in *bone scans*, allowing detection of fractures, inflammation, and metastatic cancers. ^{99m}Tc-labeled compounds that are taken up by the liver and excreted into the biliary tract are used for imaging the liver, biliary tract, and gall bladder in *cholescintigraphy*, helping in the diagnosis of cholecystitis and congenital biliary anatomical anomalies. *Parathyroid scintigraphy* with ^{99m}Tc-labeled complexes can identify an active parathyroid adenoma, allowing selective excision and retaining uninvolved glands. ^{99m}Tc or γ-emitting ^{133}Xe are used in *pulmonary scintigraphy* to evaluate ventilation and perfusion defects, helping to diagnose pulmonary embolism and other anomalous pulmonary perfusion defects. ^{99m}Tc-labeled dimercaptosuccinic acid or 1,1-ethylenedicysteine is concentrated in the renal cortex, allowing *renal scintigraphy* to identify anomalous or diseased areas of the kidney, ureters, and bladder.

Single-photon emission computed tomography (SPECT)

Rotating the γ detector around the patient to acquire images from varying vectors allows computer analysis similar to X-ray CT scan and MRI technology, and can produce a 3D image. *SPECT bone* scan produces a 3D image, allowing better localization of a defect. Cardiac *SPECT scans* use ^{201}Tl-(thallium) or ^{99m}Tc-labeled compounds to detect perfusion defects and help to differentiate between infarction and ischemia. *Brain SPECT scans* use ^{99m}Tc-labeled compounds to measure perfusion defects in the brain, helping in differentiation of stroke and dementia. Gallium $^{67}Ga^{3+}$ is concentrated in areas of inflammation and rapid cell growth, so *gallium scans* using either plain scintigraphy or SPECT technology can identify areas of inflammation, infection, and tumor growth.

Current SPECT scanners do not have very high spatial resolution, but are widely available and relatively inexpensive. The problem with resolution is the lack of a technology to focus these high-energy γ photons. Collimators are used which, by blocking all photons not traveling parallel, do give a type of focus that greatly reduces the signal and therefore sensitivity. In research investigation on animal models, higher levels of radiation are tolerated, so pinhole collimators that reduce signal strength dramatically but allow very good resolution are available. Positron emission radionuclides can be detected without collimation in *positron emission tomography (PET)*. The γ-emitting radioisotopes used in scintigraphy are, in general, heavier elements, relatively inexpensive, and have half-lives that are long enough to be commercially available, except for ^{99m}Tc, which is generated from the more stable isotope ^{99}Mo. Positron-emitting isotopes are, in general, lighter, more expensive, and short lived. No collimator is needed in PET, for focusing the image because γ radiation is detected, not from the tissue, but from the position of the positron when positron–electron annihilation occurs, giving a very small localization error but high sensitivity. With annihilation, two photons of γ energy are released in approximately opposite directions, which give vector information, in contrast to single photon emission. This allows for a greater sensitivity of 100–1000×, with better signal-to-noise ratio and higher resolution. There are positron-emitting isotopes of oxygen, carbon, nitrogen, and fluorine, allowing for easier labeling of specific biological molecules. The shorter half-lives of PET isotopes allow a higher initial dose, shorter time sequence intervals for dynamic studies, and shorter intervals for repeat studies. Glucose, labeled with ^{18}F as fluorodeoxyglucose, is able to pass freely through the blood–brain barrier and concentrates at areas of metabolic activity, allowing brain imaging and cancer detection. *Brain PET* is useful in evaluating stroke, tumor, and neurodegenerations, and has been used extensively in brain research because of the ability to measure functional changes and to identify various neuroreceptors using specific ^{18}F-labeled neurotransmitters. An *Oncology PET* is widely used in cancer diagnosis, staging and evaluating treatment effects. The fluorodeoxyglucose is rapidly taken up by growing cancer cells, enzymatically phosphated, but not metabolized because of the fluorine substitution for an oxygen, so that the radioactive label is highly concentrated in the cancer cell, allowing sensitive detection of metastatic tumors. The 110 min half-life of ^{18}F allows time for transport from a cyclotron facility to the PET facility, but the shorter half-lives of ^{11}C, ^{13}N, and ^{15}O require that the PET scanner be located close to the cyclotron. Rubidium is metabolized as if it were potassium, allowing rapid uptake of intravenously injected positron emitter ^{82}Rb by myocardium, indicating areas of normal and deficient perfusion. The problem with the short halve-life of ^{82}Rb (75 s) has been overcome by an ^{82}Rb generator using ^{82}Sr (strontium), which decays to ^{82}Rb by electron capture with a half-life of 25 days. *Cardiac PET* scans using ^{82}Rb chloride are more sensitive than are cardiac SPECT scans, but are more expensive and not as widely available.

Multiple choice questions

1 SPECT scan technology using γ emission has resolution limited mainly by
   a. The dose of the radionuclide given
   b. The quality of the γ detectors used
   c. The lack of technology to focus γ radiation
   d. The duration of the scan

2 In evaluating a patient for metastatic cancer, the most important first step is
  a. History and physical exam
  b. PET scan using ^{18}F-labeled fluorodeoxyglucose
  c. Gallium (^{67}Ga^{3+}) SPECT scan
  d. SPECT bone scan

3 The stable isotope ^{13}C can be detected by
  a. OCT
  b. Scintigraphy
  c. Nuclear magnetic resonance spectroscopy
  d. Transmission electron microscopy

4 The short half-lives of ^{11}C, ^{13}N, and ^{15}O require PET scans using these radionuclides be
  a. Very fast scanners
  b. Near the cyclotron used to make these isotopes
  c. Used only early in the week
  d. Collimated

References

1 Marmor, M.F., Fulton, A.B., Holder, G.E. *et al.* (2009) ISCEV Standard for full-field clinical electroretinography (2008 update). *Documenta Ophthalmologica*, 118, 69–77.
2 Rahman, M.T., Uddin, M.S., Sultana, R. *et al.* (2013) Polymerase chain reaction (PCR): a short review. *AKMMC Journal*, 4(1), 30–36.
3 Garrett, T.J., Menger, R.F., Dawson, W.W. *et al.* (2011) Lipid analysis of flat-mounted eye tissue by imaging mass spectrometry with identification of contaminants in preservation. *Analytical and Bioanalytical Chemistry*, 401(1), 103–113.
4 Lee, J., Kim, J.G., Mahon, S.B. *et al.* (2014) Noninvasive optical cytochrome c oxidase redox state measurements using diffuse optical spectroscopy. *Journal of Biomedical Optics*, 19(5), 055001.

SECTION II

Clinical correlations on acute and chronic diseases

CHAPTER 3

Free radicals: their role in brain function and dysfunction

Natan Gadoth[1,2]

[1] *Department of Neurology, Mayanei HaYeshua Medical Center, Bnei Brak, Israel*
[2] *The Sackler Faculty of Medicine, Tel Aviv University, Tel-Aviv, Israel*

THEMATIC SUMMARY BOX

At the end of this chapter, students should be able to :

- Describe the main features of the disease entities covered in the chapter including possible etiology and pathophysiology

- Recognize and summarize the contribution of oxidative stress to nerve cell function

- Outline the similarities in the mechanisms responsible for OS damage in various brain disorders described in the chapter

- Review the role of age in OS

- Review the therapeutic and practical opportunities of antioxidant treatment of the described CNS disorders

---

Introduction

The central nervous system (CNS), which is depended on oxygen for its continuous normal function, is protected from damaging oxidation by an endogenous antioxidant defense system, that is, enzymatic and nonenzymatic antioxidants. Glutathione, uric acid, and nicotinamide adenine dinucleotide phosphate (NADP) are endogenous while vitamin C (ascorbic acid) and vitamin E (α-tocopherol) are the major exogenous antioxidants. When an imbalance between prooxidant and antioxidant factors is present, a state of oxidative stress (OS) is formed. This can lead to generation of reactive oxygen species (ROS) and other electrophiles that are capable of either supporting or damaging cellular functions.

The brain, which requires 4×10^{12} molecules of adenosine triphosphate (ATP) each minute for its normal function, consumes therefore 20% of the inspired oxygen

---

Oxidative Stress and Antioxidant Protection: The Science of Free Radical Biology and Disease, First Edition.
Edited by Donald Armstrong and Robert D. Stratton.
© 2016 John Wiley & Sons, Inc. Published 2016 by John Wiley & Sons, Inc.

while its weight is only 2% of the total body weight. This vital high turnover of oxygen necessary for normal brain function makes it vulnerable to the interruption of oxygen supply and/or inhibition of mitochondrial production of ATP. About 5% of the oxygen consumed by cells is reduced to form ROS; however, the high oxygen consumption of the brain relatively to other organs may result in an increased amount of ROS produced. Considering the potential damage that ROS may cause by its deleterious effect on apoptosis and on neural membranes, which are particularly rich in polyunsaturated fatty acids (PUFAs), may suggest that they play a role in "normal" aging and disease-related neurodegeneration. In addition, the antioxidant defense of the brain is relatively reduced due to relatively low levels of the antioxidants such as catalase, glutathione peroxidase (GPX), and vitamin E.

The beneficial role of oxidative stress in the brain

There are "*good*" and "*bad*" ROS. All good ROS are by-products of the turnover in the mitochondrial respiratory chain. In brain tissue, ROS are generated by microglia and astrocytes and modulate synaptic and nonsynaptic communication between neurons and glial cells. ROS also interfere with increased neuronal activity by modifying the myelin basic protein, thus altering its conduction ability and possible induction of synaptic long-term potentiation. Furthermore, results from animal models suggest that the role of O_2 in modulating synaptic plasticity and potentiation is altered by age. ROS are also involved in central control of food intake and are required for hypothalamic osmoregulation.

The harmful role of oxidative stress in the brain

The deleterious effects on the brain associated with ROS appear when those reactive compounds are in excess and overcome the relatively feeble antioxidant protective mechanism of the brain. The high concentration of PUFAs (which are rich in double bonds and are responsible for their susceptibility to peroxidation) and the high concentration of brain iron, which may act as a prooxidant, are the main reasons for the vulnerability of the brain to assault by ROS. As a result of lipid peroxidation, further potential damage can be caused by the formation of toxic by-products such as reactive aldehydes, which may enhance neuronal apoptosis (programmed cell death). An additional damaging effect of ROS on brain is their potential ability to cause oxidative modification of DNA with secondary endonuclease-mediated DNA fragmentation, which leads to DNA damage.

Brain mitochondria, the energy source for essential brain activity, are targeted by ROS. Following brain ischemia, the mitochondria will take up calcium in the form of Ca^{2+}, resulting in increased production of ROS by a poorly understood mechanism(s). The increase ROS production results in increased permeability of the mitochondrial membrane followed by a cascade of events starting with Ca^{2+} which is followed by further influx of K^+. The end result is an increase in the osmolarity inside the mitochondria resulting in entry of water (osmotic swelling) and finally a bioenergetic neural tissue failure.

The role of oxidative stress (OS) in programmed neuronal death (apoptosis)

Programmed cell (neuronal) death is an actively regulated mechanism of cell death, which plays a major role in maintenance of normal tissue growth. The intracellular machinery responsible for apoptosis depends on a family of proteases that have cysteine at their active site and cleave their target proteins at specific aspartic acid site, thus they are called caspases (**C**ysteine-dependent **ASP**artate-directed prote**ASE**s). Caspases, once activated, amplify a proteolytic cascade that leads to apoptotic cell death. A variety of ROS such as the prooxidants H_2O_2, among others, induce apoptosis while antioxidants such as N-acetylcysteine (NAC) suppress apoptosis. Thus, apoptosis is strongly associated with normal mitochondrial function, in particular changes in electron transport and altered mitochondrial oxidation–reduction. Neuronal firing is highly dependent on normal mitochondrial function; thus, the mitochondrion is highly susceptible to OS via increased apoptosis.

Oxidative stress in neonatal hypoxic-ischemic encephalopathy (HIE)

Asphyxia is a term frequently used to describe the state of severe lack of oxygen due to abnormal breathing. This state is most frequent in newborns. A total of 0.5–3% of newborns worldwide will suffer from birth asphyxia severe enough to require resuscitation. Out of those, about a million will die and a similar number will be left with significant neurobehavioral disability. Many of those infants will suffer from anoxic brain damage manifested by a variety of neurological abnormal findings, seizures, and altered state of consciousness, a combination defined as "encephalopathy." Since ischemia (poor blood supply) and hypoxia (poor oxygen supply) are linked and are concomitantly present in asphyxiated newborns, the term hypoxic-ischemic encephalopathy (HIE) was coined.

Free radicals (FRs) play a significant role in induction of brain damage in newborns, in particular those who are born prematurely. The developing brain and especially the brain of premature newborns is prone to the damaging action of FRs because the rapidly developing nervous system requires constant and increased supply of oxygen to enable its enhanced aerobic metabolism to meet the enormous energy demands of the primitive brain. Unfortunately, prematurity is associated with insufficient homeostatic mechanisms and lack of an adequate and efficient antioxidant system, which matures only during the first year of life. In newborns, the concentrations of the major endogenous antioxidants, such as the enzymes GPX and superoxide dismutase (SOD), and the exogenous antioxidants, such as vitamins E and C, are reduced in the plasma and the red blood cells. Other precipitating factors are the increase in the release of free iron, increase in the production of superoxide radicals, and increase in the concentration of cell membrane PUFAs, resulting in enhanced susceptibility to OS.

The net result is a high susceptibility of full term and especially premature newborns to the increased production and damaging effect of FRs resulting in the so-called free radical-related disorder, which consists of underdeveloped bronchi and lungs (bronchopulmonary dysplasia), damage to the retina with accompanied blindness (retinopathy of prematurity), severe intestinal necrosis (necrotizing enterocolitis), kidney failure, and bleeding into the brain ventricles.

The events that lead to anoxic irreversible brain damage start with cellular energy crisis manifested by reduced formation of the adenosine (ATP, ADP, AMP) and phosphocreatine compounds. This reduction of the cellular "fuel" impairs cellular vital organelle functions. In addition, the essential process of oxidative phosphorylation is blocked and replaced by inadequate and short-lived anaerobic glycolysis. The very high demand of the immature developing brain for ATP cannot be met by this "rescue" metabolic pathway for more than 2–3 min (in rats). The vital role of ATP in keeping K^+ inside and Na^+ outside the cell (the Na/K pump) is no longer active, Na^+ followed by Cl^- leak into the cell, increasing its osmolality and leading to increased cellular water in an attempt to lower it. The accumulation of water within the cell results in cellular edema. This cascade of events leads to an increased production of FR with enhanced OS, peroxidation of the PUFAs within the cell membrane, and intracellular acidosis when the production of ATP is not rapidly reinstituted. The neurons and glial cell membranes lose their normal ion gradients leading to massive depolarization and activation of the presynaptic voltage-dependent calcium channels, yielding the release of toxic excitatory amino acids, such as aspartate and glutamate, into the extracellular space. The vital reuptake of neurotransmitters by the presynaptic receptors is also blocked. Thus, cell death is facilitated by excitatory overload that, in turn, leads to activation of lipases, proteases, and caspases, which are responsible for the final degradation of the cytoskeleton and cell membrane.

Time to therapeutic intervention will determine the final outcome of the hypoxic brain cells. Without timely successful treatment, cell necrosis will occur and the surviving cells will eventually die from delayed apoptosis when hypoxia is not completely reversed (Figure 3.1). The best form of therapy is reversal of ischemia-hypoxia within the time window of several minutes to avoid the devastating cascade of events outlined earlier. It was shown that the best way to do so is to supply the asphyxiated newborn with sufficient ambient air coupled with cooling and reducing body and brain temperature by several centigrade for up to 7–8 h. This quite recent combined approach resulted in an improved salvage of the hypoxic brain functions with better neurobehavioral outcome.

OS in inflammatory brain disease due to infection

The two common forms of CNS infection causing an inflammatory reaction in the brain are meningitis where the pathogen invades the meninges and encephalitis when the pathogen invades the brain. Many of the infectious processes will start in the meninges and subsequently invade the brain (meningoencephalitis). While infectious meningitis presents with fever, headache, and nuchal rigidity (stiff neck), encephalitis presents with reduced and impaired consciousness followed by seizures and neurological abnormalities in the form of motor and sensory deficits. In the state of health, the brain is protected from invasion of macromolecules, microorganisms, and malignant cells by the blood–brain barrier (BBB), which is a specialized

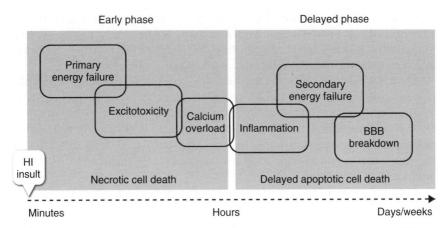

Figure 3.1 The early and late cascade of events leading to nerve cell death in neonatal HIE.

neurovascular structure with selective permeability. Infectious agents will invade the brain only if they are able to penetrate ("break") the BBB or when the BBB is damaged as happens, for example, in severe head trauma. While the majority of bacterial infections of the CNS present as meningitis, viral infections take the form of meningoencephalitis. In addition to bacteria and viruses, the CNS can be invaded by fungi and parasites, which cause a more protracted disease course with the formation of focal inflammation (cerebritis) and somewhat later brain abscess, which is a localized mass of pus surrounded by a thick fibrous membrane that grows slowly, producing marked local edema and increased intracranial pressured similar to the effect of a brain tumor.

A number of CNS resident cells and especially microglia are able to form ROS and reactive nitrogen species (RNS), a family of molecules with antimicrobial properties derived from nitric oxide (NO) and superoxide (O_2^-) that are produced via the enzymatic activity of inducible nitric oxide synthase 2 (NOS2) and nicotinamide adenine dinucleotide phosphate (NADPH) oxidase, respectively, in response to certain invading molecules. Indeed, when viruses such as herpes simplex invade the brain to cause one of the most lethal forms of viral meningoencephalitis, a rapid but low-level production of NO and its reactive products NO_2, CO_2^-, and $ONOO^-$ (peroxynitrite) are induced. These compounds play an important role in delivering immune effectors across the BBB because of their vasodilatory effect, which enhances the adherence of immune inflammatory cells such as T-lymphocytes onto the neurovasculature bed of the brain. In addition, when microglia are activated by cytokines derived from T-lymphocytes during antigen-specific inflammatory responses, the production of NO is dramatically enhanced. These ROS are considered the key antiviral agents in a variety of CNS viral encephalitis types. The antiviral activity of these radicals is manifested by inducing phagocytosis. The dual actions of radicals in permitting immune effectors to cross the BBB as well as the direct damage they cause to neurotrophic pathogens may be necessary for eliminating the infectious agent from the CNS.

The primary defense line against invasion of infectious agents are the white blood cells (leukocytes) including polymorphonuclear (PMN) neutrophils, and monocytes, which are the main cell types accumulating at the site of infection. Those phagocytic cells respond to the presence of the invading infectious agent by producing large amounts of oxygen-based radical species. In the CNS where the capacity of neurons to regenerate is quite limited, the benefit of the ability of FRs to help in eradicating infectious invaders outweighs their nonspecific tissue destructive capacity.

The PMN cells within the inflamed CNS region express high levels of three key enzymes involved in the production of radicals. Those are Nox2 isoform of NADPH oxidase that catalyzes the formation of superoxide (O_2^-); the inducible isoform of NOS2 that is responsible for the production of high level of NO, and myeloperoxidase that catalyzes the production of hypochlorous acid (HClO) from H_2O_2. The dual expression of Nox and NOS2 in PMN cells results in the formation of $ONOO^-$ from nitric acid and superoxide.

Certain cytokines play a role in the induction of NOS2 and Nox2 in PMN and monocyte. The principal cytokine is interferon γ (IFN-γ) produced mainly by the activated CD4 Th1 cells, CD8 T cells, and natural killer (NK) cells. Other cytokines, which include interleukin 1β, tumor necrosis factor-α, and interleukin 6 are produced by a variety of cells such as microglia, play an important role in modulation of the expression and activity of the enzymes mentioned earlier. Microglia are bone marrow-derived cells with high-level expression of Nox2 and NOS2, enabling them to produce significant amounts of NO, O_2^-, and their radical products. While astrocytes express a number of enzymes involved in the production of ROS and RNS, neurons express NOS1 only. The endothelial cells that are vital for the normal function of the BBB express mainly NOS3.

Out of this variety of FRs, NO, and its reactive products are the main antiviral protective agents in the CNS. In addition, they are capable of increasing the BBB permeability, enabling entrance of immune effectors that facilitates the clearing of the invading infectious agent from the CNS.

OS in neuroimmunological disorders

Multiple sclerosis (MS) is a chronic autoimmune demyelinating disease of the brain and spinal cord affecting mainly young adults. The commonest form of MS known as relapsing–remitting MS manifests with recurrent attacks of motor and sensory acute impairment due to focal or multifocal immune-mediated inflammatory damage to the central (brain and spinal cord) white matter combined with axonal damage. About 2,500,000 people around the world suffer from MS. Females are more likely to be affected. The prevalence (the number at a certain time of people with a particular condition, usually measured as cases per 100,000) of the disease varies in different parts of the world being as high as 154/100,000 in Scandinavia and as low as 1.5/100,000 in Southeast Asia. The typical neuropathological finding is the white matter MS "plaque." Plaques are spread within the white matter, frequently adjacent to the ventricular borders, and within the corpus callosum, a thick bundle of myelinated fibers connecting both hemispheres. In the acute phase of plaque formation, activated monocytes, lymphocytes, microglia, and macrophages destroy myelin and to a variable degree, oligodendrocytes. The macrophages phgocyte the degraded myelin, at first myelin fragments, somewhat later myelin proteins, and finally myelin lipids. This evolution takes a few weeks, after which a process of scarring is established. The final stage is called gliosis because it consists mainly of glial cells. The chronic plaque contains foci of BBB leakage, destroyed myelin sheets, dying oligodendrocytes (the myelin forming glial cells in the CNS), damaged and dying axons, scar fibrous tissue, and inflammatory cellular infiltrates especially monocyte-driven macrophages, which secrete cytokines, chemokines, NO, and ROS in response to inflammation. It is now well established that ROS-induced OS contribute to the damage caused by MS. Similar to their effect on infectious inflammation, BBB permeability is increased during the early phase of plaque formation due to the local damaging effect of ROS, resulting in migration of monocytes across the injured BBB. ROS contribute to the persistence of the plaque by inducing myelin phagocytosis accompanied by damage to oligodendrocytes and axons. The activated microglia and the leukocytes produce a large amount of ROS and RNS, which were shown to mediate demyelination and axonal damage in MS brains and in the brain of animals with experimental autoimmune encephalomyelitis (EAE), the classical experimental model of MS. There is also additional noninflammatory mitochondrial damage in MS believed to be caused by the increased energy demand of the axons in response to demyelination. The observed reduced concentration of antioxidants in the serum of patients with MS and within the MS plaque is considered as the result of an increase in the level of ROS.

Oligodendrocytes are the myelin-forming cells and thus, the main contributors to the existence of the central white matter, are rich in iron, a potential potent contributor to OS. These cells are, therefore, prone to oxidative damage, a risk that is enhanced by their reduced levels of the antioxidants such as superoxide dismutase, glutathione, and glutathione peroxide.

From the observations mentioned above, it is evident that ROS contribute to the persistence of MS lesions, a fact which calls for antioxidant treatment. Unfortunately such treatment, which was found beneficial in experimental models of MS, cannot be used in patients with MS due to the need to administer an extremely large amount of available exogenous antioxidants in order to achieve adequate antioxidant levels in the brain and spinal cord.

OS in cerebrovascular disease

Stroke also known as cerebrovascular accident (CVA) is the result of disruption of cerebral blood flow (CBF) with concomitant brain tissue damage. Normal cerebral circulation and cerebral blood perfusion may be impaired because of a long list of heterogeneous disorders; thus, one should refer to stroke as a syndrome rather than a single disease. Of the various etiologies, cerebrovascular disease is the most common cause of stroke while hypertension, hyperlipidemia, and diabetes are important risk factors. Age, male gender, obesity, and smoking are additional risk factors. Stroke is the third leading cause of death and a significant contributor to acute and chronic disabilities in adults. Indeed, 20% of people afflicted with stroke will die during its acute phase. The acute neurological impairment is progressive during the first hours in many cases, taking the form of motor, sensory, cognitive, and autonomic dysfunction, which are dictated by the brain area supplied by a particular artery. For example, when the basilar artery (which supplies the brain stem) lumen is blocked by a blood clot formed locally or flipped from the arterial blood supply of the brain stem, the affected patient suffers loss of consciousness severe enough to cause a state of coma accompanied by the failure of spontaneous breathing, swallowing, voluntary, and reflex eye movements as well as motor–sensory functions. There may also be erratic heart rate, brittle (jumpy) blood pressure, and impairment of one or more sensory modalities. These signs are the result of ischemia and the subsequent anoxia of the vital neural structures located or passing via the brain stem from the lower parts of the brain (cerebellum and medulla oblongata) onto the cortex, spinal cord, and vice versa. The vertebral arteries supply the basilar artery and are vital parts of the so-called posterior circulation. When the lumen gets narrowed due to arterial spasm, as may happen in children and young adults with a form of migraine called "basilar artery migraine," the blood supply becomes suddenly insufficient causing the already mentioned signs. In both situations, if the blood supply is not rapidly reestablished, the mentioned neural structures will suffer irreversible damage in the form of prolonged coma, severe functional disability, and even death. Other notable forms of brain ischemia are conditions that impair CBF by either slow pulse, low blood pressure, or the combination of both, a frequent result of cardiac failure. Other quite common conditions that lead to insufficient CBF and cerebral ischemia are loss of blood volume from either massive bleeding or dehydration.

The main causes of generalized hypoxia are due to lack of environmental oxygen, respiratory failure, or a reduced capacity of blood hemoglobin to bind and carry oxygen to the brain such as in carbon monoxide or cyanide poisoning. All of these are responsible for severe brain hypoxia, leading to a devastating cellular energy failure and cessation of vital brain functions.

As mentioned earlier in this chapter, brain anoxia and ischemia are associated with an increased production of FRs associated with OS, which are harmful to the vulnerable brain due to its high content of lipids, high energy requirement, and high oxygen consumption. Additional damage is caused by secondary inflammatory changes at the site of stroke because of the expression of several proinflammatory genes and the release of inflammatory mediators accompanied by migration of PMN and monocytes/phagocytes, which add to the OS in a similar mechanism that was described earlier for infectious inflammation. In cases where the stroke becomes hemorrhagic or is caused by primary intracerebral bleed (15% of strokes), the presence of iron from the breakdown of hemoglobin adds to the production of FRs, resulting in an increase in OS. Free iron can enhance the production of radicals via the Fenton reaction in which organic substrates are oxidized by Fe^{2+} and reduced by Fe^{3+} according to the following reactions:

1 $Fe^{2+} + H_2O_2 \rightarrow Fe^{3+} + OH^- + OH^*$

2 $Fe^{3+} + H_2O_2 \rightarrow Fe^{2+} + HO_2^* + H^+$

As described earlier regarding infectious inflammation, brain ischemia and consequent anoxia lead to the cessation of ATP production, which is dependent on adequate supply of oxygen and glucose, both deprived by brain ischemia. This leads to the inability of the cells to maintain their membrane-resting potential because of the failure of the energy-dependent Na^+/K^+ ATPase system, which is responsible for about one-third of the total intracellular energy generation. Indeed, after 2 min of the onset of stroke, neurons and glial cells depolarize as they lose their membrane potential, a process that is associated with harmful enhanced release of excitatory amino acids, as was previously noted.

Another source of damage caused by ischemia is the production of NO. Studies in experimental animals have documented the role of NO in a variety of neuronal functions including learning and memory processes, cortical arousal, nociception (pain perception), food intake, blood vessel dilatation, and immune response. The wide range of cell types affected by NO implies that it acts as a neurotransmitter. NO is synthesized in the brain upon demand, in response to the activation of N-methyl-D-aspartate (NMDA) receptors by the excitatory amino acid glutamate. Within the neuron, NO is a coproduct of the conversion of the semi-essential amino acid L-arginine to L-citrulline by the enzyme nitric oxide synthase (NOS) with calcium and calmodulin (a calcium-binding messenger protein) as cofactors. A group of NOSs is classified according to its tissue domain such as neuronal nitric oxide synthase (nNOS, aka NOS1), endothelial nitric oxide synthase (eNOS, aka NOS3), and inducible nitric oxide synthase (iNOS, aka NOS2) by proinflammatory cytokines. The synthesis of NO by eNOS during the early phase of stroke may play a protective role by causing vasodilatation and an increase in local blood supply while later in the course the accelerated synthesis of NO by nNOS and iNOS may facilitate cell death.

An additional contributing factor to brain damage caused by stroke is an oxidative modification and/or DNA fragmentation mediated by the enzyme endonuclease. These processes are active during the early phase of stroke.

OS in traumatic brain injury

Traumatic brain injury (TBI) is a major cause of physical disability, mental impairment, and significant mortality. TBI affects mainly young and elderly people and if they survive the initial trauma, then they may continue their life as disabled individuals. In the United States, the death rate from TBI in 2010 was 25.4/100,000 for men and 9.0/100,000 for women. The soft and relatively mobile brain is partly tethered to the rigid hard skull by its continuity with the spinal cord across the foramen magnum at the base of the skull and the nerves and blood vessels that either enter or leave the skull by bony apertures called "foramina." The quite rigid fibrous tissue that surrounds and protects the brain is called the dura mater (Latin – hard mother [of the brain]), which is folded into two leafs. The outer leaf that coats the inner bony skull is separated from the inner leaf that is closer to the brain by a very narrow fluid filled space. The largest meningeal arteries stem from the external carotid artery and ramify in the dura mater adjacent to the thin parts of temporal bone, which makes them prone to local skull trauma such as a depressed skull fracture where the sharp edges of the broken bone can injure the underlying meningeal artery that can bleed forcibly resulting in a rapidly expanding epidural hematoma compressing the brain, causing local compressive damage as well as life-threatening increased intracranial pressure, which may be fatal if the hematoma is not surgically evacuated promptly.

The veins that drain the blood from the brain into venous slits located between the dural leafs (called brain sinuses) may also be stretched and may rupture by the sudden unequal acceleration of the skull and brain detailed below, causing a slower forming bleed resulting in the formation of an acute and often chronic subdural hematoma. In addition to either blunt or open (when the traumatic agent penetrates the skull) brain and/or spinal injury, both epidural and subdural hematomas are common serious complication of CNS trauma. Another form of traumatic bleed within the skull is subarachnoid hemorrhage (SAH), which is confined to the space between the brain surface and a delicate mushy serous membrane which contains the cerebrospinal fluid and the major blood vessels.

The difference between the low mass of the soft brain and that of the heavy skull is responsible for another form of brain injury known as "coup contrecoup injury." The focally traumatized skull is responsible for localized brain injury adjacent to the traumatic impact (coup). Subsequently, the slower acceleration of the nervous tissue will lead to a traumatic lesion on the side opposite to the area that was impacted (contrecoup).

Another serious type of TBI is diffuse axonal injury (DAI), which is caused the shearing forces acting on the brain as the result of the brain's acceleration–deceleration detailed earlier associated with axonal stretching, damaging this delicate neuronal extension. The victims of such an injury suffer rapid and prolonged loss of consciousness, often without visible signs of significant head trauma and frequently without abnormal focal neurological signs. Early brain imaging by computerized tomography may appear normal or show generalized nonspecific brain edema while MRI will reveal multiple small bleeds. A repeated study after weeks or months will show generalized brain atrophy associated with severe neurobehavioral impairment. Many of these patients will stay in a "chronic vegetative state" or "minimal conscious state" for the rest of their life. At the site of brain trauma, many of the directly damaged cells will subsequently die. Those adjacent to the center of the impact may survive the primary injury, but will be subjected to secondary degeneration.

OS is formed almost immediately after TBI and may be the cause of mitochondrial dysfunction observed about 30 min later and sustained for at least 72 h. In the mouse model, focal TBI resulted in progressive mitochondrial damage in the form of swelling and breakage of its outer membrane lasting 3–12 h post-trauma. It was also shown that NO radicals act with superoxide radicals to form $ONOO^-$, an RNS involved in post-traumatic mitochondrial dysfunction.

Lipid peroxidation due to excessive ROS formation plays a major role in secondary CNS damage following TBI. It was shown in the rodent experimental models that following TBI, there is an increased release of glutamate as well as activation of the NMDA receptors that lead to intracellular calcium toxic excess and subsequent activation of phospholipases. The latter, in addition to the ROS formed at the site of the trauma, attack the multiple double bonds at the side chains of the PUFAs within the cell membrane. In between those double bonds lie methylene bridges ($-CH2-$) that possess reactive hydrogens, which further form lipoperoxide radicals that are unstable and attack adjacent PUFAs. This cascade of events, called "free-radical chain reaction", leads to the production and accumulation of lipid peroxides, which alter membrane permeability and cause oxidation of membrane proteins. This results in the entry of Ca^{+2} into the cell, which initiates a series of reactions starting with the activation of excitatory glutamate receptors that contribute to cell death. In addition, Ca^{+2} activates nNOS to produce excess NO that accumulates in the brain immediately after brain trauma, while additional NO is produced by eNOS that is also responsible for an additional peak of intracellular NO some days later. As mentioned in the introduction, the brain is equipped with an inadequate intrinsic antioxidant defense system. One part of this system is low molecular weight antioxidants (LMWAs), which can be measured, and reflects the ability to neutralized ROS. Measurements of temporal changes in the levels of LMWA in traumatized brains of rats have shown that the animals who were able to mobilize higher levels of antioxidants within minutes of TBI had a better recovery of brain functions. Interestingly, experimental TBI was associated with a total body increase in OS as indicated by decreased levels of LMWA in the heart, lungs, kidneys and liver observed 60 min following brain trauma. Other intrinsic antioxidants that play a role in the defense from OS induced by TBI are GPX, heme oxygenase (HO), and xanthine oxygenase (XO) as was mentioned earlier. TBI is frequently associated with cerebral hemorrhage and extravasation of blood via the damaged BBB. Heme release from the disintegrating red blood cells serves as a substrate for heme oxygenase to produce CO, Fe^{2+}, and bilirubin, adding important substances to the increased OS induced by brain trauma.

OS in neurodegenerative disorders

This is a heterogeneous group of brain disorders characterized by premature degeneration of neurons. The most common of these are known as sporadic (not autosomal dominant) Alzheimer's disease and Parkinson's disease (AD and PD, respectively). Both disorders affect aged people. In AD a progressive cognitive decline (dementia) is the hallmark of the disease, whereas PD is a chronic progressive disorder of motor control and, to a lesser extent, sensory, cognitive, and emotional impairment. Interestingly, there are patients with a combination of clinical and laboratory features of both disorders, that is, PD with dementia and AD with parkinsonism, suggesting that a common pathological process due to an unknown cause is shared by both diseases.

Alzheimer disease

AD known today as Dementia of Alzheimer type is quite frequent in the elderly affecting 1 out of 9 people aged 65 and older (11%) and a third of people aged 85 and older (32%). Currently, 5.2 million Americans suffer from AD. The disease is more frequent in women (two-thirds of the affected). The characteristic early clinical features consist of memory impairment for recent events such as remembering recent conversations and names of people and events. This is followed by impaired communicative skills, disorientation, confusion, poor judgment, and behavior and mood changes. At the advanced stage, the patients require help with all basic daily functions and need continuous close supervision. The characteristic brain pathology consists of neuronal accumulation of twisted strands of protein (tau [τ] protein) and protein fragments made of amyloid β (Aβ), the main constituent of the characteristic "amyloid plaques" located in between neurons. Amyloids are insoluble fibrous protein aggregates with a mutual structural core, which are derived from inappropriately folded versions of proteins and polypeptides normally present in the body. The plaques are composed of amyloid, amyloid precursor protein (APP), dystrophic neuronal extensions, activated microglia, and reactive astrocytes.

Tau protein is normally associated with microtubules that are vital components of the neuronal cytoskeleton made of a dimer of two globular proteins, α tubulin and β tubulin. Tau protein binds to microtubules promoting their assembly and providing stability to the neuron. The hyperphosphorylated form of this protein is the main constituent of the plaques and the characteristic neurofibrillary tangles within the neuron. The net consequence of those neurodegenerative alterations, both intra and extra neuronal, is impairment of the "cross talk" between brain cells.

From the gross structural point of view, the degenerative process in AD leads to brain atrophy especially of the hippocampus, a part of the limbic system, which is early involved in AD and plays an important role in the consolidation of information from short-term to long-term memory and spatial orientation, especially navigation. In addition, other brain regions that participate in a variety of cognitive functions are the frontal cortex, cerebellum, the deep nuclei of gray matter within each hemisphere (the basal ganglia), and the locus coeruleus, a small nucleus of adrenergic neurons located in the pons.

In recent years, the "oxidative stress hypothesis" of AD was introduced, gained scientific merit, and was supported by biochemical evidence. The essence of this theory was the vicious cycle model of continuous progressive brain degeneration due to OS. Studies on patients with AD and non-AD mild memory impairment disclosed higher levels of lipid peroxidation products in the CNS and peripheral tissues in both disorders concomitant with a decrease in the two major antioxidant enzymes GPX and SOD. The low levels of those important antioxidants could represent a state of "overconsumption" of those enzymes by the large load of ROS and RNS formed in the degenerating brain, exposing it to further OS and FR damage, which results in additional production of ROS and RNS. Heterogeneous sources of ROS, RNS, and FR such as malfunctioning mitochondria that will be most likely present with respiratory chain defect and the consequent formation of FRs with excess oxygen were suggested. Additional support for the mentioned hypothesis is the observation of increased redox-active sources, such as some transition metals, in particular iron, in the early stages of Alzheimer's disease. The Aβ extracellular deposits, typical for AD brain, can induce local inflammatory processes with activated microglia forming ROS similar to their response to invading infectious agent in meningoencephalitis or to the production of proinflammatory immune complexes such as in the acute phase of MS described above. In addition, OS can lead to intralysosomal induction of Aβ being indirectly involved in amyloid genesis. In addition, Aβ has the ability to destabilize lysosomal membranes, thus facilitating pathogenic macroautophagocytic processes.

Could aging be a central causative factor in AD? It is well known that aging affects the cellular "power plants," the mitochondria, which play important roles in apoptosis. The process of cellular aging is associated with massive synthesis of FRs concomitant with abnormal alteration of mitochondrial function and damage to mitochondrial membrane integrity. The altered mitochondrial membrane is now exposed to excess FRs, which in turn "attack" the lipid-rich membrane PUFAs to produce additional lipid peroxide radicals. Thus, the mitochondria could be the "blue-collar worker" involved in the complicated etiopathology of AD.

OS in Parkinson disease (PD)

PD is the other neurodegenerative disease closely related to aging, feared of by elderly people, their families, and friends. PD is quite common, reaching a frequency of

0.1–0.5% among people aged 55–99 years. Indeed, 52% of people aged 85 and older suffer from PD. The characteristic clinical features of PD were first reported by James Parkinson in 1817 based on six patients, three of them observed by him while walking on the street. These are slowness of voluntary movements (bradykinesia), stiffness of muscles (rigidity), resting rhythmical tremor, and postural instability. The onset may be so mild that it is often unrecognized or assured by their doctor everything is "just old age." At that stage, they are frequently referred to an orthopedic surgeon and/ or a psychiatrist.

The disease is ultimately progressive leading to severe impairment of life quality not only due to motor symptoms but frequently due to associated sensory, mental, and cognitive deficits as PD is not only a form of motor impairment. In about 10% of the patients, genetic factors are present that can cause familial PD. Those patients have a relatively young age of onset, rapid clinical deterioration associated with drug treatment unresponsiveness. The disease is the result of progressive degeneration and death of about 400,000 neurons containing neuromelanin, which renders them a dark appearance at their location in the substantia nigra pars compacta (SNPC) (from Latin – the cell-packed black section) located in the midbrain (diencephalon). Those cells produce the neurotransmitter dopamine that travels along a neural circuit called the nigrostriatal pathway to reach its destination in the corpus striatum, which consists of the deeply situated "basal nuclei," that is, the tail-shaped caudate and the lentil-shaped lentiform that is divided into the putamen and small pale appearance medial part called globus pallidus. A complicated network of central and peripheral connections and the associated excitatory and inhibitory neurotransmitters is involved in the striatal network, which is responsible for maintaining smooth coordinated voluntary motor functions as well as emotional and cognitive abilities.

The clinical picture of PD is the result of a progressive decrease in dopaminergic activity in the nigrostriatal network. Most of the symptoms and signs may be completely or partially resolved by increasing brain DA. The etiology of this degenerative neuronal process is unknown. However, OS is considered a significant causative factor stemming from the metabolic reactions involved in synthesis of dihydroxyphenylalanine (DOPA) and dopamine (DA). The levo-stereoisomer of DOPA (L-DOPA; L-3,4-dihydroxyphenylalanine) is synthesized from the amino acid L-tyrosine by the enzyme tyrosine hydroxylase and further metabolized to its centrally active DA by either the DOPA-specific decarboxylase (DOPA decarboxylase) or the unspecific aromatic amino acid decarboxylase. A portion of the produced DA that is released into the synaptic gap in the striatum is extensively metabolized by the enzyme monoamine oxidase (MAO) located at the mitochondrial outer membrane. The "surviving" DA is stored in vesicles and then following appropriate stimuli is released into the synaptic gap. Reuptake of the "leftover" DA by the protein DA transporter (DAT) enables maintenance of a stable and constant availability of DA which is vital for normal voluntary movements.

Evidence for increased OS in brains of PD patients was derived from the presence in the SNPC of a decreased ratio of GSH/GSSG (glutathione and oxidized glutathione, respectively), a known marker of OS. Both DOPA and DA contain a catechol ring and are therefore named catecholamines. The catechol ring is a very reactive compound that can explain the role of DA and DOPA in the increased OS present in PD brains.

The complex metabolic pathways involved in synthesis and breakdown of DOPA and DA are the source of a variety of ROS and RNS with a potential to oxidize DNA to form oxidized nucleic acids, which may cause harmful mutations. These ROS can act on mitochondrial membranes to decrease the ability of those organs to supply adequate ATP required by the "overworked" aging nigrostriatal system. In addition, OS may impair intracellular protein folding, resulting in the formation of abnormal protein aggregates within the cytoplasm in the form of the acidophilic intracellular inclusions known by the name Lewy bodies. These are aggregates of α-synuclein, a cytosolic protein of nerve cells, and τ protein, which are present within neurons in the SNPC of patients with PD. The presence of such protein aggregates is the expression of protein misfolding, which impairs the neuronal cytoskeleton. This α-synuclein protein is also found in AD, a fact that further supports its role in neuronal degeneration.

Therapeutic implications and opportunities

The development and introduction of any drug that is supposed to act locally on the CNS is dependent on its ability to cross the BBB to reach its target. This obstacle is bypassed in CNS disorders, since many of them are associated with damaged BBB as mentioned earlier. Once the therapeutic molecules reach their target cell and/or synapse, their affinity and solubility in fat, the main constitute of the CNS, determine their potential ability to affect the brain and its network. It can be stated, in general, that as for today, the net results of antioxidant therapy in human brain disease is dismal. Moreover, the beneficial role of OS in the CNS may be nulled by the action of drugs with antioxidant properties.

In HIE, few clinical trials such as supplementation of vitamin E that has strong antioxidant activity in preterm infants reduced the risk of intracranial hemorrhage, a frequent complication of prematurity, but increased the risk of sepsis. In premature newborns of very low birthweight, it increased the risk of sepsis but reduced the risk of severe retinopathy and blindness. It was concluded that supplementation of high-dose vitamin E in this particular population has more disadvantages than benefits. N-Acetyl Cysteine (NAC) is a another potential agent with marked antioxidant properties that showed promising results in asphyxiated newborn piglets, but in few clinical trials, which assessed its supplementation in premature newborns, there was no sufficient support to their value; in a single study, NAC supplementation to premature septic newborns resulted in increased mortality rate.

Although there are attractive theoretical possibilities on how to enhance endogenous antioxidant activity to counteract ROS-induced damage in MS, practical therapeutic convincing conclusions are still needed. The same can be said for infectious inflammation.

Lazaroids are aminosteroids with potent antioxidant properties, especially acting on lipid peroxidation but have no glucocorticoid or mineralocorticoid activity. Tirilazad mesylate is one of the Lazaroids that was chosen for clinical trials in TBI. Only a subgroup of the population study that consisted of males with severe TBI accompanied by subarachnoid hemorrhage benefited from this mode of treatment by reduced mortality and a better state of consciousness than placebo. It can be stated that in spite of the promising results from experimental models of TBI, antioxidant treatment in patients has so far failed. This statement is also valid for stroke, MS, and the neurodegenerative disorders AD and PD.

Multiple choice questions

The questions are intended to expand the knowledge of the student by encouraging her/him to reach the correct answer by additional reading and using commonsense.

1 The neuron is susceptible to OS and requires a constant rich supply of oxygen due to the following:
   a. Different from other cell types, many motor neurons in the cortex have axons that reach distal muscles, that is, those which innervate the small muscles in the feet that may be at least 2 m long in an NBA player
   b. Neurons are connected to a large number of other neurons by a network of multiple synapses reaching their dendritic spines (small membranous protrusions from the neuron's dendrite that typically receives input from a single synapse of an axon)
   c. Neurons are "factories" of vital neurotransmitters requiring constant energy supply
   d. All are correct

2 All the following are correct for stroke except:
   a. Hypoglycemia will cause a stroke localized to the most active part of the brain, which are the basal ganglia
   b. There is a particular type of very small stroke deep in the brain called "lacuna." Those are the typical findings in people with long-standing hypertension
   c. Stroke can be the result of a thrombus located on the aortic valve
   d. An autoimmune process causing inflammation in the wall of a cerebral artery (Vasculitis) can result in stroke
   e. Migraine is a risk factor for stroke

3 Choose the correct statement form of the following:
   a. The aging brain is partially immune to OS due to its reduced volume and decreased metabolic activity
   b. The immature brain is partially immune to OS due the underdeveloped connections between the immature nerve cells
   c. The developing brain of newborns is dependent on continuous energy supply because of its steep growth spurt
   d. Females are at a greater risk of stroke than males due to their lifelong periodic hormonal fluctuations increasing the risk of altered blood coagulation functions
   e. Athletes are more prone to OS-induced brain ischemia and subsequent stroke due to their high energy expenditure

Additional Reading

1 Gadoth, N. & Gobel, H.H. (eds) (2011) (Chapters 1, 2, 3, 4, 5, 6, 7, 9, 12) *Oxidative Stress and Free Radical Damage in Neurology*. Humana Press and Springer Science, NY.
2 Popa-Wagner, A., Mitran, S., Sivanesan, S. *et al.* (2013) ROS and brain disease: the good, the bad, and the ugly. *Oxidative Medicine and Cellular Longevity*, Vol. 2013, article ID 963520, 1–14.

3 St-Louis, R., Parmentier, C., Raison, D. *et al.* (2012) Reactive oxygen species are required for the hypothalamic osmoregulatory response. *Endocrinology*, 153, 1317–1329.

4 Elmore, S. (2007) Apoptosis: a review of programmed cell death. *Toxicologic Pathology*, 35, 495–516.

5 Cerio, F.G., Lara-Celador, I., Alvarez, A. *et al.* (2013) Neuroprotective therapies after perinatal hypoxic-ischemic brain injury. *Brain Science*, 3, 191–214.

6 Andersen, J.K. (2004) Oxidative stress in neurodegeneration: cause or consequence? *Nature Reviews Neuroscience*, 10 (Suppl), S18–S25.

7 Perrone, S., Negro, S., Tataranno, M.L. *et al.* (2010) Oxidative stress and antioxidant strategies in newborns. *Journal of Maternal-Fetal and Neonatal Medicine*, 23 (Suppl. 3), 63–65.

8 Raichle, M.E. (1983) The pathophysiology of brain ischemia. *Annals of Neurology*, 13, 2–10.

9 Adams, J.H., Graham, D.I., Murray, L.S. *et al.* (1982) Diffuse axonal injury due to nonmissile head injury in humans: an analysis of 45 cases. *Annals of Neurology*, 12, 557–563.

10 Chung, J.A. & Cummings, J.L. (2000) Neurobehavioral and neuropsychiatric symptoms in Alzheimer's disease: characteristics and treatment. *Neurologic Clinics*, 18, 829–846.

11 Padurariu, M., Ciobica, A., Lefter, R. *et al.* (2013) The oxidative stress hypothesis in Alzheimer's disease. *Psychiatria Danubina*, 25, 401–409.

12 Jellinger, K.A. (2010) Basic mechanisms of neurodegeneration: a critical update. *Journal of Cellular and Molecular Medicine*, 14, 457–487.

13 Jankovic, J. (2008) Parkinson's disease: clinical features and diagnosis. *Journal of Neurology, Neurosurgery, and Psychiatry*, 79, 368–376.

CHAPTER 4

Mediators of neuroinflammation

Rajiv Tikamdas, Ping Zhang, and Bin Liu

Department of Pharmacodynamics, College of Pharmacy, University of Florida, Gainesville, FL, USA

THEMATIC SUMMARY BOX

At the end of this chapter, students should be able to:

- Define neuroinflammation

- List the main cellular mediators of neuroinflammation

- Describe the characteristics of microglial activation

- List the microglial effector molecules

- List the exogenous and endogenous signals that induce microglial activation

- Differentiate between the proinflammatory and anti-inflammatory states of activated microglia

- Describe the interaction between neurons and microglia

- Illustrate the contribution of dysregulated microglial activation to the pathogenesis of various neurological disorders

- Understand how microglial activation could be used as an early-stage diagnostic biomarker for neurological disorders

- Understand how mediators/pathways of neuroinflammation could be used as potential therapeutic targets

---

Introduction

Inflammation in the brain, that is, neuroinflammation, can be triggered by the presence of foreign substances including microbial pathogens and the occurrence of neuronal injury inflicted by ischemic, traumatic, and neurotoxic insults. The inflammatory response, by design, aids to restore the disrupted homeostasis in the brain and, ideally, should subside after the eradication of the invading pathogens, removal of the cellular debris, and/or the initiation of potential repair mechanisms. However, a persistent presence of inflammatory stimuli could, on the other hand, result in a

---

Oxidative Stress and Antioxidant Protection: The Science of Free Radical Biology and Disease, First Edition.
Edited by Donald Armstrong and Robert D. Stratton.
© 2016 John Wiley & Sons, Inc. Published 2016 by John Wiley & Sons, Inc.

sustained inflammatory response, leading to the overproduction and accumulation of proinflammatory and cytotoxic effector molecules that could exacerbate neuronal damage, which, in turn, could trigger an additional inflammatory response. Therefore, neuroinflammation and neuronal damage, mediated by various inflammatory mediators, may create a reciprocal and vicious cycle that may potentially play a key role in driving the degenerative process in various neurological disorders. Mediators of this vicious cycle could serve as potential diagnostic biomarkers and/or therapeutic targets for the detection and treatment of the disorders. In this chapter, we will attempt to summarize the key features of cellular and molecular mediators of neuroinflammation and their relevance to a number of neurological disorders.

Cells mediating neuroinflammation

Neuroinflammation is primarily mediated by two non-neuronal cell types in the brain, namely microglia and astroglia. Microglia are the resident immune cells in the brain and represent around 10% of the cell population in the adult central nervous system (CNS). *Microglia*, unlike astroglia, have a mesodermal origin and arise from primitive myeloid progenitor cells, which migrate into the CNS early during development where they play a crucial role in sculpting the newly formed neuronal synapses and eliminating excess neurons.[1,2] Microglia are also important for adult neurogenesis and synaptic plasticity through its role in eliminating newborn hippocampal neurons that fail to integrate into the adult brain circuitry and continuously monitoring and pruning the synaptic connections.[2,3] The primary function of microglia is immune surveillance and maintenance of homeostasis in the brain. They constantly survey the surrounding microenvironment and become readily activated in response to various stimuli such as invading pathogens and neuronal injury and, once activated, secrete proinflammatory and anti-inflammatory factors and neurotrophic factors in an attempt to restore homeostasis.[4] *Astroglia*, important in providing structural and nutritional support to neurons, are also involved in neuroinflammation, but with a less robust response compared to microglia in terms of type and quantity of factors released.[5] Therefore, in an intact brain, microglia are the chief mediators of the inflammatory response. However, if the integrity of the blood–brain barrier is compromised, peripheral immune cells may gain access to the CNS and contribute to neuroinflammation.

Characteristics of microglial activation

Once activated, microglia undergo significant changes in morphology, number, gene expression, and secretory profiles. Activated microglia transform to an amoeboid shape with larger soma and much shorter and less ramified processes, which is functionally required for it to carry out its increased phagocytic activity. Microglia accumulate at the site of injury and significantly increase in number as a result of proliferation and chemotactic migration. Microglial activation is marked by the upregulation of cell surface markers such as CD11b, CD45, ionized calcium-binding adapter molecule 1

Table 4.1 Components of microglial activation cascade.

Amoeboid transformation
Proliferation
Chemotaxis
Phagocytosis
Production of:
Proinflammatory cytokines (TNF-α, IL-1β, IL-6)
Free radicals (ROS & NO)
Lipid metabolites (PGs)

TNF-α, tumor necrosis factor-alpha; IL-1β, interleukin-1beta; IL-6, interleukin-6; ROS, reactive oxygen species; NO, nitric oxide; PGs, prostaglandins.

(Iba1), and major histocompatibility complex II (MHC-II). Furthermore, they release a wide variety of soluble factors including cytokines, chemokines, free radicals, and lipid metabolites[6] (Table 4.1).

Proinflammatory cytokines such as tumor necrosis factor-α (TNF-α), interleukin-1 β (IL-1β), and IL-6 are among the key mediators of neuroinflammation. During the course of microglial activation, cytokines are in general synthesized as inactive precursor proteins that are proteolytically processed to the active form before release into the extracellular space. Cytokines then interact with their specific surface receptors on target cells to initiate cell signaling pathways or exert cytotoxic effects. TNF-α acts on TNFR1 and TNFR2 receptors, resulting in the activation of diverse signaling pathways that include induction of apoptotic cell death via activation of the caspase-8 dependent extrinsic pathway of apoptosis and activation of the nuclear factor-κ B (NF-κB) pathway, which results in further production of proin-flammatory cytokines including TNF-α itself in an autocrine manner.[7] IL-1β acts on the IL-1R receptor and activates multiple inflammatory signaling pathways such as NF-κB, C-Jun N-terminal kinase (JNK), and p38 mitogen-activated protein kinase (MAPK), leading to an increase in the expression of cytokines such as IL-6 and IL-1β itself,[8] in addition to increased synthesis of prostaglandin E2 (PGE2) and nitric oxide (NO).[9,10] IL-1β signaling is controlled by its natural antagonist interleukin-1 receptor antagonist (IL-1RA), which blocks the binding of IL-1β to its receptor and aborts its downstream signaling cascade.[8] IL-6 is a versatile cytokine that acts through IL-6R receptor and glycoprotein 130 (gp130), eliciting proinflammatory as well as anti-inflammatory responses.[11] Reactive oxygen species (ROS) are highly reactive oxygen-containing molecules that could damage vital cellular constituents such as proteins, nucleic acids, and lipids resulting in a myriad of detrimental effects including protein misfolding, enzyme inactivation, DNA damage, and lipid peroxidation. ROS are crucial for their role in destroying any invading pathogens or aberrant cells through the nicotinamide adenine dinucleotide phosphate (NADPH) oxidase-dependent respiratory burst in addition to the elimination of excess uninte-grated neurons during development. Oxidative stress occurs due to the imbalance between excessive ROS production and deficient antioxidant defenses leading to ROS-mediated tissue damage. Neurons are particularly vulnerable to oxidative stress due to its high energy consumption, which leads to the collateral generation of ROS from the mitochondrial respiratory chain; the relative scarcity of intracellular antioxidants such as glutathione; the presence of high levels of iron in particular brain regions such as substantia nigra, which facilitates the generation of more ROS through the Fenton reaction; and the generation of ROS as by-products of the metabolism of certain neurotransmitters such as dopamine. In addition, the brain is rich in polyunsaturated fatty acids that are highly susceptible to oxidative damage through lipid peroxidation.[12,13] *Nitric oxide (NO)* is a nitrogen free radical that could react with integral cellular molecules inducing nitrosative damage. NO is generated upon the conversion of L-arginine to L-citrulline by the action of nitric oxide synthase (NOS) that exists in three isoforms: neuronal nitric oxide synthase (nNOS/NOS1), inducible nitric oxide synthase (iNOS/NOS2), and endothelial nitric oxide synthase (eNOS/NOS3). iNOS/NOS2 is the isoform present in microglia and macrophages and normally has a low constitutive activity, but it is readily upregulated upon microglial activation.[14] NO exerts a neurotoxic effect through multiple mechanisms. NO inhibits the mitochondrial respiratory chain, resulting in decreased ATP synthesis and energy failure. It blocks glutamate uptake resulting in the accumulation of high levels of extracellular glutamate and subsequent excitotoxic cell death.[15] Importantly, NO could react with superoxide free radical to form the highly reactive peroxynitrite free radical, which wreaks havoc on cellular constituents.[16] Lipid mediators such as prostaglandin*s (PGs)* are arachidonic acid derivatives synthesized by cyclooxygenase (COX) enzyme that exists in two isoforms: the ubiquitously expressed and constitutively active COX-1 and the inducible COX-2 that is mainly expressed in immune cells.[17] PGs act on G-protein-coupled receptors in a paracrine and/or autocrine manner. PGE2 is the main proinflammatory PG that signals through four distinct receptors (EP1–4). PGE2 acts on EP1 receptors that may cause neuronal excitotoxicity via alteration of cellular Ca^{2+} levels and disruption of its homeostasis.[18] In addition to the proinflammatory factors, activated microglia also secrete anti-inflammatory cytokines such as IL-4, IL-10, and IL-13[19] as well as neurotrophic factors such as brain-derived neurotrophic factor (BDNF)[20] and insulin-like growth factor (IGF),[21] which serve to resolve the inflammatory process and promote tissue repair. The overall inflammatory response represents the net effect of the interaction of various proinflammatory and anti-inflammatory factors owing to the extensive cross talk among them.

Detrimental versus beneficial roles of microglial activation

Microglia alternate between two main phenotypes: the deactivated (resting) "ramified" phenotype that plays important roles in neurogenesis and synaptic plasticity, and the activated "amoeboid" phenotype that, depending on the gene expression and secretory profiles, could be further subdivided into the proinflammatory classically activated (M1) state and the anti-inflammatory alternatively activated (M2) state.[19] M1 state involves secretion of proinflammatory and cytotoxic factors including cytokines, free radicals, and prostaglandins. On the other hand, M2 state involves secretion of anti-inflammatory and neurotrophic factors. Classical microglial activation and the resulting inflammatory response is beneficial only if it occurs in a controlled and self-limiting manner as it helps in the destruction of invading pathogens and aberrant cells, as well as phagocytosis of apoptotic cellular debris. However, a sustained and uncontrolled inflammatory response will result in the secretion of excessive amounts of the neurotoxic proinflammatory factors leading to various neu-rological pathologies. Therefore, it is imperative to establish and maintain a proper balance between the proinflammatory (M1) and the anti-inflammatory (M2) states, where the inflammatory response is resolved once the stimulus has been eliminated and a subsequent shift toward the anti-inflammatory state occurs in order to support tissue repair.

Microglia-activating signals

Microglia are highly sensitive to changes in the CNS homeostasis as they are con-stantly surveying their microenvironment and become readily activated in response to potential threats to the CNS. Microglial activation can be triggered either by foreign substances and pathogens that invade the CNS tissue (exogenous signals) or certain factors released from injured neurons (endogenous signals). Activation is mediated through a set of microglial receptors known as pattern recognition receptors (PRRs), which recognize specific molecular patterns known as pathogen-associated molecular patterns (PAMPs) that are present in pathogens, or danger-associated molecular pat-terns (DAMPs), which are present in intracellular molecules that are released upon neuronal damage. PRRs include toll-like receptors (TLRs), scavenger receptors, recep-tor for advanced glycation end-products (RAGE), Nucleotide Oligomeriaztion domain (NOD)-like receptors (NLRs), and purinergic receptors.[22] TLRs are the major PRRs that recognize and become activated by certain PAMPs and DAMPs, triggering a cascade of downstream signaling events involving NF-κB and MAPKs that culminate in the production of various proinflammatory effector molecules.[23]

Exogenous signals

Invading pathogens such as bacteria and viruses induce a robust inflammatory response in the CNS that is primarily mediated through TLRs. Pneumococcal infection trig-gers microglial activation via TLR2 and TLR4.[24] Viral encephalitis is associated with a strong immune response mediated through TLR2 and TLR9.[25] Furthermore, the bacterial endotoxin lipopolysaccharide *(*LPS*)* induces a robust inflammatory response through activation of the microglial TLR4 and has been used in experimental animals to model pathogenic processes involving inflammation.[23] *Environmental toxicants* such as the pesticides paraquat[26] and rotenone,[27] as well as the organochlorinated pesticide dieldrin,[28] have also been reported to induce microglial activation and subsequent neurotoxicity. In addition, heavy metals such as manganese have been shown to activate microglia to produce ROS and cause subsequent neuronal loss.[29,30]

Endogenous signals

Research in the last decade or so has identified a number of factors that are capa-ble of inducing microglial activation, and the list is growing. These factors could be released from neurons injured or stressed by a variety of insults including ischemia, trauma, exposure to neurotoxins, oxidative stress, and gene mutations. The currently identified factors include *histone H1,* a nuclear protein whose main function includes packaging the nucleic acid into the highly dense chromosomes. Interestingly, it is released from ischemic neurons and induces the expression of microglial MHC-II together with promoting microglial survival and chemotaxis.[31] Another factor is *heat shock protein 60 (HSP60)* whose main function is to assist the proper folding of newly synthesized or denatured polypeptides. It is released from injured neurons and is able to induce microglial activation via TLR4 signaling and results in a significant neuronal death in culture.[32] Both the calcium-dependent cysteine protease *μ-calpain*[33] *and matrix metalloproteinase-3 (MMP3)*[34,35] have been shown to cause microglial activation and subsequent neurotoxicity. A cysteine protease inhibitor *cystatin C* has been shown to be released from neurotoxin-injured neurons and cause microglial activation and exacerbate neurotoxicity.[36]

Neuron–microglia interplay

The neuron–microglia relationship is now viewed as a reciprocal "two-way" interaction. In the healthy CNS, microglia are tightly regulated through the interaction with the surrounding neurons where neurons inhibit microglial activity through both contact-mediated and paracrine inhibition. Contact-mediated inhibition is mediated through membrane-bound neuronal signals such as the glycoprotein CD200 and the chemokine ligand fractalkine (CX3CL1), while paracrine inhibition is mediated through released soluble neuronal signals such as transforming growth factor-β (TGF-β), BDNF, and the soluble form of CX3CL1.[37] On the other hand, injured or stressed neurons release neuronal factors that would act as "emergency signals" to their surrounding milieu to activate microglia. Therefore, in a degenerating brain, sustained microglial activity occurs due to not only the release of microglia-activating signals from injured neurons but also the disinhibition of microglial activity as a result of loss of the inhibitory neuronal signals. This sets in motion a self-perpetuating cycle of neuronal loss and microglial activation, which could explain the progressive nature of neurodegenerative diseases. Intriguingly, despite the well-established association between microglial activation and neurodegeneration, it remains a hotly debated "chicken and egg" issue whether microglial activation could initiate neurodegeneration by itself or it just accelerates and propagates the already initiated neurodegenerative process.

Pathological implications of dysregulated microglial activation

Microglial activation and the resulting neuroinflammation have been implicated in a variety of neurological disorders including neurodegenerative diseases such as Alzheimer's disease (AD) and Parkinson's disease (PD), brain injuries such as stroke and traumatic brain injury, as well as psychiatric disorders such as schizophrenia and depression.

Neurodegenerative diseases
Alzheimer's disease (AD)
AD is the most common neurodegenerative disease that involves a progressive loss of memory and cognitive dysfunction. It is characterized by the presence of β-amyloid

plaques and neurofibrillary tangles in the brain.[38] Epidemiological studies have reported a reduced risk of AD development in nonsteroidal anti-inflammatory drug (NSAID) users.[39–41] Brain imaging studies suggest early-stage microglial activation in disease progression.[42] Microglial activation[43] has been observed in postmortem AD brains, and elevated levels of TNF-α,[44] IL-1β, and IL-6[45] are seen in the cerebrospinal fluid (CSF) of AD patients. In animal models of AD, robust microglial activation in proximity to β-amyloid plaques[46,47] and significant increase in proinflammatory cytokines are observed.[48] The role of oxidative stress in AD pathogenesis is supported by the neuroprotective effect of the free radical scavenger edaravone against β-amyloid neurotoxicity.[49] As a potential anti-inflammatory intervention strategy for AD, a pilot study has shown that the TNF-α inhibitor, etanercept, significantly improved the clinical outcome in AD patients.[50]

Parkinson's disease (PD)

Parkinson's disease (PD) is the second most prevalent neurodegenerative disease. The movement disorders observed in PD patients is associated with a selective and progressive degeneration of the nigrostriatal dopamine-releasing neurons. Similar to AD, microglial activation was detected in the substantia nigra of postmortem PD brains,[51] and elevated levels of TNF-α,[52] IL-1β, and IL-6[45,53] have been detected in the brain and CSF of PD patients. Epidemiological studies have shown that the use of NSAIDs reduces the incidence of PD.[54,55] Experimentally, neuroinflammation has been shown to play an important role in PD pathogenesis.[13,56] At least as a proof of concept, administration of bacterial endotoxin LPS results in a robust inflammatory response and subsequent PD-like dopaminergic neurodegeneration in the brain.[23] The proinflammatory effectors, TNF-α,[57] NADPH oxidase,[58] iNOS[59], and COX-2[60] have all been shown to be key mediators of dopaminergic neurotoxicity. Several anti-inflammatory agents have been shown to provide neuroprotection in various PD animal models.[56]

Brain injuries

Stroke (ischemic brain injury)

Stroke is one of the leading causes of death and disability. Besides hemorrhagic stroke which is caused by blood vessel rupture inside the brain, the majority of cases of stroke are ischemic and occur as a result of a blockage of blood flow to certain brain areas due to clot formation, which consequently leads to neuronal damage in the ischemic areas. Microglia become readily activated in response to the ischemic neuronal injury and are detected in both the ischemic core and the penumbral regions.[61,62] Activated microglia mediate the postischemic neuroinflammatory response via producing several proinflammatory factors, which have been implicated in worsening the clinical outcome in stroke patients by causing secondary bystander neuronal damage. Elevated levels of TNF-α[63,64] and IL-6[64] have been detected in the CSF of stroke patients. A role of TNF-α in ischemic injury is supported by the exacerbated neuronal damage following TNF-α administration and the neuroprotective effect of blocking TNF-α using either recombinant type-I-soluble TNF receptor or neutralizing anti-TNF-α antibody in a rat model of focal brain ischemia.[65,66] Role of IL-1β in ischemic injury is supported by the reduced neuronal damage observed in IL-1R1[67] or IL-1β converting enzyme-deficient mice,[68] in addition to the ability of IL-1RA to mitigate the ischemic brain injury.[69,70] Furthermore, deletion of TLR2[71] and TLR4,[72] which are major mediators of the ischemic injury-associated immune response, results in significant neuroprotection against ischemic brain injury. Minocycline,[73] a microglial activation modulator, and edaravone,[74] a free radical scavenger, have been shown to significantly reduce ischemic brain damage.

Traumatic brain injury (TBI)

Traumatic brain injury (TBI) induces a robust inflammatory response mediated through increased microglial activity, which results in the aggravation of neuronal damage and impairment of tissue repair.[75] This is supported by the chronic persistent microglial activation seen in TBI patients.[76] Elevated CSF levels of IL-1β and IL-6 have been detected in TBI patients.[77] Also, a significant increase in levels of TNF-α, IL-1, IL-6,[78–81] and superoxide free radical[82] has been observed in brains of TBI animal models. Conversely, reduced microglial activation, neuroprotection, and improved motor functions have been observed in animal models of TBI following application of TNF-α inhibitor,[78] soluble TNF receptor,[79] IL-1RA,[83] NADPH oxidase inhibitor,[84] free radical scavenger,[82] and the anti-inflammatory agent minocycline.[85]

Neuropsychiatric diseases

Schizophrenia

Accumulating evidence supports a role for microglial activation in the pathogenesis of schizophrenia. Epidemiological studies have shown that prenatal infection significantly increased the risk of developing schizophrenia.[86] Microglial activation was observed in positron emission tomography (PET) scan of brains of schizophrenic patients as well as postmortem brains of patients with chronic schizophrenia.[87–89] Furthermore, increased expression or levels of cytokines TNF-α, IL-1β and/or IL-6, and/or the microglial marker MHC-II have been detected in brains, CSF, and/or serum of schizophrenic patients.[90–93] Interestingly, addition of microglial activity modulator minocycline to antipsychotic therapeutic regimens has been shown to improve the clinical outcome in schizophrenic patients,[94,95] consistent with the observed microglial activation–inhibitory activity of several antipsychotic drugs observed *in vitro*.[96–98]

Depression

Postmortem analysis has detected robust microglial activation in brains of depressed patients who committed suicide.[99] In patients with major depressive disorders, elevated levels of IL-1β and IL-6 were found in serum and/or CSF.[91,100,101] Experimentally, involvement of IL-6 in depression is supported by the observation of depression-like symptoms in animals following IL-6 central administration and the abolished antidepressant effect of the selective serotonin reuptake inhibitor (SSRI) fluoxetine in mice overexpressing IL-6.[102] TNF-α serum levels were significantly elevated in major depressive disorder patients, but returned to normal levels after antidepressant treatment.[103] Deletion of TNFR1 or TNFR2 has been shown to suppress depression-like behaviors in mice.[104] The potent microglial activator, bacterial endotoxin LPS, induced depressive-like behaviors in mice.[105] For the same token, interleukin converting enzyme (ICE) knockout mice showed decreased

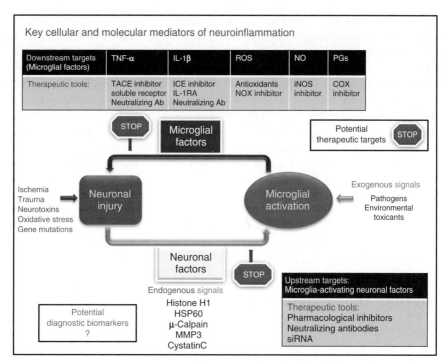

Figure 4.1 Interplay between neuronal injury and microglial activation and the various mediators of neuroinflammation. TNF-α, tumor necrosis factor alpha; TACE, TNF-α converting enzyme; IL-1β, interleukin-1 beta; ICE, IL-1β converting enzyme; IL-1RA, interleukin-1 receptor antagonist; ROS, reactive oxygen species; NOX, NADPH oxidase; NO, nitric oxide; iNOS, inducible nitric oxide synthase; PGs, prostaglandins; COX, cyclooxygenase; HSP60, heat shock protein 60; MMP3, matrix metalloproteinase 3; siRNA, small interfering RNA.

depressive-like behaviors following central LPS administration, implying a key role for IL-1β.[106] The role of microglial activation in depression pathogenesis was further supported by the antidepressant-like effects of minocycline, the microglial activity modulator, in a rat model of depression[107], in addition to the anti-inflammatory effects of several antidepressant drugs including the SSRI fluoxetine[108] and the tricyclic antidepressants (TCAs) amitriptyline and nortriptyline,[109] which have been shown to inhibit LPS-induced microglial activation and the subsequent release of microglial proinflammatory factors, suggesting that their antidepressant effects involve an anti-inflammatory component.

Microglial activation as a diagnostic biomarker and therapeutic target

Components of microglial activation cascade may hold the promise as potential early-phase diagnostic biomarkers for CNS disorders in which neuroinflammation plays a role in the initiation stage of the progressive neurodegeneration process. If a particular neurological disorder is characterized by neuronal dysfunction at a specific brain region, the spatial localization of early stage microglial activation could forecast the development of diseases unique to the brain region such as the substantia nigra region for PD. Microglial activation could be visualized, with ever increasing sensitivity due to advance in technology and probe development by magnetic resonance imaging (MRI) and PET.[42,110] Moreover, identification of certain inflammatory factors or factors specific to injured neurons in CSF or serum could also potentially be used as indicators of the symptom-causing neuronal loss. Similarly, key regulatory steps in the microglial activation pathways and critical factors that serve to initiate or modulate microglial activation could be potential targets for therapeutic intervention. Upstream targets include the microglia-activating neuronal factors. Downstream targets include the neurotoxic effector molecules produced by the activated microglia (Figure 4.1). Continued investigation into the precise role of neuroinflammation in the pathogenesis of neurological disorders will certainly help us diagnose and manage those devastating diseases.

Multiple choice questions

1 Which of the following is NOT true about microglia?
   a. They are a highly specialized type of glial cells in the brain
   b. They originate from the ectoderm during development
   c. They are the key mediator of neuroinflammation
   d. They secrete both proinflammatory and anti-inflammatory cytokines

2 Microglia can be activated by
   a. Invading pathogens (e.g., bacteria and viruses) in the brain
   b. Brain injury
   c. Certain environmental toxins
   d. All of the above

3 Microglial activation is thought to be involved in the pathogenesis of which of the following diseases?
   a. Alzheimer's disease
   b. Parkinson's disease
   c. Ischemic stroke
   d. All of the above

References

1 Alliot, F., Godin, I. & Pessac, B. (1999) Microglia derive from progenitors, originating from the Yolk Sac, and which proliferate in the brain. *Brain Research. Developmental Brain Research*, 117(2), 145–152.
2 Paolicelli, R.C., Bolasco, G., Pagani, F. *et al.* (2011) Synaptic pruning by microglia is necessary for normal brain development. *Science*, 333(6048), 1456–1458.
3 Schafer, D.P., Lehrman, E.K., Kautzman, A.G. *et al.* (2012) Microglia sculpt postnatal neural circuits in an activity and complement-dependent manner. *Neuron*, 74(4), 691–705.

4 Hanisch, U.K. & Kettenmann, H. (2007) Microglia: active sensor and versatile effector cells in the normal and pathologic brain. *Nature Neuroscience*, 10(11), 1387–1394.

5 Farina, C., Aloisi, F. & Meinl, E. (2007) Astrocytes are active players in cerebral innate immunity. *Trends in Immunology*, 28(3), 138–145.

6 Kettenmann, H., Hanisch, U.K., Noda, M. *et al.* (2011) Physiology of microglia. *Physiological Reviews*, 91(2), 461–553.

7 Wajant, H., Pfizenmaier, K. & Scheurich, P. (2003) Tumor necrosis factor signaling. *Cell Death and Differentiation*, 10(1), 45–65.

8 Weber, A., Wasiliew, P. & Kracht, M. (2010) Interleukin-1 (IL-1) pathway. *Science Signaling*, 3(105), cm1.

9 Pinteaux, E., Parker, L.C., Rothwell, N.J. *et al.* (2002) Expression of interleukin-1 receptors and their role in interleukin-1 actions in murine microglial cells. *Journal of Neurochemistry*, 83(4), 754–763.

10 Casamenti, F., Prosperi, C., Scali, C. *et al.* (1999) Interleukin-1beta activates forebrain glial cells and increases nitric oxide production and cortical glutamate and GABA release in vivo: implications for Alzheimer's disease. *Neuroscience*, 91(3), 831–842.

11 Kamimura, D., Ishihara, K. & Hirano, T. (2004) IL-6 signal transduction and its physiological roles: the signal orchestration model. *Reviews of Physiology, Biochemistry and Pharmacology*, 149, 1–38.

12 Hsieh, H.L. & Yang, C.M. (2013) Role of redox signaling in neuroinflammation and neurodegenerative diseases. *Biomedical Research International*, 2013, 484613.

13 Hald, A. & Lotharius, J. (2005) Oxidative stress and inflammation in Parkinson's disease: is there a causal link? *Experimental Neurology*, 193(2), 279–290.

14 Thiel, V.E. & Audus, K.L. (2001) Nitric oxide and blood-brain barrier integrity. *Antioxidants and Redox Signaling*, 3(2), 273–278.

15 Bal-Price, A. & Brown, G.C. (2001) Inflammatory neurodegeneration mediated by nitric oxide from activated glia-inhibiting neuronal respiration, causing glutamate release and excitotoxicity. *The Journal of Neuroscience: The Official Journal of the Society for Neuroscience*, 21(17), 6480–6491.

16 Bal-Price, A., Matthias, A. & Brown, G.C. (2002) Stimulation of the NADPH oxidase in activated rat microglia removes nitric oxide but induces peroxynitrite production. *Journal of Neurochemistry*, 80(1), 73–80.

17 Choi, S.H., Aid, S. & Bosetti, F. (2009) The distinct roles of cyclooxygenase-1 and -2 in Neuroinflammation: implications for translational research. *Trends in Pharmacological Sciences*, 30(4), 174–181.

18 Kawano, T., Anrather, J., Zhou, P. *et al.* (2006) Prostaglandin E2 EP1 receptors: downstream effectors of COX-2 neurotoxicity. *Nature Medicine*, 12(2), 225–229.

19 Colton, C.A. (2009) Heterogeneity of microglial activation in the innate immune response in the brain. *Journal of Neuroimmune Pharmacology: The Official Journal of the Society on NeuroImmune Pharmacology*, 4(4), 399–418.

20 Gomes, C., Ferreira, R., George, J. *et al.* (2013) Activation of microglial cells triggers a release of brain-derived neurotrophic factor (BDNF) inducing their proliferation in an adenosine A2A receptor-dependent manner: A2A receptor blockade prevents BDNF release and proliferation of microglia. *Journal of Neuroinflammation*, 10, 16.

21 Suh, H.S., Zhao, M.L., La, D. *et al.* (2013) Insulin-like growth factor 1 and 2 (IGF1, IGF2) expression in human microglia: differential regulation by inflammatory mediators. *Journal of Neuroinflammation*, 10, 37.

22 Block, M.L., Zecca, L. & Hong, J.S. (2007) Microglia-mediated neurotoxicity: uncovering the molecular mechanisms. *Nature Reviews. Neuroscience*, 8(1), 57–69.

23 Dutta, G., Zhang, P. & Liu, B. (2008) The lipopolysaccharide Parkinson's disease animal model: mechanistic studies and drug discovery. *Fundamental & Clinical Pharmacology*, 22(5), 453–464.

24 Klein, M., Obermaier, B., Angele, B. *et al.* (2008) Innate immunity to pneumococcal infection of the central nervous system depends on toll-like receptor (TLR) 2 and TLR4. *The Journal of Infectious Diseases*, 198(7), 1028–1036.

25 Sørensen, L.N., Reinert, L.S., Malingaard, L. *et al.* (2008) TLR2 and TLR9 synergistically control herpes simplex virus infection in the brain. *Journal of Immunology*, 181(12), 8604–8612.

26 Wu, X.F., Block, M.L., Zhang, W. *et al.* (2005) The role of microglia in paraquat-induced dopaminergic neurotoxicity. *Antioxidants & Redox Signaling*, 7(5–6), 654–661.

27 Gao, H.M., Hong, J.S., Zhang, W. *et al.* (2002) Distinct role for microglia in rotenone-induced degeneration of dopaminergic neurons. *The Journal of Neuroscience: The Official Journal of the Society for Neuroscience*, 22(3), 782–790.

28 Mao, H., Fang, X., Floyd, K.M. *et al.* (2007) Induction of microglial reactive oxygen species production by the organochlorinated pesticide dieldrin. *Brain Research*, 1186, 267–274.

29 Zhang, P., Wong, T.A., Lokuta, K.M. *et al.* (2009) Microglia enhance manganese chloride-induced dopaminergic neurodegeneration: role of free radical generation. *Experimental Neurology*, 217(1), 219–230.

30 Liu, Y., Barber, D.S., Zhang, P. *et al.* (2013) Complex II of the mitochondrial respiratory chain is the key mediator of divalent manganese-induced hydrogen peroxide production in microglia. *Toxicological Sciences: An Official Journal of the Society of Toxicology*, 132(2), 298–306.

31 Gilthorpe, J.D., Oozeer, F., Nash, J. *et al.* (2013) Extracellular histone H1 is neurotoxic and drives a pro-inflammatory response in microglia. *F1000Research*, 2, 148.

32 Lehnardt, S., Schott, E., Trimbuch, T. *et al.* (2008) A vicious cycle involving release of heat shock protein 60 from injured cells and activation of toll-like receptor 4 mediates neurodegeneration in the CNS. *The Journal of Neuroscience: The Official Journal of the Society for Neuroscience*, 28(10), 2320–2331.

33 Levesque, S., Wilson, B., Gregoria, V. *et al.* (2010) Reactive microgliosis: extracellular micro-calpain and microglia-mediated dopaminergic neurotoxicity. *Brain: A Journal of Neurology*, 133(Pt 3), 808–821.

34 Kim, Y.S. *et al.* (2005) Matrix metalloproteinase-3: a novel signaling proteinase from apoptotic neuronal cells that activates microglia. *The Journal of Neuroscience: The Official Journal of the Society for Neuroscience*, 25(14), 3701–3711.

35 Kim, Y.S., Kim, S.S., Cho, J.J. *et al.* (2007) A pivotal role of matrix metalloproteinase-3 activity in dopaminergic neuronal degeneration via microglial activation. *FASEB Journal: Official Publication of the Federation of American Societies for Experimental Biology*, 21(1), 179–187.

36 Dutta, G., Barber, D.S., Zhang, P. *et al.* (2012) Involvement of dopaminergic neuronal cystatin C in neuronal injury-induced microglial activation and neurotoxicity. *Journal of Neurochemistry*, 122(4), 752–763.

37 Biber, K., Neumann, H., Inoue, K. *et al.* (2007) Neuronal 'on' and 'off' signals control microglia. *Trends in Neurosciences*, 30(11), 596–602.

38 Lue, L.F., Brachova, L., Civin, W.H. *et al.* (1996) Inflammation, A beta deposition, and neurofibrillary tangle formation as correlates of Alzheimer's disease neurodegeneration. *Journal of Neuropathology and Experimental Neurology*, 55(10), 1083–1088.

39 Szekely, C.A., Thorne, J.E., Zandi, P.P. *et al.* (2004) Nonsteroidal anti-inflammatory drugs for the prevention of Alzheimer's disease: a systematic review. *Neuroepidemiology*, 23(4), 159–169.

40 Etminan, M., Gill, S. & Samii, A. (2003) Effect of non-steroidal anti-inflammatory drugs on risk of Alzheimer's disease: systematic review and meta-analysis of observational studies. *BMJ (Clinical Research Ed.)*, 327(7407), 128.

41 Anthony, J.C., Breitner, J.C., Zandi, P.P. *et al.* (2000) Reduced prevalence of AD in users of NSAIDs and H2 receptor antagonists: the cache county study. *Neurology*, 54(11), 2066–2071.

42 Cagnin, A., Brooks, D.J., Kennedy, A.M. *et al.* (2001) In-vivo measurement of activated microglia in dementia. *Lancet*, 358(9280), 461–467.

43 Vehmas, A.K., Kawas, C.H., Stewart, W.F. *et al.* (2003) Immune reactive cells in senile plaques and cognitive decline in Alzheimer's disease. *Neurobiology of Aging*, 24(2), 321–331.

44 Tarkowski, E., Liljeroth, A.M., Minthon, L. *et al.* (2003) Cerebral pattern of pro- and anti-inflammatory cytokines in dementias. *Brain Research Bulletin*, 61(3), 255–260.

45 Blum-Degen, D., Müller, T., Kuhn, W. *et al.* (1995) Interleukin-1 beta and interleukin-6 are elevated in the cerebrospinal fluid of Alzheimer's and de Novo Parkinson's disease patients. *Neuroscience Letters*, 202(1–2), 17–20.

46 Frautschy, S.A., Yang, F., Irrizarry, M. *et al.* (1998) Microglial response to amyloid plaques in APPsw transgenic mice. *The American Journal of Pathology*, 152(1), 307–317.

47 Stalder, M., Phinney, A., Probst, A. *et al.* (1999) Association of microglia with amyloid plaques in brains of APP23 transgenic mice. *The American Journal of Pathology*, 154(6), 1673–1684.

48 Patel, N.S., Paris, D., Mathura, V. *et al.* (2005) Inflammatory cytokine levels correlate with amyloid load in transgenic mouse models of Alzheimer's disease. *Journal of Neuroinflammation*, 2(1), 9.

49 He, F., Cao, Y.P., Che, F.Y. *et al.* (2014) Inhibitory effects of edaravone in B-amyloid-induced neurotoxicity in rats. *BioMed Research International*, 2014, 370368.

50 Tobinick, E., Gross, H., Weinberger, A. *et al.* (2006) TNF-alpha modulation for treatment of Alzheimer's disease: a 6-month pilot study. *MedGenMed: Medscape General Medicine*, 8(2), 25.

51 McGeer, P.L., Itagaki, S., Boyes, B.E. *et al.* (1988) Reactive microglia are positive for HLA-DR in the substantia nigra of Parkinson's and Alzheimer's disease brains. *Neurology*, 38(8), 1285–1291.

52 Mogi, M., Harada, M., Riederer, P. *et al.* (1994) Tumor necrosis factor-alpha (TNF-alpha) increases both in the brain and in the cerebrospinal fluid from parkinsonian patients. *Neuroscience Letters*, 165(1–2), 208–210.

53 Mogi, M., Harada, M., Kondo, T. *et al.* (1994) Interleukin-1 beta, interleukin-6, epidermal growth factor and transforming growth factor-alpha are elevated in the brain from Parkinsonian patients. *Neuroscience Letters*, 180(2), 147–50.

54 Chen, H., Zhang, S.M., Hernán, M.A. *et al.* (2003) Nonsteroidal anti-inflammatory drugs and the risk of Parkinson disease. *Archives of Neurology*, 60(8), 1059–1064.

55 Chen, H., Jacobs, E., Schwarzschild, M.A. *et al.* (2005) Nonsteroidal antiinflammatory drug use and the risk for Parkinson's disease. *Annals of Neurology*, 58(6), 963–967.

56 Liu, B. (2006) Modulation of microglial pro-inflammatory and neurotoxic activity for the treatment of Parkinson's disease. *The AAPS Journal*, 8(3), E606–E621.

57 Sriram, K., Matheson, J.M., Benkovic, S.A. *et al.* (2006) Deficiency of TNF receptors suppresses microglial activation and alters the susceptibility of brain regions to MPTP-induced neurotoxicity: role of TNF-alpha. *FASEB Journal: Official Publication of the Federation of American Societies for Experimental Biology*, 20(6), 670–682.

58 Wu, D.C., Teismann, P., Tieu, K. *et al.* (2003) NADPH oxidase mediates oxidative stress in the 1-methyl-4-phenyl-1,2,3,6-tetrahydropyridine model of Parkinson's disease. *Proceedings of the National Academy of Sciences of the United States of America*, 100(10), 6145–6150.

59 Dehmer, T., Lindenau, J., Haid, S. *et al.* (2000) Deficiency of inducible nitric oxide synthase protects against MPTP toxicity in vivo. *Journal of Neurochemistry*, 74(5), 2213–2216.

60 Feng, Z.H., Wang, T.G., Li, D.D. *et al.* (2002) Cyclooxygenase-2-deficient mice are resistant to 1-methyl-4-phenyl1, 2, 3, 6-tetrahydropyridine-induced damage of dopaminergic neurons in the substantia nigra. *Neuroscience Letters*, 329(3), 354–358.

61 Gulyás, B., Tóth, M., Schain, M. *et al.* (2012) Evolution of microglial activation in ischaemic core and peri-infarct regions after stroke: a PET study with the TSPO molecular imaging biomarker [((11))C]vinpocetine. *Journal of the Neurological Sciences*, 320(1–2), 110–117.

62 Thiel, A. & Heiss, W.D. (2011) Imaging of microglia activation in stroke. *Stroke; a Journal of Cerebral Circulation*, 42(2), 507–512.

63 Zaremba, J. & Losy, J. (2001) Early TNF-alpha levels correlate with ischaemic stroke severity. *Acta Neurologica Scandinavica*, 104(5), 288–295.

64 Vila, N., Castillo, J., Dávalos, A. *et al.* (2000) Proinflammatory cytokines and early neurological worsening in ischemic stroke. *Stroke; a Journal of Cerebral Circulation*, 31(10), 2325–2329.

65 Barone, F.C., Arvin, B., White, R.F. *et al.* (1997) Tumor necrosis factor-alpha. A mediator of focal ischemic brain injury. *Stroke; a Journal of Cerebral Circulation*, 28(6), 1233–1244.

66 Dawson, D.A., Martin, D. & Hallenbeck, J.M. (1996) Inhibition of tumor necrosis factor-alpha reduces focal cerebral ischemic injury in the spontaneously hypertensive rat. *Neuroscience Letters*, 218(1), 41–44.

67 Basu, A., Lazovic, J., Krady, J.K. *et al.* (2005) Interleukin-1 and the interleukin-1 type 1 receptor are essential for the progressive neurodegeneration that ensues subsequent to a mild hypoxic/ischemic injury. *Journal of Cerebral Blood Flow and Metabolism: Official Journal of the International Society of Cerebral Blood Flow and Metabolism*, 25(1), 17–29.

68 Schielke, G.P., Yang, G.Y., Shivers, B.D. *et al.* (1998) Reduced ischemic brain injury in interleukin-1 beta converting enzyme-deficient mice. *Journal of Cerebral Blood Flow and Metabolism: Official Journal of the International Society of Cerebral Blood Flow and Metabolism*, 18(2), 180–185.

69 Loddick, S.A. & Rothwell, N.J. (1996) Neuroprotective effects of human recombinant interleukin-1 receptor antagonist in focal cerebral ischaemia in the rat. *Journal of Cerebral Blood Flow and Metabolism: Official Journal of the International Society of Cerebral Blood Flow and Metabolism*, 16(5), 932–940.

70 Pradillo, J.M., Denes, A., Greenhalgh, A.D. *et al.* (2012) Delayed administration of interleukin-1 receptor antagonist reduces ischemic brain damage and inflammation in comorbid rats. *Journal of Cerebral Blood Flow and Metabolism: Official Journal of the International Society of Cerebral Blood Flow and Metabolism*, 32(9), 1810–1819.

71 Lehnardt, S., Lehmann, S., Kaul, D. *et al.* (2007) Toll-like receptor 2 mediates CNS injury in focal cerebral ischemia. *Journal of Neuroimmunology*, 190(1–2), 28–33.

72 Caso, J.R., Pradillo, J.M., Hurtado, O. *et al.* (2007) Toll-like receptor 4 is involved in brain damage and inflammation after experimental stroke. *Circulation*, 115(12), 1599–1608.

73 Yrjänheikki, J., Tikka, T., Keinänen, R. *et al.* (1999) A tetracycline derivative, minocycline, reduces inflammation and protects against focal cerebral ischemia with a wide therapeutic window. *Proceedings of the National Academy of Sciences of the United States of America*, 96(23), 13496–13500.

74 Zhang, N., Komine-Kobayashi, M., Tanaka, R. *et al.* (2005) Edaravone reduces early accumulation of oxidative products and sequential inflammatory responses after transient focal ischemia in mice brain. *Stroke; a Journal of Cerebral Circulation*, 36(10), 2220–2225.

75 Davalos, D., Grutzendler, J., Yang, G. *et al.* (2005) ATP mediates rapid microglial response to local brain injury in vivo. *Nature Neuroscience*, 8(6), 752–758.

76 Ramlackhansingh, A.F., Brooks, D.J., Greenwood, R.J. *et al.* (2011) Inflammation after trauma: microglial activation and traumatic brain injury. *Annals of Neurology*, 70(3), 374–383.

77 Winter, C.D., Iannotti, F., Pringle, A.K. *et al.* (2002) A microdialysis method for the recovery of IL-1beta, IL-6 and nerve growth factor from human brain in vivo. *Journal of Neuroscience Methods*, 119(1), 45–50.

78 Shohami, E., Gallily, R., Mechoulam, R. *et al.* (1997) Cytokine production in the brain following closed head injury: dexanabinol (HU-211) is a novel TNF-alpha inhibitor and an effective neuroprotectant. *Journal of Neuroimmunology*, 72(2), 169–177.

79 Knoblach, S.M., Fan, L. & Faden, A.I. (1999) Early neuronal expression of tumor necrosis factor-alpha after experimental brain injury contributes to neurological impairment. *Journal of Neuroimmunology*, 95(1–2), 115–125.

80 Shohami, E., Novikov, M., Bass, R. *et al.* (1994) Closed head injury triggers early production of TNF alpha and IL-6 by brain tissue. *Journal of Cerebral Blood Flow and Metabolism: Official Journal of the International Society of Cerebral Blood Flow and Metabolism*, 14(4), 615–619.

81 Woodroofe, M.N., Sarna, G.S., Wadhwa, M. *et al.* (1991) Detection of interleukin-1 and interleukin-6 in adult rat brain, following mechanical injury, by in vivo microdialysis: evidence of a role for microglia in cytokine production. *Journal of Neuroimmunology*, 33(3), 227–236.

82 Miyamoto, K., Ohtaki, H., Dohi, K. *et al.* (2013) Therapeutic time window for edaravone treatment of traumatic brain injury in mice. *BioMed Research International*, 2013, 379206.

83 Jones, N.C., Prior, M.J., Burden-Teh, E. *et al.* (2005) Antagonism of the interleukin-1 receptor following traumatic brain injury in the mouse reduces the number of nitric oxide synthase-2-positive cells and improves anatomical and functional outcomes. *The European Journal of Neuroscience*, 22(1), 72–78.

84 Lu, X.Y., Wang, H.D., Xu, J.G. *et al.* (2014) NADPH oxidase inhibition improves neurological outcome in experimental traumatic brain injury. *Neurochemistry International*, 69, 14–19.

85 Bye, N., Habgood, M.D., Callaway, J.K. *et al.* (2007) Transient neuroprotection by minocycline following traumatic brain injury is associated with attenuated microglial activation but no changes in cell apoptosis or neutrophil infiltration. *Experimental Neurology*, 204(1), 220–233.

86 Brown, A.S. & Derkits, E.J. (2010) Prenatal infection and schizophrenia: a review of epidemiologic and translational studies. *The American Journal of Psychiatry*, 167(3), 261–280.

87 Van Berckel, B.N., Bossong, M.G., Boellaard, R. *et al.* (2008) Microglia activation in recent-onset schizophrenia: a quantitative (R)-[11C]PK11195 Positron emission tomography study. *Biological Psychiatry*, 64(9), 820–822.

88 Doorduin, J., de Vries, E.F., Willemsen, A.T. *et al.* (2009) Neuroinflammation in schizophrenia-related psychosis: a PET study. *Journal of Nuclear Medicine: Official Publication, Society of Nuclear Medicine*, 50(11), 1801–1807.

89 Radewicz, K., Garey, L.J., Gentleman, S.M. *et al.* (2000) Increase in HLA-DR immunoreactive microglia in frontal and temporal cortex of chronic schizophrenics. *Journal of Neuropathology and Experimental Neurology*, 59(2), 137–150.

90 Fillman, S.G., Cloonan, N., Catts, V.S. *et al.* (2013) Increased inflammatory markers identified in the dorsolateral prefrontal cortex of individuals with schizophrenia. *Molecular Psychiatry*, 18(2), 206–214.

91 Sasayama, D., Hattori, K., Wakabayashi, C. *et al.* (2013) Increased cerebrospinal fluid interleukin-6 levels in patients with schizophrenia and those with major depressive disorder. *Journal of Psychiatric Research*, 47(3), 401–406.

92 Song, X.Q., Lv, L.X., Li, W.Q. *et al.* (2009) The interaction of nuclear factor-kappa B and cytokines is associated with schizophrenia. *Biological Psychiatry*, 65(6), 481–488.

93 Van Kammen, D.P., McAllister-Sistilli, C.G., Kelley, M.E. *et al.* (1999) Elevated interleukin-6 in schizophrenia. *Psychiatry Research*, 87(2–3), 129–136.

94 Miyaoka, T., Yasukawa, R., Yasuda, H. *et al.* (2008) Minocycline as adjunctive therapy for schizophrenia: an open-label study. *Clinical Neuropharmacology*, 31(5), 287–292.

95 Levkovitz, Y., Mendlovich, S., Riwkes, S. *et al.* (2010) A double-blind, randomized study of minocycline for the treatment of negative and cognitive symptoms in early-phase schizophrenia. *The Journal of Clinical Psychiatry*, 71(2), 138–149.

96 Bian, Q., Kato, T., Monji, A. *et al.* (2008) The effect of atypical antipsychotics, perospirone, ziprasidone and quetiapine on microglial activation induced by interferon-gamma. *Progress in Neuro-Psychopharmacology & Biological Psychiatry*, 32(1), 42–48.

97 Kato, T., Mizoguchi, Y., Monji, A. *et al.* (2008) Inhibitory effects of aripiprazole on interferon-gamma-induced microglial activation via intracellular Ca^{2+} regulation in vitro. *Journal of Neurochemistry*, 106(2), 815–825.

98 Zheng, L.T., Hwang, J., Ock, J. *et al.* (2008) The antipsychotic spiperone attenuates inflammatory response in cultured microglia via the reduction of proinflammatory cytokine expression and nitric oxide production. *Journal of Neurochemistry*, 107(5), 1225–1235.

99 Steiner, J., Bielau, H., Brisch, R. *et al.* (2008) Immunological aspects in the neurobiology of suicide: elevated microglial density in schizophrenia and depression is associated with suicide. *Journal of Psychiatric Research*, 42(2), 151–157.

100 Owen, B.M., Eccleston, D., Ferrier, I.N. *et al.* (2001) Raised levels of plasma interleukin-1beta in major and postviral depression. *Acta Psychiatrica Scandinavica*, 103(3), 226–228.

101 Maes, M., Bosmans, E., De Jongh, R. *et al.* (1997) Increased serum IL-6 and IL-1 receptor antagonist concentrations in major depression and treatment resistant depression. *Cytokine*, 9(11), 853–858.

102 Sukoff Rizzo, S.J., Neal, S.J., Hughes, Z.A. *et al.* (2012) Evidence for sustained elevation of IL-6 in the CNS as a key contributor of depressive-like phenotypes. *Translational Psychiatry*, 2, e199.

103 Tuglu, C., Kara, S.H., Caliyurt, O. *et al.* (2003) Increased serum tumor necrosis factor-alpha levels and treatment response in major depressive disorder. *Psychopharmacology*, 170(4), 429–433.

104 Simen, B.B., Duman, C.H., Simen, A.A. *et al.* (2006) TNF-alpha signaling in depression and anxiety: behavioral consequences of individual receptor targeting. *Biological Psychiatry*, 59(9), 775–785.

105 O'Connor, J.C., Lawson, M.A., André, C. *et al.* (2009) Lipopolysaccharide-induced depressive-like behavior is mediated by indoleamine 2,3-dioxygenase activation in mice. *Molecular Psychiatry*, 14(5), 511–522.

106 Lawson, M.A., McCusker, R.H. & Kelley, K.W. (2013) Interleukin-1 beta converting enzyme is necessary for development of depression-like behavior following intracerebroventricular administration of lipopolysaccharide to mice. *Journal of Neuroinflammation*, 10, 54.

107 Arakawa, S., Shirayama, Y., Fujita, Y. *et al.* (2012) Minocycline produced antidepressant-like effects on the learned helplessness rats with alterations in levels of monoamine in the amygdala and no changes in BDNF levels in the hippocampus at baseline. *Pharmacology, Biochemistry, and Behavior*, 100(3), 601–606.

108 Liu, D., Wang, Z., Liu, S. *et al.* (2011) Anti-inflammatory effects of fluoxetine in lipopolysaccharide(LPS)-stimulated microglial cells. *Neuropharmacology*, 61(4), 592–599.

109 Obuchowicz, E., Kowalski, J., Labuzek, K. *et al.* (2006) Amitriptyline and nortriptyline inhibit interleukin-1 release by rat mixed glial and microglial cell cultures. *The International Journal of Neuropsychopharmacology / Official Scientific Journal of the Collegium Internationale Neuropsychopharmacologicum (CINP)*, 9(1), 27–35.

110 Walberer, M., Jantzen, S.U., Backes, H. *et al.* (2014) In-vivo detection of inflammation and neurodegeneration in the chronic phase after permanent embolic stroke in rats. *Brain Research*, 1581, 80–88.

CHAPTER 5

Oxidative and nitrative stress in schizophrenia

Anna Dietrich-Muszalska

Department of Biological Psychiatry of the Chair of Experimental and Clinical Physiology, Medical University of Lodz, 92-215 Lodz, Poland

---

THEMATIC SUMMARY BOX

At the end of this chapter, students should be able to:

- Describe the main features of schizophrenia

- Outline the etiopathogenesis of this disorder

- Describe the changes in cellular biomolecules induced by oxidative stress in schizophrenia

- Analyze the contribution of oxidative and nitrative stress in schizophrenia

- Describe the changes in cellular biomolecules induced by oxidative stress in schizophrenia

- Know which enzymes are responsible for the synthesis of NO

- Explain the use of oxidative stress biomarkers in psychiatric research

- Point out the role of antioxidants in psychiatric disorders

---

Introduction

Since the discovery of the first free radical by Gomberg in 1900 and the suggestion of its role in the etiology of schizophrenia in the mid-1950s, there has been a great progress in the study of oxidative stress in neuropsychiatric disorders.[1,2] The role of oxidative stress and lipid peroxidation causing the alteration in the composition of phospholipid membrane was presented by Horrobin in 1998.[3]

Antioxidant defense deficiency in the cortex of maturational development and stressful physiological activity leads to increased concentrations of reactive oxygen species (ROS), which in turn cause cellular injury dysfunction, and potentially cell death.[4,5] A large body of evidence suggests that ROS are involved in cell membrane pathology and may play a role in schizophrenia and other psychiatric disorders.

Reactive oxygen species include both free radicals and species that are not free radicals (hydrogen peroxide, singlet oxygen, ozone, hypochlorite, and peroxynitrite).

Oxidative Stress and Antioxidant Protection: The Science of Free Radical Biology and Disease, First Edition. Edited by Donald Armstrong and Robert D. Stratton.
© 2016 John Wiley & Sons, Inc. Published 2016 by John Wiley & Sons, Inc.

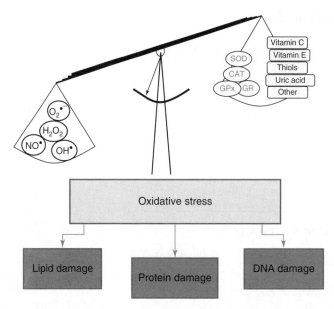

Figure 5.1 The predominance of oxidation processes caused by oxidative stress. (Figure prepared by A. Dietrich-Muszalska.)

Free radicals play an important role in cellular signaling. However, being highly unstable with unpaired electrons, they have enough oxidative potential and strength to damage cellular proteins, lipids, carbohydrates, and nucleic acids. In healthy organisms under physiological conditions, there is a balance between the production of ROS and the activities of antioxidant enzymes and other antioxidants since multiple defense mechanisms exist to protect the organism against free radicals. The balance between the prooxidative and antioxidative processes is a physiological phenomenon used in many important functions of the body. The imbalance with a predominance of oxidation, defined as oxidative stress (Figure 5.1), may be involved in the pathogenesis of many diseases including schizophrenia.

Reactive oxygen species are produced in mitochondria as a leak of electrons from the electron transport chain. Oxidative phosphorylation in the mitochondrial electron transport chain is highly effective, but incomplete reduction in oxygen during respiration produces superoxide anions ($O_2^{\bullet-}$) that are enzymatically dismutated by superoxide dismutase (SOD) to hydrogen peroxide (H_2O_2). Hydroxyl radicals, in turn, are formed via the Fenton reaction.[6] Because of these reactions, superoxide anion ($O_2^{\bullet-}$) is one of the most important ROS and is controlled via multiple enzyme systems: SOD, catalase (CAT), glutathione (GSH) transferase and thioredoxin (Trx). Nicotinamide adenine dinucleotide phosphate (NADPH) oxidase (NOX2) and xanthine oxidase (XO) are also involved in the production of free radicals.[6,7]

Since the brain with its high metabolic rate is particularly vulnerable to oxidative damage, the impairment of redox mechanisms in the brain is manifested as an imbalance in the generation and scavenging of ROS and reactive nitrogen species (RNS) as well as altered regulation of these fundamental redox mechanisms. This imbalance leads to oxidative stress, which can contribute to the pathogenesis of the brain abnormalities in schizophrenia.[6] Moreover, cognitive dysfunction in schizophrenia is associated with changed redox mechanisms.[8,9]

Oxidative stress participates in the development of many diseases and is believed to contribute to aging of cells and organisms. Oxidative stress-induced impairment of neuronal processes may be involved in neurodegeneration and also in the pathophysiology of neuropsychiatric diseases.

Oxidative stress and psychiatric disorders

Although the brain is the major organ affected by oxidative stress in neuropsychiatric disorders, the results of numerous studies indicate that the markers of oxidative cell injury in the central nervous system (CNS) correlate with the markers in peripheral cells, mainly the blood cells (erythrocytes and platelets).

The CNS presents high amount of oxidizable substrates, high oxygen tension, and relatively low antioxidant capacity, making it extremely vulnerable to oxidative damage.[10,11] Moreover, high metal content (e.g., iron, zinc, copper and manganese) in the brain can catalyze the formation of ROS/RNS. The brain utilizes more than 20% of oxygen consumed by the body, yet it comprises only 2% of the total body weight.[12,13] The high energy demand from oxidative glucose metabolism and a high concentration of polyunsaturated fatty acids (PUFAs) and relatively low levels of antioxidants are, therefore, thought to render the brain more vulnerable to oxidative damage than most organs.[12,14]

The impaired redox regulation caused by genetic and environmental factors, with alterations of antioxidant defense systems and oxidative stress in brains of schizophrenic patients, is associated with various pathophysiological processes including mitochondrial dysfunction, inflammation, epigenetic changes, impairment of cell signaling, hypoactive N-methyl-D-aspartate (NMDA) receptors, and impairment of γ-aminobutyric acid (GABA) interneurons.[15–19] Schizophrenia is characterized by mitochondrial dysfunction with oxidative damage to the mitochondrial membrane and the mitochondrial electron transport chain dysfunction.[20–22] Dopamine (DA) may also impair mitochondrial membrane potential.[18,23]

In some brain structures, the presence of significant quantities of transition metal ions such as iron (Fe), copper (Cu), and manganese (Mn) can also contribute to the formation of ROS.[6] Free radical-induced peroxidation of lipid/phospholipid in cell membranes may cause damage to the membrane and changes in its biophysical properties, such as fluidity and inactivation of receptors/enzymes associated with the membrane.[24] These processes, in turn, may lead to disturbances in the physiological function of cells, especially neurons. Neurons are particularly sensitive to oxidative stress; therefore, the efficiency of the antioxidant defense system plays an important role in their normal functioning. The low antioxidant defense system observed in schizophrenia seems to be, in part, responsible for oxidative stress in the disorder.[25,26]

All types of biological molecules such as DNA, proteins, and lipids are under the attack of ROS and RNS. Because oxygen free radicals are rapidly metabolized *in vivo*, their evaluation represents an extremely difficult task to perform. Therefore, the estimation of biomarkers of oxidative/nitrative changes in biomolecules is used.

Exposure of proteins to ROS/RNS can alter the structure of target proteins by the oxidation of side-chain groups, protein scission, backbone fragmentation, cross-linking, unfolding, and the formation of new reactive groups, causing the loss of protein function and resulting in by-products and/or protein aggregates. Markers of oxidative/nitrative changes in proteins are levels of carbonyl groups, free thiols, nitrotyrosine, and dityrosine. Biomarkers of lipids are the products of lipid peroxidation such as malondialdehyde (MDA), lipid peroxides, thiobarbituric acid reactive substances (TBARS), and isoprostanes.

The effects of oxidative modification of phospholipids, neuronal DNA, mitochondrial DNA, and proteins on the neuronal functions may lead to various altered mechanisms in the course of schizophrenia.

Schizophrenia and oxidative stress

Schizophrenia is fully recognized as a multidimensional illness, with a profound impact on behavior, perception, thinking, emotions, neurocognition, and social function.[27] It affects 0.5–1% of total world population. The disease is characterized by a chronic and often recurrent course, and causes serious deterioration in cognitive and psychosocial functioning. These findings occur, in most cases, as early as during the first 5 years of the course.[28] There are three main categories of symptoms: positive symptoms (e.g., delusions and hallucinations), negative symptoms (e.g., flat affect, lack of motivation, and deficits in social function), and cognitive deficits.[28,29]

The etiopathogenesis of schizophrenia is difficult to study because of the heterogeneity of patient populations, differing symptoms, long-term treatment, various side effects, and difficulties in diagnosis, particularly in the early stage of the disease. Numerous hypotheses of etiopathogenesis of schizophrenia have been postulated, with neurodevelopmental hypothesis being the leading one, but others include immunological, inflammatory, infectious, and neurodegenerational.[30–44] These hypotheses have been proposed in an attempt to explain the pathophysiology of schizophrenia, but no single theory seems to account for all aspects of the disease.

Neuronal maldevelopment, impaired neurotransmission, genetic factors, viral infections, environmental factors, and stressors are the main triggering causes of schizophrenia.[45–48] Moreover, mitochondrial pathology and oxidative stress may be the most critical components in the pathology of schizophrenia.[49–51] Molecular mechanisms leading to oxidative stress involved in the pathophysiology of schizophrenia are not yet fully elucidated. This diversity reflects the considerable neurobiological heterogeneity and complexity of clinical syndromes of the disease. The implication of oxidative stress in inflammatory process, neurodegeneration and neurodevelopment, is emerging as an important mechanism underlying various pathological processes. Evidence accumulated from numerous types of studies, including those of obstetric complications, facial dysmorphogenesis, genetic neuroimaging, and neuropathological studies, supports both neurodevelopmental and neurodegeneration theories. The development of new investigative techniques, especially neuroimaging, and studies of apoptotic pathways seem to prove neurodegenerative and neurodevelopmental theories, indicating that oxidative/nitrative damage may integrate and support basic mechanisms of abnormal neurodevelopment and neurodegeneration processes in schizophrenia. A unifying hypothesis has been proposed to conceptualize schizophrenia as a progressive neurodevelopmental disorder.[52]

Presently, pathophysiology of schizophrenia is regarded as the result of pathological alterations in architecture and function of the brain circuits supporting perceptual, cognitive, and emotional processes, leading to disturbances of neural coordination within these networks.[53–55]

Converging lines of evidence (including reduced neuropil) suggests that disrupted cortical synaptic circuitry is a central deficit in schizophrenia.[56] Apoptotic mechanisms may also be involved in this process and the pathophysiology of the illness.[57] The postmortem findings contribute to the hypothesis that schizophrenia has its source in altered synaptic circuitry.[4] Neuroimaging studies also indicate that a progressive loss of cortical gray matter occurs in the early course.[54,58] Mechanisms of the defects and synaptic dysfunction suggest that dysregulation of neuronal apoptosis may contribute also to its pathophysiology. Moreover, apoptotic mechanism seems to be responsible for the progressive gray matter volume loss (first onset of psychosis) when antioxidant activity against oxidative stress is low.[5,59]

Biochemical alterations in the brain, especially the DA system with the production of free radicals and oxidative damage to the brain structure, might be partly responsible for the pathogenesis of this heterogeneous disease. Oxidative stress is involved in different complicated mechanisms of the disease and might be involved in the explanation of various pathological processes, integrating some etiopathogenetic hypotheses about schizophrenia.

The hypothesis of oxidative stress with lower antioxidant defense and oxidative damage, reported by numerous studies, supports pathophysiological progression of schizophrenia and is consistent with the neurodegeneration hypothesis.[34] Moreover, inflammation has been postulated to be a factor in the pathophysiology of this disorder leading to overproduction of prostaglandins from arachidonic acid (AA), especially PGE2 and production of proinflammatory cytokines (IL-6).[60] The activity of cyclooxygenase-2 (COX-2) responsible for the synthesis of prostaglandins is also elevated.

Biomarkers of oxidative stress in schizophrenia

Specific markers of oxidative stress associated with schizophrenia may provide a key link connecting mechanisms underlying numerous processes described in the disease. Mitochondrial dysfunction triggers extensive generation of free radicals leading to oxidative/nitrative stress, abnormalities, and pathologic consequences. Patients with schizophrenia present an increase in oxidative damage to lipids, proteins, and DNA in both central and peripheral tissues with deficits of antioxidant defense, especially a deficit of GSH. Oxidative/nitrative changes induced by oxidative stress in schizophrenic patients are not restricted only to the brain, but may be observed in the blood and peripheral cells (plasma, erythrocytes, and blood platelets). These elements are used to estimate the level of oxidative damage markers that may give clues to the relation of oxidative damage to the illness onset and progression.

Direct evidence of oxidative stress, especially when tardive dyskinesia occurs, has been obtained in animal models.[61] The functional and biochemical consequences of oxidative stress are studied using animal models and postmortem brain analysis.

Common markers used to assess the extent of oxidative/nitrative stress in schizophrenic patients are the products of oxidized and changed biomolecules: lipids, proteins, nucleic acids, and also the activities of antioxidant defense system. The increase in the level of lipid peroxidation products (measured commonly as TBARS, MDA, 4-hydroxynonenal, and isoprostanes), altered proteins and amino acids (estimated as the level of generated carbonyl groups or protein peroxides), 3-nitrotyrosine, and the presence of DNA damage products (8-hydroxyguanosine, telomere shortening) as well as reduced antioxidant defense systems were observed in schizophrenia.[26,62–67]

Lipid peroxidation

Lipid peroxidation through free radical oxidation of unsaturated fatty acids (arachidonic acid and docosahexaenoic acid (DHA)) is involved in the oxidative injury in schizophrenia. Several methods have been developed for the measurement of the products of free radical-induced lipid peroxidation, such as lipid peroxides, and TBARS, to assess oxidative injury in the disorder. There are other markers of free radical-induced lipid peroxidation: 4-hydroxynonenal, ethane, and isoprostanes.[68–70] Analysis confirmed the value of these markers in the assessment of oxidative stress.[71,72] The contribution of oxidative injury to the pathophysiology of schizophrenia was shown by the increase in lipid peroxidation products in the plasma and CSF from patients, and the altered levels of both enzymatic and nonenzymatic antioxidants in chronic, drug-naïve, and first-episode patients.[72–76]

Enhanced carbonyl stress as a disease feature in a subpopulation of schizophrenics was observed.[63,77] Dicarbonyls, for example, methylglyoxal (MG), a potent protein-glycating agent, are formed from sugars, lipids, and amino acids.[78,79] Dicarbonyl accumulation modifies proteins and leads to the eventual formation of advanced glycation end-products (AGEs).[80] An increase in the level of AGE was observed in schizophrenia.[77] Carbonyl stress (dicarbonyl accumulation) was found in patients with schizophrenia and may be a key concept for clarifying some of the pathogenic and pathological mechanisms.

Isoprostanes are a group of prostaglandin isomers belonging to the family of prostaglandin derivatives synthesized mainly from AA *in vivo* through a nonenzymatic free radical-mediated mechanism.[81–83] Estimation of isoprostanes is recommended to monitor the oxidative stress in patients with schizophrenia, according to the clinical status of the disease (type of schizophrenia, predominance of positive or negative symptoms, first episode or recurrence, acute episode or remission) and can be useful for monitoring the effectiveness of treatment. The study on the assessment of increased F_2-isoprostane concentration (8-iso-PGF$_2\alpha$) in patients with schizophrenia was the first one showing that the oxidation of arachidonic acid occurs through the nonenzymatic pathway (not associated with cyclooxygenase), as a result of free radical attack on membrane structures.[70]

Cell membrane phospholipids and polyunsaturated fatty acids in schizophrenia

Membrane phospholipids such as phosphatidylserine (PS), phosphatidylethanolamine (PE), and phosphatidylinositol (PI) are highly rich in AA and DHA, which are predominantly released by phospholipase A2 (PLA2) after the stimulation of cell receptors.[84]

Polyunsaturated fatty acids (PUFAs), (such as AA and DHA) components of membrane phospholipids play an important role in cell membrane dynamics.[85,86] These defects, caused partly by oxidative stress, have been observed in schizophrenia patients during the course of illness, suggesting PUFAs dysregulation in schizophrenia.[84] The reduced level of PUFA has been reported in schizophrenia.[87,88] Phospholipids in the brain are extremely rich in PUFAs and in the neuronal membranes, constituting as high as 65% compared with other cells (15–35%).[69,89] The postmortem cortical tissue study also implicates abnormal fatty acid composition in the frontal cortex in schizophrenia patients.[26,90,91] The reduced level of secondary messengers derived from membrane arachidonic acid, caused by oxidative stress, may be a key factor in modified signal transduction and neuronal deficits.[3,84] Impairment of membrane phospholipids induced by free radicals and their dysfunction lead to disturbance of signal transduction and can also be linked to the changed function of neurotransmitters in schizophrenia, especially glutamatergic and serotonergic systems.[84,86] Dysregulation of glutamatergic mechanisms, particularly hyperactive in exacerbations of psychosis, and the related glutamate excitotoxicity may be associated with oxidative stress and changes in the composition of membrane phospholipids.[92]

The omega-6/omega-3 PUFA ratio in membrane phospholipids not only has an affect on the fluidity of the membrane but also can affect the ligand–receptor interaction, possibly by increasing the availability of surface protein receptors and/or by increasing the concentration of receptors in the membrane.[93]

Reactive nitrogen species in the central nervous system

Nitric oxide (NO) is a small gas molecule of profound importance in intercellular signaling in the CNS. NO is generated mainly enzymatically by nitric oxide synthases (NOSs) from L-arginine in the presence of NADPH, tetrahydrobiopterin, and flavin adenine nucleotides as cofactors (Figure 5.2). There are three NOS isoforms: neuronal (nNOS or NOS-1), endothelial (eNOS or NOS-3), and cytokine inducible (iNOS or NOS-2).

Nitric oxide is a widespread and multifunctional biological messenger molecule in the CNS and functions as a neurotransmitter playing roles in neurodevelopment, cell migration, and neurotransmission (long-term potentiation (LTP), neurosecretion, formation of synapses, synaptic plasticity, and release of other neurotransmitters). NO is especially important as a second messenger of the NMDA receptor activation, interacting with both dopaminergic and serotonergic pathways.[8] NO is important in tissue injury in various neuropsychiatric disorders including schizophrenia.[51,94] A

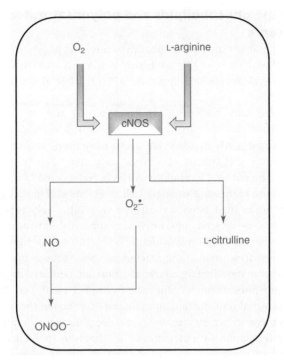

Figure 5.2 Nitric oxide (NO) and peroxynitrite (ONOO⁻) syntheses; cNOS – cellular NOS. (Figure prepared by A. Dietrich-Muszalska.)

positive correlation between the duration of schizophrenia and levels of NO could be explained by pathophysiological differences between the acute and chronic phases of schizophrenia and by the use of medications in the course of the disease.[95]

Nitric oxide is synthesized in the postsynaptic region and during the excessive activation of NMDA receptors. The excessive release of NO may be involved in free radical damage to DNA. NO can be converted into a variety of different RNS including nitrosonium cation (NO⁺), nitroxyl anion (NO⁻), peroxynitrite (ONOO⁻), S-nitrosocysteine, and S-nitrosoglutathione.[96] S-nitrosylation and subsequent further oxidation of critical cysteine residues can also lead to mitochondrial dysfunction.[97] NO may play a dual role, both neurotoxic and neuroprotective. The neurotoxic effects of NO are dependent on its rapid reaction with superoxide radical and formation of ONOO⁻.[96,98] ONOO⁻ is an anion of unstable peroxynitrous acid. It is a relatively long-lived cytotoxic oxidant and may be involved in schizophrenia due to its reaction with a wide range of biological molecules: damaging proteins and nucleic acids; initiating lipid peroxidation; and depleting cellular antioxidants, such as GSH and ascorbate. ONOO⁻ initiates lipid peroxidation in membranes of schizophrenia patients and may react directly with free thiols, leading to enzyme inhibition and autooxidation of DA cell membranes. These reactions may be especially relevant for psychiatric disorders.[99]

Peroxynitrite causes nitration of tyrosine residues in proteins to form 3-nitrotyrosine.[100] Nitration of protein, as measured by the level of 3-nitrotyrosine, was increased in the prefrontal cortex of patients with schizophrenia.[101] Elevated level of 3-nitrotyrosine was found in the plasma proteins and blood platelet proteins of patients with schizophrenia.[62,63] The presence of the lowered activity of SOD found in the platelets from patients with schizophrenia leads to higher levels of superoxide anion not being scavenged and, in the presence of NO, causes enhanced ONOO⁻ formation and increased oxidative stress.[75,102]

Nitrative stress in schizophrenia

Nitric oxide and the main enzyme-producing NO in the CNS (NOS-1) play an important role in the pathogenesis of schizophrenia. NO induces DA release in the striatum and prefrontal cortex, and this suggests that NO is an important modulator of dopaminergic activity. NO can act as neuromodulator of NMDA receptor functions while glutamate stimulation of NMDA receptors results in NO synthesis and release. The NMDA–NO–cyclic GMP pathway modulates the release of neurotransmitters such as glutamate and DA (Figure 5.3). The level of protein nitration, dependent on the formation of ONOO⁻, is increased in the disease.

Several studies have demonstrated a potential pathological link between NMDA hypofunction, enhanced neuronal production of IL-6, and oxidative stress, which may in turn be associated with the GABAergic dysfunction often observed in the

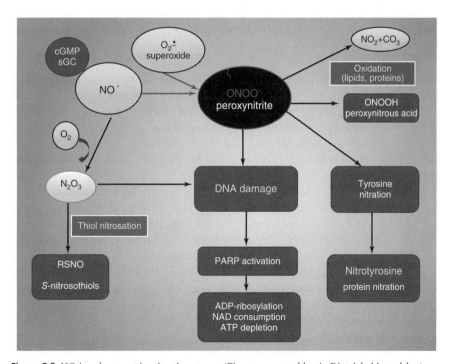

Figure 5.3 NO involvement in nitrative stress. (Figure prepared by A. Dietrich-Muszalska.)

brain of patients with schizophrenia. NO generation in the brain of schizophrenia patients may have effects on synaptogenesis and synaptic remodeling, impaired receptor expression and/or functioning, release of neurotransmitters, mitochondrial pathology, oligodendrocyte injury, and impaired myelination of outer membrane. The excessive production of RNS (NO, ONOO⁻) and reduced defense against their elimination is called nitrosative stress. Enhanced content of products of NO metabolism, increased protein nitrosylation, and nitration (especially increased content of 3-nitrotyrosine residues) belong to biomarkers of nitrative stress.[103,104]

Role of glutathione in schizophrenia

Glutathione, a tripeptide formed of glutamate, glycine, and cysteine with an active thiol group, is involved in disulfide bond formation. GSH may be oxidized to form glutathione disulfide (GSSG), giving a low GSH/GSSG ratio in this disorder.[70] GSH plays a role in detoxification and antioxidant defense.[12] As an important antioxidant in the brain, it is implicated in the pathophysiology of schizophrenia. The antioxidant function of GSH is due to the redox-active thiol group that becomes oxidized when GSH reduces target molecules. The GSH/GSSG ratio is a marker of the redox status, which is low in this disorder. An impairment of GSH synthesis may be one of the causes of oxidative stress in schizophrenia.[105] GSH is a major nonprotein antioxidant in the brain and protects the neurons against damage caused by ROS.

The decrease in low-molecular-weight thiols in schizophrenic patients is associated with a significant reduction in the antioxidant defense system.[70] The deficit of GSH can lead to peroxidation of membrane lipids and microdamage in dopaminergic terminals, causing the loss of synaptic connectivity.[106]

The decrease in plasma levels of total and reduced GSH, with altered antioxidant enzyme activities, was reported in drug-naive first-episode patients with schizophrenia.[107] Reduced GSH levels in prefrontal cortex and deficient levels of GSH and its metabolite γ-D-glutamylglycine in the cerebrospinal fluid were found in drug-naïve or drug-free patients with schizophrenia.[108] The significant decrease in GSH levels in the prefrontal cortex in patients with schizophrenia was proved by magnetic resonance spectroscopy studies *in vivo*.[91,109] GSH may exert its antioxidant effects through several different mechanisms. GSH is involved in the detoxification of drugs, storage of cysteine, and may affect gene expression and development of neurons.[12] Moreover, GSH also increases the action of glutamate at the NMDA, and the reduction in GSH concentration could also contribute to the hypofunction of NMDA receptor in the brain.[110–112] Metabolism of DA also plays a pathological role in schizophrenia. DA metabolism in brain with autoxidation is the major contributor to ROS production, and monoamine oxidase (MAO) activity is the most significant contributor to this process.[113] Because of the high iron content in the brain, it seems to be reasonable to conclude that DA autoxidation is the most important source of elevated hydrogen peroxide concentration.[113]

It was shown that DA in cultured cortical neurons decreased the GSH level by 40% due to conjugation.[106] Postmortem studies have also revealed reduction in GSH level (about 40%) in the caudate nucleus and prefrontal cortex of schizophrenia patients.[91,114] GSH acts as a cofactor for antioxidant enzymes such as glutathione peroxidase (GPx) and GSH transferase; regenerates other important antioxidants, vitamins C and E; and may directly eliminate ROS.[105,115] The significant reduction in the amount of low-molecular-weight thiols such as GSH and its precursors, cysteine and cysteinylglycine, in schizophrenia was described. A significant increase in the amount of homocysteine (HCSH) in plasma of schizophrenic patients was also found.[70]

Increased risk of schizophrenia is linked to polymorphisms of genes associated with GSH synthesis.[116,117] Oxidative stress in schizophrenia may be partly associated with polymorphisms in the gene coding the key enzyme for GSH synthesis such as glutamate cysteine ligase.[116,117] GSH deficit causes impairment of parvalbumin neurons with concomitant reduction in gamma oscillations in the hippocampus.[118] GSH seems to be a novel treatment target in schizophrenia and other psychiatric disorders, since GSH has gained dominance as the important cellular antioxidant. N-acetylcysteine (NAC), a GSH precursor, seems to be a useful agent in the treatment of various psychiatric disorders including schizophrenia.[119–121] The mechanism of NAC action is not clearly understood. NAC may act as a precursor of GSH and may be involved in the modulation of glutamatergic, neurotrophic, and inflammatory pathways.

Antioxidants

The antioxidant defense system consists of various antioxidant enzymes including SOD, CAT, GPx, Trx, and numerous nonenzymatic compounds. The antioxidants may act cooperatively and protect the biomolecules against oxidative damage.

To evaluate the antioxidant defense system in plasma reflecting the total activity of all antioxidants present in plasma, the total antioxidant status (TAS) should be measured together with the estimation of individual antioxidants. TAS in plasma was reduced in patients with chronic schizophrenia as well as in drug-naive first-episode patients with schizophrenia.[25,122] TAS in plasma depends mainly on a high concentration of albumin and the presence of ascorbic and uric acids, bilirubin, tocopherol, melatonin, and other exogenous antioxidants derived from diet, especially numerous polyphenols. Individual plasma antioxidants, albumin, bilirubin, and uric acid were also found lower in schizophrenia subjects.[123–125] Plasma and urinary vitamin C levels are also lower in chronic schizophrenia subjects, even after controlling for diet. McCreadie et al. found lower ratios of vitamin E to cholesterol in schizophrenic patients than in normal control subjects.[126] Decreased levels of GSH, ascorbic acid, and plasma vitamin E levels were also found in erythrocytes from schizophrenic patients.[127]

Uric acid, a product of purine metabolism, is an important nonenzymatic antioxidant present in plasma. The level of uric acid is relatively high and accounts for approximately half of the free radical scavenging activity in human blood.[128–130] Properties of uric acid include quenching of superoxide and singlet oxygen and protecting against oxidation of ascorbic acid through the chelation of iron.

The level of uric acid in CSF is about 10 times lower than in the serum.[131] This suggests that the purine metabolite may be generated peripherally, and its migration to CNS is limited by the blood–brain barrier (BBB). Lower levels of uric acid have been reported in schizophrenic patients.[124,132]

Nitric oxide and antipsychotics in schizophrenia

Antipsychotics used in the treatment of schizophrenia may have different effects on the generation of NO. It is unclear how antipsychotics may affect NO generation and its metabolic pathways. Plasma NO metabolite levels in schizophrenia patients, both before and after the treatment with antipsychotics, were significantly lower than in normal control subjects.[133] Lee and Kim suggested that treatment of schizophrenia patients can lead to a partial normalization of the NO metabolite deficiency, with increased NO metabolite levels paralleling to the improvement of psychiatric symptoms.[134]

Treatment with chlorpromazine, haloperidol, or clozapine normalizes platelet NOS in schizophrenic patients.[135] A significant increase in the level of NO in erythrocytes of patients with schizophrenia treated with antipsychotics was described.[136] Moreover, patients with schizophrenia presented higher nitrate levels than controls, but after olanzapine treatment, lower nitrate levels than those treated with risperidone or haloperidol were observed. Risperidone decreased the reduction of CNS MRI volume in the clinical course of schizophrenia.[137]

Concluding remarks

The investigation of the redox mechanisms involved in schizophrenia (and the potential protective effects of antioxidative therapies) remains a very active field and helps in the understanding and treatment of schizophrenia and other psychiatric disorders associated with oxidative stress.

Multiple choice questions

1 Which statement about free radicals is *not* true?
   a. Free radicals play a role in cellular signaling
   b. Free radicals have no unpaired electrons
   c. Free radicals are unstable and very short living, making measurement difficult
   d. Free radicals can damage cellular proteins, lipids, carbohydrates, and nucleic acids

2 The brain is vulnerable to oxidative stress due to all but which factor?
   a. High level of polyunsaturated fatty acids
   b. Relatively low antioxidant defense system
   c. Stable metabolism with low oxygen utilization
   d. Relatively high levels of polyvalent metal cations

3 Which biomarkers are not used for the estimation of lipid peroxidation?
  a. 4-Hydroxynonenal
  b. Isoprostanes
  c. Thiobarbituric acid reactive substances (TBARS)
  d. 3-Nitrotyrosine

4 Which statement about NO molecule is not true?
  a. NO functions as a neurotransmitter
  b. NO does not react with superoxide anion and does not form ONOO$^-$.
  c. NO modulates the release of neurotransmitters such as glutamate and DA

References

1 Gomberg, M. (1990) An instance of trivalent carbon: triphenylmethyl. *Journal of the American Chemical Society*, 22, 757–771.

2 Hoffer, A., Osmond, H. & Smythies, J. (1954) Schizophrenia: a new approach. II. Results of a year's research. *Journal of Mental Set*, 100, 29.

3 Horrobin, D.F. (1998) The membrane phospholipid hypothesis as a biochemical basis for the neurodevelopmental concept of schizophrenia. *Schizophrenia Research*, 30(3), 193–208.

4 Kasper, S. & Papadimitriou, G.N. (2009) *Schizophrenia*, 2nd edn. Informa Health Care, New York.

5 Glantz, L.A., Gilmore, J.H., Lieberman, J.A. *et al.* (2006) Apoptotic mechanisms and the synaptic pathology of schizophrenia. *Schizophrenia Research*, 8(1), 47–63.

6 Halliwell, B. (1992) Reactive oxygen species and the central nervous system. *Journal of Neurochemistry*, 59(5), 1609–1623 Erratum in: Jornal of Neurochemistry, 2012, 120(5): 850.

7 Halliwell, B. (2006) Oxidative stress and neurodegeneration: where are we now? *Journal of Neurochemistry*, 97, 1634–1658.

8 Bitanihirwe, B.K. & Woo, T.U. (2011) Oxidative stress in schizophrenia: an integrated approach. *Neuroscience and Biobehavioral Reviews*, 35, 878–893.

9 Zhang, X.Y., Chen da, C., Xiu, M.H. *et al.* (2012) Plasma total antioxidant status and cognitive impairments in schizophrenia. *Schizophrenia Research*, 39(1–3), 66–72.

10 Sies, H. (1991) Role of reactive oxygen species in biological processes. *Klinische Wochenschrift*, 69(21–23), 965–968.

11 Halliwell, B. & Whiteman, M. (2004) Measuring reactive species and oxidative damage in vivo and in cell culture: how should you do it and what do the results mean? *British Journal of Pharmacology*, 142(2), 231–255.

12 Dringen, R. (2000) Metabolism and functions of glutathione in brain. *Progress in Neurobiology*, 62(6), 649–671.

13 Berg, D., Youdim, M.B. & Riederer, P. (2004) Redox imbalance. *Cell and Tissue Research*, 318(1), 201–213.

14 Bains, J.S. & Shaw, C.A. (1997) Neurodegenerative disorders in humans: the role of glutathione in oxidative stress-mediated neuronal death. *Brain Research. Brain Research Reviews*, 25(3), 335–358.

15 Benes, F.M. & Berretta, S. (2001) GABAergic interneurons: Implications for understanding schizophrenia and bipolar disorder. *Neuropsychopharmacology*, 25, 1–27.

16 Reynolds, G.P., Abdul-Monim, Z., Neill, J.C. *et al.* (2004) Calcium binding protein markers of GABA deficits in schizophrenia–postmortem studies and animal models. *Neurotoxicity Research*, 6(1), 57–61.

17 Picchioni, M.M. & Murray, R.M. (2007) Schizophrenia. *BMJ*, 335(7610), 91–95.

18 Ben-Shachar, D. (2002) Mitochondrial dysfunction in schizophrenia: a possible linkage to dopamine. *Journal of Neurochemistry*, 83(6), 1241–1251.

19 Nakazawa, K., Zsiros, V., Jiang, Z. *et al.* (2012) GABAergic interneuron origin of schizophrenia pathophysiology. *Neuropharmacology*, 62(3), 1574–1583.

20 Ben-Shachar, D. & Karry, R. (2008) Neuroanatomical pattern of mitochondrial complex I pathology varies between schizophrenia, bipolar disorder and major depression. *Plos One*, 3(11), e3676.

21 Yao, J.K., Reddy, R.D. & van Kammen, D.P. (2001) Oxidative damage and schizophrenia: an overview of the evidence and its therapeutic implications. *CNS Drugs*, 15(4), 287–310.

22 Prabakaran S, Swatton JE, Ryan MM, et al. 2004. Mitochondrial dysfunction in schizophrenia: evidence for compromised brain metabolism and oxidative stress. *Molecular Psychiatry*, 9(7):684-697, 643.

23 Elkashef, A.M., Al-Barazi, H., Venable, D. *et al.* (2002) Dopamine effect on the mitochondria potential in B lymphocytes of schizophrenic patients and normal controls. *Progress in Neuropsychopharmacology and Biological Psychiatry*, 26(1), 145–148.

24 Perluigi, M., Coccia, R. & Butterfield, D.A. (2012) 4-Hydroxy-2-nonenal, a reactive product of lipid peroxidation, and neurodegenerative diseases: a toxic combination illuminated by redox proteomics studies. *Antioxidants and Redox Signaling*, 17(11), 1590–1609.

25 Yao, J.K., Reddy, R., McElhinny, L.G. *et al.* (1998) Reduced status of plasma total antioxidant capacity in schizophrenia. *Schizophrenia Research*, 32(1), 1–8.

26 Yao, J.K. & Reddy, R. (2011) Oxidative stress in schizophrenia: pathogenetic and therapeutic implications. *Antioxidants and Redox Signaling*, 15(7), 1999–2002.

27 O'Leary, D.S., Flaum, M., Kesler, M.L. *et al.* (2000) Cognitive correlates of the negative, disorganized, and psychotic symptom dimensions of schizophrenia. *Journal of Neuropsychiatry and Clinical Neurosciences*, 12, 4–15.

28 Tamminga, C.A. & Holcomb, H.H. (2005) Phenotype of schizophrenia: a review and formulation. *Molecular Psychiatry*, 10, 27–39.

29 Carpenter, W.T. Jr., (1994) The deficit syndrome. *The American Journal of Psychiatry*, 151(3), 327–329.

30 Murray, R. & Lewis, S. (1987) Is schizophrenia a neurodevelopmental disorder? *British Medical Journal (Clinical Research Ed.)*, 295, 681–682.

31 Lewis, D.A. & Levitt, P. (2002) Schizophrenia as a disorder of neurodevelopment. *Annual Review of Neuroscience*, 25, 409–432.

32 Weinberger, D.R. (1986) The pathogenesis of schizophrenia: a neurodevelopmental theory. In: Nasrallah, H.A.W. & Weinberger, D.R. (eds), *The Neurobiology of Schizophrenia*. Elsevier, Amsterdam, pp. 397–406.

33 Weinberger, D.R. (1987) Implications of normal brain development for the pathogenesis of schizophrenia. *Archives of General Psychiatry*, 44, 660–669.

34 Lieberman, J.A. (1999) Is schizophrenia a neurodegenerative disorder? A clinical and neurobiological perspective. *Biological Psychiatry*, 46(6), 729–739.

35 Rund, B.R. (2009) Is schizophrenia a neurodegenerative disorder? *Nordic Journal of Psychiatry*, 63(3), 196–201.

36 Kinney, D.K., Hintz, K., Shearer, E.M. *et al.* (2010) A unifying hypothesis of schizophrenia: abnormal immune system development may help explain roles of prenatal hazards, post-pubertal onset, stress, genes, climate, infections, and brain dysfunction. *Medical Hypotheses*, 74(3), 555–563.

37 Kliushnik, T.P., Kalinina, M.A., Sarmanova, Z.V. *et al.* (2009) Dynamics of immunological and clinical parameters in the treatment of childhood schizophrenia. *Zhurnal Nevrologii i Psikhiatrii Imeni S.S. Korsakova*, 109(6), 46–49.

38 O'Donnell, P. (2012) Cortical interneurons, immune factors and oxidative stress as early targets for schizophrenia. *European Journal of Neuroscience*, 35(12), 1866–1870.

39 Covelli, V., Pellegrino, N.M. & Jirillo, E. (2003) A point of view: the need to identify an antigen in psychoneuroimmunological disorders. *Current Pharmaceutical Design*, 9(24), 1951–1955.

40 Hanson, D.R. & Gottesman, I.I. (2005) Theories of schizophrenia: a genetic-inflammatory -vascular synthesis. *BMC Medical Genetics*, 6, 7.

41 Babulas, V., Factor-Litvak, P., Goetz, R. *et al.* (2006) Prenatal exposure to maternal genital and reproductive infections and adult schizophrenia. *The American Journal of Psychiatry*, 163(5), 927–929.

42 Brown, A.S., Deicken, R.F., Vinogradov, S. *et al.* (2009) Prenatal infection and cavum septum pellucidum in adult schizophrenia. *Schizophrenia Research*, 108(1–3), 285–287.

43 Yolken, R.H. & Torrey, E.F. (1995) Viruses, schizophrenia, and bipolar disorder. *Clinical Microbiology Reviews*, 8, 131–145.

44 Yolken, R.H., Karlsson, H., Yee, F. *et al.* (2000) Endogenous retroviruses and schizophrenia. *Brain Research. Brain Research Reviews*, 31(2–3), 193–199.

45 Kendler, K.S. (2003) The genetics of schizophrenia: chromosomal deletions, attentional disturbances, and spectrum boundaries. *The American Journal of Psychiatry*, 160(9), 1549–1553.

46 Thome, J., Foley, P. & Riederer, P. (1998) Neurotrophic factors and the maldevelopmental hypothesis of schizophrenic psychoses. Review article. *Journal of Neural Transmission*, 105(1), 85–100.

47 Carlsson, A., Waters, N. & Carlsson, M.L. (1999) Neurotransmitter interactions in schizophrenia–therapeutic implications. *Biological Psychiatry*, 46(10), 1388–1395.

48 Pearce, B.D. (2001) Schizophrenia and viral infection during neurodevelopment: a focus on mechanisms. *Molecular Psychiatry*, 6(6), 634–646.

49 Ben-Shachar, D. & Laifenfeld, D. (2004) Mitochondria, synaptic plasticity, and schizophrenia. *International Review of Neurobiology*, 59, 273–296.

50 Bubber, P., Tang, J., Haroutunian, V. *et al.* (2004) Mitochondrial enzymes in schizophrenia. *Journal of Molecular Neuroscience*, 24(2), 315–321.

51 Yao, J.K. & Keshavan, M.S. (2011) Antioxidants, redox signalling, and pathophysiology in schizophrenia: an integrative view. *Antioxidants and Redox Signaling*, 15(7), 2011–2035.

52 Gupta, S. & Kulhara, P. (2010) What is schizophrenia: A neurodevelopmental or neurodegenerative disorder or a combination of both? A critical analysis. *Indian Journal of Psychiatry*, 52(1), 21–27.

53 Insel, T.R. (2009) Disruptive insights in psychiatry: transforming a clinical discipline. *Journal of Clinical Investigation*, 119(4), 700–705.

54 Lewis, D.A. & Sweet, R.A. (2009) Schizophrenia from a neural circuitry perspective: advancing toward rational pharmacological therapies. *Journal of Clinical Investigation*, 119(4), 706–716.

55 Palop, J.J., Chin, J. & Mucke, L. (2006) A network dysfunction perspective on neurodegenerative diseases. *Nature*, 443, 768–773.

56 Lewis, D.A. & Lieberman, J.A. (2000) Catching up on schizophrenia: natural history and neurobiology. *Neuron*, 28(2), 325–334.

57 Jarskog, L.F., Glantz, L.A., Gilmore, J.H. *et al.* (2005) Apoptotic mechanisms in the pathophysiology of schizophrenia. *Progress in Neuropsychopharmacology and Biological Psychiatry*, 29(5), 846–858.

58 Lewis, D.A. (2012) Cortical circuit dysfunction and cognitive deficits in schizophrenia: Implications for preemptive interventions. *European Journal of Neuroscience*, 35(12), 1871–1878.

59 Glantz, L.A., Gilmore, J.H., Overstreet, D.H., Salimi, K., Lieberman, J.A. & Jarskog, L.F. (2010) Pro-apoptotic Par-4 and dopamine D2 receptor in temporal cortex in schizophrenia, bipolar disorder and major depression. *Schizophrenia Research*, 118(1-3), 292–299.

60 Pedrini, M., Massuda, R., Fries, G.R. *et al.* (2012) Similarities in serum oxidative stress markers and inflammatory cytokines in patients with overt schizophrenia at early and late stages of chronicity. *Journal of Psychiatric Research*, 46(6), 819–824.

61 Harrison, P.J. (1999) The neuropathological effects of antipsychotic drugs. *Schizophrenia Research*, 40(2), 87–99.

62 Dietrich-Muszalska, A. & Olas, B. (2009) Modifications of blood platelet proteins of patients with schizophrenia. *Platelets*, 20(2), 90–96.

63 Dietrich-Muszalska, A., Olas, B., Głowacki, R. *et al.* (2009) Oxidative/nitrative modifications of plasma proteins and thiols from patients with schizophrenia. *Neuropsychobiology*, 59(1), 1–7.

64 Miller, B., Suvisaari, J., Miettunen, J. *et al.* (2011) Advanced paternal age and parental history of schizophrenia. *Schizophrenia Research*, 133(1–3), 125–132.

65 Jorgensen, A., Broedbaek, K., Fink-Jensen, A. *et al.* (2013) Increased systemic oxidatively generated DNA and RNA damage in schizophrenia. *Psychiatry Research*, 209(3), 417–423.

66 Malaspina, D., Harlap, S., Fennig, S. *et al.* (2001) Advancing paternal age and the risk of schizophrenia. *Archives of General Psychiatry*, 58(4), 361–367.

67 Malaspina, D., Dracxler, R., Walsh-Messinger, J. *et al.* (2014) Telomere length, family history, and paternal age in schizophrenia. *Molecular Genetics & Genomic Medicine*, 2(4), 326–331.

68 Puri, B.K., Counsell, S.J., Ross, B.M. *et al.* (2008) Evidence from in vivo 31-phosphorus magnetic resonance spectroscopy phosphodiesters that exhaled ethane is a biomarker of cerebral n-3 polyunsaturated fatty acid peroxidation in humans. *BMC Psychiatry*, 8(Suppl 1), S2.

69 Ross, B.M., Maxwell, R. & Glen, I. (2011) Increased breath ethane levels in medicated patients with schizophrenia and bipolar disorder are unrelated to erythrocyte omega-3 fatty acid abundance. *Progress in Neuropsychopharmacology and Biological Psychiatry*, 35(2), 446–53.

70 Dietrich-Muszalska, A. & Olas, B. (2009) Isoprostanes as indicators of oxidative stress in schizophrenia. *World Journal of Biological Psychiatry*, 10(1), 27–33.

71 Grignon, S. & Chianetta, J.M. (2007) Assessment of malondialdehyde levels in schizophrenia: a meta-analysis and some methodological considerations. *Progress in Neuropsychopharmacology and Biological Psychiatry*, 31(2), 365–369.

72 Zhang, M., Zhao, Z., He, L. *et al.* (2010) A meta-analysis of oxidative stress markers in schizophrenia. *Science China. Life Sciences*, 53(1), 112–124.

73 Mahadik, S.P. & Scheffer, R.E. (1996) Oxidative injury and potential use of antioxidants in schizophrenia. *Prostaglandins, Leukotrienes, and Essential Fatty Acids*, 55(1–2), 45–54.

74 Khan, M.M., Evans, D.R., Gunna, V. *et al.* (2002) Reduced erythrocyte membrane essential fatty acids and increased lipid peroxides in schizophrenia at the never-medicated first-episode of psychosis and after years of treatment with antipsychotics. *Schizophrenia Research*, 58(1), 1–10.

75 Dietrich-Muszalska, A., Olas, B. & Rabe-Jablonska, J. (2005) Oxidative stress in blood platelets from schizophrenic patients. *Platelets*, 16(7), 386–391.

76 Tsai, M.C., Liou, C.W., Lin, T.K. *et al.* (2013) Changes in oxidative stress markers in patients with schizophrenia: the effect of antipsychotic drugs. *Psychiatry Research*, 209(3), 284–290.

77 Arai, M., Yuzawa, H., Nohara, I. *et al.* (2010) Enhanced carbonyl stress in a subpopulation of schizophrenia. *Archives of General Psychiatry*, 67(6), 589–597.

78 Thornalley, P.J. (2003) Glyoxalase I–structure, function and a critical role in the enzymatic defence against glycation. *Biochemical Society Transactions*, 31, 1343–1348.

79 Jaisson, S. & Gillery, P. (2010) Evaluation of nonenzymatic posttranslational modification-derived products as biomarkers of molecular aging of proteins. *Clinical Chemistry*, 56, 1401–1412.

80 Kouidrat, Y., Amad, A., Desailloud, R. *et al.* (2013) Increased advanced glycation end-products (AGEs) assessed by skin autofluorescence in schizophrenia. *Journal of Psychiatric Research*, 47, 1044–1048.

81 Morrow, J.D. & Roberts, L.J. (1997) The isoprostanes: unique bioactive products of lipid peroxidation. *Progress in Lipid Research*, 36(1), 1–21.

82 Chen, Y., Morrow, J.D. & Roberts, L.J. (1999) Formation of reactive cyclopentenone compounds in vivo as products of the isoprostane pathway. *Journal of Biological Chemistry*, 274(16), 10863–10868.

83 Brame, C.J., Salomon, R.G., Morrow, J.D. *et al.* (1999) Identification of extremely reactive γ-ketoaldehydes (isolevuglandins) as products of the isoprostane pathway and characterization of their lysyl protein adducts. *Journal of Biological Chemistry*, 274(19), 13139–13146.

84 Skosnik, P.D. & Yao, J.K. (2003) From membrane phospholipid defects to altered neurotransmission: is arachidonic acid a nexus in the pathophysiology of schizophrenia? *Prostaglandins, Leukotrienes, and Essential Fatty Acids*, 69(6), 367–384.

85 Rana, R.S. & Hokin, L.E. (1990) Role of phosphoinositols in transmembrane signaling. *Physiological Reviews*, 70(1), 115–164.

86 du Bois, T.M., Deng, C. & Huang, X.F. (2005) Membrane phospholipid composition, alterations in neurotransmitter systems and schizophrenia. *Progress in Neuropsychopharmacology and Biological Psychiatry*, 29(6), 878–888.

87 Yao, J.K., Stanley, J.A., Reddy, R.D. *et al.* (2002) Correlations between peripheral polyunsaturated fatty acid content and in vivo membrane phospholipid metabolites. *Biological Psychiatry*, 52(8), 823–830.

88 Yao, J.K., Sistilli, C.G. & van Kammen, D.P. (2003) Membrane polyunsaturated fatty acids and CSF cytokines in patients with schizophrenia. *Prostaglandins, Leukotrienes, and Essential Fatty Acids*, 69(6), 429–436.

89 Horrocks, L.A., Ansell, G.B. & Porcellati, G. (1992) *Phospholipids in the Nervous System* Vol. 1: Metabolism. Raven, New York.

90 McNamara, R.K., Jandacek, R., Rider, T. *et al.* (2007) Abnormalities in the fatty acid composition of the postmortem orbitofrontal cortex of schizophrenic patients: gender differences and partial normalization with antipsychotic medications. *Schizophrenia Research*, 91(1-3), 37–50.

91 Gawryluk, J.W., Wang, J.F., Andreazza, A.C. *et al.* (2011) Decreased levels of glutathione, the major brain antioxidant, in post-mortem prefrontal cortex from patients with psychiatric disorders. *International Journal of Neuropsychopharmacology*, 14(1), 123–130.

92 Tsai, G.E., Ragan, P., Chang, R. *et al.* (1998) Increased glutamatergic neurotransmission and oxidative stress after alcohol withdrawal. *The American Journal of Psychiatry*, 155(6), 726–732.

93 Farkas, E., de Willde, M.C., Kiliaan, A.J. *et al.* (2002) Dietary long chain PUFAs differentially affect hippocampal muscarinic 1 and serotonergic 1A receptors in experimental cerebral hypoperfusion. *Brain Research*, 954(1), 32–41.

94 Bernstein, H.G., Bogerts, B. & Keilhoff, G. (2005) The many faces of nitric oxide in schizophrenia. A review. *Schizophrenia Research*, 78(1), 69–86.

95 Oliveira, J.P., Trzesniak, C., Oliveira, I.R. *et al.* (2012) Nitric oxide plasma/serum levels in patients with schizophrenia: a systematic review and meta-analysis. *Revista Brasileira de Psiquiatria*, 34(Suppl 2), S149–S155.

96 Bartosz, G. (1996) Peroxynitrite: mediator of the toxic action of nitric oxide. *Acta Biochimica Polonica*, 43, 645–659.

97 Nakamura, T. & Lipton, S.A. (2011) Redox modulation by S-nitrosylation contributes to protein misfolding, mitochondrial dynamics, and neuronal synaptic damage in neurodegenerative diseases. *Cell Death and Differentiation*, 18(9), 1478–86.

98 Pryor, W.A. & Squadrito, G.L. (1995) The chemistry of peroxynitrite: a product from the reaction of nitric oxide with superoxide. *American Journal of Physiology*, 268, 699–722.

99 Antunes, F., Nunes, C., Laranjinha, J. *et al.* (2005) Redox interaction of nitric oxide with dopamine and its derivatives. *Toxicology*, 208, 207–212 BMC Psychiatry. 11:124.

100 Radi, R. (2004) Nitric oxide, oxidants, and protein tyrosine nitration. *Proceedings of the National Academy of Sciences of the United States of America*, 101, 4003–4008.

101 Andreazza, A.C., Shao, L., Wang, J.F. *et al.* (2010) Mitochondrial complex I activity and oxidative damage to mitochondrial proteins in the prefrontal cortex of patients with bipolar disorder. *Archives of General Psychiatry*, 67, 360–368.

102 Beckman, J.S. & Koppenol, W.H. (1996) Nitric oxide, superoxide, and peroxynitrite: the good, the bad, and ugly. *American Journal of Physiology*, 271, C1424–C1437.

103 White, P.J., Charbonneau, A., Cooney, G.J. *et al.* (2010) Nitrosative modifications of protein and lipid signaling molecules by reactive nitrogen species. *American Journal of Physiology. Endocrinology and Metabolism*, 299, E868–E878.

104 Tao, R.R., Ji, Y.L., Lu, Y.M. *et al.* (2012) Targeting nitrosative stress for neurovascular protection: new implications in brain diseases. *Current Drug Targets*, 13, 272–284.

105 Do, K.Q., Cabungcal, J.H., Frank, A. *et al.* (2009) Redox dysregulation, neurodevelopment, and schizophrenia. *Current Opinion in Neurobiology*, 19(2), 220–230.

106 Grima, G., Benz, B., Parpura, V. *et al.* (2003) Dopamine-induced oxidative stress in neurons with glutathione deficit: implication for schizophrenia. *Schizophrenia Research*, 62(3), 213–224.

107 Raffa M, Atig F, Mhalla A, et al. 2011. Decreased glutathione levels and impaired antioxidant enzyme activities in drug-naive first-episode schizophrenic patients, *BMC Psychiatry*. 11:124.

108 Woo, T.U., Kim, A.M. & Viscidi, E. (2008) Disease-specific alterations in glutamatergic neurotransmission on inhibitory interneurons in the prefrontal cortex in schizophrenia. *Brain Research*, 1218, 267–277.

109 Gawryluk, J.W., Wang, J.F., Andreazza, A.C. *et al.* (2011) Prefrontal cortex glutathione S-transferase levels in patients with bipolar disorder, major depression and schizophrenia. *International Journal of Neuropsychopharmacology*, 14(8), 1069–1074.

110 Köhr, G., Eckardt, S., Lüddens, H. *et al.* (1994) NMDA receptor channels: subunit-specific potentiation by reducing agents. *Neuron*, 12(5), 1031–1040.

111 Papadia, S., Soriano, F.X., Leville, F. *et al.* (2008) Synaptic NMDA receptor activity boosts intrinsic antioxidant defenses. *Nature Neuroscience*, 11(4), 476–487.

112 Steullet, P., Neijt, H.C., Cuénod, M. *et al.* (2006) Synaptic plasticity impairment and hypofunction of NMDA receptors induced by glutathione deficit: relevance to schizophrenia. *Neuroscience*, 137(3), 807–819.

113 Bošković, M., Vovk, T., Kores Plesničar, K.B., Bošković, M., Vovk, T. & Kores Plesničar, K.B. (2011) Oxidative stress in schizophrenia. *Current Neuropharmacology*, 9(2), 301–312.

114 Yao, J.K., Leonard, S. & Reddy, R. (2006) Altered glutathione redox state in schizophrenia. *Disease Markers*, 22(1–2), 83–93.

115 Lu, S.C. (2009) Regulation of glutathione synthesis. *Molecular Aspects of Medicine*, 30(1–2), 42–59.

116 Saadat, M., Mobayen, F. & Farrashbandi, H. (2007) Genetic polymorphism of glutathione S-transferase T1: a candidate genetic modifier of individual susceptibility to schizophrenia. *Psychiatry Research*, 153(1), 87–91.

117 Tosic, M., Ott, J., Barral, S. *et al.* (2006) Schizophrenia and oxidative stress: glutamate cysteine ligase modifier as a susceptibility gene. *American Journal of Human Genetics*, 79(3), 586–592.

118 Steullet, P., Cabungcal, J.H., Kulak, A. *et al.* (2010) Redox dysregulation affects the ventral but not dorsal hippocampus: impairment of parvalbumin neurons, gamma oscillations, and related behaviors. *Journal of Neuroscience*, 30(7), 2547–2558.

119 Matsuzawa, D. & Hashimoto, K. (2011) Magnetic resonance spectroscopy study of the antioxidant defense system in schizophrenia. *Antioxidants and Redox Signaling*, 15(7), 2057–2065.

120 Berk, M., Copolov, D., Dean, O. *et al.* (2008) N-acetyl cysteine as a glutathione precursor for schizophrenia – a double-blind, randomized, placebo-controlled trial. *Biological Psychiatry*, 64(5), 361–368.

121 Berk, M., Ng, F., Dean, O. *et al.* (2008) Glutathione: a novel treatment target in psychiatry. *Trends in Pharmacological Sciences*, 29(7), 346–351.

122 Li, X.F., Zheng, Y.L., Xiu, M.H. *et al.* (2011) Reduced plasma total antioxidant status in first-episode drug-naive patients with schizophrenia. *Progress in Neuropsychopharmacology and Biological Psychiatry*, 35(4), 1064–1067.

123 Yao, J.K., Reddy, R. & van Kammen, D.P. (2000) Abnormal age-related changes of plasma antioxidant proteins in schizophrenia. *Psychiatry Research*, 97(2, 3), 137–151.

124 Yao, J.K., Reddy, R. & van Kammen, D.P. (1998) Reduced level of plasma antioxidant uric acid in schizophrenia. *Psychiatry Research*, 80(1), 29–39.

125 Suboticanec, K., Folnegović-Smalc, V., Korbar, M. *et al.* (1990) Vitamin C status in chronic schizophrenia. *Biological Psychiatry*, 28(11), 959–966.

126 McCreadie, R.G., MacDonald, E., Wiles, D. *et al.* (1995) The Nithsdale Schizophrenia Surveys. XIV: plasma lipid peroxide and serum vitamin E levels in patients with and without tardive dyskinesia, and in normal subjects. *British Journal of Psychiatry*, 167, 610–617.

127 Surapaneni, K.M. (2007) Status of lipid peroxidation, glutathione, ascorbic acid, vitamin E and antioxidant enzymes in schizophrenic patients. *Journal of Clinical and Diagnostic Research*, 1, 39–44.

128 Korte, S., Arolt, V., Peters, M. *et al.* (1998) Increased serum neopterin levels in acutely ill and recovered schizophrenic patients. *Schizophrenia Research*, 32(1), 63–67.

129 Murr, C., Widner, B., Wirleitner, B. *et al.* (2002) Neopterin as a marker for immune system activation. *Current Drug Metabolism*, 3(2), 175–187.

130 Chittiprol, S., Venkatasubramanian, G., Neelakantachar, N. *et al.* (2010) Oxidative stress and neopterin abnormalities in schizophrenia: a longitudinal study. *Journal of Psychiatric Research*, 44(5), 310–313.

131 Bowman, G.L., Shannon, J., Frei, B. *et al.* (2010) Uric acid as a CNS antioxidant. *Journal of Alzheimer's Disease*, 19(4), 1331–1336.

132 Reddy, R., Keshavan, M. & Yao, J.K. (2003) Reduced plasma antioxidants in first-episode patients with schizophrenia. *Schizophrenia Research*, 62(3A), 205–212.

133 Lee, B.H., Lee, S.W., Yoon, D. *et al.* (2006) Increased plasma nitric oxide metabolites in suicide attempters. *Neuropsychobiology*, 53(3), 127–132.

134 Lee, B.H. & Kim, Y.K. (2008) Reduced plasma nitric oxide metabolites before and after antipsychotic treatment in patients with schizophrenia compared to controls. *Schizophrenia Research*, 104(1-3), 36–43.

135 Das, I., Khan, N.S., Puri, B.K. *et al.* (1995) Elevated platelet calcium mobilization and nitric oxide synthase activity may reflect abnormalities in schizophrenic brain. *Biochemical and Biophysical Research Communications*, 212, 375–380.

136 Herken, H., Uz, E., Ozyurt, H. *et al.* (2001) Red blood cell nitric oxide levels in patients with schizophrenia. *Schizophrenia Research*, 52(3), 289–290.

137 Kato, T., Monji, A., Hashioka, S. *et al.* (2007) Risperidone significantly inhibits interferongamma-induced microglial activation in vitro. *Schizophrenia Research*, 92(1–3), 108–115.

CHAPTER 6

The effects of hypoxia, hyperoxia, and oxygen fluctuations on oxidative signaling in the preterm infant and on retinopathy of prematurity

M. Elizabeth Hartnett

Department of Ophthalmology, Moran Eye Center, University of Utah, Salt Lake City, UT, USA

THEMATIC SUMMARY BOX

At the end of this chapter, students should be able to:

- Describe the differences in term and premature birth leading to greater oxidative stress

- Describe ROP and its appearances in high oxygen or regulated oxygen

- Describe the role of hypoxia, hyperoxia, and fluctuations in oxygenation in ROP

- Describe effects from oxidative signaling in normal and aberrant retinal vascular development

---

Introduction

Retinopathy of prematurity (ROP) is a leading cause of childhood blindness worldwide. When ROP was first identified in the 1940s, oxygen and oxygen-related oxidative stress were believed to be culprits. Regulation and monitoring of oxygen greatly reduced the occurrence of ROP, but as smaller and younger premature infants survived, it became clear that high oxygen at birth was not the only cause of ROP. Clinical studies that tested the role of various antioxidants or of different oxygenation profiles in ROP have consistently failed recently to show efficacy and safety. This chapter seeks to define ROP and review the relationship between oxygen levels and oxidative stress in preterm birth and term birth, and the role of current-day oxygen stresses

---

Oxidative Stress and Antioxidant Protection: The Science of Free Radical Biology and Disease, First Edition.
Edited by Donald Armstrong and Robert D. Stratton.
© 2016 John Wiley & Sons, Inc. Published 2016 by John Wiley & Sons, Inc.

including high oxygen and fluctuations in oxygenation in activating oxidative signaling pathways that can lead to normal or aberrant retinal vascular development. Also included is the use of animal models to study ROP.

Anatomy and physiology of the human eye in adult and development

It is important to review anatomic and histologic characteristics of the human eye (Figure 6.1). The eye is a globe, and the retina is a multilayered tissue that lines the back of the globe and consists of many types of cells, including neurons, endothelial cells, and glial cells. The optic nerve consists of approximately a million nerve fibers, the nuclei of which are ganglion cells that line the inner retina. The macula provides high-resolution visual acuity to humans and some other species. Notably, rats

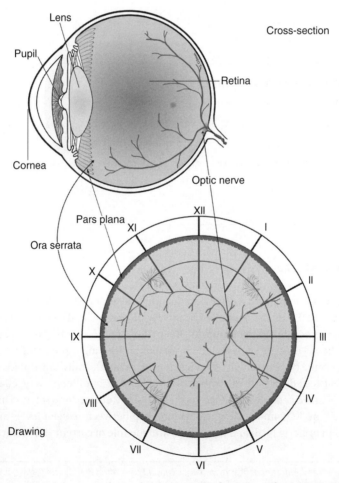

Figure 6.1 Anatomy of the eye and vasculature. (Drawing by James Gilman, CRA, FOPS.) *(See color plate section for the color representation of this figure.)*

and mice do not have maculae. Because of the spherical shape of the eye, locations on the retina are described as follows: posterior (near the optic nerve and macula); anterior (near the peripheral retina adjacent to the ora serrata or furthest extent of the retina); and superior (toward the top of the skull), inferior (toward the chin), temporal (toward the ear), and nasal (toward the nose).

Premature birth

Premature birth in humans is defined as birth before 37 weeks of gestational age. Full-term human birth occurs at 40 weeks of gestational age. Although all aspects of development can be affected in the infant born prematurely, the most extremely premature infants are those with the greatest comorbidities and highest risk of developing ROP. In the United States and in other regions throughout the world that have technologic advances and resources to provide prenatal and perinatal care to the premature infant, the extremely low gestational aged infants (<28 weeks of gestational age) and the extremely low birthweight infants (<1000 g birthweight) are those at the greatest risk.[1] In regions lacking resources for optimal prenatal and perinatal care or having poor maternal nutrition, drug use, or high infant oxygen at birth, ROP develops in premature infants older than 31 weeks of gestational age or larger than 1500 g.[2]

Vascular development (retina and hyaloid) and effects from preterm birth

The adult eye receives its blood supply from two separate circulations, the retinal vasculature, which extends through the inner retina to about the inner nuclear layer, and the choroid, which is beneath the retina but supplies the outer retina and the retinal pigment epithelium (Table 6.1). In development, the hyaloidal vasculature is also important. Most ocular development is known from studies in animals, but there are differences among species in the development of the vasculatures of the eye. Therefore, it is necessary to review the data that are available from human studies. In normal ocular development, the hyaloidal artery enters the eye through the

Table 6.1 Circulations of adult and developing eye.

Name	When develops	Function	Overlap
Hyaloid	Fetal, ~4–5 weeks of gestation	Nutrition + O_2 to developing eye	With choroid and retina until regression at ~36 weeks of gestation
Choroid	Fetal, ~6–8 weeks of gestation	Nutrition + O_2 for RPE and developing photoreceptors	Present in adult with retinal circulation
Retinal	Fetal, angioblasts ~12 weeks of gestation	Nutrition + O_2 for sensory retina	Present in adult with choroid circulation

optic fissure at about 4 weeks and becomes prominent at 5 weeks. At 8–12 weeks, the entire hyaloidal vasculature fills the eye and includes the vaso hyaloidea propria, the hyaloidal artery, the tunica vasculosa lentis, and the pupillary membrane.[3] The choroid begins to form at about 6–8 weeks of gestation and is believed to be almost completely mature by about 26 weeks of gestation. The retinal circulation begins by a process of vasculogenesis whereby precursors migrate from the deeper layers of the eye to the inner retina and become angioblasts.[4] These angioblasts give rise to endothelial cells at about 12 weeks of gestation. During the time from 12 to 22 weeks of gestation, other neural cells in the retina develop and form synapses or connections to one another or with other cells. Some of these neural or glial connections, including endothelial cells, appear important to retinal vascular development. Particularly important are the glial Müller cells that span the inner-to-outer retina and the ganglion cells whose axons form the optic nerve. In general, the maturation of retinal neurons starts in the posterior retina near the optic nerve and extends to the peripheral ora serrata. However, differentiation of the photoreceptors starts peripherally and finishes at the posterior eye in the normal full-term birth. The photoreceptors at the macular center are not fully developed until about term birth. The initial retinal vascular development is through a process of vasculogenesis at least through 22 weeks of gestation. After that, it is harder to obtain intact human eyes, and the ensuing vasculature of the retina is believed to occur through budding angiogenesis based mainly on animal models. Human retinal vascular development is largely completed by about term birth. The hyaloid undergoes regression beginning with the vaso hyaloidea propria at about 12 weeks, continues with the tunica vasculosa lentis and then with the pupillary membrane, and finishes at about 35–36 weeks of gestation.[5]

Preterm infants at the highest risk of severe ROP in the United States are often born younger than 28 weeks of gestation. At this time, the choroid is mostly mature, the hyaloid is still present and has not regressed, and the retinal vasculature is immature. In addition, retinal neurons are still undergoing differentiation and forming synaptic connections. The photoreceptors use much oxygen, and the demand can place additional stress on the retina. In normal development, the oxygen demand can even create a physiologic hypoxia that drives retinal vascularization from posterior to the ora serrata by creating a gradient of angiogenic factors that stimulate endothelial cell growth and migration. However, in the preterm infant, there can be other stresses that affect retinal vascularization and lead to pathologic signaling and aberrant angiogenesis.[5,6]

Oxidative stress and reserves in preterm compared to full-term birth

Many factors are involved in the pathogenesis of ROP, including premature birth, poor nutritional reserve, oxygenation, inflammation, and oxidative stress. There is also evidence that inherited factors are involved in ROP.[7] Several aspects of preterm birth place the infant at risk of oxidative damage and pathologic signaling of oxidative pathways. Evidence suggests that there is an imbalance of reactive species and antioxidative mechanisms in prematurity. Preterm infants are believed to consume more glutathione (GS), which creates a nonenzymatic oxidative reserve in red blood cells.[6,8] Hypoxia in complicated births may limit the ability of the infant to quench mitochondrial-produced reactive species.[6,9] Some prostaglandins, such as E2, are believed to help preserve neural function during episodes of hypoxia that occur in progressive labor, but it is unclear if this mechanism is sufficient with abrupt preterm birth.[6,10] There is also concern that preterm infants have reduced expression of antioxidative enzymes.[6]

Therefore, in the preterm infant, there may be increased levels of oxidative compounds and decreased ability to quench them. Not only can oxidative compounds directly damage tissue, but they also trigger aberrant signaling processes that can lead to pathologic biologic events.[6,11]

Oxygen and oxidative stress

Relationship of arterial oxygen and oxygen saturation and the oxyhemoglobin dissociation curve

**Retinopathy of prematurity was first described as retrolental fibroplasia (RLF) by Terry in 1942,[12] and RLF was likely the most severe form of ROP, stage 5. At that time, there was no technologic way to measure the oxygen content in the blood stream, and much of the oxygen delivered was based on the concentration coming from an external oxygen tank. Now there is understanding of the relationships among inspired oxygen level, arterial oxygen concentration, and what is now commonly measured as arterial oxygen saturation. As an illustration, the adult breathes 21% oxygen at sea level, and usually this causes arterial oxygen concentration of 100 mm Hg. The main transport protein for oxygen in the blood stream is hemoglobin; at 21% inspired oxygen, normal adult hemoglobin is 100% saturated. The relationship between arterial oxygen levels and oxygen saturation is represented by the oxyhemoglobin dissociation curve[13] (Figure 6.2). There are several considerations in the premature infant. The main oxygen transport protein is fetal hemoglobin, which has greater affinity for oxygen than adult hemoglobin and, therefore, releases oxygen at a lower tissue oxygen concentration. Fetal hemoglobin tends to shift the oxyhemoglobin dissociation curve to the left. The other consideration is that above 90% oxygen saturation, the curve flattens so that it is harder to distinguish 95 mm Hg arterial oxygen from much higher levels above 100 mm Hg, which can be damaging to endothelial cells of newly formed retinal capillaries in preterm infants. Neonatologists now strive to maintain oxygen saturations along the slope of the oxyhemoglobin dissociation curve, but despite multiple clinical trials to test the effect of different oxygen saturation targets on ROP severity, seemingly the only consensus achieved is to avoid high oxygen at birth. This is because attempts to reduce oxygen saturation targets have led to increased mortality in some studies as discussed in the Section "The role of oxygen in ROP".

Role of oxygen levels on retinal vascular development

The preterm infant is born into a relatively hyperoxic environment at birth. *In utero*, oxygen levels are about 30–40 mm Hg. When ROP was first identified as RLF in the United States, preterm infants were placed into 100% inspired oxygen (compared to room air at 21% oxygen), which can cause very high arterial oxygen levels. High

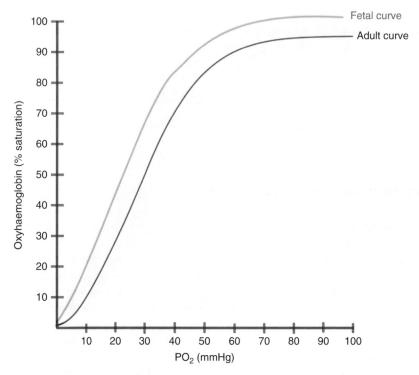

Figure 6.2 Oxyhemoglobin dissociation curve depicting oxygen saturations with fetal and adult hemoglobin. (Drawing by James Gilman, CRA, FOPS.)

oxygen can damage newly formed endothelial cells of capillaries and cause attenuation of newly formed retinal capillary beds resulting in broad areas of avascular retina. When preterm infants were then moved from high supplemental oxygen to a room air environment, the avascular retina that had been created became hypoxic and resulted in the release of "angiogenic" compounds that stimulated blood vessel growth.[14] However, the vessels grew aberrantly rather than into the oxygen-deprived retina. In the United States, attempts to avoid high oxygen at birth are made; however, in developing countries that lack the resources to monitor and regulate oxygen, high oxygen-induced retinopathy (OIR) is seen.[2] Now, it is recognized that besides high oxygen at birth, other oxygen stresses are associated with severe ROP, including fluctuations in oxygen levels and intermittent hypoxia.[15]

Oxygen levels and generation of oxidative compounds

High oxygen levels can lead to an increase in the generation of superoxide radicals in the mitochondria. It is also possible that low oxygen can lead to the generation of the superoxide radical by slowing upstream events in the electron transport chain and by increasing the concentration of oxygen donors.[6,16] Fluctuations in oxygen may also lead to an increase in the reactive oxygen species (ROS) by activating the enzyme NADPH oxidase, which is one of the leading generators of reactive oxygen in endothelial cells.[17] In addition, superoxide, in the presence of nitric oxide (NO), can lead to damaging reactive nitrogen species (RNS) including peroxynitrite.[16]

Human ROP phases

When preterm birth occurs, the retinal vasculature is incompletely developed. It is believed that after vasculogenesis, the retinal vasculature is extended to the ora serrata through a process of angiogenesis (Figure 6.3) whereby endothelial cells proliferate and migrate toward a gradient created by vascular endothelial growth factor (VEGF) and potentially other factors. The retinal tissue oxygenation in the full-term or preterm infant is not known, but it is known that the preterm infant experiences minute-to-minute fluctuations in blood oxygen concentration as measured transcutaneously.[18] The combination of fluctuations in oxygenation with periodic hyperoxia and hypoxia leads to the first phase of ROP known as delayed physiologic retinal vascular development (PRVD). Upon transfer from supplemental oxygen to ambient air, the second phase of ROP occurs with growth of blood vessels into the vitreous and is called vasoproliferation.[19] Blindness occurs from retinal detachment and scarring that result from the interaction between intravitreal angiogenesis from vasoproliferation and the vitreous. However, even less severe forms of ROP can lead to high myopia (severe near-sightedness) and strabismus (crossed eyes), which can reduce visual acuity in the developing child.

Retinopathy of prematurity is clinically defined by several parameters (zone, stage, plus disease) (Table 6.2) useful in determining the severity of ROP and risk of

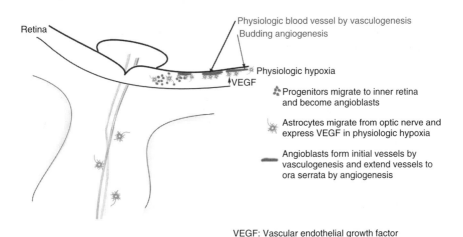

Figure 6.3 VEGF dilemma: retinal vascular development requires VEGF. (Drawing by James Gilman, CRA, FOPS.)

Table 6.2 Parameters used to classify retinopathy of prematurity.

Zone	Area of developed retinal vascularization of vascular cells	Three zones; zone I least mature and worst ROP
Stage	Severity of ROP	Stages 1–2: delayed physiologic retinal vascular development – phase 1 (NOX1, NOX2 involved in apoptosis) Stage 3: vasoproliferation – phase 2 (NOX1, NOX2, NOX4 involved) Stages 4–5: fibrovascular phase – phase 3 (unknown)
Plus disease	Dilation and tortuosity of vessels	Evidence of severe ROP and vascular activity (NOS involved)
Presence and location of retinal detachment	Not reflected in most OIR models	Stage 4a – Not involving fovea Stage 4b – Involves fovea Stage 5 – Total retinal detachment

blindness. The "phases" of ROP are not as clearly defined in preterm human infants as they are in experimental models because of variability among preterm infants and because the individual premature infant has other conditions that can affect ROP.[19]

Phase 1: delayed physiologic retinal vascular development

Delayed PRVD leads to a peripheral area of avascular retina. Clinically, the amount of peripheral avascular retina present is reflected in the zone of ROP or the region of the retina in which retinal vascularization has developed. Three ROP zones have been defined, and zone I has the smallest area of vascularized retina and, in the presence of ROP, has the greatest risk of developing into severe ROP. The causes of delayed retinal vascular development in the preterm infant are partly explained by experimental evidence and some inferred as hypothesis. There is evidence that reduced levels of insulin-like 1 growth factor (IGF-1) in the human premature infant is correlated with delayed PRVD,[20] and experimentally IGF-1 has a permissive role in angiogenesis in experimental models.[21]

Recurrent fluctuations in oxygenation increase the activation of NADPH oxidase, which is associated with apoptosis of cells important in PRVD. Recurrent fluctuations in oxygenation also downregulate erythropoietin expression in Müller cells through a mechanism mediated by VEGF.[22] Erythropoietin, apart from being important in blood cell development (i.e., hematopoiesis), can also lead to angiogenesis and facilitate PRVD. NADPH oxidase can generate superoxide, which in combination with NO, can lead to damaging nitrative compounds that lead to cell damage. Also trans-arachidonic acids in the cell membranes are generated through NO and lead to apoptosis of endothelial cells directly and in an independent mechanism through upregulation of the angiogenic inhibitor, thrombospondin, in experimental models.[23] Oxidative compounds have also been linked experimentally to delayed PRVD in phase I ROP through the use of liposomes of superoxide dismutase (SOD) and the antioxidants, apocynin, and vitamins C and E, which all improve

PRVD.[17,24] The blood fluctuations in oxygenation may directly impact the concentration of VEGF in the retinal tissue, and this upregulation and downregulation of VEGF can first facilitate and then slow angiogenesis, and can thereby theoretically slow PRVD.[25]

Phase 2: vasoproliferation

There is evidence that the same signaling pathways involved in delayed PRVD can also play a role in vasoproliferation.[19] During the phase of vasoproliferation, blood vessels grow into the vitreous rather than into the retina. The hypothesis has been that retinal hypoxia upregulates angiogenic factors that lead to blood vessel growth into the vitreous. Indeed, in several experimental models, hypoxia is increased in the retina once the animal is placed into ambient oxygen conditions either by hyperoxia-induced vaso-obliteration of newly formed capillaries[14] or by recurrent fluctuations in oxygenation that cause delayed PRVD.[19] Several angiogenic factors and inflammatory factors have been associated with intravitreal angiogenesis, but of those, VEGF has been the most recognized and studied factor associated with vasoproliferation. There is also cross talk between VEGF and oxidative pathways. In addition, oxidative pathways have been involved in vasoproliferation independent of VEGF.[11]

Phase 3: fibrovascular phase

In human preterm infants, retinal detachment can occur because of fibrovascular changes that occur between vasoproliferation and the vitreous. Contraction of the scar tissue leads to tractional pulling on the underlying retina and causes retinal detachment. Retinal detachments represent the most severe stages of ROP, stages 4 and 5.

The role of oxygen in ROP

High oxygen and ROP in the 1940s

When ROP was first described in the 1940s, methods to examine the retina had not been widely adopted and it was not clear that high oxygen could cause ROP. Therefore, newborn animals were used in experimental models to test the effects of high oxygen.[26] These newborn animals vascularized their retinas after birth, unlike in the human infant. It was discovered that high oxygen levels were associated with damage to the sensitive newly formed retinal capillaries. The hypothesis was developed that resultant areas of avascular retina stimulated the formation of blood vessels into the vitreous when the animals were removed from high oxygen to room air. This hypothesis described two phases of ROP, vaso-obliteration and vasoproliferation, but these were based on animal models and not human infants.[14] Now it is recognized that much of ROP in the United States occurs as delayed PRVD, with potentially some central vasoconstriction when infants are exposed to high oxygen, followed by later vasoproliferation upon return to ambient air.[19]

Current levels of oxygen in regions with oxygen regulation and monitoring

Historically, high oxygen at birth is known to cause ROP. However, most ROP in the United States develops in association with other oxygen stresses, because attempts are generally made to avoid high oxygen at birth. In regions throughout the developing world that lack resources for good prenatal care and to monitor and regulate oxygen, ROP is seen in larger and older infants who receive 100% oxygen.[2] Recently, in the United States, other oxygen stresses have been associated with severe ROP and these include fluctuations in transcutaneous oxygen[18] and intermittent hypoxic episodes.[27]

Clinical trials on oxygen

In the 1950s, clinical trials were performed in infants to test whether high oxygen at birth was more often associated with ROP.[26] Indeed this was found and efforts were made to restrict high oxygen at the time of birth. However, with a reduction in oxygen, there was greater incidence of cerebral palsy and other serious conditions. Later, methods were used to measure arterial oxygen in the capillaries by transcutaneous monitoring, with additional technological advances by oxygen saturation measurements. Based on retrospective studies or single institution studies, various trials were performed. The supplemental therapeutic oxygen to prevent retinopathy of prematurity (STOP-ROP) tested the hypothesis that higher oxygen saturation targets (96–99% compared to 89–94%) would reduce vasoproliferation and reduce the number of eyes reaching the "threshold" level of the severity of ROP.[28] The hypothesis was not proven although it was found that high oxygen did not increase ROP incidence and, in a subgroup, high oxygen saturation targets were associated with reduced ROP severity in a post hoc analysis. Several previous small studies suggested that infants with low oxygen saturation were less likely to develop ROP. Therefore, other clinical trials such as Surfactant Positive Airway Pressure Pulse Oximetry Randomized Trial (SUPPORT)[29] in the United States; Benefits of Oxygen Saturation Targeting Study II (BOOST II) in Australia, the United Kingdom, and New Zealand[30]; and the Canadian Oxygen Trial (COT) in Canada, the United States, Argentina, Finland, Germany, and Israel[31] were performed. In SUPPORT and BOOST II, there was increased death in the infants with low oxygen saturation targets (85–89% SaO_2) compared to infants with high oxygen saturation targets (91–95% SaO_2), but in survivors, ROP was reduced in infants with low oxygen saturation targets. In the COT, neither ROP nor survival was affected. Apart from the differences in countries in which infants were enrolled and differences in the years when enrolled, infants with pulmonary hypertension were excluded in the COT. Currently, there is no consensus as to the correct oxygen saturations to target in the preterm infant to prevent ROP and maximize survival and development of other organ systems.[32]

Use of animal models to study ROP

Animals that vascularize their retinas after birth are used in models of OIR to simulate the events seen in human ROP. All models have limitations, the common being that the animals are not premature and do not suffer from other conditions associated with prematurity (Table 6.3).

Table 6.3 Major animal models of OIR.

Species	Limitations	Strengths
Mouse	• Not premature • Oxygen levels and constant high oxygen do not reflect human ROP • Difficult to control for growth of pups and metabolism. • Vessels already developed and then die • Does not look like human ROP	• Easier to use transgenic animals to study human ROP
Rat	• Not premature • Hard to study molecular mechanisms (transgenic rats not as common as mouse), but gene therapy and new technologies permit study of molecular mechanisms	• Fluctuations in oxygen • Extrauterine growth restriction (poor postnatal growth similar to risk of human ROP) • Reflects arterial oxygen in human ROP • Looks like phases 1 and 2 human ROP
Beagle	• Not premature • High constant oxygen • Hard to study molecular mechanisms, but gene therapy and new technologies may permit this	• Size of eye more similar to infant, making it a better model for drug studies • Looks like phases 1 and 2 ROP • May share some similarities to human phase 3 ROP

The mouse OIR model[33] subjects newborn mice (with vascularization of the retina to the ora serrata) to high oxygen to create areas of vaso-obliteration in the central retina. The animals are brought into room air, and the avascular retinal area experiences a relative hypoxia that induces angiogenic factors that lead to vasoproliferation into the vitreous at the margins of the avascularized and vascularized retina. The benefit of the model is the relative ease of using transgenic animals to study molecular mechanisms. The limitations are the lack of prematurity, the high and constant oxygen levels used (which do not reflect what most preterm infants experience), the fact the animals have full retinal vascularization when exposed to oxygen, and the retinal appearance, which is dissimilar from the phases of human ROP.

The rat OIR model[34] is the most representative model of human ROP in many aspects. In the model, newborn rat pups within 6 h of birth are placed into a controlled oxygen environment that uses repeated fluctuations in oxygenation. The fluctuations recreate the extremes of transcutaneous oxygen levels that preterm infants experience. Also, similar to the preterm infant retina, physiologic retinal vascularization is incomplete. The model also recreates an appearance similar to the phases of human ROP. The limitations had been the difficulty in studying molecular mechanisms, but this issue has now been addressed using gene therapy approaches.[35] Other limitations are that the animals require long durations of fluctuations in oxygenation to develop phases similar to human ROP, and the newborn rat pups are not preterm.

The beagle OIR model[36] exposes pups to high constant oxygen for 4 days followed by room air. The pups do not have complete retinal vascularization when high oxygen is used and develop appearances similar to the phases of ROP. Limitations include high oxygen levels, newborn, but not premature pups, and difficulty in studying molecular mechanisms. However, there is an advantage in studying the effects of drugs on vasoproliferation or delayed PRVD since the puppy eyes are more similar in size to preterm infant eyes than are either newborn rat or mouse eyes. All models are useful in studying questions related to oxygen induced aberrant angiogenesis, vascular loss and ROP.

Link between oxidative stress and oxygen in ROP

Fluctuations in oxygenation and activation of endothelial NADPH oxidase

NOX isoforms

NADPH oxidase was originally recognized as the mechanism for superoxide burst generation by leukocytes to fight infection. It is now also recognized as the main mechanism of ROS generation in endothelial cells.[37] NADPH oxidase can also cause angiogenesis.[38,39] Isoforms of NADPH oxidase are NOX1–5 and Duox1 and Duox2. The isoforms are differentially expressed in tissues and cells, and evidence suggests that NADPH oxidase isoforms play different roles in physiologic and pathological responses.[38] NOX1, NOX2, and NOX4 are the most widely expressed isoforms in endothelial cells and have been implicated in pathological retinal diseases.[11,40]

Effects of NOX activation on phases of ROP

NADPH oxidase has been shown to be activated and cause vasoproliferation in experimental models. NADPH oxidase isoforms can generate different types of ROS. NOX2 is in leukocytes and endothelial cells and, when activated by repeated oxygen fluctuations in the rat OIR model, has been associated with delayed PRVD.[41] Activation of NADPH oxidase involving NOX2 generates superoxide and requires the aggregation of cytoplasmic and membrane-bound subunits (Figure 6.4). Also, NOX1 requires aggregation of cytoplasmic and membrane-bound subunits to generate superoxide radical and has been shown important in hyperoxia-induced vaso-obliteration.[42] NOX4 is activated in endothelial cells by aggregation with membrane-bound p22phox and does not require aggregation of cytoplasmic subunits. The primary ROS generated by NOX4 activation of NADPH oxidase is H_2O_2. NOX4 activation leads to vasoproliferation in both VEGF dependent and independent ways.[40]

Effects of activation in different retinal cell types in ROP

NADPH oxidase can trigger activation of STAT3,[43,44] which is a transcription factor that can translocate to the nucleus and lead to transcription of genes. It can also trigger cytoplasmic signaling of biochemical pathways. When STAT3 is activated in Müller cells, it leads to downregulation of erythropoietin in association with delayed PRVD.[45] However, when activated in endothelial cells, it can lead to vasoproliferation. VEGF can trigger activation of STAT3, but in some cases of high oxygen, VEGF may be reduced and it is unclear if its presence is needed for the biologic outcomes seen experimentally.

Subunits of NADPH oxidase

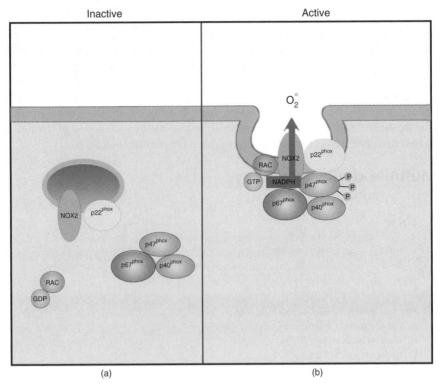

Figure 6.4 Subunits of NADPH oxidase and activation. The isoform NOX4 does not require aggregation with cytoplasmic subunits and becomes activated when it aggregates with p22phox. (Drawing by James Gilman, CRA, FOPS.)

Clinical applications

Clinical trials of antioxidants

Although the preterm infant has a predisposition to increased ROS and RNS and evidence from experimental OIR models suggests that oxidative compounds contribute to features of severe ROP, clinical trials in human preterm infants have not yielded reproducible or significant results. Vitamin E supplementation was tested in several clinical trials, but many studies were inconclusive as to the efficacy in reducing severe ROP or were stopped because of complications such as increased infections and sepsis. A meta-analysis of vitamin E studies showed that supplementation significantly reduced the risk of phase 2, vasoproliferation. In separate studies, low dose N-acetylcysteine[25] or lutein[11] was tested, but no effects were found on ROP. Although other doses or conditions may be efficacious, current studies suggest that the complexity of oxidative signaling in different cells may account for limited success in ROP. Also, oxidative signaling is important as a line of defense to invading microorganisms. The preterm infant is relatively immunosuppressed so that global inhibition of oxidative processes may also be unsafe.

Conclusions

Premature birth can lead to situations in which oxidative compounds are increased, but the ability to quench them is limited. Because of the effects of oxygen on ROP historically, it is not surprising to learn that oxidative signaling plays a role in the development of pathologic features associated with the phases of ROP, leading to greater oxidative stress. However, the complexity of oxidative signaling in different cell types limits the efficacy of broad treatment. In addition, the necessity for oxidative mechanisms in handling invading microorganisms may make global inhibition of oxidative stress unsafe in the immunosuppressed premature infant.

Multiple choice questions

1 Proposed mechanisms of the phases and causes of retinopathy of prematurity include all except:
   a. Mechanisms based mainly on studies using animals exposed to oxygen stresses
   b. Involve generation of ROS from activation of NADPH oxidase and other enzymes
   c. Only occur in infants that are born between 36 and 40 weeks of gestational age
   d. May involve potentially protective mechanisms from ROS

2 The vasculatures of the eye include all except:
   a. The choroidal vasculature
   b. The hyaloid vasculature
   c. The retinal vasculature
   d. The renal vasculature

3 Premature birth is believed to do all of the following except:
   a. Increase consumption of glutathione that provides nonenzymatic oxidative reserve
   b. Increase fluctuations in oxygenation and activation of endothelial NADPH
   c. Decrease the chance of developing myopia and strabismus
   d. Produce fewer antioxidant enzymes and more oxidative compounds

References

1 Good, W.V., Hardy, R.J., Dobson, V. *et al.* (2005) The incidence and course of retinopathy of prematurity: findings from the early treatment for retinopathy of prematurity study. *Pediatrics*, 116, 15–23.
2 Shah, P.K., Narendran, V. & Kalpana, N. (2012) Aggressive posterior retinopathy of prematurity in large preterm babies in South India. *Archives of Disease in Childhood Fetal and Neonatal Edition*, 97, F371–5.
3 Mann, I. (1964) *The Development of the Human Eye*, 1st edn. Grune and Stratton Inc., New York.
4 McLeod, D.S., Hasegawa, T., Prow, T., Merges, C. & Lutty, G. (2006) The initial fetal human retinal vasculature develops by vasculogenesis. *Developmental Dynamics*, 235, 3336–47.
5 Hartnett, M.E. (2014) Retinopathy of prematurity: a template for studying retinal vascular disease. In: Chalupa, L.M. & Werner, J.S. (eds), *The New Visual Neurosciences*. MIT Press, Cambridge MA, pp. 1483–1502.
6 Hartnett, M.E. & DeAngelis, M.M. (2012) Studies on retinal and choroidal disorders. In: Armstrong, D. (ed), *Studies on Retinal and Choroidal Disorders*, 1st edn. Humana Press, New York, pp. 559–584.

7 Bizzarro, M.J., Hussain, N., Jonsson, B. *et al.* (2006) Genetic susceptibility to retinopathy of prematurity. *Pediatrics*, 118, 1858–1863.

8 Buhimschi, I.A., Buhimschi, C.S., Pupkin, M. & Weiner, C.P. (2003) Beneficial impact of term labor: Nonenzymatic antioxidant reserve in the human fetus. *American Journal of Obstetrics and Gynecology*, 189, 181–188.

9 Sanchez-Alvarez, R., Almeida, A. & Medina, J.M. (2002) Oxidative stress in preterm rat Brain is due to mitochondrial dysfunction. [Article]. *Pediatric Research*, 51, 34–39.

10 Najarian, T., Hardy, P., Hou, X. *et al.* (2000) Preservation of neural function in the perinate by high PGE2 levels acting via EP2 receptors. *Journal of Applied Physiology*, 89, 777–784.

11 Wang, H., Zhang, S.X. & Hartnett, M.E. (2013) Signaling pathways triggered by oxidative stress that mediate features of severe retinopathy of prematurity. *JAMA Ophthalmology*, 131, 80–85.

12 Terry, T.L. (1942) Extreme prematurity and fibroblastic overgrowth of persistent vascular sheath behind each crystalline lens:(1) Preliminary report. *American Journal of Ophthalmology*, 25, 203–204.

13 Hartnett, M.E. & Lane, R.H. (2013) Effects of oxygen on the development and severity of retinopathy of prematurity. *Journal of AAPOS : The Official Publication of the American Association for Pediatric Ophthalmology and Strabismus / American Association for Pediatric Ophthalmology and Strabismus*, 17, 229–234.

14 Ashton, N., Ward, B. & Serpell, G. (1954) Effect of oxygen on developing retinal vessels with particular reference to the problem of retrolental fibroplasia. *British Journal of Ophthalmology*, 38, 397–430.

15 Hartnett, M.E. (2010) The effects of oxygen stresses on the development of features of severe retinopathy of prematurity: knowledge from the 50/10 OIR model. *Documenta Ophthalmologica*, 120, 25–39.

16 Ward, J.P.T. (2008) Oxygen sensors in context. *Biochimica et Biophysica Acta (BBA) – Bioenergetics*, 1777, 1–14.

17 Saito, Y., Geisen, P., Uppal, A. & Hartnett, M.E. (2007) Inhibition of NAD(P)H oxidase reduces apoptosis and avascular retina in an animal model of retinopathy of prematurity. *Molecular Vision*, 13, 840–853.

18 Cunningham, S., Fleck, B.W., Elton, R.A. & McIntosh, N. (1995) Transcutaneous oxygen levels in retinopathy of prematurity. *Lancet*, 346, 1464–1465.

19 Hartnett, M.E. & Penn, J.S. (2012) Mechanisms and management of retinopathy of prematurity. *New England Journal of Medicine*, 367, 2515–2526.

20 Hellstrom, A., Carlsson, B., Niklasson, A. *et al.* (2002) IGF-I is critical for normal vascularization of the human retina. *Journal of Clinical Endocrinology Metabolism*, 87, 3413–3416.

21 Smith, L.E.H., Kopchick, J.J., Chen, W. *et al.* (1997) Essential role of growth hormone in ischemia-induced retinal neovascularization. *Science*, 276, 1706–1709.

22 Wang, H., Byfield, G., Jiang, Y., Smith, G.W., McCloskey, M. & Hartnett, M.E. (2012) VEGF-mediated STAT3 activation inhibits retinal vascularization by down-regulating local erythropoietin expression. *American Journal of Pathology*, 180, 1243–1253.

23 Kermorvant-Duchemin, E., Sapieha, P., Sirinyan, M. *et al.* (2010) Understanding ischemic retinopathies: emerging concepts from oxygen-induced retinopathy. *Documenta Ophthalmologica*, 120, 51–60.

24 Niesman, M.R., Johnson, K.A. & Penn, J.S. (1997) Therapeutic effect of liposomal superoxide dismutase in an animal model of retinopathy of prematurity. *Neurochemical Research*, 22, 597–605.

25 Penn, J.S., Madan, A., Caldwell, R.B., Bartoli, M., Caldwell, R.W. & Hartnett, M.E. (2008) Vascular endothelial growth factor in eye disease. *Progress in Retinal and Eye Research*, 27, 331–371.

26 Patz, A. (1954) Oxygen studies in retrolental fibroplasia. *American Journal of Ophthalmology*, 38, 291–308.

27 Di Fiore, J.M., Kaffashi, F., Loparo, K. *et al.* (2012) The relationship between patterns of intermittent hypoxia and retinopathy of prematurity in preterm infants. *Pediatric Research*, 72, 606–612.

28 Group, T.S.-R.M.S. (2000) Supplemental therapeutic oxygen for prethreshold retinopathy of prematurity (STOP-ROP), a randomized, controlled trial. I: primary outcomes. *Pediatrics*, 105, 295–310.

29 2010) Early CPAP versus surfactant in extremely preterm infants. *New England Journal of Medicine*, 362, 1970–1979.

30 Stenson, B.J., Tarnow-Mordi, W.O., Darlow, B.A. *et al.* (2013) Oxygen saturation and outcomes in preterm infants. *New England Journal of Medicine*, 368, 2094–2104.

31 Schmidt, B., Whyte, R.K., Asztalos, E.V. *et al.* (2013) Effects of targeting higher vs lower arterial oxygen saturations on death or disability in extremely preterm infants: a randomized clinical trial. *JAMA*, 309, 2111–2120.

32 Owen, L. & Hartnett, M.E. (2014) Current concepts of oxygen management and ROP. *Journal of Ophthalmic & Vision Research*, 1, 94–100.

33 Smith, L.E.H., Wesolowski, E., McLellan, A. *et al.* (1994) Oxygen induced retinopathy in the mouse. *Investigative Ophthalmology & Visual Science*, 35, 101–11.

34 Penn, J.S., Tolman, B.L. & Lowery, L.A. (1993) Variable oxygen exposure causes preretinal neovascularization in the newborn rat. *Investigative Ophthalmology & Visual Science*, 34, 576–585.

35 Wang, H., Smith, G.W., Yang, Z. *et al.* (2013) Short hairpin RNA-mediated knockdown of VEGFA in Muller cells reduces intravitreal neovascularization in a rat model of retinopathy of prematurity. *American Journal of Pathology*, 183, 964–74.

36 McLeod, D.S., Brownstein, R. & Lutty, G.A. (1996) Vaso-obliteration in the canine model of oxygen-induced retinopathy. *Investigative Ophthalmology & Visual Science*, 37, 300–311.

37 Ushio-Fukai, M. (2006) Redox signaling in angiogenesis: role of NADPH oxidase. *Cardiovascular Research*, 71, 226–235.

38 Wilkinson-Berka, J.L., Rana, I., Armani, R. & Agrotis, A. (2013) Reactive oxygen species, Nox and angiotensin II in angiogenesis: implications for retinopathy. *Clinical Science (London)*, 124, 597–615.

39 Ushio-Fukai, M. & Nakamura, Y. (2008) Reactive oxygen species and angiogenesis: NADPH oxidase as target for cancer therapy. *Cancer Letters*, 266, 37–52.

40 Wang, H., Yang, Z., Jiang, Y. & Hartnett, M.E. (2014) Endothelial NADPH oxidase 4 mediates vascular endothelial growth factor receptor 2-induced intravitreal neovascularization in a rat model of retinopathy of prematurity. *Molecular Vision*, 20, 231–241.

41 Saito, S., Shiozaki, A., Nakashima, A., Sakai, M. & Sasaki, Y. (2007) The role of the immune system in preeclampsia. *Molecular Aspects of Medicine*, 28, 192–209.

42 Wilkinson-Berka, J., Deliyanti, D., Rana, I. *et al.* (2014) NADPH oxidase, NOX1, mediates vascular injury in ischemic retinopathy. *Antioxidants & Redox Signaling*, 20, 2726–2740.

43 Byfield, G., Budd, S. & Hartnett, M.E. (2009) The role of supplemental oxygen and JAK/STAT signaling in intravitreous neovascularization in a ROP rat model. *Investigative Ophthalmology & Visual Science*, 50, 3360–3365.

44 Saito, Y., Uppal, A., Byfield, G., Budd, S. & Hartnett, M.E. (2008) Activated NAD(P)H oxidase from supplemental oxygen induces neovascularization independent of VEGF in retinopathy of prematurity model. *Investigative Ophthalmology Visual Science*, 49, 1591–1598.

45 Wang, H., Byfield, G., Jiang, Y., Smith, G.W., McCloskey, M. & Hartnett, M.E. (2011) VEGF-mediated STAT3 activation inhibits retinal vascularization by downregulating erythropoietin expression. *American Journal of Pathology*, 2011, 180, 1243–53.

CHAPTER 7

Oxidative damage in the retina

Robert D. Stratton[1,2]

[1] Department of Ophthalmology, University of Florida College of Medicine, Gainesville, FL, USA
[2] Department of Small Animal Clinical Sciences, College of Veterinary Medicine, Gainesville, FL, 32610 USA

List of abbreviations

OS   Oxidative stress
UV   Ultraviolet
PR   Photoreceptor
RPE   Retinal pigment epithelium
AOX   Antioxidant
PVD   Posterior vitreous detachment
NSR   Neurosensory retina
SLC23A2   Sodium-dependent ascorbate transporter
RAL   Retinaldehyde
LCPUFAs   Long-chain polyunsaturated fatty acids
ROL   Retinol
RP   Retinitis pigmentosa
CHOP   CCAAT-enhancer-binding protein homologous protein
DHA   Docosahexaenoic acid
NPD1   Neuroprotectin
CEP   ω-2-carboxyethylpyrrole
4-HNE   4-Hydroxynonenal
NV   Neovascularization
ROS   Reactive oxygen species
PE   Phosphatidylethanolamine
NRPE   N-retinylidene-PE
A2E   N-retinylidene-N-retinyl-ethanolamine
BM   Bruch's membrane
LDL   Low-density lipoprotein
ONOO⁻   Peroxynitrite
8-OHdG   8-Oxo-2′-deoxyguanosine
mtBER   Mitochondrial base excision repair
nDNA   Nuclear DNA
mtDNA   Mitochondrial DNA
OGG1   8-Oxoguanine glycosylase
Pol-γ   Polymerase gamma
PGC-1α   Peroxisome proliferator-activated receptor-gamma coactivator 1 alpha

Oxidative Stress and Antioxidant Protection: The Science of Free Radical Biology and Disease, First Edition.
Edited by Donald Armstrong and Robert D. Stratton.
© 2016 John Wiley & Sons, Inc. Published 2016 by John Wiley & Sons, Inc.

mt   Mitochondria
AMD   Age-related macular degeneration
CNVM   Choroidal neovascular membrane
VEGF   Vascular endothelial growth factor
CFH   Complement factor H
ARMS2   mt age-related susceptibility protein 2
HTRA1   High-temperature-required serine protease
BMP-4   Bone morphogenetic protein 4
CFHRA   CFH-risk alleles
A2RA   ARMS2-risk alleles
CNTF   Ciliary neurotrophic factor
HGF   Hepatocyte growth factor
RBP   Retinol-binding protein
CME   Cystoid macular edema
AGEs   Advanced glycation end-products
GPx-1 RDH   Glutathione peroxidase-1 retinol dehydrogenase

THEMATIC SUMMARY BOX

At the end of this chapter, students should be able to:

• Describe the main features of age-related macular degeneration and diabetic retinopathy

• Outline the pathogenesis of these disorders

• Analyze the contribution of oxidative stress in the pathogenesis

• Analyze the contribution of polymorphism in the pathogenesis

• Analyze therapeutic targets

Introduction

Oxidative stress (OS)-induced damage is likely in the eye due to daily exposure to ultraviolet (UV) and blue light, high oxygen tension, high choroidal blood flow, dramatic oxygen tension gradient across the vitreous, maintenance of an intraocular pressure, highly unsaturated long-chain fatty acids, large surface area of the stacked cell membrane disks in the photoreceptor (PR) outer segments, metabolic load on the retinal pigment epithelium (RPE), and high metabolic rate in the PR cell layer. Major diseases of the eye, namely, age-related macular degeneration (AMD), glaucoma, cataract, diabetic retinopathy, and retinitis pigmentosa (RP), have different etiologies, but the cellular response to injury from OS is remarkably similar. Depending on the extent of damage, cells undergo the OS response, endoplasmic reticulum-folded protein response, heat-shock response, inflammatory response, apoptotic response, and necrotic response. The major endogenous enzymatic systems to protect against oxidative damage are superoxide dismutase (SOD), catalase, glutathione peroxidase, and thioredoxin reductase. The main nonenzymatic antioxidants are vitamins A, C, and E; glutathione; ubiquinone; lutein; zeaxanthin; and meso-zeaxanthin, along with a host of antioxidant (AOX) molecules from dietary sources.

The vitreous

The vitreous, filling the space between the lens and the retina, is composed of long-chain hyaluronic acid polymer that, along with a collagen framework, create a gel that is 99% water. The collagen framework is strongly attached to the posterior lens capsule, ciliary body, peripheral retina, and optic disk, and weakly attached to the rest of the retina. The vitreous is transparent because the vitreous collagen fibers are smaller than the wavelength of visible light, and there are specific vitreous proteins to keep the fibers from aggregating.[1] In disease and in aging, hyaluronic acid polymer ester bonds hydrolyze causing the gel to liquefy. Simultaneously, there is fragmentation of the collagen fibers, so that over time there is liquefaction and collapse of the collagen structure. At first, there are pockets of liquid vitreous allowing currents to flow, carrying oxygen from the highly saturated preretinal vitreous to the relatively hypoxic postlenticular vitreous leading to oxidative stress and damage to the lens. With liquefaction, the vitreous fiber structure detaches from the posterior retina, remaining tightly attached to the anterior, peripheral retina and to the retina at the optic nerve head. Eventually the vitreous attached to the optic nerve pulls off, creating a posterior vitreous detachment (PVD). In about 1% of human eyes, there are spots of tight adhesion to the collagen structure of the vitreous in the peripheral retina so that a PVD will trigger a U-shaped flap (horseshoe-shaped) tear in the retina. In about one-third of eyes with such retinal tears, the hole created by the tear leads to a rhegmatogenous (caused by a tear) retinal detachment by allowing the now liquid vitreous to enter the subretinal space. Normal neurosensory retina (NSR) is held in place by the suction force created by the pumping action of the RPE. Liquid vitreous leaking through the tear breaks this suction so that the retina detaches, not by being pulled off, but rather falling off by no longer being held in place by suction. Extracellular fluid in the normal NSR is also pumped out by the RPE pumping action, creating a state of deturgescence necessary for transparency and normal neuronal function, so the detached swollen retina is no longer transparent and there is retinal neuronal dysfunction. The retinal detachment is repaired by closing the hole in the retina and re-establishing the normal force holding the retina in place.

The intact vitreous gel has an ascorbate level that is 30–40 times higher than plasma levels, achieved by sodium-dependent ascorbate transporter (SLC23A2) in the pigmented layer of the ciliary body epithelium.[2] This is responsible for the high level of ascorbate in the aqueous humor. The ascorbate in the intact vitreous consumes oxygen by an unknown mechanism, possibly involving free iron.[3] The oxygen consumption and the convection current barrier inherent in the intact vitreous gel protect the anterior portions of the eye from the very saturated oxygen concentration at the retina. Degeneration and liquefaction of the vitreous due to inflammation, retinal microvascular leakage, or surgical removal of the vitreous lead to higher oxygen concentrations at the lens and the trabecular meshwork with development of nuclear sclerotic cataract and open angle glaucoma from the increased oxidative damage. Extracellular SOD is also present in the vitreous but in very low concentration. Attempts to prevent cataract by diet have shown that extra-cellular SOD can be increased by diet but have yet to show prevention of the changes in lens.[4] There is evidence that the AOX verbascoside, a phenylpropanoid glycoside, can increase AOX levels in vitreous and lens of the rabbit.[5] Crystallins have been found in the vitreous and inhibit inflammation by blocking the Ras–Raf–MEK–ERK pathway.[6] Under hypoxic conditions, a variety of soluble factors are secreted into the vitreous cavity including growth factors, cytokines, and chemokines.[7]

The retina

The retina is especially susceptible to OS-induced damage because of the retina's exposure to light; the high production rate of protein in the PR cell; the complex visual cycle of visual pigments; high oxygen tension; high choroidal blood flow; main-tenance of a significant intraocular pressure; preponderance of highly unsaturated long-chain fatty acids in the PR cell; polymorphism in very complex protein path-ways; and oxidative, metabolic, and phagocytic loads on the RPE.

OS in the visual cycle

Phototransduction of light into neural activity in the PR involves absorption of light by the visual pigment 11-*cis*-retinaldehyde (RAL)-opsin rhodopsin for rods and a similar pigment for cones. Initial production of the visual pigment rhodopsin occurs in the inner segment of the PR where a multistage protein production process occurs with folding and multiple postproduction alterations. Once this inner segment disk membrane-activated 11-*cis*-RAL-opsin protein absorbs light, there is a conforma-tional change in the 11-*cis*-RAL chromophore to all-*trans*-RAL, releasing it from the opsin protein and triggering a cyclic GMP pathway to signal neural transmission (see Figure 7.1). The all-*trans*-RAL is then metabolized back to the activated 11-*cis*-RAL chromophore by the visual cycle, with conversion of the all-*trans*-RAL aldehyde to the alcohol, transport to the RPE with specific binding proteins, esterification of the all-*trans*-retinol (ROL) (esterification allows self-aggregation into an inert retinosome for storage), isomerization of the all-*trans*-retinyl ester to 11-*cis*-ROL, oxidation to 11-*cis*-RAL, and transportation back to the PR outer segment where it is covalently bonded to the outer segment disk membrane-bound apo-opsin protein to make the activated visual pigment. This complex process has many protein components including chaperone and transport proteins.

Recycling of the long-chain polyunsaturated fatty acids (LCPUFA)

The main mechanism for replacement of oxidized lipoproteins is the diurnal phagocy-tosis of about one-tenth of the PR outer segment membranes per day by the RPE. The renewal process is carried out by a very active lipoprotein production in the PR inner segment ER and Golgi apparatus. Not only does an extraordinary amount of visual pigment need to be made, but also recycling of the long-chain polyunsaturated fatty acids (LCPUFAs) in the shed outer segments has to be done, and new cell membrane

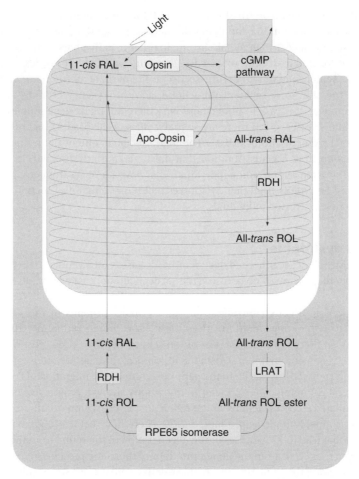

Figure 7.1 The rod photoreceptor cell (PRC) outer segment is the top figure and the retinal pigment epithelial (RPE) cell is the bottom figure. The active form of rhodopsin visual pigment is an outer segment disk membrane-bound protein that absorbs light causing a conformational shift in the 11-*cis*-retinal (11-*cis*-RAL) moiety to all-*trans*-retinal (all-*trans*-RAL), resulting in cleavage of all-*trans*-RAL and opsin protein that stimulates the photoreceptor to trigger a neuronal impulse through a cyclic GMP pathway. The all-*trans*-RAL aldehyde is reduced to all-*trans*-retinol (all-*trans*-ROL) by an NADPH-dependent retinal dehydrogenase (RDH). The all-*trans*-ROL is transported through the interphotoreceptor matrix and into the RPE through a partially understood mechanism involving specific retinal-binding proteins in the interphotoreceptor matrix and in the RPE microsome. The all-*trans*-ROL is esterified by lecithin retinol acyltransferase to a fatty acid all-*trans*-retinyl ester, which then aggregates in a specialized microsome called a retinosome. RPE65 isomerase then hydrates the all-*trans*-retinyl ester and changes the ROL to the 11-*cis*-isomer and a free fatty acid. An NAD^+-dependent retinal dehydrogenase oxidizes the 11-*cis*-ROL to 11-*cis*–RAL, which is then transported back to the membrane-bound apo-opsin where the two are covalently joined to form the activated rhodopsin, 11-*cis*-RAL-opsin.

lipoprotein has to be made. To effect this, the PR inner segment has a large number of very active mitochondria supplying the needed energy for this renewal process. The RPE also has a very high metabolic rate in order to daily phagocytize such a large lipoprotein load; detoxify the oxidized lipoproteins and lipid peroxides; regenerate the visual pigments; maintain deturgescence in the neural environment; and supply the numerous PR inner segment mitochondria with oxygen, nutrients, and waste product removal. It is easy to understand why there are many gene mutations leading to retinal diseases, given there are so many steps in the production and recycling of the visual pigment and LCPUFAs. The high metabolic rate of the retina will lead to dysfunction from a gene mutation that does not cause dysfunction in other tissues. For the same reason, some individuals that have more OS from environmental sources and less AOXs from dietary sources will have retinal dysfunction that does not occur in less stressed individuals.

Retinal dystrophies

Retinal dystrophies are inherited degenerative diseases of the retina. An ever increasing number of gene mutations causing retinal degeneration have been identified (see Table 7.1).

Mutations in a gene controlling one of the many proteins in the visual cycle may lead to retinal degeneration. RP is the most common retinal dystrophy phenotype. In RP, there is progressive loss of rod cells earlier than cone cells, with loss of night-time vision and peripheral visual field. Clinical findings are attenuation of the retinal blood vessels, loss of RPE in the peripheral retina, and migration of the pigment into a typical clumping pattern. Electroretinography shows a progressive decrease in light-stimulated electrical activity of the retina, first in the scotopic dark-adapted ERG and then in the photopic light-adapted ERG. There have been 140 gene mutations identified that lead to RP.[8] Causes of the degeneration of the retina from the errors in metabolism that occur from the gene mutations of the many proteins involved in the visual cycle share the same stresses, with oxidative damage, ER stress, mitochondrial DNA (mtDNA) damage, and triggering of the apoptotic pathways.

Table 7.1 Mapped and identified retinal dystrophy genes.[8]

Disease category	Mapped only	Gene identified
Cone dystrophy autosomal dominant	3	10
Cone dystrophy autosomal recessive	1	17
Macular dystrophy AD	7	12
Macular dystrophy AR	0	3
Retinitis pigmentosa AD	1	24
Retinitis pigmentosa AD with syndrome	1	12
Retinitis pigmentosa AR	3	45
Retinitis pigmentosa AR with syndrome	4	42
Retinitis pigmentosa AR with Usher syndrome	3	12
Retinitis pigmentosa X-linked recessive	3	3
Retinitis pigmentosa X-linked recessive with syndrome	0	2

Despite knowing the exact genetic cause of a particular form of RP, we understand the pathogenesis of the retinal degeneration incompletely.[9] Many different gene defects in the rhodopsin production pathway lead to RP. Rhodopsin is a protein that is processed abundantly in the PR so that a misfolded protein can lead to a buildup, causing the ER to fill up; undergo stress; trigger the unfolded protein response; and release caspase, signaling apoptosis, or indirectly signal apoptosis by stimulating CCAAT-enhancer-binding protein homologous protein (CHOP).[10,11] Visual cycle inhibitors and CHOP inhibitors are potential treatments for RP that have some promising results.[12,13] A particularly severe form of RP is caused by mutations in the retinol dehydrogenase RDH12 with early and severe loss of vision, clinically identified as Leber's congenital amaurosis. RDH12 is not just involved in the reduction of all-*trans*-RAL to all-*trans*-retinol (ROL) but also enzymatically detoxifies a variety of other aldehydes, which are oxidation products of LCPUFAs.[14] Docosahexaenoic acid (DHA) is the main PUFA in the PR and is retained in the PR even in omega-3 FA deficiency through a highly effective recycling process that is similar to the recycling of visual pigment. RPE cells make DHA-derived neuroprotectin (NPD1) in response to OS. NPD1 provides the PR with many protective, anti-inflammatory, and prosurvival repair signals through the induction of antiapoptotic proteins and inhibition of pro-apoptotic proteins.[15] DHA is an omega-3 PUFA with six double bonds that oxidizes rapidly without AOX protection. A unique oxidation product of DHA oxidation is ω-(2-carboxyethyl)pyrrole (CEP). CEP protein adducts accumulate in the RPE and BM and have been found to promote neovascularization (NV) in a pathway independent of vascular endothelial growth factor (VEGF).[16]

A large number of mitochondria in the PR inner segment produce an abundance of reactive oxygen species (ROS). The PR also contains an abundance of LCPUFAs, the source material for forming new outer segment disk membranes. With the abundant ROS and abundant LCPUFA, lipid peroxidation occurs extensively. A common and toxic oxidation product of LCPUFA oxidation is 4-hydroxynonenal (4-HNE). RDH12 appears to detoxify 4-HNE by reducing this aldehyde to an alcohol.[14] A defective RDH12 results in a buildup of the very reactive 4-HNE, resulting 4-HNE adducts, which are the products of 4-HNE reactions with protein. The buildup of 4-HNE adducts appears to be the cause of early and severe retinal degeneration in this form of RP.[14]

Mutations in the ABCA4 gene lead to some forms of Stargardt disease, cone–rod dystrophy and RP in homozygous individuals, and increase the chance of AMD in heterozygous individuals.[17] OS occurs when all-*trans*-RAL readily reacts with phosphatidylethanolamine (PE) found in the disk membranes to make N-retinylidene-phosphatidylethanolamine (NRPE), which is usually transported by ABCA4 protein out of the disk membrane, where NRPE is reduced by RDH, freeing RAL to be transported to the RPE where it reenters the visual cycle.[18] With defective ABCA4 transport, there is buildup of the NRPE, which may then lead to adduction of a second RAL to form N-retinylidene-N-retinyl-phosphatidylethanolamine (A2PE), a toxic bisretinoid, A2 being two vitamin A moieties (RAL) and PE being one phosphatidylethanolamine.[17] When the phosphate group is removed by a phosphatase, A2E results. When there is deficient ABCA4 activity, the buildup of A2E and other oxidants leads to OS and apoptosis. The importance of this bisretinoid is that there are no known detoxifying chemical reactions, so A2E and its oxidation products, adducts, and dimers build up in the retina even when there is normal ABCA4 activity, as part of normal aging. A2E damages the cell through its two RAL chromophores that absorb blue light and produces superoxide and peroxyl radicals. The two hydrophobic retinal side chains and the hydrophilic ethanolamine of A2E make it a detergent, giving A2E more ability to damage cell membranes. The amine group in A2E complexes with lysosome enzymes and interferes with normal degradation processes.[19] A2E also damages the mitochondria causing the release of cytochrome c leading to apoptosis. Through shedding of the PR disks and phagocytosis by the RPE, bisretinoids are removed from the PR, but build up in the RPE lysosomes, and end up in residual bodies with pigment granules called lipofuscin. Many bisretinoids strongly absorb blue light and produce ROS, including superoxide. Photooxidation of A2E and related compounds can involve the incorporation of as many as nine oxygen atoms into the molecule resulting in multiple toxic epoxides, endoperoxides, and furanoid compounds.[19] Some of these are stable, polar, and can diffuse to cause oxidative damage widely in the RPE and even diffuse out of the RPE stimulating pattern recognition receptors RAGE and TLR3 that signal inflammation.[20]

Lipofuscin age pigment

Lipofuscin is the autofluorescent collection of aging pigments found in the RPE residual bodies. So lipofuscin is the accumulation of phagocytosed lysosomal-oxidized cell membranes that could not be completely degraded. Lipofuscin accumulates rapidly in children, possibly because of more photooxidative damage due to lower filtering of blue light by the immature lens nucleus. The rate of lipofuscin accumulation decreases in adulthood, but continues to accumulate until plateauing in the seventh decade, reaching up to 20% of the volume of the RPE cytoplasm.[19] Fundus photography can detect the greenish yellow autofluorescence of lipofuscin. A2E and its derivatives make up most of the material, along with divalent metals and some protein, but the exact composition of RPE lipofuscin has not yet been determined. Within lipofuscin are pigment granules that appear to have little protein, yet have all of the phototoxicity and ROS-producing activity, along with carboxyethylpyrrole adducts.[21] Tandem mass spectroscopy studies show that bisretinoids in lipofuscin self-react to produce higher molecular weight species that become more hydrophobic.[19] A2E has long been thought to be the main fluorescent chromophore in lipofuscin, but it appears that A2E may self-quench and fluorescence may actually occur from A2E self-reacting adducts.[19] It also appears that the phototoxicity of lipofuscin may prove to be from one or more of these A2E adducts.[19] PR cell loss occurs with aging, but in Caucasians, PR loss is significantly and directly correlated with lipofuscin concentration in the adjacent RPE.[17]

RPE melanosomes

RPE melanosomes synthesize melanin from L-DOPA during fetal development and then synthesis diminishes, so little is made in the adult RPE.[22] Melanosomes are located in the apical RPE close to the PR where this dark brown pigment absorbs light that passes through the PR cells, protecting the RPE from phototoxicity. Melanin quenches singlet oxygen and scavenges ROS and metal ions.[22] Melanosomes take up some of the lipofuscin granules, gradually changing into melanolysosomes with aging, first in the macula and then in the peripheral retina. The apical location shielding effect of the melanosomes not only reduces the need for phagocytosis by protecting the PR outer segment (which protrudes deeply into the RPE) but also aids the degradation of the ROS in the phagosomes.[23] As the RPE ages and there are fewer pure melanosomes and more lipofuscin-laden melanolysosomes, there is less apical localization so the shielding effect and the aid in ROS degradation become less effective.[22] The RPE continues to fill with lipofuscin and lipofuscin-laden melanolysosomes.

Bruch's membrane

The RPE is separated from the choroidal capillaries (the choriocapillaris) by a permeable Bruch's membrane (BM), which consists of permeable extracellular matrix material interposed between the basement membrane of the RPE and the fenestrated basement membrane of the choriocapillaris. The large quantity of lipoprotein needed for producing PR cell membranes is transported as low-density lipoprotein (LDL) particles through the choriocapillaris and BM to the RPE where lipid is transported to the PR. Oxidized lipoprotein begins to accumulate in the choriocapillaris and BM early in life, and with aging it may reach a level that creates an obstruction to transport of lipoprotein through the BM and RPE. There is also accumulation of oxidized lipoprotein in the lipofuscin and melanolysosomes. The BM deposit of oxidized lipoproteins leads to a barrier for transport of phagocytized lipoproteins out of the RPE and into the choriocapillaris, causing cellular debris to accumulate between the RPE and the choriocapillaris.

Oxidative damage to the DNA

OS causes DNA damage, and with the highly oxidizing environment of the PR and RPE, the nuclear deoxyribonucleic acid (nDNA) repair mechanisms need to be very effective. The nDNA has robust protection against environmental insults as well as a very reliable single-strand break DNA repair mechanism. The mtDNA does not have such a robust protection, leading to mtDNA damage and dysfunction as a significant contributing factor to aging from oxidative damage in the retina. Mitochondria are at the same time highly vulnerable to oxidative damage and the source of significant production of ROS that cause the damage. Superoxide is formed as an unavoidable reaction product during the mitochondrial oxidative phosphorylation. Superoxide is detoxified by the mitochondrial SOD2 and catalase, but some ROS are nevertheless formed, including hydroxyl radical $HO^{\bullet}$ and peroxynitrite ($ONOO^-$). These are very reactive and can attack the iron in the cytochrome as well as oxidize PUFAs, leading to lipid peroxidation chain reaction. The hydroxyl radical is the main vector of oxidative damage to base-pairs in the mtDNA, and because guanine is most easily oxidized, 8-oxo-2′-deoxyguanosine (8-OHdG) is the main oxidation product. 8-OHdG pairs with not only cytosine but also adenine, so a G–C base pair may become an 8-OHdG–A pair and then a T–A transversion mutation.[24]

Mitochondrial base excision repair (mtBER) pathway

Oxidized nDNA and mtDNA bases are repaired through similar base excision and repair pathways by proteins that are encoded in the nucleus and not the mt.

8-Oxoguanine glycosylase (OGG1) is the main enzyme that cleaves the 8-OHdG from the mtDNA. The main DNA polymerase in the mitochondrial short-patch mtBER pathway is DNA polymerase gamma (Pol-γ). Although most mtBER proteins are splice variants of the nuclear counterparts, Pol-γ is unique to the mt.[24] Pol-γ is sensitive to OS, and this vulnerability in the mtBER may be a major reason why the mtDNA is more susceptible to OS than is nDNA.[25] It appears that OS-induced RPE mtDNA damage activates mtBER enzymes but then the mtDNA repair capacity can be overwhelmed due to Pol-γ dysfunction, resulting in decreased mtDNA repair.

The mt are dynamic, move along the cytoskeleton, divide, and fuse together. Mt fusion and fission are part of the cellular response to stress. Fusion allows some protection from mtDNA mutations by combining the mitochondrial genomes, so the OPA1 (optic atrophy 1 autosomal dominant) gene defect that leads to faulty mitochondrial fusion thus leads to increasing mtDNA damage, which is the cause of about half of the cases of dominant optic nerve atrophy, and some cases of glaucoma with normal intraocular pressure. Polymorphism in the DNM1L gene involved in mitochondrial fission is suspected in a variety of neurodegenerative diseases but has not yet been implicated in retinal disease.[26] Mutations in the mtDNA gene encoding the mt NADH dehydrogenase involved in oxidative phosphorylation lead to underperforming mt, causing Leber's hereditary optic neuropathy, a form of optic nerve atrophy.[24]

The PR and RPE do have a mt renewal process of mt biogenesis and autophagy of damaged mt marked for removal. Biogenesis of the mt is a very complex process involving over 100 genes in both the nuclear and the mt genomes and includes fusion and fission processes.[24] Despite the complexity of mitochondrial biogenesis, it appears peroxisome proliferator-activated receptor-gamma coactivator 1 alpha (PGC-1α) is a major regulator.[27] Aging causes a decrease in mitochondrial biogenesis through reduced PGC-1α activity, thought to be due in part to inactivation of 5'-adenosine monophosphate-activated protein kinase through an unknown process.[28] Stimulation of PGC-1α occurs through caloric restriction, a therapy that is difficult to use in humans.[29] Caloric restriction mimetics resveratrol, polyphenols, SRT1720, and SRT501, and metformin stimulate mitochondrial biogenesis through mechanisms that upregulate PGC-1α activity and are possible aids in treating age-related diseases.[29] A cautionary note: high doses of vitamin C can inhibit mitochondrial biogenesis, so indiscriminant use of antioxidants may cause harm.

The nDNA protection can be upregulated by prior treatment with nonlethal OS. Catalase, glutathione peroxidase, and SOD1 (cytosol SOD) increase, but SOD2 (mitochondrial SOD) does not, so it appears that there is limited adaptive response for the protection of the mitochondria when the cell is chronically stressed.[30] VEGF is neuroprotective in part due to the upregulation of SOD2. VEGF inhibition reduces this mtDNA protective mechanism. Another source of OS to the mt is the mt cytochrome, which is a photosensitizer, so light can induce ROS and it may be this property that causes accumulative damage in the mitochondria in all light-exposed cells.[24]

Age-related macular degeneration

Age-related macular degeneration is an acquired disease of the central area of the retina called the macula lutea, a term derived from Latin which means yellow spot.

The macula is where the acuity, color vision, and sensitivity are highest, the cone PR predominates and where the retinal carotenoid pigments are located. The center of the macula is the fovea where the neurosensory retina is thinnest, being only the layer of specialized foveal cone cells, with all of the other neuronal elements pushed radially away from these cones so that neural elements do not interfere with the quality of the image on the center of the fovea (the foveola). The disease of AMD causes a gradual accumulation of debris under the RPE basement membrane of the macula and in Bruch's membrane (located between the RPE and the choroidal capillaries). Early in the development of AMD, the debris is not visible on biomicroscopy, but as the material builds and microglia and macrophages migrate in, small crystal-like deposits are most often seen first, called drusen. Central vision is most often preserved, but there is delay in recovery of the central vision to bright light bleaching as the PR and RPE functions decline and the permeable BM accumulates hydrophobic material. As cellular debris accumulates, it coalesces with the small hard drusen to form larger white deposits called soft drusen. Drusen contain phospholipids, glycolipids, cholesterol and oxidized cholesterol (7-keto cholesterol), oxidized fatty acids, and various proteins, including misfolded amyloid oligomers, C-reactive protein, adducts of the carboxyethylpyrrole protein, bisretinoids, immunoglobulins, acute phase molecules, as well as the complement-related proteins C3a, C5a, C5, C5b-9, complement factor H (CFH), CD35, and CD46.[31]

Progressive apoptotic loss of PR cells and RPE causes focal geographic areas of atrophy and depigmentation to occur. When the fovea is involved, there is loss of central vision. Geographic atrophy can cause a profound loss of central vision but does not extend into the peripheral retina, so that peripheral vision is preserved. This "dry" form of AMD is call nonexudative AMD with the more advanced form called geographic atrophy. When choroidal vessels break through the chronically inflamed BM as NV, plasma and blood leak out, elevating the RPE into a dome-shaped detachment. Often the neovascular vessels break through the RPE layer and enter the subretinal space, causing a localized detachment of the retina as the leaking and bleeding vessels overwhelm the RPE pump. This process produces a choroidal neovascular membrane (CNVM). Sudden detachment of the retina causes distortion in vision, called metamorphopsia, due to the displacement of the PR cells. Over time, as leaked plasma is reabsorbed by the retinal vessels and the RPE pump, large and hydrophobic molecules accumulate while water, electrolytes, and small hydrophilic molecules are easily reabsorbed, giving rise to pale yellow lipid deposits in the retina and subretinal space called hard exudate. This "wet" form of the disease is called exudative AMD. Typically, exudative AMD can be very rapid in onset and can progress rapidly. Peripheral vision is usually preserved except when there is massive bleeding in the presence of a bleeding diathesis, such as warfarin anticoagulation. Laser cautery of a CNVM preserves central vision if the fovea is not yet involved, but over 95% of patients present with foveal involvement. A clinical trial showed that laser treatment of a subfoveal CNVM was beneficial, but a sudden loss of central vision from treatment was hard to justify to patients for a small gain in the preservation of vision later, so most patients with subfoveal membranes were treated with a variety of other treatments including steroids, ionizing radiation, surgical removal of the CNVM, 360° peripheral retinotomy with rotation and reattachment of the fovea on RPE that did not have a CNVM (macular translation surgery), and photodynamic therapy of the CNVM using verteporfin activated with a diode laser to cauterize the CNVM with the induced ROS. Exudative AMD had no treatment that consistently improved vision until the availability of VEGF inhibitors.

VEGF in AMD

VEGF-A is an angiogenesis regulating protein as well as a survival factor for the ganglion cells, PR, and RPE that is produced by the RPE, Müller cells, astrocytes, and pericytes. The isoform, VEGF165, is proangiogenic, and the splice variant VEGF165b is antiangiogenic.[32] Serine-rich protein kinase-1 (SRPK1) is involved in the control of the splice variance. Topically applied inhibitors of SRPK1, such as disubstituted furan-morphonyl (SPHINX), piperidinyl-isonicotinamide (SRPIN340), and morphonyl-nicotinamide (MVRL09), reduced choroidal NV in a rodent model of exudative AMD by decreasing the VEGF165 isoform, but not the VEGF165b isoform.[33] A short RNA aptamer that bound to VEGF165 with high affinity gave some benefit to AMD patient vision, but when the humanized mouse monoclonal antibody bevacizumab and its Fab fragment ranibizumab, with affinity to all isoforms of VEGF-A, became available, great success was found in controlling loss of vision from the exudative form of AMD.[34] A soluble decoy VEGF receptor fusion protein aflibercept has been shown to have similar benefit.[34] Because treatment requires repeated intravitreous injections every 4–6 weeks, investigation into longer lasting VEGF inhibitors continues.[35] The European Society of Retinal Specialists paper is an excellent review of the current treatment of exudative AMD.[34]

Induction of VEGF in the retina leads to NV, but conversely also protects the retina from OS and apoptosis through upregulation of mt SOD2; AOX enzymes and anti-inflammatory heme oxygenase-1; and prosurvival NF-κB and antiapoptotic PI3-K/Akt.[36] By preventing VEGF protection of the neural retina, chronic treatment of exudative AMD with these very effective VEGF inhibitors may lead to OS and apoptosis in the retina.[36]

Parainflammation and complement in AMD

With chronic OS and the buildup of intra- and extracellular debris, a low-grade inflammatory process is triggered. The macrophage-like retinal microglia migrate out of the neurosensory retina into the subretinal space and BM. As the disease progresses, macrophages are recruited to remove the debris. Often the macrophages phagocytize not only the debris but also the extracellular matrix material leading the macrophage to trigger autoantibodies as well as antibodies to CEP adducts, and the inflammation becomes chronic. This low-level chronic inflammatory process in AMD has been called parainflammation.[37] Complement activation occurs in the inflamed BM and in the overlying RPE and PR cells through both the antibody initiated classical pathway and the more predominant spontaneously initiated alternate pathway. Apoptosis of PR cells and RPE occurs due to a combination of factors discussed, such as the ER-folded protein stress, A2E, lipofuscin, DHA oxidation with CEP adducts, mtDNA damage, reduced mt biosynthesis, poor SOD2 response to chronic stress, mt dysfunction, and activation of microglia and macrophages, with resulting complement activation. Rapamycin can prevent the irreversible change to senescence induced by OS, demonstrating an interesting protection from OS by suppressing the immune system.[38]

Polymorphism and complement factors

The complement cascade is elaborate with promoting factors and inhibiting factors, allowing discovery of many polymorphisms of genes that lead to either an increased risk of AMD, such as genes related to CFH, complement component 3, complement factor 1, or polymorphisms that reduce risk, such as genes related to complement factor B, complement component 2, and some factor H related genes.[39] Patients with CFH Y402H polymorphism have a deficient inhibitor of the alternate complement pathway, factor H, and an increased risk of AMD, and loss of vision from AMD. ARMS2 (mt age-related susceptibility protein 2) and related HTRA1 (high-temperature-required serine protease) genes have polymorphisms that increase the risk for AMD through an unknown mechanism.[40] Polymorphism in xeroderma pigmentosum complementation group D (XPD codon 751) may be associated with protection from the development of AMD.[41]

Polymorphism in the mtDNA glycosylases has been found. One study found that c.977C>G-hOGG1 (OGG1 cleaves 8-OHdG from oxidized mtDNA) polymorphism may be associated with nonexudative AMD.[42] Uracil-DNA glycosylase (UNG) and single-strand-selective monofunctional uracil-DNA glycosylase (SMUG1) polymorphisms are also associated with AMD, affecting the severity.[43]

Iron in AMD

Iron is a potent free radical generator through the Fenton reaction so that hydroxyl and hydroperoxyl radicals are formed. Ceruloplasmin acts as an AOX by oxidizing the more toxic Fe^{2+} to Fe^{3+}. The amount of iron in the retina increases with age and is increased even more in AMD affected eyes, especially in the RPE and BM, even in early AMD.[44] The iron-binding glycoprotein transferrin is also increased in AMD patients.[44] Ceruloplasmin deficiency causes iron overload and severe retinal degeneration.

AMD treatment targets

VEGF inhibition is the current treatment of choice for exudative AMD. There are multiple trials reported attempting to clarify treatment regimes, which are summarized in reference (34). In advanced nonexudative AMD, decreased levels of the endoribonuclease DICER1 occur, leading to accumulation of small fragments of DICER1 substrate, *Alu* RNA (small double-stranded RNA). This suggests a possible target for the treatment of nonexudative AMD.[45] Bone morphogenetic protein-4 (BMP-4) is elevated in the RPE and BM of patients with nonexudative AMD and is absent in RPE of active exudative AMD, but BMP-4 increases once the exudative AMD becomes fibrotic.[46] Cultures of RPE cells exposed to chronic sublethal OS express high levels of BMP-4, which mediates RPE senescence through activation of p53, so the cells become senescent or undergo apoptosis if p53-mediated DNA repair is not adequate.[46] BMP-4 may be the molecular switch determining the type of AMD that develops, and tissue necrosis factor may regulate the BMP-4 levels.[46]

OS in AMD etiology

Age-related macular degeneration is a multifactorial disease, with retinal degeneration resulting from a combination of an age-related increase in chronic OS, ER stress, chronic parainflammation, mt dysfunction, apoptotic loss of PR cells and RPE, BM disruption, iron overload, and complement activation. These multiple factors interact with a vast array of polymorphisms in the genome to give the phenotype we recognize as AMD. The importance of this disease has led to a rapidly growing field of research, with each identified etiologic factor explored for possible treatment targets.

Evidence that OS is involved in the development of AMD is extensive. Epidemiologic evidence shows there is a 2.54-fold increase in the risk of nonexudative AMD and a 4.55-fold increase in the risk of exudative AMD in current smokers versus never smokers.[47] A correlation between systemic OS biomarkers and AMD as well as a correlation between systemic AOX depletion and AMD also supports the role of OS in AMD.[48] Success of AOX therapy is also an evidence. The correlation between PR cell loss and accumulation of lipofuscin in the underlying RPE also supports the OS role in AMD. OS-induced increase in mtDNA damage and mutations and a decrease in the efficacy of DNA repair have been correlated with the severity of AMD.[49] Damage to nDNA accumulates in AMD although not as much as in mtDNA, so the mtBER proteins that are nucleus encoded are affected by the nDNA damage, and thus mutagenesis in nDNA affects the ability of the mtDNA repair.[49] Lymphocytes from AMD patients displayed a higher amount of oxidative DNA damage, exhibited a higher sensitivity to hydrogen peroxide and UV radiation, and repaired the lesions induced by these factors less effectively than did cells from control individuals.[49]

Although OS is an important factor in the development of AMD, it appears that the most important factor is aging, a process of slow accumulation of cellular damage from all causes including OS and inflammation, with OS leading to inflammation. The extensive variation in the human genome and in the environments experienced leads to a heterogeneous presentation of age-related diseases. Such heterogeneity presents opportunities for treatment.

Treatment of AMD
Antioxidants
The Age-Related Eye Disease Study (AREDS) showed that in a large group of moderately advanced nonexudative AMD, AOX vitamins of 500 mg of vitamin C, 400 IU of vitamin E, 15 mg of β-carotene, 80 mg zinc oxide, and 2 mg of copper reduced the progression to severe visual loss by 25% at 5 years.[50] The Age-Related Eye Disease Study 2 (AREDS2) showed that the substitution of a combination of lutein 10 mg and zeaxanthin 2 mg for β-carotene improved the outcome with a hazard ratio of 0.9 for progression to severe visual loss.[51] The AREDS2 also showed no apparent benefit of DHA and EPA on the AMD progression. In a recent study, it was found that AMD patients with two CFH-risk alleles (CFHRA) and no ARMS2-risk alleles (A2RA) had more progression of the AMD with zinc treatment than placebo whether alone or in combination with AOXs. For those with two CFHRAs and one or two A2RAs, there was no difference in placebo.[52] For AMD patients with none or one CFHRAs and no A2RAs, zinc gave no benefit while AOX did decrease AMD progression.[52] Those with none or one CFHRAs and one or two A2RAs are the ones that did benefit from zinc, either alone or in combination with AOXs.[52] The chapter on clinical trials by Palacio and Mooradian points out the danger of recommending AOX treatment to the general population based of the benefit demonstrated in a specific patient pool.

A reasonable approach would be to determine the AOX status of patients at risk for AMD, and give supplements to those found to be deficient. DHA is antioxidant in low concentration but a significant prooxidant in high concentration. Great care is needed in treating with DHA, and perhaps OS monitoring of blood is needed to get the right DHA dosage.

Amyloid β

An antibody to amyloid β removed amyloid β from soft drusen and protected the RPE and PR cells in a mouse model.[97] Upregulation of apolipoprotein E may lead to proteolytic degradation of amyloid β, an area for future investigation.

Anti-inflammatory treatment

Steroids are potentially useful to decrease the inflammatory component of AMD. The vitreous cavity is a very useful treatment approach for a variety of drug containers that can be sutured to a pars plana sclerotomy, allowing for long-duration delivery to the whole eye. Encapsulated cell technology can potentially deliver a variety of biologics for prolonged periods. Many inhibitors and modulators of complement are proposed to retard AMD progression, including inhibitors of factor B, factor D, C1, C3, C3 convertase and C5, as well as CFH recombinant and fusion proteins.[47] A small prospective, double-masked, randomized clinical trial on intravenous eculizumab, a C5 inhibitor, failed to retard progression of geographic atropy.[53]

LDL modulation

Lowering LDL with statin drugs to retard the progression of AMD has not yet been proven. Some evidence is hopeful, but a large randomized clinical trial has not been done.[54] Modulation of LDL with long-chain PUFAs may also be helpful. 7-Keto cholesterol is a common component of oxidized LDL and is a very strong stimulator of macrophages and inflammation, while DHA decreases this stimulated inflammation.[55]

Emerging treatments for AMD

Recent evidence that the neuroprotective protein DJ-1 protects the RPE from OS may lead to a way to manipulate this pathway to protect the RPE.[56] The metalloproteinase inhibitor TIMP1 can inhibit the destruction of BM, making it a possible treatment pathway. Polymorphism in monocyte chemotactic protein 1 (MCP-1) has recently been shown to be correlated with AMD, suggesting a possible treatment pathway. Pigment epithelium-derived factor (PEDF) is both neurotrophic and antiangiogenic, and levels are decreased in patients with AMD, making PEDF an attractive treatment for both types of AMD. Ciliary neurotrophic factor (CNTF) is neurotrophic and an encapsulated cell technology to treat RP found some preservation of retinal thickness, but no benefit in vision compared to control.[57] A preliminary trial of encapsulated cell technology CNTF to treat geographic atrophy showed success in slowing vision loss.[58]

Platelet-derived growth factor (PDGF) is proangiogenic. An anti-PDGF aptamer Fovista is undergoing trials as a supplement to anti-VEGF therapy. Hepatocyte growth factor (HGF) and its membrane receptor, mesenchymal–epithelial transition (MET) factor, are potential targets for AMD treatment. HGF protects various cells from oxidative stress-induced apoptosis mainly via the phosphorylation of phosphoinositide 3-kinase/Akt pathway and the upregulation of AOX enzymes. HGF is made in the RPE, and it is upregulated in exudative AMD where it is proangiogenic.

Visual cycle inhibitors

N-(4-hydroxyphenyl) retinamide (fenretinide) displaces ROL from retinol-binding protein (RBP) and, together with dietary restriction of vitamin A, can reduce the amount of ROL available for the visual cycle, thereby reducing the amount of A2E and lipofuscin produced.[59] A study of oral fenretinide in AMD patients with GA showed that the group of patients who had reduction of RBP from treatment did have a reduction in the growth of GA lesions as well as a reduction in the development of CNVM.[60] A1120 is a potent binder of RBP and is effective at reducing lipofuscin in a mouse model.[61] Emixustat is a nonretinoid inhibitor of RPE65 isomerase, which also inhibits the recycling of ROL and can reduce the accumulation of A2E and lipofuscin.[62] A phase 1b study is completed and phase 2 trial is ongoing.[62]

Diabetic retinopathy

Pathogenesis of diabetic retinopathy

The chronic hyperglycemia of diabetes leads to abnormally thickened capillary basement membranes, OS-induced cell stress, cell dysfunction, and ultimately loss through apoptosis of the retinal capillary pericytes first and then the retinal capillary endothelial cells. This gradual damage to the retinal vessels is similar to the microangiopathy that occurs systemically in diabetes. The deturgescence of the retina is maintained by the blood retinal barrier that is similar to the blood–brain barrier. When pericytes are lost due to apoptosis, focal areas of weakness develop that eventually dilate into microaneurysms. As long as the endothelial cells can stretch and cover the dilated areas and maintain tight junctions, the blood retinal barrier is maintained. Once there are too many microaneurysms and there is too much dysfunction or loss of endothelial cells, the microaneurysms leak and the barrier is compromised. The chronic hyperglycemia can also weaken the RPE pump and further compromise the blood retinal barrier. When the macular retina becomes thickened, called cystoid macular edema (CME), vision is compromised. What looks like cysts in the retina, are actually dilated extracellular spaces filled with extravasated plasma. With chronic leakage, exudative lipid and large molecular moieties collect as the extravasated water, electrolytes, and smaller molecule are reabsorbed by what is left of the blood retinal barrier and RPE pump. Microaneurysms, hard exudates, and CME are the hallmarks of nonproliferative diabetic (NPD) retinopathy. As capillaries lose pericytes first and then endothelial cells, capillaries become acellular, maintaining a structure because the thickened shared basement membranes of the missing pericytes and endothelium remain. These acellular capillaries leak diffusely, but are very prone to closure, creating areas of retinal vascular nonperfusion. The choroidal capillaries undergo a similar microangiopathy, but have a density of capillaries that is about 20 times greater than the retinal circulation, so significant vascular nonperfusion of the choroid is rarely seen in diabetes. The choroidal vessels supply the nonperfused retina through the RPE, preventing infarction of the retina, but not preventing ischemia. The ischemic retina responds by producing angiogenic factors, including high levels of soluble VEGF. Unfortunately the neural structure of the retina appears to not allow vessels to grow into the ischemic retina, so NV proliferates into the overlying vitreous. If a posterior vitreous detachment has occurred, NV will not typically proliferate on or in the retina. Rather, the soluble VEGF diffuses into the anterior chamber where it stimulates NV on the iris and trabecular meshwork, scarring over the aqueous outflow channels and causing neovascular glaucoma.

Treatment of diabetic retinopathy

Control of the hyperglycemia slows the progression of NPDR and helps to prevent NV. Although laser photocoagulation of focal areas of leakage found by fluorescein angiography combined with a mild grid pattern treatment of thickened areas was found to help preserve vision in patients with diabetic retinopathy and CME, VEGF inhibitors have proven to be the better treatment choice. Thorough pan-retinal laser treatment of the peripheral retina to ablate the ischemic retina was found to be very effective to prevent further proliferation of NV in patients with early proliferative diabetic retinopathy. Vitrectomy surgery to remove vitreous and fibrotic NV can help prevent traction retinal detachment.

Advanced glycation end-products (AGEs)

Hyperglycemia leads to covalent bonding of glucose adducts to plasma proteins through an autoxidative mechanism.[63] Glycation of proteins interferes with their normal functions by changing conformations, altering enzymatic activity, and interfering with receptor functioning. Advanced glycation end-products (AGEs) form intra- and extracellular cross linking not only with proteins but also with many other molecules including lipids and nucleic acids to promote the development of diabetic microangiopathy.[63] Recent studies suggest that AGEs interact with plasma membrane localized receptors for AGEs to alter intracellular signaling, gene expression, and release of proinflammatory molecules and free radicals.[63]

OS in diabetic microangiopathy

Hyperglycemia also appears to be the cause of metabolic insult to the retinal vasculature in other ways. Hyperglycemia damages retinal vessels by increasing activity of enzymes nitric oxide synthase (NOS) and NADH/NADPH oxidase. Mitochondria are sensitive to the increased superoxide and nitric oxide (NO) that are made, and mtDNA damage occurs, eventually leading to apoptosis in the pericytes and endothelial cells. Superoxide quickly reacts with the endothelial nitric oxide synthase (eNOS)-produced NO, making $ONOO^-$ that adds to the OS damage. When superoxide reduces the circulating NO, inducible nitric oxide synthase (iNOS) is triggered, explaining the increase in NOS and creating a positive feedback loop for producing more ROS. Excessive superoxide leads to caspase-3 activation and apoptosis. Animal models of upregulation of SOD2 inhibited diabetes-induced increases in mitochondrial superoxide, restored mitochondrial function, and prevented vascular pathology both *in vitro* and *in vivo*.[24] Hyperglycemia-induced superoxide also leads to inflammation, which can be blocked by AOXs (*N*-acetylcysteine).[64] Remember that VEGF also upregulates SOD2. The OS/AOX imbalance that is caused by hyperglycemia reduces the capillary repair system of the circulating endothelial progenitor cell (EPCs). EPC dysfunction leads to failure to replace damaged and apoptotic retinal endothelial cells, which then leads to the breakdown of the blood retinal barrier and eventually to nonperfusion. Attempts to prevent hyperglycemic cellular damage with AOX therapy have had mixed results. Glutathione peroxidase-1 (GPx-1) enzymatically detoxifies H_2O_2. Upregulation of GPx-1 in mice reduces intracellular H_2O_2, interfering with cellular pathways and attenuating normal insulin-mediated Akt signaling causing the development of insulin resistance, hyperinsulinemia, and obesity.[65] In mice, GPx-1 deficiency protects against high-fat-induced insulin resistance, enhances insulin-mediated ROS production, increases insulin-mediated Akt signaling, and preserves glucose uptake in muscle.[66] This surprising correlation of decreased ROS with insulin resistance may be due to mitochondrial dysfunction from the lost ROS-growth factor-mediated signaling.[65,67] These findings show the perils of indiscriminant AOX therapy and point out the need for a careful investigation. This is a very important field of investigation for students to consider as a career.

Multiple choice questions

1 The most effective treatment of neovascularization in exudative AMD is
   a. Focal laser photocoagulation of the choroidal neovascular membrane
   b. Powerful antioxidant therapy
   c. Intravitreous anti-VEGF therapy
   d. Steroid therapy

2 The visual cycle recycles which molecule
   a. Docosahexaenoic acid (DHA)
   b. All-*trans*-retinal
   c. Complement H factor
   d. A2E

3 Which are products of oxidation?
   a. ω-(2-carboxyethyl)pyrrole (CEP)
   b. A2E
   c. 4-Hydroxynonenal (4-HNE)
   d. All of the above

4 Hyperglycemia leads to
   a. Decreased activity of NOS
   b. Decreased activity of NADH/NADPH oxidase
   c. Decreased advanced glycation end-products
   d. Increased activity of NOS, NADH/NADPH oxidase, and advanced glycation end-products

References

1 Los, L.I., van der Worp, R.J., van Luyn, M.J. *et al.* (2003) Age-related liquefaction of the human vitreous body: LM and TEM evaluation of the role of proteoglycans and collagen. *Investigative Ophthalmology & Visual Science*, 44(7), 2828–2833.

2 Holekamp, N.M. (2010) The vitreous gel: more than meets the eye. *American Journal of Ophthalmology*, 149(1), 32–36.

3 Shui, Y.B., Holekamp, N.M., Kramer, B.C. *et al.* (2009) The gel state of the vitreous and ascorbate-dependent oxygen consumption: relationship to the etiology of nuclear cataracts. *Archives of Ophthalmology*, 127(4), 475–482.

4 Rokicki, W., Zalejska-Fiolka, J., Mrukwa-Kominek, E. *et al.* (2013) Effect of selected dietary compounds on extracellular superoxide dismutase in the vitreous of chinchillas. *Ophthalmic Research*, 50(1), 54–58.

5 Mosca, M., Ambrosone, L., Semeraro, F. *et al.* (2014) Ocular tissues and fluids oxidative stress in hares fed on verbascoside supplement. *International Journal of Food Sciences and Nutrition*, 65(2), 235–240.

6 Hong, S.M. & Yang, Y.S. (2012) A potential role of crystallin in the vitreous bodies of rats after ischemia-reperfusion injury. *Korean Journal of Ophthalmology*, 26(4), 248–254.

7 Dell'Omo, R., Semeraro, F., Bamonte, G. *et al.* (2013) Vitreous mediators in retinal hypoxic diseases. *Mediators of Inflammation*, 2013, 935301.

8 https://sph.uth.edu/retnet/sum-dis.htm.

9 Wright, A.F., Chakarova, C.F., Abd El-Aziz, M.M. *et al.* (2010) Photoreceptor degeneration: genetic and mechanistic dissection of a complex trait. *Nature Reviews Genetics*, 11(4), 273–284.

10 Dou, G., Kannan, R. & Hinton, D.R. (2012) Endoplasmic reticulum response to oxidative stress in RPE. In: Stratton, R.D., Hauswirth, W.W. & Gardner, T.W. (eds), *Oxidative Stress in Applied Basic Research and Clinical Practice*. Vol. 11. Humana Press, New York, pp. 241–258.

11 Xiong, B. & Bellen, H.J. (2013) Rhodopsin homeostasis and retinal degeneration: lessons from the fly. *Trends in Neurosciences*, 36(11), 652–660.

12 Maeda, A., Maeda, T., Golczak, M. *et al.* (2008) Retinopathy in mice induced by disrupted all-trans-retinal clearance. *Journal of Biological Chemistry*, 283(39), 26684–26693.

13 Griciuc, A., Aron, L., Roux, M.J. *et al.* (2010) Inactivation of VCP/ter94 suppresses retinal pathology caused by misfolded rhodopsin in Drosophila. *PLoS Genetics*, 6(8), e1001075.

14 Saadi, A., Ash, J.D. & Ngansop, T.N. (2012) Role of photoreceptor retinol dehydrogenases in detoxification of lipid oxidation products. In: Stratton, R.D., Hauswirth, W.W. & Gardner, T.W. (eds), *Oxidative Stress in Applied Basic Research and Clinical Practice*. Vol. 8. Humana Press, New York, pp. 165–180.

15 Bazan, N.G. & Halabi, A. (2012) The role of mitochondrial oxidative stress in retinal dysfunction. In: Stratton, R.D., Hauswirth, W.W. & Gardner, T.W. (eds), *Oxidative Stress in Applied Basic Research and Clinical Practice*. Vol. 7. Springer Science + Business Media, LLC, New York, pp. 141–163.

16 Ebrahem, Q., Renganathan, K., Sears, J. *et al.* (2006) Carboxyethylpyrrole oxidative protein modifications stimulate neovascularization: Implications for age-related macular degeneration. *Proceedings of the National Academy of Sciences of the United States of America*, 103(36), 13480–13484.

17 Sparrow, J.R. (2012) Bisretinoid lipofuscin in the retinal pigment epithelium: oxidative processes and disease implications. In: Stratton, R.D., Hauswirth, W.W. & Gardner, T.W. (eds), *Oxidative Stress in Applied Basic Research and Clinical Practice*. Vol. 5. Humana Press, New York, pp. 95–111.

18 Molday, R.S., Beharry, S., Ahn, J. *et al.* (2006) Binding of N-retinylidene-PE to ABCA4 and a model for its transport across membranes. *Advances in Experimental Medicine and Biology*, 572, 465–470.

19 Murdaugh, L.S., Dill, A.E., Dillon, J. *et al.* (2012) Age-related changes in RPE lipofuscin lead to hydrophobic polymers. In: Stratton, R.D., Hauswirth, W.W. & Gardner, T.W. (eds), *Oxidative Stress in Applied Basic Research and Clinical Practice*. Vol. 6. Humana Press, New York, pp. 113–139.

20 Crabb, J.W. (2012) Oxidative modifications as triggers of AMD pathology. In: Stratton, R.D., Hauswirth, W.W. & Gardner, T.W. (eds), *Oxidative Stress in Applied Basic Research and Clinical Practice*. Vol. 3. Humana Press, New York, pp. 65–84.

21 Liu, J., Itagaki, Y., Ben-Shabat, S. *et al.* (2000) The biosynthesis of A2E, a fluorophore of aging retina, involves the formation of the precursor, A2-PE, in the photoreceptor outer segment membrane. *Journal of Biological Chemistry*, 275(38), 29354–29360.

22 Sparrow, J.R., Hicks, D. & Hamel, C.P. (2010) The retinal pigment epithelium in health and disease. *Current Molecular Medicine*, 10(9), 802–823.

23 Sarangarajan, R. & Apte, S.P. (2005) Melanization and phagocytosis: implications for age related macular degeneration. *Molecular Vision*, 11, 482–90.

24 Jarrett, S.G., Lewin, A.S. & Boulton, M.E. (2012) The role of mitochondrial oxidative stress in retinal dysfunction. In: Stratton, R.D., Hauswirth, W.W. & Gardner, T.W. (eds), *Oxidative Stress in Applied Basic Research and Clinical Practice*. Vol. 10. Humana Press, New York, pp. 203–239.

25 Graziewicz, M.A., Day, B.J. & Copeland, W.C. (2002) The mitochondrial DNA polymerase as a target of oxidative damage. *Nucleic Acids Research*, 30(13), 2817–2824.

26 Reddy, P.H., Reddy, T.P., Manczak, M. *et al.* (2011) Dynamin-related protein 1 and mito-chondrial fragmentation in neurodegenerative diseases. *Brain Research Reviews*, 67(1–2), 103–118.

27 López-Lluch, G., Irusta, P.M., Navas, P. *et al.* (2008) Mitochondrial biogenesis and healthy aging. *Experimental Gerontology*, 43(9), 813–819.

28 Reznick, R.M., Zong, H., Li, J. *et al.* (2007) Aging-associated reductions in AMP-activated protein kinase activity and mitochondrial biogenesis. *Cell Metabolism*, 5, 151–156.

29 Martin-Montalvo, A. & de Cabo, R. (2013) Mitochondrial metabolic reprogramming induced by calorie restriction. *Antioxidants and Redox Signaling*, 19(3), 310–320.

30 Jarrett, S.G., Albon, J. & Boulton, M. (2006) The contribution of DNA repair and antiox-idants in determining cell type-specific resistance to oxidative stress. *Free Radical Research*, 40, 1155–1165.

31 Nita, M., Grzybowski, A., Ascaso, F.J. *et al.* (2014) Age-related macular degeneration in the aspect of chronic low-grade inflammation (pathophysiological parainflammation). *Mediators of Inflammation*, 2014, 930671.

32 Klettner, A. & Roider, J. (2012) Mechanisms of pathological VEGF production in the retina and modification with VEGF-antagonists. In: Stratton, R.D., Hauswirth, W.W. & Gardner, T.W. (eds), *Oxidative Stress in Applied Basic Research and Clinical Practice*. Vol. 13. Humana Press, New York, pp. 277–305.

33 Gammons, M.V., Fedorov, O., Ivison, D. *et al.* (2013) Topical antiangiogenic SRPK1 inhibitors reduce choroidal neovascularization in rodent models of exudative AMD. *Investigative Ophthalmology & Visual Science*, 54(9), 6052–6062.

34 Schmidt-Erfurth, U., Chong, V., Loewenstein, A. *et al.* (2014) Guidelines for the manage-ment of neovascular age-related macular degeneration by the European Society of Retina Specialists (EURETINA). *British Journal of Ophthalmology*, 98(9), 1144–1167.

35 Souied EH, Devin F, Mauget-Faÿsse M, et al. 2014. Treatment of exudative age-related mac-ular degeneration with a designed ankyrin repeat protein that binds vascular endothelial growth factor: a phase I/II study***. *American Journal of Ophthalmology* 158(4):724-732.

36 Brar, V.S. & Chalam, K.V. (2012) VEFG inhibitor induced oxidative stress in retinal ganglion cells. In: Stratton, R.D., Hauswirth, W.W. & Gardner, T.W. (eds), *Oxidative Stress in Applied Basic Research and Clinical Practice*. Vol. 29. Humana Press, New York, pp. 585–593.

37 Medzhitov, R. (2008) Origin and physiological roles of inflammation. *Nature*, 454(7203), 428–435.

38 Demidenko, Z.N., Zubova, S.G., Bukreeva, E.I. *et al.* (2009) Rapamycin decelerates cellular senescence. *Cell Cycle*, 8, 1888–1895.

39 Francis, P.J. & Klein, M.L. (2011) Update on the role of genetics in the onset of age-related macular degeneration. *Clinical Ophthalmology*, 5, 1127–1133.

40 Nita, M., Grzybowski, A., Ascaso, F.J. *et al.* (2014) Age-related macular degeneration in the aspect of chronic low-grade inflammation (pathophysiological parainflammation). *Mediators of Inflammation*, 2014, 930671.

41 Görgün, E., Güven, M., Unal, M. *et al.* (2010) Polymorphisms of the DNA repair genes XPD and XRCC1 and the risk of age-related macular degeneration. *Investigative Ophthalmology & Visual Science*, 51(9), 4732–4737.

42 Synowiec, E., Blasiak, J., Zaras, M. *et al.* (2012) Association between polymorphisms of the DNA base excision repair genes MUTYH and hOGG1 and age-related macular degeneration. *Experimental Eye Research*, 98, 58–66.

43 Synowiec, E., Wysokinski, D., Zaras, M. *et al.* (2014) Association between polymorphism of the DNA repair SMUG1 and UNG genes and age-related macular degeneration. *Retina*, 34(1), 38–47.

44 Mehta, S. & Dunaief, J.L. (2012) The role of iron in retinal diseases. In: Stratton, R.D., Hauswirth, W.W. & Gardner, T.W. (eds), *Oxidative Stress in Applied Basic Research and Clinical Practice*. Vol. 12. Humana Press, New York, pp. 259–276.

45 Xu, J., Zhu, D., He, S. *et al.* (2011) Transcriptional regulation of bone morphogenetic protein 4 by tumor necrosis factor and its relationship with age-related macular degeneration. *The FASEB Journal*, 25(7), 2221–2233.

46 Zhu, D.H., Deng, X., Xu, J. *et al.* (2009) What determines the switch between atrophic and neovascular forms of age related macular degeneration? – The role of BMP4 induced senescence. *Aging (Albany NY)*, 1(8), 740–745.

47 Zarbin, M. & Rosenfeld, P.J. (2012) Review of emerging treatments for age-related macular degeneration. In: Stratton, R.D., Hauswirth, W.W. & Gardner, T.W. (eds), *Oxidative Stress in Applied Basic Research and Clinical Practice*. Vol. 1. Humana Press, New York, pp. 1–46.

48 Brantley, M.A., Osborn, M.P., Cai, J. *et al.* (2012) Oxidative stress and systemic changes in age-related macular degeneration. In: Stratton, R.D., Hauswirth, W.W. & Gardner, T.W. (eds), *Oxidative Stress in Applied Basic Research and Clinical Practice*. Vol. 18. Humana Press, New York, pp. 367–397.

49 Blasiak, J., Glowacki, S., Kauppinen, A. *et al.* (2013) Mitochondrial and nuclear DNA damage and repair in age-related macular degeneration. *International Journal of Molecular Sciences*, 14(2), 2996–3010.

50 Chew, E.Y. (2013) Nutrition effects on ocular diseases in the aging eye. *Investigative Ophthalmology & Visual Science*, 54(14), ORSF42–ORSF47.

51 Chew, E.Y., Clemons, T.E., Sangiovanni, J.P. *et al.* (2014) Secondary analyses of the effects of lutein/zeaxanthin on age-related macular degeneration progression: AREDS2 report no. 3. Age-Related Eye Disease Study 2 (AREDS2) Research Group. *JAMA Ophthalmology*, 132(2), 142–149.

52 Awh, C.C., Hawken, S. & Zanke, B.W. (2015) Treatment response to antioxidants and zinc based on CFH and ARMS2 genetic risk allele number in the age-related eye disease study. *Ophthalmology*, 122, 162–9.

53 Yehoshua, Z., de Amorim Garcia Filho, C.A., Nunes, R.P. *et al.* (2014) Systemic complement inhibition with eculizumab for geographic atrophy in age-related macular degeneration: the COMPLETE study. *Ophthalmology*, 121(3), 693–701.

54 Peponis, V., Chalkiadakis, S.E., Bonovas, S. *et al.* (2010) The controversy over the association between statins use and progression of age-related macular degeneration: a mini review. *Clinical Ophthalmology*, 4, 865–869.

55 Rodriguez, I.R. (2012) Deposition and oxidation of lipoproteins in Bruch's membrane and choriocapillaris are "Age-Related" risk factors with implications in age-related macular degeneration. In: Stratton, R.D., Hauswirth, W.W. & Gardner, T.W. (eds), *Oxidative Stress in Applied Basic Research and Clinical Practice*. Vol. 15, Humana Press, New York, pp. 321–335.

56 Shadrach, K.G., Rayborn, M.E., Hollyfield, J.G. *et al.* (2013) DJ-1-dependent regulation of oxidative stress in the retinal pigment epithelium (RPE). *Plos One*, 8(7), e67983.

57 Birch, D.G., Weleber, R.G., Duncan, J.L. *et al.* (2013) Randomized trial of ciliary neurotrophic factor delivered by encapsulated cell intraocular implants for retinitis pigmentosa. *American Journal of Ophthalmology*, 156(2), 283–292.

58 Zhang, K., Hopkins, J.J., Heier, J.S. *et al.* (2011) Ciliary neurotrophic factor delivered by encapsulated cell intraocular implants for treatment of geographic atrophy in age-related macular degeneration. *Proceedings of the National Academy of Sciences of the United States of America*, 108(15), 6241–6245.

59 Querques, G., Rosenfeld, P.J., Cavallero, E. *et al.* (2014) Treatment of dry age-related macular degeneration. *Ophthalmic Research*, 52, 107–115.

60 Mata, N.L., Lichter, J.B., Vogel, R. *et al.* (2013) Investigation of oral fenretinide for treatment of geographic atrophy in age-related macular degeneration. *Retina*, 33(3), 498–507.

61 Dobri, N., Qin, Q., Kong, J. *et al.* (2013) A1120, a nonretinoid RBP4 antagonist, inhibits formation of cytotoxic bisretinoids in the animal model of enhanced retinal lipofuscinogenesis. *Investigative Ophthalmology & Visual Science*, 54(1), 85–95.

62 Kubota, R., Al-Fayoumi, S., Mallikaarjun, S. *et al.* (2014) Phase 1, dose-ranging study of emixustat hydrochloride (ACU-4429), a novel visual cycle modulator, in healthy volunteers. *Retina*, 34(3), 603–609.

63 Singh, V.P., Bali, A., Singh, N. *et al.* (2014) Advanced glycation end products and diabetic complications. *Korean Journal of Physiology and Pharmacology*, 18, 1–14.

64 Lin, Y., Berg, A.H., Iyengar, P. *et al.* (2005) The hyperglycemia-induced inflammatory response in adipocytes: the role of reactive oxygen species. *Journal of Biological Chemistry*, 280(6), 4617–4626.

65 Handy, D.E., Lubos, E., Yang, Y. *et al.* (2009) Glutathione peroxidase-1 regulates mitochondrial function to modulate redox-dependent cellular responses. *Journal of Biological Chemistry*, 284(18), 11913–11921.

66 Loh, K., Deng, H., Fukushima, A. *et al.* (2009) Reactive oxygen species enhance insulin sensitivity. *Cell Metabolism*, 10(4), 260–272.

67 Lubos, E., Loscalzo, J. & Handy, D.E. (2011) Glutathione peroxidase-1 in health and disease: from molecular mechanisms to therapeutic opportunities. *Antioxidants and Redox Signaling*, 15(7), 1957–1997.

CHAPTER 8

The role of oxidative stress in hearing loss

Colleen G. Le Prell[1] and Josef M. Miller[2]

[1] Callier Center for Communication Disorders, School of Behavioral and Brain Sciences, University of Texas at Dallas, Dallas, TX, USA

[2] Department of Otolaryngology, Kresge Hearing Research Institute, University of Michigan, Ann Arbor, MI, USA

THEMATIC SUMMARY BOX

At the end of this chapter, students should be able to:

- Describe the functional impact of hearing loss

- Describe different insults and disorders that cause hearing loss

- Outline the pathogenesis of hearing loss

- Analyze the contribution of oxidative stress to hearing loss

- Infer new therapeutic targets

Introduction

Sound waves are a physical stimulus; much like the ripples in a pond that spread from the point of impact to the water, there is a wave-like deflection of air molecules that travels from a sound source (vibrating vocal folds, a speaker, metal parts striking each other, etc.) outwards. A human, or any animal standing in that spreading sound field, has the potential to detect "sound" when the systematically mobile air molecules enter the ear canal. The sound waves travel down, and are amplified in, the ear canal as a function of ear canal resonances, before arriving at and impinging on the tympanic membrane. The sound waves deflect the tympanic membrane (or the ear drum), which results in motion of three connected bones, termed the middle ear ossicles. Motion of these bones results in pressure concentrated at a second smaller membrane, termed the oval window. The oval window is the point at which the third middle ear ossicle, the stapes, contacts the cochlea. The cochlea is filled with fluid, and when the stapes exerts pressure on the fluid-filled cochlea, in wave-like patterns reflecting the sound waves striking the tympanic membrane, then that mechanical signal is transferred to the inner ear as waves in the fluid inside the

---

Oxidative Stress and Antioxidant Protection: The Science of Free Radical Biology and Disease, First Edition.
Edited by Donald Armstrong and Robert D. Stratton.
© 2016 John Wiley & Sons, Inc. Published 2016 by John Wiley & Sons, Inc.

cochlea. The pressure waves in the fluid inside the cochlea displaces a membrane, termed the basilar membrane, which serves as the supporting base for the sensory cells inside the cochlea, which are the outer hair cells (OHCs) and the inner hair cells (IHCs).

The OHCs are essential for normal hearing thresholds. The OHCs sit between the basilar membrane and the tectorial membrane. With basilar membrane motion, a shearing force is exerted on the OHCs. Until a decade ago, it was generally thought that most, if not all, noise-induced hearing loss (NIHL) occurred when this shearing force directly caused a mechanical destruction of hair cells and their supporting structures, with perhaps some contribution of noise-induced reductions in blood flow to the cochlea. What we now know, however, is that cell damage after noise insult can be largely a by-product of oxidative stress.[1–5] In addition to loss of OHCs, there can be loss of IHCs and a secondary loss of auditory nerve fibers. Finally, the structure associated with cochlear blood supply, the stria vascularis, critical to maintain homeostasis of the organ of Corti, can be damaged, and with that the generation of the endocochlear potential, which is essential for maintained transduction sensitivity. Oxidative stress has been implicated in the damage that occurs to all of these structures. In fact, these structures are vulnerable not only to noise but also to a variety of other insults, and they are lost during the normal process of aging, in the absence of any environmental insult. As the loss of cells to different insults and to aging continues, hearing loss develops and continually progresses with continuing loss of cells in the inner ear.

Hearing loss is a significant clinical, social, and economic issue; the third most prevalent disability, it affects some 278 million people, including more than 10% of the adult male population.[6,7] In the United States, hearing loss in the speech frequency region affects some 29 million Americans aged 20–69 years; when the higher frequencies of 3, 4, and 6 kHz are included, the number of affected individuals doubles.[8] In this chapter, we will review some of the major causes of hearing loss, the underlying pathology, the role of oxidative stress in that pathology, and the potential that antioxidant therapeutics can prevent the onset or progression of hearing loss. For more detailed information on all of the pathologies reviewed here, the role of oxidative stress, and the current status of therapeutic research, readers are referred to the upcoming edition: *Free Radicals in ENT Pathology*.[9]

Age-related hearing loss

Age is one of the major risk factors for hearing loss.[10] Age-related hearing loss (ARHL), also termed presbycusis, begins at higher frequencies and progresses to lower frequencies with increasing age. According to the National Institutes of Health, approximately one-third of Americans aged between 65 and 74 and nearly half of those aged over 75 have hearing loss.[11] One of the earliest and most systematic efforts to identify causes of ARHL was that of Schuknecht, who proposed four ARHL "phenotypes" based on different patterns of pathology detected during dissection of donated human temporal bones. These early papers are reviewed in detail elsewhere.[12,13] Here, we briefly note that the four proposed patterns of pathology include neural, strial, sensory, and mixed origins, reflecting degeneration of the afferent neurons, stria vascularis, organ of Corti, or a mixed pathology with degeneration of multiple structures. A cochlear conductive ARHL has also been proposed, with stiffening of the basilar membrane being the proposed mechanism, but this remains unproven.[13]

One of the key challenges in the identification of "pure" ARHL in humans is that hearing loss in humans is significantly associated with a variety of health-related factors. Cardiovascular disease, diabetes, smoking, nutritional deficiencies, poorer socioeconomic status, anticancer drugs such as cisplatin and antibiotics from the aminoglycoside category, and noise exposure are all associated with poorer hearing outcomes during aging. In humans, Gates and colleagues specifically described different rates of progression of ARHL in human participants that had different noise exposure histories.[14] These reports stimulated a systematic analysis in which mice were exposed to noise inducing a temporary (reversible) change in hearing at different ages than allowed to age for different durations, and differences in age-related pathology and ARHL were detected as a function of that early noise history despite the lack of noise-induced injury at the time of exposure.[15] Data from additional studies suggest that the spiral ganglion cell loss observed late in life in those mice with early noise exposure was likely the result of an immediate and lasting loss of the synaptic connections between the IHCs and the auditory nerve receptors.[16] Taken together, the data clearly indicate that a variety of pathological changes occur in aging ears, and the type and extent of pathology is influenced by the various "insults" that accrue over the life span.

There has been significant interest in the extent to which oxidative stress underlies ARHL and the extent to which antioxidant treatments might reduce or prevent it. The notion of free radical formation across the life span contributing to age-related pathology is well developed.[17–20] Much of the early work establishing a potential role of oxidative stress was completed in mouse models, taking advantage of the ability to manipulate individual genes that mediate the neutralizing of free radicals. When genes controlling the production of the antioxidant enzyme superoxide dismutase (SOD1) were knocked out, there was an increase in ARHL and age-related cochlear pathology reported.[21–23] Similar effects have been observed in mice that have been genetically manipulated so that they cannot synthesize vitamin C; there is an increase in ARHL and age-related cochlear pathology.[24] Other studies have taken the opposite approach and induced an overexpression of antioxidant enzymes. With overexpression of catalase, less ARHL is observed.[25] Perhaps somewhat surprisingly, however, the overexpression of SOD1 had no benefit.[26] There are significant ongoing efforts by multiple groups to screen a larger number of antioxidant genes for potential changes during aging in hope of identifying new therapeutic targets.[27,28]

In rodents, the potential contribution of free radical formation to hearing loss has largely been driven by studies assessing the potential prevention of ARHL using free radical scavengers ("antioxidants"). Early work in rats demonstrated reduced ARHL in rats maintained with supplemented levels of vitamins C or E, melatonin, or lazaroid, compared to rats maintained on a nutritionally complete *ad libitum* diet with no supplements.[29] Interestingly, a fifth experimental group, maintained on 30% calorie restriction, had the best outcomes of all groups, with less hearing loss than the antioxidant-supplemented groups as well as the control group. The benefits of calorie restriction have been confirmed by Someya and colleagues in mice; they provided the novel mechanistic insight that calorie restriction turns down apoptotic gene expression.[30] Additional data show mixed success across different single-agent supplements with lipoic acid, N-acetylcysteine (NAC), and coenzyme Q_{10} (CoQ_{10}) all reducing hearing loss, with no benefits from other supplements at the doses tested (including acetyl-L-carnitine, β-carotene, L-carnosine, curcumin, dl-tocopherol, epigallocatechin gallate, gallic acid, lutein, lycopene, melatonin, proanthocyanidin (grape-seed extract), quercetin dehydrate, and resveratrol).[25] The studies by Someya and colleagues used C57BL/6J mice as subjects; this is important in that this mouse loses its hearing quickly and early, as a consequence of a progressive stereocilia defect. Data from this strain are mixed, with no benefit of NAC-supplemented drinking water in this strain, whereas robust protection against ARHL was achieved using a combination of six antioxidant agents including L-cysteine–glutathione mixed disulfide, ribose-cysteine, NW-nitro-L-arginine methyl ester, vitamin B_{12}, folate, and ascorbic acid.[31,32]

There are considerable and significant differences in age of onset and rate of progression of changes in hearing with age across mouse strains.[33,34] The extent to which the C57BL/6J mouse is an appropriate subject for studies modeling human ARHL and the extent to which such protection will translate to humans is a topic of considerable debate.[35] There have also been significant efforts to assess ARHL in CBA/J and CBA/CaJ mice, which have slower progressive hearing loss beginning in the middle of the life span. Using the CBA/J strain, Sha *et al.* reported no benefits of a diet including supplemented levels of vitamins A, C, and E; L-carnitine; and α-lipoic acid.[36] At this point, it is not clear whether the discrepancies across studies are related to the rodent model selected, the specific agent or agent combination selected, the age at which dosing started, the amount of the supplemented agent in the modified diet, or some combination of all of the above. A second key issue is the control group. Benefits of a supplement may be more evident in nutritionally deficient populations.[37–39] *Ad libitum* feeding may also result in an inherently "unhealthy" control group in which supplement benefits may be more readily apparent, potentially leading to challenges in the translation of benefits to populations that include healthy human volunteers.[40]

The control group issue, and its potential impact on translation to humans, is not trivial. Two different ongoing long-term studies on the effects of caloric restriction on life span are being conducted in macaque monkeys, with one colony maintained by the National Institute on Aging (NIA) and a second colony maintained by the University of Wisconsin (UW) primate center. The control monkeys in the UW study are fed *ad libitum* diets, whereas the control monkeys in the NIA study are maintained on a mildly restricted diet in an effort to avoid obesity in the largely sedentary laboratory animals. In reports on the preliminary outcomes, improved survival outcomes have been reported for subjects in the UW study, but not the NIA study. It has been suggested that the mild caloric restriction for control animals in the NIA study might make it more difficult to detect any additional accrued benefit with more significant caloric restriction in the experimental subjects.[41] This is clearly an important issue, and additional work in rodent models may help to clarify these key questions and set the stage for human translational research.

Translation of effective interventions to humans in prospective studies will be challenging, given the slow progression of ARHL and the requirement for participant compliance with a long-term intervention. Longitudinal data from older adults generally suggest age-related changes in hearing on the order of 0.7–1 dB/year, although the rate of change varies with frequency (more rapid change at higher frequencies), age (more rapid change at older ages), and sex (more rapid change in men).[42–49] A variety of retrospective human epidemiological data are available, and our teams have reviewed much of this literature in recent publications.[50–52] In brief, there are a number of studies that have assessed the "contribution" of different single agents to hearing health by comparing either nutrient intake (measured using food frequency questionnaires) or nutrient status (measured levels in blood samples) and threshold sensitivity. These studies are plagued by the issue of covariation across micronutrients and macronutrients. A diet that is healthy with respect to one nutrient is often healthy with respect to other nutrients as well, making it difficult to identify specific causal relationships. Two approaches used to resolve these covariation issues are multiple regression techniques, which seek to weight the contributions of each nutrient measured, and the use of overall dietary quality metrics such as the healthy eating index (HEI). New data on other nutrients continue to emerge.[53]

To summarize the data to date, ARHL is a major problem worldwide, with significant economic and humanitarian (quality-of-life) costs. ARHL is difficult to distinguish from hearing loss driven by insults across the life span, such as a slowly accruing effects of cardiovascular disease, diabetes, or nutritional deficiency. Genetic knockout models suggest an important role for free radical formation and endogenous antioxidant defense, and data from intervention studies have provided mixed support for potential therapeutic intervention. Much of the literature implicates either increased free radical formation during aging or decreased antioxidant defenses during aging in the development of ARHL and age-related cochlear pathology. (For recent review, see Ref. 54.) For more detailed information on ARHL, the role of oxidative stress, and the current status of therapeutic research, readers are referred to: *Free Radicals in ENT Pathology*; the chapters by RD Frisina and DR Frisina, Juiz and colleagues, and Yamasoba discuss these topics in detail. This is an important area for further research and development.

Noise-induced hearing loss

For many people, hearing loss is caused by an injury induced by exposure to a single loud sound (acoustic trauma). More commonly, noise induces hearing loss via the repeat exposure of the individual to moderately loud sound on a repeat basis. There is a significant debate over the precise level and the duration of exposure at which hazard begins, resulting in different noise limits around the world. The relationship between high-level sound and hearing loss is the foundation for the hearing conservation regulations in the United States (i.e., 29 CRF 1910.95) and elsewhere (for summary of various international standards, see Ref. 55). The federal noise regulations for occupational exposure, 29 CFR 1910.95, are enforced by the Occupational Safety & Health Administration (OSHA). OSHA mandates use of hearing protection devices (HPDs) for any worker exposed to sounds that exceed the permissible exposure limit (PEL) of 90 dBA for 8 h or more per day. Sound levels are, of course, rarely constant, and fluctuating levels throughout the work day are accounted for using the "time-weighted-average" (TWA). TWA estimates the "equivalent" sound-level exposure accrued over the work day if sound levels had been constant. Those calculations use a 5-dB "exchange rate," meaning that for every 5-dB increase in sound level the allowed exposure time is cut in half. Thus, 8 h of 90-dBA exposure is considered equivalent to 4 h at 95 dBA or 2 h at 100 dBA; each of these are considered a 100% dose. In the case of changing levels over the course of the day, for example, 4 h of 90-dBA exposure plus 2 h of 95-dBA exposure, exposure would also be defined as a 100% dose as each of the two components generates 50% of the daily allowed exposure. Per 29 CFR 1910.95, HPDs are required for anyone exceeding 100% daily noise dose, and HPDs must be provided as an option for any worker exceeding a 50% dose.

Even within the context of occupational exposure, there is no uniform consensus regarding noise risk. Some data suggest that, throughout a 40-year career, there is some 10–15% excess risk of hearing loss for those exposed to 85-dBA sound levels during their working career and some 21–29% excess risk of hearing loss for those exposed to 90-dBA sound levels during their working career (see Ref. 56; especially Table 9-1 on p. 187). Given the potential that some individuals will be at an increased risk for developing hearing loss, even if they do not exceed the 100% dose as defined by 29 CFR 1910.95, the National Institute on Occupational Health & Safety (NIOSH) and others have advocated that the PEL be reduced to an 85-dBA TWA with daily exposure calculated using a 3-dB "exchange rate." In other words, for every 3-dB increase in sound level, the allowed exposure time would be cut in half. Thus, a 100% dose is defined as: 8 h of 85-dBA exposure, 4 h of 88-dBA exposure, 2 h of 91-dBA exposure, and so on. The empirical evidence supporting specific exposure limits and exchange rates is limited, and the "best" standard remains an issue of significant debate.[57] The point here is not to advocate one set of standards or another but rather to acknowledge that, despite debate regarding the specific sound levels expected to increase the risk of hearing loss over a 40-year working career, there is generally unanimous agreement that there is a dose–response relationship in which higher levels and longer durations of sound exposure are increasingly hazardous.

Hearing loss attributed to noise exposure typically is observed as a "notched" configuration of the patient's audiogram in combination with a positive history of noise exposure. Not all individuals identified as having an audiometric notch report a history of noise exposure, and not all individuals reporting a history of noise have an audiometric notch. Nonetheless, the "notched" audiogram in combination with the noise history is the most used clinical metric for assessing potential NIHL. The notch is most frequently observed at frequencies ranging from 3 to 6 kHz in humans. This is in part related to the spectral distribution of energy within the acoustic stimulus, and it is also related to the robust amplification of sound at these frequencies, which can be as much as 20 dB based on ear canal resonance properties. Finally, it appears that these higher frequency regions of the cochlea are more metabolically active, and inherently more vulnerable to noise damage as well as other insults.

As NIHL was long assumed to result from direct mechanical destruction of the sensory structures inside the cochlea, mechanical devices (ear plugs, ear muffs) that reduce sound coming into the ear were assumed to be the only strategies for reducing NIHL. Impulse noise and other very loud sounds clearly do cause mechanical damage; however, oxidative stress is an important contributing insult in many cases.[58,59] The timeline for free radical formation in the inner ear is well characterized, beginning with an immediate noise-induced free radical production, well-documented increases in free radical production during the first 1–2 h postexposure, and maximum accumulations of free radicals and their by-products at 7–10 days postnoise.[60,61]

As discussed earlier for ARHL, much of the early work establishing a potential role of oxidative stress in NIHL was completed in mouse models. When genes controlling the production of either SOD1 or glutathione peroxidase (GPx) were knocked out, there was an increase in NIHL.[62,63] In contrast, treatment with agents that increase SOD1 or glutathione levels resulted in less NIHL when guinea pigs were treated prior to noise exposure.[64,65] These data directly stimulated research studies in laboratories around the world, all assessing potential therapeutic agents that might reduce or prevent NIHL through the reduction of noise-induced oxidative stress and the corresponding apoptotic cascade of events, resulting in cell death in the cochlea, with the OHCs and the stria vascularis appearing to be the two structures in which the greatest oxidative stress was produced and where the greatest cell loss occurred.

A wide variety of agents have had at least some benefit in a number of rodent models, which is encouraging for the likely ultimate success of this category of agents being translated into human use. However, the wide variety of agents, models, and conditions of noise exposure and assessments make it challenging to integrate the results into a unified theory. Agents have been tested in different species, with guinea pigs, chinchillas, mice, and rats being the most common. Agents have been tested in different noise models, with some insults inducing 20-dB permanent NIHL in control animals, and some insults inducing 50–60 dB permanent NIHL. Dosing has started at different prenoise times and continued for variable durations. Dosing has also varied, with some agents delivered orally, some injected, and some placed directly into the inner ear using surgical delivery procedures. In most cases, there is little dose–response data; studies have focused on a single dose that "worked" in some initial investigation. Overall, these methodological differences make it difficult, if not impossible, to make predictions about the "best" agent, the "most effective" agent, or the "most likely to succeed" agents. (For review see Ref. 4.) Nonetheless, we reiterate the significant enthusiasm that comes from the pattern of consistently positive outcomes across most studies, and we stress the need for the clinical trials that are urgently needed in order to establish efficacy in humans. A small number of trials have been completed or are in progress now. (For recent review, see Ref. 66.[66]) However, as of the time of the writing of this chapter, there are no drugs that have been approved by the US Food and Drug Administration (FDA) for the prevention of hearing loss.

With respect to both animal studies and clinical trials, there are considerable and significant differences in the total amount of change and the rate of progression for changes in hearing as a function of the specific noise insult used in the study. Some noise insults induce a temporary threshold shift (TTS), whereas others induce a permanent threshold shift (PTS). Recent data revealed that a robust TTS, that is, a large change in hearing lasting at least 24 h but completely recovering within 2 weeks of the exposure, results in an immediate and permanent loss of synaptic connections between the IHCs and the auditory nerve fibers.[15,16,67,68] Oxidative stress is almost surely implicated, as it plays a role in both NIHL and ARHL. Assessing this phenomenon in human ears will be complicated, however, for two reasons. First, primary loss of the auditory afferent nerve fibers occurs as a function of age in animals that have never been exposed to investigational noise; thus, we cannot attribute all primary afferent loss to previous noise alone.[69–71] Second, a TTS is by definition an insult that resolves. The delayed hearing loss that is observed later in life can be reliably attributed to that single TTS experience in an animal that has had no other environmental insult. However, human subjects rarely have a single noise insult. They attend multiple concerts and go to bars, clubs, movies, or sporting events, on a fairly regular basis, with the specific insult varying across individuals. They may be exposed to noise at work, be a recreational shooter that hunts or shoots target at a firing range, or have military service that was accompanied by noise. They may be exposed to chemicals or drugs that are toxic to the ear. They may have health histories that predispose them to hearing loss or genetic backgrounds that render them more vulnerable. There are indeed very significant individual variability issues. Susceptibility to noise damage varies across strains of mice and across rodent species.[72–74] Human vulnerability surely differs from mice, as guinea pigs appear to be less vulnerable than mice and rats, and humans have been suggested to be less vulnerable than guinea pigs.[74,75] Even within species, the variability in individual vulnerability to NIHL is well known in both animal models and human participants.[59,76–83]

It is reasonable to assume that both robust TTS and aging, or aging alone, would have neural consequences in humans similar to those observed in animals. However, early data directly assessing changes in the auditory brainstem response (ABR – auditory brainstem evocated potentials) Wave I amplitude with TTS in humans failed to reveal consistent Wave I changes with TTS.[84] A recent cross-sectional study assessed this topic using subject recall of noise exposure during the previous year, with sound exposures estimated based on commonly reported measurements, and provided some evidence consistent with a relationship between noise history and ABR amplitude in humans.[85] Specifically, there was a statistically significant association between noise history and ABR Wave I amplitudes collected with a mastoid electrode for 70–90 dB nHL click stimuli and 4 kHz pure-tones. The relationship at lower sound pressure levels was not statistically significant, which was not surprising. However, the relationship disappeared when the recording electrode was placed against the tympanic membrane – a recording condition that improves signal-to-noise ratio. Finally, there was no relationship between noise history and ABR Wave V amplitude for either the mastoid or tympanic membrane recording electrode configuration, a result that the authors suggested might imply central compensation for reduced peripheral input. In contrast to this recent report, other physiological data from humans have not shown corresponding deficits. There were no deficits in ABR amplitude in either veterans with known noise exposure[86] or professional pop/rock musicians[87] compared to their respective control subjects. Moreover, the Institute of Medicine report on Noise and Military Service specifically concluded, "The committee's understanding of the mechanisms and processes involved in the recovery from noise exposure suggests, however, that a prolonged delay in the onset of noise-induced hearing loss is unlikely."[88]

Evidence supporting a role of oxidative stress in human NIHL largely comes from three data sets. First, there have been a variety of genetic studies that implicate genes related to oxidative stress in human NIHL.[89,90] Second, there have been a small number of intervention studies in which TTS was reduced in humans using an agent that has antioxidant actions (although this may not be its only mechanism of action).[91–93] Promising outcomes were also reported for a subset of presumably more vulnerable subjects treated with a combination therapy, although the lack of a reliable TTS precluded any conclusions about statistically reliable group benefits.[94] In a different study with a more significant noise insult, a reduction in the rate of PTS after weapon-generated impulse noise was reported.[95] Finally, epidemiological data suggest better hearing at higher frequencies in noise-exposed populations as a function of better dietary quality, with fruit and vegetable intake being the primary mediators of NIHL.[96] For more detailed information on NIHL, the role of oxidative stress, and the current status of therapeutic research, readers are referred to: *Free Radicals in ENT Pathology*; the chapters by Altschuler and Dolan, Yamashita, and Le Prell and Lobarinas discuss these topics in detail. More work in this area is urgently needed to better understand the phenomena of NIHL in humans, and also to identify more completely the role of oxidative stress in NIHL in humans.

Drug-induced hearing loss

Hearing loss is often caused by the use of drugs that are harmful to the auditory system; some of the most ototoxic drugs that are required to be used for lifesaving purposes are the aminoglycoside antibiotics and the chemotherapeutic cisplatin (for review, see Ref. 97). Some drugs are cochleotoxic, meaning they damage sensory cells in the cochlea and induce hearing loss; other drugs are vestibulotoxic, meaning they damage cells in the semicircular canals and induce balance deficits. Most aminoglycoside antibiotics are both cochleotoxic and vestibulotoxic. Drugs such as neomycin, tobramycin, kanamycin, streptomycin, amikacin, and gentamicin can affect hearing, balance, or both.[98–100] Multiple factors influence observed prevalence of hearing loss after aminoglycoside antibiotics. Specifically, hearing loss can develop slowly, sometimes arising weeks after aminoglycoside treatment ends; thus, testing during or immediately after the aminoglycoside treatments end could miss late-developing hearing loss in some patients. Risks may increase as a function of previous aminoglycoside treatment based on cumulative drug dose. The nutritional health of the patient may also influence vulnerability, with reduced protein intake increasing aminoglycoside ototoxicity in animal subjects, as a function of reduced endogenous glutathione (GSH) levels.[101] Finally, the tests selected for use in monitoring patient auditory function will influence prevalence, as some tests are more sensitive to the earliest changes. Because the basal cochlea is affected first, hearing loss prevalence increases when extended high frequency (EHF) threshold changes are considered. The use of otoacoustic emission tests also increases prevalence, as these tests reveal subtle changes in OHC function.[102,103]

Despite the potential adverse side effects of ototoxicity and nephrotoxicity, aminoglycosides remain a drug of choice for use against Gram-negative bacteria and are widely used to treat multidrug-resistant tuberculosis (MDR-TB). Tuberculosis (TB) is an airborne infectious disease caused by the bacteria *Mycobacterium tuberculosis*, and it is the leading infectious cause of morbidity and mortality in adults worldwide, killing about 2 million people every year. Approximately 1.6 billion people are infected worldwide; of these, about 15 million have active disease at any given time. Case rates vary widely by country, age, race, sex, and socioeconomic status. According to the US Center for Disease Control (CDC), the number of TB cases is slowly decreasing in the United States; however, there were still almost 13,000 cases reported in the United States in 2008.[104] All 50 states and the District of Columbia continue to report TB cases, and four states (California, Florida, New York, and Texas) reported more than 500 cases each for 2008 (approximately half of all TB cases in 2008). HIV infection is the single greatest medical risk factor for TB as HIV impairs cell-mediated defense against TB. The problem of MDR-TB is growing throughout the world as a consequence of poorly defined treatment regimens (there is no single "gold standard" for care), poor treatment supervision and inadequate resources for prolonged treatment, and HIV coinfection. Tuberculosis is the leading AIDS-related killer, responsible for perhaps half of all AIDS-related deaths. In some parts of Africa, 75% of people with HIV have TB.[105]

The first-line drugs for TB treatment include isoniazid, rifampin, pyrazinamide, and ethambutol. These drugs have good penetration and generally minimal side effects, but drug resistance is increasing. The recommended length of therapy for most types of TB is 6–9 months, and many patients, particularly those in developing countries, do not complete the full course of treatment. When symptoms re-emerge, the disease is often resistant to the first-line treatments. MDR-TB requires "second-line" antibiotic treatments, commonly including aminoglycoside antibiotics (such as streptomycin, kanamycin, and amikacin), and the polypeptide antibiotic capreomycin. Each of these antibiotics carries a risk of adverse side effects, and hearing loss is a common adverse side effect of each of these drug treatments. The availability of drugs or other agents that preserve hearing sensitivity during treatment for MDR-TB using aminoglycoside antibiotics would be a major advance in clinical care. The challenges of drug-induced hearing loss for TB-positive patients are acute, particularly for those patients exposed to loud sound as part of their occupation.

Treatment with d-methionine or α-tocopherol reduces hearing loss in guinea pigs treated with the aminoglycoside antibiotics gentamicin and amikacin.[106–108] Confirming the potential for translation to human patients, aspirin, which also has potent antioxidant properties, reduced gentamicin-induced hearing loss in human patients in China.[109] The major dose-limiting factor for chronic use of aspirin to protect the human inner ear is gastric side effects. The ideal antioxidant strategy will be inexpensive, safe, and readily available, and will have minimal side effects. We have shown compelling reductions in gentamicin-induced hearing loss in guinea pigs using a combination of β-carotene, vitamins C and E, and magnesium.[110] For more detailed information on aminoglycoside-induced hearing loss, the role of oxidative stress, and the current status of therapeutic research, readers are referred to: *Free Radicals in ENT Pathology*; the chapter by Rybak and Brenner discusses this topic in detail.[9] Cisplatin-induced hearing loss is reviewed in the chapters by Laurell and Pierre, and Campbell and Anderson.

Summary and conclusions

The role of oxidative stress in acquired hearing loss is not limited to aging, noise, and ototoxic drugs; these insults were highlighted in this chapter because they induce hearing loss in a significant number of individuals. Acquired hearing loss is a compelling public health issue, and these are among the primary causes. However, there is clearly a role for oxidative stress in sudden idiopathic sensorineural hearing loss, and the loss of residual hearing that can occur after the surgical insertion of a cochlear implant (a device that restores some auditory function using direct electrical stimulation of the auditory nerve). Oxidative stress has been widely implicated in ear, nose, and throat pathologies. The development of novel therapeutic agents to reduce hearing loss would increase the potential protection of workers exposed to occupational noise as well as noise-exposed military populations. Adolescents and young adults may benefit as well with the popularity of clubs, concerts, sporting events, and personal music players. The development of novel therapeutic agents to reduce hearing loss would perhaps protect the hearing of patients that require lifesaving treatment with cisplatin, to kill cancer cells and preserve life, a benefit that all too often comes at great cost to their hearing. It would be a great advance indeed if patients did not need to choose between their hearing and their life.

A better understanding of oxidative stress during and after noise, and activation of other mechanisms of cellular and molecular events that lead to cell death subsequent to noise insult, has advanced the potential to identify and develop novel therapeutic agents. Identification of oxidative stress and improved knowledge as to how cells die has been particularly significant for the development of novel therapeutic agents to reduce NIHL. Widespread clinical acceptance of any novel therapeutic will be driven by demonstration that the agent reduces noise-induced PTS in randomized, placebo-controlled, prospective human clinical trials. Identification and access to populations that develop PTS despite the use of traditional HPDs (which are ethically required in such studies) is challenging. Moreover, such studies are necessarily slow, requiring years of data collection from each individual subject, given that NIHL is generally slow to develop. Given these and other obstacles, many groups are turning to TTS models for initial human proof-of-concept testing.

The last decade has seen a significant increase in our understanding of the mechanisms of stress-induced hearing impairment, and with that windows of opportunity have opened to define interventions that may reduce this disability and improve the quality-of-life form many millions worldwide. It is an exciting and optimistic time.

Multiple choice questions

1 Most, if not all, noise-induced hearing loss occurs when a shearing force directly causes a mechanical destruction of hair cells and their supporting structures.
   a. True
   b. False

2 Studies in mice show all except:
   a. Inhibition of vitamin C increases age-related hearing loss (ARHL)

   b. Knockout of SOD1 increases ARHL

   c. Overexpression of catalase decreases ARHL

   d. Overexpression of SOD1 decreases ARHL dramatically

3 Thirty percent calorie restriction in rats decreases ARHL.

   a. True

   b. False

References

1 Le Prell, C.G. *et al.* (2007) Mechanisms of noise-induced hearing loss indicate multiple methods of prevention. *Hearing Research*, 226(1–2), 22–43.

2 Henderson, D. *et al.* (2006) The role of oxidative stress in noise-induced hearing loss. *Ear and Hearing*, 27(1), 1–19.

3 Abi-Hachem, R.N., Zine, A. & Van De Water, T.R. (2010) The injured cochlea as a target for inflammatory processes, initiation of cell death pathways and application of related otoprotectives strategies. *Recent Patents on CNS Drug Discovery*, 5(2), 147–163.

4 Le Prell, C.G. & Bao, J. (2012) Prevention of noise-induced hearing loss: potential therapeutic agents. In: Le Prell, C.G., *et al.* (eds), *Noise-Induced Hearing Loss: Scientific Advances, Springer Handbook of Auditory Research*. Springer Science+Business Media, LLC, New York, pp. 285–338.

5 Poirrier, A.L. *et al.* (2010) Oxidative stress in the cochlea: an update. *Current Medicinal Chemistry*, 17(30), 3591–3604.

6 Stevens, G. *et al.* (2013) Global and regional hearing impairment prevalence: an analysis of 42 studies in 29 countries. *European Journal of Public Health*, 23(1), 146–152.

7 Tucci, D., Merson, M.H. & Wilson, B.S. (2010) A summary of the literature on global hearing impairment: current status and priorities for action. *Otology & Neurotology*, 31(1), 31–41.

8 Agrawal, Y., Platz, E.A. & Niparko, J.K. (2008) Prevalence of hearing loss and differences by demographic characteristics among US adults: data from the National Health and Nutrition Examination Survey, 1999–2004. *Archives of Internal Medicine*, 168(14), 1522–1530.

9 Miller, J.M. *et al.* (eds) (2015) *Free Radicals in ENT Pathology*. Springer Science+Business Media, New York.

10 Cruickshanks, K.J. *et al.* (1998) Prevalence of hearing loss in older adults in Beaver Dam, Wisconsin. The Epidemiology of Hearing Loss Study. *American Journal of Epidemiology*, 148(9), 879–886.

11 National Institute on Deafness and Other Communication Disorders. (2014) *Quick Statistics*. 3 October 2013 [cited 1 November 2014]. URL http://www.nidcd.nih.gov/health/statistics/Pages/quick.aspx [accessed on 9 September 2015].

12 Ohlemiller, K.K. (2004) Age-related hearing loss: the status of Schuknecht's typology. *Current Opinion in Otolaryngology. Head and Neck Surgery*, 12(5), 439–443.

13 Ohlemiller, K.K. & Frisina, R.D. (2008) Age-related hearing loss and its cellular and molecular bases. In: Schacht, J., Popper, A.N. & Fay, R.R. (eds), *Auditory Trauma, Protection, and Repair: Springer Handbook of Auditory Research*. Vol. 31. Springer Science+Business Media, LLC, New York, pp. 145–194.

14 Gates, G.A. *et al.* (2000) Longitudinal threshold changes in older men with audiometric notches. *Hearing Research*, 141(1–2), 220–228.

15 Kujawa, S.G. & Liberman, M.C. (2006) Acceleration of age-related hearing loss by early noise exposure: evidence of a misspent youth. *Journal of Neuroscience*, 26(7), 2115–2123.

16 Kujawa, S.G. & Liberman, M.C. (2009) Adding insult to injury: cochlear nerve degeneration after "temporary" noise-induced hearing loss. *Journal of Neuroscience*, 29(45), 14077–14085.

17 Pacifici, R.E. & Davies, K.J. (1991) Protein, lipid and DNA repair systems in oxidative stress: the free-radical theory of aging revisited. *Gerontology*, 37(1–3), 166–180.

18 Davies, K.J. (1995) Oxidative stress: the paradox of aerobic life. *Biochemical Society Symposia*, 61, 1–31.

19 Mattson, M.P. (2000) *Apoptosis in neurodegenerative disorders. Nature Reviews Molecular and Cellular Biology*, 1, 120–129.

20 Mattson, M.P. (2006) Neuronal life-and-death signaling, apoptosis, and neurodegenerative disorders. *Antioxid Redox Signal*, 8(11–12), 1997–2006.

21 McFadden, S.L. *et al.* (1999) Cu/Zn SOD deficiency potentiates hearing loss and cochlear pathology in aged 129,CD-1 mice. *Journal of Comparative Neurology*, 413(1), 101–112.

22 McFadden, S.L. *et al.* (1999) Age-related cochlear hair cell loss is enhanced in mice lacking copper/zinc superoxide dismutase. *Neurobiology of Aging*, 20(1), 1–8.

23 Keithley, E.M. *et al.* (2005) Cu/Zn superoxide dismutase and age-related hearing loss. *Hearing Research*, 209(1–2), 76–85.

24 Kashio, A. *et al.* (2009) Effect of vitamin C depletion on age-related hearing loss in SMP30/GNL knockout mice. *Biochemical and Biophysical Research Communications*, 390(3), 394–398.

25 Someya, S. *et al.* (2009) Age-related hearing loss in C57BL/6J mice is mediated by Bak-dependent mitochondrial apoptosis. *Proceedings of the National Academy of Sciences of the United States of America*, 106(46), 19432–19437.

26 Coling, D.E. *et al.* (2003) Effect of SOD1 overexpression on age- and noise-related hearing loss. *Free Radical Biology and Medicine*, 34(7), 873–880.

27 Tadros, S.F. *et al.* (2014) Gene expression changes for antioxidants pathways in the mouse cochlea: relations to age-related hearing deficits. *PLoS One*, 9(2), e90279.

28 Tanaka, C. *et al.* (2012) Expression pattern of oxidative stress and antioxidant defense-related genes in the aging Fischer 344/NHsd rat cochlea. *Neurobiology of Aging*, 33(8), 1842.e1–1842.e14.

29 Seidman, M.D. (2000) Effects of dietary restriction and antioxidants on presbyacusis. *Laryngoscope*, 110(5 Pt 1), 727–738.

30 Someya, S. *et al.* (2007) Caloric restriction suppresses apoptotic cell death in the mammalian cochlea and leads to prevention of presbycusis. *Neurobiology of Aging*, 28(10), 1613–1622.

31 Heman-Ackah, S.E. *et al.* (2010) A combination antioxidant therapy prevents age-related hearing loss in C57BL/6 mice. *Otolaryngology – Head and Neck Surgery*, 143(3), 429–434.

32 Davis, R.R. *et al.* (2007) N-acetyl L-cysteine does not protect against premature age-related hearing loss in C57BL/6J mice: a pilot study. *Hearing Research*, 226(1–2), 203–208.

33 Erway, L.C. *et al.* (1993) Genetics of age-related hearing loss in mice: I. Inbred and F1 hybrid strains. *Hearing Research*, 65(1–2), 125–132.

34 Willott, J.F. & Erway, L.C. (1998) Genetics of age-related hearing loss in mice. IV. Cochlear pathology and hearing loss in 25 BXD recombinant inbred mouse strains. *Hearing Research*, 119(1–2), 27–36.

35 Schacht, J. *et al.* (2012) Alleles that modulate late life hearing in genetically heterogeneous mice. *Neurobiology of Aging*, 33(8), 1842.e15–1842.e29.

36 Sha, S.H. *et al.* (2012) Antioxidant-enriched diet does not delay the progression of age-related hearing loss. *Neurobiology of Aging*, 33(5), 1010.e15–1010.e6.

37 Kang, W.S. *et al.* (2012) Effects of a zinc-deficient diet on hearing in CBA mice. *Neuroreport*, 23(4), 201–205.

38 Durga, J. *et al.* (2007) Effects of folic acid supplementation on hearing in older adults: a randomized, controlled trial. *Annals of Internal Medicine*, 146(1), 1–9.

39 Saposnik, G. (2011) The role of vitamin B in stroke prevention: a journey from observational studies to clinical trials and critique of the VITAmins TO Prevent Stroke (VITATOPS). *Stroke*, 42(3), 838–842.

40 Martin, B. *et al.* (2010) "Control" laboratory rodents are metabolically morbid: why it matters. *Proceedings of the National Academy of Sciences of the United States of America*, 107(14), 6127–6133.

41 Cava, E. & Fontana, L. (2013) Will calorie restriction work in humans? *Aging (Albany NY)*, 5(7), 507–514.

42 Echt, K.V. *et al.* (2010) Longitudinal changes in hearing sensitivity among men: the Veterans Affairs Normative Aging Study. *Journal of the Acoustical Society of America*, 128(4), 1992–2002.

43 Kiely, K.M. *et al.* (2012) Cognitive, health, and sociodemographic predictors of longitudinal decline in hearing acuity among older adults. *The Journals of Gerontology. Series A, Biological Sciences and Medical Sciences*, 67(9), 997–1003.

44 Ostri, B. & Parving, A. (1991) A longitudinal study of hearing impairment in male subjects – an 8-year follow-up. *British Journal of Audiology*, 25(1), 41–48.

45 Lee, F.S. *et al.* (2005) Longitudinal study of pure-tone thresholds in older persons. *Ear and Hearing*, 26(1), 1–11.

46 Morrell, C.H. *et al.* (1996) Age- and gender-specific reference ranges for hearing level and longitudinal changes in hearing level. *Journal of the Acoustical Society of America*, 100(4 Pt 1), 1949–1967.

47 Cruickshanks, K.J. *et al.* (2003) The 5-year incidence and progression of hearing loss: the epidemiology of hearing loss study. *Archives of Otolaryngology – Head and Neck Surgery*, 129(10), 1041–1046.

48 Brant, L.J. & Fozard, J.L. (1990) Age changes in pure-tone hearing thresholds in a longitudinal study of normal human aging. *Journal of the Acoustical Society of America*, 88(2), 813–820.

49 Pearson, J.D. *et al.* (1995) Gender differences in a longitudinal study of age-associated hearing loss. *Journal of the Acoustical Society of America*, 97(2), 1196–1205.

50 Le Prell, C.G. & Spankovich, C. (2013) Noise-induced hearing loss: detection, prevention and management. In: Kirtane, M.V., *et al.* (eds), *Otology and Neurotology*. Thieme, Stuttgart, Germany, pp. 268–284.

51 Spankovich, C. & Le Prell, C.G. (2013) Healthy diets, healthy hearing: National health and nutrition examination survey, 1999–2002. *International Journal of Audiology*, 52(6), 369–276.

52 Choi, Y.H. *et al.* (2014) Antioxidant vitamins and magnesium and the risk of hearing loss in the US general population. *American Journal of Clinical Nutrition*, 99(1), 148–155.

53 Kang, J.W. *et al.* (2014) Dietary vitamin intake correlates with hearing thresholds in the older population: the Korean National Health and Nutrition Examination Survey. *American Journal of Clinical Nutrition*, 99(6), 1407–1413.

54 Yamasoba, T. *et al.* (2013) Current concepts in age-related hearing loss: epidemiology and mechanistic pathways. *Hearing Research*, 303, 30–38.

55 Suter, A.H. (2007) Development of standards and regulations for occupational noise. In: Crocker, M. (ed), *Handbook of Noise and Vibration Control*. John Wiley and Sons, Inc., Hoboken.

56 Pelton, H.K. (2001) Hearing conservation. In: Dobie, R.A. (ed), *Medical–Legal Evaluation of Hearing Loss*, 2nd edn. Singular Publishing, San Diego, pp. 184–208.

57 Dobie, R.A. & Clark, W.W. (2014) Exchange rates for intermittent and fluctuating occupational noise: a systematic review of studies of human permanent threshold shift. *Ear and Hearing*, 35(1), 86–96.

58 Ohlemiller, K.K. (2008) Recent findings and emerging questions in cochlear noise injury. *Hearing Research*, 245, 5–17.

59 Wang, Y., Hirose, K. & Liberman, M.C. (2002) Dynamics of noise-induced cellular injury and repair in the mouse cochlea. *Journal of the Association for Research in Otolaryngology*, 3(3), 248–268.

60 Ohlemiller, K.K., Wright, J.S. & Dugan, L.L. (1999) Early elevation of cochlear reactive oxygen species following noise exposure. *Audiology and Neuro-Otology*, 4(5), 229–236.

61 Yamashita, D. *et al.* (2004) Delayed production of free radicals following noise exposure. *Brain Research*, 1019, 201–209.

62 Ohlemiller, K.K. *et al.* (1999) Targeted deletion of the cytosolic Cu/Zn-superoxide dismutase gene (Sod1) increases susceptibility to noise-induced hearing loss. *Audiology and Neuro-Otology*, 4(5), 237–246.

63 Ohlemiller, K.K. *et al.* (2000) Targeted mutation of the gene for cellular glutathione peroxidase (Gpx1) increases noise-induced hearing loss in mice. *Journal of the Association for Research in Otolaryngology*, 1(3), 243–254.

64 Cassandro, E. *et al.* (2003) Effect of superoxide dismutase and allopurinol on impulse noise-exposed guinea pigs – electrophysiological and biochemical study. *Acta Oto-Laryngologica*, 123(7), 802–807.

65 Ohinata, Y. *et al.* (2000) Glutathione limits noise-induced hearing loss. *Hearing Research*, 146(1–2), 28–34.

66 Le Prell, C.G. & Lobarinas, E. (2015) Strategies for assessing antioxidant efficacy in clinical trials. In: Miller, J.M., *et al.* (eds), *Oxidative Stress in Applied Basic Research and Clinical Practice: Free Radicals in ENT Pathology*. Springer, New York, pp. 163–192.

67 Lin, H.W. *et al.* (2011) Primary neural degeneration in the guinea pig cochlea after reversible noise-induced threshold shift. *Journal of the Association for Research in Otolaryngology*, 12(5), 605–616.

68 Wang, Y. & Ren, C. (2012) Effects of repeated "benign" noise exposures in young CBA mice: shedding light on age-related hearing loss. *Journal of the Association for Research in Otolaryngology*, 13(4), 505–515.

69 Sergeyenko, Y. *et al.* (2013) Age-related cochlear synaptopathy: an early-onset contributor to auditory functional decline. *Journal of Neuroscience*, 33(34), 13686–13694.

70 Boettcher, F.A. *et al.* (1995) Age-related changes in auditory evoked potentials of gerbils. III. Low-frequency responses and repetition rate effects. *Hearing Research*, 87(1–2), 208–19.

71 Schmiedt, R.A., Mills, J.H. & Boettcher, F.A. (1996) Age-related loss of activity of auditory-nerve fibers. *Journal of Neurophysiology*, 76(4), 2799–2803.

72 Davis, R.R. *et al.* (2001) Genetic basis for susceptibility to noise-induced hearing loss in mice. *Hearing Research*, 155(1–2), 82–90.

73 Yoshida, N. *et al.* (2000) Acoustic injury in mice: 129/SvEv is exceptionally resistant to noise-induced hearing loss. *Hearing Research*, 141(1–2), 97–106.

74 Duan, M. *et al.* (2008) Susceptibility to impulse noise trauma in different species: guinea pig, rat, and mouse. *Acta Otolaryngologica*, 128(3), 277–283.

75 Liang, Z. (1992) Parametric relation between impulse noise and auditory damage. In: Dancer, A.L., *et al.* (eds), *Noise-Induced Hearing Loss*. Mosby Year Book, St. Louis, pp. 325–335.

76 Maison, S.F. & Liberman, M.C. (2000) Predicting vulnerability to acoustic injury with a noninvasive assay of olivocochlear reflex strength. *Journal of Neuroscience*, 20(12), 4701–4707.

77 Yoshida, N. & Liberman, M.C. (2000) Sound conditioning reduces noise-induced permanent threshold shift in mice. *Hearing Research*, 148(1–2), 213–219.

78 Mills, J.H. *et al.* (2001) A comparison of age-related hearing loss and noise-induced hearing loss. In: Henderson, D., *et al.* (eds), *Noise Induced Hearing Loss: Basic Mechanisms, Prevention and Control*. Noise Research Network, London, pp. 497–511.

79 Strasser, H., Irle, H. & Legler, R. (2003) Temporary hearing threshold shifts and restitution after energy-equivalent exposures to industrial noise and classical music. *Noise Health*, 5(20), 75–84.

80 Le Prell, C.G. *et al.* (2012) Digital music exposure reliably induces temporary threshold shift (TTS) in normal hearing human subjects. *Ear and Hearing*, 33(6), e44–e58.

81 Ward, W.D. (1970) Temporary threshold shift and damage-risk criteria for intermittent noise exposures. *Journal of the Acoustical Society of America*, 48(2), 561–574.

82 Spankovich, C. *et al.* (2014) Temporary threshold shift after impulse-noise during video game play: Laboratory data. *International Journal of Audiology*, 53(Suppl 2), S53–S65.

83 Ward, W.D. (1960) Recovery from high values of temporary threshold shift. *Journal of the Acoustical Society of America*, 32(4), 497–500.

84 Klein, A.J. & Mills, J.H. (1981) Physiological and psychophysical measures from humans with temporary threshold shift. *Journal of the Acoustical Society of America*, 70(4), 1045–1053.

85 Stamper, G.C. & Johnson, T.A. (2014) Auditory function in normal-hearing, noise-exposed human ears. Ear & Hearing, 36(2):172–84; see additional information in 36(6):738–40.

86 Konrad-Martin, D. *et al.* (2012) Age-related changes in the auditory brainstem response. *Journal of the American Academy of Audiology*, 23(1), 18–35 quiz 74–75.

87 Samelli, A.G. *et al.* (2012) Audiological and electrophysiological assessment of professional pop/rock musicians. *Noise Health*, 14(56), 6–12.

88 Institute of Medicine (2005) Humes, L.E., Joellenbeck, L.M. & Durch, J.S. (eds), *Noise and Military Service: Implications for Hearing Loss and Tinnitus*. The National Academies Press, Washington, DC.

89 Abreu-Silva, R.S. *et al.* (2011) The search of a genetic basis for noise-induced hearing loss (NIHL). *Annals of Human Biology*, 38(2), 210–8.

90 Gong, T.W. & Lomax, M.I. (2011) Genes that influence susceptibility to noise-induced hearing loss. In: Le Prell, C.G., *et al.* (eds), *Noise-Induced Hearing Loss: Scientific Advances*. Springer, New York, pp. 179–203.

91 Quaranta, A. *et al.* (2004) The effects of 'supra-physiological' vitamin B12 administration on temporary threshold shift. *International Journal of Audiology*, 43(3), 162–165.

92 Quaranta, N. *et al.* (2012) The effect of alpha-lipoic acid on temporary threshold shift in humans: a preliminary study. *Acta Otorhinolaryngologica Italica*, 32(6), 380–385.

93 Attias, J. *et al.* (2004) Reduction in noise-induced temporary threshold shift in humans following oral magnesium intake. *Clinical Otolaryngology*, 29(6), 635–641.

94 Le Prell, C.G. *et al.* (2011) Increased vitamin plasma levels in Swedish military personnel treated with nutrients prior to automatic weapon training. *Noise & Health*, 13, 432–443.

95 Attias, J. *et al.* (1994) Oral magnesium intake reduces permanent hearing loss induced by noise exposure. *American Journal of Otolaryngology*, 15(1), 26–32.

96 Spankovich, C. & Le Prell, C.G. (2014) Associations between dietary quality, noise, and hearing: data from the National Health and Nutrition Examination Survey, 1999-2002. *International Journal of Audiology*, 53(11), 796–809.

97 Campbell KC, & Le Prell CG. (2011) Potential therapeutic agents. *Seminars in Hearing*, 32(3):281–296.

98 Clark, C.H. (1977) Toxicity of aminoglycoside antibiotics. *Modern Veterinary Practice*, 58(7), 594–598.

  99 Xie, J., Talaska, A.E. & Schacht, J. (2011) New developments in aminoglycoside therapy and ototoxicity. *Hearing Research*, 281(1–2), 28–37.
 100 Ahmed, R.M. *et al.* (2012) Gentamicin ototoxicity: a 23-year selected case series of 103 patients. *Medical Journal of Australia*, 196(11), 701–704.
 101 Lautermann, J., McLaren, J. & Schacht, J. (1995) Glutathione protection against gentamicin ototoxicity depends on nutritional status. *Hearing Research*, 86(1–2), 15–24.
 102 Constantinescu, R.M. *et al.* (2009) Otoacoustic emissions analysers for monitoring aminoglycosides ototoxicity. *Romanian Journal of Internal Medicine*, 47(3), 273–278.
 103 Stavroulaki, P. *et al.* (1999) Otoacoustic emissions – an approach for monitoring aminoglycoside induced ototoxicity in children. *International Journal of Pediatric Otorhinolaryngology*, 50(3), 177–184.
 104 Pratt, R., Robison, V. & Navin, T. (2009) Trends in Tuberculosis – United States, 2008. *Morbidity and Mortality Weekly Survey Surveillance Reports*, 58(10), 249–253.
 105 Rosenberg T. (2004) Necessary treatments. *The Wall Street Journal* [cited 2 November 2009]. URL http://www.nytimes.com/2004/09/19/magazine/19IDEALAB.html?_r=1&pagewanted=all&position=.
 106 Sha, S.H. & Schacht, J. (2000) Antioxidants attenuate gentamicin-induced free radical formation in vitro and ototoxicity in vivo: D-methionine is a potential protectant. *Hearing Research*, 142(1–2), 34–40.
 107 Campbell, K.C.M. *et al.* (2007) Prevention of noise- and drug-induced hearing loss with D-methionine. *Hearing Research*, 226, 92–103.
 108 Fetoni, A.R. *et al.* (2004) alpha-Tocopherol protective effects on gentamicin ototoxicity: an experimental study. *International Journal of Audiology*, 43(3), 166–171.
 109 Sha, S.H., Qiu, J.H. & Schacht, J. (2006) Aspirin to prevent gentamicin-induced hearing loss. *New England Journal of Medicine*, 354(17), 1856–1857.
 110 Le Prell, C.G. *et al.* (2014) Assessment of nutrient supplement to reduce gentamicin-induced ototoxicity. *Journal of the Association for Research in Otolaryngology*, 15(3), 375–393.

CHAPTER 9

Disorders of children

Hirokazu Tsukahara and Masato Yashiro

Department of Pediatrics, Okayama University Graduate School of Medicine, Dentistry and Pharmaceutical Sciences, Okayama, 700-8558, Japan

List of abbreviations

ADMA   Asymmetric dimethylarginine
BMI   Body mass index
BW   Body weight
CNS   Central nervous system
CO   Carbon monoxide
Cr   Creatinine
CSF   Cerebrospinal fluid
ELISA   Enzyme-linked immunosorbent assay
GA   Gestational age
GST   Glutathione S-transferase
HNE   4-Hydroxy-2-nonenal
IAE   Influenza-associated acute encephalopathy
L-FABP   L-type fatty acid binding protein
L-NAME N^G-nitro-L-arginine methyl ester
NO   Nitric oxide
NOS   Nitric oxide synthase
8-OHdG   8-Hydroxy-2′-deoxyguanosine
OSI   Oxidative stress index
ROS   Reactive oxygen species
SOD   Superoxide dismutase
TAC   Total antioxidative capacity
TH   Total hydroperoxides
TRX   Thioredoxin

THEMATIC SUMMARY BOX

At the end of this chapter, students should be able to:

- Describe the main features of acute and chronic diseases in prenatal, neonatal, and pediatric populations

---

Oxidative Stress and Antioxidant Protection: The Science of Free Radical Biology and Disease, First Edition.
Edited by Donald Armstrong and Robert D. Stratton.
© 2016 John Wiley & Sons, Inc. Published 2016 by John Wiley & Sons, Inc.

- Outline the pathogenesis of these disorders
- Analyze the contribution of oxidative stress and antioxidant defense systems
- Review biomarkers that are available to measure oxidative stress and antioxidant treatment outcomes.
- Summarize redox modulation strategies for acute influenza encephalopathy

Introduction – reactive oxygen species, antioxidative systems, and oxidative stress

Reactive oxygen species (ROS), such as superoxide anion (O_2^-), hydroxyl radical (OH), hydrogen peroxides (H_2O_2), and nitric oxide (NO), serve in cell signaling as messenger molecules of the autocrine and paracrine systems and serve in host defense.[1] However, the generation of ROS can also engender damage to multiple cellular organelles and processes, which can ultimately disrupt normal physiology. Excess NO can injure tissues and does so mainly by its rapid reaction with O_2^-, thereby producing peroxynitrite anion ($ONOO^-$).[2]

As countermeasures, cells, tissues, and organs possess various antioxidant systems for the elimination of ROS. Under physiological conditions, a well-managed balance prevails between the formation and elimination of ROS by these systems. "Oxidative stress" can occur when ROS production is accelerated or when the mechanisms for maintaining the normal reductive environment are impaired. Oxidative damage to the cellular components can be deleterious and concomitant in the body.[2,3]

The following sections present discussion of the rapidly accruing data linking oxidative events as critical participants in prenatal, neonatal, and pediatric medicine. The initial three sections present the fundamental scope. The subsequent seven sections present specific examination of this topic in prenatal, neonatal, and pediatric fields while incorporating a discussion of our recent findings. The last section outlines the therapeutic value of available antioxidative drugs in the treatment of acute encephalopathy of severe type.

Biomarkers for oxidative stress

Oxidative molecular damage is unavoidable: every biological molecule is placed at risk by persistent oxidative stress. Lipids, proteins, nucleic acids, and carbohydrates are modified in a manner that is characteristic of oxidizing species. Direct measurement of ROS *in vivo* is difficult because the half-lives of ROS are usually short. "Oxidative stress biomarkers" are often measured using stable adducts that are produced as a result of the oxidative processes that occur *in vivo*.[3]

Measurement of these specific biomarkers in body fluids enables repeated monitoring of the oxidative stress status *in vivo*, which is otherwise not possible with invasive tests. Analysis of oxidative stress in various pathological processes is now per-formed predominantly using enzyme-linked immunosorbent assay (ELISA) because improved antibody production technologies enable the production of specific anti-bodies against antigenic proteins or peptides that receive chemical alteration reactions by oxidative damage. This technique has proved to be particularly suitable for clinical medicine.[4]

Oxidative stress biomarkers are separable into two categories: (i) formation of modified molecules by ROS and (ii) consumption or induction of enzymes or antiox-idants. The first category (i) includes molecules generated in a reaction with ROS (including ONOO⁻). Molecules are subjected to scission, cross-linking, or covalent modification in these reactions. These molecules are increased when ROS are gener-ated. Some are removed rapidly or are repaired rapidly, but others remain for a long time in intracellular or extracellular compartments.

Major targets of ROS in the cell molecular components are membrane lipids, proteins, nucleic acids, and carbohydrates. Clinically applicable biomarkers include 4-hydroxy-2-nonenal (HNE), malondialdehyde, acrolein, F2-isoprostane (markers of lipid oxidation), 8-hydroxy-2'-deoxyguanosine (8-OHdG), 8-nitroguanine (markers of oxidative DNA damage), carboxymethyl lysine, pentosidine (markers of glycoxi-dation), 2-pyrrolidone, 3-nitrotyrosine (markers of protein oxidation), nitrite/nitrate (a marker of nitro-oxidation), and bilirubin oxidative metabolites (markers of heme oxygenase activity).

The second category (ii) includes antioxidative enzymes and molecules associated with ROS metabolism. In most cases, these molecules are destroyed or modified. They exhibit decreased activity or quantity after ROS exposure. Conversely, they often show an overshooting response for a matter of hours, days, or weeks.

Very recently, compact machines by which serum/plasma total hydroperoxides (TH) and "total antioxidative capacity (TAC)" or urinary 8-OHdG can be measured are available.[5] These machines quickly provide highly reproducible results.

Furthermore, assessment of oxidative stress markers in exhaled breath has proven useful for managing airway inflammatory diseases (Figure 9.1).[6,7] These "lung biomarkers" might be helpful in making diagnoses, defining specific phenotypes of diseases, monitoring exacerbations, and evaluating drug effects in airway diseases.

Nitric oxide system blockade, endothelial dysfunction, and oxidative stress

The NO radical is synthesized from L-arginine and molecular oxygen in every cell type by nitric oxide synthases (NOSs), which have three isoforms. They are neuronal NOS (NOS1), inducible NOS (NOS2), and endothelial NOS (NOS3). Actually, NOS1 and NOS3, which are present constitutively in cells of various types, are activated by transient increases in intracellular calcium. The third isoform, NOS2, is induced in response to inflammatory and immunological stimuli in myriad cells such as vascular endothelial cells, smooth muscle cells, and activated immune cells. The output of NO from NOS2 is about 1000 times that of other constitutive isoforms.[3]

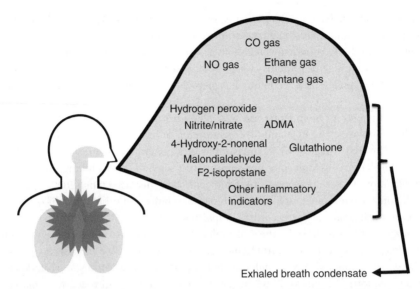

CO gas

NO gas Ethane gas
Pentane gas

Hydrogen peroxide
Nitrite/nitrate ADMA
4-Hydroxy-2-nonenal Glutathione
Malondialdehyde
F2-isoprostane
Other inflammatory
indicators

Exhaled breath condensate ◄

Figure 9.1 Oxidative stress biomarkers in exhaled breath ("lung biomarkers"). Abbreviations: ADMA, asymmetric dimethylarginine; CO, carbon monoxide; NO, nitric oxide.

The vascular endothelium, rather than being a mere barrier between intravascular and interstitial compartments, is a widely distributed organ that is responsible for the regulation of hemodynamics; angiogenic vascular remodeling; and metabolic, synthetic, anti-inflammatory, and antithrombogenic processes. Understanding of the interrelationship of NO deficiency, endothelial dysfunction, and oxidative stress will enable the delineation of a rational therapeutic strategy in conditions that are linked to oxidative damage.

Our experimentally obtained results related to this topic, which are presented below, are illustrative. We examined the effects of endogenous NO blockade on oxidative stress status and renal function in young rats.[8,9] Two NOS inhibitors were used: N^G-nitro-L-arginine methyl ester (L-NAME) as a nonselective inhibitor and aminoguanidine as a selective inhibitor of NOS2 (Figure 9.2). Oral administration of L-NAME, but not aminoguanidine, for 4 weeks induced systemic hypertension, significant reduction in urinary nitrite/nitrate, and a significant increase in urinary 8-OHdG compared with nontreated animals. Combining all the data revealed a significant negative correlation between urinary nitrite/nitrate and 8-OHdG. The L-NAME-treated rats also developed proteinuria and tubular enzymuria. The effects of L-NAME on blood pressure and urinary parameters were restored by a large dose of L-arginine. These observations highlight the importance of continuous generation of NO by constitutive NOS (especially NOS3) in the control of vascular tone, renal function, and antioxidative capacity in young animals.

Furthermore, rats receiving chronic, nonselective NOS inhibitor treatment are reported to exhibit various parenchymal lesions.[10] Genetic disruption of all three NOS isoforms exhibited markedly reduced survival in mice, possibly caused by spontaneous myocardial infarction associated with multiple cardiovascular risk factors of

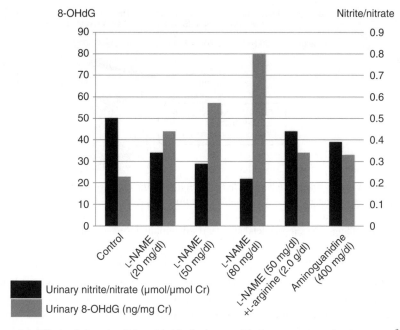

Figure 9.2 Effects of chronic nitric oxide blockade on oxidative stress status in young rats.[8,9] Abbreviations: Cr, creatinine; L-NAME, N^G-nitro-L-arginine methyl ester; 8-OHdG, 8-hydroxy-2′-deoxyguanosine. Presented data are mean values of the markers. Oral administration of L-NAME (20, 50, and 80 mg/dl of drinking water), but not aminoguanidine (400 mg/dl), for 4 weeks of induced systemic hypertension and a significant reduction in urinary excretion of nitrite/nitrate. Rats treated with L-NAME also showed a significant increase in urinary 8-OHdG excretion compared with the control animals. The above effects were dependent on the dosage of L-NAME. The effects of L-NAME (50 mg/dl) on blood pressure and urinary nitrite/nitrate and 8-OHdG were restored by a large dose of L-arginine (2.0 g/dl), a precursor for nitric oxide synthesis.

metabolic origin.[11] Enhanced oxidative stress is likely to participate in the development of such organ damage in animals with chronic NO deficiency and endothelial dysfunction.

Pregnancy as a state of oxidative stress

Pregnancy *per se* is a state of oxidative stress arising from increased placental metabolic activity and increased production of ROS, in addition to reduced total antioxidant capacity (Figure 9.3).[12] Concentrations of oxidative stress biomarkers such as blood lipid peroxides, oxidized LDL, and 8-isoprostane and urinary 8-OHdG became higher toward the third trimester of pregnancy than in nonpregnant women.[13–15] Regarding antioxidants, erythrocyte activities of superoxide dismutase (SOD) and glutathione peroxidase increased toward the third trimester of pregnancy.[15] In contrast, the TAC values determined by the ferric reducing ability of plasma test were lower in pregnant women than in nonpregnant women.[15] The serum TAC values determined using

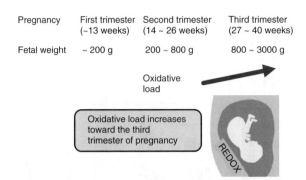

Figure 9.3 Oxidative stress status in the fetoplacental unit. Increased generation of reactive oxygen species during growth of the fetoplacental unit is a prominent feature of pregnancy. Further enhancement of oxidative stress is likely to promote several pregnancy-related disorders including preeclampsia, fetal growth restriction, preterm labor, and low birthweight.

2,2′-azino-di-(3-ethylbenzothiazoline sulfonate) decreased gradually as pregnancy advanced.[14] Collectively, these results indicate that the total antioxidant capacity in plasma or serum is not enhanced in spite of the presence of excessive ROS during later pregnancy.

Very recently, we measured the oxidative stress status in 60 healthy pregnant women at the early third trimester.[16] The age of the subjects was 31 ± 5 years (range: 21–39 years), and venous blood was sampled at the gestational age (GA) of 28 ± 1 weeks (27–29 weeks). THs and TAC were measured using the Free Radical Analytical System (Diacron International, Grosseto, Italy).[5] In healthy Japanese adults, serum TH values are 275 ± 48 U.CARR for those aged 20–29 years ($n = 80$) and 283 ± 50 U.CARR ($n = 118$) for those aged 30–39 years, where one U.CARR (unit of TH value) is equivalent to 0.08 mg/dl of H_2O_2.[17] The ratio of TH to TAC was calculated and designated as "oxidative stress index (OSI)." The rough estimation of OSI is around 0.1 in healthy Japanese adults.

Thioredoxin (TRX) is a ubiquitously expressed, multifunctional protein (12 kDa) that has a redox-active dithiol-disulfide within the conserved -Cys-Gly-Pro-Cys-sequence.[18,19] This defensive protein plays a crucial role in ROS detoxification and transcription factor regulation, each of which is crucially important for normal cellular function. Reportedly, TRX-1 plays a role in reproduction as a component of the "early pregnancy factor."[20] Target disruption of the mouse TRX-1 gene results in early embryonic lethality.[21] In the 60 pregnant women, serum TRX-1 concentrations were also measured using a sensitive sandwich ELISA system (Redox Bioscience Inc., Kyoto, Japan). In healthy Japanese adults aged 32 ± 7 years ($n = 13$), serum TRX-1 concentrations are 20 ± 17 ng/ml (5–53 ng/ml).

In the pregnant women, the serum TH was 471 ± 105 U.CARR (193-708 U.CARR), TAC was 2142 ± 273 µmol/l (1430–2601 µmol/l), OSI was 0.23 ± 0.08 (0.09–0.45), and TRX-1 was 90 ± 42 ng/ml (11–205 ng/ml). A significant negative correlation was found between TH and TAC in the subjects ($r = -0.46$, $p = 0.0002$). For correlations between oxidative stress biomarkers (TH, TAC, OSI, TRX-1) and clinical data (body weight, BW; height; body mass index, BMI), statistical significance was found only

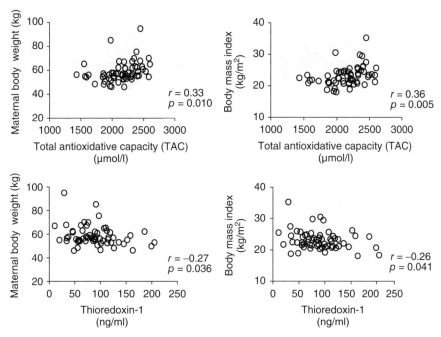

Figure 9.4 Correlations between oxidative stress biomarkers (total antioxidative capacity (TAC), thioredoxin-1) and clinical data (maternal body weight, body mass index).[16]

for TAC versus BW ($r=0.33$, $p=0.010$), TAC versus BMI ($r=0.36$, $p=0.005$), TRX-1 versus BW ($r=-0.27$, $p=0.036$), TRX-1 versus BMI ($r=-0.26$, $p=0.041$) (Figure 9.4).

Results showed that, compared with healthy adult values, the TH values were higher, but the TAC values were lower in the pregnant women; moreover, the TRX-1 concentrations were several times higher than those of healthy adults. The TAC values were found to have significant negative correlation with the TH values in the subjects, thereby implying that antioxidant buffering capacity is attenuated as the oxidative load increases. The high concentrations of TRX-1 are likely to be linked to physiologically high oxidative stress status and reduced antioxidant capacity in pregnant women.

Among the correlations analyzed statistically, TAC and TRX-1, respectively, showed significant positive and negative correlations with each of BW and BMI in these pregnant women. Although the causal relation remains unclear, these results suggest that TAC in serum and systemic release of TRX-1 are related closely to the maternal body size.

Prenatal disorders

In the study described earlier, all 60 women gave birth to healthy infants. When correlations between oxidative stress biomarkers and neonatal birth weight were tested,

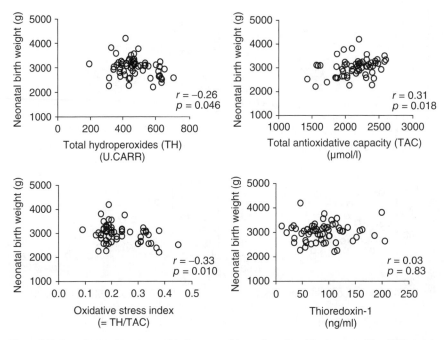

Figure 9.5 Correlations between oxidative stress biomarkers (total hydroperoxides (TH), total antioxidative capacity (TAC), oxidative stress index (OSI), thioredoxin-1) and neonatal birthweight.[16]

statistical significance was found for TH versus BW ($r = -0.26$, $p = 0.046$), TAC versus BW ($r = 0.31$, $p = 0.018$), and OSI versus BW ($r = -0.33$, $p = 0.010$) (Figure 9.5).[16] Results showed that TAC correlated significantly and positively and TH correlated significantly and negatively with neonatal birth weight. Although the study design is limited by its cross-sectional nature, the results suggest that neonatal birthweight is affected by the maternal oxidative condition during later pregnancy.

Increased generation of ROS during growth of the fetoplacental unit is a prominent feature of pregnancy.[12] Further enhancement of oxidative stress is likely to promote several pregnancy-related disorders including preeclampsia, diabetes mellitus, fetal growth restriction, preterm labor, low birthweight (as suggested earlier), and other pregnancy-related disorders.[22–24] Previous histological examinations revealed that the concentrations of HNE, TRX, glutaredoxin, and protein disulfide isomerase were higher in the placenta in preeclampsia than in uncomplicated pregnancy and that levels of 8-OHdG and TRX were higher in the placenta in cases of preeclampsia or fetal growth restriction than in uncomplicated pregnancy during the third trimester.[25,26] These results indicate that TRX might be induced adaptively against oxidative stress in the placenta in preeclampsia or fetal growth restriction.

Pentosidine is accepted as a satisfactory marker for glycoxidation *in vivo*.[3] Previously, we measured pentosidine concentrations in umbilical cord blood from newborns using the high-performance liquid chromatography method.[27] Results showed that the umbilical pentosidine concentrations were considerably lower than normal adult values, but that they were significantly elevated in newborns of mothers with preeclampsia compared with those of mothers without preeclampsia. Our findings suggest that accumulation of pentosidine and oxidative stress occur in fetal tissues and organs *in utero*, and that oxidative stress is augmented *in utero* during preeclampsia.

According to these findings, strategies to reinforce the total antioxidant capacity are expected to be beneficial for sick pregnant women as an adjuvant therapy, but this inference requires additional fundamental investigations.

Oxidative stress in fetal-to-neonatal transition

As air-breathing organisms, we inhale an atmosphere containing 20–21% oxygen. In contrast, fetal development occurs in a much more hypoxic environment. Consequently, at birth, neonates are exposed suddenly to remarkably higher concentrations of inspired oxygen. The energy metabolism efficiency increases rapidly after birth because all aerobic organisms require oxygen for energy production and for the maintenance of cellular functions. Neonates must also withstand the associated generation of ROS, which can oxidize critical macromolecules. The input of ROS depends not only on the ambient oxygen being inspired but also on its conversion to ROS (e.g., by activated polymorphonuclear cells in inflammatory conditions, during resuscitation after hypoxia by damaged mitochondria or activated xanthine oxidase enzyme).[28]

A protective mechanism occurs during the third trimester of fetal life in the form of increased antioxidative enzyme levels. This mechanism adequately prepares the neonate to withstand higher levels of oxygen and ROS generated concurrently after birth. However, sick preterm neonates frequently suffer from oxidative injury because of their insufficient ability to protect themselves against oxidative insult.

Oxidative stress appears to play important roles in the respective pathogeneses of various diseases in neonates, such as neonatal asphyxia, respiratory distress syndrome, bronchopulmonary dysplasia, intraventricular hemorrhage, necrotizing enterocolitis, and retinopathy of prematurity – so-called oxygen radical diseases of the newborn.[28]

Evaluation of oxidative stress status in neonates using specific biomarkers

Previously, we measured urinary levels of acrolein-lysine, 8-OHdG, and nitrite/nitrate in 1-month-old neonates to examine the status of oxidative stress and its relation to the degree of prematurity and clinical condition.[29] They received mixed feeding with 30–70% of their intake as breast feeding. Study subjects consisted of three groups: healthy term neonates ($n = 10$), stable preterm neonates requiring no supplemental oxygen ($n = 21$), and sick preterm neonates requiring supplemental oxygen and ventilator support ($n = 16$).

Urinary levels of acrolein-lysine and 8-OHdG were significantly higher in sick preterm neonates than those of stable preterm and healthy term neonates. In the sick preterm group, neonates developing active retinopathy showed significantly higher levels of acrolein-lysine than other neonates without retinopathy (Figure 9.6). No

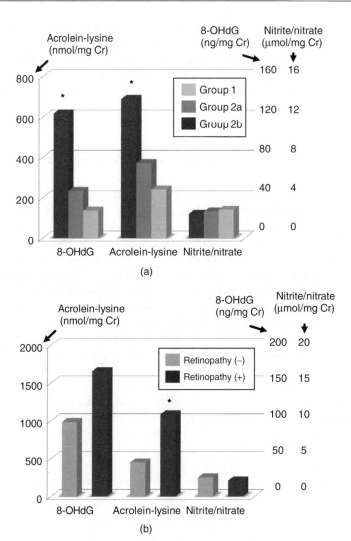

Figure 9.6 Urinary levels of acrolein-lysine, 8-hydroxy-2′-deoxyguanosine, and nitrite/nitrate in 1-month-old term and preterm neonates.[29] Abbreviations: Cr, creatinine; 8-OHdG, 8-hydroxy-2′-deoxyguanosine. Presented data are mean values of the markers. (a) Group 1: healthy term neonates ($n = 10$); Group 2a: stable preterm neonates ($n = 21$); Group 2b: sick preterm neonates ($n = 16$). *$p < 0.05$ versus Group 1, Group 2a. (b) In Group 2b, neonates developing active retinopathy exhibited significantly higher levels of acrolein-lysine than the other neonates without retinopathy did. *$p < 0.05$ versus sick preterm neonates without retinopathy.

significant differences were found between urinary markers in stable preterm and healthy term neonates. Urinary nitrite/nitrate levels were not significantly different among the three groups, suggesting no difference in endogenous NO formation. These results provide evidence of augmentation of oxidative damage to DNA, lipids, and proteins, especially in clinically sick preterm neonates.

Other investigations revealed elevated levels of oxidative stress biomarkers (e.g., tracheal aspirate protein carbonyls, plasma allantoin, urinary orthotyrosine, plasma heptanal, 2-nonenal, and HNE), and decreased levels of antioxidants (plasma sulfhydryls, red blood cell glutathione, and tracheal aspirate glutathione) in oxygen-treated preterm infants. It appears likely that exposure to high concentrations of inspired oxygen can contribute to excessive ROS production in neonates relying on intensive clinical care.

Reactive oxidative metabolites increase endothelial permeability and activate leukocytes, endothelial cells, and other cells with secretion of cytokines and growth factors.[3] Those factors consequently accelerate inflammation and enhance oxidative stress. Particularly, our findings suggest an important role for enhanced oxidative stress in the pathogenesis of respiratory distress, bronchopulmonary dysplasia, and active retinopathy in sick preterm infants.

In another part of the study, we measured the urinary excretion of L-type fatty acid binding protein (L-FABP) and glutathione S-transferase (GST)-pi in neonates to examine the oxidative stress status in the kidney and its relation to the degree of prematurity and clinical condition.[30,31] Sick preterm neonates were found to excrete significantly more L-FABP than healthy term neonates during the late neonatal period. Urinary L-FABP levels showed significant positive correlation with those of 8-OHdG. Sick preterm neonates treated with supplemental oxygen and mechanical ventilation also showed significantly higher levels of GST-pi and of 8-OHdG than clinically stable neonates at 1 month of age.

Considering this accumulated evidence, it appears likely that preterm neonates have two major antioxidant enzymes in the kidney (i.e., L-FABP and GST-pi in the proximal and distal tubules, respectively) through which potentially oxidative endogenous and exogenous substances should be detoxified. Further studies must be undertaken to explore the activity and expression regulation of these enzymes in the neonatal kidney.

Breast milk – a rich source of antioxidants

Breast milk is recognized as an ideal and natural food for the first 6 months of age. It is a complex biological fluid that provides infants with both nutritional and nonnutritional factors.[32] Breast milk contains various enzymatic and nonenzymatic antioxidant constituents such as SOD, catalase, glutathione peroxidase, lactoferrin, TRX, vitamins C and E, β-carotene, coenzyme Q_{10}, albumin, and NO. These antioxidant factors are likely to play an important role in protecting the high level of potentially oxidizable lipids in breast milk.

Ezaki and coworkers measured "TAC" in breast milk using the Free Radical Analytical System (Diacron International, Grosseto, Italy).[33] They analyzed 56 breast milk samples collected from mothers of preterm infants born with GA of 34 (mean) weeks (24–37 weeks) and at postnatal age 39 (mean) days (4–145 days). The mean TAC value for breast milk was 3807 μmol/l. The value in colostrum was about 4200 μmol/l and about 3500 μmol/l in mature milk; the TAC values were significantly negatively correlated with the days after delivery. The values for breast milk were significantly higher than those of standard infant formulas (about 2671 µmol/l).

In the study described earlier, the TAC of breast milk decreased during the course of lactation, which might be a natural result of decline in the antioxidant storage capacity of lactating women. High antioxidant capacity in colostrum is likely to be effective for preventing neonates from exposure to an oxygen-rich environment after birth, four to five times as much as intrauterine environment. Preterm neonates are potentially vulnerable to oxidative stress because of the deficiency of their antioxidant defense system and increase in ROS production.[28] Therefore, neonatal feeding with breast milk, especially colostrum, can be useful to improve antioxidant system and to suppress oxidative stress in such susceptible infants. It is also noteworthy that the antioxidant capacity of breast milk in lactating women was related to maternal plasma TAC values or antioxidant (pro-) vitamin intakes from their diets. Special attention must be devoted to women's dietary habits during pregnancy and lactation to optimize the TAC of their breast milk.

Shoji and coworkers used 8-OHdG as a marker of oxidative stress status in neonates.[34,35] In healthy 1-month-old term neonates, urinary 8-OHdG levels of the breast-fed group ($n = 10$, 39 (mean) ng/mg Cr) were significantly lower than that of formula-dominant, mixed-fed group ($n = 11$, 161 ng/mg Cr) or the formula-fed group ($n = 10$, 204 ng/mg Cr). In preterm infants (born at about 29 weeks of gestation), urinary 8-OHdG levels of the breast-fed group were also significantly lower than that of formula-fed group at 14 and 28 days of age.

It is also important to note the results of *in vitro* experiments conducted by Shoji and coworkers.[36] Cultured Intestinal Epithelial Cells-6 were preincubated with 100-fold dilutions of defatted human breast milk, bovine milk, or three infant formulas for 24 h, followed by a 30-min 0.5-mmol/l H_2O_2 challenge to induce oxidative stress. Results showed that human milk treatment maintained the highest cell survival rate (50%) when compared with no pretreatment (27%), bovine milk treatment (6%), or formula treatment (13–16%) of cells. The results support the contention that human milk is a rich source of antioxidants and reduces oxidative stress in a cell culture model representative of the intestinal mucosa.

It is assumed that human milk exerts antioxidant properties in the gastrointestinal tracts of infants. Proteins of many kinds from human milk remain almost intact in the neonatal gut. Therefore, the antioxidant proteins and enzymes are likely to be absorbed substantially to provide systemic protection against oxidative stress in neonates.

Oxidative stress biomarkers in pediatric medicine

Clinical results of studies indicate that an imbalance between oxidative and antioxidative activities in favor of the former contributes to the pathogeneses of many diseases in the field of pediatric medicine (Table 9.1).[3] In most studies, oxidative stress biomarkers were determined in samples of blood (such as serum, plasma, erythrocytes, granulocytes, or lymphocytes) or urine. In other studies, the parameters were measured using different body fluids (such as cerebrospinal fluid (CSF), bronchoalveolar lavage fluid, joint fluid, nasal lavage fluid, and middle-ear fluid), tissues, or exhaled breath, either alone or in combination with samples of blood or urine.

Table 9.1 Pediatric diseases possibly associated with enhanced oxidative stress.

---

1 *Allergic/inflammatory:*
   Allergic rhinitis, atopic dermatitis, bronchial asthma, burn, Kawasaki disease, systemic lupus erythematosus
2 *Cardiovascular:*
   Cardiac transplantation, cardiopulmonary bypass, congenital cardiac defects, primary hypertension
3 *Endocrinologic:*
   Hyperthyroidism, iodine-deficient goiter, thyroiditis
4 *Environmental/toxicologic:*
   Air pollution, carcinogenic metal exposure, exercise, ozone exposure, passive smoking
5 *Gastrointestinal/hepatologic:*
   Autoimmune hepatitis, inflammatory bowel disease, live failure, nonalcoholic fatty liver disease, viral hepatitis
6 *Genetic:*
   Alagille syndrome, Down syndrome
7 *Hematologic/neoplastic:*
   Acute leukemia, bone marrow transplantation, sickle cell anemia, solid tumors, thalassemia major
8 *Infectious:*
   Acute bronchiolitis, acute otitis media/tonsillitis, chronic otitis media/tonsillitis, encephalitis HIV infection, malaria, meningitis, pandemic influenza (H1N1), sepsis
9 *Metabolic:*
   Citrin deficiency, diabetes mellitus, glycogen storage disease, phenylketonuria, urea cycle enzyme defects, Wilson disease
10 *Neonatal:*
   Asphyxia, hypoxic ischemic encephalopathy, maternal chorioamnionitis, maternal preeclampsia, neonatal sepsis, premature birth, respiratory distress syndrome, retinopathy
11 *Neurologic/muscular:*
   Autism spectrum disorders, cerebral palsy, congenital muscular dystrophy, developmental brain disorders, epilepsy, mitochondrial encephalopathy, psychosis, traumatic brain injury
12 *Nutritional:*
   Hypercholesterolemia, hyperlipidemia, Kwashiorkor, multimetabolic syndrome, obesity
13 *Pharmacologic/therapeutic:*
   Analgesics, anticancer drugs, immunosuppressive drugs, total body irradiation
14 *Renal:*
   Glomerulonephritis, hemolytic uremic syndrome, nephrotic syndrome, renal insufficiency/failure, urinary tract infection
15 *Respiratory:*
   Chronic pulmonary disease, cystic fibrosis, obstructive sleep apnea

---

Pediatric diseases in which enhanced oxidative stress might be involved are listed and divided into 15 categories. Note that it remains to be clarified yet whether excessive formation of reactive oxygen species is a primary cause or a downstream consequence of the pathological process in each disease.

Investigation of the role of oxidative stress in pediatric diseases requires information about the oxidative stress status of young populations. It will be possible to evaluate the contribution of oxidative stress to various pediatric diseases and to establish better approaches for each disease when we ascertain the reference normal levels of oxidative stress in children and adolescents.

Our data for the evaluation of oxidative stress levels in healthy young subjects were of 113 healthy Japanese people of a broad age range (1.5–21.0 years).[37] Early morning void urine samples were obtained for analyses of biomarkers reflecting oxidative damage to lipids (i.e., acrolein-lysine), DNA (i.e., 8-OHdG), carbohydrates (i.e., pentosidine), and NO formation (i.e., nitrite/nitrate). Our subjects were classified into the following five groups to verify the influence of age on the oxidative stress parameters: 1–6 years ($n = 33$), 6–11 years ($n = 34$), 11–16 years ($n = 20$), 16–21 years ($n = 13$), and 21–30 years ($n = 13$) (Figure 9.7).

The concentrations of urinary acrolein-lysine, 8-OHdG, pentosidine, and nitrite/nitrate were highest in the youngest age group (1–6 years). They decreased with age to reach constant levels by early adolescence. No significant difference was found between males and females for any oxidative stress parameter. The physiological meaning and mechanisms for the high levels of oxidative stress and NO formation in younger subjects warrant further research, but they remain unclear.

Therapeutic interventions that decrease exposure to ROS or which augment antioxidative defenses are expected to be beneficial as adjunctive therapies for oxidative-stress-related diseases. Antioxidative strategies such as administration of pharmacological or dietary agents are based on two main mechanisms: enhancement of ROS elimination and inhibition of ROS generation.[3] The following results of our clinical experiments will provide an illustrative and useful example.

Infectious and inflammatory disorders (especially acute encephalopathy)

Various infectious and inflammatory disorders appear to be linked to oxidative damage attributable to ROS (including $ONOO^-$) in their pathogenesis and progression (Table 9.1).[38] These disorders are common and often severe in young generations. Severe forms of these disorders are occasionally fatal or leave severe sequelae, for which effective treatment is currently either insufficient or unavailable. The excessive host response and enhanced oxidative stress are thought to play an important role in the progression and deterioration of the disorders.

Acute encephalopathy is a severe central nervous system (CNS) complication of common infections (such as influenza, exanthem subitum, and acute viral gastroenteritis) that can result in sudden death or development of neurological sequelae. Computed tomography and magnetic resonance imaging are useful to evaluate brain injury severity, but it is often difficult to perform such radiological examinations during the critical period when key therapeutic decisions are made. Therefore, assessing ongoing brain injury and predicting outcomes using CSF samples are extremely valuable in these patients.

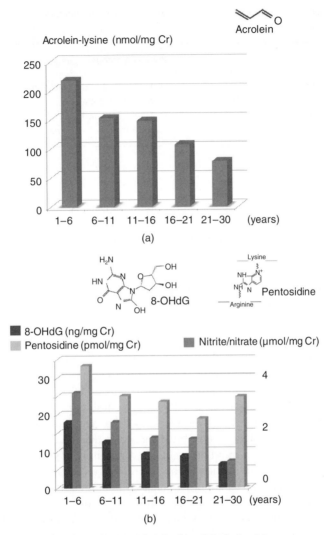

Figure 9.7 Age-related changes of urinary levels of acrolein-lysine (a), 8-hydroxy-2′-deoxyguanosine, pentosidine, and nitrite/nitrate (b) in healthy children.[37] Abbreviations: Cr, creatinine; 8-OHdG, 8-hydroxy-2′-deoxyguanosine. Presented data are mean values of the markers. Note that younger subjects exhibit higher levels of urinary markers.

The brain is vulnerable to free radical damage because of high oxygen consumption, in addition to its consequent generation of high levels of NO and ROS, high contents of unsaturated lipids and cellular iron, and weakened antioxidant defense systems.[38] Previous results for the CSF analyses showed enhanced production of ROS and NO in the CNS of children with acute encephalopathy compared with those of nonencephalopathy subjects. In these studies, the pathogenic viruses were influenza, enterovirus, respiratory syncytial virus, and other viruses (Table 9.2).

Table 9.2   Increased cerebrospinal fluid levels of oxidative stress biomarkers in acute encephalopathy.[38]

Diseases	Primary findings
Influenza encephalopathy	Higher hydroperoxides,[a] higher nitrite/nitrate
Respiratory syncytial virus encephalopathy	Higher nitrite/nitrate
Enterovirus encephalopathy	Higher hydroperoxides[a]
"Clinically mild encephalopathy"	Higher 8-hydroxy-2′-deoxyguanosine, hexanoyl-lysine

[a]Rapid analytical method.[5]

Strategies for severe encephalopathy should be established in the near future. Prevention (or modulation) of excessive host inflammatory response such as ROS/NO release, a possible therapeutic approach, is explained in the following section.

Redox modulation strategy for severe influenza encephalopathy

Influenza-associated acute encephalopathy (IAE) is an abrupt disorder of the CNS triggered by influenza virus infection, often engendering severe sequelae or death.[39,40] The Centers for Disease Control and Prevention have designated IAE as an important public health problem at least since 2003. The 2009 pandemic influenza A (H1N1) virus emerged in Mexico in April 2009, thereafter spreading rapidly worldwide. In June 2009, The World Health Organization declared that the spreading novel influenza virus constituted a global pandemic.

Because influenza virus infection occurs predominantly in younger generations, great concern has arisen in relation to the severity of complications, such as IAE or severe pneumonia, among children. Pathological findings such as the lack of viral antigen and sparse (if any) inflammatory infiltrates in the brain imply that direct viral invasion and subsequent inflammation are unlikely to cause encephalopathy. A prevailing theory is that excessive host inflammatory response characterized by massive production of proinflammatory cytokines/chemokines and ROS/NO and excessive apoptosis exacerbate IAE.[38] Widespread vascular endothelial activation, dysfunction, and damage occur, ultimately resulting in multiple organ failure and death (Figure 9.8).

In Japan, the guideline for diagnosis and management of pediatric IAE was formulated in 2005 by the collaborating study group on IAE, which was organized by the Japanese Ministry of Health, Labour, and Welfare. The guideline has been used widely among general and pediatric hospitals in our country. Neuraminidase inhibitors, pulse steroid therapy, high-dose immunoglobulin, antioxidative agent (edaravone), coagulation modifying agent (thrombomodulin), plasma exchange, and hypothermia are listed as selectable treatments for severe IAE.[41] These cocktail treatments are expected to impede excessive inflammatory host response and enhanced oxidative stress in the patients (Figure 9.8).

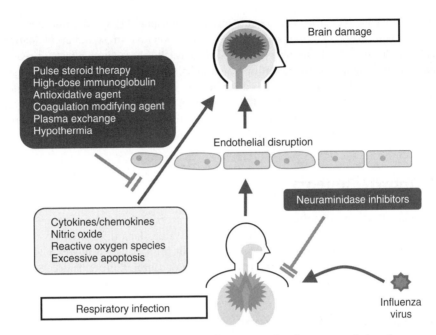

Figure 9.8 Mechanisms of brain damage in influenza-associated acute encephalopathy (IAE).[38] The findings presented in recent reports suggest that, in cases of severe IAE, either seasonal or 2009 pandemic, pathological manifestations similarly result from complex biological phenomena including overproduction of cytokines/chemokines and nitric oxide/reactive oxygen species, apoptosis induction, and vascular endothelial disruption. Additional exploration of these pathways is expected to contribute to the development of more effective adjunctive strategies in IAE.

For pediatric patients with IAE, the mortality rate was about 30% in the preguideline era in Japan when no efficient strategy had been proposed. Thereafter, the mortality rate decreased to about the one-fourth along with the nationwide distribution of this practical guideline. However, the incidence of poor outcomes of pediatric IAE has not been ideally low: 7% for death and about 20% for neurological sequelae. More global studies with sequential monitoring of oxidative stress biomarkers must be conducted to identify more effective strategies for IAE of severe type.

Summary and conclusions

Oxidative damage affecting tissues and organs contributes to the development and progression of acute and chronic health problems in the prenatal, neonatal, and pediatric fields. This chapter has briefly presented the (patho) physiology of ROS and antioxidative defense systems, the clinical application of oxidative stress biomarkers, and the interrelationship of NO system blockade, endothelial dysfunction, and oxidative stress. The chapter has also discussed prenatal, neonatal, and pediatric disorders in which enhanced oxidative stress is likely to be involved. The discussion emphasizes that many good biomarkers are readily measurable using ELISA and other methodologies. Our recent and interesting data related to oxidative stress and antioxidative defenses in these fields are also presented. Finally, the therapeutic effects of redox modulation strategies for acute influenza encephalopathy of severe type are summarized. However, most of the studies described above are preliminary in nature and, therefore, warrant further studies. It is hoped that further elucidation of the clinical and pathological correlation between oxidative stress biomarkers and disease course will spur new therapeutic approaches to alleviating many human health problems.

Acknowledgments

We thank Professor Mitsufumi Mayumi (Fukui University, Fukui, Japan) and Professor Tsuneo Morishima (Okayama University, Okayama, Japan) for their invaluable help with these studies. This work was supported by the Japanese Ministry of Health, Labour and Welfare and the Japanese Ministry of Education, Culture, Sports, Science and Technology.

Multiple choice questions

1 In young rats, compared to a selective NOS2 inhibitor, administration of the nonselective NOS inhibitor N^G-nitro-L-arginine methyl ester (L-NAME) caused all except which dysfunction?
   a. Hypertension
   b. Proteinuria
   c. Enzymuria
   d. Decrease oxidative DNA damage

2 Pregnancy is a state of oxidative stress caused by a combination of all but which factor?
   a. Increased thioredoxin (TRX) levels
   b. Increased placental metabolic activity
   c. Increased production of ROS
   d. Reduced total antioxidant capacity

3 All these statements are true except:
   a. Glycoxidation, as measured by pentosidine levels is lower in healthy neonates compared to adult levels
   b. Glycoxidation, as measured by pentosidine levels is lower in neonates of healthy mothers compared to neonates of mothers with preeclampsia
   c. In preeclampsia, oxidative stress increases *in utero*
   d. Antioxidant therapy in preeclampsia is a well-established treatment and has virtually eliminated preeclampsia as a risk to pregnancy

4 Preterm neonates suffer oxidative stress from a combination of all but which factor?
   a. Increase ambient inspired oxygen levels
   b. Increased ROS production by inflammatory reaction to hypoxia
   c. Natural increase in antioxidant levels in the third trimester
   d. Depletion of antioxidants

References

1 Dröge, W. (2002) Free radicals in the physiological control of cell function. *Physiological Reviews*, 82, 47–95.

2 Auten, R.L. & Davis, J.M. (2009) Oxygen toxicity and reactive oxygen species: the devil is in the details. *Pediatric Research*, 66, 121–127.

3 Tsukahara, H. (2014) Oxidative stress biomarkers: current status and future perspective. In: Tsukahara, H. & Kaneko, K. (eds), *Oxidative Stress in Applied Basic Research and Clinical Practice – Pediatric Disorders*. Springer, Berlin, Germany, pp. 87–113.

4 Noiri, E. & Tsukahara, H. (2005) Parameters for measurement of oxidative stress in diabetes mellitus: applicability of enzyme-linked immunosorbent assay for clinical evaluation. *Journal of Investigative Medicine*, 53, 167–175.

5 Kaneko, K. (2014) Rapid diagnostic tests for oxidative stress status. In: Tsukahara, H. & Kaneko, K. (eds), *Oxidative Stress in Applied Basic Research and Clinical Practice – Pediatric Disorders*. Springer, Berlin, Germany, pp. 137–148.

6 Popov, T.A. (2011) Human exhaled breath analysis. *Annals of Allergy, Asthma, and Immunology*, 106, 451–456.

7 Dodig, S., Richter, D. & Zrinski-Topić, R. (2011) Inflammatory markers in childhood asthma. *Clinical Chemistry and Laboratory Medicine*, 49, 587–599.

8 Tsukahara, H., Imura, T., Tsuchida, S. *et al.* (1996) Renal functional measurements in young rats with chronic inhibition of nitric oxide synthase. *Acta Paediatrica Japonica*, 38, 614–618.

9 Tsukahara, H., Hiraoka, M., Kobata, R. *et al.* (2000) Increased oxidative stress in rats with chronic nitric oxide depletion: measurement of urinary 8-hydroxy-2'-deoxyguanosine excretion. *Redox Report*, 5, 23–28.

10 Zatz, R. & Baylis, C. (1998) Chronic nitric oxide inhibition model six years on. *Hypertension*, 32, 958–964.

11 Nakata, S., Tsutsui, M., Shimokawa, H. *et al.* (2008) Spontaneous myocardial infarction in mice lacking all nitric oxide synthase isoforms. *Circulation*, 117, 2211–2223.

12 Myatt, L. & Cui, X. (2004) Oxidative stress in the placenta. *Histochemistry and Cell Biology*, 122, 369–382.

13 Toescu, V., Nuttall, S.L., Martin, U. *et al.* (2002) Oxidative stress and normal pregnancy. *Clinical Endocrinology*, 57, 609–613.

14 Belo, L., Caslake, M., Santos-Silva, A. *et al.* (2004) LDL size, total antioxidant status and oxidize LDL in normal human pregnancy: a longitudinal study. *Atherosclerosis*, 177, 391–399.

15 Hung, T.H., Lo, L.M., Chiu, T.H. *et al.* (2010) A longitudinal study of oxidative stress and antioxidant status in women with uncomplicated pregnancies throughout gestation. *Reproductive Sciences*, 17, 401–409.

16 Nakatsukasa, Y., Tsukahara, H., Tabuchi, K. *et al.* (2013) Thioredoxin-1 and oxidative stress status in pregnant women at early third trimester of pregnancy: relation to maternal and neonatal characteristics. *Journal of Clinical Biochemistry and Nutrition*, 52, 27–31.

17 Nojima, J., Miyakawa, M., Kodama, M. *et al.* (2010) Measurement of the oxidation stress degree by the automated analyzer JCA-BM 1650. *Japanese Journal of Medical Technology*, 59, 199–207(Japanese).

18 Nakamura, H., Hoshino, Y., Okuyama, H. *et al.* (2009) Thioredoxin 1 delivery as new therapeutics. *Advanced Drug Delivery Reviews*, 61, 303–309.

19 Yashiro, M., Tsukahara, H. & Morishima, T. (2014) Thioredoxin therapy: challenges in translational research. In: Tsukahara, H. & Kaneko, K. (eds), *Oxidative Stress in Applied Basic Research and Clinical Practice – Pediatric Disorders*. Springer, Berlin, Germany, pp. 233–252.

20 Clarke, F.M., Orozco, C., Perkins, A.V. *et al.* (1991) Identification of molecules involved in the 'early pregnancy factor' phenomenon. *Journal of Reproduction and Fertility*, 93, 525–539.

21 Matsui, M., Oshima, M., Oshima, H. *et al.* (1996) Early embryonic lethality caused by targeted disruption of the mouse thioredoxin gene. *Developmental Biology*, 178, 179–185.

22 Rogers, M.S., Wang, C.C., Tam, W.H. *et al.* (2006) Oxidative stress in midpregnancy as a predictor of gestational hypertension and pre-eclampsia. *BJOG*, 113, 1053–1059.

23 Potdar, N., Singh, R., Mistry, V. *et al.* (2009) First-trimester increase in oxidative stress and risk of small-for-gestational-age fetus. *BJOG*, 116, 637–642.

24 Min, J., Park, B., Kim, Y.J. *et al.* (2009) Effect of oxidative stress on birth sizes: consideration of window from mid pregnancy to delivery. *Placenta*, 30, 418–423.

25 Shibata, E., Ejima, K., Nanri, H. *et al.* (2001) Enhanced protein levels of protein thiol/disulphide oxidoreductases in placentae from pre-eclamptic subjects. *Placenta*, 22, 566–572.

26 Takagi, Y., Nikaido, T., Toki, T. *et al.* (2004) Levels of oxidative stress and redox-related molecules in the placenta in preeclampsia and fetal growth restriction. *Virchows Archiv*, 444, 49–55.

27 Tsukahara, H., Ohta, N., Sato, S. *et al.* (2004) Concentrations of pentosidine, an advanced glycation end-product, in umbilical cord blood. *Free Radical Research*, 38, 691–695.

28 Saugstad, O. (2014) Oxygen and oxidative stress in the newborn. "*Oxidative Stress in Applied Basic Research and Clinical Practice – Pediatric Disorders*", Tsukahara H, Kaneko K, pp. 3-13, Springer, Berlin, Germany.

29 Tsukahara, H., Jiang, M.Z., Ohta, N. *et al.* (2004) Oxidative stress in neonates: evaluation using specific biomarkers. *Life Sciences*, 75, 933–938.

30 Tsukahara, H., Sugaya, T., Hayakawa, K. *et al.* (2005) Quantification of L-type fatty acid binding protein in the urine of preterm neonates. *Early Human Development*, 81, 643–646.

31 Tsukahara, H., Toyo-Oka, M., Kanaya, Y. *et al.* (2005) Quantitation of glutathione S transferase-pi in the urine of preterm neonates. *Pediatrics International*, 47, 528–531.

32 Tsukahara, H. (2013) Redox modulatory factors of human breast milk. In: Zibadi, S., Watson, R.R. & Preedy, V.R. (eds), *Handbook of Dietary and Nutritional Aspects of Human Breast Milk*. Wageningen Academic Publishers, Wageningen, Netherlands, pp. 599–614.

33 Ezaki, S., Ito, T., Suzuki, K. *et al.* (2008) Association between total antioxidant capacity in breast milk and postnatal age in days in premature infants. *Journal of Clinical Biochemistry and Nutrition*, 42, 133–137.

34 Shoji, H., Oguchi, S., Shimizu, T. *et al.* (2003) Effect of human breast milk on urinary 8-hydroxy-2'-deoxyguanosine excretion in infants. *Pediatric Research*, 53, 850–852.

35 Shoji, H., Shimizu, T., Shinohara, K. *et al.* (2004) Suppressive effects of breast milk on oxidative DNA damage in very low birthweight infants. *Archives of Disease in Childhood. Fetal and Neonatal Edition*, 89, F136–F138.

36 Shoji, H., Oguchi, S., Fujinaga, S. *et al.* (2005) Effects of human milk and spermine on hydrogen peroxide-induced oxidative damage in IEC-6 cells. *Journal of Pediatric Gastroenterology and Nutrition*, 41, 460–465.

37 Tamura, S., Tsukahara, H., Ueno, M. *et al.* (2006) Evaluation of a urinary multi-parameter biomarker set for oxidative stress in children, adolescents and young adults. *Free Radical Research*, 40, 1198–1205.

38 Tsukahara, H., Yashiro, M., Nagaoka, Y. *et al.* (2014) Infectious and inflammatory disorders. In: Tsukahara, H. & Kaneko, K. (eds), *Oxidative Stress in Applied Basic Research and Clinical Practice – Pediatric Disorders*. Springer, Berlin, Germany, pp. 371–386.

39 Morishima, T., Togashi, T., Yokota, S. *et al.* (2002) Collaborative Study Group on Influenza-Associated Encephalopathy in Japan (2002) Encephalitis and encephalopathy associated with an influenza epidemic in Japan. *Clinical Infectious Diseases*, 35, 512–517.

40 Okumura, A., Nakagawa, S., Kawashima, H. *et al.* (2013) Severe form of encephalopathy associated with 2009 pandemic influenza A (H1N1) in Japan. *Journal of Clinical Virology*, 56, 25–30.

41 Yamashita, T. & Abe, K. (2014) Edaravone therapy: from bench to bedside. In: Tsukahara, H. & Kaneko, K. (eds), *Oxidative Stress in Applied Basic Research and Clinical Practice – Pediatric Disorders*. Springer, Berlin, Germany, pp. 211–218.

CHAPTER 10

Oxidative stress in oral cavity: interplay between reactive oxygen species and antioxidants in health, inflammation, and cancer

Maurizio Battino[1,2], Maria Greabu[3], and Bogdan Calenic[3]

[1] Department of Dentistry and Specialized Clinical Sciences, Biochemistry Section, Università Politecnica delle Marche, Ancona, Italy
[2] Centre for Nutrition & Health, Universidad Europea del Atlantico (UEA), Santander, Spain
[3] Department of Biochemistry, Faculty of Dental Medicine, University of Medicine and Pharmacy 'CAROL DAVILA', Bucharest, Romania

---

THEMATIC SUMMARY BOX

At the end of this chapter, students will be able to:

- Describe the significance of oxidative stress and its involvement in major oral and general diseases
- Describe the damage inflicted by oxidative stress to cellular structures
- Define reactive oxygen species and explain the reactions that generate them
- Show how the balance between oxidative stress and antioxidants influences general and oral diseases
- Describe exogenous antioxidants with particular significance to oral cavity affections
- Understand the antioxidant roles of saliva and describe its major antioxidant systems
- Enumerate main methods for the determination of oxidative stress

---

Oxidative stress – significance for oral and general environment

Out of the total oxygen amount inspired by humans, approximately 10% is used in different nonenzymatic chemical reactions while the majority (around 90%) is used for energy production by mitochondria.[1] Normal oxygen metabolism releases a group of compounds collectively named reactive oxygen species (ROS). Most often they

---

Oxidative Stress and Antioxidant Protection: The Science of Free Radical Biology and Disease, First Edition.
Edited by Donald Armstrong and Robert D. Stratton.
© 2016 John Wiley & Sons, Inc. Published 2016 by John Wiley & Sons, Inc.

are produced in the mitochondria following either electron leakage during respiratory chain or chemical reactions of transition metals (usually iron or copper ions). ROS are reactive molecules due the presence of one or more unpaired electrons that occupy an orbital alone. The most common ROS molecules include hydroxyl radicals, hydrogen peroxide, and superoxide anion radicals.[2] O_2 can also be considered a free radical but its reactivity is low due to a specific arrangement of the two unpaired electrons it contains (triplet oxygen). Some ROS, such as hydroxyl radical, are extremely active, short-lived, and act at close distance of their production. Others, more long-lived, such as H_2O_2 (which is freely diffusible across biomembranes), are suggested to be responsible for the signaling properties of ROS. In normal conditions, ROS have beneficial biological effects being involved in cellular homeostasis and key molecular mechanisms such as modulation of cellular metabolism and cellular redox state, cell signaling, inhibition or activation of different gene transcription factors, inhibition of bacterial growth, inactivation of viruses, and release of inflammatory cytokines.[3] However, as a result of intense environmental stress stimuli, ROS levels can increase significantly. Oxidative stress (OS) can be defined as a redox imbalance between antioxidant systems (AOs) and endogenous or exogenous prooxidants in favor of the prooxidants (Figure 10.1). Damage as a result of free radicals accumulates over time and can represent a major underlining cause for human diseases. A solid body of research literature shows that this loss of balance is in many cases closely associated with initiation and development of a wide range of systemic or organ-specific diseases including cancer, diabetes, metabolic syndrome, cardiovascular pathology, pulmonary diseases, or oral conditions (Figure 10.2).[4]

At a cellular level, increased concentrations of ROS are potent inducers of cellular damage targeting structures that include (but are not limited to) lipids, proteins, and DNA.[5] These further affect cellular components by causing cytoskeleton deorganization, protein dysfunction, and membrane function impairment with decreased membrane permeability. Increased levels of OS lead to cellular death either by necrosis or

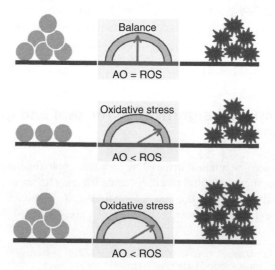

Figure 10.1 Interplay between reactive oxygen species and antioxidants.

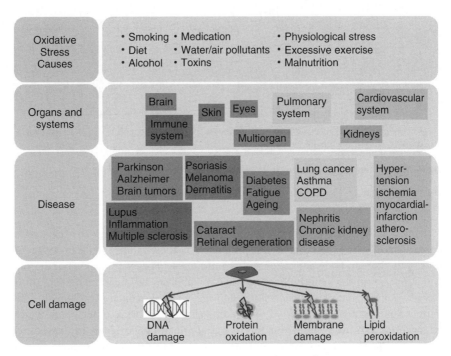

Figure 10.2 Oxidative stress and its involvement in several major diseases.

programmed cell death through intrinsic or extrinsic apoptotic pathways. Different ROS have different "affinities" for their target substrates: thus DNA or DNA deoxyribose can be damaged by the hydroxyl radical; lipid residues can be easily oxidized by metal-induced formation of ROS while protein damage is usually acquired following exposure to superoxide/hydroxyl radicals. One important issue in free radical biology research is the need for accurate identification and quantification of biomarkers that identify and quantify OS. Few methods can detect free radicals directly due to their very short life span. However, one technique is electron spin resonance where the free radical reacts with a molecule that stabilizes the end-product and allows for identification and monitoring. Alternatively, indirect methods have been devised where the tests measure the extent of OS damage in different cellular components (lipids proteins or DNA).

Reactive oxygen species – general outline

Atmospheric oxygen presents two unpaired electrons in the outer electron shell. This particular structure makes atomic oxygen prone to radical formation. Oxygen reduction takes place in a sequential addition of electrons that leads to ROS formation among which are superoxide, hydrogen peroxide, and hydroxyl radical (Figure 10.3, Table 10.1). These free radical species and reactive molecules are capable of radical

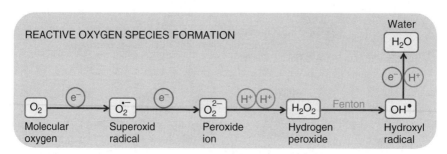

Figure 10.3 Generation of different reactive oxygen species.

formation in the extra- and intracellular environments and can cause oxidative stress damage in cell tissues and organs. The production of these radicals tends to be ubiquitous in the human body and takes place continuously. For example, even in normal conditions, around 5% of the liver's oxygen metabolism is transformed into various oxygen radicals that can cause cellular damage. ROS of exogenous origins include various sources such as radiation or cigarette smoking. On the other hand, typical endogenous ROS are generated in physiological or pathological body processes such as aging, inflammation, or cancer and can have multiple origins: cell organelles such as mitochondria or peroxisomes or neutrophils and other cells of the immune system. Particular to the oral environment, other sources of ROS production encompass dental materials such as metals, alloys, composite fillings or cements, dental implants, or bleaching agents.

It is important to remember that in normal concentrations, ROS play key beneficial roles and are implicated in major biological processes, such as the release of proinflammatory cytokines and apoptosis signaling by nuclear factor-κB (NF-κB) activation and initiation of gene transcription factors and by intracellular thiol depletion. ROS are also involved in cell signaling, regulation of bacterial proliferation, and modulation of cellular redox state and cell metabolism.[6]

Oxidative stress – damage to cellular structures

Lipids are a group of compounds with important biological functions in the human body such as key constituents of cell membranes, hormones, or source of energy. Lipid peroxidation (LPO) represents the oxidative degradation of lipids usually containing double carbon bonds.[7] Three main mechanisms are described as being involved in LPO: enzymatic oxidation, ROS-independent nonenzymatic oxidation and ROS-mediated oxidation. Each type is characterized by specific end-products and requires certain antioxidants to suppress OS activity. These end-products play different roles depending on their concentration and local conditions: pro- or anti-apoptotic effects, pro- or anti-inflammatory effects, and cytotoxic or cytoprotective roles. Several end-products are also known to be toxic with a proven mutagenic and carcinogenic potential. To date, it is well documented that LPO is intimately

Table 10.1   Reactive oxygen species and their main characteristics.

Reactive oxygen species	Chemical formula	Observations
Superoxide anion	$O_2^{\bullet-}$	• Does not have a high reactivity, so is able to act at long distances from its site of production • Produced continuously by processes such as mitochondrial electron chain transport system; oxidases such as xanthine dehydrogenase oxidase; immune system through NADPH oxidase • Plays an important role in removal of bacterial species • Is converted into hydrogen peroxide by superoxide dismutase residing in cellular cytoplasm and mitochondria • Can form complex reactive species such as peroxynitrite in combination with nitric oxide • Can occur in intra- or extracellular sites
Hydrogen peroxide	H_2O_2	• It is not a radical molecule • Produced mainly from superoxide anion in a reaction catalyzed by superoxide dismutase • Has the ability to diffuse through membranes and can act at long distances from its site of production • Can occur in intra- or extracellular sites
Hydroxyl radical	$HO^{\bullet}$	• Very reactive molecule • Key molecule in organ or tissue toxicity due to reactive oxygen species • Reacts with biological macromolecules close to its site of production; therefore, the damage is site specific
Peroxyl radical	$RO_2^{\bullet}$	• One prominent example is ozone or trioxygen • High levels of ozone induce programmed cell death in different cell types • Increased concentrations of ozone contribute to inflammatory responses in many tissues
Hydroperoxyl radical	$HO_2^{\bullet}$	• Formed as a result of hydrogen atom transfer to molecular oxygen • Molecule with increased reactivity playing an important role in degrading atmospheric organic pollutants • Important biological role in lipid peroxidation of polyunsaturated fatty acids of the cell membrane
Alkoxyl radical	$RO^{\bullet}$	• Oxygen-centered radicals formed usually indirectly from carbohydrate derivatives • Important intermediates in the atmospheric oxidation of hydrocarbons

Table 10.1   *(Continued)*

Reactive oxygen species	Chemical formula	Observations
Hypochlorous acid	HOCl	• An important precursor of free radical in the mammalian body • High reactivity that makes the molecule highly effective in the bactericidal defense system • Its high reactivity is also the basis for its cytotoxic effects • Can be associated with cancer, neurodegenerative, or cardio-vascular diseases

connected with the development of a wide range of pathological and physiological processes: cancer, diabetes, cardiovascular and neurological diseases, inflammation, and aging.

Point and extended alterations of genetic material as a result of oxidative damage are among the first events involved in the initiation and further development of cancer. In this aspect, DNA bases such as pyrimidine and purine as well as DNA deoxyribose can be damaged by the hydroxyl radical or as a result of metal-induced formation of oxygen species.[8]

Protein oxidation and its underlying mechanisms were studied in experiments involving proteins, peptides, or aminoacids being exposed to ROS such as superoxide/hydroxyl radicals formed as a result of ionizing radiation. ROS-mediated protein oxidation can potentially target any aminoacid side chain, but it was shown to preferentially damage methionine and cysteine.[9]

Oxidative stress and antioxidants – implications in general and oral diseases

It is well established that the control of ROS production and antioxidant mechanisms are intimately connected with the initiation and development of major organ-specific or systemic diseases. Thus, oral conditions, cancer, cardiovascular diseases, pulmonary diseases, and kidney diseases are initiated by a discontinuity in the redox control and signaling, that is, when the redox balance between prooxidants and antioxidants is broken in favor of the prooxidants.

Periodontal diseases, which encompass gingivitis and periodontitis, are the most widespread chronic inflammatory conditions and affect a large group of the population worldwide. Periodontitis, a nonreversible inflammatory disease, has as an underlying cause, the presence of the oral microorganisms in the dental plaque eliciting an immune response in the tissue surrounding the teeth. Tissue destruction involves apical migration of the oral epithelia, loss of the periodontal collagen fibers, and resorption of the alveolar bone leading to pathological tooth mobility and subsequent tooth loss. Current research shows that the pathogenesis of periodontitis is a direct consequence of the inflammatory response together with a progressive accumulation of ROS in both oral tissues and surrounding biological fluids.[10] As described earlier, OS is triggered by an imbalance between AO defense mechanisms and ROS levels. Thus, ROS can influence directly the behavior and metabolism of cells from oral tissues. To date, positive correlations have been described between oral OS and systemic markers of inflammation. At the same time, OS may represent a logical link between initiation and development of periodontal diseases and systemic disorders such as cardiovascular diseases, cancer, diabetes mellitus, and metabolic syndrome.[11] For researchers, the oral environment represents an ideal accessible location where OS-mediated tissue destruction and antioxidant response to oxidative stress can be studied.

Cancer is the second cause of death in developing countries and the main cause of death in the economically developed countries. Head and neck squamous cell carcinomas are the sixth most common malignancies in the world, with cancers of the oral cavity and pharynx being the most common. Oral cavity cancers are twice more common in men than in women with major risk factors including smoking, alcohol use, and human papilloma virus infections. There is solid evidence that high concentrations of free radicals are directly connected to the development of oral cancer.[8] The main mechanisms of OS involvement in carcinogenesis involve overexpression of proto-oncogenes, inactivation or mutations of tumor suppressor genes, and DNA damage at different levels: DNA base oxidation, DNA strand breaks, and DNA point-mutations.[12,13] OS damage of lipids and proteins (discussed earlier in the chapter) are also of particular importance in cancer development.

The human body has developed highly complex antioxidant mechanisms that can function in synergy in order to prevent tissue or organ damage generated by ROS. Antioxidants are molecules that have the ability to donate electrons and to reduce the damage inflicted by free radicals to biological structures.

Some antioxidants such as uric acid or glutathione are continuously produced during normal body metabolism while others such as vitamins E and C or β-carotene have to be supplied in the diet as micronutrients.

Historically, the term antioxidant was coined to define a general chemical that has the capacity to prevent oxygen consumption. Early research in the beginning of the 20th century focused on antioxidant characteristics important for industrial process. Thus, antioxidants were used for vulcanization of rubber, prevention of metal surfaces corrosion, and fuel polymerization.[14] Studies on the roles of antioxidants in the biochemistry of living organisms first explored their antirancid properties, that is, their ability to prevent the oxidation of unsaturated fatty acids. So the field was revolutionized with the discovery that certain vitamins such as A, C, and E are important antioxidants in biology. There are four general mechanisms through which AO can counteract the harmful effects of ROS and they can be systematized as follows: prevention of ROS formation, elimination (scavenger) of reactive radicals once formed, repair of damaged biomolecules, and potential to adapt and participate with other AO in ROS depletion.[15] The first mechanism implies hydrogen peroxide and hydroperoxide reduction to water before they generate free radicals. Several examples include phospholipid hydroperoxide glutathione peroxidase, glutathione peroxidase and glutathione S-transferase. The second mechanism involves antioxidants with the capacity to scavenge free radicals, such as vitamin E (a very potent lipophilic radical-scavenging AO), vitamin C, or bilirubin. The repair defense mechanism is characteristic of different proteolytic enzymes, peptidases, and DNA repair enzymes and limits the accumulation of oxidized proteins. The last important mechanism relies on the ability of living organisms to produce site-specific antioxidants following ROS formation. Antioxidants can be defined as molecules that have the potential to prevent or inhibit the oxidation process of other molecules. Depending on their origin, AO can be divided into two major categories being synthesized by the body (endogenous) or obtained from the diet or supplements (exogenous).[16]

One example of endogenously secreted AO with particular importance for oral tissues is the antioxidant system present in saliva. Within the oral cavity, saliva is one body fluid that can precisely mirror OS status of a particular disease as well as important markers of oral or general pathologies.[17] Among other functions, saliva acts as a defense mechanism and is fully equipped with several antioxidant mechanisms that, in normal conditions, counteract the effects of ROS, canceling the harmful effects that OS can have on oral tissues.[18–20] The most important antioxidants include the following: uric acid, albumin, glutathione, and ascorbic acid. Uric acid, with a salivary concentration of 40–240 µM, accounts for almost 90% of the total antioxidant activity and represents one of the major salivary AOs. Its scavenger mechanism lies in its ability to react with hydroxyl radicals and metal ions such as iron and copper. Uric acid concentration is usually lower in patients with chronic periodontitis than in healthy subjects. Albumin with a concentration of ~10 µM can act as a "replacement" for uric acid, binding metal ions when needed. Just as for uric acid, albumin levels are also decreased in patients with periodontal conditions. Ascorbic acid or vitamin C plays a similar role and may reduce oxidative damage induce by smoking cigarettes. Its concentration is significantly higher in the gingival crevicular fluid than in plasma. Another important AO with a particular importance in inflammation is glutathione. Although found in low concentrations, 2 µM, glutathione acts as a scavenger for metal ions released from dental materials and regulates cytokines involved in inflammation such as IL-8 or IL-6. Several enzymes such as salivary peroxidase, gamma glutamyl transferase, glutathione peroxidase, superoxide dismutase, and lactate dehydrogenase also have important antioxidant significance.

Exogenous antioxidants in the dental field can be usually incorporated in mouth rinses or toothpastes and include compounds presented in Table 10.2.[31,32] Several

Table 10.2 Exogenous antioxidants with particular significance to oral cavity conditions.

Oral condition	Exogenous antioxidants	Observations	References
Gingivitis/periodontitis	Green vegetables, flavonoids, vitamin E	Reduction of inflammation molecules	(21, 22)
Caries	Epigallocatechin-3-gallate, cranberries – type A oligomers, grape seed	Dental caries prevention, antibacterial effect	(23–25)
Implants	Grape seed, caffeic acid phenethyl ester	Treatment of peri-implantitis, stimulation of bone healing	(26, 27)
Orthodontics	Vitamin C, resveratrol, propolis	Bone formation and expansion	(28, 29)
Cancer	Proanthocyanidins from flavonoids	Reduce proliferation of oral cancer cells	(30)

trials have been undertaken to test the efficacy of various antioxidants on patients with different forms of periodontitis or gingivitis in clinical settings. Recent examples include the following: vitamin E supplementation together with periodontal treatment in patients with chronic periodontitis; dietary supplements of different combinations of fruits, vegetables, and berries together with nonsurgical periodontal treatments; intake of dietary antioxidants such as C and E vitamins and α- and β-carotene by patients over 75 years old; topical applications of coenzyme Q_{10} in patients with chronic periodontitis; and systemic administration of micronutrient antioxidants in postmenopausal women.[33–37] The general conclusion of these trials is that adjunctive antioxidant supplements improve both the periodontal healing and the antioxidant defense system.

Scientific literature describes a wide variety of methods and protocols that assess ROS presence and their effects on biological structures.[38–40] It is not our purpose to provide a complete outline of these methods but rather to provide a general overview of the approaches used to determine and quantify OS. The methodologies can be generally grouped along three lines: direct analysis of free radicals, assessment of the damage inflicted to biomolecules, and quantification of antioxidant systems or redox levels. Direct measurement of ROS would of course be the method of choice. Unfortunately ROS are very unstable and difficult to measure directly; therefore, scientists prefer to measure OS indirectly by assessing damage to protein, lipids, or DNA, or quantifying specific antioxidant enzymes levels (Figure 10.4).

In conclusion, there is solid evidence that many oral diseases such as periodontitis or oral cancer are directly linked to loss of balance between antioxidant systems and endogenous or exogenous prooxidants. As a result, the accumulated free radicals target cell structures such as lipids, DNA, or proteins, and promote cellular damage. Moreover, oxidative stress is directly associated with several general pathological conditions, and in many cases, it represents the systemic link between the initiation of general diseases and the development of various oral conditions. These observations thoroughly justify the continuous search for new antioxidants as well as antioxidant usage in the prophylaxis and treatment of oral diseases.

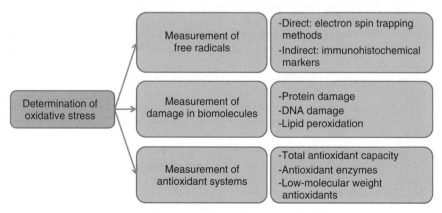

Figure 10.4 Methods for determination of oxidative stress.

Multiple choice questions

1 Electron spin trapping is a direct method of measuring:
   a. Free radicals
   b. Protein damage
   c. Antioxidant levels

2 Saliva is one body fluid that can precisely mirror OS status of a particular disease as well as important markers of oral or general pathologies:
   a. True
   b. False

3 The most important antioxidants in saliva include all except:
   a. Uric acid
   b. Albumin
   c. Glutathione
   d. Ascorbic acid
   e. Docosahexaenoic acid

References

1 Sena, L.A. & Chandel, N.S. (2012) Physiological roles of mitochondrial reactive oxygen species. *Molecular Cell*, 48, 158–167.
2 Ray, P.D., Huang, B.-W. & Tsuji, Y. (2012) Reactive oxygen species (ROS) homeostasis and redox regulation in cellular signaling. *Cellular Signalling*, 24, 981–990.
3 Finkel, T. (2011) Signal transduction by reactive oxygen species. *The Journal of Cell Biology*, 194, 7–15.
4 Brieger, K., Schiavone, S., Miller, F.J. Jr., *et al.* (2012) Reactive oxygen species: from health to disease. *Swiss Medical Weekly*, 142, w13659.
5 Winterbourn, C.C. (2008) Reconciling the chemistry and biology of reactive oxygen species. *Nature Chemical Biology*, 4, 278–286.
6 Murphy, M.P., Holmgren, A., Larsson, N.-G. *et al.* (2011) Unraveling the biological roles of reactive oxygen species. *Cell Metabolism*, 13, 361–366.
7 Niki, E. (2009) Lipid peroxidation: physiological levels and dual biological effects. *Free Radical Biology and Medicine*, 47, 469–484.
8 Kryston, T.B., Georgiev, A.B., Pissis, P. *et al.* (2011) Role of oxidative stress and DNA damage in human carcinogenesis. *Mutation Research/Fundamental and Molecular Mechanisms of Mutagenesis*, 711, 193–201.
9 Luo, S. & Levine, R.L. (2009) Methionine in proteins defends against oxidative stress. *The FASEB Journal*, 23, 464–472.
10 Chung, H.Y., Lee, E.K., Choi, Y.J. *et al.* (2011) Molecular inflammation as an underlying mechanism of the aging process and age-related diseases. *Journal of Dental Research*, 90, 830–840.
11 Bullon, P., Morillo, J., Ramirez-Tortosa, M. *et al.* (2009) Metabolic syndrome and periodontitis: is oxidative stress a common link? *Journal of Dental Research*, 88, 503–518.
12 Sosa, V., Moliné, T., Somoza, R. *et al.* (2013) Oxidative stress and cancer: an overview. *Ageing Research Reviews*, 12, 376–390.
13 Choudhari, S.K., Chaudhary, M., Gadbail, A.R. *et al.* (2014) Oxidative and antioxidative mechanisms in oral cancer and precancer: a review. *Oral Oncology*, 50, 10–18.

14 Lobo, V., Patil, A., Phatak, A. *et al.* (2010) Free radicals, antioxidants and functional foods: impact on human health. *Pharmacognosy Reviews*, 4, 118.

15 Birben, E., Sahiner, U.M., Sackesen, C. *et al.* (2012) Oxidative stress and antioxidant defense. *World Allergy Organization Journal*, 5, 9–19.

16 Carocho, M. & Ferreira, I.C. (2013) A review on antioxidants, prooxidants and related controversy: natural and synthetic compounds, screening and analysis methodologies and future perspectives. *Food and Chemical Toxicology*, 51, 15–25.

17 Miricescu, D., Totan, A., Calenic, B. *et al.* (2014) Salivary biomarkers: relationship between oxidative stress and alveolar bone loss in chronic periodontitis. *Acta Odontologica Scandinavica*, 72, 42–47.

18 Bullon, P., Newman, H.N. & Battino, M. (2014) Obesity, diabetes mellitus, atherosclerosis and chronic periodontitis: a shared pathology via oxidative stress and mitochondrial dysfunction? *Periodontology*, 2000(64), 139–153.

19 Greabu, M., Battino, M., Mohora, M. *et al.* (2009) Saliva-a diagnostic window to the body, both in health and in disease. *Journal of Medicine and Life*, 2, 124–132.

20 Greabu, M., Totan, A., Battino, M. *et al.* (2008) Cigarette smoke effect on total salivary antioxidant capacity, salivary glutathione peroxidase and gamma-glutamyltransferase activity. *Biofactors*, 33, 129–136.

21 Carnelio, S., Khan, S. & Rodrigues, G. (2008) Definite, probable or dubious: antioxidants trilogy in clinical dentistry. *British Dental Journal*, 204, 29–32.

22 Carvalho Rde, S., de Souza, C.M., Neves, J.C. *et al.* (2013) Vitamin E does not prevent bone loss and induced anxiety in rats with ligature-induced periodontitis. *Archives of Oral Biology*, 58, 50–58.

23 Schmidt, M.A., Riley, L.W. & Benz, I. (2003) Sweet new world: glycoproteins in bacterial pathogens. *Trends in Microbiology*, 11, 554–561.

24 Berger, S.B., De Souza Carreira, R.P. *et al.* (2013) Can green tea be used to reverse compromised bond strength after bleaching? *European Journal of Oral Sciences*, 121, 377–381.

25 Vidhya, S., Srinivasulu, S., Sujatha, M. *et al.* (2011) Effect of grape seed extract on the bond strength of bleached enamel. *Operative Dentistry*, 36, 433–438.

26 Shrestha, B., Theerathavaj, M.L., Thaweboon, S. *et al.* (2012) In vitro antimicrobial effects of grape seed extract on peri-implantitis microflora in craniofacial implants. *Asian Pacific Journal of Tropical Biomedicine*, 2, 822–825.

27 Ucan, M.C., Koparal, M., Agacayak, S. *et al.* (2013) Influence of caffeic acid phenethyl ester on bone healing in a rat model. *Journal of International Medical Research*, 41, 1648–1654.

28 Uysal, T., Gorgulu, S., Yagci, A. *et al.* (2011) Effect of resveratrol on bone formation in the expanded inter-premaxillary suture: early bone changes. *Orthodontics and Craniofacial Research*, 14, 80–87.

29 Altan, B.A., Kara, I.M., Nalcaci, R. *et al.* (2013) Systemic propolis stimulates new bone formation at the expanded suture: a histomorphometric study. *Angle Orthodontist*, 83, 286–291.

30 King, M., Chatelain, K., Farris, D. *et al.* (2007) Oral squamous cell carcinoma proliferative phenotype is modulated by proanthocyanidins: a potential prevention and treatment alternative for oral cancer. *BMC Complementary and Alternative Medicine*, 7, 22.

31 Singh, N., Niyogi, R.G., Mishra, D. *et al.* (2013) Antioxidants in oral health and diseases: future prospects. *IOSR Journal of Dental and Medical Sciences*, 10(3), 36–44.

32 Sharma, A. & Sharma, S. (2011) Reactive oxygen species and antioxidants in periodontics: a review. *International Journal of Dental Clinics*, 3(2), 44–47.

33 Singh, N., Chander Narula, S., Kumar Sharma, R. *et al.* (2014) Vitamin E supplementation, superoxide dismutase status, and outcome of scaling and root planing in patients with chronic periodontitis: a randomized clinical trial. *Journal of Periodontology*, 85(2), 242–249.

34 Chapple IL, et al. 2012. "Adjunctive daily supplementation with encapsulated fruit, vegetable and berry juice powder concentrates and clinical periodontal outcomes: a double-blind RCT." *Journal of Clinical Periodontology* 39.1:62-72.

35 Iwasaki, M., Moynihan, P., Manz, M.C. *et al.* (2013) Dietary antioxidants and periodontal disease in community-based older Japanese: a 2-year follow-up study. *Public Health Nutrition*, 16(02), 330–338.

36 Hans, M., Prakash, S. & Gupta, S. (2012) Clinical evaluation of topical application of perio-Q gel (Coenzyme Q_{10}) in chronic periodontitis patients. *Journal of Indian Society of Periodontology*, 16(2), 193–199.

37 Daiya, S., Sharma, R.K., Tewari, S. *et al.* (2014) Micronutrients and superoxide dismutase in postmenopausal women with chronic periodontitis: a pilot interventional study. *Journal of Periodontal and Implant Science*, 44(4), 207–213.

38 Niki, E. (2011) Antioxidant capacity: which capacity and how to assess it? *Journal of Berry Research*, 1, 169–176.

39 Takashima, M., Horie, M., Shichiri, M. *et al.* (2012) Assessment of antioxidant capacity for scavenging free radicals in vitro: a rational basis and practical application. *Free Radical Biology and Medicine*, 52, 1242–1252.

40 Bompadre, S., Leone, L., Politi, A. *et al.* (2004) Improved FIA-ABTS method for antioxidant capacity determination in different biological samples. *Free Radical Research*, 38, 831–838.

CHAPTER 11

Oxidative stress and the skin

Christina L. Mitchell

Department of Dermatology, University of Florida College of Medicine, Gainesville, FL 32606, USA

THEMATIC SUMMARY BOX

At the end of this chapter, students should be able to:

- Describe the main mechanisms of oxidative stress (OS) in the skin

- List the main OS-related diseases of the skin

- Outline the pathogenesis of these disorders

- Analyze the contribution of oxidative stress in carcinogenesis

- Review the available topical treatments

- Contrast the mechanistic basis of chronological aging and photoaging of the skin

---

Introduction

The skin is the largest organ of the body. It functions as the first-line defense from external insults, protecting all other organs from the environment. The skin protects the body from radiation and acts as a barrier against pathogens and desiccation. The skin also functions as a thermoregulator, both acting as an insulator and, by way of sweating, a coolant. It is a major sensory organ as well, containing a variety of nerve endings. The skin is also important for vitamin D production and immune function as it plays an active part in immune functions.

The skin is a complex organ system with many different cell types and adnexal structures. It can be subdivided into three major layers: the epidermis, the dermis, and the subcutaneous tissue. The epidermis is mainly composed of keratinocytes, melanocytes (pigment-producing cells), and Langerhans cells (antigen-presenting cells). The dermis is primarily composed of extracellular proteins made by fibroblasts. Collagens, elastin, proteoglycans, and fibronectin are among the proteins found in the dermis, with type-I collagen being the most abundant protein among these.[1] The dermis is important for skin strength and resiliency and is directly altered by oxidative imbalance.

Oxidative Stress and Antioxidant Protection: The Science of Free Radical Biology and Disease, First Edition.
Edited by Donald Armstrong and Robert D. Stratton.
© 2016 John Wiley & Sons, Inc. Published 2016 by John Wiley & Sons, Inc.

Since it functions as the major barrier to the environment, the skin is constantly exposed to chemical and physical insults. The integrity and health of the skin thus depends on its ability to defend and protect itself from insult. Oxidative stress and imbalance contribute greatly to the health, appearance, and function of the skin. Many environmental pollutants act as oxidants or catalyze the production of reactive oxygen species (ROS), which are able to promote proliferation and cell survival, signaling, and altering apoptotic pathways. The uncontrolled formation of ROS is implicated in a number of skin disorders including inflammatory dermatoses and nonmelanoma skin cancer development.[2] Sunlight and other forms of ultraviolet (UV) radiation significantly contribute to environmental stress on the skin. Smoking and other chemical pollutants, such as automobile exhaust, contribute to oxidative imbalance as well. Compounding oxidative stressors hastens photoaging and contributes to the development of skin cancer.

Mechanisms of oxidative stress in the skin

Oxidative stress is an intrinsic part of aerobic anabolism and catabolism in the skin. Under normal circumstances, low levels of oxygen-free radicals (superoxide anion, hydroxyl radicals, hydrogen peroxide, and molecular oxygen) are produced by various cellular processes. These ROS play a role in cell proliferation, differentiation, and apoptosis as well as in immune responses. Oxidative stress occurs as a result of high metabolic demands and the introduction of external sources such as ultraviolet radiation, pollution, tobacco smoke, microorganisms, and xenobiotics.[3] Damage by free radicals results when protective mechanisms become overwhelmed and depleted during oxidative imbalance. While there are antioxidant (AOx) systems in the skin, they can be overpowered by the generation of excess ROS. Cellular damage takes the form of lipid peroxidation, DNA damage, and protein oxidation resulting from oxidative imbalance.[4] The uncontrolled production of ROS plays a major role in a range of skin diseases including skin cancer.

Reactive oxygen species

The majority of oxygen in the skin is used for cellular metabolism. Most of the oxygen is used by the mitochondria to produce energy. O_2 is altered by a series electron-transfer reactions and energy-transfer reactions. In electron-transfer reactions, single-electron subtractions form superoxide anions, hydrogen peroxide, hydroxyl radicals, and then water. In energy-transfer reactions, a molecule known as a sensitizer absorbs energy upon irradiation and transfers the energy to O_2, forming singlet oxygen. Mitochondria produce most of the intrinsic ROS. However, xanthine oxidase enzyme systems and synthesis of prostaglandins produce ROS as well.[5] Porphyrins, which are intermediate products of heme metabolism, are another example of sensitizers/oxidizers in the skin. While environmental insults and normal cellular metabolism generate ROS, activated leukocytes are another source of oxygen radicals in the skin. Inflammation-activated leukocytes also generate ROS via myeloperoxidase and nitric oxide synthase activity, and also generate superoxide, nitric oxide, and hypochlorite anion. The ROS thus produced serve to kill pathogenic microorganisms and remove damaged tissue.[2] Adjacent normal tissues, however, are damaged by the inexact targeting of the inflammatory process.

Protective cellular enzymes and nonenzymatic AOxs are able to keep oxidative cellular damage to a minimum in normal, healthy settings. The skin and other tissues have specific enzymatic defenses such as catalase (CAT), glutathione reductase, glutathione peroxidase, and superoxide dismutase (SOD), which destroy hydrogen peroxide, lipid hydroperoxides, and superoxide, respectively. Nonenzymatic AOxs such as vitamins C and E, ubiquinone, and uric acid neutralize ROS.[6]

Skin aging

Photodamage and photoaging

There are two types of skin aging: chronological aging and photoaging. Chronological or intrinsic aging of the skin is inevitable and has been implicated in the development of fine wrinkles or rhytids and benign growths such as seborrheic keratoses and cherry angiomas. Chronological skin aging has not been implicated in causing deep wrinkles/rhytids or significant dyspigmentation as has photoaging. While chronological and intrinsic skin aging is independent of sun exposure, photoaging of the skin is mainly dependent on two factors: the amount of pigment or melanin in the skin, and the aggregate amount of sun exposure. People who live in sunny environments and have a history of intense sun exposure acquire the largest amount of solar radiation and , therefore, experience the most significant photoaging. Ultraviolet A (UVA) rays, which have lower energy but deeper penetration into the skin than the radiation of ultraviolet B (UVB) spectrum, are implicated as the major causative agent of photoaging.

Exposure to environmental factors such as sun, smoking, and air pollution increases oxidative stress on skin by the generation of ROS. These ROS are capable of damaging DNA, oxidizing proteins, altering intracellular calcium levels, and activating cell surface receptors. ROS damage the dermis by activating cell surface receptors and kinases, which increase transcription factors AP-1 and NF-k2. Increased activity of AP-1 decreases the expression of collagens I and III fibroblasts in the skin, thereby reducing collagen synthesis. AP-1 also triggers the production of matrix metalloproteinases by keratinocytes and fibroblasts, which degrade and destroy preexisting mature collagen. UV light also induces the production of transcription factor NF-κB. This proinflammatory cytokine increases collagen degradation by the process of neutrophil chemotaxis and neutrophil collagenases.[7]

The role of chronic UVA exposure is demonstrated in Figure 11.1. This patient presented was with unilateral wrinkling of the skin. He had been a truck driver for 28 years and had extensive exposure to UVA light, which passed through the window of his truck.[4] UVB rays are filtered by windowpanes; thus, this is a pictorial example of UVA effects on the skin: thickening of the epidermis and destruction of collagen and elastin.

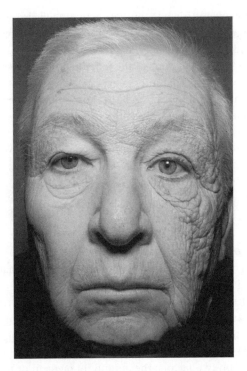

Figure 11.1 A 69-year-old man presented with a 25-year history of gradual, asymptomatic thickening and wrinkling of the skin on the left side of his face. The physical examination showed hyperkeratosis with accentuated ridging, multiple open comedones, and areas of nodular elastosis. Histopathological analysis showed an accumulation of elastolytic material in the dermis and the formation of milia within the vellus hair follicles. Findings were consistent with the Favre–Racouchot syndrome of photodamaged skin, known as dermatoheliosis. The patient reported that he had driven a delivery truck for 28 years. Ultraviolet A (UVA) rays transmit through window glass, penetrating the epidermis and upper layers of dermis. Chronic UVA exposure can result in thickening of the epidermis and stratum corneum, as well as destruction of elastic fibers. This photoaging effect of UVA is contrasted with photocarcinogenesis. Although exposure to ultraviolet B (UVB) rays is linked to a higher rate of photocarcinogenesis, UVA has also been shown to induce substantial DNA mutations and direct toxicity, leading to the formation of skin cancer. The use of sun protection and topical retinoids and periodic monitoring for skin cancer were recommended for the patient. (Image from *The New England Journal of Medicine*, Jennifer Gordon and Joaquin Brieva, Unilateral dermatoheliosis, 366; 16 Copyright ©2012 Massachusetts Medical Society. Reprinted with permission from Massachusetts Medical Society.)

Ultraviolet light from the sun is a well-recognized source of stress to the skin. Damage from sunlight causes premature aging of the skin resulting in rhytids, dyspigmentation, telangiectasias, and xerosis. The skin can have greater laxity and a leathery appearance. Photodamage also induces skin cancer. The sunlight-induced cascade of damage begins when ultraviolet light is absorbed by a target chromophore. This sets off a series of photochemical reactions that may cause skin aging and cancer.[8]

These photo-induced reactions can alter DNA and oxidize nucleic acids. ROS generated by ultraviolet light can also destroy proteins and lipids, thereby changing a cell's ability to function.

This photo-induced cascade of damage begins when UV radiation from the sun contacts the skin. UV light can be subdivided into groups by wavelength. UVA and UVB are the clinically relevant wavelengths of light, since they are able to penetrate into the skin. Ultraviolet light is absorbed into the skin by target chromophores, mainly DNA and urocanic acid, a product of filaggrin degradation. The primary chromophore in the skin is DNA. The keratinocyte's DNA absorbs UVB (290–320 nm) and develops damaged DNA in the form of thymine dimers and (6–4)-photoproducts.[9] This solar-induced DNA damage must be corrected for a cell to function properly. A specific repair pathway, the nucleotide excision repair pathway of DNA repair, is essential for correcting this photo-induced DNA damage. This damage results in an abnormal expression of various gene products, which may be important for apoptosis, genomic repair, or growth arrest.[10] The most studied of these genes is *p53*, which is involved in regulating the cell cycle. Once *p53* alleles are irreparably mutated by ultraviolet light-induced DNA damage, squamous cell carcinoma develops. This relationship also holds for basal cell carcinoma, but the effect is not as striking. Mutations in the patched gene or other parts of the hedgehog signaling pathway result in the formation of basal cell carcinoma.[11]

Urocanic acid, a product of filaggrin degradation, is found in high concentrations in the epidermis. Urocanic acid is a target chromophore for UV light as well. Urocanic acid undergoes a *trans*-to-*cis* isomerization when exposed to UVB. When *trans*-urocanic acid absorbs one photon of UV light, singlet oxygen is formed. This extremely active ROS can trigger a cascade of cell membrane damage and generate further ROS.[12] *cis*-Urocanic acid has also been implicated as a mediator of immunosuppression in the skin through the suppression of delayed type of hypersensitivity reactions, reduction in Langerhans cells, suppression of natural killer lymphocytes, and polymorphonuclear leukocytes.[10]

Smoking and skin aging

In 1969, Harry Daniell noticed that smokers looked older than nonsmokers.[13] He subsequently developed a wrinkle-scoring system to more objectively analyze the association between smoking and wrinkles. Now the association between smoking and skin wrinkling has been reproduced in multiple studies.[14]

Skin cancer

Overview

There is strong evidence that oxidative stress following ultraviolet radiation contributes to skin cancer.[15] Following oxidative stress, cells may suffer DNA damage including DNA mutations, DNA repair and replication dysfunction, and cell cycle dysregulation. ROS also induce the production of immunosuppressive cytokines that contribute to carcinogenesis (Figure 11.2).[3]

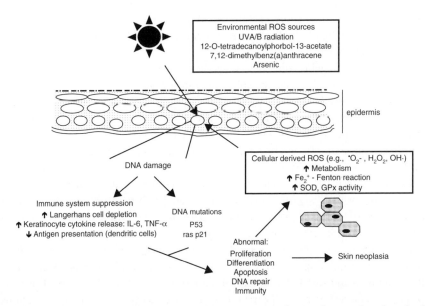

Figure 11.2 DNA damage: a central event in skin cancer. DNA damage is central to altered cell proliferation, differentiation, DNA repair, cell death, and immune system function required for skin cancer development. Exogenously and endogenously derived ROS (e.g., OH−. and H_2O_2) induce mutations in growth regulatory genes (e.g., p21 ras, p53) and can disrupt normal immune system function. The effects of ROS result in abnormal cellular physiology that contribute to elevated ROS (e.g., increased SOD activity), thus maintaining the DNA damage cycle, and the potential for cancer-causing events to occur.

The ultraviolet spectrum of an electromagnetic radiation spans wavelengths from 100 to 400 nm. Ultraviolet light is divided by wavelength into UVA (315–400 nm), UVB (280–315 nm), and ultraviolet C (UVC) (100–280 nm). UVC is not clinically relevant to skin damage because it is completely absorbed by ozone. UVB is greatly absorbed by ozone and accounts for 5% of UV exposure in the environment. UVA, which is not absorbed by ozone, accounts for the remaining 95% of UV exposure.

Carcinogenesis of the skin is a multistep process that involves three stages: initiation, promotion, and progression. Oxidative imbalance plays a significant role in skin carcinogenesis. The development of cancer in the skin, as that in any other organ, occurs in stages and is accompanied by numerous biochemical and molecular alterations. ROS are involved in all three stages of skin cancer development.

The first step, initiation, involves structural change in DNA that results in DNA mutations. Particular mutations in either tumor suppressor genes or proto-oncogenes cause dysfunction in terminal differentiation in cells. ROS are known to cause DNA damage via DNA base alteration, single- and double-stranded breaks, and cross-linking DNA with proteins.[16] UV light has an important role in initiation. Carcinogenic compounds that produce ROS are able to induce thymine glycol formation. Thymine glycol is the main product of thymidine residue damage in DNA.[17] 8-Hydroxyguanosine, a mutated DNA by-product, is found in increased concentration in the urine of animals exposed to oxidative stress.[16] Mouse skin has been shown to develop skin cancer when exposed to free oxygen radicals in tobacco smoke condensation.[18] Medications have also been implicated in skin cancer tumor induction. Azathioprine, a purine analogue, is an antimetabolite immuno-suppressant. The use of azathioprine has been linked to an increased incidence of skin cancer. Azathioprine is a UVA sensitizer and causes the accumulation of 6-thioguanine DNA, thereby acting as an inducer of skin cancer.[19]

Promotion, the second stage of carcinogenesis, involves clonal expansion of mutated cells. Oxidative stress as a mechanism for tumor promotion is widely accepted now because of the evidence from mouse model skin cancer studies. For example, peroxides and free radical generators promote tumor formation in mouse skin, and conversely, AOxs inhibit tumor promotion and carcinogenesis in mouse models.[20] A number of tumor promoters have been shown to generate ROS. These tumor promoters can decrease the levels of ROS scavengers and antioxidants that prevent the growth of tumors.[2] ROS can also mimic the biochemical effects of tumor promoters.[21]

The third stage of carcinogenesis is tumor progression. This occurs when benign papillomas develop into malignant tumors. Athar *et al.* showed papilloma-bearing mice developed an increased number of skin carcinomas when treated with free radical-generating compounds, thus showing that ROS contributes to tumor progression.[22] Oxidative stress has been shown to induce formation of ornithine decarboxylase, thus increasing polyamines. This stimulates cellular proliferation and carcinogenesis. Conversely, it is known that diethyl maleate, which promotes the degradation of reduced glutathione, slows tumor progression.[23]

UVB and UVA radiations play different roles in the development of skin cancer. UVB radiation is a complete carcinogen with the ability to generate squamous cell carcinomas.[22, 24] UVB is absorbed by DNA, resulting in the signature UV-induced thymidine dimers. These signature mutations are important for tumor initiation. UVA, on the other hand, is important for tumor promotion.[25] UVB directly damages DNA while UVA alters DNA through ROS intermediaries. UVA causes more oxidative stress than UVB and reacts with photosensitive molecules in the skin such as cytochromes, porphyrins, riboflavins, and heme, which act as sensitizers. These compounds absorb energy from the UVA and become excited and unstable. They transfer this energy to O_2, generating singlet oxygen and other ROS.[26] The ROS cause damage to cellular DNA, proteins, and lipids. ROS can cause nonspecific oxidation of DNA by causing single-stranded breaks and oxidized pyrimidine bases.[27] The most specific and characteristic mutation is generated by oxidative stress through the modified guanine nucleotide, 8-hydroxyguanine. This generates a G:C-to-T:A transversion during replication that promotes carcinogenesis.[28] UVA also mutates mitochondrial DNA through the generation of singlet oxygen. It has been implicated as a cause of the 4977-base pair mitochondrial DNA deletion, also known as "the common deletion."[29]

ROS generated from UVA can target cellular phospholipid membranes and proteins. Lipid peroxidation occurs when a hydroxyl radical steals a hydrogen atom from an unsaturated fatty acid. This forms unstable lipid molecules with extra electrons that can generate peroxyl radicals, setting off a chain reaction.[30] The UVA from sunlight is 10 times more efficient at lipid peroxidation than UVB, which destroys cell membranes through the lipoperoxide chain reaction.[31] Proteins are altered by oxidation when polypeptide chains are altered by ROS, generating carbonyl derivatives.

This protein alteration of the skin is most noticeable in the dermis through the cumulative and persistent degradation of collagen and elastin.[32]

Immunosuppression and skin cancer development

Consistent UV radiation contributes to skin cancer development through two major mechanisms: (i) direct DNA damage leading to mutations, and (ii) inhibition of normal immune system function. It is generally accepted that skin cancer development requires a loss of immune surveillance for neoplasia in the skin. In other words, skin cancer develops in the setting of immunosuppression.

The immune system of the skin is composed of different cell populations: keratinocytes, monocytes, epidermotropic T cells, free nerve endings, dermal macrophages, and Langerhans cells. These various cells interact through an intricate network of prostaglandins and cytokines.[10] In 1974, Kripke demonstrated that UV radiation caused systemic immunosuppression through the induction of UV-induced T-suppressor cells.[33] Kripke's work became the basis for modern photoimmunology. The skin's immune system is a complex organization that is greatly altered by UV exposure. Oxidative stress from ultraviolet light alters the normal immune response via DNA damage and *cis*-urocanic acid. UV production of *cis*-urocanic acid is associated with suppression of cellular-mediated responses, alteration of antigen-presenting cells, reduction of Langerhans cell numbers, and suppression of natural killer lymphocytes and polymorphonuclear leukocytes. The UV-induced formation of cyclobutane pyrimidine dimers in the skin causes suppression of cell-mediated immunity, increases formation of T-suppressor cells, alters antigen-presenting cells, and increases expression of IL-10 and TNF-α in the skin. UV light also greatly diminishes the number of Langerhans cells, which are the main antigen-presenting cells in the skin. These cells also undergo UV-induced morphological changes as well as alterations in growth, differentiation, and migration, consequently altering function. UV radiation both suppresses the systemic and cutaneous immune systems and inhibits immune response to skin tumors.[34]

Oxidative stress and melanoma

Epidermal melanocytes are susceptible to excessive ROS production due to their role in melanin synthesis. Melanin synthesis is stimulated by sun exposure and inflammation; it requires oxidation reactions and generates superoxide anion and hydrogen peroxide, thereby increasing oxidative stress on the melanocyte. Oxidative stress in turn can result in melanocyte apoptosis or lead to malignant transformation.[35]

Oxidative stress appears to be important for the initiation and progression of melanoma. Mutations in several melanoma-associated genes develop as a result of oxidative stress. Some mutations associated with melanoma exacerbate oxidative stress as well. A common mutation in melanoma, the V600E BRAF mutation, may be induced by oxidative stress.[36] P16 mutations and deletions are associated with the development of various cancers, but are more commonly found in melanoma and familial melanoma syndrome. p16 is a tumor suppressor protein, but until recently, it was not understood why mutations in p16 seemed to be more associated with development of melanomas than other malignancies. It was found that melanocytes are more sensitive to p16 depletion than other cell types. p16 was found to be an important regulator of oxidative stress, as it modulates intracellular oxidative stress in a cell cycle-independent manner separate from its control of the retinoblastoma pathway. p16 expression is greatly upregulated in melanocytes exposed to UV light and oxidative stress. Depletion of p16 in cultured melanocytes greatly increases ROS levels.[37] This oxidative imbalance may also contribute to the initiation and progression of melanoma. Loss of PTEN, a tumor suppressor protein, is also associated with melanoma progression. This is likely due to an increased level of superoxide anion resulting from sustained activation of Akt.[38] Mutations in GSTM1 and GSTT1, both of which belong to the glutathione *S*-transferase family of AOx genes, are associated with a high risk of melanoma. This is especially true for such subjects with a history of childhood sunburns.[39] Melanoma tumors have an abnormal redox state, and melanoma tumor cells have higher intracellular levels of ROS than normal cells.[40] Melanoma tumor cells also express abnormally high levels of nitric oxide, and this increase in nitric oxide correlates with the disease stage.[41] The role of oxidative stress in melanoma is also supported by the evidence that the AOx *N*-acetylcysteine inhibits melanoma tumor formation in a melanoma mouse model.[42] For this reason, AOxs are being considered for the prevention and treatment of melanoma.

Vitiligo

Vitiligo is an acquired disorder of skin depigmentation. It is characterized by loss/destruction of melanocytes in the epidermis. Two main theories of vitiligo pathogenesis exist: oxidative stress-mediated melanocytic toxicity and autoimmune antibody-mediated melanocyte destruction. Vitiligo is a skin disease of multifactorial etiology. It involves genetic susceptibility and autoimmune, biochemical, oxidant–AOx, neural, and viral mechanisms.[43] Vitiligo has a prevalence of 0.5% worldwide. While the exact mechanism of vitiligo pathogenesis is unknown, oxidative stress has been shown to play a major role in the initiation of white patches by the destruction of melanocytes from the accumulation of ROS.[44] In addition, hypersensitivity of epidermal melanocytes to oxidative stress is a known contributor to vitiligo pathogenesis.[45]

Some patients with vitiligo have a slightly blue or yellow/green fluorescence of the affected areas with Wood's lamp exam. This finding led to the discovery of oxidized pteridines in the skin. This overproduction of pteridines in the skin was due to a metabolic defect in tetrahydrobiopterin homeostasis. This imbalance results in the accumulation of hydrogen peroxide, which is toxic to melanocytes.[46]

Several studies have been conducted to evaluate the role of oxidative stress as the initial pathologic event in melanocyte destruction. Alterations in AOx patterns have been observed in various tissues of vitiligo patients. In addition, elevated levels of SOD and low levels of CAT activity have been found in the skin of vitiligo patients.[47] As previously mentioned, H_2O_2 has also been found to accumulate in the epidermis of patients with vitiligo.[48] During oxidative stress, SOD increases to scavenge the increased levels of superoxide anions, and CAT levels decrease.[49] Sravani *et al.* set out to evaluate the role of oxidative stress in vitiligo by measuring the levels of AOx enzymes in normal patients and patients with vitiligo.[47] They concluded that oxidative stress is increased in vitiligo because of the high levels of SOD and low levels of CAT in the skin of vitiligo patients.[47]

Intrinsic defenses against free radicals

The skin has an intricate defense network guarding against ROS damage. An enzymatic AOx system and a nonenzymatic AOx system work in conjunction to neutralize ROS. Notably, both the concentration and activity of AOxs are higher in the epidermis than the dermis.[30] The main AOx enzymes are SOD, CAT, and glutathione peroxidase. SOD converts superoxide radicals into H_2O_2 and O_2. Then, CAT and glutathione peroxidase contribute to the reduction of H_2O_2 to water and O_2. Nonenzymatic AOxs are found in both the lipid-soluble and water-soluble compartments of the cells and include vitamins C and E, glutathione, and ubiquinol. These work in a coordinated and self-perpetuating manner to neutralize ROS. For example, glutathione reductase can reduce the level of oxidized glutathione disulfide, generating glutathione once again. Then glutathione perpetuates the AOx action by restoring oxidized vitamins C and E to the normal state so that they can scavenge ROS again. There is a new interest in NF-E2-related factor 2 (Nrf2). This transcription factor activates the production of AOx enzymes. Studies have shown that Nrf2 protects the epidermis and fibroblasts against UVA-induced oxidative damage.[50,51]

Topical antioxidants

Topical AOxs are now incorporated into sunscreens and cosmeceuticals to replenish the skin's natural reservoirs. Topical AOxs reduce UVA-induced oxidative damage and act as photoprotectants. Product stability and product penetration are important for the effectiveness of topically applied AOxs: Product stabilization is important because AOxs are unstable and may become oxidized and inactive before they can reach their target. Topical AOxs must also be properly absorbed into the skin to reach the target tissue.

Vitamin C
Vitamin C is the predominant water-soluble AOx in the skin. On a molar basis, vitamin C is the predominant AOx in the skin. Its concentration is 15-fold greater than glutathione, 200-fold greater than vitamin E, and 1000-fold greater than ubiquinol.[52] It neutralizes ROS in the aqueous compartments of the skin and also aids in regenerating the AOx, vitamin E.[53] Vitamin C plays other roles in antiaging. It is a cofactor for enzymes required for collagen synthesis. Topical 1% vitamin C increases collagen synthesis and decreases collagenase expression.[54] Subsequently, a nutritional deficiency of vitamin C results in scurvy. ʟ-Ascorbic acid also inhibits elastin biosynthesis, which may decrease the elastin accumulation that occurs in aging skin.[55] Vitamin C can aid in photo-induced hyper/dyspigmentation by inhibiting tyrosinase. It can also

aid in skin hydration through protection of the epidermal barrier.[56] Topical use of vitamin C also has demonstrated photoprotective effects including the reduction of erythema and sunburn cell formation.[57] Because ʟ-ascorbic acid is unstable in a vehicle that can penetrate the stratum corneum, esterified substitutes such as magnesium ascorbyl phosphate and ascorbyl-6-palmitate are used in cosmeceutical formulations. Unfortunately, these esterified, stable substitutes do not demonstrate the same AOx activity *in vivo* as that of ʟ-ascorbic acid and are inferior AOxs.[56]

Vitamin E

Vitamin E is a lipid-soluble AOx. Vitamin E protects cell membranes from oxidative stress. When a ROS attacks a cell membrane, a lipid peroxyl radical is formed that can set off a chain reaction of oxidation, creating more lipid peroxyl radicals and threatening the cell structure.[58] Tocopherols and tocotrienols stop the chain reaction by scavenging the peroxyl radical. α-Tocopherol is the most abundant form of vitamin E although it also exists as eight compounds, four tocopherols, and four tocotrienols. Vitamin E is more abundant in the stratum corneum of the epidermis than in other parts of the skin.[59] The highest concentrations of vitamin E are found in the lower levels of the stratum corneum. Vitamin E is naturally delivered to the epidermis via sebum from the sebaceous gland.[30] Numerous studies have demonstrated the photoprotective effects of topical vitamin E in animal skin. Topical α-tocopherol was found to protect rabbit skin against UV-induced erythema.[60] Topical application of vitamin E to mouse skin resulted in a reduction in lipid peroxidation, photoaging, immunosuppression, and photocarcinogenesis.[61–63] Vitamin E prevents photocarcinogenesis by inhibiting UV-induced cyclopyrimidine dimer formation in the epidermal p53 gene.[64] α-Tocopherol has modest absorption of UV light (near 290 nm), which may add to its photoprotective character.[65]

The lipophilic character of vitamin E makes it a good compound for application to the skin. The hydroxyl group on the chromanol ring is essential for AOx activity. However, α-tocopherol is not stable in topical formulations unless that hydroxyl group is esterified. In order for vitamin E to become active again, the ester must be hydrolyzed. While this easily occurs after oral ingestion, this hydrolysis occurs very slowly after application to the skin.[66] In mouse studies, the esterified α-tocopherol succinate and α-tocopherol acetate were less effective AOxs for protecting against UV-induced erythema, immunosuppression, photoaging, and carcinogenesis.[67]

As previously mentioned, vitamins C and E function in conjunction to maintain AOx stores in the skin. The functions of these AOxs are dependent on one another for providing AOx defense: vitamin C regenerates oxidized vitamin E, glutathione regenerates oxidized vitamin C, and so on in a self-perpetuating manner. The oral combination of vitamins C and E in high doses provides protection against UV-induced erythema in humans, whereas either vitamin alone was ineffective.[68] In porcine skin, the topical combination of 15% ʟ-ascorbic acid and 1% α-tocopherol provided fourfold protection against UV-induced erythema of thymine dimer formation.[69] Topical combinations of vitamins C and E have also been shown to inhibit tanning in humans.[70]

Vitamin A

Retinoids and carotenoids are the two forms of vitamin A used in topical preparations. Carotenoids such as β-carotene and lycopene act as AOxs by scavenging singlet oxygen and inhibiting lipid peroxidation.[71] Retinoids, on the other hand, increase collagen production and epidermal thickness by binding to nuclear receptors, RARs, and RXRs, inhibiting AP-1 and MMP-1 expression.[55] Retinoids are commonly found in cosmeceutical products. While the safety of retinoids in cosmetic products has been in question since mouse studies suggested the carcinogenic effects of UV radiation, there is a long-standing evidence from use in clinical medicine that topical retinoids are safe.[72] Retinol, tretinoin, and tazarotene are all marketed for their antiaging properties.

Drug-induced skin photosensitization

Some drugs are capable of inducing skin photosensitivity by triggering a ROS-mediated inflammatory response.[73] Modern medicine has developed therapeutics from the destructive nature of porphyrins by the process of photodynamic therapy, which is a common treatment for actinic keratoses. Photodynamic therapy employs porphyrins as photosensitizing agents. These agents are applied topically or ingested. Exposure to specific wavelengths of red or blue light then causes the porphyrins to generate ROS that destroy tumor cells.[74] In the practice of dermatology, aminolevulinic acid or methyl-amino-levulinic acid are the porphyrins used to treat AKs and other skin tumors such as squamous cell carcinoma, basal cell carcinoma, and benign viral warts. Generation of singlet oxygen and various other ROS is required in cutaneous porphyrin photosensitization.[74] Both aminolevulinic acid and methyl-amino-levulinic acid are mainly converted into protoporphyrin IX intracellularly. Protoporphyrin IX is activated by blue or red light to generate ROS. The enzyme xanthine oxidase generates the free radical, $O_2^{\bullet-}$. Inhibition of xanthine oxidase with the gout medication allopurinol blocks porphyrin-mediated photosensitivity responses by blocking the synthesis of $O_2^{\bullet-}$.[22]

Conclusion

Oxygen-free radicals from endogenous and exogenous sources are known to damage the skin and contribute to aging and skin disease. While the skin has a complex defense system in place to prevent damage by ROS, this innate system can be overwhelmed, which leads to oxidative stress that may result in skin aging, immunosuppression, and skin cancer formation. A better understanding of the sources of ROS and the mechanisms of damage from these radicals would allow better, more targeted, therapies for prevention and treatment of oxidative stress in the skin.

Multiple choice questions

1 The first step in carcinogenesis of the skin is:
   a. Tumor promotor generation of ROS

   b. Polyamine stimulation of cellular proliferation through ROS induction of ornithine decarboxylase

   c. Oxidative damage to DNA

2 The type of UV radiation that causes direct DNA damage by absorption and induction of thymidine dimers is:

   a. UVA

   b. UVB

3 UV radiation alters normal skin immune response by:

   a. DNA damage

   b. *cis*-Urocanic acid immunosuppression

   c. UV-induced T-suppressor cells

   d. All the above

References

1 Mukherjee, S., Date, A., Patravale, V. *et al.* (2006) Retinoids in the treatment of skin aging: An overview of clinical efficacy and safety. *Clinical Interventions in Aging*, 4, 327–348.

2 Bickers, D. & Athar, M. (2006) Oxidative stress in the pathogenesis of skin disease. *Journal of Investigative Dermatology*, 126, 2565–2575.

3 Trouba, K.J., Mamadeh, H.K., Amin, R.P. *et al.* (2002) Oxidative stress and its role in skin disease. *Antioxidants and Redox Signalling*, 4, 665–673.

4 Gordon, J. & Brieva, J. (2012) Unilateral dermatoheliosis. *The New England Journal of Medicine*, 366, 16.

5 Podhaisky, H.P., Riemschneider, S. & Wohlrab, W. (2002) UV light and oxidative damage in the skin. *Pharmacie*, 57, 30–33.

6 Young, I.S. & Woodside, J.V. (2001) Antioxidants in health and disease. *Journal of Clinical Pathology*, 54, 176–186.

7 Pandel, R.E. (2013) Skin photoaging and the role of antioxidants in its prevention. *ISRN Dermatology*, 2013:930164, 1–11.

8 Trautinger, F. (2001) Mechanisms of photodamage of the skin and its functional consequences for skin aging. *Clinical and Experimental Dermatology*, 26, 573–577.

9 Young, A.R., Potten, C.S., Nikaido, O. *et al.* (1998) Human melanocytes and keratinocytes exposed to UVB or UVA in vivo show comparable levels of thymine dimers. *Journal of Investigative Dermatology*, 111, 936–940.

10 Amerio, P.E. (2009) UV-induced skin immunosuppression. *Anti-inflammatory and Anti-allergy Agents in Medicinal Chemistry*, 8(1), 3–13.

11 Bale, A. & Yu, K.P. (2001) The hedgehog pathway and basal cell carcinomas. *Human Molecular Genetics*, 10, 757–762.

12 Hanson, K. & Simon, J.D. (1998) Epidermal trans-urocanic acid and the UVA-induced photoaging of the skin. *Proceedings of the National Academy of Sciences of the United States of America*, 95, 10576–10578.

13 Daniell, H. (1969) Smooth tobacco and wrinkled skin. *New England Journal of Medicine*, 280, 53.

14 Vierkotter, A. & Krutmann, K. (2012) Environmental influences on skin aging and ethnic-specific manifestations. *Dermato-Endocrinology*, 4, 227–231.

15 Kensler, T. & Trush, M.A. (1984) Role of oxygen radicals in tumor promotion. *Environmental Mutagenesis*, 6, 593–616.

16 Athar, M. (2002) Oxidative stress and experimental carcinogenesis. *Indian Journal of Experimental Biology*, 40, 656–667.

17 Nishigori, C., Hatttori, Y. & Toyokuni, S. (2004) Role of reactive oxygen species in skin carcinogenesis. *Antioxidants and Redox Signaling*, 6, 561–570.

18 Curtin, G., Hanausek, M., Walaszec, Z. *et al.* (2004) Short-term in vitro and in vivo analyses for assessing the tumor-promoting potentials of cigarette smoke condensates. *Toxicological Sciences*, 81, 185–193.

19 O'Donovan, P., Perrett, C.M., Zhang, X. *et al.* (2005) Azathioprine and UVA light generate mutagenic oxidative DNA damage. *Science*, 309, 1871–1874.

20 Perchellet, J.P., Perchellet, E.M., Gali, H.U. *et al.* (1995) Oxidative stress and multistage skin carcinogenesis. In: Mukhtar, H. (ed), *Skin Cancer: Mechanisms and Human Relevance*. CRC Press, Boca Raton, FL, pp. 145–180.

21 Nakamura, Y., Colburn, N.H. & Gindhart, T.D. (1985) Role of reactive oxygen in tumor promotion: Implication of superoxide anion in promotion of neoplastic transformation in JB-6 cells by TPA. *Carcinogenesis*, 6, 229–235.

22 Athar, M., Elmets, C.A., Bickers, D.R. & Mukhtar, H. (1989) A novel mechanism for the generation of superoxide anions in hematoporphyrin derivative-mediated cutaneous photosensitization. *Activation of the xanthine oxidase pathway. The Journal of Clinical Investigation*, 83, 1137–1143.

23 Sander, C.S., Chang, H., Hamm, F. *et al.* (2004) Role of oxidative stress and the antioxidant network in cutaneous carcinogenesis. *International Journal of Dermatology*, 43, 326–335.

24 Black, H., deGruijl, F.R., Forbes, P.D. *et al.* (1997) Photocarcinogenesis: an overview. *Photochemistry and Photobiology*, 40, 29–47.

25 de Gruijl, F. (2000) Photocarcinogenesis: UVA vs UVB. Singlet oxygen, UVA, and ozone. *Methods in Enzymology*, 319, 359–366.

26 Cadet, J. & Douky, T. (2009) Sensitized formation of oxidatively generated damage to cellular DNA by UVA radiation. *Photochemical and Photobiological Sciences*, 8, 903–911.

27 Gaboriau, F., Demoulins-Giacco, N., Tiracche, I. *et al.* (1995) Involvement of singlet oxygen in ultraviolet A-induced lipid peroxidation in cultured human fibroblasts. *Archives for Dermatological Research*, 287, 338–340.

28 Parsons, P.G. & Hayward, I.P. (1985) Inhibition of DNA repair synthesis by sunlight. *Photochemistry and Photobiology*, 42, 287–293.

29 Berneburg, M., Grether-Beck, S., Kürten, V. *et al.* (1999) Singlet oxygen mediates the UVA-induced generation of the photoaging-associated mitochondrial common deletion. *Journal of Biological Chemistry*, 274, 15345–15349.

30 Chen, L., Hu, J.Y. & Wang, S.Q. (2012) The role of antioxidants in photoprotection: a critical review. *Journal of the American Academy of Dermatology*, 67, 1013–1024.

31 Morlière, P., Moysan, A. & Tirache, I. (1995) Action spectrum for UV-induced lipid peroxidation in cultured human skin fibroblasts. *Free Radical Biology and Medicine*, 19, 365–371.

32 Sander, C.S., Chang, H., Salzmann, S. *et al.* (2002) Photoaging is associated with protein oxidation in human skin in vivo. *Journal of Investigative Dermatology*, 118, 18–25.

33 Kripke, M.L. (1974) Antigenicity of murine skin tumors induced by ultraviolet light. *Journal of the National Cancer Institute*, 53, 1333–1336.

34 Kripke, M.L. & Fisher, M.S. (1976) Immunologic parameters of ultraviolet carcinogenesis. *Journal of the National Cancer Institute*, 57, 211–215.

35 Denat, L., Kadekaro, A., Marot, L. *et al.* (2014) Melanocytes as instigators and victims of oxidative stress. *Journal of Investigative Dermatology*, 134, 1512–1518.

36 Landi, M.T., Bauer, J., Pfeiffer, R.M. *et al.* (2006) MC1R germline variants confer risk for BRAF-mutant melanoma. *Science*, 313, 521–522.

37 Jenkins, N.C., Liu, T., Cassidy, P. *et al.* (2011) The p16INK4A tumor suppressor regulates cellular oxidative stress. *Oncogene*, 30, 265–274.

38 Govindarajan, B., Slight, J., Vincent, B. *et al.* (2007) Overexpression of Akt converts radial growth of melanoma to vertical growth melanoma. *Journal of Clinical Investigation*, 117, 719–729.

39 Fortes, C., Mastroeni, S., Boffetta, P. *et al.* (2011) Polymorphisms of GSTM1 and GSTT1, sun exposure and the risk of melanoma: a case–control study. *Acta Dermato-Venereologica*, 91, 284–289.

40 Meyskens, F., McNulty, S.E., Buckmeier, J.A. *et al.* (2001) Aberrant redox regulation in human metastatic melanoma cells compared to normal melanocytes. *Free Radical Biology and Medicine*, 31, 799–808.

41 Yang, Z., Misner, B., Ji, H. *et al.* (2013) Targeting nitric oxide signaling with nNOS inhibitors as a novel strategy for the therapy and prevention of human melanoma. *Antioxidants and Redox Signaling*, 19, 433–447.

42 Cotter, M.A., Thomas, J., Cassidy, P. *et al.* (2007) *N*-acetylcysteine protects melanocytes against oxidative stress/damage and delays onset of ultra-violet-induced melanoma in mice. *Clinical Cancer Research*, 13, 5952–5958.

43 Halder, R.M. & Chappel, J.L. (2009) Vitiligo update. *Seminars in Cutaneous Medicine and Surgery*, 28, 86–92.

44 Rokos, H., Beazley, W.D. & Schallreuter, K.U. (2002) Oxidative stress in vitiligo: photo-oxidation of pterins produces H_2O_2 and pterin-6-carboxylic acid. *Biochemical and Biophysical Research Communications*, 292, 805–811.

45 Qui, L., Song, Z. & Vijayasaradhi, S. (2014) Oxidative stress and vitiligo: the Nrf2-ARE signalling connection. *Journal of Investigative Dermatology*, 134, 2074–2076.

46 Ortonne, J.P. (2008) Vitiligo and other disorders of hypopigmentation. In: Bolognia, J.J. (ed), *Dermatology*. Elsevier, Spain, pp. 65.

47 Sravani, P.V., Babu, N.K., Gopal, K.V. *et al.* (2009) Determination of oxidative stress in vitiligo by measuring SOD and CAT levels in vitiliginous and non-vitiliginous skin. *Indian Journal of Dermatology, Venereology and Leprology*, 75(3), 268–271.

48 Schallreuter, K.E. (1999) In vivo and in vitro evidence for hydrogen peroxide accumulation in the epidermis of patients with vitiligo and its successful removal by a UVB-activated pseudocatalase. *Journal of Investigative Dermatology Symposium Proceedings*, 4, 91–96.

49 Koca, R., Armutcu, F., Altinyazar, H.C. *et al.* (2004) Oxidant-antioxidant enzymes and lipid peroxidation in generalized vitiligo. *Clinical and Experimental Dermatology*, 29, 406–409.

50 Tian, F.F., Zhang, F.F., Lai, X. *et al.* (2011) Nrf2-mediated protection against UVA radiation in human skin keratinocytes. *Bioscience Trends*, 5, 23–29.

51 Hirota, A., Kawachi, Y., Itoh, K. *et al.* (2005) Ultraviolet A irradiation induces NF-E2-related factor 2 activation in dermal fibroblasts. *Journal of Investigative Dermatology*, 124, 825–832.

52 Shindo, Y., Witt, E., Han, D. *et al.* (1994) Enzymatic and non-enzymatic antioxidants in the epidermis and dermis of human skin. *Journal of Investigative Dermatology*, 102, 122–124.

53 Thiele, J.J., Trabler, M.G. & Packer, L. (1998) Depletion of human stratum corneum vitamin E: an early and sensitive in vivo marker of UV-induced photo-oxidation. *Journal of Investigative Dermatology*, 110, 756–761.

54 Varani, J., Spearman, D., Perone, P. *et al.* (2001) Inhibition of type I procollagen synthesis by damaged collagen in photoaged skin and by collagenase-degraded collagen in vitro. *American Journal of Pathology*, 158, 931–942.

55 Fisher, G., Datta, S.C., Talwar, H.S. *et al.* (1996) Molecular basis of sun-induced premature skin aging and retinoid antagonism. *Nature*, 379, 335–339.

56 Campos, P., Gonçalves, G.M. & Gaspar, L.R. (2008) In-vitro antioxidant activity and in-vivo efficacy of topical formulations containing vitamin C and its derivatives studied by non-invasive methods. *Skin Research and Technology*, 14, 376–380.

57 Darr, D., Combs, S., Dunston, S. *et al.* (1992) Topical vitamin C protects porcine skin from ultraviolet radiation-induced damage. *British Journal of Dermatology*, 127, 247–253.

58 Munne-Bosch, S. & Alegre, L. (2002) The function of tocopherols and tocotrienols in plants. *Critical Reviews in Plant Sciences*, 21, 31–57.

59 Podda, M., Weber, C., Traber, M.G. *et al.* (1996) Simultaneous determination of tissue tocopherols, tocotrienols, ubiquinols, and ubiquinones. *Journal of Lipid Research*, 37, 893–901.

60 Roshchupkin, D.I., Pistsov, M.Y. & Potapenko, A.Y. (1979) Inhibition of ultraviolet light-induced erythema by antioxidants. *Archives for Dermatological Research*, 266, 91–94.

61 Lopez-Torres, M., Thiele, J.J., Shindo, Y. *et al.* (1998) Topical application of alpha-tocopherol modulates the antioxidant network and diminishes ultraviolet-induced oxidative damage in murine skin. *British Journal of Dermatology*, 138, 207–15.

62 Bissett, D., Chatterjee, R. & Hannon, D.P. (1990) Photoprotective effect of superoxide-scavenging antioxidants against ultraviolet radiation-induced chronic skin damage in the hairless mouse. *Photodermatology, Photoimmunology and Photomedicine*, 7, 56–62.

63 Gensler, H. & Magdaleno, M. (1991) Topical vitamin E inhibition of immunosuppression and tumorigenesis induced by ultraviolet irradiation. *Nutrition and Cancer*, 15, 97–106.

64 Chen, W., Barthelman, M., Martinez, J. *et al.* (1997) Inhibition of cyclobutane pyrimidine dimer formation in the epidermal p53 gene of UV-irradiated mice by alpha-tocopherol. *Nutrition and Cancer*, 29, 205–211.

65 Sorg, O., Tran, C. & Saurat, J.H. (2001) Cutaneous vitamins A and E in the context of ultraviolet- or chemically-induced oxidative stress. *Skin Pharmacology and Applied Skin Physiology*, 14, 363–372.

66 Pinnell, S. (2003) Cutaneous photodamage, oxidative stress, and topical antioxidant protection. *Journal of the American Academy of Dermatology*, 48, 1–19.

67 Gensler, H., Aickin, M., Peng, Y.M. *et al.* (1996) Importance of the form of topical vitamin E for prevention of photocarcinogenesis. *Nutrition and Cancer*, 26, 183–191.

68 Eberlein-Konig, B., Plaszek, M. & Przybilla, B. (1998) Protective effect against sunburn of combined systemic ascorbic acid vitamin C and d-alpha-tocopherol vitamin E. *Journal of the American Academy of Dermatology*, 38, 45–48.

69 Lin, J.Y., Selim, M.A., Shea, C.R. *et al.* (2003) UV photoprotection by combination topical antioxidants vitamin C and E. *Journal of the American Academy of Dermatology*, 48, 866–874.

70 Quevedo, W.C., Holstein, T.J., Dyckman, J. *et al.* (2000) The responses of the human epidermal melanocyte system to chronic erythemal doses of UVR in skin protected by topical applications of a combination of vitamins C and E. *Pigment Cell Research*, 13, 190–192.

71 Sies, H. & Stahl, W. (1995) Vitamins E and C, beta-carotene, and other carotenoids as antioxidants. *American Journal of Clinical Nutrition*, 62, 1315s–1321s.

72 Wang, S.Q., Dusza, S.W. & Lim, H.W. (2010) Safety of retinyl palmitate in sunscreens: a critical analysis. *Journal of the American Academy of Dermatology*, 63, 903–906.

73 Briganti, S. & Picardo, M. (2003) Antioxidant activity, lipid peroxidation and skin diseases. What's new. *The European Academy of Dermatology and Venereology*, 17, 663–669.

74 Athar, M., Mukhtar, H., Elmets, C.A. *et al.* (1988) In situ evidence for the involvement of superoxide anions in cutaneous porphyrin photosensitization. *Biochemical and Biophysical Research Communications*, 151, 1054–1059.

CHAPTER 12

Oxidative stress in osteoarticular diseases

María José Alcaraz[1], Sergio Portal-Núñez[2], Juan A. Ardura[2], and Pedro Esbrit[2]

[1] Department of Pharmacology and IDM, University of Valencia, Valencia, Spain
[2] Bone and Mineral Metabolism Laboratory, Instituto de Investigación Sanitaria (IIS)-Fundación Jiménez Díaz and UAM, Madrid, Spain

---

THEMATIC SUMMARY BOX

At the end of this chapter, students should be able to:

- Describe the main features of rheumatoid arthritis, osteoarthritis, and osteoporosis

- Outline the pathogenesis of these disorders

- Analyze the contribution of oxidative stress

- Revise the mechanisms involved in immune alterations and cartilage degradation

- Point out the mechanistic basis of age-related bone loss

- Infer new therapeutic targets

- Evaluate the impact of skeletal diseases

Introduction

A wide range of evidence supports the contribution of reactive oxygen species (ROS) and reactive nitrogen species (RNS) to osteoarticular disorders. These chronic conditions have a high prevalence and pose an important public health problem as they are a major cause of disability. Their pathogenesis is poorly understood although a number of risk factors have been revealed in recent studies. A better understanding of the mechanisms that contribute to the initiation and progression of disease can help to find ways of preventing or treating these disorders.

Oxidative Stress and Antioxidant Protection: The Science of Free Radical Biology and Disease, First Edition.
Edited by Donald Armstrong and Robert D. Stratton.
© 2016 John Wiley & Sons, Inc. Published 2016 by John Wiley & Sons, Inc.

Rheumatoid arthritis

Rheumatoid arthritis (RA) is a chronic autoimmune disease that affects 0.5–1% of the adult population worldwide. RA is characterized by synovial angiogenesis, hyperplasia of the synovial membrane, and infiltration of immune (e.g., CD4+ and CD8+ T cells, B cells) and inflammatory cells (e.g., neutrophils, monocytes, and macrophages). Synovial cell transformation and production of inflammatory mediators and degradative enzymes result in chronic inflammation and degradation of cartilage and adjacent bone. Synovial fibroblasts exhibit a proliferative phenotype, invade the cartilage, release proinflammatory cytokines and chemokines, and produce ROS and catabolic enzymes. In addition, proinflammatory cytokines activate chondrocytes and osteoclasts to release further inflammatory mediators, ROS, RNS, and degradative enzymes sustaining chronic inflammation and joint destruction.[1]

Oxidative stress markers

Redox processes are involved in the regulation of the immune system, as well as in the metabolism and survival of different cell types in the joint. Nevertheless, an imbalance between the production of ROS and/or RNS and the activity of the antioxidant defense systems results in oxidative stress, which contributes to RA pathology. Increased levels of oxidative stress markers such as nitrite (derived from nitric oxide, NO) and oxidatively modified proteins, lipids, extracellular matrix components, and nucleic acids accompanied by a reduction in antioxidant molecules have been detected in serum and synovial fluid from RA patients. Therefore, IgG aggregates modified by oxygen radicals (chlorinated IgG [Cl-IgG]) and peroxynitrite (nitrated IgG [N-IgG]) have been reported in the synovial fluid of RA patients. In addition, glycoxidation of proteins form advanced glycation end-products (AGE) that are related to endothelial dysfunction in RA. AGE receptor is upregulated in RA synovial fibroblasts, and its stimulation by AGE or ligands such as S100 proteins or high mobility group box-1 induces oxidative stress contributing to the amplification and chronification of the inflammatory response.[2]

Oxidative stress production

A number of factors can lead to the generation of oxidative stress in the joints. There is an increased intra-articular pressure in RA joints, which could result in ischemia–reperfusion episodes. Leukocytes and other cells can become activated by extracellular matrix components released upon tissue damage, by inflammatory cytokines, or by lipopolysaccharides in the presence of infection. As a consequence, NADPH oxidase activation and inducible NO synthase (iNOS) expression lead to the generation of high levels of ROS and RNS, which exceed the ability of the antioxidant defense systems including enzymatic antioxidants (superoxide dismutase [SOD], catalase, glutathione peroxidase, peroxiredoxin), and nonenzymatic antioxidants such as vitamins C and E, β-carotene, and glutathione. Mitochondrial dysfunction leading to an increased ROS production can also contribute to the pathology. In RA, mitochondrial alterations are present in key cell types such as T cells and synoviocytes, leading to an excessive leakage of electrons from the electron transport chain and increased superoxide anion ($O_2^{\bullet-}$) production. RNS such as peroxynitrite radical ($ONOO^-$) are generated by the reaction between $O_2^{\bullet-}$ and NO and induce cell damage by the oxidation and nitration of proteins, lipids, and DNA. $ONOO^-$ depletes thiol groups and modifies glutathione balance toward oxidative stress. Moreover, tissue injury releases iron and copper ions and heme proteins that catalyze free radical reactions. Some environmental factors such as smoking could also be involved in the generation of oxidative stress.[2]

Oxidative stress and the immune system

The functioning of immune cells is influenced by alterations in the intracellular redox balance. ROS have a physiological role in priming the immune system as second messengers. Mild oxidative conditions lead to changes in the intracellular thiol pool of T cells and can mediate the normal immune response. In addition, exposure of T cells to ROS from activated macrophages amplifies the TCR-mediated signal transduction and the defense response against pathogens. Nevertheless, the situation is different in chronic conditions such as RA, where long-term exposure to oxidants might lead to chronic changes in the redox status of immune cells. T cells from synovial fluid of RA patients have decreased reduced glutathione and increased NO levels. These cells exhibit hyporesponsiveness to mitogenic and death stimuli with a reduced proliferation and a long-term persistence at the inflammation site. It is likely that early exposure to ROS results in the upregulation of antioxidant systems such as thioredoxin reductase 1, which protect cells against ROS-driven apoptosis. RNS contribute to T-lymphocyte dysfunction. After a short exposure, RNS induce tyrosine phosphorylation of CD3ζ (zeta) chain of the TCR complex and cell activation. In contrast, when the exposure to RNS is prolonged, T cells downregulate membrane receptors, such as CD4, CD8, and chemokine receptors, and become refractory to stimulation. Furthermore, oxidative stress leads to posttranslational modifications of proteins involved in autoimmunity, for example, type-II collagen (the main component of cartilage), increasing its antigenicity.[3,4]

Oxidative stress signaling

ROS are second messengers in intracellular signal transduction upon activation of cytokine or growth factor receptors. Proinflammatory cytokines such as tumor necrosis factor-alpha (TNF-α) play an important role in the pathogenesis of RA. They induce the production of ROS and the synthesis of iNOS and NO generation in different cell types. Anti-TNF-α therapy has greatly improved RA treatment. Interestingly, this type of agent may partly act through the reduction of oxidative stress.[5]

Proinflammatory cytokines upon interaction with their specific receptors on the cell membrane activate nuclear factor-kappa B (NF-κB), phosphatidylinositol 3-kinase/Akt-1, and signal transducer and activator of transcription-3 leading to the transcription of inflammatory genes, including the cytokines themselves in a positive feedback loop, and matrix metalloproteinases (MMPs) that degrade the components of cartilage. Oxidative stress results in IκB kinase activation leading to IκB phosphorylation and the release of NF-κB, which translocates into the cell nucleus to activate gene transcription. In addition, ROS activate mitogen-activated protein kinases (extracellular signal-regulated kinase [ERK]1/2, Jun-NH_2-terminal kinase [JNK] and p38) and may directly regulate the activity of transcription factors (e.g., NF-κB, activator protein-1, p53, early growth response-1, and hypoxia inducible factor-1) through oxidative modifications of cysteine residues.[2]

Toll-like receptors (TLRs) in cell membranes are activated by endogenous molecules released during inflammation or tissue injury in the joints of RA patients, namely fibrinogen, heat shock proteins, or breakdown products of hyaluronic acid, leading to the upregulation of proinflammatory cytokines and chemokines. ROS are mediators of signal transduction after activation of TLR4, leading to the phosphorylation of ERK1/2, Akt, and p38, as well as to NF-κB activation. An abnormal activation of TLR4 can lead to RA progression.[6]

Synovial hyperplasia

Synovial hyperplasia and chronic inflammation are associated with a lack of apoptosis of immune cells and synovial fibroblasts, which depends on the activation of NF-κB and other signaling pathways inducing the expression of antiapoptotic molecules. Activation of NF-κB also results in the transcription of cell-growth-promoting factors such as cyclin D1 and c-Myc. Some antioxidant enzymes such as peroxiredoxin or thioredoxin reductase may protect synovial cells against apoptosis induced by ROS, whereas oxidative stress may induce mutations of *p53* that contribute to synovial cell transformation and excessive production of cytokines, MMPs, ROS, and RNS.[2]

Cartilage damage

An excessive ROS and/or RNS production in RA joints can potentiate cartilage damage as well as osteoclast activation and bone resorption. In addition, ROS inactivate antiproteinases and depolymerize hyaluronic acid, altering the properties of synovial fluid.

Chondrocytes express endothelial NOS (eNOS) and iNOS, and produce NO and $O_2^{\bullet-}$ that generate $ONOO^-$ and hydrogen peroxide (H_2O_2). In addition, chondrocytes possess an antioxidant enzyme system formed by SOD, catalase, glutathione peroxidase, and glutathione reductase, which regenerates reduced glutathione from the oxidized glutathione at the expense of NADPH.

Cartilage degradation by ROS may result from several mechanisms such as the inhibition of matrix synthesis and the induction of matrix degradation either directly by oxidative modification of matrix components or indirectly by inducing the expression of matrix-degrading enzymes such as MMPs and aggrecanases (Figure 12.1). In addition, posttranscriptional activation of MMPs is driven by ROS. The latter may directly oxidize different intracellular and extracellular components including nucleic acids, transcriptional factors, and membrane phospholipids. As a result, cellular membranes and extracellular matrix components such as proteoglycans and collagens are degraded.[2]

Osteoarthritis

Osteoarthritis (OA) is the most prevalent condition leading to functional limitation and physical disability among population aged 65 and older. Therefore, OA represents a heavy economic burden on healthcare systems as the number of older people

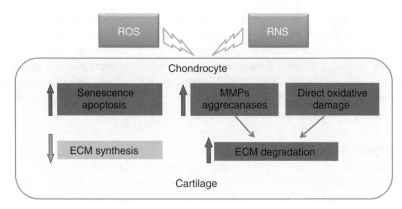

Figure 12.1 Effects of oxidative stress on chondrocytes and cartilage. ECM, extracellular matrix.

increases. Main risk factors are genetics, aging, obesity, injury, and biomechanical forces. This condition is associated with progressive hyaline articular cartilage loss, low-grade synovitis, and alterations in subchondral bone and periarticular tissues. In OA, there is an imbalance between anabolic and catabolic processes in the joint with a predominance of the latter. Different mechanisms may contribute to this imbalance such as reduced repair ability, increased production of proinflammatory mediators, and oxidative stress. Degradation of joint components in OA is thought to be largely elicited by mechanical stress and proinflammatory cytokines inducing ROS production and iNOS expression. Both ROS and RNS can play a role in the suppression of glycosaminoglycan and collagen synthesis, expression of MMPs, and activation of proenzymes leading to cartilage breakdown. In turn, products derived from altered tissues induce further synovial inflammation and joint degradation in a vicious circle, resulting in the progression of disease and the exacerbation of clinical symptoms. Chondrocytes show a hypertrophic phenotype and mitochondrial alterations that may be induced by proinflammatory cytokines and would contribute to cell death. Therefore, ROS suppress mitochondrial oxidative phosphorylation and ATP formation, which results in a reduced synthesis of matrix components such as collagen and proteoglycan. Oxidative stress may also induce telomere genomic instability and senescence of chondrocytes, leading to cartilage aging. ROS may cooperate with NO to induce chondrocyte apoptosis and also with other factors such as AGEs to modify proteins and alter cartilage homeostasis.[7–9] The increased ROS in aging and OA would interfere with normal insulin-like growth factor I (IGF-I) signaling in chondrocytes, leading to a defective repair ability that may shift the balance toward an increased catabolism and OA progression.[10]

Osteoporosis

Osteoporosis, classically defined by a decreased bone mass per unit volume of bone, is a skeletal disease characterized by a low bone mineral density (BMD) and deterioration of bone microarchitecture, related to accelerated bone resorption and/or low bone formation. This leads to bone loss and fragility and ultimately increase in the risk of fractures. Quantification of the fracture risk by the so-called FRAX score takes into consideration BMD and other clinical risk factors, such as age, race, exercise, calcium intake, and the associated morbidities such as RA and diabetes mellitus (DM). The high risk of fracture figures for men and all women aged >50 years (about 20% and 50%, respectively) makes osteoporosis a global health issue with an enormous socioeconomic impact.

Pathogenesis of osteoporosis

The maintenance of adult bone mass depends on the coordinated actions of matrix-resorbing osteoclasts, deriving from the hematopoietic lineage, and bone matrix producing osteoblasts of mesenchymal strain. This involves a process (bone remodeling) whereby old or fatigued bone is replaced by new bone in each basic multicellular unit (BMU) on the bone surfaces. BMUs are constituted of osteoclast and osteoblast lineage cells that act at specific times during the bone remodeling cycle. This is initiated by the activation of bone resorption, and bone formation normally ensues in a coupled manner. The activity of both cell types is balanced when bone formation by osteoblasts matches the amount of osteoclast-dependent bone resorption within each BMU. On the other hand, uncoupled or unbalanced bone remodeling associated with an increased number of BMUs per bone surface area as observed in primary osteoporosis leads to profound alterations in bone mass and/or structure.

Clinical conditions such as hyperparathyroidism and diabetes mellitus (DM), particularly type-I DM, and long-term corticosteroid use are frequently associated with osteoporotic bone loss. In these scenarios, the main pathogenetic factor responsible for osteoporosis is clearly delineated (e.g., the deficit of insulin in diabetic patients, a well-characterized osteogenic factor). Osteoporosis is most often due to postmenopausal estrogen deficiency combined with aging. Indeed, osteoporosis can now be envisioned as a comorbidity entity encompassing old age.[11] In both sexes, trabecular bone loss begins immediately after reaching peak bone mass, in the setting of sex steroid sufficiency; but osteopenia (at both trabecular and cortical levels) dramatically increases in women after menopause. Thus, estrogen deficiency is thought to be a key factor accelerating bone loss but other age-related changes, such as glucocorticoid overproduction, renal insufficiency (causing deficit of calcitriol), and sarcopenia, significantly contribute to involutional osteoporosis in aging subjects.

Age-related alterations of bone tissue and osteoblastic function

In rodents, as in humans, bone loss occurs with age. In fact, the bulk of current knowledge about the development of senile osteoporosis mostly relies on experimental rodent (mainly mouse) models. Of note though, in a difference from humans, mice do not undergo acute loss of estrogens with age, providing a suitable experimental tool for dissecting the effects of aging from those of sex steroid deficiency. Mice also display other particular features such as continuous modeling at the growth plate and absence of cortical remodeling and lack of Haversian system in contrast to humans. Notwithstanding these considerations, the main histological finding associated with aged bone in rodents and humans is a decline in bone thickness, likely a consequence of a deficit in the number of osteoblasts and/or a poor osteoblastic function and low bone remodeling. However, mechanical properties of skeletal tissue are relatively conserved with aging, likely related to a maintained subperiosteal apposition that increases momentum of inertia.

A deficit of bone formation is the major culprit of age-related bone loss. It is the current belief that osteoblasts are unable to balance osteoclast activity in each BMU of old bone. This is accounted for, in part, by a diminished number of osteoprogenitors at the expense of an increased adipogenesis in the bone marrow. Moreover, the bone content of osteocalcin, a marker of osteoblast maturation, is decreased in aging humans. *Ex vivo* osteoblast-like cell cultures from human bone biopsies show a diminished proliferative capacity and a poor response to calcitriol related to donor age. The underlying molecular mechanisms of age-related bone loss have remained poorly understood until recently.[12] A decrease in osteoprotegerin/receptor activator of NF-κB ligand ratio, an important modulator of bone remodeling, has been reported in mice in this setting: a fact probably related to a decrease in osteoclast precursors in the bone marrow of aged subjects. Aging mice exhibit an increased activity of 11 β-hydroxysteroid dehydrogenase type I leading to the activation of endogenous glucocorticoids, which contribute to a reduction in bone cell viability and bone angiogenesis. Mice with a deficit of telomerase exhibit cell senescence and osteopenia related to diminished osteoblast function and bone formation.[13] A decreased abundance of osteogenic factors, such as IGF-I and parathyroid hormone-related protein (PTHrP), has also been reported to occur in aged bone.

Oxidative stress and bone loss

Accumulated evidence in recent years indicates that oxidative stress is a pivotal pathogenetic culprit in aging and other osteoporosis-related conditions (e.g., DM). Augmented oxidative stress has been detected in both old mouse bone and primary osteoblasts from osteoporotic patients. Oxidative stress markers have been detected in plasma of subjects with reduced bone mass, and a positive association has been found between antioxidant vitamin C intake and BMD in postmenopausal women.[14] In fact, the bone-sparing effect of sex steroids seems to likely result from their antioxidant effects. A major consequence of reduced sex steroid levels is bone loss associated with an increased bone remodeling; most likely because estrogen or androgen deficiency induces both osteoblastogenesis and osteoclastogenesis, the latter exceeding the former due to the prolonged life span of osteoclasts and reduced life span of osteoblasts. In this scenario, a decrease in oxidative stress defenses has been described in osteoporotic bone. Oxidative stress also appears to contribute to diabetic osteopenia. In a mouse model of this condition, high levels of 8-hydroxydeoxyguanosine, a marker of oxidative DNA damage, have been detected in osteoblasts and in urine whose levels were inversely associated with histomorphometric parameters of bone formation.[15]

In both aging and diabetic subjects, mitochondria produce an excess of ROS, namely $O_2^{\bullet-}$, hydroxyl radical ($^{\bullet}OH$), and H_2O_2. The role of ROS in aging has been reconsidered in recent years. Thus, ROS can be regarded as pro-survival effectors, which are primarily triggered for homeostatic purposes; as ROS accumulate beyond a certain threshold, as occurs in elder and diabetic subjects, they contribute to deterioration of the cell status (Figure 12.2).

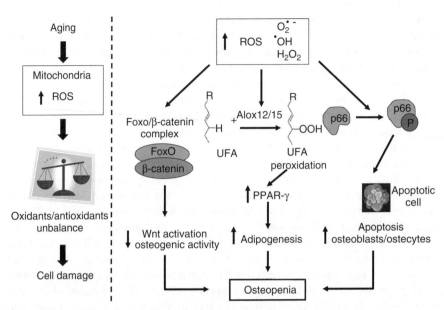

Figure 12.2 General mechanism of aging and bone damage produced by oxidative stress. UFA, unsaturated fatty acid.

H_2O_2 is by far the most stable and thus abundant ROS, which can induce osteoblast apoptosis through a mechanism involving the adaptor protein p66[Shc]. In response to H_2O_2, this protein is released from an inhibitor complex in the inner mitochondrial membrane. Free monomeric p66[Shc] induces the opening of the mitochondrial permeability transition pore leading to the rupture of mitochondrial integrity and releasing proapoptotic factors. Moreover, a feedback loop occurs based on further accumulation of ROS by p66[Shc]-dependent activation of membrane NADPH oxidases and downregulation of ROS scavenging enzymes by inhibiting Forkhead box O (FOXO) transcription factors.[12] The latter factors are cell defense proteins triggered by oxidative stress that exhibit a winged-helix DNA-binding domain named Forkhead box and have different functions at distinct stages of the osteoblast lineage. In mature osteoblasts, oxidative stress promotes FOXOs translocation into the nucleus upon posttranslational modifications by stress-related kinases including JNK. In the nucleus, FOXOs regulate the transcription of antioxidant enzymes such as SOD and catalase as well as DNA repair genes, decreasing oxidative stress and osteoblast apoptosis. On the other hand, FOXO activation in osteoprogenitors results in the inhibition of bone formation through interaction with Wnt/β-catenin signaling. This pathway promotes the progression of osteoblast precursor cells to bone-producing osteoblasts through the association of β-catenin with the T-cell factor (TCF)/lymphoid-enhancer binding factor family of transcription factors that regulates the expression of osteogenic genes. Binding of β-catenin to FOXOs diverts the limited pool of β-catenin from TCF- to FOXO-mediated transcription and thus decreases osteoblastogenesis. Consistent with these concepts, aging mice display oxidative stress associated with an increase in the FOXO-dependent antioxidant compensatory response. As chronological age advances and ROS increase, FOXO may eventually aggravate rather than alleviate the age-associated damage to bone by diminishing osteogenic Wnt signaling, thus decreasing bone formation.

Lipid oxidation induced by oxidative stress also seems to have an important role in age-related osteopenia. Lipid-oxidizing activity of Alox12 and Alox15 lipoxygenases and 4-hydroxynonenal, a product of lipid peroxidation, increase in old mouse bone associated with osteoblast apoptosis. It has also been reported that urinary excretion levels of 8-iso-prostaglandin $F_{2\alpha}$, a major prostaglandin derivative generated by nonenzymatic free radical-catalyzed oxidation of arachidonic acid and a reliable lipid peroxidation biomarker, were negatively associated with BMD in a wide age-range population of men and women. Moreover, lipid oxidation has been reported to inhibit the osteogenic action of factors such as bone morphogenetic protein-2 and PTH but activate the restraining effects of peroxisome proliferator-activated receptor-γ (PPAR-γ) on osteoblastogenesis. Current data also indicate that the product of Klotho gene, a coreceptor for fibroblastic growth factor (FGF)-23 produced by osteocytes, may act as a modulator of oxidative stress and cell aging.[16] The extracellular domain of Klotho protein can be secreted and exerts endocrine actions, including protection against oxidative stress through FOXO activation. Mice deficient in either Klotho or FGF-23 present premature aging and osteopenia with reduced cortical thickness and low bone formation and remodeling, as occurs in involutional osteoporosis.

Acknowledgments

RETICEF RD12/0043/0013, RETICEF RD12/0043/0008, PI11/00449 (MINECO, ISCIII-FEDER), SAF2013-48724-R (MINECO, FEDER) and PROMETEOII/2014/071 (Generalitat Valenciana).

Multiple choice questions

1 Rheumatoid arthritis is characterized by
   a. Mitochondrial homeostasis
   b. Downregulation of advanced glycation end-products
   c. Increased expression of antioxidant systems
   d. Chronic changes in the redox status of immune cells
   e. Apoptosis of synovial cells

2 What statement is correct concerning osteoarthritis?
   a. Aging is an important risk factor
   b. Inflammation does not contribute to disease
   c. Collagen synthesis is enhanced by oxidative stress
   d. The repair ability of chondrocytes is normal
   e. There is a predominance of anabolic processes in the joint

3 The major mechanism causing osteoblastic dysfunction in age-related osteopenia is:
   a. Increased osteoclastogenesis
   b. Oxidative stress

   c. Augmented production of endogenous glucocorticoids
   d. Immobilization
   e. Fractures

References

1 Firestein, G.S. (2003) Evolving concepts of rheumatoid arthritis. *Nature*, 423(6937), 356–361.

2 Filippin, L.I., Vercelino, R., Marroni, N.P. *et al.* (2008) Redox signalling and the inflammatory response in rheumatoid arthritis. *Clinical and Experimental Immunology*, 152(3), 415–422.

3 Bennett, S.J. & Griffiths, H.R. (2013) Regulation of T cell functions by oxidative stress. In: Alcaraz, M.J., Gualillo, O. & Sanchez-Pernaute, O. (eds), *Studies on Arthritis and Joint Disorders*. Oxidative Stress in Applied Basic Research and Clinical Practice Armstrong, D. Ed. Humana Press, New York, pp. 33–48.

4 Kasic, T., Colombo, P., Soldani, C. *et al.* (2011) Modulation of human T-cell functions by reactive nitrogen species. *European Journal of Immunology*, 41(7), 1843–1849.

5 Kageyama, Y., Takahashi, M., Ichikawa, T. *et al.* (2008) Reduction of oxidative stress marker levels by anti-TNF-alpha antibody, infliximab, in patients with rheumatoid arthritis. *Clinical and Experimental Rheumatology.*, 26(1), 73–80.

6 Asehnoune, K., Strassheim, D., Mitra, S. *et al.* (2004) Involvement of reactive oxygen species in toll-like receptor 4-dependent activation of NF-kappaB. *Journal of Immunology*, 172(4), 2522–2529.

7 Abramson, S.B., Attur, M., Amin, A.R. *et al.* (2001) Nitric oxide and inflammatory mediators in the perpetuation of osteoarthritis. *Current Rheumatology Reports*, 3(6), 535–541.

8 Goldring, M.B. & Marcu, K.B. (2009) Cartilage homeostasis in health and rheumatic diseases. *Arthritis Research and Therapy*, 11(3), 224.

9 Henrotin, Y.E., Bruckner, P. & Pujol, J.P. (2003) The role of reactive oxygen species in homeostasis and degradation of cartilage. *Osteoarthritis and Cartilage*, 11(10), 747–755.

10 Yin, W., Park, J.I. & Loeser, R.F. (2009) Oxidative stress inhibits insulin-like growth factor-I induction of chondrocyte proteoglycan synthesis through differential regulation of phosphatidylinositol 3-Kinase-Akt and MEK-ERK MAPK signaling pathways. *Journal of Biological Chemistry*, 284(46), 31972–31981.

11 Khosla S, Melton LJ III,, Riggs B. Lawrence. 2010. The unitary model for estrogen deficiency and the pathogenesis of osteoporosis: is a revision needed?. *Journal of Bone and Mineral Research* 26(3):441–451.

12 Manolagas, S.C. (2010) From estrogen-centric to aging and oxidative stress: a revised perspective of the pathogenesis of osteoporosis. *Endocrine Reviews*, 31(3), 266–300.

13 Saeed, H., Abdallah, B.M., Ditzel, N. *et al.* (2011) Telomerase-deficient mice exhibit bone loss owing to defects in osteoblasts and increased osteoclastogenesis by inflammatory microenvironment. *Journal of Bone and Mineral Research*, 26(7), 1494–1505.

14 Cervellati, C., Bonaccorsi, G., Cremonini, E. *et al.* (2013) Bone mass density selectively correlates with serum markers of oxidative damage in post-menopausal women. *Clinical Chemistry Laboratory Medicine*, 51(2), 333–338.

15 Hamada, Y., Kitazaw, S., Kitazawa, R. *et al.* (2007) Histomorphometric analysis of diabetic osteopenia in streptozotocin-induced diabetic mice: a possible role of oxidative stress. *Bone*, 40(50), 1408–1414.

16 Kuro-o, M. (2009) Klotho and aging. *Biochimica Biophysica Acta*, 1790(10), 1049–1058.

CHAPTER 13

Gene therapy to reduce joint inflammation in horses

Patrick Colahan[1], Rachael Watson[2], and Steve Ghivizzani[2]

[1] Department of Large Animal Clinical Sciences, College of Veterinary Medicine, University of Florida, Gainesville, FL, USA

[2] Department of Orthopedics and Rehabilitation, College of Medicine, University of Florida, Gainesville, FL, USA

THEMATIC SUMMARY BOX

At the end of this chapter, the student should be able to:

- Know the limitations of rodent models

- Understand an alternative large animal model for orthopedic research

- Understand the process of investigating a potential therapeutic agent

- Describe how gene therapy can be used to modify the degenerative processes of arthritis

- Understand the current status of the application of gene therapy to the treatment of arthritis

---

Introduction

The study of gene therapy for the treatment of arthritis has been conducted using horses in the animal model. The rationale for the selection of the animal model, the decisions guiding the investigation of cDNA gene introduction for therapy, and the current status of the development of the therapy are described as an example of an investigatory process used to develop a new therapy.

Animal model considerations

The development of a therapy for a human condition obviously requires the investigation of that therapy in an animal model. Most of such studies are conducted in

---

Oxidative Stress and Antioxidant Protection: The Science of Free Radical Biology and Disease, First Edition.
Edited by Donald Armstrong and Robert D. Stratton.
© 2016 John Wiley & Sons, Inc. Published 2016 by John Wiley & Sons, Inc.

rodents or rabbits. These species offer many advantages. They are relatively inexpensive to produce. They can be bred or, particularly in the case of rodents, genetically manipulated to produce very specific phenotypes and genotypes. Research animal care facilities have extensive experience housing and caring for these rodents and rabbits. Society does not highly value these species, so that their use in research does not provoke public reaction.

The shortcomings of rodent and rabbit models have become increasingly obvious over decades of research, however. High-profile studies taken from rodents to clinical trials in humans have demonstrated a number of serious failures. Even given appropriate animal numbers and research design, the metabolism, physiology, and biomechanics of these small animals may not sufficiently mimic the attributes of human beings to reliably permit the direct application of therapeutic responses developed in rodent and rabbit animal models to the treatment of human beings.[1,2]

The use of animals other than rodents and rabbits in the development of therapeutic agents is not new, but occurs in a small portion of research using animal models. Using large domestic species as research models poses a number of problems. Animal care facilities are not accustomed to housing or caring for these species, particularly cattle and horses. They are individually more expensive than smaller species. There is also some public resistance to the use of horses as research animals.

The large domestic species, such as sheep, cattle, and pigs, have all been used as animal models. Horses have been used rarely, but offer excellent animal models for the study of orthopedic problems especially arthritis. Unlike rodents and rabbits and even the smaller food animal species, sheep and pigs, horses have joints of similar size to humans and unlike cattle can be handled without extensive specialized facilities. The dense connective tissues of the joint, the articular surfaces, and the volume of the joint spaces approximate those of the supporting limb joints of humans more closely than any other readily available species. They suffer arthritic degeneration as a naturally occurring condition. The equine genome shows nearly 50% conserved synteny with human chromosomes.[3]

The nature and origins of horses

Horses are the most athletic domestic animal. The cardiopulmonary capacity of horses (160 ml/kg/min VO_2 max) is considerably greater than other athletic species such as humans (69–85 ml/kg/min VO_2 max), dogs (100 ml/kg/min), and racing camels (51 ml/kg/min VO_2 max). Horses are also the fastest of the domestic species capable of speeds up to 19 m/s besting people (10–11 m/s), dogs (16.6 m/s), and camels (10–11 m/s).[4]

There are hundreds of breeds of horses. The athletic abilities of the light breeds of horses, those bred for speed and not heavy draft work, are a result of evolution to occupy open grasslands where survival required the ability to travel distances and to outrun predators. Other breeds, those selected for draft work, originate from the European forest horse. There is tremendous variation among the breeds of domestic horses because of these diverse origins, human selection for traits, and interbreeding between breeds. The phenomic diversity between breeds is considerable, and basic traits such as muscle fiber type predominance, type I to II, can vary by 20% to 80%.[5] However, the variation within each breed is much less so that a specific breed can provide phenomic and genomic consistencies. For instance, 80% of the genome of Thoroughbred horses is derived from just 31 horses.[6]

As a model for joint disease, Thoroughbred horses or related breeds have been used. The Thoroughbred breed was developed in the 17th and 18th centuries and the lineage of each individual has been recorded by the breed registry beginning in 1791. Selective breeding has biomechanically and physiologically adapted the breed to high-speed exercise. The selection for speed has reduced the size of the extremities, thus reducing the energy required to move them and increasing the velocity of their movement. The loads imposed by the repetitive movement of locomotion on the anatomical components of the limbs are also increased and contribute to the prevalence of traumatic arthritis. Because the joints of horses are similar in size to the major joints, the hip and knee, of people and they suffer traumatic arthritis, they are becoming increasingly accepted as an appropriate model for traumatic arthritis in people.

Pathobiology of joint disease

The properly functioning joint has supple, smooth cartilage surfaces that minimize friction between the bones of the articulation. The maintenance of the optimized frictionless surface and the cartilage tissue requires an intricate organ system that is the joint.

Cartilage is a complex tissue composed primarily of ground substance, collagen, and chondrocytes. Proper function of cartilage depends upon the healthy chondrocytes to maintain the resilient ground substance and the smooth low friction surface of the tissue. The cartilage is supported by the underlying subchondral bone and nourished by the joint fluid produced by the synoviocytes lining the joint capsule.

The subchondral bone mechanically supporting the cartilage determines the geometry of the joint surface and provides cushioning of the forces imposed on the articular surfaces. Proper functioning of subchondral bone aids in protecting the cartilage from the compressive forces of locomotion. The cartilage is attached to the underlying bone by collagen fibers that arc through the ground substance of the cartilage and anchor into the bone. These fibers support the ground substance and attach the cartilage to bone countering the shear forces imposed by the articulating bones.

The synoviacytes of the synovial membrane produce a complex viscoelastic fluid that provides lubrication and nourishes the chondrocytes. Cartilage is avascular so that the joint fluid produced by the synovial fluid is the only source of nutrition for cartilage. The synovial membrane is a very metabolically active tissue playing a critical immunological role and producing synovial fluid.

Traumatically induced arthritis is initiated by mechanical injury to the bone, joint capsule, or ligaments of the joint. This can occur in an almost infinite number of ways and most frequently involves multiple joint components. Disruption of the mechanical integrity and stimulation of an inflammatory response within the joint leads to the biochemical and cellular processes described in the previous chapter.

On the tissue level, the synovial membrane becomes thickened initially with edema and inflammatory cell invasion. Overtime, the synovial membrane becomes fibrotic and the synoviacytes are lost. Consequently, the synovial fluid becomes less viscous and loses its nutritional value to the chondrocytes. The subchondral and periarticular bone respond to chronic mechanical or inflammatory insult. Subchondral bone typically becomes more dense with the addition of more mineral, losing flexibility and hence the ability to protect the articular cartilage from excessive compressive loading. The bone on the margins of the joint where the fibrous joint capsule attaches becomes proliferative and produces spur-like bone growth.

The cartilage being the least metabolically active tissue in the organ is at a great risk of damage as a result of the changes induced by mechanical stress and inflammation in the organ as a whole. Damage to the cartilage is, for the most part, irreversible. The proinflammatory cytokines, produced in response to the injury, induce reactive oxygen species (ROS) production and iNOS expression. Both ROS and reactive nitrogen species (RNS) lead to the suppression of glycosaminoglycan and collagen synthesis. Both of these are essential to the maintenance of cartilage ground substance. The expression of matrix metalloproteinases (MMPs) and the activation of proenzymes lead to cartilage breakdown. Loss of nutrition leads to injured chondrocytes and cell apoptosis. Continuing release of proinflammatory cytokines and enzymatic substances from damaged tissues leads to more synovial inflammation and joint degradation. This becomes a cyclic process in a progressive degenerative condition with ever worsening clinical signs.

Current therapy for traumatic arthritis

The stimulus for the development of a gene therapy is the inadequacy of current pharmaceutical therapies. Nonsteroidal anti-inflammatory drugs (NSAIDs) are the standard therapy in medicine and veterinary medicine. These drugs block the cyclooxygenase isoenzymes (COX I and COX II) and prevent the production of prostaglandins, blocking some of the signs of inflammation, vasodilation, and pain. Intra-articular injection of hyaluronan and the oral administration of chondroitin sulfate and glycosamine are common treatments and may help reduce inflammation or contribute in some incompletely described ways to the return to homeostasis in the joint. However, they produce short-lived therapeutic effects and are not listed as recommended therapies by the American Academy of Orthopedic Surgeons (AAOS).[7,8] Intra-articular administration of corticosteroid has some chondroprotective effect and is recommended by the AAOS, but has a limited duration of effect. All the available therapies are palliative. They do not effectively interdict the long-term, cyclic degenerative process in the joint.

Attempts to develop pharmacologic agents that disrupt the action of the proteases that breakdown the extracellular matrix of cartilage, such as stromelysins, aggrecanases, and collagenases, have not been successful. Advanced clinical trials of a drug to inhibit iNOS have been halted because of lack of efficacy. An effective therapy to end the degenerative processes of arthritis remains elusive.

Development of gene therapy in horses

The goal of gene therapy is to insert cDNA into the cells of the joint to induce the production of anti-inflammatory cytokines that would stop the cyclic inflammatory/degenerative processes of arthritis.[7]

A useful gene therapy has a number of essential properties. It must be safe. Neither the cDNA nor the transfecting agent can induce a noxious response, particularly a long-term or life-threatening response such as cancer or genetic mutation. It must be efficacious. It must induce long-term production, by a large population of cells, of high concentrations of an anti-inflammatory cytokine that interdicts the inflammatory/degenerative cycle. To create a therapy that meets these requirements, both the cDNA and transfecting agent must be tested and refined. To this end, a number of studies have been conducted primarily in horses to develop a gene therapy for traumatic arthritis.

The safety and efficacy of the cDNA depends on the selection of the appropriate anti-inflammatory mediator and the accurate replication of the coding sequence of the species. The genomes of both horses and humans have been sequenced so that species-specific cDNAs can be generated.[3]

Gene selection

Considerable research into the role of cytokines in the inflammatory process indicates that interleukin-1 (IL-1) concentrations are increased in osteoarthritic joints following trauma, and are produced within the joint by chondrocytes and synovial cells. The increased IL-1 concentrations drive the progression of osteoarthritis (OA).[8] IL-1 is known to be the most potent physiological inducer of chondrocytic chondrolysis.[9] At picogram concentrations, IL-1 inhibits extracellular matrix production in cartilage by blocking collagen type II and proteoglycan synthesis and by enhancing the rate of chondrocyte apoptosis.[10] Slightly higher concentrations induce proteolytic enzyme synthesis in chondrocytes, driving enhanced production of MMPs and aggrecanases that degrade the cartilaginous matrix.[11] IL-1 is a primary mediator of the inflammatory cascade and stimulates articular cells to produce numerous downstream OA effector molecules including cyclooxygenases I and II; nitric oxide (NO); phospholipase A2; prostaglandin E2 (PGE2); proinflammatory cytokines such as IL-6, IL-8, IL-15, and IL-18; and chemokines such as CXC-2, CXC-5, CXC-10 and CCL-3, CCL-5, CCL-7.[12] Release of these agents further stimulates cartilage matrix degradation and causes bone erosion, synovitis, fibrosis, and pain sensitivity.[12–15]

The pivotal role of IL-1 in the development of OA has been demonstrated *in vivo* in human cell culture models. Transection of the anterior cruciate ligament in dogs induces IL-1 synthesis by the synovium and articular chondrocytes, and, predictably, OA.[16–19] Studies conducted by Amin, Abramson, Attur and colleagues provide the most compelling evidence of the role of IL-1 in human OA.[20–22] Human articular cartilage recovered from OA knees and grown in cell culture demonstrated elevated NO,

PGE2, MMPs, and IL6 dependent on the exposure to IL-1. IL-1 is the critical inflammatory mediator, the prime target for interdiction of the inflammatory/degenerative cascade.

In the homeostatic joint, the proinflammatory action of IL-1 is blocked equally effectively by the anti-inflammatory action of IL-1 receptor antagonist (IL-1Ra) or soluble IL-1 type-II receptor antagonist (sIL-1RII). In the transected ACL model in dogs, repeated injection of IL-1Ra beginning shortly after transection effectively blocked MMP synthesis, suppressed early degenerative changes in the articular cartilage of the tibial plateau and femoral condyles, and reduced the number and size of periarticular osteophytes compared to the control joints.[23] Clinical trials of a recombinant IL-1Ra (anakinra, Kineret®) produced relief of clinical signs, but in a double-blinded trial against a placebo, the effect only lasted 4 days postinjection.[24]

IL-1Ra does not have the signaling activity provided by recruitment of IL-1R accessory protein (IL-1R-AcP) and has a half-life in the joint of less than 1 h. IL-1 has a potent spare signaling effect so that IL-1Ra must be maintained at 10–100 fold molar excess to block IL-1 effectively. Conversely, sIL-1RII binds and inactivates IL-1. However, the cDNA for sIL-1RII is larger than that of IL-1Ra. The cDNA for IL-1Ra is easily expressed at high concentrations and recombinant IL-1Ra is used therapeutically. Because it can be used to create a smaller transfecting packet that will readily enter cells and induce the synthesis of high concentrations of a protein that is know to be therapeutically effective, the gene selected for therapeutic development is that coding for IL-1Ra.

Delivery system

The ideal delivery system effectively carries the genetic material into the target cells, does not move outside the target organ, produces no local or systemic immune response, and carries no viral genetic material that would cause disease directly or induce genetic changes leading to disease. Both nonviral and viral options are available for use as a transfecting agent to carry the cDNA into the cells of the joints. Nonviral delivery systems are not effective, but a viral system using recombinant adeno-associated virus (AAV) has been shown to provide an effective, stealthy delivery system.

AAV vectors provide several advantages. (i) In people, the wild-type virus is not associated with any pathologic conditions. (ii) The recombinant form contains no native viral coding sequences, reducing the immunogenicity of transfected cells. (iii) AAV vectors can infect both dividing and quiescent cells. (iv) In many applications *in vivo*, persistent transgenic expression has been observed. (v) The recombinant form does not integrate into the genome of the target cell with significant frequency.[25] (vi) The small size of the viral particle (20–30 nm) permits penetration into then cells of tissue with dense extracellular matrix.

Refinement of the AAV vector was necessary to improve the slow onset of gene expression. A self-complementary, double-strand vector is much more effective, producing a rapid onset and a 20-fold increase in gene expression.[26] Packaging or pseudotyping the vector in different capsid serotypes alters the tropism of the vector and often improves transduction. It also varies the antigenicity and enables selection of least antigenic types for the initial treatment, allowing alternative capsid variants for repeat treatments. The refinement of the delivery system over 15 years has led to an AAV system that has an acceptable level of biosafety and a high degree of gene delivery efficiency.[27]

Studies developing gene therapy for arthritis

One model of traumatically induced arthritis in horses has been used for the studies of gene therapy. In this model, a small osteochondral fragment is created on the radial carpal bone using arthroscopic surgical techniques. Controlled exercise on a treadmill is initiated as soon after the surgery as the arthroscopic portals are healed. The carpus of the horse is the major joint in the forelimb approximately midway between the foot and the body. As the name suggests, the equine carpus is analogous anatomically to the human wrist and is composed of multiple bone and three major articulations. The model mimics a naturally occurring injury in horses that occurs with the stresses of high-speed exercise and results in osteoarthritis.[28–32]

The first study used a first-generation adenovirus vector and gene coding for human IL-1Ra. A clear chondroprotective effect was observed at 8 weeks, but a cell-mediated immune response to both the adenovirus and the human protein abbreviated the duration of IL-1Ra production.[33] While obvious problems limited the production of IL-1Ra, the feasibility of using direct viral-mediated IL-1Ra gene delivery was demonstrated.

A series of cell culture studies using human and equine fibroblasts identified several AAV capsid serotypes that were effective transducers (AAV1, AAV2, AAV3, AAV5) and some (AAV7, AAV8, AAV9) that were not. To test serotype efficiency *in vivo*, scAAV vector containing human IL-1Ra was packaged into AAV2, AAV5, and AAV8. Vector preparations containing ~2×10^{11} viral genomes of each serotype were injected into the intercarpal and metacarpal joints of several Thoroughbred horses. Human cDNA was selected because the reagents to measure human Il-1Ra were commercially available and those for equine IL-1Ra were not.

Measurement of the concentrations of human IL-1Ra in the synovial fluid using ELISA over a 10-week period showed that each serotype elevated the steady-state concentrations in synovial fluids to between 500 and 1500 pg/ml for at least 1 month. The AAV5 serotype produced about 2× higher expression than AAV2 or AAV8. However, all three serotypes generated functional levels of IL-1Ra even at the conservative dose administered.

To determine the cell types transfected, scAAV was packaged with genome for fluorescent green protein and injected into healthy forelimb joints of Thoroughbred horses. Following euthanasia 10 days after injection, fluorescence was clearly obvious in synovial cells particularly in the villi and even some chondrocytes. In later studies, injection of the same package into arthritic joints produced greatly increased transfection of all tissues but particularly cartilage in damaged areas. The increased transfection is likely due to reduced ECM and greater exposure of the chondrocytes to

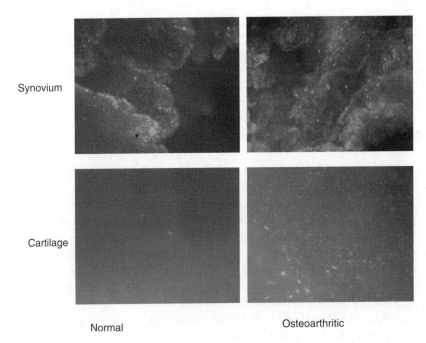

Figure 13.1 Normal and osteoarthritic synovium and cartilage transfected with scAAV packaged with genome coding for fluorescent green protein 10 days postinjection. *(See color plate section for the color representation of this figure.)*

the transfecting agent, as well as inflammation-induced activation of the regulatory sequences driving gene expression (Figure 13.1).

To determine the duration of increased IL-1Ra production and the effect of transfecting dose, optimized equine cDNA was packaged in AAV and injected into the joints of normal horses. Dose rates 1×10^{11} to 1×10^{13} viral genomes were used, and sampling was continued for more than 6 months. At dose rates of 1×10^{12} vg and above, IL-1Ra concentration in the joint fluid samples was greater than that previously demonstrated to be chondroprotective and persisted for more than 6 months after injection.[34]

Furthermore, no effect was noted in any control joints indicating that a treatment effect was contained in the injected joints. Subsequent studies indicated that the cDNA could not be found outside the fibrous joint capsule of the injected joint. No other organs including regional lymph nodes or spleen contained meaningful amounts of the injected cDNA.

Because the optimized equine cDNA packaged in a minimally antigenic AAV capsid effectively increased IL-1Ra concentrations to assumed therapeutic concentrations, produced no adverse responses locally or systemically, and all synthetic DNA material was contained to the treated joints, studies of the efficacy of the therapy in controlling the degenerative processes of arthritis were undertaken. A short-term efficacy study was conducted in 20 horses using the carpal chip model. Treatment was given 2 weeks following the creation of the osteochondral fragment. Monitoring of the joints with magnetic resonance imaging, radiographs, sequential sampling of joint fluid, subjective lameness evaluation, kinematic analysis, and repeat arthroscopic examination indicated that increased IL-1Ra concentrations in the joint fluid does occur and that the signs of osteoarthritis are effectively reduced during the 8 weeks following injection. A long-term study is currently underway to determine if gene therapy for arthritis using cDNA for IL-1Ra will effectively control the development of osteoarthritis following joint trauma.

Multiple choice questions

1 Cartilage is nourished by:
   a. Capillaries in the cartilage
   b. Synovial fluid
   c. Bone marrow

2 NSAIDs do not control the proteases that breakdown the extracellular matrix of cartilage:
   a. True
   b. False

3 The goal of gene therapy for arthritis is to inhibit the inflammatory cytokine:
   a. IL-1
   b. VEGF
   c. Retinaldehyde

References

1 Hackam, D.G. & Redelmeier, D.A. (2006) Translation of research evidence from animals to humans. *JAMA*, 296(14), 1727–1732.
2 van der Worp, H.B., Howells, D.W., Sena, E.S. *et al.* (2010) Can animal models of disease reliably inform human studies? *PLoS Medicine*, 7(3), e1000245.
3 Wade, C.M., Giulotto, E., Sigurdsson, S. *et al.* (2009) Genome sequence, comparative analysis, and population genetics of the domestic horse. *Science*, 326(5954), 865–7.
4 Derman, K.D. & Noakes, T.D. (1994) Comparative aspects of exercise physiology. In: Hodgeson, D.R. & Rose, R.J. (eds), *The Athletic Horse*. WB Saunders Co, Philadelphia, pp. 13–25.
5 Snow, D.H. & Guy, P.S. (1976) Percutaneous needle muscle biopsy in the horse. *Equine Veterinary Journal*, 8, 150.
6 Gaffney, B. & Cunningham, E.P. (1988) Estimation of genetic trend in racing performance of Thoroughbred horses. *Nature*, 332, 722–4.
7 Ghivizzani, S.C., Gouze, E., Gouze, J.N. *et al.* (2008) Perspectives on the use of gene therapy for chronic joint diseases. *Current Gene Therapy*, 8, 273–286.
8 Goldring, M.B., Otero, M., Tsuchimochi, K. *et al.* (2008) Defining the roles of inflammatory and anabolic cytokines in cartilage metabolism. *Annals of the Rheumatic Diseases*, 67(Suppl 3), iii75–iii82.
9 Hubbard, J.R., Steinberg, J.J., Bednar, M.S. *et al.* (1988) Effect of purified human interleukin-1 on cartilage degradation. *Journal of Orthopaedic Research*, 6, 180–187.
10 Dinarello, C.A. (2000) Proinflammatory cytokines. *Chest*, 118, 503–508.
11 Smith, R.L. (1999) Degradative enzymes in osteoarthritis. *Frontiers in Bioscience*, 4, D704–D712.

12 Gouze, J.N., Gouze, E., Popp, M.P. *et al.* (2006) Exogenous glucosamine globally protects chondrocytes from the arthritogenic effects of IL-1beta. *Arthritis Research and Therapy*, 8, R173.

13 Geng, Y., Blanco, F.J., Cornelisson, M. *et al.* (1995) Regulation of cyclooxygenase-2 expression in normal human articular chondrocytes. *Journal of Immunology*, 155, 796–801.

14 Mengshol, J.A., Vincenti, M.P., Coon, C.I. *et al.* (2000) Interleukin-1 induction of collagenase 3 (matrix metalloproteinase 13) gene expression in chondrocytes requires p38, c-Jun N-terminal kinase, and nuclear factor kappa B: differential regulation of collagenase 1 and collagenase 3. *Arthritis and Rheumatism*, 43, 801–811.

15 Stadler, J., Stefanovic-Racic, M., Billiar, T.R. *et al.* (1991) Articular chondrocytes synthesize nitric oxide in response to cytokines and lipopolysaccharide. *Journal of Immunology*, 147, 3915–3920.

16 Pond, M.J. & Nuki, G. (1973) Experimentally-induced osteoarthritis in the dog. *Annals of the Rheumatic Diseases*, 32, 387–388.

17 Fernandes, J.C., Martel-Pelletier, J. & Pelletier, J.P. (2002) The role of cytokines in osteoarthritis pathophysiology. *Biorheology*, 39, 237–246.

18 Pelletier, J.P., Faure, M.P., DiBattista, J.A. *et al.* (1993) Coordinate synthesis of stromelysin, interleukin-1, and oncogene proteins in experimental osteoarthritis. An immunohistochemical study. *American Journal of Pathology*, 142, 95–105.

19 Sadouk, M.B., Pelletier, J.P., Tardif, G. *et al.* (1995) Human synovial fibroblasts coexpress IL-1 receptor type I and type II mRNA. The increased level of the IL-1 receptor in osteoarthritic cells is related to an increased level of the type I receptor. *Laboratory Investigation*, 73, 347–355.

20 Attur, M.G., Dave, M., Cipolletta, C. *et al.* (2000) Reversal of autocrine and paracrine effects of interleukin 1 (IL-1) in human arthritis by type II IL-1 decoy receptor. Potential for pharmacological intervention. *Journal of Biological Chemistry*, 275, 40307–40315.

21 Attur, M.G., Dave, M.N., Leung, M.Y. *et al.* (2002) Functional genomic analysis of type II IL-1beta decoy receptor: potential for gene therapy in human arthritis and inflammation. *Journal of Immunology*, 168, 2001–2010.

22 Attur, M.G., Patel, I.R., Patel, R.N. *et al.* (1998) Autocrine production of IL-1 beta by human osteoarthritis-affected cartilage and differential regulation of endogenous nitric oxide, IL-6, prostaglandin E2, and IL-8. *Proceedings of the Association of American Physicians*, 110, 65–72.

23 Caron, J.P., Fernandes, J.C., Martel-Pelletier, J. *et al.* (1996) Chondroprotective effect of intraarticular injections of interleukin-1 receptor antagonist in experimental osteoarthritis. Suppression of collagenase-1 expression. *Arthritis and Rheumatism*, 39, 1535–1544.

24 Chevalier, X., Goupille, P., Beaulieu, A.D. *et al.* (2009) Intraarticular injection of anakinra in osteoarthritis of the knee: a multicenter, randomized, double-blind, placebo-controlled study. *Arthritis and Rheumatism*, 61, 344–352.

25 Choi, V.W., McCarty, D.M. & Samulski, R.J. (2005) AAV hybrid serotypes: improved vectors for gene delivery. *Current Gene Therapy*, 5, 299–310.

26 Li, C., Diprimio, N., Bowles, D.E. *et al.* (2012) Single amino acid modification of adeno-associated virus capsid changes transduction and humoral immune profiles. *Journal of Virology*, 86(15), 7752–7759.

27 Bowles, D.E., McPhee, S.W., Li, C. *et al.* (2012) Phase 1 gene therapy for Duchenne muscular dystrophy using a translational optimized AAV vector. *Molecular Therapy*, 20, 443–455.

28 Frisbie, D.D., Al-Sobayil, F., Billinghurst, R.C. *et al.* (2008) Changes in synovial fluid and serum biomarkers with exercise and early osteoarthritis in horses. *Osteoarthritis and Cartilage*, 16, 1196–1204.

29 Frisbie, D.D., Kawcak, C.E., McIlwraith, C.W. *et al.* (2009) Evaluation of polysulfated glycosaminoglycan or sodium hyaluronan administered intra-articularly for treatment of

horses with experimentally induced osteoarthritis. *American Journal of Veterinary Research*, 70, 203–209.

30 Frisbie, D.D., McIlwraith, C.W., Kawcak, C.E. *et al.* (2009) Evaluation of topically administered diclofenac liposomal cream for treatment of horses with experimentally induced osteoarthritis. *American Journal of Veterinary Research*, 70, 210–215.

31 Kawcak, C.E., Frisbie, D.D., Werpy, N.M. *et al.* (2008) Effects of exercise vs experimental osteoarthritis on imaging outcomes. *Osteoarthritis and Cartilage*, 16, 1519–1525.

32 Kawcak, C.E., Norrdin, R.W., Frisbie, D.D. *et al.* (1998) Effects of osteochondral fragmentation and intra-articular triamcinolone acetonide treatment on subchondral bone in the equine carpus. *Equine Veterinary Journal*, 30, 66–71.

33 Frisbie, D.D., Ghivizzani, S.C., Robbins, P.D. *et al.* (2002) Treatment of experimental equine osteoarthritis by in vivo delivery of the equine interleukin-1 receptor antagonist gene. *Gene Therapy*, 9, 12–20.

34 Gouze, E., Gouze, J.N., Palmer, G.D. *et al.* (2007) Transgene persistence and cell turnover in the diarthrodial joint: implications for gene therapy of chronic joint diseases. *Molecular Therapy*, 15, 1114–1120.

CHAPTER 14

Muscle and oxidative stress

Reza Ghiasvand and Mitra Hariri

Department of Community Nutrition, School of Nutrition and Food Sciences, Isfahan University of Medical Sciences, Isfahan, Iran

---

THEMATIC SUMMARY BOX

At the end of this chapter, students should be able to:

- Understand how free radicals are produced during exercise

- Be able to describe the effect of free radicals on muscles

- Be able to discuss the role of vitamins C and E in oxidative stress reduction

- Understand which antioxidant nutrients can help in myopathy and muscle dystrophy treatment

- Understand which factors may cause muscle cramps and how nutrition helps to relieve it

- Be able to give nutritional recommendations to athletes and patients for preventing muscle disease

- Search and find new diets and supplements for oxidative stress reduction

---

The well-documented benefits of regular physical exercise include reduced risk of cardiovascular disease, cancer, osteoporosis, and diabetes.[1] Beneficial effects caused by exercise are decreased adipose tissue, altered lipid and hormonal profiles, receptor and transport-protein adaptations, improved mitochondrial coupling, and alterations to antioxidant defenses.[2-4] Aerobic organisms increase the production of reactive oxygen species (ROS) during normal respiration and inflammation so antioxidant defenses are necessary.[5] Exercise can create an imbalance between oxidant and antioxidant levels, a situation known as oxidative stress.[6,7] The generation of oxygen-free radicals and lipid peroxides increases during exercise.[7,8] Produced free radicals in an unconditioned individual will induce oxidative damage and result in muscle injury.[8] New studies show increased oxygen consumption in prolonged exercise increases the production of superoxide radicals that results in lipid peroxidation by-products in muscle and whole body; therefore, these mediators are associated with deleterious effects in host.[9,10] Recent *in vivo* and *in vitro* animal and human studies provide evidence that the effect of free radical generation during

---

Oxidative Stress and Antioxidant Protection: The Science of Free Radical Biology and Disease, First Edition. Edited by Donald Armstrong and Robert D. Stratton.
© 2016 John Wiley & Sons, Inc. Published 2016 by John Wiley & Sons, Inc.

and after exercise depends on the rate of oxygen consumption and presence of cellular antioxidant defense mechanisms.[11,12] Free radicals may cause muscle protein break down, elevation of lipid peroxidation, and depletion of muscle antioxidants, but the exact mechanisms of this oxidative stress has not been discovered yet and are under investigation in laboratories around the world.

Loss of Ca^{2+} homeostasis may be an important event in increasing exercise-caused free radical generation. Decrease in cellular thiols and enhancement of intracellular Ca^{2+} may raise free radical generation, membrane lipid peroxidation, and exhaustion of intracellular enzymes.[13] Cannon *et al.* discovered a significant positive correlation between superoxide radicals and plasma creatine kinase activity by using a single bout of downhill running in men and women.[13] In addition, exhaustive running among animals may cause decreased oxidative capacity in adipose tissue mitochondria.[14]

Several sources may produce free radicals during or after exercise: (i) the mitochondrial oxygen radicals that avoided antioxidant enzymes present in the mitochondria may enter the sarcoplasm; (ii) the capillary endothelium, because hypoxia or reoxygenation process during exercise may create free radicals; and (iii) muscle or tissue damage during exercise causes inflammation in cells and create an oxidative burst from inflammatory cells.[15]

The results of different studies show that skeletal muscle fibers contain a high antioxidant capacity and a large amount of antioxidant enzymes to reduce the damaging effects of ROS.[16] This reflects the fact that free radical formation occurs in a large amount of muscle.

Oxidative stress does not just cause pathological conditions. Recent evidence suggests ROS also act as important signaling molecules in muscle contraction and adaptation.[17] The effects of ROS are dose dependent and, at high levels, develop toxic effects on the cell and exert profound changes in gene expression. ROS can reduce lean body mass by stimulating the expression and activity of skeletal muscle protein degradation pathways.[18] These compounding factors of oxidative stress may ultimately lead to muscle wasting and may cause muscular diseases such as dystrophy, myopathy, soreness, and cramps. Multiple enzymatic and nonenzymatic antioxidant defense systems are present in cells to protect the membranes and other cell organelles from the damaging effects of free radical reactions. These include vitamins C and E, coenzyme Q_{10}, superoxide dismutase (SOD), and glutathione peroxidase. Many studies have been done on the effect of antioxidants and muscle disorders, but most are on vitamins C and E.[19,20] The purpose of this chapter is to describe muscular disease that may be caused by oxidative stress and how nutrition can help in the treatment of these diseases. We also discuss the most popular antioxidants in relation to muscle.

Vitamins E and C and oxidative stress in muscle

All cell membranes have vitamin E (α-tocopherol) as an antioxidant. The inner mitochondrial membrane is the main store place for vitamin E. Studies show the deficiency of vitamin E increases susceptibility to free radical damage during exercise in rats and leads to premature fatigue (40% decline in endurance capacity) because vitamin E deficiency depresses respiratory control of muscle mitochondria and it cause vulnerable of lysosomal membranes.[21,22] The vitamin E content of skeletal muscle is 50% of that seen in liver, heart, and lung tissue (20–30 nmol/g). Meydani *et al.* measured the vitamin E concentration in skeletal muscle after vitamin E supplementation.[23] All participations took 800 IU α-tocopherol/day (800 mg/day) for 4 weeks. After 15 days, the plasma concentration of α-tocopherol increased 300% but γ-tocopherol concentration decreased 74% and was maintained at this plateau until the end of study. Muscle biopsies indicated a significant increase in α-tocopherol (53%) and decrease in γ-tocopherol after supplementation in comparison with baseline values (from 37.6 ± 7.0 to 57.3 ± 12.1 nmol/g, $P < 0.0001$). Scientists believe a significant inverse relation between plasma α-tocopherol concentration and the percentage of type I muscle fibers; therefore, this inverse relation may prove that to reduce oxidative stress, a person with high physical activity and a high percentage of type I fibers may have a greater requirement for vitamin E than those with more type II glycolytic fibers.[15] Type I fibers, also called slow twitch or slow oxidative fibers, contain a large amount of myoglobin, many mitochondria, and many blood capillaries. Type I fibers are red, use ATP at a slow rate, have a slow contraction velocity, are very resistant to fatigue, and have a high capacity to generate ATP by oxidative metabolic processes. Such fibers are found in a large number in the postural muscles of the neck. Type IIA fibers, also called fast twitch or fast oxidative fibers, contain a very large amount of myoglobin, many more mitochondria, and many more blood capillaries. Type IIA fibers are red, have a very high capacity for generating ATP by oxidative metabolic processes, split ATP at a very rapid rate, have a fast contraction velocity, and are resistant to fatigue. Such fibers are infrequently found in humans. Type IIB fibers, also called fast twitch or fast glycolytic fibers, contain a low content of myoglobin, relatively few mitochondria, relatively few blood capillaries, and a large amount of glycogen. Type IIB fibers are white, geared to generate ATP by anaerobic metabolic processes, not able to supply skeletal muscle fibers continuously with sufficient ATP, fatigue easily, use ATP at a fast rate, and have a fast contraction velocity. Such fibers are found in a large number in the muscles of the arms.

During oxidative stress, vitamin E in cell membranes scavenges ROS, becoming a radical. Cytosolic water-soluble antioxidants such as vitamin C reduce vitamin E radicals by donating an electron.[24] The results of human and animal studies show that after-exercise antioxidants such as plasma vitamins C and E and uric acid increase, but acute submaximal exercise has been shown to decrease vitamin E concentrations in skeletal muscle.[25–27] However, only a few human studies have examined the interaction between vitamin E and exercise. The results suggest that vitamin E supplementation decreases oxidative stress and rates of lipid peroxidation, and vitamin E requirements may increase with exercise.[28] The amount of lipid peroxidation is increased after exercise, and vitamin E supplementation can reduce this increased lipid peroxidation. Sumida *et al.* examined the effect of vitamin E supplementation on oxidative stress and found that supplementation with vitamin E decreases the increased amount of circulating aspartate transaminase, β-glucuronidase, and the rate of lipid peroxidation caused by exercise.[29] The beneficial effect of vitamin E on oxidative stress after exercises was shown by Meydani *et al.* in young and older men and in women.[30] These scientists showed vitamin E supplementation 800 IU

(800 mg) β-tocopherol/day for 48 days decreased oxidative injury caused by exercise in all participants. Data of this study verify the result of other studies about increasing concentrations of vitamins E and C after supplementation, but after correction for changes in plasma volume, the differences were no longer significant. These data suggest that previously observed increases in these vitamins (none of which were corrected for plasma volume changes) were most likely a result of exercise-induced hemoconcentration. Vitamin E supplementation may help in reducing the amount of membrane damage because of oxidative stress in untrained subjects, but there is little evidence that vitamin E benefits high-performance athletes.[31,32] Yet, there have not been any well-controlled studies about vitamin E supplementation and oxidative stress that can prove ergogenic effects of vitamin E supplementation among athletics. A review of the effect of vitamin E on athletics' performance, such as maximal oxygen uptake, muscle strength, swimming endurance, or blood lactate concentrations, and cardiorespiratory fitness tests showed no beneficial effect.[33] Other studies showed a potential ergogenic effect of vitamin E, for it was found that vitamin E supplementation was associated with an increased anaerobic threshold and decreased expired pentane production during cycle ergometry in high-altitude environments.[34] The result of a relatively large number of well-conducted investigations is that vitamin C has no ergogenic effect in athletes without vitamin C deficiency. Therefore, vitamin supplements may have beneficial effects on improvement in time to run a specific distance or increased maximal oxygen uptake among athletics especially with antioxidant deficiency.[35]

Muscular disease and nutrition

Muscular dystrophy

Muscular dystrophy (MD) is a family of muscle diseases that weaken the muscles responsible for movement. Muscular dystrophies are defined as very progressive disease that make skeletal muscles very weak and are responsible for the death of muscle cells and tissue, so these diseases cause deficiency in muscle proteins. National Institute of Neurological Disorders and Stroke explains muscular dystrophy as a genetic sickness that causes progressive weakness and degeneration of the musculoskeletal system.[35]

Duchenne muscular dystrophy (DMD) is the most common form of muscular dystrophy. In this disease, dystrophin is the gene responsible for producing the affected protein. Studies show there is no treatment for muscular dystrophy, but a diet with high amount of protein, vitamin and mineral supplements, and herbs may perhaps improve muscle strength.[36] Scientists believe oxidative stress is a main pathologic factor in DMD because skeletal muscles of these patients show increased oxidative damage markers such as by-products of lipid peroxidation and carbonyls.[37] Several symptoms of DND, including muscle fatigue and cardiomyopathy, may be the result of increased oxidative stress in the muscles. One potential therapy is coenzyme Q_{10} (CoQ_{10}).[38] CoQ_{10} is a hydrophobic molecule that binds to the inner mitochondrial membrane and acts as an electron acceptor molecule for complexes I and II of the respiratory chain. CoQ_{10} supplementation and incorporation into mitochondria can fortify the antioxidative activity of NADH and create metabolic support to muscle.[38] The antioxidative activity of CoQ_{10} is able to reduce the excessive oxidative radicals and calcium overload in DMD muscles. In addition, CoQ_{10} can decrease calcium accumulation in mitochondria and regulates the mitochondrial transition pore.[39] Animal studies in exercised X-linked muscular dystrophy (mdx) mice showed that CoQ_{10} treatment increases the strength by 42% compared to that of controls. CoQ_{10} is presumably so important to muscle cells; contributes to growth control, cellular energy production, and other vital functions; and it deserve special attention for persons with muscular dystrophy.[38] A small quantity of CoQ_{10} is found in different foods. CoQ_{10} is made in most young people's body but a person with muscular dystrophy makes too little and has higher requirements because of the illness. Two placebo-controlled trials indicated that 100 mg CoQ_{10} daily in patient with muscular atrophies and dystrophies caused improved physical performance. Similar to CoQ_{10}, vitamin E as an antioxidant may help in the treatment of myopathy.[40] There are a lot of studies about the plasma level of vitamin E and effect of this vitamin in these patients. Recent studies show decreased blood levels of vitamin E and selenium in patients with muscular dystrophy, but until now scientists have not found significant result in vitamin E supplementation in muscular dystrophy treatment. Animal studies show vitamin E with selenium may be effective in improving major symptoms of muscular dystrophy. According to new studies, green tea polyphenols can act as antioxidants in myopathy disease and may be able to reduce the inflammation leading to degradation of muscles. Scientists suggest 250–500 mg/day of green tea can reduce inflammation[41] and oxidative stress in myopathy.[42]

Protein for myopathy patients is very essential because protein is needed for muscle growth, repair, and regeneration; therefore, protein is important in their diet. They should eat protein sources such as fish, lean meats, beans, and fewer fatty meats.[43,44] Leucine is a branched amino acid and is suitable for human consumption without any side effects. New evidence shows that leucine can be helpful in muscle regeneration in these patients.[45] They should avoid refined foods, such as whitened breads, sugars and pasta, and avoid coffee, alcohol, tobacco, and other stimulants. They should also avoid foods containing additives such as artificial colorings and preservatives and should completely remove potential food allergens, including dairy, wheat (gluten), corn, and soy.[46] Evidence shows inflammatory mediators are high in these patients; therefore, anti-inflammation supplement such as omega-3 fatty acids may be effective in this disease. Scientists recommend eating cold water, oily fish at least twice a week or take DHA and EPA supplement at least 1–2 g/day.[47] Other supplements such as calcium, magnesium, and vitamin D may be beneficial for preventing muscle and skeletal weakness.

Myopathy

Myopathy is a disease without any disorder of innervation or the neuromuscular junction, but muscles are in a pathological condition and cannot function normally.[48] Different forms of myopathy have been determined such as metabolic myopathy, inflammatory myopathy, and mitochondrial myopathy. Different factors may cause these disorders such as inflammation, endocrine, hereditary, metabolic, and toxic causes. Some conditions such as chronic inflammation during strenuous exercise or disease states may increase oxidative stress that causes muscle weakness leading to pathology of myopathy.[49] Muscle contraction is very sensitive to oxidative stress because proteins troponin, tropomyosin, myosin, and actin are sensitive to oxidation.[49] Some scientists believe the main cause of muscular damage in the myopathies is oxidative stress. Many studies prove the role of oxidative stress in magnifying the problems in various myopathies. Some authors even suggest oxidative stress as a causative basis.[50] The results of some studies show that as oxidative stress increases in these patients, the activity of oxidative enzymes such as superoxide dismutase, glutathione reductase, and superoxide dismutase decreases, so in order to reduce some complications, consuming antioxidants such as vitamin E, selenium, and CoQ_{10} may be effective. But patients should not substitute nutritional supplement for their medications.[51] Nutritionists recommend adding antioxidant supplements to the present medical regimen. One study showed that if patients consume antioxidants and vitamin D supplement, they will tolerate their medications better and nutritional supplements could even increase the pharmaceutical effect of the drugs.[52] Glutathione (GSH) is the most abundant thiol and endogenous antioxidant in the human body. Recent investigations have focused on GSH. GSH has three amino acids in its structure, and cysteine is a critical amino acid. Some evidence shows that the requirement for cysteine increases during periods of catabolic illness. In critically ill myopathy patients, the levels of GSH and glutamine decrease in muscle compared to control; therefore, such patients may benefit from supplementation with GSH or its precursors for myopathy treatment or prevention.[53] But no trials to date have examined the efficacy of this compound in treating or preventing myopathies.

In metabolic myopathies, energy-generating processes are disturbed and cause blocking of energy production, so muscle cells cannot work properly. There are 10 metabolic diseases of muscle, each one getting its name from the substance that it is lacking.

In addition to antioxidant supplementation, patients with metabolic myopathy may benefit from dietary changes.[53] If there are problems with carbohydrate processing, high-protein diet may be helpful, while those with difficulty processing fats may do well on a diet high in carbohydrates and low in fat.[54] Carnitine supplements can be very effective in reversing heart failure in this disorder especially when patient suffers from carnitine deficiency.[55] In other conditions, carnitine supplementation has not been therapeutic. The result of one case report study showed that a low-fat diet with nighttime uncooked cornstarch meal may be effective in the treatment of a patient with mutation in the carnitine phosphatetransferase 1A (CPT1A) enzyme.[56] Alternatively, the triheptanoin (anaplerotic) diet appears promising. One study evaluated the efficiency of triheptanoin in seven patients (10–55 years old) with CPT II deficiency. The daily percent for macronutrients in this diet for carbohydrate, fat, protein, and triheptanoin was 37, 20, 13, and 30, respectively. Patients received an extra dose of triheptanoin 30 min before strenuous exercise. The duration of study was 61 months, and just two patients experienced mild muscle pain with exercise after 61 months and two patients experienced rhabdomyolysis with exercise and none experienced rhabdomyolysis or hospitalizations after 61 months intervention. All patients had normal physical activities including strenuous sports.[57]

If the immune system attacks muscles, inflammatory myopathy may occur, making muscles weak and damaging muscle tissue. Systemic inflammation may trigger a catabolic/anabolic imbalance that results in skeletal muscle wasting and reduced muscle strength. Cytokine-mediated pathways, particularly TNF-α, can have a negative impact on muscle protein catabolism.[58] Some studies have proved that chronic inflammation may result in increased oxidative stress. Laboratory-based evidence suggests that purified polyphenol compounds such as resveratrol, epigallocatechin gallate, and curcumin that are found in red grapes, green tea, and turmeric, respectively, might be useful for the treatment of inflammatory myopathy.[42] The result of one study by Bonnefont-Rousselot *et al.* showed that plasma glutathione peroxidase (GSH-Px) activity, selenium, and vitamin E concentrations in patients with inflammatory myopathy were significantly lower than in controls.[59] Fedacko *et al.* studied the effect of 3 months intervention with selenium and CoQ_{10} on 60 patients with myopathy.[60] The results showed that plasma levels of CoQ_{10} increased in intervention groups compared with the placebo. Also, the symptoms of myopathy significantly improved in the active group, such as the intensity of muscle pain, muscle weakness, muscle cramps, and tiredness.

Apart from antioxidants, other dietary supplementations such as creatine may be effective in the treatment of inflammatory myopathies. New evidence indicated that creatine may have a potential role in the treating this disorder.[7,61] In one study, myopathy patients that received creatine supplement for 8 days in conjunction with exercise based on functional performance times and changes on magnetic resonance spectroscopy improved significantly in comparison with exercise alone.[62] Scientists believe myopathy may have an immune-mediated pathogenesis with gluten possibly responsible in the pathogenesis. Therefore, a diet without gluten may be as a useful therapeutic target.[63] Studies showed a gluten-free diet with antioxidant supplements can improve all abnormalities in muscle biopsy in these patients. If inflammatory myopathy is an autoimmune disease, vitamin D can be an important factor in the treatment of this disease because several autoimmune diseases such as type I diabetes mellitus, multiple sclerosis (MS), inflammatory bowel disease, systemic lupus erythematosus (SLE), and rheumatoid arthritis (RA) are associated with the low levels of vitamin D. Also, some studies show people with myopathy suffer from vitamin D deficiency.[64] Therefore, vitamin D supplementation may be helpful in the treatment and prevention of inflammatory myopathy. In this disease, patients may be under chronic corticosteroid therapy and need to make dietary modifications in order to minimize some side effects, such as fluid retention, hyperglycemia, and weight gain. Patients on prednisone therapy should consume a low-fat, low-carbohydrate, and low-salt diet. For the prevention of osteopenia, they should also take calcium (1 g/day) and vitamin D (400–800 IU/day) supplementation.[64–66]

Mitochondrial myopathy is a type of myopathy associated with mitochondrial disease. New studies show that antioxidants such as CoQ_{10}, selenium, vitamins C and E can treat mitochondrial myopathies. In one study by Sacconi, it was demonstrated the level of muscle CoQ_{10} in patients with mitochondrial myopathy was decreased compared to a control group.[67] Caso *et al.* indicated that CoQ_{10} for 30 days can decrease pain severity by 40% and pain interference with daily activities by 38%.[68] In one case-report study of a myopathy patient, after taking riboflavin (100 mg, three times per day), vitamin C (500 mg, twice daily), and ubiquinone (300 mg, three times per day) for 3 months the patient felt better, and the authors concluded that antioxidant supplements may replace statin drugs in the future.[69] Probiotics have been used therapeutically to modulate immunity, improve digestive processes, lower cholesterol, treat rheumatoid arthritis, and prevent cancer. Recent studies show that probiotics can improve inflammation, oxidative stress, and muscle damage in athletics, so according this evidence probiotics can be effective in improving myopathy.[70,71]

Ketogenic diet (KD) has been suggested as a possible treatment of mitochondrial disorders. This diet causes ketone body production by stimulating lipid utilization because carbohydrate content is very low but lipid content is very high in this diet. Ketone bodies can be used as an energy source by the brain, heart, and skeletal muscle.[45,72] Increased production of ketone bodies causes an increase in the transcript levels of genes associated with Krebs cycle and glycolysis, and are associated with mitochondrial biogenesis and increases the number of mitochondria in muscles.

Muscle cramps

Sometimes during or immediately after exercise, people, including athletes, may suffer from exercise-associated muscle cramping (EAMC). Statistics shows up to 95% of general population may be affected by EAMC, yet the cause remains unknown.

Muscles are the most dynamic organ in the body and the heart and arteries supply blood for them, but blood cannot come out of muscles unless muscles itself contract. Therefore, muscle contraction and relaxation are largely responsible for blood circulation through the muscle. When muscles are used, toxins such as lactic acid are produced in them. Therefore, blood circulation is essential for removing toxins from muscles. If too many toxins accumulate in muscles, the muscle response is to go into spasm.[73] The problem for people with muscle cramps is that blood supply is not adequate in muscles and toxins accumulate, and the reflex response of that muscle is to go into spasm. Sometimes, a muscle goes into spasm and remains in spasm for a few minutes (i.e., a cramp), so the blood supply is further impaired and there is a sudden and quick buildup of toxic metabolites, which causes more pain and spasm. Secondary muscle damage may happen by free radicals if a muscle starts to become painful and toxic metabolites increase in them. High levels of free radicals develop from different sources such as strenuous exercise, unhealthy diets, toxins, air, water, food pollution, and stress. Free radicals promote muscle cramps because of the damage to muscle fibers, and therefore having good antioxidant status helps protect against this secondary damage.[74]

During exercise lactic acid is produced in the muscles. When lactic acid increases in muscle in high amount, it can cause soreness, spasm, painful muscle cramps, and increase in recovery time needed between workouts.[75] Raven et al. compared the effect of four capsules of a dietary silicate mineral antioxidant supplement, Microhydrin (MH), in male bicyclists before racing with bicyclist who took a placebo.[76] Their result showed a statistically significant decrease in lactic acid among athletes compared to the placebo group, so these scientists suggested that the antioxidant may be helpful in reducing muscle cramps and hastening muscle recovery. In another study, authors studied the effect of 9.6 g of MH or placebo over 48 h in a randomized, double-blind, crossover design. Their results showed that MH did not have any effect on measured oxygen uptake, respiratory exchange ratio, and carbohydrate and fat oxidation rates; all blood parameters (lactate, glucose, and free fatty acids) were unaffected by MH supplementation. The volume of expired CO_2 and ventilation were significantly greater with MH supplementation. New evidence has shown that selenium deficiency is associated with muscle cramps in athletes.[77] Fedacko et al. showed that intake of CoQ_{10} and selenium for 8 weeks could reduce muscle cramps, muscle weakness, and pain in patients with statin-associated myopathy.[60] In one systematic review and meta-analysis by El-Tawil et al. on 23 articles about quinine (as an antioxidants) and vitamin E found that quinine (300 mg/day) compared to placebo significantly reduce cramp numbers over 2 weeks by 28% and cramp intensity by 70% and cramp days by 20%, but cramp duration was not significantly affected.[78] A quinine–vitamin E combination and vitamin E alone were not significantly different from quinine across at outcomes. Khajehdehi et al. showed that intake of vitamin E (400 mg/day and vitamin C (250 mg/day) for 8 weeks can reduce muscle cramps.[79]

Free radicals and other factors like dehydration, electrolyte imbalance, calcium and magnesium deficiency, and low carbohydrate stores may cause muscle cramps.

The dehydration/electrolyte imbalance

Fluid imbalances, dehydration, and serum electrolyte abnormalities are often related to EAMC. One of the popular explanations for EAMCs is a dehydration/electrolyte imbalance theory for it seems that fluid and ion shift from the extracellular space cause EAMCs.[80] New studies have indicated the serum electrolyte changes that occur with endurance exercise are related to EAMCs. In certain clinical conditions, serum electrolyte and fluid disturbances have been associated with the development of muscle cramps. Scientists believe that sodium (Na) losses and potassium (K) imbalances (e.g., hypokalemia, hyperkalemia) are the main factors for EAMCs.[81] However, the results of several studies have shown no differences in plasma K concentration between athletes with and without EAMCs.[82] In these studies, investigators often compared hematologic characteristics between crampers and noncrampers postexercise, but it is very important how quickly postexercise blood sampling was performed, because plasma K can return to normal concentration before 5 min postexercise. Potassium deficiency is very rare because it is an available mineral in common foods. Eating five or more servings of fruit and/or vegetables a day is the easiest way to be ensured of having abundant potassium. Preparing fat-free soup with fresh or frozen vegetables by boiling them very lightly makes a source loaded with potassium. Apples, grapes, and carrots have potassium in high level in comparison with other foods.

One prospective study of marathon runners did not show any association between EAMC and serum concentration of sodium, potassium, calcium, phosphate, bicarbonate, urea, or creatinine concentrations.[83]

Magnesium

Magnesium has been shown to play an important role in muscle and nerve function, and it is the most important electrolyte supplement for preventing skeletal muscle cramping in athletes.[84] If a person eats fresh fruit, vegetables, and legumes every-day, magnesium deficiency will not develop, but people with western diet that lacks fresh fruit, vegetables, legumes, and unprocessed grains and cereals may have magnesium deficiency. Therefore, athletes with muscle cramps should increase the intakes of green, leafy vegetables such as spinach, cabbage, lettuce, and broccoli.

Calcium

Some exercise experts believe calcium cannot play a role in muscle cramps because calcium could be released from muscles if dietary calcium intake were low; however, calcium deficiency is known as a responsible factor for impaired muscle contraction and muscle cramps.[85,86] Some evidence shows temporary imbalance of calcium in muscle during exercise that may cause muscle cramps.[73,87] A recent study presented that calcium supplementation may not be effective in treating leg cramps in pregnancy.[88] However, studies on athletes show that calcium supplementation can reduce muscle cramps. Two case report studies showed a hiker and a ballet dancer suffering from muscle cramps whose disorder disappeared after taking foods with high amount of calcium. Some athletes are vegan and vegetarian, and the amount of calcium is very low in these diets.[89–92]

Inadequate carbohydrate stores

Inadequate glycogen stores are also known to be a potential factor of muscle cramps. Scientists believe cramping may occur after long duration exercise because of glycogen store depletion.[93–95] Therefore, athletes with a history of cramping who want to participate in endurance sports should consume high-carbohydrate meals during exercise in days before and days following event.

Multiple choice questions

1 Which mineral may cause the increase in oxidative stress?
   a. Loss of Mg
   b. Loss of K
   c. Loss of Ca
   d. Loss of Zn

2 What is the most important amino acid for the muscle dystonia treatment?
   a. Arginine
   b. Leucine
   c. Phenylalanine
   d. Glycine

3 Which mineral may reduce in the diet of people with western diet and what it may cause?
   a. Calcium, cramps
   b. Potassium, cramps
   c. Potassium, myopathy
   d. Calcium, myopathy

References

1 Lee, I.M. & Paffenbarger, R.S. Jr., (2000) Associations of light, moderate, and vigorous intensity physical activity with longevity. The Harvard Alumni Health Study. *American Journal of Epidemiology*, 151(3), 293–299.

2 Leeuwenburgh, C., Hansen, P.A., Holloszy, J.O. & Heinecke, J.W. (1999) Hydroxyl radical generation during exercise increases mitochondrial protein oxidation and levels of urinary dityrosine. *Free Radical Biology & Medicine*, 27(1–2), 186–192.

3 Bejma, J., Ramires, P. & Ji, L.L. (2000) Free radical generation and oxidative stress with ageing and exercise: differential effects in the myocardium and liver. *Acta Physiologica Scandinavica*, 169(4), 343–351.

4 Alessio, H.M., Hagerman, A.E., Fulkerson, B.K., Ambrose, J., Rice, R.E. & Wiley, R.L. (2000) Generation of reactive oxygen species after exhaustive aerobic and isometric exercise. *Medicine and Science in Sports and Exercise*, 32(9), 1576–1581.

5 Uttara, B., Singh, A.V., Zamboni, P. & Mahajan, R.T. (2009) Oxidative stress and neurodegenerative diseases: a review of upstream and downstream antioxidant therapeutic options. *Current Neuropharmacology*, 7(1), 65–74.

6 McKenzie, M.J., Goldfarb, A., Garten, R.S. & Vervaecke, L. (2014) Oxidative stress and inflammation response following aerobic exercise: role of ethnicity. *International Journal of Sports Medicine*, 35, 822–7.

7 Canale, R.E., Farney, T.M., McCarthy, C.G. & Bloomer, R.J. (2014) Influence of acute exercise of varying intensity and duration on postprandial oxidative stress. *European Journal of Applied Physiology*, 114, 1913–24.

8 Isner-Horobeti, M.E., Rasseneur, L., Lonsdorfer-Wolf, E. *et al.* (2014) Effect of eccentric vs concentric exercise training on mitochondrial function. *Muscle & Nerve*, 50, 803–11.

9 Kanda, K., Sugama, K., Sakuma, J., Kawakami, Y. & Suzuki, K. (2014) Evaluation of serum leaking enzymes and investigation into new biomarkers for exercise-induced muscle damage. *Exercise Immunology Review*, 20, 39–54.

10 Faes, C., Balayssac-Siransy, E., Connes, P. *et al.* (2014) Moderate endurance exercise in patients with sickle cell anaemia: effects on oxidative stress and endothelial activation. *British Journal of Haematology*, 164(1), 124–130.

11 Watson, T.A., Callister, R., Taylor, R., Sibbritt, D., MacDonald-Wicks, L.K. & Garg, M.L. (2003) Antioxidant restricted diet increases oxidative stress during acute exhaustive exercise. *Asia Pacific Journal of Clinical Nutrition*, 12(Suppl), S9.

12 Watson, T.A., Callister, R., Taylor, R.D., Sibbritt, D.W., MacDonald-Wicks, L.K. & Garg, M.L. (2005) Antioxidant restriction and oxidative stress in short-duration exhaustive exercise. *Medicine and Science in Sports and Exercise*, 37(1), 63–71.

13 Cannon, J.G., Meydani, S.N., Fielding, R.A. *et al.* (1991) Acute phase response in exercise. II. Associations between vitamin E, cytokines, and muscle proteolysis. *The American Journal of Physiology*, 260(6 Pt 2), R1235–R1240.

14 Gohil, K., Henderson, S., Terblanche, S.E., Brooks, G.A. & Packer, L. (1984) Effects of training and exhaustive exercise on the mitochondrial oxidative capacity of brown adipose tissue. *Bioscience Reports*, 4(11), 987–993.

15 Evans, W.J. (2000) Vitamin E, vitamin C, and exercise. *The American Journal of Clinical Nutrition*, 72(2 Suppl), 647s–652s.

16 Requena-Mendez, A., Lopez, M.C., Angheben, A. *et al.* (2013) Evaluating chagas disease progression and cure through blood-derived biomarkers: a systematic review. *Expert Review of Anti-infective Therapy*, 11(9), 957–976.

17 Lam, T., Chen, Z., Sayed-Ahmed, M.M., Krassioukov, A. & Al-Yahya, A.A. (2013) Potential role of oxidative stress on the prescription of rehabilitation interventions in spinal cord injury. *Spinal Cord*, 51(9), 656–662.

18 Droge, W. (2005) Oxidative stress and ageing: is ageing a cysteine deficiency syndrome? *Philosophical Transactions of the Royal Society of London Series B. Biological Sciences*, 360(1464), 2355–2372.

19 Askari, G., Hajishafiee, M., Ghiasvand, R. *et al.* (2013) Quercetin and vitamin C supplementation: effects on lipid profile and muscle damage in male athletes. *International Journal of Preventive Medicine*, 4(Suppl 1), S58–S62.

20 Daneshvar, P., Hariri, M., Ghiasvand, R. *et al.* (2013) Effect of eight weeks of quercetin supplementation on exercise performance, muscle damage and body muscle in male badminton players. *International Journal of Preventive Medicine*, 4(Suppl 1), S53–S57.

21 Davies, K.J., Quintanilha, A.T., Brooks, G.A. & Packer, L. (1982) Free radicals and tissue damage produced by exercise. *Biochemical and Biophysical Research Communications*, 107(4), 1198–1205.

22 Venditti, P., Napolitano, G., Barone, D. & Di Meo, S. (2014) Vitamin E supplementation modifies adaptive responses to training in rat skeletal muscle. *Free Radical Research*, 48(10), 1179–1189.

23 Meydani M., Evans W.J., Handelman G. *et al.* (1993) Protective effect of vitamin E on exercise-induced oxidative damage in young and older adults. *The American Journal of Physiology* 264: R992–8.

24 Ji, L.L. (1995) Exercise and oxidative stress: role of the cellular antioxidant systems. *Exercise and Sport Sciences Reviews*, 23, 135–166.

25 Bailey, D.M., Evans, K.A., McEneny, J. *et al.* (2011) Exercise-induced oxidative-nitrosative stress is associated with impaired dynamic cerebral autoregulation and blood-brain barrier leakage. *Experimental Physiology*, 96(11), 1196–1207.

26 Draeger, C.L., Naves, A., Marques, N. *et al.* (2014) Controversies of antioxidant vitamins supplementation in exercise: ergogenic or ergolytic effects in humans? *Journal of the International Society of Sports Nutrition*, 11(1), 4.

27 Paulsen, G., Cumming, K.T., Holden, G. *et al.* (2014) Vitamin C and E supplementation hampers cellular adaptation to endurance training in humans: a double-blind, randomised, controlled trial. *The Journal of Physiology*, 592(Pt 8), 1887–1901.

28 Nikolaidis, M.G., Kerksick, C.M., Lamprecht, M. & McAnulty, S.R. (2012) Does vitamin C and E supplementation impair the favorable adaptations of regular exercise? *Oxidative Medicine and Cellular Longevity*, 2012, 707941.

29 Sumida, S., Tanaka, K., Kitao, H. & Nakadomo, F. (1989) Exercise-induced lipid peroxidation and leakage of enzymes before and after vitamin E supplementation. *The International Journal of Biochemistry*, 21(8), 835–838.

30 Meydani, M., Evans, W.J., Handelman, G. *et al.* (1993) Protective effect of vitamin E on exercise-induced oxidative damage in young and older adults. *The American Journal of Physiology*, 264(5 Pt 2), R992–R998.

31 Taghiyar, M., Ghiasvand, R., Askari, G. *et al.* (2013) The effect of vitamins C and E supplementation on muscle damage, performance, and body composition in athlete women: a clinical trial. *International Journal of Preventive Medicine*, 4(Suppl 1), S24–30.

32 Martinovic, J., Dopsaj, V., Kotur-Stevuljevic, J. *et al.* (2011) Oxidative stress biomarker monitoring in elite women volleyball athletes during a 6-week training period. *Journal of Strength and Conditioning Research/National Strength & Conditioning Association*, 25(5), 1360–1367.

33 Clarkson, P.M. & Thompson, H.S. (2000) Antioxidants: what role do they play in physical activity and health? *The American Journal of Clinical Nutrition*, 72(2 Suppl), 637s–646s.

34 Gomes, E.C., Allgrove, J.E., Florida-James, G. & Stone, V. (2011) Effect of vitamin supplementation on lung injury and running performance in a hot, humid, and ozone-polluted environment. *Scandinavian Journal of Medicine & Science in Sports*, 21(6), e452–60.

35 Braakhuis, A.J. (2012) Effect of vitamin C supplements on physical performance. *Current Sports Medicine Reports*, 11(4), 180–184.

36 Davidson, Z.E. & Truby, H. (2009) A review of nutrition in Duchenne muscular dystrophy. *Journal of Human Nutrition and Dietetics : The Official Journal of the British Dietetic Association*, 22(5), 383–893.

37 Terrill, J.R., Radley-Crabb, H.G., Iwasaki, T., Lemckert, F.A., Arthur, P.G. & Grounds, M.D. (2013) Oxidative stress and pathology in muscular dystrophies: focus on protein thiol oxidation and dysferlinopathies. *The FEBS Journal*, 280(17), 4149–4164.

38 Spurney, C.F., Rocha, C.T., Henricson, E. *et al.* (2011) CINRG pilot trial of coenzyme Q10 in steroid-treated Duchenne muscular dystrophy. *Muscle & Nerve*, 44(2), 174–178.

39 Giorgi, C., Agnoletto, C., Bononi, A. *et al.* (2012) Mitochondrial calcium homeostasis as potential target for mitochondrial medicine. *Mitochondrion*, 12(1), 77–85.

40 Folkers, K. & Simonsen, R. (1995) Two successful double-blind trials with coenzyme Q10 (vitamin Q10) on muscular dystrophies and neurogenic atrophies. *Biochimica et Biophysica Acta*, 1271(1), 281–286.

41 Davoodi, J., Markert, C.D., Voelker, K.A., Hutson, S.M. & Grange, R.W. (2012) Nutrition strategies to improve physical capabilities in Duchenne muscular dystrophy. *Physical Medicine and Rehabilitation Clinics of North America*, 23(1), 187–99 xii–xiii.

42 Fuller, H.R., Humphrey, E.L. & Morris, G.E. (2013) Naturally occurring plant polyphenols as potential therapies for inherited neuromuscular diseases. *Future Medicinal Chemistry*, 5(17), 2091–2101.

43 Chen, Z., Holland, W., Shelton, J.M. *et al.* (2014) Mutation of mouse Samd4 causes leanness, myopathy, uncoupled mitochondrial respiration, and dysregulated mTORC1 signaling. *Proceedings of the National Academy of Sciences of the United States of America*, 111(20), 7367–7372.

44 Yatsuga, S. & Suomalainen, A. (2012) Effect of bezafibrate treatment on late-onset mitochondrial myopathy in mice. *Human Molecular Genetics*, 21(3), 526–535.

45 Ahola-Erkkila, S., Carroll, C.J., Peltola-Mjosund, K. *et al.* (2010) Ketogenic diet slows down mitochondrial myopathy progression in mice. *Human Molecular Genetics*, 19(10), 1974–1984.

46 Peretti, N., Sassolas, A., Roy, C.C. *et al.* (2010) Guidelines for the diagnosis and management of chylomicron retention disease based on a review of the literature and the experience of two centers. *Orphanet Journal of Rare Diseases*, 5, 24.

47 Carvalho, S.C., Apolinario, L.M., Matheus, S.M., Santo Neto, H. & Marques, M.J. (2013) EPA protects against muscle damage in the mdx mouse model of Duchenne muscular dystrophy by promoting a shift from the M1 to M2 macrophage phenotype. *Journal of Neuroimmunology*, 264(1–2), 41–7.

48 Apostolakis, E., Papakonstantinou, N.A., Baikoussis, N.G. & Papadopoulos, G. (2014) Intensive care unit-related generalized neuromuscular weakness due to critical illness polyneuropathy/myopathy in critically ill patients. *Journal of Anesthesia*, 29, 112–21.

49 Canton, M., Menazza, S. & Di Lisa, F. (2014) Oxidative stress in muscular dystrophy: from generic evidence to specific sources and targets. *Journal of Muscle Research and Cell Motility*, 35(1), 23–36.

50 Shin, J., Tajrishi, M.M., Ogura, Y. & Kumar, A. (2013) Wasting mechanisms in muscular dystrophy. *The International Journal of Biochemistry & Cell Biology*, 45(10), 2266–2279.

51 Fairclough, R.J., Perkins, K.J. & Davies, K.E. (2012) Pharmacologically targeting the primary defect and downstream pathology in Duchenne muscular dystrophy. *Current Gene Therapy*, 12(3), 206–244.

52 Urso, M.L. & Clarkson, P.M. (2003) Oxidative stress, exercise, and antioxidant supplementation. *Toxicology*, 189(1–2), 41–54.

53 Pastore, A., Petrillo, S., Tozzi, G. *et al.* (2013) Glutathione: a redox signature in monitoring EPI-743 therapy in children with mitochondrial encephalomyopathies. *Molecular Genetics and Metabolism*, 109(2), 208–214.

54 Vorgerd, M. (2008) Therapeutic options in other metabolic myopathies. *Neurotherapeutics: The Journal of the American Society for Experimental NeuroTherapeutics*, 5(4), 579–582.

55 Vishwanath, S., Abdullah, M., Elbalkhi, A. & Ambrus, J.L. Jr., (2011) Metabolic myopathy presenting with polyarteritis nodosa: a case report. *Journal of Medical Case Reports*, 5, 262.

56 Campos Filho Wde, O., Nicolini, E., Auxiliadora Martins, M., Zucoloto, S. & Basile Filho, A. (2002) Respiratory and renal dysfunctions due to lipid storage metabolic myopathy: case report. *Arquivos de Neuro-psiquiatria*, 60(3-a), 647–650.

57 Roe, C.R., Yang, B.Z., Brunengraber, H., Roe, D.S., Wallace, M. & Garritson, B.K. (2008) Carnitine palmitoyltransferase II deficiency: successful anaplerotic diet therapy. *Neurology*, 71(4), 260–264.

58 Langhans, C., Weber-Carstens, S., Schmidt, F. *et al.* (2014) Inflammation-induced acute phase response in skeletal muscle and critical illness myopathy. *PloS One*, 9(3), e92048.

59 Bonnefont-Rousselot, D., Chantalat-Auger, C., Teixeira, A., Jaudon, M.C., Pelletier, S. & Cherin, P. (2004) Blood oxidative stress status in patients with macrophagic myofasciitis. *Biomedicine & Pharmacotherapy = Biomedecine & Pharmacotherapie*, 58(9), 516–519.

60 Fedacko, J., Pella, D., Fedackova, P. *et al.* (2013) Coenzyme Q(10) and selenium in statin-associated myopathy treatment. *Canadian Journal of Physiology and Pharmacology*, 91(2), 165–170.

61 Kley, R.A., Tarnopolsky, M.A. & Vorgerd, M. (2013) Creatine for treating muscle disorders. *The Cochrane Database of Systematic Reviews*, 6, Cd004760.

62 Tarnopolsky, M.A., Parshad, A., Walzel, B., Schlattner, U. & Wallimann, T. (2001) Creatine transporter and mitochondrial creatine kinase protein content in myopathies. *Muscle & Nerve*, 24(5), 682–688.

63 Albany, C. & Servetnyk, Z. (2009) Disabling osteomalacia and myopathy as the only presenting features of celiac disease: a case report. *Cases Journal*, 2(1), 20.

64 Crescioli, C. & Vitamin, D. (2014) Receptor agonists: suitable candidates as novel therapeutic options in autoimmune inflammatory myopathy. *Biomed Research International*, 2014, 949730.

65 Ahmed, W., Khan, N., Glueck, C.J. *et al.* (2009) Low serum 25 (OH) vitamin D levels (<32 ng/mL) are associated with reversible myositis-myalgia in statin-treated patients. *Translational Research : The Journal of Laboratory and Clinical Medicine*, 153(1), 11–16.

66 Riphagen, I.J., van der Veer, E., Muskiet, F.A. & DeJongste, M.J. (2012) Myopathy during statin therapy in the daily practice of an outpatient cardiology clinic: prevalence, predictors and relation with vitamin D. *Current Medical Research and Opinion*, 28(7), 1247–1252.

67 Sacconi S., Trevisson E., Salviati L., *et al.* (2010) Coenzyme Q10 is frequently reduced in muscle of patients with mitochondrial myopathy. *Neuromuscular Disorders*, 20(10):44–8.

68 Caso, G., Kelly, P., McNurlan, M.A. & Lawson, W.E. (2007) Effect of coenzyme q10 on myopathic symptoms in patients treated with statins. *The American Journal of Cardiology*, 99(10), 1409–1412.

69 Patchett, D. & Grover, M. (2011) Mitochondrial myopathy presenting as rhabdomyolysis. *The Journal of the American Osteopathic Association*, 111, 404–405.

70 Ghoneim, M.A. & Moselhy, S.S. (2013) Antioxidant status and hormonal profile reflected by experimental feeding of probiotics. *Toxicology and Industrial Health*. 2013 Nov 20. [Epub ahead of print]

71 Lu, H.K., Hsieh, C.C., Hsu, J.J., Yang, Y.K. & Chou, H.N. (2006) Preventive effects of *Spirulina platensis* on skeletal muscle damage under exercise-induced oxidative stress. *European Journal of Applied Physiology*, 98(2), 220–226.

72 Hassani, A., Horvath, R. & Chinnery, P.F. (2010) Mitochondrial myopathies: developments in treatment. *Current Opinion in Neurology*, 23(5), 459–465.

73 Drouet, A. (2013) Management of muscle cramp: what's to be done? *La Revue du Praticien*, 63(5), 619–623.

74 Noakes, T.D. (1999) Postexercise increase in nitric oxide in football players with muscle cramps. *The American Journal of Sports Medicine*, 27(5), 688–689.

75 Miles, M.P. & Clarkson, P.M. (1994) Exercise-induced muscle pain, soreness, and cramps. *The Journal of Sports Medicine and Physical Fitness*, 34(3), 203–216.

76 Purdy Lloyd, K.L., Wasmund, W., Smith, L. & Raven, P.B. (2001) Clinical effects of a dietary antioxidant silicate supplement, Microhydrin((R)), on cardiovascular responses to exercise. *Journal of Medicinal Food*, 4(3), 151–159.

77 Brahme-Isgren, M. & Stenhammar, L. (2007) Muscular symptoms common in selenium deficiency. Association with growth pain, restless legs and calf cramps. *Lakartidningen*, 104(4), 214.

78 El-Tawil, S., Al Musa, T., Valli, H., Lunn, M.P., El-Tawil, T. & Weber, M. (2010) Quinine for muscle cramps. *The Cochrane Database of Systematic Reviews*, 2010 (12), CD005044.

79 Khajehdehi, P., Mojerlou, M., Behzadi, S. & Rais-Jalali, G.A. (2001) A randomized, double-blind, placebo-controlled trial of supplementary vitamins E, C and their combination for treatment of haemodialysis cramps. *Nephrology, Dialysis, Transplantation: Official Publication of the European Dialysis and Transplant Association – European Renal Association*, 16(7), 1448–1451.

80 Schwellnus, M.P. (2009) Cause of exercise associated muscle cramps (EAMC) – altered neuromuscular control, dehydration or electrolyte depletion? *British Journal of Sports Medicine*, 43(6), 401–408.

81 Maquirriain, J. & Merello, M. (2007) The athlete with muscular cramps: clinical approach. *The Journal of the American Academy of Orthopaedic Surgeons*, 15(7), 425–431.

82 Lagueny, A. (2005) Cramp-fasciculation syndrome. *Revue Neurologique*, 161(12 Pt 1), 1260–1266.

83 Schwellnus, M.P., Allie, S., Derman, W. & Collins, M. (2011) Increased running speed and pre-race muscle damage as risk factors for exercise-associated muscle cramps in a 56 km ultra-marathon: a prospective cohort study. *British Journal of Sports Medicine*, 45(14), 1132–1136.

84 Garrison, S.R., Allan, G.M., Sekhon, R.K., Musini, V.M. & Khan, K.M. (2012) Magnesium for skeletal muscle cramps. *The Cochrane Database of Systematic Reviews*, 9, Cd009402.

85 Katzberg, H.D., Khan, A.H. & So, Y.T. (2010) Assessment: symptomatic treatment for muscle cramps (an evidence-based review): report of the therapeutics and technology assessment subcommittee of the American academy of neurology. *Neurology*, 74(8), 691–696.

86 Loprinzi, C.L., Qin, R., Dakhil, S.R. *et al.* (2014) Phase III randomized, placebo-controlled, double-blind study of intravenous calcium and magnesium to prevent oxaliplatin-induced sensory neurotoxicity (N08CB/Alliance). *Journal of Clinical Oncology: Official Journal of the American Society of Clinical Oncology*, 32(10), 997–1005.

87 Rabe, E., Jaeger, K.A., Bulitta, M. & Pannier, F. (2011) Calcium dobesilate in patients suffering from chronic venous insufficiency: a double-blind, placebo-controlled, clinical trial. *Phlebology/Venous Forum of the Royal Society of Medicine*, 26(4), 162–168.

88 Young, G.L. & Jewell, D. (2002) Interventions for leg cramps in pregnancy. *The Cochrane Database of Systematic Reviews*, 2002 (1), CD000121.

89 Venderley, A.M. & Campbell, W.W. (2006) Vegetarian diets : nutritional considerations for athletes. *Sports Medicine (Auckland, NZ)*, 36(4), 293–305.

90 Leitzmann, C. (2005) Vegetarian diets: what are the advantages? *Forum of Nutrition*, 57, 147–156.

91 Leischik, R. & Spelsberg, N. (2014) Vegan triple-ironman (raw vegetables/fruits). *Case Reports in Cardiology*, 2014, 317246.

92 Barr, S.I. & Rideout, C.A. (2004) Nutritional considerations for vegetarian athletes. *Nutrition (Burbank, Los Angeles County, Calif.)*, 20(7–8), 696–703.

93 Haller, R.G., Wyrick, P., Taivassalo, T. & Vissing, J. (2006) Aerobic conditioning: an effective therapy in McArdle's disease. *Annals of Neurology*, 59(6), 922–8.

94 Jung, A.P., Bishop, P.A., Al-Nawwas, A. & Dale, R.B. (2005) Influence of hydration and electrolyte supplementation on incidence and time to onset of exercise-associated muscle cramps. *Journal of Athletic Training*, 40(2), 71–75.

95 Vissing, J., Quistorff, B. & Haller, R.G. (2005) Effect of fuels on exercise capacity in muscle phosphoglycerate mutase deficiency. *Archives of Neurology*, 62(9), 1440–1443.

CHAPTER 15

Role of oxidants and antioxidants in male reproduction

Ashok Agarwal[1], Hanna Tadros[1,2], Aaron Panicker[1,3], and Eva Tvrdá [1,4]

[1] American Center for Reproductive Medicine, Cleveland Clinic, Cleveland, OH 44195, USA

[2] College of Medicine The Royal College of Surgeons in Ireland-Bahrain, Bahrain

[3] Department of Urology, Wayne State University, Detroit, MI, USA

[4] Department of Animal Physiology, Slovak University of Agriculture, Nitra, Slovakia

THEMATIC SUMMARY BOX

At the end of this chapter, students should be able to:

- Define the concept of oxidative stress and its sources in the body

- Outline some methods used to detect oxidative stress

- Outline environmental and lifestyle factors associated with seminal oxidative stress

- Describe the physiological roles of reactive oxygen species in male reproduction

- Describe the pathological roles of oxidative stress in the male reproductive system

- List the antioxidants used for male factor infertility

- Discuss the efficacy of antioxidants as a therapeutic approach for infertility

Introduction

Oxidative stress (OS) is highly damaging to most, if not all, systems of the body. This stress, caused by increased levels of reactive oxygen species (ROS) and reactive nitrogen species (RNS), is associated with multiple pathologies such as atherosclerosis, Parkinson's disease, cancer, and motor neuron disease.[1] Nonetheless, OS is a complicated concept because both ROS and RNS are essential for normal physiological cell function. Hence, while they play a vital role in normal spermatogenesis, they can also lead to pathologies of the male reproductive system.

Oxidative Stress and Antioxidant Protection: The Science of Free Radical Biology and Disease, First Edition.
Edited by Donald Armstrong and Robert D. Stratton.
© 2016 John Wiley & Sons, Inc. Published 2016 by John Wiley & Sons, Inc.

Male infertility: an oxidative role

Infertility is defined as a lack of clinical pregnancy after 12 months or more of unprotected sex.[2] It affects 15% of all couples in the United States, and up to 50% of cases are due to male factors.[1] An estimated 9–14% of American men of reproductive age will experience difficulty in initiating a pregnancy.[2]

Male infertility can stem from multiple factors, including defects in the hypothalamic–pituitary–gonadal axis either due to pretesticular, testicular, or posttesticular issues (the latter of which includes inflammation and obstruction).[2] Some specific causes include Y-chromosome microdeletions, Klinefelter's syndrome, direct trauma, hypogonadotropic hypogonadism, and Kallmann syndrome.[2] The etiology of male factor infertility still remains largely idiopathic and multifactorial. Nonetheless, above all, ROS-induced oxidative stress has been highly implicated.

In the male reproductive system, spermatozoa face the oxygen paradox.[1] On the one hand, a small amount of ROS are needed to reinforce sperm-fertilizing capability. In fact, hyperactivation, acrosome reaction, capacitation, and sperm–oocyte interaction depend on the presence of physiological levels of ROS.[1] On the other hand, an excess of ROS will lead to oxidative stress and subsequent damage. In fact, oxidative stress-induced damage, especially with respect to sperm motility, has been observed since the 1940s.[3] Today, we know that oxidative stress is positively correlated with poor sperm parameters.[2]

Background of oxidative stress and ROS

Oxidative stress is caused by an imbalance of antioxidants and prooxidants.[2] This imbalance stems from excessive ROS production or diminished antioxidant capacity. OS is a compartmentalized phenomenon and is proportional to reductive stress.[4]

ROS levels are inversely related to sperm motility and concentration.[5] As a result of increased ROS levels, there is a greater probability of poor sperm parameters due to structural and functional damage. Spermatozoa membranes are particularly vulnerable to oxidative stress damage due to their high levels of polyunsaturated fatty acids (PUFAs) on the lipid bilayer, which has kinks and susceptible double bonds.[3] ROS can also cause damage DNA at the nuclear and mitochondrial levels, which can be extremely dangerous if passed on to offspring.[3]

Redox reactions

In general, when a molecule gains an electron, it is reduced and the molecule that reduced it is called a reductant. Conversely, when a molecule loses an electron, it is oxidized and the molecule that oxidizes it is considered as an oxidant. Furthermore, molecules containing orbitals with unpaired electrons are unstable and have a tendency to act as an oxidant. Free radicals are produced, for example, when the stable oxygen molecule gains an electron (reduction), a highly reactive oxygen radical, superoxide anion is formed.[6] When ROS such as the superoxide anion are unable to obtain a stable octet configuration, the radical will try to get rid of the extra unpaired electron, making them highly reactive.[6]

It was initially proposed that the Haber–Weiss reaction could generate more toxic radicals from the less reactive superoxide and hydrogen peroxide, which are generated enzymatically.[4] Nonetheless, this reaction was found to have a second-order rate constant of zero in aqueous solution and to be thermodynamically unstable.[4] Hence, it was hypothesized that a catalyst was needed in order for the Haber–Weiss reaction to proceed. This introduces the Fenton reaction in which metal ions such as iron and copper act as the catalysts and generate the hydroxyl radical as an end-product.[4] In the example of the Fenton reaction below, the presence of $Fe^{2+/3+}$ induces ROS by ultimately generating OH^- as an end-product. Along with $Fe^{2+/3+}$, Cu is another molecule that is considered a vital metal cation in the following reaction.[7]

Physiological processes:

$$Fe^{3+} + O_2^{\bullet-} \rightarrow Fe^{2+} + O_2 \quad \text{reduction–oxidation reaction}$$

$$Fe^{2+} + H_2O_2 \rightarrow Fe^{3+} + HO^{\bullet} + OH^- \quad \text{Fenton reaction}$$

The net reaction:

$$O_2^{\bullet-} + H_2O_2 \rightarrow O_2 + OH^- + OH \quad \text{Haber–Weiss reaction}$$

Other sources of ROS include the electron transport chain (ETC) – a cascaded reduction–oxidation reaction. In the cytochrome oxidase complex of the ETC, four electrons interact with dioxygen to generate H_2O. During this process, electrons may leak out of the chain.[7] These unpaired electrons continuously try to bond with free radicals or stable molecules to increase levels of ROS in the system. Thus, cellular respiration is also a significant source of ROS.

Reactive oxygen species

A free radical refers to any species capable of independent existence with at least one unpaired electron in its outer orbit.[8] These are very unstable, short-lived species that react rapidly with adjacent molecules.[8] In the human biological system, oxygen is the most abundant element by mass. Since O_2 is a diradical, it reacts rapidly with other reactive species.[4] In fact, oxygen is the source of most radicals because partially reduced species are generated through normal metabolic processes requiring oxygen.[4]

ROS are prominent toxicological intermediates involved in oxidative stress.[4] They are considered a class of free radicals as they contain oxygen molecules with one or more unpaired electrons. With this highly reactive state, they tend to react and induce radical formation.[2] The different forms of ROS (Figure 15.1) include the primary superoxide anion radical, which can in turn react to form secondary forms such as hydrogen peroxide, peroxyl radicals, and hydroxyl radicals.[2] Tertiary forms include the reactive nitrogen species, which will be detailed later.[2]

Another means of categorizing ROS is into free radicals (such as the hydroxyl radical, $HO^{\bullet}$), combined free radicals (such as superoxide anion, $O_2^{\bullet-}$), nonradical molecules (such as hydrogen peroxide), and ions (such as the hypochlorite ion, ClO^-).[8] Hydroxyl radicals are of particular significance because they can alter

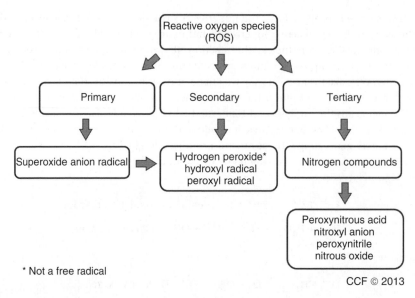

* Not a free radical

CCF © 2013

Figure 15.1 Primary, secondary, and tertiary types of reactive oxygen species, including the reactive nitrogen species.

purines and pyrimidines in DNA strands.[7] Superoxide anions, on the other hand, are common by-products of electron transfer reactions.[7]

A basic understanding of organic chemistry and biochemistry helps explain the sources of ROS. Oxygen, other free radicals, and peroxides contribute to ROS production and form an autocatalytic cycle.[6] This explains why ROS sources include, but are not limited to, hydroxyl radicals, superoxide, peroxyl, and hydrogen peroxide (H_2O_2), all of which are also reactive species in themselves[3] (Table 15.1).

Reactive nitrogen species

Although the focus of this chapter is on ROS, RNS are equally important. Of the nitrogen-derived radicals, nitrogen dioxide ($NO_2^{\bullet}$) and nitric oxide ($NO^{\bullet}$) are of particular interest.[7] Similar to ROS, RNS cause lipid peroxidation (LPO) and nitrosation.

Table 15.1 Common oxygen-free radicals in male reproduction.

Reactive oxygen species	
Hydrogen peroxide	H_2O_2
Superoxide anion	$O_2^{\bullet-}$
Singlet oxygen	$^1O_2$
Hydroxyl radical	$OH^{\bullet}$
Peroxyl radical	$ROO^{\bullet}$
Alkoxyl radical	$RO^{\bullet}$
Alkyl radical	$R^{\bullet}$

Nonetheless, a small concentration of NO• can be beneficial for motility, capacitation, and acrosome reaction in spermatozoa.

In mammals, most RNS are derived from NO•. In fact, NO• production occurs through nitric oxide synthase (NOS) catalyzing L-arginine and oxygen reactions. Nicotinamide adenine dinucleotide phosphate (NADPH) is used as an electron donor. There are three forms of NOS: neuronal nitric oxide synthase (nNOS), inducible nitric oxide synthase (iNOS), and endothelial nitric oxide synthase (eNOS).[7]

The NOS system depends on oxygen and several cofactors, which include NADPH, flavin adenine dinucleotide (FAD), calmodulin, flavin mononucleotide (FMN), and calcium. The result is the formation of NO• with a by-product of L-citrulline.[9] The NOS system takes on several forms depending on the exact physiological role. For instance, nNOS plays a vital role in the testes as there is a testis-specific subclass known as TnNOS – which is a major contributor to the formation of NO• in the male reproductive tract.[9] TnNOS is found solely in the Leydig cells, suggesting it may be involved in steroidogenesis.[9] Furthermore, eNOS and iNOS associate with structural proteins in tight junctions in the testis and may play a role in germ cell apoptosis.[9] eNOS, specifically, plays a role in degenerating germ cell lines while iNOS has been associated with maintaining germ cell lines in the seminiferous epithelium.[9] In fact, this was postulated because iNOS in the testes, rather than being induced by immunological means, has been shown to be induced by factors released from round spermatids.[9] Thus, germ cells may regulate NOS function in Sertoli–Leydig cells.[9] All in all, with the variety of NOS systems in the male reproductive tract, the importance of RNS in the physiological processes of male reproduction is evident.

The product of the NOS reaction is nitric oxide. NO• is a free radical with vasodilatory properties and is an important cellular signaling molecule involved in many physiological and pathological processes.[7] Furthermore, NO• can be therapeutic as a vasodilator.[7] Nonetheless, NO• and its actions depend on several factors within the cell that include its concentration, the amount of thiols, proteins, and metals within the cell, and the redox reactions in the cell.[7] Thus, NO• roles vary among different cells at different concentrations.

Regardless, it has been shown that cyclic guanosine monophosphate (cGMP) may mediate NO-associated signal transduction as a second messenger at low concentrations of NO•.[7] On the other hand, when RNS levels become excessive, damaging effects may occur. These effects are capable of influencing protein structure and function and catalytic enzyme activity, altering cytoskeletal organization and impairing cell signal transduction.[7]

Peroxynitrite (ONOO−), another RNS, is derived from NO• reacting with the superoxide anion.[7] Peroxynitrite can induce lipid peroxidation and nitrosation as other RNS, but its effects are focused on tyrosine molecules that normally act as mediators of enzyme function and signal transduction[7] (Table 15.2).

Sources of ROS and RNS in semen

As oxygen is abundant in our bodies, so is the production of radicals. In regard to the male reproductive system, several sources of radicals can be identified. Figure 15.2 showcases the endogenous and exogenous sources of oxidative stress along with its

Table 15.2 Common nitrogen-free radicals in male reproduction.

Reactive nitrogen species	
Nitrogen dioxide radical	$NO_2{}^\bullet$
Nitric oxide radical	$NO^\bullet$
Peroxynitrite	$ONOO^-$

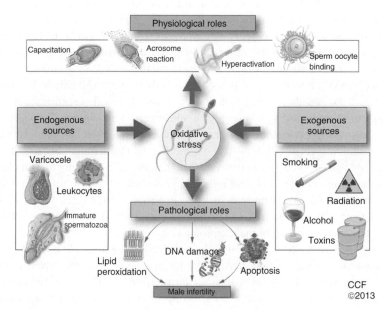

Figure 15.2 Potential generators of reactive oxygen species leading to oxidative stress in the male comprise endogenous and exogenous sources. Physiological levels of reactive oxygen species play a role in sperm capacitation, acrosome reaction, hyperactivation, and sperm–oocyte binding. However, at pathological levels, reactive oxygen species causes lipid peroxidation, DNA damage, and apoptosis, which lead to detrimental effects on male fertility. *(See color plate section for the color representation of this figure.)*

physiological and pathological roles in the male reproductive system, which will be elaborated upon in this section.

Sources of RNS include mature and immature spermatozoa in seminal ejaculate; accessory glands such as the coagulating gland; and glands in the urethra, penis, neck of the bladder, prostate, and seminal vesicles.[9] In the penis, RNS are produced in the pelvic plexus, corpus cavernous, cavernous nerves and their terminal endings within the corporeal erectile tissue, and branches of the dorsal penile nerves, as well as in nerve plexuses in the adventitia of the deep cavernous arteries.[9] In the testes, Leydig cells in adolescent males produce RNS along with round cells, infiltrating leukocytes, epithelial cells, endothelial cells, smooth muscles cells, and macrophages in the intertubular area of the seminiferous tubules.[9] The ejaculatory duct and vas deferens also produce RNS.[9]

ROS has been shown to originate from electron leakage from actively respirating spermatozoa, mediated by intracellular redox reactions.[10] Generation of ROS by spermatozoa has been proposed to occur through two means. One is through the NADPH oxidase system at the level of the sperm plasma membrane and the second through NADPH-dependent oxidoreductase (diaphorase) at the mitochondrial level.[1] The latter seems to be the main source of ROS production. Furthermore, it should be noted that spermatozoa are rich in mitochondria because they need a constant supply of energy for their motility.[10] Sources of ROS can be categorized as endogenous and exogenous. The former includes immature spermatozoa, leukocytes, and varicocele.

Endogenous sources of ROS

Immature spermatozoa are an important source of ROS. During spermatogenesis, developing spermatozoa extrude their cytoplasm in order to prepare for fertilization. In dysfunctional sperm, the cytoplasm is not extruded, leading to excess residual cytoplasm (ERC). Cytoplasmic droplets (excess residual cytoplasm or ERC) explain the missing link between poor sperm quality and increased ROS generation.[1] In fact, retention of residual cytoplasm is positively correlated with ROS generation. ERC leads to the activation of the NADPH system, bringing about an excess of electrons that can contribute to elevated ROS levels.[11] Furthermore, cytosolic enzyme glucose-6-phosphate dehydrogenase (G6PD) activation has been shown to also play a significant role.[1] Defective spermiogenesis is the cause of these changes and elevates ROS production in spermatozoa, which in turn affects its mitochondrial function.[10] Hence, ERC ultimately affects sperm motility, morphology, and fertilization potential, which may lead to male infertility.[10]

Another endogenous source includes peroxidase-positive leukocytes such as polymorphonuclear leukocytes, neutrophils, and the less common macrophages.[10] Many of these originate from the seminal vesicles and prostate.[10] In fact, when activated, excessive stimulation of the hexose monophosphate shunt and the subsequent heightened NADPH production may lead to ROS production 100 times higher than normal.[10] This, for example, becomes exaggerated in leukocytospermia, characterized by a leukocyte concentration greater than 1×10^6/ml.[12] Both exogenous and endogenous factors can play a role in this condition. One of the most important factors includes infection. With infection and inflammation, ROS levels become more exacerbated due to cytokine release. With leukocytospermia, controversy exists about its clinical significance. Sperm parameters such as poor quality, decreased hyperactivation, and defective sperm function have been attributed to leukocytospermia.[1] However, no correlation has been found between seminal leukocyte concentrations and impaired sperm quality or function in other studies.[1]

Varicocele is the excessive dilation of the pampiniform venous plexus around the spermatic cord and this endogenous condition is highly correlated with OS. Its role in male infertility is well researched, as 40% of male partners of infertile couples are diagnosed with varicocele.[10] Furthermore, higher grades of varicocele have been correlated with higher ROS levels.[10] In addition, research has shown that spermatozoa from varicocele patients tend to have high levels of DNA damage induced by oxidative stress.[13] The most common management option is varicocelectomy, which has been proven to reduce ROS levels in males although this does not necessarily result in increased pregnancy rates.[12]

Sertoli cells have also shown to play a significant role in ROS production during spermatogenesis.[14] In one study, it was found that Sertoli cells produce ROS at various levels mediated by certain antioxidants.[15] This led to the hypothesis that ROS potentially play a role in the regulation of spermatozoa production.

Environmental and lifestyle factors contributing to ROS-related infertility

Fertility rates can be affected by a number of variables including obesity, alcohol/illicit drug/tobacco use (the age of cigarette smoking initiation continues to fall), exposure to ionizing radiation via electronic devices, and pollution.[16] It is no coincidence that these changes are occurring at a time when sperm parameters in men are rapidly decreasing. Declining semen parameters come in the form of diminished sperm count, viability, and quality of spermatozoa. As a result, the following section sets out to outline these factors and how they affect sperm parameters.

Industrial components

Certain industrial compounds have been associated with several adverse health effects, many of which are related to male infertility. These chemicals have been shown to increase the production of reactive species such as the superoxide anion and hydrogen peroxide in the testes.[17] For instance, phthalate, a compound in plastics and beauty products, has been shown to damage sperm DNA damage and impair spermatogenesis.[17] Heavy metals and pesticides may also lead to OS, which is a serious issue as they are common and persistent in the environment. Workers exposed to these pollutants were more likely to have decreased sperm quality, count, volume, and density.[10]

Diet

Obesity rates in the United States have grown consistently over the past 30 years with some groups being affected more than others due to various factors including race, income, age, gender, and geographic location. The connection of obesity to male infertility has not been fully elucidated. However, diet can affect semen parameters. For example, a diet high in saturated fat has been shown to reduce sperm quality. In a study of young Danish men, those with the highest saturated fat intake had a 38% lower sperm concentration and a 41% lower total sperm count than those with the lowest saturated fat intake.[18] These results were confirmed in a study of dairy food intake, which found that a low-fat dairy diet resulted in higher sperm concentration.[19] However, omega-3 fatty acids and omega-6 fatty acids were shown to improve morphology, motility, and sperm count.[18,20]

With regard to obesity and its relation to semen parameters, data are conflicting. In a study of Dutch men, it was found that overweight men tended to have lower sperm counts.[20] However, another study reported that underweight men had lower sperm counts than normal and overweight individuals. In the latter study, the researchers concluded that being overweight may protect against a low sperm concentration.[21] Furthermore, a study on Tunisian men showed that motility, morphology, and sperm count did not vary across different BMI levels.[22]

Alcohol

By directly affecting the liver, ethanol metabolism increases ROS production while simultaneously decreasing the antioxidant capacity of the human body.[17] When acetaldehyde, one of the by-products of ethanol metabolism, interacts with proteins and lipids, ROS is formed.[10] This damages proteins, lipids, and DNA.[17] Although alcohol has been associated with OS, its effect on semen parameters has not been explored to a large extent. Overall, studies have failed to find any adverse effects on semen parameters in males who drink regularly.[23] In a study of 8344 males, moderate alcohol consumption did not negatively affect semen parameters.[23] In fact, testosterone levels were actually higher in those men.[23] However, it was noted that chronic drinkers had reduced levels of testosterone, possibly due to an impaired hypothalamic–pituitary axis and damaged Leydig cells (via Bax-dependent caspase-3 activation and subsequent apoptosis).[23,24] Alcohol blocks gonadotropin-releasing hormone (GnRH), reducing LH levels and testosterone synthesis. Furthermore, consumption of alcohol has been shown to increase ROS production when consumed by someone who is malnourished.[17]

Radiation

Radiation is a natural source of energy and has significant clinical effects on humans. Cell phones are becoming more accessible to the general population, especially to males entering puberty and to those of reproductive age. Cell phones release radiofrequency electromagnetic radiation (RF-EMR).[25] With the advent of blue tooth and hands-free calling, more users are leaving phones in their pockets, resulting in direct radiation transmission to the testes. Studies have shown that exposure to RF-EMR increases the risk of asthenozoospermia and teratozoospermia.[25] Furthermore, increased daily exposure to radiation reportedly leads to a decrease in general semen parameters.

In addition, *in vitro* studies have demonstrated that electromagnetic radiation induces ROS production and DNA damage in human spermatozoa, which further decreases the motility and vitality of sperm cells as well as their concentration depending on the duration of radiation.[10]

The electron flow along the internal membranes of a cell is disrupted as a result of numerous charged molecules within the cytosol, which in turn negatively affects normal cellular and organelle function.[10]

Tobacco use

Various components of cigarette smoke have been associated with OS. Cigarettes contain a plethora of free radical-inducing agents including nicotine, cotinine, hydroxycotinine, alkaloids, and nitrosamines.[12] The prime component of tobacco is nicotine, which has been known to induce ROS production in sperm, reducing sperm motility and concentration and altering sperm morphology.[17] When compared with nonsmokers, smokers had lower motility and hypo-osmotic swelling (HOS) test percentages – the latter indicating weak plasma membrane integrity.

LPO results in the loss of membrane integrity. It also leads to malondialdehyde (MDA) production that can combine with various molecules and lead to cytotoxic effects.[26] In a comparative study, asthenoteratozoospermic nonsmokers had higher motility rates than asthenoteratozoospermic smokers.[26] This study also showed that motility and morphology were positively correlated with total antioxidant capacity (TAC) and negatively correlated with MDA levels.[27]

Smoking increases free radical production by increasing leukocyte concentration in the seminal plasma.[17] A study solidifying these facts was done by Saleh *et al.*, which showed that in smokers, the seminal ROS and total antioxidant capacity score was increased – a direct indication that ROS production had also increased.[28]

Furthermore, a different study showed that levels of seminal plasma antioxidants such as vitamins E and C were diminished in smokers. This was confirmed by the presence of increased levels of 8-hydroxy-2'-deoxyguanosine (8-OHdG).[10] Furthermore, spermatozoa from smokers were significantly more sensitive to acid-induced DNA denaturation than those of nonsmokers and resulted in higher levels of DNA strand breaks.[10]

Methods used to measure OS and TAC

Various methods can be used to measure ROS levels and antioxidant capacity within the body. The most common, however, are chemiluminescence and flow cytometry. Table 15.3 outlines these methods along with other less frequently used procedures.

Table 15.3 Overview of major methods of ROS and RNS detection.

Chemiluminescence assay	• Chemiluminescence assays are specific and sensitive when used for samples with high sperm concentration ($>1 \times 10^6$/ml).[29] The process of chemiluminescence highly depends on the emission of light after exciting molecules of a certain substance. The light emitted by these molecules is measured via a luminometer. Luminometers use a photomultiplier tube to detect photons.[30] The method with which these photons are counted varies based on the luminometer. *Photon-counting luminometers* keep track of individual photons while *direct current (DC) luminometers* measure the electric current induced by the photon flux that passes through a photomultiplier tube. It is of utmost importance to be consistent with reagent measurement – this helps produce scientifically accurate results that do not deviate from experiment to experiment.[29]
	• Commercially, two variants are used. The single- or double-tube luminometer is used in small laboratories, whereas the multiple tube variant is used in larger commercial laboratories as it can measure multiple samples simultaneously.[30] With regard to measurements, two approaches can be taken. The first is more prone to error as it measures the reaction as a predetermined time after the reaction has occurred.[30] Thus, variability exists as to when the predetermined time begins after mixing. With the second approach, the luminescence signal is measured within a predetermined interval, making it the preferred method.[30]
	• However, one disadvantage is that the test is unable to differentiate between different types of ROS. According to Ref. 30, the chemical lucigenin can be used specifically to detect the extracellular superoxide anion. Chemiluminescence also cannot detect the source of ROS. This can be problematic as clinicians and researchers are unable to determine whether spermatozoa or leukocytes contributed to potentially elevated ROS levels in the sample of interest. Furthermore, error may be introduced if the sample volume, temperature change, reagent, and reactant concentration are not accurate.[30]

Table 15.3   (*Continued*)

Flow cytometry	• Flow cytometry detects individual intracellular reactive oxygen radicals by measuring fluorescence emitted by ROS-binding probes. Cells are suspended at a density of 10^5–10^7/ml and an average of 10,000 events is recorded.[31] The flow cytometer utilizes two different dyes: dichlorofluorescein (DCFH) for intracellular H_2O_2 and dihydroethidium (DHE) for intracellular detection of $O_2^{\bullet-}$.[32] The latter dye, hydroethidium, targets intracellular super. A red fluorescein color is emitted when the sample is positive, which indicates that hydroethidine was oxidized into ethidium bromide.[31] In a comparison between the mentioned procedures, flow cytometry is considered a better detector of ROS. • As noted earlier, chemiluminescence requires a sperm count higher than 10^6/ml. However, flow cytometry can detect ROS in samples with a lower sperm count.[32] It also is more sensitive, accurate, and specific than chemiluminescence.[32] The disadvantage lies in the fact that cytometry is more expensive. • *TUNEL assays* can be used alongside flow cytometry to assess sperm DNA damage. In these assays, terminal deoxynucleotidyl transferase (TdT) adds deoxyribonucleotides to 3′ hydroxyls of DNA molecules.[33] Deoxyuridine triphosphate (dUTP) is the substrate that attaches to the free 3′-OH breaks. These substrates can serve as markers that can be detected by a flow cytometer. Similar assays measuring DNA integrity include COMET and Sperm Chromatin Structure Assay (SCSA).[34] These processes are relatively inexpensive, but current threshold levels are unclear, and the protocols vary. • When a quantifiable measurement of antioxidants is necessary, a colorimetric assay can be used. One assay of interest is the *ferric reducing antioxidant power* (FRAP) *assay*. FRAP can be utilized to measure the total antioxidant capacity (TAC) of a semen sample.[35] The advantage of FRAP is in its ability to measure all available antioxidants in a sample. By scoring an ROS–TAC metric in a clinical setting, a clearer diagnosis can be offered to infertile patients.
Microscopy and stain	• General microscopy is a commonly used tool with restricted capabilities. To measure ROS, nitroblue tetrazolium is added to the sample. Its reduction to the blue–black compound called formazan indicates that ROS is present within a sample.[31] • To prepare the stain and slide, a sample of 1–5×10^6/ml in concentration mixed in Kreb's buffer is created.[31] A total of 10 µl is then loaded onto a glass slide and is incubated for 20 min at 37 °C.[31] This is followed by a gentle rinse with 0.154 M NaCl and the addition of an equal volume nitroblue tetrazolium and phorbol 12-myristate 13-acetate in Kreb's buffer with 5 mM glucose.[31] After 15 min of incubation, slides are washed with 0.154 mM NaCl, fixed for 1 min in absolute methanol, and then counter-stained with 1% or 2% safranin.[31] Exhibiting the strenuous task of microscopy, observation of 100 consecutive cells under oil immersion is then necessary. Grades of formazan infiltration of cells are then given being either heavy formazan density, intermediate formazan density, scattered, or no formazan presence.[31] • Both the *nitroblue tetrazolium staining* and an additional *cytochrome c reduction* used to detect extracellular ROS are less expensive tests compared to many of the other methods. • Another means of using microscopy includes *electron spin resonance spectroscopy* which measures unpaired electrons in order to find free radicals.[29] However, this means that oxidative stress caused by certain species such as H_2O_2 cannot be adequately measured.

Table 15.3   (*Continued*)

Epifluorescence microscopy	• Epifluorescence microscopy is a specific type of microscopy vital in that the end-product is detectable via fluorescence. This process is similar to the usage of one of the aforementioned dyes in flow cytometry, except it is now applied to general microscopy. As explained earlier, when hydroethidine is added to a sample, it will react with superoxide anions in the vicinity. When a sample is positive, a red fluorescence will emit from the oxidized form of hydroethidine, ethidium bromide.[31] The equipment needed for this process is quite simple and inexpensive and, hence, is more commonly used than many of the other methods.[31]
Enzymatic antioxidants	• Both nonenzymatic and enzymatic antioxidants exist in the male reproductive tract. Though different in concentration and nature, they are both good indicators of ROS. Since solutions can be made to react with certain enzymes, enzymatic antioxidants can be measured.
	• For instance, superoxide dismutase (SOD) is a vital enzymatic antioxidant that can be detected using a tetrazolium salt. Through means of xanthine oxidase and hypoxanthine, SOD activity may be measured.[31] In fact, SOD will react and produce a chromophore, which then can be measured at its maximal absorbance of 525 nm.[31] One unit of SOD is the amount of enzyme needed to exhibit 50% dismutation of a superoxide radical.[31] Glutathione peroxidase is also a detectable and vital antioxidant. Using a kinetic colorimetric assay, this enzyme's activity is detectable by an indirect glutathione reductase coupled reaction.[31] This reaction causes glutathione oxidation via NADPH and glutathione reductase. When NADPH is subsequently oxidized, a decrease in the absorbance at 340 nm is detectable and, hence, is proportional to glutathione peroxidase activity in that sample.[31] The catalase enzyme is also important. Using a CAT assay, when the enzyme in the solution reacts with methanol in the presence of hydrogen peroxide, formaldehyde is formed.[31] Purpald chromogen, which reacts with aldehydes to form a heterocycle, is added to detect the formaldehyde spectrophotometrically as a purple color.[31]
Total antioxidants	• Though measuring enzymatic antioxidants is a possible approach, it is more logical to look at a person's full capability to fight off oxidative stress. This can be measured in the form of total antioxidant measurement, being either nutrient derived or endogenous to the body. All measurement approaches to detect total antioxidants include the capability of antioxidants in a certain sample to inhibit oxidation of 2,2'-azino-di-ethylbenzthiazoline sulfonate by metmyoglobin.[31] This is then compared with a control solution, usually Trolox, a water-soluble tocopherol analog.[31] Results are presented as the equivalence to micromoles of Trolox.[31]
ELISA, immunohistochemistry, and western blotting	• Enzyme-linked immunosorbent assay (ELISA) utilizes detection of antibodies or antigens present in the sample produced via ROS reactions. If any of the antigens or antibodies are detected, they will appear when analyzed using spectrophotometer. Detection of these antigens is based on using antibodies that target certain antigens. Hence, ELISA can also be utilized to detect antioxidants and their antigens.

Table 15.3   (*Continued*)

	• Two other viable methods of detection include immunohistochemistry and western blotting. Oxidative DNA adducts may be measured via immunohistochemistry. In fact, immunohistochemistry has been used to detect oxidative stress molecules such as 8-hydroxy-2'-deoxyguanosine (8-OHdG), thioredoxin (TRX), 4-hydroxynonenal (4-HNE), and redox factor-1 (ref-1).[36] The specifics of how these molecules are detectable include the idea of incubation of samples with enzyme-linked secondary antibodies and a substrate of which will react with the enzyme if binding occurs.[31] Western blotting is a process of detecting certain proteins and is rarely used for ROS detection.
Other methods and approaches	• Other methods exist to detect ROS, though most are infrequently used. These include using nitrate reductase and the Griess reactions to detect nitrate and nitrite. Total NO levels can also be detected via rapid response of chemiluminescence assay.[31] Other measurable products of ROS reactions include protein oxidation levels, lipid peroxides, total plasma lipid hydroperoxides, and total 8-F2-isoprostane.[31]

Physiological role of ROS in reproductive system

ROS play a role in a variety of physiological processes that are crucial in the male reproductive system. In fact, a certain amount of ROS are required for sperm maturation, capacitation, hyperactivation, acrosome reaction, and sperm–oocyte fusion.[37]

Sperm maturation

Maturation occurs in the epididymis and results in the remodeling (release, attachment, and rearrangement) of surface proteins at the membrane level.[38] This allows for signal transduction machinery to be assembled in order for proper hyperactivation and capacitation to take place. ROS are important during maturation because they participate in redox reactions.[39] They bring stability through disulfide bond formation in cysteine residues. The protamines formed during spermiogenesis possess disulfide bonds in the majority of their cysteine residues.[17,40] In fact, antioxidant therapy has a variety of adverse effects at the nuclear level – for example, it interferes with the formation of disulfide bonds, which in turn decreases sperm DNA condensation.[17,41] Thus, it is implied that ROS increase chromatin stability through their redox activity.

Capacitation

During capacitation, sperm motility changes from a progressive state to a highly energetic one. Many receptors on the sperm head become activated, which gives the strength it needs to penetrate the zona pellucida.[42] Hence, capacitation sets up the path necessary for hyperactivation and acrosome reactions. Without this mechanism, the fertilizing capacity of a spermatozoa is lost.

ROS are necessary for capacitation to occur.[10] A combination of superoxide anion and nitric oxide forms peroxynitrite ($ONOO^-$). This allows oxysterol to be produced. Oxysterol, which removes cholesterol from the lipid bilayer, inhibits tyrosine phosphate and promotes cyclic adenosine 3',5'-monophosphate (cAMP) production.[43]

This is vital because cAMP must increase in concentration for capacitation to occur.[10] cAMP and its subsequent pathway involves protein kinase A, which goes on to phosphorylate its substrates. Throughout the pathway, phosphorylation of MEK (extracellular signal-regulated kinase) -like proteins and threonine-glutamate-tyrosine occur, as well as tyrosine phosphorylation of fibrous sheath proteins.[10] Regardless, the cAMP pathway is vital as it is involved in enzyme activation and gene expression.

With the enzyme activation and gene expression alteration, membrane fluidity of the spermatozoa is altered to enhance fertilization.[42] All these factors allow for the continuation of sperm maturation. In fact, when ROS scavengers and NO synthase inhibitors are nearby, cAMP has been seen to decrease.[17] This exhibits the significance of ROS and NO in the regulation of cAMP in the capacitation process.

Although ROS production is necessary for capacitation, an overproduction excites the apoptotic pathway.[42] When oxysterols and lipid aldehydes continue to be produced, cell-mediated suicide or apoptosis occurs, where mitochondrial superoxide production is enhanced along with lipid peroxidation, cytochrome c release, and subsequent caspase activation.[42]

Hyperactivation

Hyperactivation follows the capacitation process and is essential for fertilization; it is considered a subcategory of capacitation. With hyperactivity, the amplitude of the flagella movements increases. Nonhyperactive spermatozoa have slow, linear movements.[17] Hyperactive cells, on the other hand, have asymmetric flagella movement, side-to-side head movement, and a nonlinear motility.[17] There is less stagnation in the oviductal epithelium and surrounding mucus. Just as capacitation requires a certain amount of ROS, an exogenous amount of ROS are necessary to promote hyperactivation. Specifically, ROS promote phosphorylation of flagellar proteins.[44]

Because elevated superoxide dismutase (SOD) levels decrease hyperactivity, it is implied that extracellular superoxide anions may play a vital role during hyperactivation. *In vitro* experiments have also shown that superoxide anions are crucial to the hyperactive motility of the spermatozoa.[17] Across other mammals, NO was also seen to regulate hyperactivation in the epididymis.[17] Hydrogen peroxide is reported to play a role as well, but its role may go hand-in-hand with NO regulation. This is postulated because catalase (CAT), which is vital as an antioxidant against hydrogen peroxide, prevents NO-induced capacitation and hyperactivity processes.[17]

Acrosome reaction (AR)

The spermatozoa head contains a compact nucleus that is covered by the acrosome. Hydrolytic enzymes such as acrosin and hyaluronidase, which are necessary for zona pellucida penetration, are located here.[10] Compared to the slow, reversible process of capacitation, this is a permanent, fast-acting step in sperm maturation.[44] Nonetheless, the reactions that occur substantially overlap with those of capacitation. During acrosomal reaction, an extensive phosphorylation, Ca^{2+} influx, and generation of ROS occur.[44]

In vivo acrosome reactions show that ROS are vital for zona pellucida penetration via the phosphorylation of plasma membrane proteins.[10] This is indicated as ROS are necessary for the phosphorylation of tyrosine proteins in the plasma membrane located in the apex of the sperm head; these tyrosine proteins include fertilin beta, P47, and the spermadhesin family.[10,45] Regardless, in *in vitro* experiments, the administration of superoxide, hydrogen peroxide, and nitric oxide in the seminal plasma induced acrosome reactions.[10]

In other studies, it has been noted that increases in H_2O_2 and decreases in CAT activity positively affect the acrosome reaction.[10] Ca^{2+} influx is necessary for the binding process, and ROS have shown to promote such an influx.[46] Furthermore, as indicated in capacitation, superoxide ions activate adenyl cyclase, a molecule involved in the biochemical cascade that is necessary for the acrosome reaction.[10,47]

Sperm–oocyte fusion

Once the zona pellucida and corona radiata are fully penetrated by a sperm, the oocyte prevents other male gametes from fusing with it by turning the vitelline layer into a hardened envelope.[37] o,o-Dityrosine cross-links allow for the formation of a single macromolecular structure to act as the envelope.[37] Ovoperoxidase is an enzyme utilized to catalyze the "oxidative hardening reaction." H_2O_2 serves as the substrate to ovoperoxidase to provide envelope formation. With our understanding of ROS and its spermicidal effect, the increase in hydrogen peroxide levels prove to be an effective spermicide against polyspermy.[48] With the balance of CATs, glutathione (GSH), and glutathione peroxidase (GPx) weighing against H_2O_2 levels, toxic oxidative stress levels are avoided.[37]

Sperm–oocyte fusion rates seem to depend on both hydrogen peroxide and superoxide anions.[17,47] This goes hand-in-hand with studies showing that SOD and CAT decrease sperm–oocyte fusion rates.[17]

Pathological roles of ROS in male reproduction

ROS play vital roles in pathological as well as physiological processes. Of particular interest is the understanding of what ROS do to induce lipid peroxidation, DNA damage, and apoptosis.

Lipid peroxidation

The lipid bilayer of sperm consists of PUFAs that encompass the cytoplasm. In comparison with somatic cells, spermatozoa have large amounts of PUFAs in their membrane.[10] As a result, the lipid bilayer of sperm has a higher sensitivity to ROS.[49] The resulting oxidation of fatty acids increases membrane permeability, decreases its fluidity, and inactivates critical membrane receptors and enzymes, leading to cell damage.[10]

DNA damage

During spermiogenesis, DNA fragmentation is quite common due to strand breaks, which occur to relieve torsional stress during the packaging process of large amounts of DNA.[43] The process is usually corrected with phosphorylation and activation of nuclear poly (ADP-ribose) polymerase and topoisomerase.[43] If DNA strand breaks cannot be restored, sperm with poorly remodeled chromatin still move on to the epithelium.

However, most DNA damage in spermatozoa is a result of oxidative stress. This can be concluded based on research, indicating that immature spermatozoa create ROS.[11] Furthermore, DNA damage that is related to ROS can be measured by the presence of 8-OHdG, a marker of DNA damage.[3] As a result of oxidation, the nitrogenous base guanine produces 8-OHdG.[3] 8-OHdG is an adduct that labilizes a glycosyl bond intended for the nearby ribose unit.[3] Hence, the higher the levels of 8-OHdG, the higher is the amount of oxidative damage to DNA in sperm.[43]

Apoptosis

Apoptosis is the mechanism by which germ cell death is needed for proper spermatogenesis to occur. It is a regulatory system used to maintain the germ cell: Sertoli cell ratio.[50] As a result of keeping the spermatozoa levels in check, Sertoli cells only nurse a specific number of sperm at one time. With an excess of damaged germ cells, the apoptotic pathway is activated. Such activation assures a proper maintenance of semen parameters in the reproductive tract. Along with this, a clear formation of the blood–testis barrier between the Sertoli cells can only result with apoptosis induction. This is indicated as the barrier is formed by tight junctions because suppression of the apoptosis-inducing gene called Bax prevents creation of the tight junctions that are necessary for barrier formation.[51]

The ROS-induced capacitation–apoptosis highway should also be examined. Initially, as mentioned, ROS/RNS production is necessary for normal physiology. However, uninterrupted generation of RNS and ROS brings about the apoptotic cascade, which includes the Fas/FasL system. FasL initiates apoptosis when it interacts with the cell surface receptor Fas.[42,51] Other important apoptotic regulator proteins are p53, caspases, c-Myc, cyclic adenosine monophosphate responsive element modulator (CREM), and the collective members of the Bcl-2 family.[51] With high levels of ROS, the mitochondrial membranes are strained, the apoptotic cascade is hyperactive, and the release of cytochrome c activates caspases to induce apoptosis. Increased apoptosis is obviously detrimental and is correlated with infertility as infertile semen samples were found to contain larger amounts of apoptotic spermatozoa.[12]

Antioxidants

Antioxidants are ROS antagonists and may be intrinsic or extrinsic in nature. Antioxidants play a significant role in the body, as an increase in ROS must be well balanced to avoid oxidative stress (Figure 15.3). In fact, a decrease in antioxidant levels may contribute to male factor infertility. This section will define and analyze enzymatic and nonenzymatic antioxidants while discussing other proven molecules and compounds with antioxidant properties.

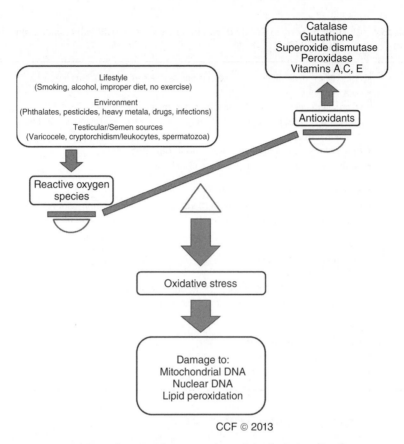

Figure 15.3 Accumulation of reactive oxygen species and the depletion of endogenous antioxidants bring about a state of oxidative stress, which could result in lipid peroxidation and damaged mitochondrial and nuclear DNA.

Enzymatic antioxidants

Enzymatic antioxidants are the main protectants in the reproductive system, serving as a counterbalance to oxidants (Table 15.4). The most important enzymatic antioxidants include SOD and CAT, which work in conjunction to remove or neutralize superoxide and hydrogen peroxide (Figure 15.4). SOD can be found in three isoforms, split between the mitochondrial matrix, intermembrane space, and cytosol.[52] The cytosolic SOD, named Cu/Zn-SOD or SOD 1, is the most common. However, the conversion of superoxide anion to a more stable form does not remove the potential for oxidative stress generation. In fact, stable H_2O_2 can still damage lipids, proteins, and DNA.[14] This is where the role of CAT comes into play. CAT converts hydrogen peroxide to H_2O and $O_2.$ Therefore, SOD and CAT together remove molecules responsible for oxidative stress and reduce oxidative stress damage (e.g., lipid peroxidation) (Figure 15.4).

Table 15.4 Glutathione peroxidase (GPx), glutathione reductase (GR), superoxide dismutase (SOD), catalase (CAT).[6,44,52–57]

GPx	• *Role*: Utilizes glutathione to reduce peroxides at the plasma membrane level.
	• *Location*:
	(a) In sperm, found in nucleus, mitochondria, and cytosol.[54]
	(b) Found in testis, prostate, seminal vesicles, vas deferens, epididymis, seminal plasma, and spermatozoa.[6]
	• Family of selenium-dependent antioxidants (except GPx5).[53]
	• Synthesized at the epithelial cells of the epididymis.[53]
GR	• *Role*: Maintains homeostasis of GSH.[56]
	• *Location*: Found in epididymis, Sertoli cells of testis, vas deferens, seminal vesicle, genital tract and epithelium, and prostate gland.[56]
SOD	• *Role*:
	(a) Scavenges superoxide anion.[52]
	(b) Converts $O_2^{\bullet -}$ to H_2O_2 and O_2.
	• *Location*: Found also in midpiece mitochondrial intermembrane space and cytosol.
	• Synthesized at the epithelial cells of the epididymis.[53]
CAT	• *Role*:
	(c) Scavenges superoxide anion.[52]
	(d) Converts $O_2^{\bullet -}$ to H_2O_2 and O_2.
	• *Location*: Found in peroxisomes.

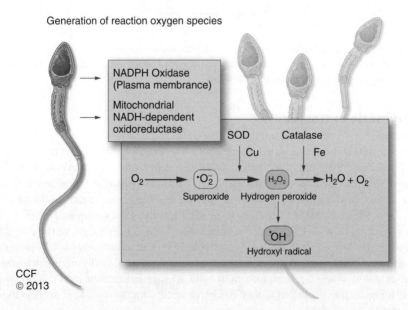

Figure 15.4 Reactive oxygen species (superoxide anion, hydrogen peroxide, and hydroxyl radical) are generated from oxidative processes in the plasma membrane and mitochondria of the male gamete. These reactions involve the SOD and catalase antioxidant enzymes along with copper and iron, respectively. *(See color plate section for the color representation of this figure.)*

GPx and glutathione reductase are two other enzymatic antioxidants important to male reproduction. GPx is a family of selenium-dependent antioxidants, with the exception of the GPx5 isoform.[53] Similar to the actions of SOD and CAT, GPx reduces H_2O_2 to water.[54] A nonenzymatic antioxidant known as GSH is concurrently converted to its oxidized form, GSSG.[54] The most vital of the glutathione peroxidase isoforms are GPx1, which is found in sperm and the genital tract, and GPx4, which is found in testicular tissue.[52] Phospholipid hydroperoxide glutathione peroxidase (PHGPx) is another name for GPx4; it is the only member of the GPx protein family that uses phospholipid hydroperoxides as the substrate of interest.[54] As a result, along with reacting to hydrogen peroxide, GPx4 can reduce lipid hydroperoxides.

Glutathione reductase (GR) is another vital enzymatic antioxidant. It is a dimeric flavoprotein that maintains homeostasis of GSH.[55] It serves to reduce GSSG – the oxidized form of glutathione – back to GSH, the reduced form of glutathione, with the aid of a reducing cofactor, NADPH.[54] It is found primarily in the Sertoli cells of the testis. Kaneko *et al.* found that Sertoli cells require GR for a successful glutathione supplementation to spermatogenic cells, thus indicating the significance of this enzymatic antioxidant.[56] Furthermore, GR in the epithelial tract serves as an antioxidant to protect unsaturated fatty acids from lipid peroxidation during maturation.[56]

Nonenzymatic antioxidants

Nonenzymatic antioxidants also play a vital role in oxidative stress balance. These are of particular interest because they are currently being used *in vivo* as the treatment for male infertility due to their ability to break up the chain reaction of lipid peroxidation.[57] Furthermore, it should be noted that this group of antioxidants is different from enzymatic antioxidants in that they are consumed immediately after interaction takes place.[57] Regardless, many forms of nonenzymatic antioxidants exist and are detailed below.

1 *Ascorbic acid:* This is more commonly known as vitamin C. It has frequently been utilized as an antioxidant due to its scavenging properties. Ascorbic acid is a water-soluble vitamin, and this is significant in distinguishing where its primary action takes place. Due to its hydrophilic nature, ascorbic acid has more effective scavenging properties at the plasma level than at the lipid bilayer.[58] This is vital because various studies have found that higher vitamin C concentrations in plasma directly result in a lower DNA fragmentation index.[59] Another study showed that sperm DNA were less affected by oxidative stress as a result of ascorbic acid supplementation.[60] Furthermore, ascorbic acid increases normal spermatozoa and improves sperm quality and motility.[58]

2 *α-Tocopherol:* The common name for α-tocopherol is vitamin E. In studies, the effect of vitamin E on semen parameters has been less pronounced than that of ascorbic acid. Vitamin E supplements (300–1200 mg) administered for 3 weeks increased seminal plasma levels but not significantly so.[58] α-Tocopherol, unlike vitamin C, is a fat-soluble vitamin. This is important when considering its effect on MDA levels. Men who took 200 mg of vitamin E had lower MDA levels in a longitudinal study.[58]

3 *Carnitine:* This is another water-soluble antioxidant. It is synthesized in the liver and acts primarily in the epididymis to assist in sperm maturation. Studies have

shown that it prevents DNA damage and apoptosis during sperm maturation.[61] Moreover, studies have found that carnitine supplementation improves sperm motility and concentration.[62]

L-Carnitine and L-acetyl carnitine are particularly important due to their effects on the utilization of energy in mitochondria. These antioxidants can shuttle long-chain lipids across the mitochondrial bilayer and start the process of ß-oxidation.[14] The effect of this process is twofold: the reduced forms NADH and FADH$_2$ are generated along with acetyl-CoA.[14] To be more specific, the significance of acetyl-CoA generation is the extra energy that can be produced once it is sent through the Krebs cycle.

4 *Lycopene:* This is another potent scavenger of free radicals. As a member of the carotenoid family, lycopene is involved in immune reactions, gap junction signals, cell growth regulation, and gene expression.[63] However, in male reproduction, lycopene is found primarily in semen and the testes. It is of clear significance here because it is thought to have one of the greatest ROS reduction rates.[64] This is supported by the fact that lycopene supplementation has consistently been noted to improve various semen parameters.[64]

5 N-*acetyl-cysteine (NAC):* This has the ability to reduce free radicals by acting with thiols and hydroxyl radicals.[65] However, it is also plays a role as a precursor to glutathione.[66] Its roles are significant within the body and are supported by the fact that NAC improves sperm concentration and acrosome reaction and reduces ROS and sperm DNA damage.[64] Furthermore, when combined with selenium, NAC also has a positive impact on sperm concentration and acrosome reaction.[64]

6 *Glutathione (GSH):* This is one of the most abundant nonenzymatic antioxidants found in the body. Being an endogenous source, it is synthesized by the liver but can also be derived from dietary sources such as fresh meat, fruits, and vegetables.[67] This molecule has three precursors: cysteine, glutamic acid, and glycine.[67] Glutathione contains a thiol group, which is the key to its antioxidant properties.[68] As explained previously, glutathione plays a vital role in enzymatic antioxidants but also helps maintain exogenous vitamins C and E.[67]

7 *Selenium (Se):* Many parts of the world have been known to have patient populations with depleted amounts of selenium (Se).[69] Adequate Se levels have been shown to positively correlate with increased levels of sperm concentration, motility, and morphology.[70] More than an antioxidant, selenium is a cofactor of phospholipid hydroxyperoxide glutathione peroxidase, GPX4.[71] This is important in condensing of the nucleus in the very small sperm head.[71] The negative effects of selenium supplementation will be discussed later in this chapter.

Other antioxidants

1 *Pentoxifylline:* Being a xanthine derivative, pentoxifylline is a phosphodiesterase inhibitor that raises intracellular cAMP levels and reduces leukotriene synthesis.[66] Pentoxifylline may improve motion characteristics of sperm but it has less of an effect on motility.[66] Furthermore, it has been shown to decrease ROS levels while higher doses also improved sperm motion parameters without changing fertilizing capability.[66]

2 *Zinc (Zn):* This, a trace element, plays a vital role and is increasingly found to be deficient in the Caucasian population.[71] In the male reproductive system, it is a cofactor for SOD and metallothioneins, assisting in scavenging superoxide and hydroxyl radicals.[71] It also helps remove ROS. Zn serves as a cofactor to dihydrofolate reductase and methionine synthase, two molecules needed for homocysteine recycling.[71] Homocysteine is a molecule found in the reproductive system that promotes oxidative stress and inhibits DNA methylation; hence, its recycling is vital.[71] Furthermore, zinc has membrane-stabilizing effects and inhibits DNases.[72]

Seminal plasma has a greater concentration of zinc than serum. This may be due to the importance of zinc-scavenging properties. Omu *et al.*'s study highlights zinc's ability to reduce antisperm antibodies.[72] Furthermore, it has been shown that zinc is vital in maintaining the quaternary structure of the DNA.[72] A reduction of zinc, therefore, can negatively impact semen parameters.

3 *Resveratrol (RSV):* This is primarily found in peanuts, grapes, and wine.[73] It garnered interest following an understanding of wine's effect on reducing cardiovascular disease. Its effects have been studied more in rats, but the antioxidant properties seem to increase spermatogenesis. This is hypothesized to occur via RSV and its actions on the hypothalamic–pituitary–gonadal axis by stimulating luteinizing hormone (LH) production.[74] This in turn has a positive impact on testosterone and sperm production.

Antioxidants: a therapeutic approach

Antioxidant treatment has proven to be effective both *in vivo* and *in vitro*.[58] However, the scientific literature does not contain strong data showing their efficacy in infertile patients. Hence, we present the most appropriate doses for the treatment of male infertility based on studies that explored antioxidant supplementation and their correlation with sperm parameters. We also emphasize that antioxidants are not always beneficial. In fact, they can have harmful effects on the male reproductive system.

Recommended dosage and regimen
Individual antioxidant treatments
1 *Glutathione (GSH):* Little information is available on the significance and efficacy of glutathione supplementation with specific dosages. Nonetheless, 600 mg of GSH per day given over 2 months to 11 men lead to an improvement in sperm kinetics and a higher sperm count.[58]

2 *Vitamin C:* It has been noted that saturation of vitamin C in plasma occurs simply with a dosage of 1 g/day.[58] Nonetheless, Lanzafame noted enough administration of ascorbic acid in order to generate a 2.2-fold rise in the system improving sperm quality.[58] In mice, on the other hand, an equivalent human therapeutic dose of 10 mg/kg of body weight lead to a decrease in MDA levels.[58] As MDA is a marker of oxidative stress, this is a clear indication of reduced oxidative stress. Furthermore, Lanzafame *et al.* noted that in a comparison of 200 versus 1000 mg/day of vitamin C given over a period of 4 months to heavy smokers, the group taking 1000 mg/day had a greater increase in semen parameters.[58]

3 *Vitamin E:* In a 1996 study by Suleiman *et al.*, 110 asthenozoospermic patients received 300 mg of vitamin E daily over 26 weeks.[75] Motility increased while lipid peroxidation levels decreased.[75]

4 *Carnitine:* Supplementation at 3 mg/day over 3 months positively affected motility.[76] A study by Vitali *et al.* reported an increase in motility to 80%.[77] Lenzi *et al.* and Cavallini *et al.* found that 2 g of L-carnitine and 1 g of acetyl L-carnitine improved sperm concentration and motility.[78,79] Another study provided 3 g of carnitine in various forms (L-carnitine completely, acetyl L-carnitine completely, or a combination of 2 g L-carnitine and 1 g acetyl L-carnitine) to 60 asthenozoospermic men over a period of 26 weeks. An improvement was found in straight progressive velocity and ROS-scavenging capacity.[61] Hence, support behind carnitine supplementation is evident.

   Nonetheless, with carnitine specifically, the dosage is highly correlated with effect. A lower dose (1 g of L-carnitine and 500 mg of acetyl L-carnitine) given over a span of 16 weeks did not improve volume, motility, concentration, forward progression, or seminal plasma, and did not increase semen levels.[61] Hence, the correct dose must be used in order to achieve maximal results.

5 *Selenium:* Studies have shown that selenium intake increases GPx activity. For instance, an older study showed that 50 and 60 ng/ml of selenium in semen was correlated with the highest motility and sperm count in 125 men from couples with infertility issues.[80]

   However, contradictory reports exist as well. Small studies that provided 200–300 mg/day supplements through selenite, selenium-enriched yeast, or diets with high selenium content increased seminal selenium levels but did not affect sperm characteristics.[80] More research is needed to discern the true impact of selenium on sperm parameters.

6 N-*acetyl-cysteine (NAC):* It has been proven to be very effective at increasing semen parameters. In a study by Ciftci *et al.*, 120 patients with idiopathic infertility received 600 mg of N-acetyl-cysteine for 13 weeks.[81] In patients receiving NAC treatment, motility, volume, viscosity, and liquefaction time improved.[81] Furthermore, there was a clear reduction in oxidative stress as measured by total antioxidant levels.[81]

7 *Zinc sulfate (ZnSO$_4$):* In a study conducted by Omu *et al.*, 100 patients with asthenozoospermia were split into two groups – a control group and an experimental group that received 250 mg of zinc sulfate twice daily for 3 months.[72] Zinc supplementation leads to an overall improvement in sperm parameters in the experimental group. A 10.5% difference between the two groups in sperm count was seen along with a 12.8% difference in sperm membrane integrity and a 11.9% difference between the two groups in asthenozoospermia.[72] In the same study, the authors found that ZnSO$_4$ also played an immunological role as T-helper cytokines and interleukin-4 levels increased in the experimental group and TNF-α and anti-sperm antibodies decreased.[72] Another study that assessed the effects of 220 mg of zinc sulfate in 14 infertile patients reported 3 pregnancies and a general increase in sperm count and motility.[72]

Combined antioxidant treatments

1 *Vitamins C and E:* As these are the two most commonly used supplements for antioxidant treatment, their combined effect has been extensively studied. In one study, 8 weeks' supplementation consisting of 1000 mg of vitamin C and 800 mg of vitamin E did not affect concentration, motility, morphology, volume, or 24-h sperm survival.[61] In a study by Greco *et al.*, 500 mg of vitamin A and 500 mg of vitamin C given twice a day improved DNA fragmentation (as assessed with the TUNEL assay).[82] All 32 men in the study showed a decrease in DNA fragmentation.[82]

However, adding 1 mg of vitamin A and 100 μg of selenium to a combined 10 mg of vitamin C and 15 mg of vitamin E improved motility rates.[82] In another study, vitamin A, *N*-acetyl-cysteine, and zinc were given to men postvaricocelectomy along with vitamins C and E.[76] This was a complicated regimen and included, over the course of 13 weeks, daily administration of 0.06 IU/kg of vitamin A, 3 mg/kg of vitamin C, 0.2 mg/kg of vitamin E, 10 mg/kg of NAC, and 0.01 mg/kg of zinc.[76] Sperm count increased by 20-fold, and, of the 20 subjects, 6 of the originally infertile men had sperm counts greater than 20 million/ml posttreatment.[76]

2 *Selenium and vitamin E:* Extensive studies have shown some support for this combination of supplementation. In a 1996 study, 100 μg/day of selenium and 400 mg of vitamin E was given to infertile men over a 1-month period.[83] Later on, selenium doses were increased by 200 μg/day over the next 4 months while the 400 mg of vitamin E remained constant.[83] Selenium levels increased by 40% in semen and by 23% in plasma.[83] Motility increased by 19% and sperm viability increased by approximately 30%. Nonetheless, in the small sample size of 9, no pregnancies were reported.[83] Furthermore, no noticeable increase in sperm concentration occurred, and ejaculate volume actually *decreased*.[83] In another study, a dose of vitamin E (400 mg) and selenium (225 μg) was compared with a control of vitamin B. MDA concentration, an oxidative stress biomarker, decreased and motility improved with the combination therapy.[84]

In a study of infertile men from Iran, researchers administered 200 μg of selenium and 400 units of vitamin E over a period of 100 days.[85] The results showed a 52.6% total improvement in sperm characteristics (362 cases) and a 10.8% increase in spontaneous pregnancies in comparison with the no-treatment group.[85] However, the authors were unable to determine whether the improvement in sperm parameters was a result of the vitamin E rather than selenium.

3 *Selenium and N-acetyl-cysteine:* In a 2009 study, 116 patients were randomized to receive 200 μg oral selenium daily, 118 patients to receive 600 mg oral *N*-acetyl-cysteine daily, 116 patients to receive 200 μg selenium plus 600 mg *N*-acetyl-cysteine daily, and 118 patients to receive a placebo, all for 26 weeks followed by no treatment over a period of 30 weeks.[86] Serum testosterone and inhibin levels increased but FSH decreased in all of the treatment groups.[86] Furthermore, average sperm concentration, motility, and morphology all increased in the treated patients compared to those receiving placebo.[86] In fact, 5–10% of each treatment group had a 50% or greater increase in sperm

concentration.[86] Hence, it is indicated that this combined treatment may a viable treatment for male infertility.

4 *Menevit:* this drug is a combination of 100 mg of vitamin C, 400 IU of vitamin E, 500 µg of folate, 1000 mg of garlic, 6 mg of lycopene, 26 µg of selenium, and 25 mg of zinc. Tremellen *et al.* conducted a study on the efficacy of Menevit during *in vitro* fertilization or intracytoplasmic sperm injection treatment. Their study did not find any effect on the oocyte fertilization rate and embryo quality.[87] However, pregnancy rates improved by 38.5% compared to 16% in the control group.[87] Hence, it can be concluded that Menevit may be effective.

5 *Vitamins C E and ZnSO₄:* In a study by Omu *et al.* in 2008 on asthenozoospermic men with normal sperm concentrations, patients were separated into three treatment groups: 200 mg $ZnSO_4$ only, 200 mg $ZnSO_4$ and 10 mg vitamin E only, and 200 mg $ZnSO_4$, 10 mg vitamin E, and 5 mg vitamin C.[88] In all three treatment groups, MDA and Bax expression decreased, and TAC increased; however, there were no statistically significant differences between the groups.[88] Thus, the combination of these compounds may be unnecessary.

However, it is interesting to note that Bcl-2, an antiapoptotic antioxidant pathway, increased, which could slow sperm apoptosis.[88] Furthermore, there was a decrease in IgG antibodies that served as antisperm antibodies.[88]

Harmful effect of antioxidants

Because there are no established antioxidant treatment guidelines or doses, negative effects can occur. Furthermore, an increase in serum concentrations following antioxidant intake does not correlate with an increase at the testicular level.[71] Hence, it is difficult to pinpoint exact dosages for patients. The following section will detail the known and proposed harmful effects of antioxidants.

1 *Vitamin C:* Few studies address the effectiveness of vitamin C in excessive amounts. Moreover, these studies lacked a solid design and were subject to bias. In fact, many observational studies, for instance, found that vitamin C decreased cardiovascular (CV) disease.[68] Nonetheless, most randomized controlled trials reported no correlation between antioxidants and lowered CV risk.[68] Thus, study design clearly impacts outcomes.

Doses of vitamin C as high as 1000 mg may decrease mitochondrial production and reduce endurance, thus leading to muscle recovery delay.[68] The vitamin C saturation of an average human body is 1 g/day. Thus, with higher levels of ingestion, excretion of the oxalate metabolite may crystallize to form kidney stones, and hence, patients present with typical symptoms of back pain and dysuria.[58] Other side effects of an oversaturated vitamin C transport system include diarrhea and indigestion.[68]

2 *Zinc and selenium:* Selenium and zinc are nutrients that are not themselves antioxidants, but are needed to promote the activity of certain antioxidant enzymes. In excess amounts, they may cause certain adverse effects:

(a) *Selenium:* Normal serum selenium levels range from 70 to 90 ng/ml as its maximal effect takes place in this specified range.[68] In fact, most people receive adequate selenium in their diets, and hence, its deficiency is rare. The Selenium and Vitamin E Cancer Prevention Trial (SELECT) was a large randomized, placebo-controlled trial set to evaluate the potential benefit of selenium and vitamin E for the prevention of prostate cancer.[68] In this study, 35,000 men who qualified for the study were divided into four groups, one of which was selenium alone. After a mean of follow-up of 5.46 years, the selenium group exhibited a statistically insignificant increased risk of type-2 diabetes, with a lack of improvement in prostate cancer prevention.[68] Although statistically insignificant, it raised concerns.[68] Furthermore, evidence also suggests that values higher than normal may contribute to asthenozoospermia and DNA methylation, causing genetic issues.[71]

(b) *Zinc*: The upper limit of tolerability of zinc is 40 mg/day in adults.[68] Above that level, adverse effects can occur. The most common effects include vomiting, abdominal cramping, diarrhea, urinary tract infections, and taste distortion.[68] Long-term excessive intake may even impair the immune system, decrease high-density lipoproteins, and hence, cause microcytic anemia or even copper deficiency.[68]

A study done by Letizmann *et al.* evaluated zinc intake and its effects on prostate cancer prevention. Among 46,974 adult men, those who were supplemented with zinc higher than 100 mg/day saw their risk for prostate cancer increase 2.3 times.[89] In those taking less than 100 mg/day, no increased risk was seen.[89] These results could be biased, however, due to the fact that calcium was given along with the zinc, so it was not possible to establish a strong cause–effect relationship. Zinc has also been shown to alter metabolism of certain drugs and other vitamins and minerals. Zinc, for example, interferes with absorption of tetracyclines, quinolones, and penicillamine.[68]

3 *Vitamin E:* This refers to a set of eight fat-soluble compounds. Vitamin E is vital in boosting the immune system and reducing CV risk.[68] Nonetheless, supplementation in excess will lead to side effects such as heightened bleeding and even death.[68] This was indicated in a meta-analysis done in 2005 by Miller *et al.*, which consisted of 135,967 adults in 19 placebo-controlled studies over a 1-year duration.[90] A total of 60% of these subjects already had heart disease or risk factors for it. The meta-analysis, though not entirely conclusive, found that 400 IU or more of vitamin E given daily increased the likelihood of death.[90] Nonetheless, the patient cohort was older, and thus, the results may not apply to younger patients. Furthermore, studies also included multivitamins rather than pure vitamin E supplements. In November 2004, the American Heart Association stated that large amounts of vitamin E (>400 IU) may be harmful.[68]

Other trials included the HOPE and HOPE-TOO trials where 400 IU of vitamin E were administered daily for a median of 7 year. [68] Supplementation did not prevent fatal and nonfatal cancers, major CV events, or death. However, the risk for heart failure increased. In fact, a regression analysis showed that vitamin E was an independent predictor of heart failure, and echography studies found that vitamin E does in fact lower left ventricular ejection fraction.[68]

A double-blind, placebo-controlled trial by Hemila *et al.* consisting of Dutch subjects aged 60 years or older found that respiratory infections were more severe in those taking 200 mg vitamin E daily.[91] Vitamin E supplementation could also cause harm in certain patient groups.[68]

A systematic review of published studies of antioxidants for male partners of couples undergoing assisted reproduction techniques by Showell *et al.* reported that miscarriages were unlikely.[92]

4 β-*Carotene:* This is an important source of vitamin A. Carotenoids can be converted into retinol in the human body. However, excessive levels can be detrimental. The ATBC trial and the CARET trial found that β-carotene study groups had increased incidences of lung cancer, and that the supplements did not help prevent other malignancies.[68] The latter of these trials could not determine a true cause–effect relationship, as vitamin A and β-carotene were both administered. The ATBC trial, however, reported a direct correlation between β-carotene and heightened mortality in the study groups.[68] Lack of β-carotene in preventing neoplasms was also exhibited in a large randomized, double-blind study over 12 years that consisted of 22,071 male physicians who received 50 mg of β-carotene supplements on alternate days.[68]

β-Carotene supplements may cause yellowing of the skin, known as hypercarotenemia, and GI disturbances, which are reversible.[68] Other studies, as that of the Women's Health Study (a large cohort of American women older than 45 years of age), reported a correlation between β-carotene and heightened risk of stroke over a period of 4.1 years.[68] This result was statistically insignificant, but could still indicate a serious adverse event that must be noted.

Other antioxidants associated with side effects include the following.
(a) *Carnitine:* Doses greater than 4 mg/day may result in GI irritability, body odor, and seizures.[93]
(b) *Lycopene:* Excess intake can bring about GI irritability and changes in skin color.[67]
(c) N-*acetyl-cysteine:* Caused GI irritability along with rash, fever, headaches, drowsiness, and hepatic toxicity.[94]
(d) *Pentoxifylline:* Caused headache, general pain, diarrhea, and abnormal stool.[95]
(e) *Resveratrol:* Excess amounts lead to nausea, abdominal pain, and diarrhea.[96]

Conclusion and key points

The significance of this chapter lies in its ability to provide a novel understanding of the current literature on oxidants and antioxidants. Furthermore, it provides a systematic review of antioxidant regimens and provides the efficacy levels of each. This was written in hopes that researchers will better understand the most novel effects of oxidative stress in the pathophysiology of male reproduction and that clinicians will have a reference to help them better treat their patients with male infertility. Finally, clinicians must be careful when determining doses of antioxidants as in excess quantity they can cause adverse effects.

Multiple choice questions

1 Physiological levels of reactive oxygen species are required for
   a Acrosome reaction

    b. Apoptosis
    c. Capacitation
    d. Hyperactivation

2 In the spermatozoa, oxidative stress causes
    a. Lipid peroxidation
    b. DNA damage
    c. Apoptosis
    d. Excessive residual cytoplasm

3 Factors that may induce oxidative stress in the male include
    a. Alcohol intake
    b. Radiation
    c. Smoking
    d. Environmental toxins

References

1 Agarwal, A., K. Makker, R. Sharma. 2008. "Clinical relevance of oxidative stress in male factor infertility: an update." *American Journal of Reproductive Immunology* 59(1):2–11.

2 Agarwal, A., D. Durairajanayagam, J. Halabi, J. Peng, M. Vazquez-Levin. 2014. "Proteomics, oxidative stress and male infertility." *Reproductive Biomedicine Online* 29(1):32–58.

3 Aitken, R. J., K. T. Jones, S. A. Robertson. 2012. "Reactive oxygen species and sperm function-in sickness and in health." *Journal of Andrology* 33(6):1096–1106.

4 Kehrer, J. P. 2000. "The Haber–Weiss reaction and mechanisms of toxicity." *Toxicology* 149(1):43–50.

5 Benedetti, S., M. C. Tagliamonte, S. Catalani, M. Primiterra, F. Canestrari, S. De Stefani, S. Palini, C. Bulletti. 2012. "Differences in blood and semen oxidative status in fertile and infertile men, and their relationship with sperm quality." *Reproductive Biomedicine Online* 25(3):300–306.

6 Tremellen, K. 2008. "Oxidative stress and male infertility – a clinical perspective." *Human Reproduction Update* 14(3):243–258.

7 Agarwal, A., A. Aponte-Mellado, B. J. Premkumar, A. Shaman, S. Gupta. 2012. "The effects of oxidative stress on female reproduction: a review." *Reproductive Biology and Endocrinology* 10:49.

8 Salway, J.G. (2012) *Medical Biochemistry at a Glance*, 3rd edn. Wiley-Blackwell.

9 Doshi, S. B., K. Khullar, R. K. Sharma, A. Agarwal. 2012. "Role of reactive nitrogen species in male infertility." *Reproductive Biology and Endocrinology* 10:109.

10 Agarwal, A., G. Virk, C. Ong, S. S. du Plessis. 2014. "Effect of oxidative stress on male reproduction." *World Journal of Men's Health* 32(1):1–17.

11 Rengan, A. K., A. Agarwal, M. van der Linde, S. S. du Plessis. 2012. "An investigation of excess residual cytoplasm in human spermatozoa and its distinction from the cytoplasmic droplet." *Reproductive Biology and Endocrinology* 10:92.

12 Said, T.; Gokul S.; Agarwal, A. 2012. "Clinical consequences of oxidative stress in male infertility" *Studies on Men's Health and Fertility*, R. J. Aitken, A. Agarwal, J. G. Alvarez, 535–549. Humana Press.

13 Shiraishi, K., H. Matsuyama, H. Takihara. 2012. "Pathophysiology of varicocele in male infertility in the era of assisted reproductive technology." *International Journal of Urology* 19(6):538–550.

14 Chen, Shu-jian, Jean-Pierre Allam, Yong-gang Duan, Gerhard Haidl. 2013. "Influence of reactive oxygen species on human sperm functions and fertilizing capacity including therapeutical approaches." *Archives of Gynecology and Obstetrics* 288(1):191–199.

15 Hipler, UC, M Görnig, B Hipler, W Römer, and G Schreiber. 1999. "Stimulation and scavestrogen-induced inhibition of reactive oxygen species generated by rat Sertoli cells." *Archives of Andrology* 44(2):147–154.

16 Thun, M. J., B. D. Carter, D. Feskanich, N. D. Freedman, R. Prentice, A. D. Lopez, P. Hartge, S. M. Gapstur. 2013. "50-Year trends in smoking-related mortality in the United States." *New England Journal of Medicine* 368(4):351–364.

17 Kothari, S., A. Thompson, A. Agarwal, S. S. du Plessis. 2010. "Free radicals: their beneficial and detrimental effects on sperm function." *Indian Journal of Experimental Biology* 48(5):425–435.

18 Jensen, T. K., B. L. Heitmann, M. B. Jensen, T. I. Halldorsson, A. M. Andersson, N. E. Skakkebaek, U. N. Joensen, M. P. Lauritsen, P. Christiansen, C. Dalgard, T. H. Lassen, N. Jorgensen. 2013. "High dietary intake of saturated fat is associated with reduced semen quality among 701 young Danish men from the general population." *American Journal of Clinical Nutrition* 97(2):411–418.

19 Afeiche, M. C., N. D. Bridges, P. L. Williams, A. J. Gaskins, C. Tanrikut, J. C. Petrozza, R. Hauser, J. E. Chavarro. 2014. "Dairy intake and semen quality among men attending a fertility clinic." *Fertility and Sterility* 101(5):1280–1287.

20 Barazani, Y., B. F. Katz, H. M. Nagler, D. S. Stember. 2014. "Lifestyle, environment, and male reproductive health." *The Urologic Clinics of North America* 41(1):55–66.

21 Qin, Dan-Dan, Wei Yuan, Wei-Jin Zhou, Yuan-Qi Cui, Jun-Qing Wu, Er-Sheng Gao. 2007. "Do reproductive hormones explain the association between body mass index and semen quality?" *Asian Journal of Andrology* 9(6):827–834.

22 Hadjkacem Loukil, L., H. Hadjkacem, A. Bahloul, H. Ayadi. 2014. "Relation between male obesity and male infertility in a Tunisian population." *Andrologia*. 47(3):282–5.

23 Jensen, T. K., S. Swan, N. Jorgensen, J. Toppari, B. Redmon, M. Punab, E. Z. Drobnis, T. B. Haugen, B. Zilaitiene, A. E. Sparks, D. S. Irvine, C. Wang, P. Jouannet, C. Brazil, U. Paasch, A. Salzbrunn, N. E. Skakkebaek, A. M. Andersson. 2014. "Alcohol and male reproductive health: a cross-sectional study of 8344 healthy men from Europe and the USA." *Human Reproduction*. 29(8)1801–9.

24 Jang, M. H., M. C. Shin, H. S. Shin, K. H. Kim, H. J. Park, E. H. Kim, C. J. Kim. 2002. "Alcohol induces apoptosis in TM3 mouse Leydig cells via bax-dependent caspase-3 activation." *European Journal of Pharmacology* 449(1–2):39–45.

25 La Vignera, S., R. A. Condorelli, E. Vicari, R. D'Agata, and A. E. Calogero. 2012. "Effects of the exposure to mobile phones on male reproduction: a review of the literature." *Journal of Andrology* 33(3):350–356.

26 Chari, Maryam Gholinezhad, and Abasalt Hosseinzadeh Colagar. 2011. "Seminal plasma lipid peroxidation, total antioxidant capacity, and cigarette smoking in asthenoteratospermic men." *Journal of Men's Health* 8(1):43–49.

27 Taha, E. A., A. M. Ez-Aldin, S. K. Sayed, N. M. Ghandour, T. Mostafa. 2012. "Effect of smoking on sperm vitality, DNA integrity, seminal oxidative stress, zinc in fertile men." *Urology* 80(4):822–825.

28 Saleh, R.A., A. Agarwal, R.K. Sharma, et al. 2002. Effect of cigarette smoking on levels of seminal oxidative stress in infertile men: a prospective study. *Fertillity and Sterility* 78(3):491–499.

29 Deepinder, F., M. Cocuzza, A. Agarwal. 2008. "Should seminal oxidative stress measurement be offered routinely to men presenting for infertility evaluation?" *Endocrine Practice* 14(4):484–491.

30 Agarwal, A., S. S. Allamaneni, T. M. Said. 2004. "Chemiluminescence technique for measuring reactive oxygen species." *Reproductive Biomedicine Online* 9(4):466–468.

31 Agarwal, A., Aziz, N. & Rizk, B. (2013) *Studies on Women's Health*. Humana Press, New York, pp. 33–60.

32 Mahfouz, R., R. Sharma, J. Lackner, N. Aziz, A. Agarwal. 2009. "Evaluation of chemiluminescence and flow cytometry as tools in assessing production of hydrogen peroxide and superoxide anion in human spermatozoa." *Fertility and Sterility* 92(2):819–827.

33 Mitchell, LA, GN De Iuliis, and R John Aitken. 2011. "The TUNEL assay consistently underestimates DNA damage in human spermatozoa and is influenced by DNA compaction and cell vitality: development of an improved methodology." *International Journal of Andrology* 34(1):2–13.

34 Sharma, R., J. Masaki, A. Agarwal. 2013. "Sperm DNA fragmentation analysis using the TUNEL assay." *Methods in Molecular Biology* 927:121–136.

35 Pahune, P. P., A. R. Choudhari, P. A. Muley. 2013. "The total antioxidant power of semen and its correlation with the fertility potential of human male subjects." *Journal of Clinical and Diagnostic Research* 7(6):991–995.

36 Takagi, Y., T. Nikaido, T. Toki, N. Kita, M. Kanai, T. Ashida, S. Ohira, and I. Konishi. 2004. "Levels of oxidative stress and redox-related molecules in the placenta in preeclampsia and fetal growth restriction." *Virchows Archiv* 444(1):49–55.

37 Pourova, J., M. Kottova, M. Voprsalova, M. Pour. 2010. "Reactive oxygen and nitrogen species in normal physiological processes." *Acta Physiologica* 198(1):15–35.

38 Vernet, P, RJ Aitken, JR Drevet. 2004. "Antioxidant strategies in the epididymis." *Molecular and Cellular Endocrinology* 216(1):31–39.

39 Sabeur, K, BA Ball. 2007. "Characterization of NADPH oxidase 5 in equine testis and spermatozoa." *Reproduction* 134(2):263–270.

40 Rousseaux, J, R Rousseaux-Prevost. 1995. "Molecular localization of free thiols in human sperm chromatin." *Biology of Reproduction* 52(5):1066–1072.

41 Ménézo, Yves JR, André Hazout, Gilles Panteix, François Robert, Jacques Rollet, Paul Cohen-Bacrie, François Chapuis, Patrice Clément, Moncef Benkhalifa. 2007. "Antioxidants to reduce sperm DNA fragmentation: an unexpected adverse effect." *Reproductive Biomedicine Online* 14(4):418–421.

42 Aitken, R. J., M. A. Baker. 2013. "Causes and consequences of apoptosis in spermatozoa; contributions to infertility and impacts on development." *International Journal of Developmental Biology* 57(2–4):265–272.

43 Aitken, R. J. 2011. "The capacitation-apoptosis highway: oxysterols and mammalian sperm function." *Biology of Reproduction* 85(1):9–12.

44 de Lamirande, E., .C. O'Flaherty. 2008. "Sperm activation: role of reactive oxygen species and kinases." *Biochimica et Biophysica Acta (BBA) – Proteins and Proteomics* 1784(1):106–115.

45 de Lamirande, Eve, and Claude Gagnon. 1998. "Paradoxical effect of reagents for sulfhydryl and disulfide groups on human sperm capacitation and superoxide production." *Free Radical Biology and Medicine* 25(7):803–817.

46 Griveau, J. F., and D. Le Lannou. 1994. "Effects of antioxidants on human sperm preparation techniques." *International Journal of Andrology* 17(5):225–231.

47 Aitken, R John. 1997. "Molecular mechanisms regulating human sperm function." *Molecular Human Reproduction* 3(3):169–173.

48 Wong, J. L., R. Creton, G. M. Wessel. 2004. "The oxidative burst at fertilization is dependent upon activation of the dual oxidase Udx1." *Developmental Cell* 7(6):801–814.

49 Makker, K., A. Agarwal, R. Sharma. 2009. "Oxidative stress & male infertility." *Indian Journal of Medical Research* 129(4):357–367.

50 Sharma R and Agarwal A. 2011. "Spermatogenesis: an overview." *Sperm Chromatin*, Zini A, Agarwal A. Springer, New York. 19–44.

51 El-Fakahany, Hasan M, Denny Sakkas. 2011. "Abortive apoptosis and sperm chromatin damage." *Sperm Chromatin*, 295–306. Springer.

52 Tvrda, E., Z. Knazicka, L. Bardos, P. Massanyi, N. Lukac. 2011. "Impact of oxidative stress on male fertility – a review." *Acta Veterinaria Hungarica* 59(4):465–484.

53 Taylor, A., A. Robson, B. C. Houghton, C. A. Jepson, W. C. Ford, J. Frayne. 2013. "Epididymal specific, selenium-independent GPX5 protects cells from oxidative stress-induced lipid peroxidation and DNA mutation." *Human Reproduction* 28(9):2332–2342.

54 Imai, H., and Y. Nakagawa. 2003. "Biological significance of phospholipid hydroperoxide glutathione peroxidase (PHGPx, GPx4) in mammalian cells." *Free Radical Biology and Medicine* 34(2):145–169.

55 Fujii, J., J. I. Ito, X. Zhang, T. Kurahashi. 2011. "Unveiling the roles of the glutathione redox system in vivo by analyzing genetically modified mice." *Journal of Clinical Biochemistry and Nutrition* 49(2):70–78.

56 Kaneko, T., Y. Iuchi, T. Kobayashi, T. Fujii, H. Saito, H. Kurachi, J. Fujii. 2002. "The expression of glutathione reductase in the male reproductive system of rats supports the enzymatic basis of glutathione function in spermatogenesis." *European Journal of Biochemistry* 269(5):1570–1578.

57 Mancini, A., S. Raimondo, M. Persano, C. Di Segni, M. Cammarano, G. Gadotti, A. Silvestrini, A. Pontecorvi, E. Meucci. 2013. "Estrogens as antioxidant modulators in human fertility." *International Journal of Endocrinology* 2013:607939.

58 Lanzafame, F., S. La Vignera, A. E. Calogero. 2012. "Best practice guidelines for the use of antioxidants." *Male Infertility*, 487–497. Springer.

59 Song, G. J., E. P. Norkus, and V. Lewis. 2006. "Relationship between seminal ascorbic acid and sperm DNA integrity in infertile men." *International Journal of Andrology* 29(6):569–575.

60 Fraga, C. G., P. A. Motchnik, M. K. Shigenaga, H. J. Helbock, R. A. Jacob, B. N. Ames. 1991. "Ascorbic acid protects against endogenous oxidative DNA damage in human sperm." *Proceedings of the National Academy of Sciences of the United States of America* 88(24):11003–11006.

61 Ross, C, A Morriss, M Khairy, Y Khalaf, P Braude, A Coomarasamy, T El-Toukhy. 2010. "A systematic review of the effect of oral antioxidants on male infertility." *Reproductive Biomedicine Online* 20(6):711–723.

62 Bornman, M. S., T. D. du, B. Otto, Muller I. I. P. Hurter, D. J. du Plessis. 1989. "Seminal carnitine, epididymal function and spermatozoal motility." *South African Medical Journal* 75(1):20–21.

63 Rao, A. V., M. R. Ray, and L. G. Rao. 2006. "Lycopene." *Advances in Food and Nutrition Research* 51:99–164.

64 Sekhon, L.H., Kashou A.H., Agarwal A. 2012. "Oxidative stress and the use of antioxidants for idiopathic OATs" In *Studies on Men's Health and Fertility*, R. J. Aitken, A. Agarwal, J. G. Alvarez, 485–516. Humana Press.

65 Ustundag, S., S. Sen, O. Yalcin, S. Ciftci, B. Demirkan, M. Ture. 2009. "L-Carnitine ameliorates glycerol-induced myoglobinuric acute renal failure in rats." *Renal Failure* 31(2):124–133.

66 Mora-Esteves, C., and D. Shin. 2013. "Nutrient supplementation: improving male fertility fourfold." *Seminars in Reproductive Medicine* 31(4):293–300.

67 Ko, E. Y., E. S. Sabanegh, Jr., 2012. "The role of over-the-counter supplements for the treatment of male infertility – fact or fiction?" *Journal of Andrology* 33(3):292–308.

68 Stewart, Adam F, and Edward D Kim. 2012. "Harmful effects of antioxidant therapy." In *Male Infertility*, 499–506. Springer.

69 Türk, Silver, Reet Mändar, Riina Mahlapuu, Anu Viitak, Margus Punab, Tiiu Kullisaar. 2014. "Male infertility: decreased levels of selenium, zinc and antioxidants." *Journal of Trace Elements in Medicine and Biology* 28(2):179–185.

70 Eroglu M., Sahin S., Durukan B., Ozakpinar O.B., Erdinc N., Turkgeldi L., Sofuoglu K., and Karateke A. 2014. "Blood serum and seminal plasma selenium, total antioxidant capacity and coenzyme levels in relation to semen param eters in men with idiopathic infertility." *Biological Trace Element Research* 159(1–3):46–51.

71 Menezo, Yves, Don Evenson, Marc Cohen, Brian Dale. 2014. "Effect of antioxidants on sperm genetic damage." In *Genetic Damage in Human Spermatozoa*, 173–189. Springer.

72 Omu, A. E., H. Dashti, and S. Al-Othman. 1998. "Treatment of asthenozoospermia with zinc sulphate: andrological, immunological and obstetric outcome." *European Journal of Obstetrics, Gynecology, and Reproductive Biology* 79(2):179–184.

73 Collodel, G, MG Federico, M Geminiani, S Martini, C Bonechi, C Rossi, N Figura, E Moretti. 2011. "Effect of trans-resveratrol on induced oxidative stress in human sperm and in rat germinal cells." *Reproductive Toxicology* 31(2):239–246.

74 Wang, H-J, Q Wang, Z-M Lv, C-L Wang, C-P Li, Y-L Rong. 2014. "Resveratrol appears to protect against oxidative stress and steroidogenesis collapse in mice fed high-calorie and high-cholesterol diet." *Andrologia* 47(1):59–65.

75 Suleiman S.A., M.E Ali, Z.M. Zaki, et al. 1996. Lipid peroxidation and human sperm motility: protective role of vitamin E. *Journal of Andrology* 17(5):530–537.

76 Paradiso Galatioto, G., G. L. Gravina, G. Angelozzi, A. Sacchetti, P. F. Innominato, G. Pace, G. Ranieri, C. Vicentini. 2008. "May antioxidant therapy improve sperm parameters of men with persistent oligospermia after retrograde embolization for varicocele?" *World Journal of Urology* 26(1):97–102.

77 Vitali, G., R. Parente, C. Melotti. 1995. "Carnitine supplementation in human idiopathic asthenospermia: clinical results." *Drugs under Experimental and Clinical Research* 21(4):157–159.

78 Lenzi, A., P. Sgrò, P. Salacone, et al. 2004. A placebo-controlled double-blind randomized trial of the use of combined l-carnitine and l-acetyl-carnitine treatment in men with asthenozoospermia. *Fertility and Sterility* 81(6):1578–1584.

79 Cavallini, G., A.P. Ferraretti, L. Gianaroli, et al. 2004. Cinnoxicam and L-carnitine/acetyl-L-carnitine treatment for idiopathic and varicocele-associated oligoasthenospermia. *Journal of Andrology* 25(5):761–772.

80 Mistry, HD., FB Pipkin, CW G Redman, L Poston. 2012 "Selenium in reproductive health." *American Journal of Obstetrics & Gynecology* 206(1):21–30.

81 Ciftci, H., A. Verit, M. Savas, et al. 2009. Effects of N-acetylcysteine on semen parameters and oxidative/antioxidant status. *Urology* 74(1):73–76.

82 Greco, E., M. Iacobelli, L. Rienzi, et al. 2005. Reduction of the incidence of sperm DNA fragmentation by oral antioxidant treatment. *Journal of Andrology* 26(3):349–353.

83 Vezina, D., F. Mauffette, K. D. Roberts, G. Bleau. 1996. "Selenium-vitamin E supplementation in infertile men. Effects on semen parameters and micronutrient levels and distribution." *Biological Trace Element Research* 53(1–3):65–83.

84 Keskes-Ammar, L., N. Feki-Chakroun, T. Rebai, Z. Sahnoun, H. Ghozzi, S. Hammami, K. Zghal, H. Fki, J. Damak, A. Bahloul. 2003. "Sperm oxidative stress and the effect of an oral vitamin E and selenium supplement on semen quality in infertile men." *Archives of Andrology* 49(2):83–94.

85 Moslemi, M. K., S. Tavanbakhsh. 2011. "Selenium-vitamin E supplementation in infertile men: effects on semen parameters and pregnancy rate." *International Journal of General Medicine* 4:99–104.

86 Safarinejad, MR, S Safarinejad. 2009. "Efficacy of selenium and/or N-acetyl-cysteine for improving semen parameters in infertile men: a double-blind, placebo controlled, randomized study." *The Journal of Urology* 181(2):741–751.

87 Tremellen, K., G. Miari, D. Froiland, and J. Thompson. 2007. "A randomised control trial examining the effect of an antioxidant (Menevit) on pregnancy outcome during IVF-ICSI treatment." *Australian and New Zealand Journal of Obstetrics and Gynaecology* 47(3):216–221.

88 Omu, A. E., M. K. Al-Azemi, E. O. Kehinde, J. T. Anim, M. A. Oriowo, T. C. Mathew. 2008. "Indications of the mechanisms involved in improved sperm parameters by zinc therapy." *Medical Principles and Practice* 17(2):108–816.

89 Leitzmann, M.F., M.J. Stampfer, K. Wu, et al. 2003. Zinc supplement use and risk of prostate cancer. *Journal of the National Cancer Institute* 95(13):1004–1007.

90 Miller, E.R. 3rd, R. Pastor-Barriuso, D. Dalal, et al. 2005. Meta-analysis: high-dosage vitamin E supplementation may increase all-cause mortality. *Annals of Internal Medicine* 142(1):37–46.

91 Hemilä H, Kaprio J, Albanes D, et al. 2002. Vitamin C, vitamin E, and beta-carotene in relation to common cold incidence in male smokers. *Epidemiology.* 13(1):32–7.

92 Showell, M. G., J. Brown, A. Yazdani, M. T. Stankiewicz, and R. J. Hart. 2011. "Antioxidants for male subfertility." *Cochrane Database of Systematic Reviews* 19(1):Cd007411.

93 Alpers, DH, WF Stenson, BE Taylor, and DM Bier. 2008. *Manual of Nutritional Therapeutics.* 5th ed. Philadelphia, PA: Lippincott Williams & Wilkins.

94 Holdiness, M. R. 1991. "Clinical pharmacokinetics of N-acetylcysteine." *Clinical Pharmacokinetics* 20(2):123–134.

95 Dawson, D. L., B. S. Cutler, W. R. Hiatt, R. W. Hobson, 2nd,, J. D. Martin, E. B. Bortey, W. P. Forbes, D. E. Strandness, Jr., 2000. "A comparison of cilostazol and pentoxifylline for treating intermittent claudication." *American Journal of Medicine* 109(7):523–530.

96 Patel, K. R., E. Scott, V. A. Brown, A. J. Gescher, W. P. Steward, K. Brown. 2011. "Clinical trials of resveratrol." *The Annals of the New York Academy of Sciences* 1215:161–169.

CHAPTER 16

Role of oxidants and antioxidants in female reproduction

Ashok Agarwal[1], Hanna Tadros[1,2], and Eva Tvrdá[1,3]

[1] American Center for Reproductive Medicine, Cleveland Clinic, Cleveland, OH 44195, USA
[2] College of Medicine, The Royal College of Surgeons in Ireland-Bahrain, Muharraq, Bahrain
[3] Department of Animal Physiology, Slovak University of Agriculture, Nitra, Slovakia

---

THEMATIC SUMMARY BOX

At the end of this chapter, students should be able to:

- Define the term oxidative stress

- Describe the importance of reactive oxygen species and reactive nitrogen species in oxidative stress

- Describe the general physiological roles of reactive oxygen species

- Discuss the significance of antioxidants in our body and their therapeutic role in female infertility

- Describe how endogenous and exogenous free radicals stimulate and initiate disease

- List the different methods used to detect reactive oxygen species

- Discuss how reactive oxygen species play a role in female reproductive physiology

- List the factors that contribute to oxidative stress in the female

- Describe the pathological roles of oxidative stress in the female reproductive tract

---

Introduction

Reactive oxygen species (ROS) and reactive nitrogen species (RNS) are naturally produced in the human body. In fact, they are key by-products in oxygen-utilizing metabolic processes such as oxidative phosphorylation and play essential roles as secondary messengers in intracellular signaling pathways.[1,2] Levels of ROS and RNS are kept within physiologic limits by antioxidants. Thus, normal cell function is a balanced interplay of antioxidant defenses and reactive oxygen and nitrogen species production. As with any concept of balance in the human system, it is assumed that

---

Oxidative Stress and Antioxidant Protection: The Science of Free Radical Biology and Disease, First Edition.
Edited by Donald Armstrong and Robert D. Stratton.

a graded response will occur when it is disrupted, with small and immediate changes manageable by homeostatic mechanisms.[2] Major changes, on the other hand, either due to aggressive increases in ROS or RNS or inadequate antioxidants, lead to a condition called oxidative stress.[3] Oxidative stress is highly pathogenic and associated with reproductive diseases such as polycystic ovary syndrome and endometriosis as well as pregnancy complications including spontaneous abortion and preeclampsia – these will be further outlined in this chapter.

Reactive oxygen species

Characteristics of reactive oxygen species

The production of ROS requires the presence of oxygen. Oxygen is necessary as a final electron acceptor in the production of adenosine-5-triphosphate (ATP) during mitochondrial oxidative phosphorylation to yield water. Nonetheless, what makes oxygen necessary is also what makes it potentially damaging. Oxygen (O_2), by nature, is a highly reactive, diradical species that accepts free electrons readily. When oxygen consumption occurs during electron transfer, radicals or oxygen ions are produced, which are highly reactive due to unpaired electrons in their outermost shell.[4] These short-lived, toxic ROS are produced due to electron leakage from the electron transport chain process toward the end of oxidative phosphorylation.[3] The main type of ROS that is produced includes the superoxide anion – a product of NADPH oxidase reactions and the hydroxyl radical – by spontaneous degeneration of hydrogen peroxide.[5] There are two redox reactions highly implicated in ROS production:

1 The Haber–Weiss reaction:

$$^{\bullet}O_2^- + H_2O_2 \rightarrow {}^{\bullet}OH + OH^- + O_2$$

The Haber–Weiss reaction produces the highly reactive hydroxyl radical ($^{\bullet}OH$), which can also facilitate interactions between the superoxide anion ($^{\bullet}O_2^-$) and hydrogen peroxide (H_2O_2). Hydrogen peroxide is not a free radical *per se*, but is highly implicated in their generation and breakdown.[3] This reaction is thermodynamically unstable and, hence, requires a catalyst to proceed – this is known as the Fenton reaction.

2 The Fenton reaction:

$$Fe^{3+} + {}^{\bullet}O_2^- \rightarrow Fe^{2+} + O_2$$

$$Fe^{2+} + H_2O_2 \rightarrow Fe^{3+} + OH^- + {}^{\bullet}OH$$

In the Fenton reaction, a metal ion acts as a catalyst and produces a net product of hydroxyl radicals ($^{\bullet}OH$). Thus, metallic cations such as copper or iron ions contribute to ROS generation by acting as catalysts.[3]

Other systems contributing to superoxide production include NADPH oxidase, which is a core enzyme in cellular respiration. It is found in the form of NADPH oxidase 1 (NOX1), NADPH oxidase 2 (NOX2), and NADPH oxidase 3 (NOX3) in smooth muscle and vascular endothelium, dual oxidases 1 and 2 (DUOX1 and DUOX2), and NADPH oxidase 4 (NOX4) in epithelial cells.[4]

Furthermore, the short electron chain in the endoplasmic reticulum, cytochrome P450 of the liver, and the enzyme xanthine oxidase also produce ROS.[3]

General physiological roles

ROS are produced as a leukocyte defense mechanism. When foreign microbes are detected, phagocytic leukocytes such as neutrophils and macrophages ingest and kill microbes via ROS-releasing, respiratory bursts in the phagolysosome. This is facilitated by a phagosome membrane enzyme called myeloperoxidase, which produces superoxide ($^\bullet O_2^-$).[5] Subsequently, hydrogen peroxide (H_2O_2), followed by a highly reactive compound called hypochlorite (ClO^-), is generated.[5]

Beyond the immune system, ROS are also vital secondary messengers that activate downstream signal transducing pathways that influence expression and activity of cytokines, ions, and growth factors. As a matter of fact, ROS have been specifically implicated in core apoptotic pathways. The final product of this pathway is the release of cytochrome c and other proapoptotic cell mediators. This is achieved by activating c-Jun N-terminal kinase (a member of the mitogen-activated protein (MAP) kinases family) in response to stress stimuli, which in turn phosphorylates and releases Bcl-2-related proteins.[4] This initiates the uncoupling of protein Bax and its translocation to oligomerize inside the mitochondria.[4] Bax in the mitochondria catalyzes the production of the final proapoptotic mediators. Both Bax (proapoptotic) and Bcl-2 (antiapoptotic) are proteins implicated in the mitochondrial apoptotic pathway (Figure 16.1).

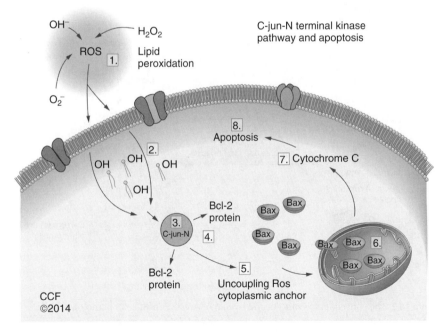

Figure 16.1 c-Jun N terminal kinase pathway and apoptosis. ROS act as secondary messengers that activate core apoptotic pathways via the activation of the c-Jun N-terminal kinase. *(See color plate section for the color representation of this figure.)*

ROS and potential means of damage

When the effects of ROS are either exaggerated or left unchecked, cellular injury occurs. The extent of damage depends on rates of production and removal of ROS.[5] Cell injury manifests as (Figure 16.2):

1 *Lipid peroxidation of membranes:* The cell membrane is a compilation of a variety of macromolecules. However, lipids in the form of polyunsaturated fatty acids (PUFAs) make up the majority of the cell membrane. Thus, being unsaturated, exposed double bonds that form kinks (turns) in the fatty acid chain are susceptible to oxidation.[5] Specifically, at these kinks, PUFAs and ROS react to produce peroxide (O_2^{2-}), which is highly implicated in inducing an autocatalytic chain reaction.[5] Furthermore, excess peroxide may induce apoptosis. It should be noted that lipid peroxidation depends on the enzyme sphingomyelinase and on the subsequent ceramide release.[4]

2 *DNA injury:* Because reactive species are highly diffusible, both nuclear DNA and mitochondrial DNA are susceptible to attack. Highly damaging single-strand breaks occur via thymine reactions.[5] DNA injury is vastly detrimental and produces defects in several downstream processes including future transcription and translation of proteins and even gene expression and cellular adaption. Premature cellular aging, cell death, and malignant transformation could also occur.[5]

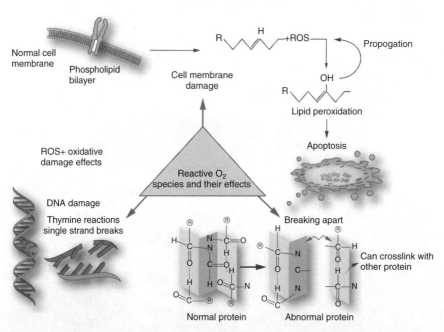

Figure 16.2 The consequences of ROS and oxidative stress. Exposure to pathological levels of ROS and oxidative stress causes cellular membrane and DNA damage along with protein manipulation. *(See color plate section for the color representation of this figure.)*

3 *Protein manipulation:* Proteins, regardless of their degree of structural complexity, pose a viable site of attack for ROS. Free radical interaction may cross-link sulfhydryl molecules, leading to conformational change of the polypeptide and, thus, loss of function or enhanced protein degradation.[5] Direct oxidation or nitrosylation of proteins could also produce carbonyl, nitration, or nitrotyrosine, all of which are capable of disrupting enzyme function as well.[4]

Characteristics of reactive nitrogen species, physiological roles, and mechanisms of damage

Reactive nitrogen species are also vital contributors to oxidative stress. These include both nitric oxide (NO) and nitrogen dioxide (NO_2) and other less reactive species such as nitrosamines and peroxynitrite ($ONOO^-$).[3] Nitric oxide is physiologically fundamental in human vasoregulation and also plays a role in signaling at the cellular level. Nonetheless, although RNS exist at some level in humans, excess sources of RNS in mammals also lead to oxidative stress. RNS are produced mainly through a reaction of oxygen and L-arginine and their reaction with superoxide anions, which ultimately leads to peroxynitrite formation.[3]

To understand how RNS can cause damage, a general overview of the physiological attributes of RNS must be discussed. Nitric oxide production and function depend on its location within the body and nitric oxide synthase (NOS) enzymes present in tissue. For example, NO acts as a neurotransmitter in the central nervous system and is produced by the neuronal nitric oxide synthase (nNOS) enzyme system.[3] NO is also produced in macrophages by inducible nitric oxide synthase (iNOS) and plays a vital role in the initial vasodilation and immune cell recruitment in inflammation.[3]

In the female reproductive system, the endothelial nitric oxide synthase (eNOS) enzyme produces NO at an elevated level in response to luteinizing hormone (LH) and human chorionic gonadotropin (hCG).[3] This is because eNOS activity is sensitive to intracellular calcium, and in a normal pregnancy, sustained or capacitive calcium entry occurs to maintain eNOS activity.[3] Lack thereof will lead to inadequately vasodilated uterine vessels and hypertensive-related complications.[3] On the other hand, overproduction of RNS may lead to an exaggerated response.

Other effects of high levels of RNS include protein structure anomalies, which in turn compromise enzymatic activity, cytoskeletal organization, and cell signal transduction.[3]

Antioxidants

Throughout our evolutionary adaptations, the human body has developed several defense mechanisms to balance levels of reactive species in our bodies. There are repair mechanisms, preventative mechanisms, physical defenses, and antioxidant defenses. The latter can be divided into enzymatic and nonenzymatic molecules.

Nonenzymatic antioxidants

Nonenzymatic antioxidants are generally exogenous and must be ingested in the form of nutrients.[6] These low molecular weight molecules, nonetheless, are necessary.

The most vital molecules include ascorbic acid (vitamin C) and α-tocopherol (vitamin E). They act as dual cofactors, with vitamin C needed to regenerate vitamin E.[2] Other molecules include thiol compounds such as thioredoxin, which breaks down H_2O_2.[2] Once oxidized, thioredoxin reductase in the vicinity will reduce and recycle the thiol compounds, maintaining viable thiol levels.[2] Other nonenzymatic antioxidants include ceruloplasmin and transferrin, which aim to sequester free iron ions – a vital catalyst for the previously mentioned Fenton reaction.[2]

Although these antioxidants are vital, the glutathione redox system is the most significant and abundant. Arguably the decisive antioxidant in oxidative balance, glutathione is a tripeptide, thiol antioxidant, and redox buffer. Glutathione, in its reduced form and in the presence of ROS, becomes oxidized to glutathione disulphide.[7] Unlike other antioxidant patterns in the body, this peptide is found in different concentrations within the cell and is a major soluble antioxidant within cell compartments. Its presence in different cellular compartments suggests that glutathione is involved in an intricate, intracellular transport system. Glutathione, a combination of L-glutamate, L-cysteine, and glycine, is synthesized by glutamate–cysteine ligase and glutathione synthetase – enzymes found only in the cytosol.[7] A concentration gradient drives the transport of glutathione into the mitochondria.[7]

Moreover, as an example of the interplay of enzymatic and nonenzymatic antioxidants, glutathione can act as a cofactor for several detoxifying enzymes, which include glutathione peroxidase (GPx) and glutathione transferase.[7] Glutathione also participates in amino acid transport through the plasma membrane and can take up and neutralize hydroxyl radicals ($^{\bullet}OH$) and singlet oxygen molecules ($^1O_2$) directly.[7] Glutathione regenerates vitamins C and E back to their active forms.[7] More specifically, through the subsequent reduction of semi-dehydroascorbate to ascorbate, glutathione is capable of reducing α-tocopherol radicals to vitamin E.[7] A lack of any of these antioxidants will leave the body, especially the reproductive system, susceptible to oxidative damage.

Enzymatic antioxidants

On the other hand, enzymatic antioxidants are endogenous and are generally more efficient in their role. The core of these antioxidants is a transition metal capable of different valences, which is necessary for the transfer of electrons during detoxification reactions.[2] These antioxidants help maintain homeostatic oxidative balance.

They include the following:

1 *Superoxide dismutases (SOD):* This enzyme catalyzes the dismutation of the superoxide anion – a free radical produced after oxygen reduction and radical chain propagation thereafter.[6] There are many isoforms but the main isoforms are SOD1, SOD2, and SOD3. SOD1 (Cu–Zn SOD), a mainly cytosolic form, contains a core of two metal cofactors forming a copper–zinc enzyme.[6] A specific mutation in this enzyme alone can produce a phenotype of female infertility.[6] SOD2 (Mn-SOD), on the other hand, is the manganese counterpart, encoded by nuclear DNA and restricted in activity to the mitochondria.[2] SOD2 is inducible under various levels

of oxidative stress and inflammatory conditions.[6] SOD3 encodes an extracellular form of this enzyme and is structurally similar to SOD1 but has not been linked to fertility or reproduction.[6] Therefore, SOD enzymes vary in type, location, and makeup.

2 *Peroxidases:* These are a large family of enzymes that preferably use hydrogen peroxide as a substrate. One of the most vital of these is GPx. GPx is a tetrameric selenoprotein that depends on reduced glutathione as a hydrogen molecule donor.[2] With a main role of detoxifying peroxides, the selenocysteine core is critical.[6] In fact, four isoforms exist in mammals, and it is a multisystem antioxidant defense system.

Another vital peroxidase is catalase, which detoxifies hydrogen peroxide but does not require an electron donor.[6] Its role is that of an antiapoptotic and will be detailed in a later section.

3 *Thioredoxin (Trx) system:* The thioredoxin system regulates various enzymes and transactivating factors for genes and is substantially involved in cell growth, differentiation, and death.[6] It is also a repair system rather than just a protective mechanism. Trx is a protein disulfide isomerase that corrects errors such as disulfide bridges and is a cofactor for the DNA-synthesizing ribonucleotide reductase enzyme.[6] Trx is not only necessary to maintain an oxidative balance and limit oxidative damage but is also critical for continued cell survival. This is highly evident, as Trx-deficient mice are embryonically lethal.[6]

Antioxidant treatment for female infertility

The global vitamin and supplement market is an enormous and growing industry, worth \$68 billion.[8] However, it is unregulated and supplements can be purchased at retail stores. This could pose issues of unnecessary supplementation, overdose, and ineffective treatment.

Antioxidants are chemical or biological compounds that include vitamins, minerals, and PUFAs, the latter of which includes omega-3, omega-6, and omega-9.[8] These can be taken as oral supplements. Other antioxidants given as supplements and studied in female infertility include melatonin, folic acid, myo-inositol, zinc, selenium, *N*-acetyl-cysteine, and vitamins A, C, and E, which can be used individually or as a combination therapy.[8] Other commonly used antioxidant supplements include pentoxifylline, a trisubstituted xanthine derivative, and L-arginine.[8] Apart from synthetic, nonenzymatic antioxidants, antioxidant mimetic molecules are currently being developed, including porphyrinic, peptidylic, and phenolic structures of zinc, copper, and manganese complexes that mimic SOD and its enzymatic role.[9]

In present-day medicine, antioxidants are utilized clinically. For example, women take antioxidant supplements to improve their fertility before undergoing assisted reproductive techniques (ART).[8] Theoretically, replacing deficient antioxidants should be effective as a means of treatment. Nevertheless, evidence of the efficacy of antioxidant supplementation is extremely limited, and the literature further demonstrates the controversy behind this topic (Figure 16.3).

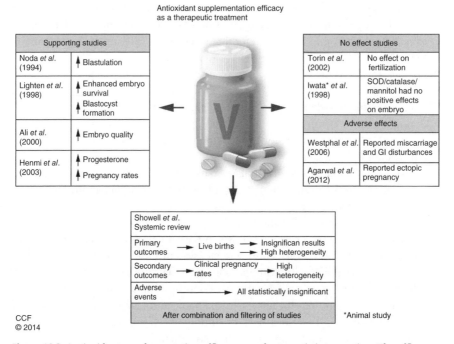

Antioxidant supplementation efficacy as a therapeutic treatment

Supporting studies	
Noda *et al.* (1994)	↑ Blastulation
Lighten *et al.* (1998)	↑ Enhanced embryo survival ↑ Blastocyst formation
Ali *et al.* (2000)	↑ Embryo quality
Henmi *et al.* (2003)	↑ Progesterone ↑ Pregnancy rates

No effect studies	
Torin *et al.* (2002)	No effect on fertilization
Iwata* *et al.* (1998)	SOD/catalase/ mannitol had no positive effects on embryo
Adverse effects	
Westphal *et al.* (2006)	Reported miscarriage and GI disturbances
Agarwal *et al.* (2012)	Reported ectopic pregnancy

Showell *et al.* Systemic review

Primary outcomes	→ Live births →	Insignifican results High heterogeneity
Secondary outcomes	Clinical pregnancy rates	High heterogeneity
Adverse events		All statistically insignificant

After combination and filtering of studies

*Animal study

CCF
© 2014

Figure 16.3 Antioxidant supplementation efficacy as a therapeutic intervention. The efficacy of antioxidant supplementation as a therapeutic intervention remains inconclusive at this stage. The outcomes of studies involving intervention with oral antioxidant supplementation either support its use in alleviating the damaging effects of oxidative stress or show no effects on the reproductive parameters studied (selected studies are shown here).

Although controversial and contradictory, this topic has received some support. A 1994 study by Noda *et al.*, incorporating low oxygen tension and low illumination, showed enhanced blastulation.[10] A 1998 study by Lighten *et al.* used human insulin-like growth factor-I (IGF-1) ligand and reported enhanced embryo survival and blastocyst formation.[11] Other studies, including that of Ali *et al.* in 2000, reported improved embryo quality with a combination of antioxidant regimens.[12] A study done by Henmi *et al.* in 2003 assessed vitamin C supplementation in patients with a luteal phase defect and found that the treatment increased pregnancy rates and progesterone levels.[13] Another study found that vitamin C levels in follicular fluid were higher in supplemented individuals than in controls, but the pregnancy rate between the groups was not statistically significant.[14]

These types of studies shine a bright light on the potential of antioxidant treatments. However, not all studies have found supplementation to be beneficial to fertilization, including a 2002 study by Tarin *et al.*, which used 62.5 μmol/l of ascorbate in a human tubal fluid medium and found no correlation with increased probability of fertilization.[15] Moreover, antioxidants can have side effects. Indeed, vitamin C is a prooxidant at high doses and may cause further reproductive damage.[14]

With such an abundant amount of studies with contradicting results, a systematic review with predefined inclusion and exclusion criteria was performed by Showell *et al.* and published in the Cochrane Library in 2013.[16] This systematic review was significant because it excluded quasi-randomized trials and studies that tested antioxidants alone against fertility drugs as controls, as well as studies that exclusively included fertile women attending fertility clinics due to male factor infertility. Hence, the review included randomized controlled trials and crossover trials, which were limited to only first-phase data usage. Participants in these studies also were either subfertile women who had been referred to a fertility clinic and who may or may not have undergone ART. The participants were randomized to treatment groups taking oral antioxidant supplementation versus control groups (placebo or no treatment/standard treatment). Other interventions that were included are individual or combined oral antioxidants versus any antioxidant (head-to-head trials), pentoxifylline versus control (no or standard treatment), and antioxidants versus fertility drugs such as metformin and clomiphene citrate. The desired primary outcomes included live birth rate per woman, and secondary outcomes included clinical pregnancy rate per women and any adverse effects reported by the trials.

Primary outcomes showed that antioxidants were not associated with an increased live birth rate. In fact, reports on these studies revealed that among subfertile women with a pretreatment live birth rate of 37%, those on antioxidant therapy exhibited live birth rates between 10% and 83%.[8] This indicates the heterogeneity of the results. Even in the two specific trials that reported a live birth, it was shown that the results were overestimated. In trials comparing antioxidants, both the odds ratio and confidence intervals indicated a possibility of not having an effect. This included a study comparing supplementation with myo-inositol plus folic acid and melatonin versus just myo-inositol and folic acid, as well as a study that compared supplementation with myo-inositol versus D-chiro-inositol.[8] Hence, these studies had statistically insignificant results.

Secondary outcomes included clinical pregnancy rates, and even with a random-effects model, the heterogeneity remained high and, thus, no relation was shown. Although none of the single antioxidants, including melatonin, vitamins E and C, L-arginine, and *N*-acetyl-cysteine, led to clinical pregnancy; combined antioxidant therapy was associated with increased clinical pregnancy rates. Yet, heterogeneity remained high. Only one study assessed the use of vitamin E, ascorbic acid, and L-arginine as combination therapy, which made it impossible to pool results. In that study, the authors found no effect on clinical pregnancy rates.

In regard to another secondary outcome measures, the systematic review found no association between antioxidants (*N*-acetyl-cysteine, zinc, biotin, selenium, etc.) and adverse events such as multiple pregnancies and miscarriage. A few studies reported gastrointestinal disturbances and ectopic pregnancies, all of which were statistically insignificant.

From the findings of this systematic review, we can conclude that as of now, there is no viable indication or evidence suggesting that antioxidant efficacy in the clinical setting is an effective treatment for subfertility. Nonetheless, as much as there is a lack of evidence for their efficacy, there is also a lack of evidence suggesting that antioxidant intake causes adverse events. Treatment with antioxidants may not be effective, but it is also highly unlikely to be harmful.

Methods of detection of ROS in the female

Oxidative stress biomarkers such as ROS and the end-products of ROS reactions can be measured and used to assess oxidative damage in tissues and biological fluids. Prime locations of these biomarkers in the female reproductive system include the peritoneal, amniotic, and follicular fluids. The major methods used to measure biomarkers of oxidative stress are listed and detailed in Table 16.1.

Physiological roles and sources of ROS

Although ROS are dangerous in excess, their presence in moderate amounts is critical for normal cellular function. They play a physiological role in nearly every human system, including the female reproductive system.

Hence, this section will outline how ROS play a vital role in the menstrual cycle, a process in which a single ovum is released monthly and the uterine endometrium is prepared for implantation. A total of 6–12 primary follicles are formed per cycle. Initial folliculogenesis begins when the ovum and other layers of the granulosa layers grow, with the formation of ovary interstitium-derived spindle cells in the form of theca interna (responsible for steroidogenesis) and the capsule, theca externa. Granulosa cells secrete follicular fluid, and the antrum appears.

Oogenesis and folliculogenesis

In the ovaries, parenchymal steroidogenic cells, endothelial cells, and phagocytic macrophages produce ROS.[17] Initially, it was thought that ROS were present in the ovaries and played a significant role there because ovarian tissue expressed markers of oxidative stress. Also, adverse effects were thought to occur in the absence of ROS.

An example of oxidative stress biomarkers includes antioxidants such as Cu-SOD, Zn-SOD, and Mn-SOD, as well as GPx glutamyl synthetase and lipid peroxides.[17] Decreased antioxidant levels were associated with negative effects on fertilization.[17] Further studies have also suggested that GPx and Mn-SOD can be used as markers to detect oocyte maturation, thus implying that ROS are involved in this process as well. Moreover, aldose reductase and aldehyde reductase are found in granulosa cells and epithelia of the genital tract.[17] NO also has a physiological significance. High levels of NO have adverse effects on cleavage rates, implantation rates, and overall embryo quality.[17] High levels of NO have been positively correlated with peritoneal factor infertility.[17] Therefore, it is apparent that ROS/RNS are present in the female reproductive tract and that they play important roles. The rest of the section will discuss the specifics of their physiological roles in folliculogenesis.

Each month, a cohort of oocytes begins to grow and develop in the ovary, but only meiosis I resumes in one of the embryos, which becomes the dominant oocyte. This initial process is promoted via ROS and is inhibited by antioxidants.[3] However, ROS inhibits and antioxidants promote this process when progressing into meiosis II.[3] During growth, steroid production in the growing follicle causes an increase in P450, which leads to ROS production. ROS is also generated by the post-LH surge inflammatory precursors.[3] Thus, physiological levels of ROS regulate oocyte maturation.

Table 16.1 Overview of the major methods of ROS and RNS detection.

Chemiluminescence assay and luminometers	• *Overview:* This method excites molecules and then measures the amount of light they emit in a selected amount of time. The oxidative end-products produced by an *in vitro* reaction between a reagent and ROS are detected via light emission as measured by a luminometer.[18] Luminometers measure light intensity using a photomultiplier tube to detect photons on a luminol-stained sample.[18] Intracellular and extracellular radicals can be measured.[18] The reaction is of a short half-life, and subsequently the probe-sample reaction is rapid.[18] Nonetheless, it is not possible to differentiate between the types of ROS detected. Other reagents can be used for this task such as lucigenin, which is specific for extracellular superoxide anions.[18]
	• *Instrumentation:*
	  (a) *Photon counting luminometer* – measures individual photons
	  (b) *Direct current luminometer* – measures electric currents that are proportional to the photon flux in the photomultiplier tube.[18]
	• *Means of error:* Although it is a commonly used method, several factors can affect its accuracy including reactant concentration, sample volume, temperature regulation, reagent administration, and background luminescence.[18] The most common operator issue is consistency with sample and reagent volumes, which has been eliminated via automation.[18] Nonetheless, the operator should be aware of these factors.

Flow cytometry	• *Overview:* This method can be used to detect changes in cells via fluorescence emitted by ROS-binding probes. Flow cytometry detects individual intracellular reactive oxygen radicals.[17] What follows is oxidation of 2,7-dichlorofluorescein diacetate by these radicals and subsequent fluorescence emission.[17] Using hydroethidine to measure intracellular superoxide is possible, and it will emit a red fluorescein color when oxidized.[17] • *Advantages:* Studies evaluating oxidative stress and damage in sperm found that flow cytometry is more sensitive, accurate, and specific than chemiluminescence.[19]
General microscopy and stain	• *Overview:* This method is a simple and cost-effective but limiting tool. In order to detect ROS, a simple histochemical method called nitroblue tetrazolium staining is used. This compound is reduced in the presence of ROS to form a bluish-black insoluble compound, dubbed formazan.[17] • *Method:* Microscopy usage is strenuous as the operator must observe 100 consecutive cells under oil immersion. Cells are then graded based on density of formazan granules inside the cells.[17]
Epifluorescence microscopy	• *Overview:* This is another method of microscopy where an end-product is detected via fluorescence. A similar method was mentioned in flow cytometry. Hydroethidine is added to react with superoxide, and a red-fluorescence emitting product called ethidium bromide is produced.[17] This technique is more commonly used because the equipment is less complex and expensive.[17]
Enzymatic antioxidants	• *Overview:* In this method, both nonenzymatic and enzymatic antioxidants are measurable and are good indicators of ROS in a sample. The latter can be measured depending on the enzyme that is present.

(continued)

Table 16.1 (*Continued*)

	• *Measurement enzymes measured:* Superoxide dismutase activity is detected via a tetrazolium salt, and the formed chromophore is then measured at its maximal absorbance of 525 nm.[17] Glutathione peroxidase can be measured via a kinetic colorimetric assay that will measure the indirect glutathione reductase-coupled reaction at its maximal wavelength absorbance.[17] Catalase can be detected via a CAT assay based on the reaction of this enzyme with methanol in the presence of hydrogen peroxide, which produces formaldehyde.[17]
Immunohistochemistry and Western blotting	• *Overview:* This method can measure oxidative DNA adducts via immunohistochemistry, as exhibited in a study done by Takagi *et al.* where oxidative stress molecules such as 8-hydroxy-2′-deoxyguanosine (8-OHdG), 4-hydroxynonenal (4-HNE), thioredoxin (Trx), and redox factor-1 (ref-1) were detected via similar methods.[20] Incubation with enzyme-linked secondary antibodies and diaminobenzidine staining could reveal the presence of these molecules.[17] As an alternative, Western blotting may be used to as a means of detecting specific proteins in a sample.
Total antioxidant capacity (TAC)	• *Overview:* This method assesses the actual antioxidant power present in the sample, as both endogenous and nutrient-derived antioxidants are measured. Total antioxidant assays are similar, in that they all rely on the antioxidants in the sample to prohibit oxidation of 2,2′-azino-di-ethylbenzthiazoline sulfonate by metmyoglobin.[17] This is then compared with a control solution.[17]
Enzyme-linked immunosorbent assay (ELISA)	• *Overview:* This method can be used to detect certain antigens or antibodies that are produced as by-products of ROS interactions or production. Fluid is analyzed for targeted molecules via a spectrophotometer. Antibodies for certain antioxidants can also be used.
Nitrate reductase and Griess reaction	• *Overview:* In this method, nitrate and nitrite can be detected via nitrate reductase and the Griess reaction as well as total nitric oxide levels in serum via rapid response chemiluminescence assay.
Other methods	• Other methods include measuring nitric oxide levels, using enzyme-linked immunosorbent assays (ELISA) to detect damaged proteins, and measuring lipid peroxides, protein oxidation, total plasma lipid hydroperoxides, total 8-F2-isoprostane, and fat-soluble antioxidants.[17] Further methods include the Raman analysis, near-infrared spectroscopy, and protein nuclear magnetic resonance.[17]

Similarly, ROS have a dominant presence in the follicle and act as the main inducers of ovulation in the preovulatory follicle. Oxygen deprivation promotes the angiogenesis that is necessary for adequate growth and development of the ovarian follicle, and ROS may play a role in this process as well.[3] Another physiological role is the induction of apoptosis. Follicular GSH and FSH counterbalance this action in the growing follicle.[3] This is shown in the dominant follicle, which experiences a surge of estrogen due to FSH, which in turn triggers the generation of catalase in the dominant follicle, lowers ROS levels, and thus, avoids apoptosis.[3]

Atresia accounts for up to 99% of germ cell depletion in the female.[21] In folliculogenesis, only one of the 6–12 follicles proceeds to mature. This is not regulated by hormones but by an intrinsic positive feedback, which studies suggest is ROS related.[21,22]

Although the apoptosis pathway of follicular atresia can be hormonally influenced, molecular mechanisms regarding the initiator of apoptosis remained vague until not long ago. Recently, key studies have elucidated the specific processes that involve ROS. A study by Tilly's group used Northern blotting analysis to study equine chorionic gonadotropin (eCG)-induced follicular growth and survival. They reported steady counts of Bcl-2 and Bcl-x expression with lower levels of Bax mRNA.[21] Bcl-x long was also dominant in granulosa cells in eCG primed ovaries.[21] With Bcl-2 and Bcl-x both being present and having primary antioxidant properties in the surviving follicle, we can assume that ROS play a role in follicular atresia. Researchers thereafter went on to develop a study with preovulatory follicles isolated from ovaries of immature rats. The follicles were separated into a group with medium alone, FSH and medium, and FSH and buthionine sulfoximine (BSO) – a specific inhibitor of glutathione synthesis.[22] As FSH is known to inhibit apoptosis, the study's most significant discovery was that the antiapoptotic effect of FSH was reversed in the group in which glutathione synthesis was inhibited (FSH and BSO group).[22] In fact, confocal microscopy showed that ROS were present in apoptotic follicles and that FSH significantly suppressed their production.[22] Thus, the antiapoptotic FSH works by inducing glutathione production, and oxidative stress plays a role in the remaining follicles that undergo apoptosis. In another study, researchers dissected goat ovarian follicles and found that the large follicles exhibited greater catalase activity than other granulosa cells from small- or medium-sized follicles.[17] Accordingly, ROS and oxidative stress are proven to play a role in follicular atresia.

Corpus luteum apoptosis and steroidogenesis

After ovulation, the corpus luteum is formed and secretes progesterone, which is vital to conception and pregnancy. In fact, when pregnancy does not occur, the corpus luteum disintegrates and regresses. During degeneration, ROS are produced via the monooxygenase reaction due to steroid hormone synthesis.[23] These P450 systems, coupled with normal levels of ROS in electron transport chain leakage reactions, cause inevitable damage.

Here, ROS play a role in progression and regression of the corpus luteum. In progression, progesterone levels decline and levels of Cu-SOD and Zn-SOD antioxidants increase during the early to mid-luteal phase.[3] Conversely, they decrease if regression occurs.[3] Lipid peroxide levels increase during luteal regression and decrease during luteal progression.[3] Levels of Cu-SOD and Zn-SOD decrease due to prostaglandin F2-α stimulating accumulation of the superoxide anion by luteal cells and nearby phagocytic leukocytes.[3] Nonetheless, Mn-SOD is directly associated with inflammatory reactions induced by oxidative stress.[3] Confirmatory to what was previously mentioned, studies have found that ROS concentrations in the rat ovarian corpus luteum increased during the regression phase, and that the prime identified antioxidants are SOD.[14]

Endometrium and endometrial cycle

The endometrium is a layer of epithelium lining the uterus that experiences cyclical degradation and regeneration. The stages of the endometrial cycle include the proliferative, secretory, and menstrual phase. ROS play a role in these stages and their progression.[24]

This was first discovered when variations of antioxidant SOD levels in the endometrial cycle were reported, as fluctuating SOD levels and elevated lipid peroxide levels were reported specifically in the late secretory phase, directly before the onset of menstruation.[20] Thus, ROS could play a regulatory role in endometrial shedding. The source or reason for heightened ROS production remains unknown. Nonetheless, a study by Serviddio *et al.* observed how varying hormone levels influenced redox reactions and lipid peroxidation in human endometrial cells.[25] They found that hormonal patterns were able to regulate GSH levels and GSH metabolism.[25] Thus, hormone changes across the endometrial cycle could mediate antioxidant activity. This was later supported by several literature reports, suggesting that estrogen and progesterone withdrawal could be the prime instigator in ROS elevation. When these hormones were removed from endometrial cells *in vitro*, antioxidant SOD activity dropped.[24,25] Thus, ROS are key players in the endometrial cycle. Their role is based on their ability to activate nuclear factor-kappa B (NF-κB), which leads to increased COX-2 mRNA expression and prostaglandin F2-α synthesis.[24] This physiological impact is detrimental to the endometrial cycle.

In other states, ROS play a potentially pathogenic but defensive role. This is indicated as vascular endothelial growth factor (VEGF, proangiogenic factor) and Angiopoietin-2 (Ang-2, angiogenic antagonist) regulate endometrial vasculature formation and are regulated by both hypoxia and ROS.[24] When, for instance, long-term progestin contraceptives are used and subsequent hypoxia and ROS production ensue, abnormal angiogenesis occurs as VEGF and Ang-2 are upregulated in an attempt to maintain homeostasis.[24]

With RNS, NO has also been implicated in the regulation of endometrial microvasculature. This was touched on earlier in the chapter and NO's vital role in vasomotor functions and regulation was explained. Endothelial NOS mRNA was observed specifically in mid-secretory and late secretory phases, implying its significance in decidualization in the preparation for pregnancy and menstruation.[24]

Factors contributing to oxidative stress in the female

Several factors that generate oxidative stress play a role in female fertility and pregnancy (Figure 16.4). In general, the first trimester is a vital point in development and carries the highest risk for miscarriage.[3] Obesity, malnutrition, drug and/or alcohol use, smoking, and exposure to environmental toxins have their greatest effects in the second trimester and can lead to poor fetal viability and influence outcomes in the third trimester.[3]

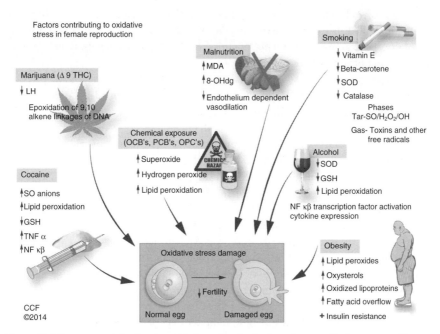

Figure 16.4 Factors contributing to oxidative stress in the female reproductive system. Factors that contribute to the generation of oxidative stress in the female reproductive system include obesity, malnutrition, alcohol intake and smoking, misuse of drugs such as marijuana and cocaine, and exposure to environmental toxins.

Obesity and overnutrition

Today, obesity has become epidemic-like and its negative effects on human health have been documented extensively. Naturally, the number of obese women of reproductive age has also increased. More specifically, two-thirds of the female population in the United States who are within the reproductive age are considered either obese or overweight.[3] In fact, obesity and overnutrition are vital when it comes to female infertility as obese women take longer to conceive and have higher risks of miscarriage than women with a normal BMI.[3] Therefore, it is important for clinicians to educate their patients about obesity and its link to reproductive diseases and complications.

Linking obesity to oxidative stress has been a sequential process. Visceral fat is associated with disordered metabolism and insulin resistance.[3] Moreover, centrally stored fat deposits are more likely to exhibit fatty acid overflow.[3] Thus, preferential storage of fat in the abdomen and visceral areas leads to lipotoxicity.[17] Lipotoxicity is the main culprit in oxidative stress, which in turn can lead to endothelial dysfunction.[17] In addition, obesity is also linked to ROS generation as intracellular fat deposits disrupt mitochondrial function.[3] This disruption leads to electron accumulation and leakage from the electron transport chain, which is a prime pathway for ROS generation. Lipid peroxides, oxidized low-density lipoproteins (oxLDL), and oxysterols are thus produced as a combined reaction between high lipid levels and oxidative stress.[3] It should be noted that in the ovaries, the mitochondrial energy production is vital to the embryonic metabolism of the oocyte.[3] Thus, mitochondrial dysfunction will directly influence embryonic metabolism. Another means of production of ROS is through increased plasma nonesterified fatty acid levels, which prompt the formation of the nitroxide radical.[3]

Oxidative stress can also be linked to obesity due to adipose tissue physiology. Adipose tissue has a bilateral relationship with the gonads, as fat secretes adipokines such as leptin, ghrelin, and resistin.[26] Of these, leptin plays a regulatory role in early embryo cleavage and development; it also stimulates the Hypothalamic-pituitary-adrenal axis and inhibits developing ovarian follicles.[26] Obese individuals also have impaired lipid and glucose metabolism due to GLUT-4 translocations, insulin–insulin receptor binding, and postreceptor signaling, along with decreased lipoprotein lipase activity.[27] With these impairments, mitochondrial dysfunction occurs, energy metabolism is impaired, and oxidative stress ensues.

The processing of macronutrients is associated with ROS production as well. When consumed and processed, don't need that there, but you can keep it. High carbohydrate or fat intake is proportional to an increase in oxidative stress biomarkers. Hence, overnutrition is also an issue. This type of oxidative stress is dubbed postprandial and is directly correlated with both postprandial glycemia and lipemia.[27] Its severity is immensely increased in diseased individuals such as those with diabetes and heart disease. Obese patients experience higher rates of disease than their healthy cohorts and are thus targets of higher postprandial oxidative stress. In fact, obese individuals have been shown to have higher resting and fasting levels of oxidative stress biomarkers compared to nonobese individuals.[27] This also is true in regard to exercise-induced oxidative stress.[27] Regardless, systemic oxidative stress has implications in all systems in the circulation, including the female reproductive system.

As mentioned earlier, endothelial dysfunction can be increased by obesity. Insulin resistance and visceral adiposity increase inflammatory reactants, diminish blood flow to skeletal muscle, and further increase ROS levels.[28] A vicious circle of endothelial dysfunction and formation of OS ensues.

Obesity also plays a role in pregnancy. Pregnancy entails an increased state of metabolic demand, which is necessary to support both maternal hormonal physiology and normal fetal development. It should be noted that a healthy pregnancy is associated with mobilization of lipids as well as increased lipid peroxides, insulin resistance, and enhanced endothelial function.[3] Obese women, on the other hand, experience increased lipid peroxide levels and limited progression of endothelial function during their pregnancies, along with an additive innate tendency for central fat storage.[3] Normally, increases in total body fat peak during the second trimester. Excessive weight gain in pregnancy can cause further complications. It also should be noted that both obese and lean women gain a similar amount of fat during pregnancy. Nonetheless, lean women tend to gain excess fat in their lower extremity or thighs, whereas obese women gain fat in their trunk and visceral areas (central obesity).[28] Preferential storage of fat in these areas leads to lipotoxicity, and its oxidative effects can be devastating. In pregnancy, oxidized lipids can inhibit trophoblast invasion and influence placental development, lipid metabolism and transport, and fetal developmental pathways.[28]

Malnutrition

In pregnancy, undernutrition leads to impaired or stunned fetal growth as well as higher rates of low birthweight and potential endothelial dysfunction.[3] *In utero*, NO levels are diminished and endothelium-dependent vasodilation is impaired.[3] In studies of undernourished dams, the offspring exhibited decreased SOD activity and increased superoxide anions, which increases NO scavenging.[3] Undernutrition during critical periods of fetal or embryonic growth can lead to overall increases in oxidative stress in the female offspring's ovaries.[3] In fact, oxidative stress coupled with mitochondrial antioxidant defense impairment may decrease primordial, secondary, and antral follicles in the offspring.[3] A study by Bernal *et al.* found that ovarian follicle numbers and mRNA levels of regulatory genes in the offspring of malnourished rats were decreased, and that this was mediated by a mechanism of oxidative stress in the ovaries and a diminished antioxidant system.[29] This is relatable to human maternal undernutrition.

Nutrition is also a time-dependent factor. Pregnant adolescents experience two states of high metabolic demand because both puberty and pregnancy require additional nutrition to sustain growth. Thus, a tug-of-war scenario ensues.

Alcohol

Alcohol (ethanol) and mechanisms of ethanol breakdown are correlated with female fertility. Specifically, alcohol can increase apoptosis, damage tissue, and alter cell structures by intensifying oxidative stress both directly and indirectly.

Hepatic oxidative metabolism plays a prime role in the primary elimination of ethanol.[3] Specifically, ethanol is dehydrogenated to produce acetyl aldehyde and is further metabolized to produce a combination of acetic acid and acetyl and methyl radicals.[3] Naturally, these are ROS, and oxidative stress is thus a component of ethanol metabolism. Nonetheless, occasional or minimal alcohol ingestion is balanced by the endogenous antioxidant systems. Regular alcohol ingestion, on the other hand, leads to an overproduction of ROS, excessive SOD and GSH antioxidant depletion, and excess lipid peroxidation.[3] Further studies have shown that acetyl aldehyde is highly implicated in ROS production, and that its presence may potentially "propagate redox cycling and catalytic generation of OS."[3]

More specifically, ethanol metabolism is linked to three metabolic pathways: alcohol dehydrogenase system, microsomal ethanol oxidation system (MEOS), and catalase system.[30] MEOS, specifically, aggravates oxidative stress directly as well as indirectly by impairing defense systems.[30] In fact, this metabolic pathway involves alkylation of hepatic proteins, and hydroxyethyl radicals are produced as coproducts.[30] The primary elimination of ethanol has been shown to potentiate the oxidation seen in the Mallard reaction, leading to a overproduction of advanced glycation end-products (AGE), which can be toxic in high numbers.[3] In fact, a study by Kalousova *et al.* found that serum levels of AGE are more abundant in people who are chronic alcoholics than in those who are nonalcoholics.[31] This is supported by the fact that AGE production in chronic alcoholics may be high due to a lack of antioxidant systems, with malnutrition, cachexia, and vitamin deficiencies being common in this patient group.[3] The physiology behind AGE products is that they bind with the receptor for advanced glycation end-products (RAGE) responsible for activating the NF-κB transcription factor and the subsequent cytokine expression.[3] Several studies have also found that antioxidant supplementation in form of vitamin B derivatives and vitamins A, C, or E may decrease production of AGE.[3] Thus, accumulation of AGE has been linked to an inflammatory state and a reflex upregulation of antioxidant systems, which insinuates that ROS are involved in the process.[3] Moreover, weakened or strained antioxidant systems may spur free radical formation in light of minimal AGE elevation.

As explained earlier, NO is vital in homeostasis of the body when it comes to regulation of vascular tone. Nonetheless, it may have cytotoxic effects in excess. Stable metabolites of nitrate and nitrite are increased in alcoholics.[30] High concentrations of NO have also been linked to vascular and endothelial dysfunction.[30] In the female reproductive system, NO in excess will disrupt the normal vascular tone of uterine vessels. Alcoholics also exhibited increased levels of oxidized LDL, which is pathogenic for atherogenesis and can lead to decreases in enzymatic antioxidants such as SOD and GPx, possibly due to their exhaustion.[30]

A Danish study found that alcohol consumption is related to delayed conception.[3] Women older than 30 years who consumed seven or more beverages per week experienced a higher rate of infertility.[3] Thus, alcohol may hasten age-related infertility in women.

Maternal alcohol ingestion undoubtedly negatively affects the fetus *in utero*. Alcohol's effects on the fetus are amplified in pregnancy, in part, due to the low levels of alcohol dehydrogenase in the fetal liver, with higher rates of low birthweight babies, congenital anomalies, and intrauterine growth retardation.[3] In fact, spontaneous abortion and early pregnancy loss are also hastened by alcohol consumption in pregnancy.[3] Alcohol metabolism leads to a prooxidant environment and thus, the antioxidant protective systems of the placenta will be depleted, and oxidative placental damage plays a large role in the pathophysiology.[3] Gauthier *et al.* reported that in pregnant women, drinking more than three alcoholic beverages at a time produces stressful systemic oxidative stress.[32] In fact, women who drank during pregnancy exhibited significant depletion of GSH, with higher rates of oxidized glutathione molecule (GSSG), during the postpartum period. Thus, alcohol plays a significant role in inducing oxidative stress.

Smoking

The negative effects of cigarette smoking are well documented and researched. The most addictive and toxic component of cigarettes is nicotine. One-third of women of reproductive age smoke cigarettes. Nicotine receptors and their actions have been linked with female reproductive pathologies. Only recently has oxidative stress become a prime focus.

Prooxidants and toxic chemicals in cigarette smoke are abundant and can be found in the two phases of smoking. The first phase is known as the tar or particulate phase.[3] Water-soluble constituents of tar, when inhaled, react to form ROS, which include the superoxide anion, hydrogen peroxide, and hydroxyl radicals – the latter of which is notorious for its role in DNA damage.[3] The gas phase, on the other hand, includes free radicals and toxins as well.[3] NO is also produced via smoking, and overproduction of NO will lead to production of the RNS known as peroxynitrite.[3]

This process begins when nicotine is inhaled and is oxidized into nicotine iminium and monoamine iminium, both of which have high reduction potentials.[3] This high free radical state will lead to the depletion of antioxidant systems such as vitamin E, β-carotene, SOD, and catalase.[3]

The effects of cigarette smoking on female fertility have been assessed in a number of studies. A meta-analysis based on 12 studies reported that smokers were 1.6 times more likely to experience infertility and increased time to conception compared to nonsmokers, in a dose-dependent manner with the number of cigarettes smoked.[33,34]

Assisted reproductive techniques are less successful in smokers as well, perhaps because ROS of exogenous origin have direct effects on the follicular microenvironment and are correlated with decreased β-carotene antioxidant systems.[33] Chelchowska *et al.* found that cigarette smoking depleted other antioxidants such as vitamin A, while other studies reported high lipid peroxidation in the follicular environment.[35] Passive smoking also had similar effects.

As for smoking during pregnancy, complications and embryo damage are common. In fact, smokers have higher rates of fetal loss, preterm deliveries, decreased fetal growth, and spontaneous abortions because placental transfer of nicotine and carbon monoxide lead to placental hypoxia.[3]

Recreational drugs

Marijuana is the most commonly used recreational drug worldwide and many leaders have called for its legalization. Nonetheless, its effects on female reproduction are evident and in pregnancy, it poses several dangers.

The main active constituents of marijuana are the cannabinoids. Cannabinoids generate free radicals when metabolized, directly affecting the peripheral and central nervous systems.[3] The fundamental component of cannabinoid is delta-9-tetrahydrocannabinol or THC, which is also responsible for producing the psychological effects attributable to this drug.[3] Cannabinoids bind to cannabinoid receptors, which include CB1 and CB2, both of which are a superfamily of G-protein-coupled receptors.[36] The first indication that marijuana may play a role in reproduction occurred when endocannibinoid receptors were localized to female reproductive organs such as the ovaries and the uterine endometrium.[36] Thus, exogenous administration of cannabinoid agonists may alter the reproductive processes within the female. THC, administered acutely, has also exhibited an ability to suppress LH.[36] In fact, this is a time-dependent process, depending on the stage of the menstrual cycle.[36] During the luteal phase, 30% of LH production is suppressed with marijuana exposure.[36] In the follicular phase, however, this effect does not occur.[36]

THC has been shown to disrupt the menstrual cycle by directly inhibiting oogenesis.[36] The mechanism of action is hypothesized to be ROS related as the negative effects of marijuana can be counterbalanced by antioxidant (i.e., vitamin E) supplementation.[3] Also, a study by Sarafian *et al.* in 1999 found that marijuana's effects are dose dependent, and that a group of controls did not exhibit increased ROS production.[37] In fact, it was later shown that generation of ROS through THC occurs via epoxidation of the 9, 10 alkene linkages in DNA.[3]

During pregnancy, THC can inhibit initial implantation and disturb embryo development.[3] The placenta is also capable of transporting THC, which accounts for its buildup in reproductive fluids.[3] Moreover, THC-exposed fetuses had low birthweights, prematurity, congenital abnormalities, and intrauterine growth retardation.[36]

Another drug of importance to fertility is cocaine. Cocaine is a potent stimulant that is highly addictive and highly detrimental to the body both psychologically and physically. When taken, cocaine is immediately oxidized into several metabolites that lead to lipid peroxidation and production superoxide anions and lipid peroxyl radicals.[3] Formaldehyde and norcocaine are oxidative metabolites, the latter of which could produce NO or peroxynitrite.[3] Oxidative stress thus leads to GSH depletion and further oxidative damage such as apoptosis. The antioxidants thiol and deferoxamine were found to inhibit apoptosis during cocaine use, suggesting that ROS play a role in this process.[3] Cocaine also has an inherent vasoconstrictive nature that can affect the uterine and placental vasculature, leading to hypoxia and further decreased GSH levels and heightened GSSG levels.[3] In fact, Lee *et al.* found a dose-dependent reduction of GSH levels as well as an inflammatory state with heightened expression of TNF-α and NF-κB in cocaine users.[38]

In pregnancy, cocaine use can cause adverse outcomes such as intrauterine growth retardation, miscarriage, and low birthweight by causing peroxidative damage to underdeveloped fetal membranes.[39] Cocaine use and subsequent ROS and RNS production also lead to limb defects, making it teratogenic as well.[39]

Chemical compound exposure

Female infertility can occur due to environmental and occupational exposure to toxins and chemicals. These substances are capable of deregulating a balanced oxidative environment, and our continued exposure to them via air, soil, contaminated food, and water poses a danger. Organochlorine pesticides (OCP), for example, accumulate in the body over time and are toxic to nerve fibers. OCPs are lipophilic and hydrophobic and thus are broken down slowly.[3] They also can accumulate in the embryo, fetus, blood, placenta, and other parts of the female reproductive system.[3] Hexachlorocyclohexane (HCH), a type of OCP, damages cell membranes via lipid peroxidation and increases superoxide radical and hydrogen peroxide production, leaving mitochondria and microsomes more susceptible to damage.[40] In a study by Pathak *et al.* (2011), a xenobiotic (insecticidal isomer in OCPs known as γ-HCH) was positively correlated with markers of oxidative stress, which included 8-OHdG, MDA, and protein carbonyl levels.[40] Moreover, as it has a tendency to conjugate with GSH, this antioxidant system was depleted, and the ferric reducing ability of plasma (FRAP, a measure of total antioxidant power) was low.[40]

An organochlorine insecticide commonly used in the past was 1,1,1-trichloro-2,2-bis (4-chlorophenyl)-ethane or DDT. Acute or transient DDT exposure is nontoxic, but continuous exposure negatively affects several systems in the body, including the reproductive system. DDT is lipophilic, and so it can remain stored in the body fat and even follicular fluid for up to two decades.[3] Its effects were outlined in a study by Jirosova *et al.* (2010), which linked DDT to decreases in diploid oocyte numbers.[41] Other studies reported correlations with miscarriages and spontaneous abortions, especially in spouses of DDT-exposed agricultural workers.[3] DDT exposure has also been correlated with intrauterine growth retardation.[3] Furthermore, another OCP associated with intrauterine growth retardation is HCH. In fact, a study by Pathak *et al.* exhibited a statistically significant elevation of gamma-HCH in maternal and cord blood of intrauterine growth retardation cases compared to controls.[40]

Polychlorinated biphenyls (PCBs) are another group of compounds implicated in female reproductive issues. PCBs are commonly found in pesticides and even cosmetics. Like OCPs, they are highly lipophilic and are slow to degrade and therefore accumulate in the body over time. Humans can be exposed to PCB via contaminated occupational environments and air and through the ingestion of meats, dairy, or fish with PCB traces. PCB has been found in follicular fluid, ovaries, uteri, and even in fetuses and embryos.

PCB exposure is associated with several pathologies although several studies contradict each other. Studies on Native American women who consumed fish containing PCBs had shorter than normal menstrual cycles.[42] Other studies reported decreased fecundability.[42] Studies by Meeker *et al.* (2011) and Toft *et al.* (2010) found that women exposed to PCBs had a heightened risk of IVF failure.[43,44]

Other studies have assessed the effects of PCBs on pregnancy. High exposure had no effect on the mean number of pregnancies in a study by Taylor, but birthweight and gestational age were decreased with higher PCB blood levels.[42,45] Furthermore, high PCB blood levels were found in women who had three or more miscarriages.[42] PCBs may induce oxidative stress via endothelial dysfunction and membrane damage with subsequent free radical formation.[3] Regardless, connections with oxidative stress are implied as antioxidant systems such as vitamin E decrease with PCB exposure.[3]

Another highly implicated group of chemicals in female reproductive oxidative stress are the organophosphorus pesticides (OPCs). OPCs have been linked to oxidative stress because GSH levels are depleted in exposed fetuses; other effects occur via lipid peroxidation.[3] In fact, Samarawickrema *et al.* found that low-grade, long-term exposure to OPCs elevated levels of oxidative markers.[46] These researchers also found elevated MDA in cord blood, DNA fragmentation in the fetus, as well as OPC accumulation in the placental-fetal compartment, suggesting the presence of fetal oxidative stress.[46] Nonetheless, oxidative biomarkers in the mothers or females in general are unaltered, and this could be due to lower maternal metabolic detoxification capacities (with continued accumulation of OPCS) or diminished conversion to toxic metabolites.[3]

Pathological effects and associations of oxidative stress

Menopause and OS

Menopause is a gradual event that occurs in the female and is marked by decline and cessation of reproductive capability due to diminished hormone production. It occurs at a mean age of 51 years in the United States but it can also occur prematurely, as early as 40 years.[47] Symptoms of menopause vary but are generally typical: breast tenderness, vaginal dryness, irregular menses, hot flashes, and osteoporosis.[47] These symptoms occur due to diminished estrogen levels as well as heightened oxidative stress. This is hypothesized to be a result of a deficient antioxidant system. In fact, as women age, they have higher serum concentrations of TNF-α, IL-4, IL-10, and IL-12 than their younger counterparts.[47]

In regard to TNF-α, it is a highly proinflammatory mediator and the subsequent increase in IL-4 leads to an inflammatory state in menopause. Reactive oxygen species are highly implicated in inflammation and thus, heightened inflammatory markers also imply heightened ROS levels. Moreover, a study by Signorelli *et al.* reported elevated oxidative stress markers such as malondialdehyde (MDA), 4-hydroxynonenal (4-HNE), oxLDL in postmenopausal women.[48] GPx, a vital antioxidant, was also found to be diminished.[48]

Differentiated cells in the female reproductive system are quite susceptible to oxidative damage, including estrogen-producing ovary cells.[49] These specific cells already have high cellular respiration levels, which imply that a large amount of ROS are being produced via electron leakage.[49] Hence, with ROS already present in a chronic state, further accumulation of ROS during menopause will damage mitochondrial structures and genes, with consequent homeostatic, bioenergetics, and functional decline.[49]

Other menopause-related research shows that low NO concentrations may lead to vasomotor defects in the uterus and in the body as a whole. Moreover, vitamin C has been shown to decrease oxidized LDL concentrations, leading to improved parameters of cardiovascular and vascular health.[47]

Reproductive diseases and oxidative stress
Endometriosis

Endometriosis is a chronic inflammatory disease characterized by the invasion of the endometrial layer into extrauterine sites such as the ovaries and other pelvic sites with a few cases found in abdominal viscera, lungs, and the urinary tract.[3] Studies disagree on whether oxidative stress plays a role in this condition.

Positive studies found higher concentrations of oxidative stress biomarkers such as MDA, proinflammatory cytokines (IL-6, TNF-α, and IL-β), angiogenic factors (IL-8 and VEGF), monocyte chemoattractant protein 1 (MCP-1), and oxLDL.[3] Phagocytic cells are also highly implicated in ROS production in endometriosis. Oxidative stress may occur via the nonenzymatic peroxidation of arachidonic acid, which produces F2-isprostanes.[3] Lipid peroxidation, in fact, leads to 8-iso-prostaglandin F2-α. This molecule is not only a vasoconstrictor and initiator of necrosis in endothelial cells but also plays a role in mediating immune cell adhesion.[3]

Another class of proteins called heat shock proteins (HSP) is also highly implicated. HSP-70B has been singled out and is a chaperone for protein metabolism and production.[3] Under heavy stress and heightened protein misfolding, levels are elevated.[3] Hence, HSP-70B is an oxidative stress biomarker. Furthermore, it supports the production of inflammatory cytokines, further worsening oxidative stress.[3]

The most viable hypothesis at the moment regarding endometriosis and its pathophysiology is retrograde menstruation. This is built on the idea that highly prooxidant factors are carried from the endometrium into the peritoneal cavity and ovaries.[50] These factors include iron and heme, as well as apoptotic endometrial cells.[50] Iron has no direct link to ROS *per se* but may be associated with catalyzing reactions of free radicals.[50]

Polycystic ovarian syndrome

Polycystic ovarian syndrome (PCOS) is a relatively common condition in women of reproductive age. In fact, its prevalence is around 18% in the female population.[3] PCOS is a combination of hyperandrogenism, ovulatory dysfunction, and polycystic ovaries. Clinical symptoms include amenorrhea or menorrhagia and dermatological manifestations such as acne and skin darkening.[3] Furthermore, 90% of sufferers are infertile.[3] The true cause of PCOS is hypothesized to be due to insulin resistance, which is accompanied by hypertension, central fat distribution, and general obesity, the latter two being proportional to oxidized LDL serum concentrations.[3] In fact, metabolic syndrome, sleep apnea, and other diseases are intimately related to obesity and, thus, PCOS.[3]

In conditions that predispose women to PCOS, oxidative stress has been found to be abundant. However, oxidative stress also plays a role in the essence of PCOS, as antioxidant levels are diminished. Furthermore, mitochondrial dysfunction could be present as mitochondrial oxygen consumption decreases along with GSH levels, but ROS production remains elevated.[3] Studies also indicate that oxidative stress alters steroidogenesis in the ovaries, disturbs follicular development, and leads to infertility and increased androgen production.[51] Furthermore, this is an inflammatory state. Inflammation is a process by which the body reacts to an unfavorable environment, which in this case is hyperglycemia.[3] Mononuclear cells and C-reactive protein are increased.[3] Hence, ROS produced by these processes are potentiated.

Unexplained infertility

Unexplained infertility is diagnosed via a process of exclusion. In other words, when a couple fails to conceive after 1 year of unprotected sex and all other infertility-related conditions are excluded, unexplained infertility is diagnosed.[3] It affects up to 15% of couples in the United States.[3] Unfortunately, unexplained infertility is a vague term. In fact, the pathophysiology behind it is still unclear. Nonetheless, it has been shown that oxidative stress may play a role in the etiology of unexplained infertility. In patients with unexplained infertility, oxidative stress markers such as MDA are somewhat elevated in the peritoneal cavity.[3] Antioxidant defenses and ROS are in an unbalanced state.[9] Levels of vitamin E and GSH are especially low.[3]

Other causes proposed include genetic polymorphism of folate metabolism. Polymorphism in the methylenetetrahydrofolate reductase (MTHFR) enzyme can be detrimental to folate metabolism.[3] A study by Altmae *et al.* in 2010 found that genetic polymorphisms in this enzyme led to the accumulation of homocysteine, an inducer of apoptosis, and oxidative stress.[52] Other studies found a low concentration of *N*-acetyl cysteine (NAC) enzyme, leading to heightened levels of homocysteine.[3]

Pregnancy complications and oxidative stress

As explained previously, pregnancy is a state of heightened metabolic activity and requirements. Thus, it is a delicate balance that can be easily disrupted by oxidative

Table 16.2   Overview of oxidative stress and pregnancy complications.

Spontaneous abortions	• *Definition:* An unintentional termination of pregnancy either before 20 weeks of gestation or when fetal weight is below 500 g.[3] This condition is attributed most commonly to chromosomal aberrations that inhibit further viable fetal development and, hence, abortion. • *Pathology:* Oxidative stress has been associated with spontaneous abortions. At the 10th or 12th week of gestation, an oxidative burst occurs.[3] Antioxidant activity usually keeps this in check. However, in certain women, premature maternal intraplacental circulation develops early in the 8th and 9th week.[3] This development overwhelms the body. Early in the pregnancy, antioxidant development is still premature and, thus, is unable to offset the buildup. Placental development is impaired and syncytiotrophoblast degradation is heightened.[3]
Recurrent pregnancy loss	• *Definition:* Three or more consecutive pregnancy losses. It occurs in 1–3% of women in their childbearing years.[3] • *Pathology:* Although a cause–effect relationship has not been established, oxidative stress seems to play a role. Women with recurrent pregnancy loss (RPL) exhibit heightened uterine natural killer cells in preimplantation angiogenesis leading to premature maternal intraplacental circulation.[3] Thus, oxidative stress in early pregnancy may lead to pregnancy loss. It should be noted that patients with RPL have also been found to have high plasma lipid peroxides and low levels of carotene and vitamin E.[3] Genetic polymorphisms, especially those of antioxidant genes, may also predispose women to this condition.[3]
Preeclampsia	• *Definition:* A disorder occurs during pregnancy and affects multiple systems in the body. This condition can occur in all women, even those who lack a hypertensive history. It is the leading cause of morbidity and mortality during pregnancy in the mother and child, with a prevalence of 3–14%.[3] Preeclampsia usually occurs after 20 weeks of gestation and is diagnosed after two blood pressure measurements, taken 6 h apart, of over 140/90 mm Hg.[3] • *Pathology:* Focal vasospasm occurs leading to placental ischemia and hypoxia.[3] With ischemia, ROS is heightened and oxidative stress ensues. This is implied as studies show elevated levels of carbonyls, lipid peroxides, and other oxidative markers, both in placental and maternal serum, and decreased levels of antioxidants such as vitamin E.[50]

stress. Table 16.2 briefly discusses pregnancy complications associated with oxidative stress, which are depicted in Figure 16.5.

Conclusion and key points

Reactive species are necessary and play important physiological roles in the human body, but are pathological when their levels become too high and lead to a state of oxidative stress. Thus, the body attempts to maintain healthy levels of ROS/RNS with the concomitant production of antioxidants. In the female reproductive system, this balance is highly delicate, and any failure of homeostasis will lead to damaging effects. This includes the aforementioned reproductive diseases and complications. Studies of

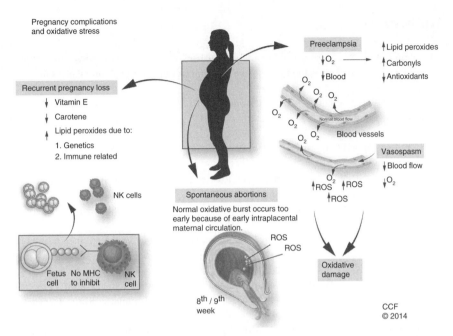

Figure 16.5 Pregnancy complications and oxidative stress. Complications that may arise from oxidative stress conditions include recurrent pregnancy loss, spontaneous abortions, and preeclampsia.

antioxidant supplementation have conflicting results. Nevertheless, further research is needed, and new approaches should be considered.

Multiple choice questions

1 In the female, physiological levels of reactive oxygen species
   a. Regulate oocyte maturation
   b. Are the main inducer of ovulation
   c. Suppresses apoptosis
   d. Plays a role in follicular atresia

2 Oxidative stress plays a role in the pathology of
   a. Polycystic ovarian syndrome
   b. Spontaneous abortion
   c. Menopause
   d. Preeclampsia

3 Factors that may induce oxidative stress in the female include
   a. Regular alcohol ingestion
   b. Abdominal fat deposition
   c. Nicotine inhalation
   d. Cocaine use

References

1 Salmon, A.B., Richardson, A. & Perez, V.I. (2010) Update on the oxidative stress theory of aging: does oxidative stress play a role in aging or healthy aging? *Free Radical Biology & Medicine*, 48(5), 642–655.

2 Burton, G.J. & Jauniaux, E. (2011) Oxidative stress. *Best Practice & Research Clinical Obstetrics & Gynaecology*, 25(3), 287–299.

3 Agarwal, A., Aponte-Mellado, A., Premkumar, B.J., Shaman, A. & Gupta, S. (2012) The effects of oxidative stress on female reproduction: a review. *Reproductive Biology and Endocrinology: RB&E.*, 10, 49.

4 Auten, R.L. & Davis, J.M. (2009) Oxygen toxicity and reactive oxygen species: the devil is in the details. *Pediatric Research*, 66(2), 121–127.

5 Kumar, V.A.A.K.A.J.C.R.S.L. (2013) *Robbins basic pathology*. Elsevier/Saunders, Philadelphia, PA.

6 Fujii, J., Iuchi, Y. & Okada, F. (2005) Fundamental roles of reactive oxygen species and protective mechanisms in the female reproductive system. *Reproductive Biology and Endocrinology: RB&E.*, 3, 43.

7 Valko, M., Leibfritz, D., Moncol, J., Cronin, M.T.D., Mazur, M. & Telser, J. (2007) Free radicals and antioxidants in normal physiological functions and human disease. *International Journal of Biochemistry and Cell Biology*, 39(1), 44–84.

8 Showell, M.G., Brown, J., Clarke, J. & Hart, R.J. (2013) Antioxidants for female subfertility. *The Cochrane Database of Systematic Reviews*, 2013(8), Cd007807.

9 Agarwal, A., Krajcir, N., Chowdary, H. & Gupta, S. (2008) Female infertility and assisted reproduction: impact of oxidative stress. *Current Women's Health Reviews*, 4(1), 9–15.

10 Noda, Y., Y. Goto, Y. Umaoka, M. Shiotani, T. Nakayama, T. Mori. (1994). Culture of human embryos in alpha modification of Eagle's medium under low oxygen tension and low illumination. *Fertility and Sterility* 62(5):1022–1027.

11 Lighten, A.D., G.E. Moore, R.M. Winston, K. Hardy. (1998). Routine addition of human insulin-like growth factor-I ligand could benefit clinical in-vitro fertilization culture. *Human Reproduction* 13(11):3144–3150.

12 Ali, A.A., J.F. Bilodeau, M.A Sirard. 2003. Antioxidant requirements for bovine oocytes varies during in vitro maturation, fertilization and development. *Theriogenology*, 59(3–4):939–949.

13 Henmi, H., T. Endo, Y. Kitajima, K. Manase, H. Hata, R. Kudo. (2003). Effects of ascorbic acid supplementation on serum progesterone levels in patients with a luteal phase defect. *Fertility and Sterility*, 80(2):459–461.

14 Agarwal, A., Allamaneni, S.S. (2004) Role of free radicals in female reproductive diseases and assisted reproduction. *Reproductive Biomedicine Online*, 9(3), 338–347.

15 Tarin, J.J., J. Brines, A. Cano. (1998). Is antioxidant therapy a promising strategy to improve human reproduction? *Human Reproduction* 13(6):1415–1424.

16 Showell M.G., Brown J, Clarke J, Hart RJ. 2013. Antioxidants for female subfertility. *Cochrane Database Syst Rev.* 2013 Aug 5;8:CD007807.

17 Agarwal, A., Aziz, N., Rizk, B. (2013) *Studies on Women's Health*. Humana Press, New York, pp. 33–60.

18 Agarwal, A., Allamaneni, S.S., Said, T.M. (2004) Chemiluminescence technique for measuring reactive oxygen species. *Reproductive Biomedicine Online*, 9(4), 466–468.

19 Mahfouz, R.Z., Sharma, R.K., Said, T.M., Erenpreiss, J., Agarwal, A. (2009) Association of sperm apoptosis and DNA ploidy with sperm chromatin quality in human spermatozoa. *Fertility and Sterility*, 91(4), 1110–1118.

20 Takagi, Y., Nikaido, T., Toki, T. *et al.* (2004) Levels of oxidative stress and redox-related molecules in the placenta in preeclampsia and fetal growth restriction. *Virchows Archiv: an International Journal of Pathology*, 444(1), 49–55.

21 Tilly, J.L., Tilly, K.I., Kenton, M.L. & Johnson, A.L. (1995) Expression of members of the bcl-2 gene family in the immature rat ovary: equine chorionic gonadotropin-mediated inhibition of granulosa cell apoptosis is associated with decreased bax and constitutive bcl-2 and bcl-xlong messenger ribonucleic acid levels. *Endocrinology*, 136(1), 232–241.

22 Tsai-Turton, M. & Luderer, U. (2006) Opposing effects of glutathione depletion and follicle-stimulating hormone on reactive oxygen species and apoptosis in cultured preovulatory rat follicles. *Endocrinology*, 147(3), 1224–1236.

23 Agarwal, A., Aziz, N. & Rizk, B. (2013) *Studies on Women's Health*. Humana Press, New York, pp. 75–94.

24 Agarwal, A., Gupta, S., Sekhon, L. & Shah, R. (2008) Redox considerations in female reproductive function and assisted reproduction: from molecular mechanisms to health implications. *Antioxidants & Redox Signaling*, 10(8), 1375–1403.

25 Serviddio, G., G. Loverro, M. Vicino, et al. (2002). Modulation of endometrial redox balance during the menstrual cycle: relation with sex hormones. *Journal of Clinical Endocrinology & Metabolism* 87(6):2843–2848.

26 Metwally, M., Li, T.C. & Ledger, W.L. (2007) The impact of obesity on female reproductive function. *Obesity Reviews: An Official Journal of the International Association for the Study of Obesity*, 8(6), 515–523.

27 Bloomer, R.J. & Fisher-Wellman, K.H. (2009) Systemic oxidative stress is increased to a greater degree in young, obese women following consumption of a high fat meal. *Oxidative Medicine and Cellular Longevity*, 2(1), 19–25.

28 Agarwal, A., Aziz, N. & Rizk, B. (2013) *Studies on Women's Health*. Humana Press, New York, pp. 131–141.

29 Bernal, A.B., Vickers, M.H., Hampton, M.B., Poynton, R.A. & Sloboda, D.M. (2010) Maternal undernutrition significantly impacts ovarian follicle number and increases ovarian oxidative stress in adult rat offspring. *PloS One*, 5(12), e15558.

30 Zima, T., Fialova, L., Mestek, O. *et al.* (2001) Oxidative stress, metabolism of ethanol and alcohol-related diseases. *Journal of Biomedical Science*, 8(1), 59–70.

31 Kalousová, M., T. Zima, P. Popov, et al. (2004. Advanced glycation end-products in patients with chronic alcohol misuse. *Alcohol and Alcoholism* 39(4):316–320.

32 Gauthier T.W., J.A. Kable, L. Burwell et al. (2010). Maternal alcohol use during pregnancy causes systemic oxidation of the glutathione redox system. *Alcoholism: Clinical and Experimental Research* 34(1):123–130.

33 Ruder, E.H., Hartman, T.J. & Goldman, M.B. (2009) Impact of oxidative stress on female fertility. *Current Opinion in Obstetrics & Gynecology*, 21(3), 219–222.

34 Augood, C., Duckitt, K. & Templeton, A.A. (1998) Smoking and female infertility: a systematic review and meta-analysis. *Human Reproduction (Oxford, England)*, 13(6), 1532–1539.

35 Chelchowska, M., J. Ambroszkiewicz, J. Gajewska, et al. (2011). The effect of tobacco smoking during pregnancy on plasma oxidant and antioxidant status in mother and newborn. *European Journal of Obstetrics, Gynecology and Reproductive Biology* 155:132–136.

36 Park, B., McPartland, J.M. & Glass, M. (2004) Cannabis, cannabinoids and reproduction. *Prostaglandins, Leukotrienes, and Essential Fatty Acids*, 70(2), 189–897.

37 Sarafian, T.A., J.A. Magallanes, H. Shau, et al. (1999). Oxidative stress produced by marijuana smoke. An adverse effect enhanced by cannabinoids. *American Journal of Respiratory Cell and Molecular Biology* 20(6):1286–1293.

38 Lee, Y.W., B. Hennig, M. Fiala, et al. (2001. Cocaine activates redox-regulated transcription factors and induces TNF-alpha expression in human brain endothelial cells. *Brain Research* 920(1–2):125–133.

39 Kovacic, P. (2005) Role of oxidative metabolites of cocaine in toxicity and addiction: oxidative stress and electron transfer. *Medical Hypotheses*, 64(2), 350–356.

40 Pathak, R., Mustafa, M.D., Ahmed, T. *et al.* (2011) Intra uterine growth retardation: association with organochlorine pesticide residue levels and oxidative stress markers. *Reproductive Toxicology (Elmsford, NY)*, 31(4), 534–539.

41 Jirsova, S., J. Masata, L. Jech, et al. (2010. Effect of polychlorinated biphenyls (PCBs) and 1,1,1-trichloro-2,2,-bis (4-chlorophenyl)-ethane (DDT) in follicular fluid on the results of in vitro fertilization-embryo transfer (IVF-ET) programs. *Fertility and Sterility* 93:1831–1836.

42 Faroon, O.M., Keith, S., Jones, D. & de Rosa, C. (2001) Effects of polychlorinated biphenyls on development and reproduction. *Toxicology and Industrial Health*, 17(3), 63–93.

43 Meeker JD, Maity A, Missmer SA. (2011). Serum concentrations of polychlorinated biphenyls in relation to in vitro fertilization outcomes. *Environ Health Perspect* 119(7):1010–6.

44 Toft, G., A.M. Thulstrup, B.A. Jönsson, et al. (2010. Fetal loss and maternal serum levels of 2,2',4,4',5,5'-hexachlorbiphenyl (CB-153) and 1,1-dichloro-2,2-bis(p-chlorophenyl)ethylene (p,p'-DDE) exposure: a cohort study in Greenland and two European populations. *Environmental Health* 9:22.

45 Taylor, P.R., J.M. Stelma, C.E. Lawrence. (1989. The relation of polychlorinated biphenyls to birth weight and gestational age in the offspring of occupationally exposed mothers. *American Journal of Epidemiology* 129(2):395–406.

46 Samarawickrema, N., A. Pathmeswaran, R. Wickremasinghe, et al. (2008. Fetal effects of environmental exposure of pregnant women to organophosphorus compounds in a rural farming community in Sri Lanka. *Clinical Toxicology (Philadelphia)* 46(6):489–495.

47 Agarwal, A., Aziz, N. & Rizk, B. (2013) *Studies on Women's Health*. Humana Press, New York, pp. 181–203.

48 Signorelli, S.S., S. Sciacchitano, M. Anzaldi, et al. (2011). Effects of long-term hormone replacement therapy: results from a cohort study. *Journal of Endocrinological Investigation* 34(3):180–184.

49 Miquel, J., Ramirez-Bosca, A., Ramirez-Bosca, J.V. & Alperi, J.D. (2006) Menopause: a review on the role of oxygen stress and favorable effects of dietary antioxidants. *Archives of Gerontology and Geriatrics*, 42(3), 289–306.

50 Sharma, R. & Agarwal, A. (2004) Role of reactive oxygen species in gynecologic diseases. *Reproductive Medicine and Biology*, 3(4), 177–199.

51 Yeon Lee, J., Baw, C.-K., Gupta, S., Aziz, N. & Agarwal, A. (2010) Role of oxidative stress in polycystic ovary syndrome. *Current Women's Health Reviews*, 6(2), 96–107.

52 Altmae, S., A. Stavreus-Evers, J.R. Ruiz, et al. (2010). Variations in folate pathway genes are associated with unexplained female infertility. *Fertility and Sterility* 94(1):130–137.

CHAPTER 17

Reactive oxygen species, oxidative stress, and cardiovascular diseases

Fatemeh Sharifpanah and Heinrich Sauer

Department of Physiology, Faculty of Medicine, Justus Liebig University, Giessen, Germany

---

THEMATIC SUMMARY BOX

At the end of this chapter, students should be able to:

- Describe sources of ROS production in the cardiovascular system

- Describe the impact of oxidative stress in the pathogenesis of hypertension

- Describe the development of atherosclerosis and the effect of oxidative stress in this process

- Describe the effect of oxidative stress on progression of ischemia/reperfusion injury

- Describe the pathogenesis of cardiac arrhythmia and the impact of oxidative stress on development of this disease

---

Introduction

The free radical theory of development was established by Allen and Balin in 1989, and described that metabolic gradients exist in embryos and may influence developmental processes.[1] Most decisive amongst these metabolic gradients are those of oxygen, which could influence the expression and activity of reactive oxygen species (ROS) and nitric oxide generating enzymes like NADPH oxidase or nitric oxide synthase (NOS).[2] Free radicals as highly reactive atoms or molecules containing unpaired electrons in their orbital are generally categorized into four groups, including oxygen free radicals, nitrogen free radicals, lipid free radicals, and other types of radicals such as chlorine radicals, and methyl radicals. Oxygen free radicals comprise the superoxide anion, the hydroxyl radical, and singlet oxygen. Nitrogen free radical includes nitric oxide.[2,3] Hydrogen peroxide and peroxynitrite are formed from free radicals, and are very reactive, but are not themselves free radicals.

Free radicals, ROS, and nitric oxide are generated during normal aerobic metabolism of organisms by various endogenous systems. Regarding the intrinsic

---

Oxidative Stress and Antioxidant Protection: The Science of Free Radical Biology and Disease, First Edition.
Edited by Donald Armstrong and Robert D. Stratton.
© 2016 John Wiley & Sons, Inc. Published 2016 by John Wiley & Sons, Inc.

ROS production in physiological states, cells evolved various enzymatic and nonenzymatic antioxidant mechanisms to scavenge ROS for keeping their concentration at physiological levels and maintain a redox balance state.[4] The enzymatic antioxidant events comprise catalases, superoxide dismutases, and glutathione peroxidases, and the nonenzymatic antioxidant molecules include glutathione and vitamins.[3,5] This balance between free radicals and antioxidants is necessary for proper physiological conditions. During pathophysiological conditions, the redox homeostasis shifts to an imbalanced state due to enhancement of ROS production or reduction of activity or expression of antioxidant events, and provides an oxidative stress environment which promotes the progression of various developmental disorders.[3] Less production of antioxidant events impairs physiological function through reduction of the repair system and also decreased adaptive responses to stimuli, resulting in the pathogenesis of different vascular and myocardial diseases. Moreover, enhancement of ROS production impairs the physiological function of the cardiovascular system due to activation of specific signaling pathways leading to apoptosis and tissue injury as well as reduction of cell function, which finally promote progression of cardiovascular disease.[3,4] In states of oxidative stress, free radicals adversely alter lipids, proteins and DNA leading to cell injury and cytotoxicity, which prompt progression of different diseases.[4–6]

Oxygen free radicals or ROS are generated in different cellular organelles comprising NADPH oxidases, mitochondria, xanthine oxidases, lipoxygenases, cytochrome P450, uncoupled endothelial NOS, myeloperoxidases, and peroxisomes.[7] The most important sources of ROS production in the cardiovascular system are the NADPH oxidase, mitochondria, and xanthine oxidase.[5] It has been shown that in the cardiovascular system, ROS are mainly produced by nonphagocytic NADPH oxidase, which influences various signaling pathways (e.g., mitogen-activated protein (MAP) kinases, tyrosine kinases, Rho kinase, protein tyrosine phosphatases, and some transcription factors (such as NF-κB, and HIF1α) to modulate cardiovascular cell function.[8–11] Furthermore, ROS increase the intracellular Ca^{2+} concentration due to ion channel activation, upregulate gene expression and enhance the activity of some proinflammatory factors.[12] Nitric oxide is another important free radical molecule to maintain physiological cardiovascular function. Nitric oxide is produced through three different types of nitric oxide synthase (NOSs) including endothelial nitric oxide synthase (eNOS), neuronal nitric oxide synthase (nNOS), and inducible nitric oxide synthase (iNOS), from which eNOS is the most important source of nitric oxide production in the cardiovascular system.[7] Nitric oxide produced by eNOS regulates blood flow and blood pressure in the body. Furthermore, nitric oxide has important antiatherogenic and antithrombotic effects on platelets, vascular smooth muscle, and endothelium.[13,14] It is also reported that the antioxidant properties of nitric oxide (nitric oxide scavenges superoxide and can also terminate the lipid peroxide chain reaction) can be intensified due to activation of some signal transduction pathways, leading to enhancement of endogenous antioxidant synthesis or downregulation of responses to proinflammatory stimuli.[14,15]

ROS are not only functioning in adult cell physiology and pathophysiology, but also play a critical role during embryonic development and cell differentiation processes. It was reported that ROS and nitric oxide under various physiological (e.g., wound healing) and pathophysiological (e.g., neovascularization of tumor nodes) conditions will be released, and so induce differentiation of vascular cells.[16,17] In embryonic stem cells the levels of ROS or nitric oxide production are enhanced upon various treatments, including cardiotrophin-1, peroxisome proliferator-activated receptor-α agonists, platelet-derived growth factor-BB, AMP-activated protein kinase agonist, α2-macroglobulin, and mechanical strain, which induces the differentiation toward cardiomyocytes, smooth muscle cells, and endothelial cells.[18–23] Moreover, it is reported that vasculogenesis in embryonic stem cells was attenuated upon treatment with β-adrenergic receptor antagonists due to reduction of nitric oxide production.[24] These observations underscore the impact of ROS and nitric oxide in redox balance and regulation of embryogenesis, and differentiation processes in the cardiovascular system. In regard to the design of new drugs, the understanding of the role of oxidative stress and the mechanism of redox signaling in initial steps of cardiovascular disease may open new avenues to develop new and more effective therapies against heart and blood vessel disorders.

Impact of oxidative stress on pathogenesis of hypertension

Hypertension or high blood pressure is a chronic disorder in which the blood pressure of arteries is elevated to or above 140 mm Hg systolic and 90 mm Hg diastolic.[25] In healthy or normal persons, systolic blood pressure is up to 120 mm Hg and diastolic pressure is up to 80 mm Hg. Hypertension is categorized to four different groups on basis of blood pressure levels.[26] (i) *Pre-hypertension* in which systolic blood pressure is between 120 and 139 mm Hg and/or diastolic blood pressure is between 80 and 89 mm Hg. (ii) *Hypertension stage 1* with 140–159 mm Hg systolic and/or 90–99 mm Hg diastolic blood pressure. (iii) *Hypertension stage 2* with 160 mm Hg or higher systolic blood pressure and/or 100 mm Hg or higher diastolic blood pressure. (iv) *Hypertensive crisis* with higher than 180 mm Hg systolic and/or higher than 110 mm Hg diastolic blood pressure. Patients with hypertensive crisis or last stage of hypertension need emergency care. Furthermore, there are two types of hypertension including primary or essential hypertension, and secondary hypertension. In essential hypertension, there is no particular reason for high blood pressure, and it usually tends to develop gradually over many years. In contrary, the secondary hypertension appears suddenly, and various conditions or medications lead to secondary hypertension such as kidney problems, adrenal gland tumors, thyroid problems, certain congenital defects of blood vessels, alcohol abuse, illegal drugs (e.g., cocaine and amphetamine), and certain medicaments such as birth control pills. Hypertension is an important and common public health problem in both developing and developed countries. According to the World Health Organization (WHO) report, hypertension with global prevalence of nearly 1 billion individuals (in 2008) causes 7.6 million deaths each year, about 13.5% of worldwide total deaths.[27–30] Furthermore, it is reported that prevalence of hypertension increases with age, and overall high-income countries have a lower prevalence of hypertension than other groups.[31]

Generally during the early stage of essential hypertension, cardiac output is increased but total peripheral resistance stays normal. Later, the cardiac output drops to normal levels and the peripheral resistance increases.[32,33] The enhancement of peripheral resistance is one of the key characteristics of hypertension and occurs due to a reduced lumen diameter of resistance vessels. According to Hagen–Poiseuille's law, very small changes in the diameter of vessels can markedly change vascular resistance.[34,35] In the sequence of hypertension, the small arteries that determine the peripheral resistance undergo structural and functional changes such as vascular remodeling, increased vasoreactivity, endothelial dysfunction, and vascular inflammation.[36–38]

Although, the exact etiology of hypertension is still unclear, it is well known that hypertension is a multi-factorial complex disorder which involves different organs. The pro-hypertensive factors which are important in the development of hypertension comprise enhancement of renin–angiotensin–aldosterone system activity, hyperactivity of the sympathetic nervous system, aberrant G-protein-coupled receptor signaling, endothelial dysfunction, and inflammation.[39–42] A common point between all these pro-hypertensive processes is oxidative stress. Data from the literature indicate that ROS cause hypertension via structural and functional changes of vessels which relate to reduction of vasodilation, enhancement of vasocontraction, and vascular remodeling. These structural and functional changes finally lead to increased peripheral resistance and elevated blood pressure.[43]

In the 1960s, Romanowski *et al.* has suggested for the first time a relation between the ROS generation level and blood pressure.[44] Some decades later in the early 2000s, it was reported that in an animal model of angiotensin-II-induced hypertension, levels of vascular ROS generation increased via nonphagocytic NADPH oxidase activation.[45,46] Since this time, a relation between oxidative stress and increased blood pressure has been reported in various experimental models of hypertension including spontaneously hypertensive rat (SHR), spontaneously hypertensive stroke prone rat (SHRSP), surgically-induced (2-kidney 1-clip (2K1C), aortic banding), hormone-induced (angiotensin-II, endothelin-1, aldosterone, deoxycorticosterone acetate (DOCA)-salt), and diet-induced hypertension (salt, fat) as well as some clinical studies.[38,47–49]

Among different sources of ROS in vascular cells, particularly ROS arising from NADPH oxidase may be involved in the development of hypertension. For instance, the sustained high blood pressure induced by angiotension-II was abolished in Nox1$^{-/-}$ knockout mice due to reduced vascular superoxide production, while Nox1 overexpressing transgenic mice show enhanced superoxide production levels and blood pressure in response to angiotensin-II.[50–52] In angiotensin-II-induced hypertensive rat models, the expression and activity of NADPH oxidase were increased, and inhibition of NADPH oxidase with a chimeric peptide (gp91ds-tat) against the p47(phox) subunit reduced the level of superoxide production and attenuated angiotensin-II-induced systolic blood pressure.[46,53,54] A decreased hypertensive response toward angiotension-II was reported in p47phox-deficient mice.[55,56] Likewise increased hypertension was observed in a mouse model with smooth muscle cell-specific p22phox overexpression.[57] Furthermore, Nox1 expression and activity were increased in small arteries of hypertensive animal models, which may also involve mitochondria, possibly through a calcium-dependent mechanism.[58]

Also other isoforms of NADPH oxidase are involved in hypertension. Of note Nox2 is highly regulated by angiotensin-II and mechanical strain, and is upregulated in animal models of hypertension.[59–62] Moreover, Nox4 which contributes to basal ROS production, upregulated the level of ROS generation when induced by angiotensin-II.[63,64] The recently identified Nox isoform Nox5 is calcium-sensitive and possesses a calmodulin-like domain with a Ca^{2+} binding site.[65–68] Vascular Nox5 is also activated by angiotensin-II and other pro-hypertensive factors such as endothelin-1, and generates ROS in response to increases in intracellular Ca^{2+}.[67,69–72] Likewise, various genetic and clinical studies support the importance of Nox isoforms in pathogenesis of hypertension. It was shown in different races that various polymorphisms in Nox subunits are associated with hypertension, and they can affect the level of blood pressure in hypertensive patients.[73–75] It was also reported that a low dose of angiotensin-II treatment by continuous intravenous infusion in human increased levels of arterial blood pressure along with reduction of nitric oxide concentration and increase in levels of F_2-isoprostane, a marker of oxidative stress.[76] These studies and investigations on patients with hypertension and coronary artery disease demonstrate that each Nox isoform of NADPH oxidase plays an individual role in cardiovascular disorders and suggests that Nox1, Nox2, and Nox5 promote endothelial dysfunction, inflammation, and apoptosis in the vessel wall, whereas Nox4 exerts a vasoprotective effect by increasing nitric oxide bioavailability and suppressing cell death pathways.[77]

The above mentioned studies support the importance of ROS-induced NADPH oxidase in hypertension upon induction by exogenous angiotensin-II treatment. However, the endogenous renin–angiotensin–aldosterone system may also participate in pathophysiologic ROS generation. In the 2K1C hypertensive animal model and Dahl rat (salt-sensitive hypertensive model) which are associated with increased activation of the renin–angiotensin–aldosterone system, levels of superoxide production were increased which partly contributed to the enhancement of blood pressure.[78] Treatment of Dahl rats with the angiotensin receptor inhibitor candesartan, and gp91phox inhibitor resulted in reduced salt-induced superoxide production, and prevented the expression of pro-inflammatory molecules.[79] However, none of these inhibitors attenuated the systolic blood pressure.[79] Furthermore, expression of p67phox subunit of NADPH oxidase was upregulated in response to a high salt diet in Dahl rats which was not observed for the other NADPH oxidase subunits. Significant reduction of salt-sensitive hypertension and oxidative stress occurred after disruption of p67phox using zinc-finger nucleases which suggests p67phox as a target for salt-sensitive hypertension therapy.[80]

Oxidative stress can also induce hypertension via an angiotensin-II independent manner. Rats with DOCA-salt-induced hypertension are a well-established animal model of oxidative and inflammatory stress in the cardiovascular system. It has been shown that NADPH oxidase is upregulated in DOCA-salt rats, which is responsible for the enhancement of superoxide production and elevation of blood pressure.[38,81] In DOCA-salt-induced hypertension, endothelin-1 has been reported as one of the key mediators of superoxide generation.[82] Endothelin-1 is mainly produced by endothelial cells, and contributes to blood pressure elevation due to its vasoconstriction effect.[83] Overexpression of human endothelin-1 in mice induced vascular remodeling, endothelial dysfunction and hypertension via NADPH oxidase activation.[84,85] It has been reported that the plasma level of endothelin-1 was increased in patients with familial hypertension, as well as its mRNA level in the endothelium of subcutaneous resistance arteries of patients with moderate to severe hypertension.[86] In another clinical study, it was observed that the plasma levels of endothelin-1 along with oxidative stress were increased upon a continuous intravenous infusion of angiotensin-II.[76] In patients with refractory hypertension, blood pressure was lowered upon treatment with the endothelin(A)-receptor inhibitor darusentan, which underscores the impact of endothelin-1 in hypertension.[87] The DOCA-salt-treated SHR rats, as a well-known animal model of hypertension, also show enhancement of endothelin-1 *mRNA* expression in the aorta in comparison to normotensive control rats.[88]

Increasing evidence suggests that NADPH oxidase is the main but not only source of ROS in hypertension. A cross talk between Nox and mitochondrial sources of ROS has been discussed in animal models of angiotensin-II-induced and DOCA-salt-induced hypertension. In this respect, treatment with a mitochondria-targeted antioxidant (mitoTEMPO) reduced blood pressure and also decreased mitochondrial and total cellular superoxide as well as cellular NADPH oxidase activity, and also restored bioavailability of nitric oxide.[89,90] Moreover, overexpression of mitochondrial superoxide dismutase 2 (SOD2) reduced angiotensin-II-induced hypertension.[89] In the mouse aorta, the mitochondrial monoamine oxidase-A and mitochondrial monoamine oxidase-B were induced upon exogenous treatment with angiotensin-II, producing significant amounts of hydrogen peroxide. This process attenuated eNOS release and aggravated endothelial dysfunction.[91] Moreover, the adaptor protein p66(Shc) contributes to ROS production in mitochondria, which occurs in stretch- or angiotensin-II-induced hypertension models.[92,93] Furthermore, eNOS uncoupling may result in hypertension and lead to ROS production. In the DOCA-salt-induced hypertension model, eNOS is uncoupled due to reduced tetrahydrobiopterin (BH4) which is an essential cofactor of eNOS. Uncoupled eNOS produces superoxide instead of nitric oxide and leads to endothelial dysfunction. Treatment of DOCA-salt-induced hypertensive rats with BH4 supplementation reduced vascular superoxide production, enhanced nitric oxide bioavailability and normalized the blood pressure.[94]

Another facet of redox signaling during physiological and pathological vascular remodeling are mechanical forces exerting shear stress on vascular walls. The blood vessels are constantly subjected to biomechanical forces which are important for vascular remodeling and significantly influence the phenotype of endothelial cells. Endothelial cells adapt to sustained shear stress, and either an increase or decrease from normal shear leads to distinct signaling events.[38] Biomechanical stress in blood vessels comprises two main forces which are acting on the blood vessel wall. The first one is steady or normal shear stress (also named laminar shear stress) which is generated by physiological blood flow within lumen of vessels. The steady shear stress acts primarily on the endothelium and induces a normal production of nitric oxide through increased expression and activation of eNOS. Laminar sheer stress also induces prostacyclin, thrombomodulin, plasminogen activators, and TGF-β, which function as anti-thrombotic agents and improve organ blood flow, thereby managing vasodilation, angiogenesis, vascular protection, and physiological vascular remodeling.[95] The other type of mechanical force in blood vessels is mechanical stretch or strain which is caused by nonlaminar flow or turbulence flow. This type of stress is also called oscillatory shear stress and arises in injured vessels and affects both endothelium and vascular smooth muscles. Oscillatory shear stress is associated with increased vessel wall thickness, reduced vessel lumen, and hypertension. Furthermore, oscillatory shear stress induces enhancement of pro-hypertensive factors such as angiotensin-II and endothelin-1, and also stimulates an acute increase of nitric oxide production and upregulation of eNOS activity. Moreover, oscillatory shear stress promotes oxidative stress via enhancement of superoxide and hydrogen peroxide production as well as inflammatory responses and leads to decrease in the endothelial function and finally vascular damage.[95] Notably, p66(Shc) protein is involved in oscillatory shear stress-induced hypertension. Silencing of p66(Shc) in oscillatory shear stress-induced hypertension blunted the effect of stretch on superoxide production and NADPH oxidase activation, and restored nitric oxide bioavailability.[93]

Highly reactive peroxynitrite, the byproduct of superoxide and nitric oxide interaction, is an important signaling molecule in shear stress-dependent activation of MAP kinases (e.g., c-Jun N-terminal kinase (JNK)), matrix metalloproteinases, and adhesion molecules.[38] The downstream signaling events depend on the concentration of produced peroxynitrite. Low concentration of peroxynitrite is associated with laminar shear stress and plays a protective role by inhibiting activation of adhesion molecules. Laminar shear stress also stimulates and enhances expression of intra- and extracellular superoxide dismutases and intracellular glutathione peroxide, which scavenge superoxide and hydrogen peroxide, respectively. The laminar shear stress via these processes protects the vessels against vascular injury.[43] In contrast, oscillatory shear stress is associated with sustained superoxide production, which, in the presence of nitric oxide, enhances peroxynitrite formation and protein nitration. These processes may contribute to vascular damage and atherosclerotic lesion formation.[96] Moreover, oscillatory shear stress in human endothelial cells over 24 h increased NADPH oxidase activity and expression of p22phox, gp91phox, and Nox4 and caused vascular inflammation through activation of the Nox1 isoform of NADPH oxidase.[38]

Impact of oxidative stress on pathogenesis of atherosclerosis

Atherosclerosis is a chronic and slowly progressive multi-factorial disease with a long asymptomatic phase, which is characterized by thickening and stiffness of the arterial wall. Besides aging, some factors such as hyperlipidemia, diabetes, hypertension, obesity, smoking, changes in arterial blood flow shear stress, and familial history of early heart disease increase the risk of atherosclerosis incidence. Atherosclerosis refers to accumulation of cholesterol deposits in macrophages (foam cells) in the intima layer of large and medium arteries to form plaque. These plaques or atheroma contain a central hypocellular lipid core including crystals of cholesterol, carbohydrate, blood and blood products, fibrous tissue, and calcium deposits which are built up in the necrotic foam cells. The lipid core is separated from the arterial lumen by a fibrous cap and myeloproliferative tissue that consists of extracellular matrix and smooth muscle cells.[97] Generally speaking, progression of atherosclerosis is categorized into six steps.[98] (i) *Initial lesion or adaptive intimal thickening* is accompanied by macrophage infiltration and focal accumulation of extracellular lipid. (ii) *Fatty streak or intimal xanthomas* with mainly intracellular lipid accumulation in foam cells without a necrotic core or fibrous cap, and reduction of arterial wall elasticity. (iii) *Intermediate lesion or pre-atheroma* step accompanied by small extracellular lipid pool buildup beside of intracellular lipid accumulation, and overlaying plaque with a thin fibrous cap. (iv) *Atheroma (plaque)* formation along with still intracellular lipid accumulation starts to build extracellular lipid cores and narrowing of the vessel lumen. (v) *Fibroatheroma* with various changes: making multiple lipid cores, covering the plaque with a fibrocalcific layer, and enhancement of smooth muscle and collagen in the affected area. During this step, plaques can sometimes grow sufficiently large to block blood flow. (vi) *Complicated lesion* or last stage of atherosclerosis accompanied by fissure and hematoma in the affected area, and also rupture of fibrous cap and release of thrombogenic materials into the blood vessels, which induce thrombosis in the vessel lumen and lead to vessel occlusion, resulting in infarction or stroke.[99]

Although, the etiology of atherosclerosis is not completely elucidated, a huge number of studies are implicating a pivotal role of oxidative stress in the pathogenesis of atherosclerosis (see Figure 17.1).[100–102] Atherosclerosis initiates with inflammatory responses of endothelial cells to low-density lipoprotein (LDL) particles, which

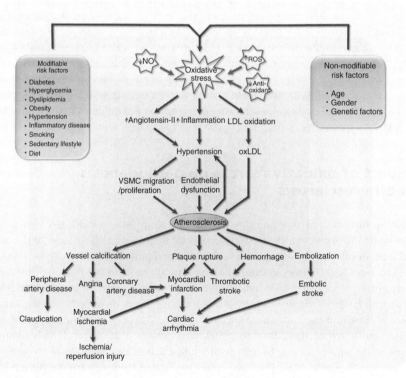

Figure 17.1 Pathogenesis of atherosclerosis.

tend to pass between endothelial cells and stay inside the vessel wall and behind the endothelium monolayer. LDL particles carry various types of fat molecules such as cholesterol, phospholipids, cholesterol esters and triglycerides, and they are susceptible to be oxidized by free radicals, forming oxidized low-density lipoprotein (oxLDL). The oxLDL is then taken up by specialized receptors on endothelial cells and induces their activation. This occurs by release of some bioactive phospholipids which stimulate activated endothelial cells to upregulate surface integrin receptors that interact with intracellular adhesion molecule (ICAM-1), vascular cell-adhesion molecule (VCAM-1), and endothelial cell-leukocyte adhesion molecule (ELAM-1), thereby providing a signaling cue for the recruitment of monocytes and facilitating their migration to subendothelial layers.[103] These migrated monocytes will be transformed into macrophages for ingestion of oxLDL. The macrophages absorb the oxLDL and form specialized foam cells. These foam cells process oxLDL and recruit functional high-density lipoprotein (HDL) to ingest cholesterol. Moreover they attract further white blood cells to the injury site. Over the time, the cholesterol and other cellular products in necrotic foam cells build up fatty deposits or plaques at the injury site, which cause the atherosclerotic lesion.

As mentioned above oxLDL, which is regarded as a major immunogen, is one of the key mediators of atherosclerosis.[104] Unlike native LDL, oxLDL causes cholesterol ester accumulation in macrophages, and also exerts cytotoxic or chemotactic actions on monocytes and inhibits the motility of macrophages.[105] It was demonstrated that ROS peroxidize lipid components in LDL particles and cause oxLDL formation.[106] In accordance with the literature, the level of ROS generation is elevated in atherosclerosis due to different mechanisms such as enhancement of ROS production or reduction of the antioxidant capacity.[107] ROS may increase upon treatment of monocytes and macrophages with oxLDL which stimulates the expression of manganese superoxide dismutase (Mn-SOD), thus leading to increase of hydrogen peroxide concentration and disturbance of the steady state levels of ROS.[108] Moreover, a markedly inverse correlation between Mn-SOD activity and glutathione concentration with plaque size was observed in rabbits with heritable hyperlipidemia.[109] The above mentioned process is enhanced by cytokines, such as TNF-α, interleukin-1 (IL-1), angiotensin-II, and interferon-γ, which induce superoxide production via NADPH oxidase in endothelial cells.[110]

In atherosclerotic lesions NADPH oxidase seems to be the primary source of ROS.[111] In apolipoprotein E$^{-/-}$ hypercholesterolemic mice (ApoE$^{-/-}$) which develop atherosclerosis spontaneously, superoxide levels were reduced upon treatment with the NADPH oxidase inhibitor apocynin, which also attenuated progression of atherosclerosis.[111] Moreover, human aortic endothelial cells exposed to hyperglycemic conditions showed enhancement of Nox1 expression, and oxidative stress.[112] Notably, Nox1$^{-/y}$/ApoE$^{-/-}$ double knockout mice displayed reduced atherosclerosis, correlating with decreased superoxide production, attenuation of chemokine expression, decreased vascular adhesion of leukocytes and macrophage infiltration as well as reduced expression of pro-inflammatory and pro-fibrotic markers.[112,113] It was observed by other research groups that levels of Nox2 were increased in ApoE$^{-/-}$ mice, while Nox4 showed no significant difference to wild type mice.[114] Moreover, endothelial-targeted Nox2 overexpression in ApoE$^{-/-}$ mice enhanced vascular superoxide production, endothelial cell activation, and increased macrophage recruitment, but no alteration on progression of atherosclerosis.[59] Furthermore, a reduction in total atherosclerotic lesion area was found in p47phox$^{-/-}$ mice crossed with ApoE$^{-/-}$ mice, suggesting a likely involvement of the Nox2 subunit of NADPH oxidase in atherogenesis.[115] Recently, Judkins *et al.* highlighted Nox2-NADPH oxidase as a potential target for future therapies of atherosclerosis on the basis of a 50% reduction of atherosclerotic lesion area observed in Nox2$^{-/y}$/ApoE$^{-/-}$ double knockout mice.[114] In humans, the genetic hereditary deficiency of Nox2 which occurs in chronic granulomatous disease is associated with reduction of vascular aging markers and oxidative stress markers. It has been described that the atherosclerotic burden was diminished in these patients, which emphasizes the possible role of Nox2 oxidase in atherosclerosis.[116,117] Regarding Nox4, it is reported that it regulates vascular smooth muscle cell migration and differentiation, which is important for neo-intima formation. Accordingly enhancement of Nox4 expression was observed in intimal lesions of coronary arteries of atherosclerotic patients.[118] It was also demonstrated that enhancement of Nox4 expression in atherosclerosis may change the smooth muscle cell phenotype, leading to instability and rupture of atherosclerotic plaques.[119] Of note Nox5 has been shown to be a calcium-dependent source of ROS in atherosclerosis and may be a new target for novel drugs to treat atherosclerosis.[120]

Nitric oxide produced by eNOS plays an anti-atherosclerotic role due to inhibition of leukocyte adhesion to the vessel wall. Leukocyte adherence is an early event in the development of atherosclerosis, and nitric oxide prevents this process via interfering with the ability of leukocyte adhesion molecule CD11/CD18 to bind to the endothelial cell surface or by suppressing CD11/CD18 expression on leukocytes.[121] It is demonstrated that hypercholesterolemia, which causes atherosclerosis, leads to a reduction in vascular nitric oxide bioavailability.[122] This process is attributed to dysfunction of the eNOS enzyme and its conversion to uncoupled eNOS. The uncoupled eNOS produces superoxide instead of nitric oxide, resulting in reduction of nitric oxide bioavailability and enhancement of pre-existing oxidative stress, which contributes to atherosclerosis. Various mechanisms lead to conversion of eNOS to uncoupled eNOS, such as deficiency of tetrahydrobiopterin (BH4) which is the essential cofactor of eNOS, depletion of L-arginine, which is the main substrate of eNOS, accumulation of asymmetrical dimethylarginine (ADMA), which is an endogenous eNOS inhibitor, and eNOS s-glutathionylation.[121] Deficiency of BH4 is presumably representing the major cause of eNOS uncoupling.[102] Under oxidative stress mediated by NADPH oxidase, superoxide modestly and peroxynitrite strongly oxidize BH4 to dihydrobiopterin (BH2), leading to BH4 deficiency. The dihydrofolate reductase (DHFR) enzyme can recycle BH4 from its BH2 oxidized form.[123] Furthermore, hypercholesterolemia may upregulate angiotensin-II due to enhancement of angiotensin-II receptor 1 (AT1R) expression, which in turn causes BH4 deficiency by hydrogen peroxide-dependent down regulation of DHFR.[124–126]

In the presence of metal ions or nitric oxide, superoxide can promote formation of lipid peroxyl radicals. Lipid peroxyl radicals are highly reactive and initiate a chain reaction by reacting with nearby lipid molecules to yield additional lipid peroxyl radicals and lipid hydroperoxides.[127] This lipid peroxidation within the vessel wall leads to oxLDL formation, and thereby atherosclerotic lesion formation.[102] It is demonstrated that oxLDL, as a main mediator of atherosclerosis, activates NADPH oxidase and xanthine oxidase, and also enhances expression of NADPH oxidase, leading to oxidative stress.[102] Furthermore, oxLDL upregulates arginase-I, which contributes to endothelial dysfunction by reducing L-arginine availability for eNOS to produce nitric oxide.[128] It is demonstrated that oxLDL stimulates synthesis of collagen IV in fibroblasts, to form a fibrous cap on top of the fatty streak.[129] Moreover, oxLDL enhances generation, release and activity of matrix metalloproteinases that have collagenase activity, which contributes to thinning of the fibrous cap, causing plaque instability and rupture.[130,131]

Myeloperoxidase (MPO), a member of the peroxidase family, has been suggested as one of the major keys in the initiation and propagation of atherosclerosis. MPO is present in neutrophils, monocytes, and macrophages, cells usually involved in host defense mechanisms through reactive oxidant production by utilizing hydrogen peroxide and chloride to generate the powerful oxidant hypochlorous acid. It is demonstrated that the MPO-derived oxidants contribute to tissue damage and are involved in progression of atherosclerosis by induction of endothelial dysfunction, modification of functional HDL into dysfunctional HDL, conversion of native LDL into modified oxLDL, and induction of endothelial cell death.[132,133]

Impact of oxidative stress on pathogenesis of ischemia/reperfusion injury

Ischemia/reperfusion injury divides to two different steps as ischemia and reperfusion, leading to myocardial injury, which is a serious complication in transplantation, heart infarction, and stroke. *Cardiac ischemia* occurs when blood flow to the heart muscle is reduced as a result of partial or complete obstruction of one or more coronary arteries, which can damage heart muscle and reduce its ability to pump efficiently.[134] Cardiac ischemia may occur slowly as a result of narrowing of the coronary arteries over time during atherosclerosis, or it may occur quickly by blockage of perfusion as a result of thrombosis or development of coronary artery spasm. Moreover, cardiac ischemia may cause serious heart arrhythmia and also heart infarction. It is reported that approximately one million people per year suffer from heart infarction in the United States.[135] Various conditions besides age increase risk of cardiac ischemia development in patients, such as diabetes, hypertension, hypercholesterolemia, obesity, smoking, and familial history. *Reperfusion* as second step of this injury is re-flowing of blood in coronary arteries following ischemia. Reperfusion itself may cause some reversible or irreversible damage in the ischemic heart, if the coronary flow is not restored within a critical period of time, and is referred to as reperfusion injury.[134,136] The reperfusion injury may cause myocardial stunning, no-reflow phenomenon, and myocardial necrosis.

In 1960, Jennings *et al.* have for the first time described myocardial reperfusion injury after ischemia in canine hearts which were subjected to coronary ligation.[137,138] In parallel, Bing *et al.* have described detailed metabolic changes during myocardial ischemia and reperfusion.[13] Over these decades of investigation on the pathology of ischemia/reperfusion injury, it has become clear that immense production of ROS during the reperfusion phase is an important causative factor. It was evidenced that oxidative stress may increase as much as 100-fold during ischemia-reperfusion conditions.[139] The major source of ROS in the ischemic heart comprises the mitochondrial electron transport chain, xanthine oxidase, and NADPH oxidase.[140] Presumably three main causes of tissue injury after ischemia/reperfusion are enhancement of free radicals, overload of intracellular calcium, and activation of neutrophils.[135–141]

In the ischemic state, the affected tissue suffers a low oxygen tension and substrate supply in the aftermath of blood flow restriction. This leads to reduction of ATP production due to impaired mitochondrial electron transport chain. The diminished mitochondrial respiration capacity then causes an impaired mitochondrial function in the ischemic human heart muscle.[142] Furthermore, production of oxygen free radicals due to disturbance of the mitochondrial electron transport chain increases in the ischemic heart, which develops an oxidative stress condition and intracellular Ca^{2+} overload.[139,143] In parallel, a metabolic change toward anaerobic glycolysis occurs to increase the levels of ATP production in the ischemic heart. However, the amount of energy generated through glycolysis is still insufficient for the maintenance of cardiac function and within 60–90 min of ischemia myocyte hyper-contracture develops in the affected area.[135,144,145] During ischemia, anaerobic glycolysis causes intracellular accumulation of lactate and protons. In the reperfusion phase following ischemia, protons are transported into the extracellular space via the Na^+–H^+ exchanger to normalize pH, resulting in increase of intracellular sodium. This condition causes activation of sarcolemmal Na^+–Ca^{2+} exchanger, leading to intracellular Ca^{2+} overload due to exchange of intracellular sodium with extracellular calcium.[139,146–148] It is supposed that the magnitude of intracellular Ca^{2+} overload depends to the duration of ischemia and degree of oxidative stress.[139] Furthermore, oxidative stress in conditions of ischemia/reperfusion detrimentally affects Ca^{2+}-handling proteins in the sarcoplasmic reticulum including the Ca^{2+} pump, the sarcoplasmic reticulum Ca^{2+}-ATPase, Ca^{2+} release channels, and also in sarcolemma (e.g., sarcolemmal Ca^{2+} pump and the L-type Ca^{2+} channels), resulting in intracellular Ca^{2+} overload.[141] It is also known that intracellular Ca^{2+} overload influences the opening of mitochondrial K^+_{ATP} channels and mitochondrial permeability transition pores (mPTP) in the reperfusion state.[143,149] The mPTP is a nonselective channel of the inner mitochondrial membrane. Upon opening of this channel the mitochondrial membrane potential collapses and uncouples oxidative phosphorylation, leading to ATP depletion and cell death. During ischemia the mPTP stays closed. During the first minutes of reperfusion the mPTP becomes open in response to Ca^{2+} overload, oxidative stress, restoration of a physiologic pH, and ATP depletion. In reperfusion injury, mitochondrial damage due to oxidative stress and long lasting opening of mPTP is a determining factor leading to loss of cardiomyocyte function and viability. Therefore, the mPTP is suggested as a new target for cardioprotection in reperfusion.[150,151] It is reported that treatment of infarcted rats as well as humans with pyridoxal 5′-phosphate (PLP), a metabolite of vitamin B6, reduced ischemia/reperfusion injury and improved cardiac function.[152,153] Administration of PLP may exert cardioprotective effects in ischemic heart by attenuating the intracellular Ca^{2+} overload due to inhibition of purinergic receptors.[154]

It is reported that elevated cytosolic calcium in ischemic tissue may activate a calcium-dependent protease, which catalyzes conversion of xanthine dehydrogenase to xanthine oxidase.[155,156] Notably, levels of xanthine oxidase activity are remarkably increased in blood of patients with myocardial infarction.[157] Furthermore, suppression of xanthine oxidase activity with its potent inhibitor allopurinol ameliorates the myocardial inefficiency in cardiac ischemic experimental models.[158] In the reperfusion state following ischemia, xanthine oxidase starts to use oxygen instead of NAD^+ as substrate. On the other hand, the oversized ATP consumption during ischemia leads to accumulation of purine metabolites, such as hypoxanthosine and hypoxanthine. Therefore, the xanthine oxidase using oxygen during reperfusion starts to catabolize hypoxanthine to xanthine and uric acid, and produces enormous amounts of superoxide and hydrogen peroxide, leading to tissue injury.[143]

Besides the impaired mitochondrial electron transport chain and xanthine oxidase, NADPH oxidase is another source of ROS production leading to oxidative stress in ischemia/reperfusion injury. It is reported that Nox2 and Nox4 among the members of the Nox family are extensively expressed in cardiomyocytes.[159] It is reported that both of them are upregulated in conditions of ischemia/reperfusion. Genetic suppression of Nox2 or Nox4 of NADPH oxidase exacerbates ischemia/reperfusion injury in mice via inadvertent down-regulation of HIF1α and upregulation of PPAR-alpha.[160] Furthermore, it is mentioned that expression of Nox2 is enhanced in rat as well as human cardiomyocytes after heart infarction.[161–163]

Oxidative stress-mediated myocardial reperfusion causes activation of the pro-inflammatory factor NF-κB. The activated form of NF-κB was also observed in experimental cardiac ischemia/reperfusion.[164] Activated NF-κB triggers gene expression of various pro-inflammatory cytokines such as TNF-α, and interleukins, leading to stimulation of cardiac remodeling.[143,165] It is demonstrated that hydrogen peroxide in the infarcted area can directly induce production of TNF-α via the p38 MAPK pathway.[166] The release of these chemokines attracts neutrophils to the ischemic area within the first 6 h of reperfusion and activates them.[143,167] These activated neutrophils produce ROS due to enhancement of cellular respiration via NADPH oxidase.[143] Furthermore, they can adhere to endothelial cells and particularly affect the vasodilation ability of vessels in response to nitric oxide. This event causes vasoconstriction and diminishes microvascular perfusion (no-reflow phenomenon), leading to apoptosis in the infarcted area. Moreover, the activated neutrophils release inflammatory factors such as leukotrienes and prostaglandins, resulting in enhancement of endothelial cell permeability. This process facilitates infiltration of neutrophils to the interstitium where they start to release proteases and elastases to destroy cell membranes and cause cell death. It is reported that inhibition of neutrophil adhesion to the endothelium or administration of antioxidants attenuates the process of reperfusion injury.[168] Treatment of infarcted animals with a nonpeptidyl mimic of superoxide dismutase ameliorates ischemia/reperfusion injury.[169,170]

Impact of oxidative stress on pathogenesis of cardiac arrhythmia

Cardiac arrhythmia occurs when the cardiac electrical activity, which coordinates the rate of heartbeats, is not working properly. This problem leads to a condition that the heart beats too fast, or too slowly, or erratically.[171] In cardiac arrhythmia, the heart is not able to pump blood effectively to all organs, resulting in low oxygen tension and may cause tissue damage in these organs. Various factors besides aging increase the risk of cardiac arrhythmia development, including coronary artery disease, hypertension, congenital heart disease, hyper- or hypothyroidism, drug abuse, diabetes, electrolyte imbalance, and obstructive sleep apnea. Generally, cardiac arrhythmia is classified into tachycardia and bradycardia.[171] *Tachycardia* refers to a fast heartbeat with over 100 bpm, which can originate from atria or ventricle.[171,172] Tachycardia in atria causes different diseases including atrial fibrillation (caused by irregular electrical impulses in atria), atrial flutter (as same as atrial fibrillation with more-organized and more-rhythmic electrical impulses in atria), supraventricular tachycardia (a kind of junctional arrhythmia with various arrhythmia forms originating above the ventricles in the atria or the atrioventricular node), and Wolff–Parkinson–White syndrome (a kind of supraventricular tachycardia with electrical impulses going between atria and ventricles without passing through the atrioventricular node). Tachycardia in ventricles also causes various diseases comprising of ventricular tachycardia (a rapid regular heartbeat with abnormal electrical impulses in the ventricles), ventricular fibrillation (a rapid irregular electrical impulse leading to ineffective trembling of ventricles instead of pumping), and long QT syndrome (a heart disorder with a rapid irregular heartbeat due to changes in the heart electrical system).

Bradycardia is a sluggish rhythm of heartbeats with less than 60 bpm, due to a slow electrical impulse from the sinoatrial node, or blockage of the electrical impulse from atria to ventricle. According to the site of conduction disturbance, bradycardia is classified to sinus node dysfunction and atrioventricular conduction disturbance, each of which contributes to various types of cardiac diseases.[171]

Among different types of cardiac arrhythmia, atrial fibrillation is the most commonly sustained arrhythmia in the clinic, and its prevalence is constantly increasing during past decades. The estimated number of patients with atrial fibrillation globally in 2010 was 33.5 million individuals.[173] Atrial fibrillation is a complex multi-factorial arrhythmia with a rapid heartbeat caused by erratic electrical impulses in the atria, leading to rapid, uncoordinated, and weak contraction of atria. This erratic electrical impulse assaults the atrioventricular node, resulting in an irregular, rapid rhythm of ventricles with a rate of 100–160 bpm, which is referred to as fibrillation. The atrial fibrillation may lead to serious complications such as stroke, and therefore requires medical attention. According to the literature, development of atrial fibrillation depends on some structural and electrical remodeling of the atria. The structural remodeling refers to abnormalities of the atrial architecture such as atrial dilatation, fibrosis, apoptotic phenomena, and tissue dedifferentiation.[174] On the other hand, the electrophysiological remodeling mainly comprises shortening of the effective refractory period, loss of rate adaptation of the refractory period, prolongation of conduction, and increased dispersion of refractoriness and conduction.[175] Moreover, it is suggested that myocardial fibrosis, inflammation, and oxidative stress, amongst all reported etiological factors, play a central role in the pathogenesis of atrial fibrillation.[171,176,177]

Myocardial fibrosis is characterized by enhancement of fibroblast proliferation and collagen deposition in the interstitial matrix of cardiomyocytes.[178] It is reported that ROS indirectly influence atrial fibrillation development by promoting myocardial fibrosis through modulation of fibroblast proliferation as well as enhancement of collagen expression.[178] Furthermore, increased serum levels of the inflammatory marker C-reactive protein and other inflammatory markers in atrial fibrillation patients confirm the role of inflammation in pathogenesis of this disease.[176,179] Moreover, a steady rise of systemic inflammatory factors as well as oxidative stress markers was observed in patients with persistent atrial fibrillation.[180] It is well established that ROS induce inflammation responses due to regulation of inflammatory mediators such as NF-κB and increased gene expression levels of inflammatory factors, which also confirms the indirect effect of ROS on atrial fibrillation development.[171–181]

Besides the indirect effects of ROS on atrial fibrillation, cumulative evidence implicates a direct pivotal role of oxidative stress in the pathogenesis of atrial fibrillation.[171,182,183] Various animal as well as clinical studies have reported an enhancement of oxidative stress markers during atrial fibrillation. For the first time, Mihm *et al.* have mentioned an association between oxidative stress and atrial fibrillation. They have observed remarkable oxidative stress in samples of the right atrial appendage obtained from atrial fibrillation patients undergoing the maze procedure.[184] Furthermore, activity of myofibrillar creatine kinase (MM-CK) was decreased in these patients in comparison to healthy individuals, which was inversely associated with the amount of 3-nitrotyrosine (a peroxynitrite biomarker), implicating the importance of increased peroxynitrite formation in atrial fibrillation development.[184] The maze procedure is a surgical treatment for ablation of atrial fibrillation. In the maze procedure, one scar will be created in the atrial tissue via small incision, ultrasound energy, radio waves, freezing, or microwave energy. This scar tissue conducts no electrical impulse, blocks the abnormal electrical signals causing arrhythmia, and leads the electrical impulses through a controlled path, or maze, to the ventricles.[185] In another study, a significant increase in atrial NADPH oxidase activity was observed in biopsies of the right atrial appendage from 170 patients who developed atrial fibrillation after coronary artery bypass surgery.[186] Genetic analysis of atrial tissue from 26 patients with atrial fibrillation displayed gene expression enhancement for the oxidative stress markers monooxygenase 1, monoamine oxidase-B, ubiquitin specific protease-8, tyrosinase-related protein-1, and tyrosine 3-monooxygenase and reduction of glutathione peroxidase-1 and heme oxygenase-2 mRNA levels.[187] This data confirmed the association between oxidative stress and pathogenesis of atrial fibrillation. Moreover, Neuman *et al.* reported a significant association between oxidative stress markers and atrial fibrillation in patients with or without persistent or permanent atrial fibrillation.[188] Moreover, Youn *et al.* have concluded that oxidative stress in the atrium causes various changes including hypertrophy via enhancement of fibroblast proliferation, extracellular matrix remodeling via enhancement of collagen and matrix metalloproteinase expression, and enhancement of inflammation due to increased expression and activity of TNF-α, interleukin-1β, interleukin-6, and inducible NOS. All these events result in structural and electrical remodeling, leading to atrial fibrillation.[183]

NADPH oxidase amongst other sources of ROS production (e.g., xanthine oxidase, uncoupled eNOS, and mitochondria) has emerged as a major enzymatic source for oxidative stress leading to atrial fibrillation.[183,186,189] Presumably NADPH oxidase activation is important for initiation of atrial fibrillation development, whereas the other sources of ROS production are involved in the maintenance of permanent atrial fibrillation.[190] It is implicated that suppressing Nox2 expression by shRNA gene based targeting methods significantly reduced levels of interstitial fibrosis as well as atrial fibrillation formation in a Nox2 shRNA animal model.[191] Overexpression of Nox4 has also been reported in patients with atrial fibrillation.[189] Furthermore, overexpression of Nox4 in animal models increased phosphorylation levels of calcium/calmodulin-activated protein kinase II (CaMKII), leading to an arrhythmic phenotype, which is suggesting a causal role of Nox4 in cardiac arrhythmia mediated by activation of the pro-arrhythmic enzyme CaMKII.[192] Nox2 and Nox4 isoforms of NADPH oxidase are the specific sources of superoxide and hydrogen peroxide production involved in atrial fibrillation. ROS-derived from NADPH oxidase activate myocardial fibrosis, atrial inflammation, and electrophysiological remodeling following activation by known upstream triggers of atrial fibrillation including angiotensin-II and atrial stretch.[171,183]

It is demonstrated that oxidative stress-induced atrial fibrillation reduced the total Na^+ current, enhanced late Na^+ current and L-type Ca^{2+} current, the open probability of the sarcoplasmic reticulum ryanodine receptor via CaMKII, and impaired sarco/endoplasmic reticulum Ca^{2+}-ATPase (SERCA) function.[171,183] These events induce intracellular Ca^{2+} overload which may facilitate early after depolarization and delayed after depolarization mechanisms in myocardial tissue. This pathway is important in arrhythmogenesis of atrial fibrillation.[171,193] It is reported that reduction of atrial Ca^{2+} current through inhibition of glutathione synthase with buthionine sulfoximine impaired the force-frequency response of isolated atrial trabeculae, and also diminished its contractility response to adrenergic agonists.[194] These data underscored the impact of oxidative stress on atrial ion channel activity and contractile function.

Multiple choice questions

1 A heartbeat of less than 60 bpm is called
   a. Tachycardia
   b. *Bradycardia*
   c. Arrhythmia
   d. Fibrillation

2 The most serious and life threatening arrhythmia of the heart is
   a. Tachycardia
   b. Bradycardia
   c. *Fibrillation*
   d. Flutter

3 Which event is NOT involved in pathogenesis of atrial fibrillation?
   a. Overexpression of Nox isoforms
   b. Intracellular calcium overload
   c. Enhancement of systemic inflammatory factors
   d. *Production of oxidized LDL*

4 The following alterations occur in the ischemic state except of
   a. *Enhancement of NADPH oxidase activity*
   b. Reduction of ATP production
   c. Disturbance of mitochondrial electron transport chain
   d. Metabolic switch toward anaerobic glycolysis

5 During progression of atherosclerosis development, oxidized LDL mediates following changes in the vessel wall occur except
   a. Upregulation of arginase-I activity
   b. Activation of NADPH oxidase and xanthine oxidase
   c. *Metabolic switch toward anaerobic glycolysis*
   d. Increased expression and activity of matrix metalloproteinase

6 Explain these steps in the etiology of atherosclerosis development and plaque rupture:

   a. Inflammatory responses of endothelial cells to low-density lipoprotein (LDL) in hypercholesterolemia

   b. ROS-produced by NADPH oxidase decrease nitric oxide bioavailability, leading to reduction of antiatherogenic functions of the endothelium

   c. LDL enters the subendothelial space

   d. LDL becomes oxidized

   e. oxLDL is then taken up by endothelial cells and induces their activation

   f. Activated endothelial cells recruit monocytes into the lesion area

   g. Transformation of monocytes into macrophages for ingestion of oxLDL

   h. Macrophages absorb oxLDL, form foam cells and are trapped in the space

   i. Foam cells attract further white blood cells to the injury site

   j. Building up atherosclerotic plaques at the injury site

   k. oxLDL induce synthesis of collagen IV in fibroblasts, to form a fibrous cap lining the plaques

   l. oxLDL enhance NADPH oxidase activity, leading to oxidative stress

   m. ROS-produced by NADPH oxidase decrease nitric oxide bioavailability, leading to reduction of antiatherogenic functions of endothelium

   n. oxLDL enhances metalloproteinases with collagenase activity

   o. Activated matrix metalloproteinases thin the fibrous cap, resulting in plaque instability and rupture

References

1 Allen, R.G. & Balin, A.K. (1989) Oxidative influence on development and differentiation: an overview of a free radical theory of development. *Free Radical Biology and Medicine*, 6, 631–661.

2 Sauer, H. & Wartenberg, M. (2010) Reactive oxygen and nitrogen species in cardiovascular differentiation of stem cells. In: *Studies on cardiovascular disorders*. Springer, pp. 61–85.

3 Madamanchi, N.R. & Runge, M.S. (2013) Redox signaling in cardiovascular health and disease. *Free Radical Biology and Medicine*, 61C, 473–504.

4 Valko, M., Leibfritz, D., Moncol, J. *et al.* (2007) Free radicals and antioxidants in normal physiological functions and human disease. *International Journal of Biochemistry and Cell Biology*, 39, 44–84.

5 Sugamura, K. & Keaney, J.F. Jr., (2011) Reactive oxygen species in cardiovascular disease. *Free Radical Biology and Medicine*, 51, 978–992.

6 Elahi, M.M., Kong, Y.X. & Matata, B.M. (2009) Oxidative stress as a mediator of cardiovascular disease. *Oxidative Medicine and Cellular Longevity*, 2, 259–269.

7 Taverne, Y.J., De Beer, V.J., Hoogteijling, B.A. *et al.* (2012) Nitroso-redox balance in control of coronary vasomotor tone. *Journal of Applied Physiology*, 112, 1644–1652.

8 Paravicini, T.M., Montezano, A.C., Yusuf, H. *et al.* (2012) Activation of vascular p38MAPK by mechanical stretch is independent of c-Src and NADPH oxidase: influence of hypertension and angiotensin II. *Journal of the American Society of Hypertension*, 6, 169–178.

9 Burger, D., Montezano, A.C., Nishigaki, N. *et al.* (2011) Endothelial microparticle formation by angiotensin II is mediated via Ang II receptor type I/NADPH oxidase/ Rho kinase pathways targeted to lipid rafts. *Arteriosclerosis, Thrombosis, and Vascular Biology*, 31, 1898–1907.

10 Meng, F.G. & Zhang, Z.Y. (2013) Redox regulation of protein tyrosine phosphatase activity by hydroxyl radical. *Biochimica et Biophysica Acta*, 1834, 464–469.

11 Knock, G.A. & Ward, J.P. (2011) Redox regulation of protein kinases as a modulator of vascular function. *Antioxidants and Redox Signaling*, 15, 1531–1547.

12 Cioffi, D.L. (2011) Redox regulation of endothelial canonical transient receptor potential channels. *Antioxidants and Redox Signaling*, 15, 1567–1582.

13 Patel, R.P., Levonen, A., Crawford, J.H. *et al.* (2000) Mechanisms of the pro- and anti-oxidant actions of nitric oxide in atherosclerosis. *Cardiovascular Research*, 47, 465–474.

14 Walford, G. & Lascalzo, J. (2003) Nitric oxide in vascular biology. *Journal of Thrombosis and Haemostasis*, 1, 2112–2118.

15 Föstermann, U. (2008) Oxidative stress in vascular disease: causes, defenses mechanisms and potential therapies. *Nature Clinical Practice. Cardiovascular Medicine*, 5, 338–349.

16 Sen, C.K. & Roy, S. (2008) Redox signals in wound healing. *Biochimica et Biophysica Acta*, 1780, 1348–1361.

17 Ushio-Fukai, M. & Nakamura, Y. (2008) Reactive oxygen species and angiogenesis: NADPH oxidase as target for cancer therapy. *Cancer Letters*, 266, 37–52.

18 Ateghang, B., Wartenberg, M., Gassmann, M. *et al.* (2006) Regulation of cardiotrophin-1 expression in mouse embryonic stem cells by HIF-1alpha and intracellular reactive oxygen species. *Journal of Cell Science*, 119, 1043–1052.

19 Sharifpanah, F., Wartenberg, M., Hannig, M. *et al.* (2008) Peroxisome proliferator-activated receptor alpha agonists enhance cardiomyogenesis of mouse ES cells by utilization of a reactive oxygen species-dependent mechanism. *Stem Cells*, 26, 64–71.

20 Lange, S., Heger, J., Euler, G. *et al.* (2009) Platelet-derived growth factor BB stimulates vasculogenesis of embryonic stem cell-derived endothelial cells by calcium-mediated generation of reactive oxygen species. *Cardiovascular Research*, 81, 159–168.

21 Padmasekar, M., Sharifpanah, F., Finkensieper, A. *et al.* (2011) Stimulation of cardiomyogenesis of embryonic stem cells by nitric oxide downstream of AMP-activated protein kinase and mTOR signaling pathways. *Stem Cells and Development*, 20, 2163–2175.

22 Sauer, H., Ravindran, F., Beldoch, M. *et al.* (2013) α2-Macroglobulin enhances vasculogenesis/angiogenesis of mouse embryonic stem cells by stimulation of nitric oxide generation and induction of fibroblast growth factor-2 expression. *Stem Cells and Development*, 22, 1443–1454.

23 Schmelter, M., Ateghang, B., Helmig, S. *et al.* (2006) Embryonic stem cells utilize reactive oxygen species as transducers of mechanical strain-induced cardiovascular differentiation. *FASEB Journal*, 20, 1182–1184.

24 Sharifpanah, F., Saliu, F., Bekhite, M.M. *et al.* (2014) β-Adrenergic receptor antagonists inhibit vasculogenesis of embryonic stem cells by downregulation of nitric oxide generation and interference with VEGF signaling. *Cell and Tissue Research*, 358, 443–453.

25 James, P.A., Oparil, S., Carter, B.L. *et al.* (2014) 2014 Evidence-based guideline for the management of high blood pressure in adults. *JAMA*, 311, 507–520.

26 Chobanian, A.V., Bakris, G.L., Black, H.R. *et al.* (2003) Seventh report of the Joint National Committee on Prevention, Detection, Evaluation, and Treatment of High Blood Pressure. *Hypertension*, 42, 1206–1252.

27 Arima, H., Barzi, F. & Chalmers, J. (2011) Mortality patterns in hypertension. *Journal of Hypertension*, 29, S3–S7.

28 Deaton, C., Froelicher, E.S., Wu, L.H. *et al.* (2011) The global burden of cardiovascular disease. *Journal of Cardiovascular Nursing*, 26, S5–S14.

29 Kearney, P.M., Whelton, M., Reynolds, K. *et al.* (2005) Global burden of hypertension: analysis of worldwide data. *Lancet*, 365, 217–223.

30 Tulman, D.B., Staeicki, S.P., Papadimos, T.J. *et al.* (2012) Advances in management of acute hypertension: a concise review. *Discovery Medicine*, 13, 375–383.

31 World Health Organization (2010). Global status report on noncommunicable diseases 2010.

32 Schiffrin, E.L. (2004) Vascular stiffening and arterial compliance: implications for systolic blood pressure. *American Journal of Hypertension*, 17, 39S–48S.

33 Sedeek, M., Hebert, R.L., Kennedy, C.R. *et al.* (2009) Molecular mechanisms of hypertension: role of Nox family NADPH oxidases. *Current Opinion in Nephrology and Hypertension*, 18, 122–127.

34 Schiffrin, E.L. & Touyz, R.M. (2004) From bedside to bench: role of rennin-angiotensin-aldosterone system in remodeling of resistance arteries in hypertension. *American Journal of Physiology. Heart and Circulatory Physiology*, 287, H435–H446.

35 Tuna, B.G., Bakker, E.N. & Van Bavel, E. (2012) Smooth muscle biomechanics and plasticity: relevance for vascular calibre and remodelling. *Basic and Clinical Pharmacology and Toxicology*, 110, 35–41.

36 Cave, A.C., Brewer, A.C., Narayanapanicker, A. *et al.* (2006) NADPH oxidases in cardiovascular health and disease. *Antioxidants and Redox Signaling*, 8, 691–727.

37 Lyle, A.N. & Griendling, K.K. (2006) Modulation of vascular smooth muscle signaling by reactive oxygen species. *Physiology (Bethesda)*, 21, 269–280.

38 Paravicini, T.M. & Touyz, R.M. (2006) Redox signaling in hypertension. *Cardiovascular Research*, 71, 247–258.

39 Harris, D.M., Cohn, H.I., Pesant, S. *et al.* (2008) GPCR signaling in hypertension: role of GRKs. *Clinical Science (London)*, 115, 79–89.

40 Touyz RM. 2005. "Molecular and cellular mechanisms in vascular injury in hypertension: role of angiotensin-II". *Current Opinion in Nephrology and Hypertension* 14: 125-131.

41 Brandes, R.P. (2014) Endothelial dysfunction and hypertension. *Hypertension*, 64, 924–928.

42 Harrison, D.G., Vinh, A., Lob, H. *et al.* (2010) Role of the adaptive immune system in hypertension. *Current Opinion in Pharmacology*, 10, 203–207.

43 Harrison, D.G., Widder, J., Grumbach, I. *et al.* (2006) Endothelial mechanotransduction, nitric oxide and vascular inflammation. *Journal of Internal Medicine*, 259, 351–363.

44 Romanowski, A., Murray, J.R. & Huston, M.J. (1960) Effects of hydrogen peroxide on normal and hypertensive rats. *Pharmaceutica Acta Helvetiae*, 35, 354–357.

45 Montezano, A.C. & Touyz, R.M. (2014) Reactive oxygen species, vascular Noxs, and hypertension: focus on translational and clinical research. *Antioxidants and Redox Signaling*, 20, 164–182.

46 Rajagopalan, S., Kurz, S., Münzel, T. *et al.* (1996) Angiotensin-II-mediated hypertension in the rat increases vascular superoxide production via membrane ADH/NADPH oxidase activation: contribution to alterations of vasomotor tone. *Journal of Clinical Investigation*, 97, 1916–1923.

47 Chen, D.D., Dong, Y.G., Yuan, H. *et al.* (2012) Endothelin-1 activation of endothelin A receptor/NADPH oxidase pathway and diminished antioxidants critically contribute to endothelial progenitor cell reduction and dysfunction in salt-sensitive hypertension. *Hypertension*, 59, 1037–1043.

48 Huang, B.S., Zheng, H., Tan, J. *et al.* (2011) Regulation of hypothalamic rennin-angiotensin system and oxidative stress by aldosterone. *Experimental Physiology*, 96, 1028–1038.

49 Schupp, N., Kolkhof, P., Quiesser, N. *et al.* (2011) Mineralocorticoid receptor-mediated DNA damage in kidneys of DOCA-salt hypertensive rats. *FASEB Journal*, 25, 968–978.

50 Gavazzi, G., Banfi, B., Deffert, C. *et al.* (2006) Decreased blood pressure in Nox1-deficient mice. *FEBS Letters*, 580, 497–504.

51 Matsuno, K., Yamada, H., Iwata, K. *et al.* (2005) Nox1 is involved in angiotensin-II-mediated hypertension: a study in Nox1-deficient mice. *Circulation*, 112, 2677–2685.

52 Dikalova, A., Clempus, R., Lassegue, B. *et al.* (2005) Nox1 overexpression potentiates angiotensin-II-induced hypertension and vascular smooth muscle hypertrophy in transgenic mice. *Circulation*, 112, 2668–2676.

53 Fukui, T., Ishizaka, N., Rajagopalan, S. *et al.* (1997) p22phox mRNA expression and NAD(P)H oxidase activity are increased in aortas from hypertensive rats. *Circulation Research*, 80, 45–51.

54 Rey, F.E., Cifuentes, M.E., Kiarash, A. *et al.* (2001) Novel competitive inhibitor of NADPH oxidase assembly attenuates vascular $O_{(2)}^{(-)}$ and systolic blood pressure in mice. *Circulation Research*, 89, 408–414.

55 Harrison, C.B., Drummond, G.R., Sobey, C.G. *et al.* (2010) Evidence that nitric oxide inhibits vascular inflammation and superoxide production via a p47(phox)-dependent mechanism in mice. *Clinical and Experimental Pharmacology and Physiology*, 37, 429–434.

56 Landmesser, U., Cai, H., Dikalov, S. *et al.* (2002) Role of p47(phox) in vascular oxidative stress and hypertension caused by angiotensin-II. *Hypertension*, 40, 511–515.

57 Weber, D.S., Rocic, P., Mellis, A.M. *et al.* (2005) Angiotensin-II-induced hypertrophy is potentiated in mice overexpressing p22(phox) in vascular smooth muscle. *American Journal of Physiology. Heart and Circulatory Physiology*, 288, H37–H42.

58 Rathore, R., Zheng, Y.M., Niu, C.F. *et al.* (2008) Hypoxia activated NADPH oxidase to increase [ROS]$_i$ and [Ca^{2+}]$_i$ through the mitochondrial ROS-PKCepsilon signaling axis in pulmonary artery smooth muscle cells. *Free Radical Biology and Medicine*, 45, 1223–1231.

59 Douglas, G., Bendall, J.K., Crabtree, M.J. *et al.* (2012) Endothelial-specific Nox2 over-expression increases vascular superoxide and macrophage recruitment in ApoE$^{-/-}$ mice. *Cardiovascular Research*, 94, 20–29.

60 Nguyen Dinh Cat, A., Montezano, A.C., Burger, D. *et al.* (2013) Angiotensin-II, NADPH oxidase, and redox signaling in the vasculature. *Antioxidants and Redox Signaling*, 19, 1110–1120.

61 Gupte, A.S., Kaminski, P.M., George, S. *et al.* (2009) Peroxide generation by p47Phox-Src activation of Nox2 has a key role in protein kinase C-induced arterial smooth muscle contraction. *American Journal of Physiology. Heart and Circulatory Physiology*, 296, H1048–H1057.

62 Touyz, R.M., Mercure, C., He, Y. *et al.* (2005) Angiotensin-II-dependent chronic hypertension and cardiac hypertrophy are unaffected by gp91phox-containing NADPH oxidase. *Hypertension*, 45, 530–537.

63 Ismail, S., Sturrock, A., Wu, P. *et al.* (2009) Nox4 mediates hypoxia-induced proliferation of human pulmonary artery smooth muscle cells: the role of autocrine production of transforming growth factor-(beta)1 and insulin-like growth factor binding protein-3. *American Journal of Physiology. Lung Cellular and Molecular Physiology*, 296, 489–499.

64 Manea, A., Tanase, L.I., Raicu, M. *et al.* (2010) Jak/STAT signaling pathway regulates Nox1 and Nox4-based NADPH oxidase in human aortic smooth muscle cells. *Arteriosclerosis, Thrombosis, and Vascular Biology*, 30, 105–112.

65 Fulton, D.J. (2009) Nox5 and the regulation of cellular function. *Antioxidants and Redox Signaling*, 11, 2443–2452.

66 Jay, D.B., Papaharalambus, C.A., Seidel-Rogol, B. *et al.* (2008) Nox5 mediates PDGF-induced proliferation in human aortic smooth muscle cells. *Free Radical Biology and Medicine*, 45, 329–335.

67 Montezano, A.C., Burger, D., Paravicini, T.M. *et al.* (2010) Nicotinamide adenine dinucleotide phosphate reduced oxidase 5 (Nox5) regulation by angiotensin-II and endothelin-a is mediated via calcium/calmodulin-dependent, rac-1-independent pathways in human endothelial cells. *Circulation Research*, 106, 1363–1373.

68 Serrander, L., Jaquet, V., Bedard, K. *et al.* (2007) Nox5 is expressed at the plasma membrane and generates superoxide in response to protein kinase C activation. *Biochimie*, 89, 1159–1167.

69 Al Ghouleh, I., Khoo, N.K., Knaus, U.G. *et al.* (2011) Oxidases and peroxidases in cardiovascular and lung disease: new concept in reactive oxygen species signaling. *Free Radical Biology and Medicine*, 51, 1271–1288.

70 Al-Shebly, M.M. & Mansour, M.A. (2012) Evaluation of oxidative stress and antioxidant status in diabetic and hypertensive women during labor. *Oxidative Medicine and Cellular Longevity*, 2012, 32974–32976.

71 Banfi, B., Tirone, F., Durussel, I. *et al.* (2004) Mechanism of Ca^{2+} activation of the NADPH oxidase 5 (Nox5). *Journal of Biological Chemistry*, 279, 18583–18591.

72 Pandey, D., Gratton, J.P., Rafikov, R. *et al.* (2011) Calcium/calmodulin-dependent kinase II mediates the phosphorylation and activation of NADPH oxidase 5. *Molecular Pharmacology*, 80, 407–514.

73 Kim, H.J. & Vaziri, N.D. (2010) Contribution of impaired Nrf2-Keap1 pathway to oxidative stress and inflammation in chronic renal failure. *American Journal of Physiology. Renal Physiology*, 298, F662–F671.

74 Moreno, M.U., Jose, G.S., Fortuno, A. *et al.* (2006) The C242T CYBA polymorphism of NADPH oxidase is associated with essential hypertension. *Journal of Hypertension,* 24, 1299–1306.

75 Schreiber, R., Ferreira-Sae, M.C., Ronchi, J.A. *et al.* (2011) The C242T polymorphism of the p22-phox gene (CYBA) is associated with higher left ventricular mass in Brazilian hypertensive patients. *BMC Medical Genomics,* 12, 114–118.

76 Romero, J.C. & Reckelhoff, J.F. (1999) Role of angiotensin and oxidative stress in essential hypertension. *Hypertension,* 34, 943–949.

77 Drummond, G.R. & Sobey, C.G. (2014) Endothelial NADPH oxidases: which Nox to target in vascular disease? *Trends in Endocrinology and Metabolism,* 25, 452–463.

78 Jung, O., Schreiber, J.G., Geiger, H. *et al.* (2004) gp91phox-containing NAD(P)H oxidase mediates endothelial dysfunction in renovascular hypertension. *Circulation,* 109, 1795–1801.

79 Zhou, M.S., Hernandez Schulman, I., Pagano, P.J. *et al.* (2006) Reduced NADPH oxidase in low renin hypertension: link among angiotensin-II, atherogenesis, and blood pressure. *Hypertension,* 47, 81–86.

80 Feng, D., Yang, C., Geurts, A.M. *et al.* (2012) Increased expression of NADPH oxidase subunit p67phox in the renal medula contributes to excess oxidative stress and salt-sensitive hypertension. *Cell Metabolism,* 15, 201–208.

81 Iyer, A., Chan, V. & Brown, L. (2010) The DOCA-salt hypertensive rat as a model of cardiovascular oxidative and inflammatory stress. *Current Cardiology Reviews,* 6, 291–297.

82 Li, L., Fink, G.D., Watts, S.W. *et al.* (2003) Endothelin-1 increases vascular superoxide via endothelin(A)-NADPH oxidase pathway in low-renin hypertension. *Circulation,* 107, 1053–1058.

83 Shreenivas, S. & Oparil, S. (2007) The role of endothelin-1 in human hypertension. *Clinical Hemorheology and Microcirculation,* 37, 157–178.

84 Amiri, F., Virdis, A., Neves, M.F. *et al.* (2004) Endothelium-restricted overexpression of human endothelin-1 causes vascular remodeling and endothelial dysfunction. *Circulation,* 110, 2233–2240.

85 Barhoumi, T., Briet, M., Kasal, D.A. *et al.* (2014) Erythropoietin-induced hypertension and vascular injury in mice overexpressing human endothelin-1: exercise attenuated hypertension, oxidative stress, inflammation, and immune response. *Journal of Hypertension,* 32, 784–794.

86 Schiffrin, E.L. (1998) Endothelin: role in hypertension. *Biological Research,* 31, 199–208.

87 Rautureau, Y. & Schiffrin, E.L. (2012) Endothelin in hypertension: an update. *Current Opinion in Nephrology and Hypertension,* 21, 128–136.

88 Sventek, P., Li, J.S., Grove, K. *et al.* (1996) "Vascular structure and expression of endothelin-1 gene in L-NAME treated spontaneously hypertensive rats". *Hypertension,* 27, 49–55.

89 Dikalov, S.I. & Ungvari, Z. (2013) "Role of mitochondrial oxidative stress in hypertension". *American Journal of Physiology. Heart and Circulatory Physiology,* 305, H1417–H1427.

90 Dikalova, A.E., Bikineyeva, A.T., Budzyn, K. *et al.* (2010) Therapeutic targeting of mitochondrial superoxide in hypertension. *Circulation Research,* 107, 106–116.

91 Sturza, A., Leisegang, M.S., Babelova, A. *et al.* (2013) Monoamine oxidases are mediators of endothelial dysfunction in the mouse aorta. *Hypertension,* 62, 140–146.

92 Graiani, G., Lagrasta, C., Migliaccio, E. *et al.* (2005) Genetic deletion of the p66Shc adaptor protein protects from angiotensin-II-induced myocardial damage. *Hypertension,* 46, 433–440.

93 Spescha, R.D., Glanzmann, M., Simic, B. *et al.* (2014) Adaptor protein p66(Shc) mediates hypertension-associated, cyclic stretch-dependent, endothelial damage. *Hypertension,* 64, 347–353.

94 Xie, H.H., Zhou, S., Chen, D.D. *et al.* (2010) GTP cyclohydrolase-I/BH4 pathway protects EPCs via suppressing oxidative stress and thrombospondin-1 in salt-sensitive hypertension. *Hypertension*, 56, 1137–1144.

95 Raaz, U., Toh, R., Maegdefessel, L. *et al.* (2014) Hemodynamic regulation of reactive oxygen species: implications for vascular disease. *Antioxidants and Redox Signaling*, 20, 914–928.

96 Cai, H., Mc Nally, J.S., Weber, M. *et al.* (2004) Oscillatory shear stress upregulation of endothelial nitric oxide synthase requires intracellular hydrogen peroxide and CaMKII. *Journal of Molecular and Cellular Cardiology*, 37, 121–125.

97 Stocker, R. & Keaney, J.F. Jr., (2004) Role of oxidative modifications in atherosclerosis. *Physiological Reviews*, 84, 1381–1478.

98 Stary, H.C., Chandler, A.B., Dinsmore, R.E. *et al.* (1995) A definition of advanced types of atherosclerotic lesions and a histological classification of atherosclerosis. A report from the Committee on Vascular Lesions of the Council on Arteriosclerosis, American Heart Association. *Arteriosclerosis, Thrombosis, and Vascular Biology*, 15, 1512–1531.

99 Bui, Q.T., Prempeh, M. & Wilensky, R.L. (2009) Atherosclerotic plaque development. *International Journal of Biochemistry and Cell Biology*, 41, 2109–2113.

100 Bonomini, F., Tengattini, S., Fabiano, A. *et al.* (2008) Atherosclerosis and oxidative stress. *Histol Histopath*, 23, 381–390.

101 Cipollone, F., Fazia, M. & Mezzetti, A. (2007) Oxidative stress, inflammation and atherosclerotic plaque development. *International Congress Series*, 1303, 35–40.

102 Li, H. & Föstermann, U. (2013) Uncoupling of endothelial NO synthase in atherosclerosis and vascular disease. *Current Opinion in Pharmacology*, 13, 161–167.

103 Amberger, A., Maczek, C., Jürgens, G. *et al.* (1997) Co-expression of ICAM, VCAM-1, ELAM-1 and Hsp60 in human arterial and venous endothelial cells in response to cytokines and oxidized low-density lipoproteins. *Cell Stress & Chaperones*, 2, 94–103.

104 Wilensky, R.L. & Hamamdzic, D. (2007) The molecular basis of vulnerable plaque: potential therapeutic role for immunomodulation. *Current Opinion in Cardiology*, 22, 545–551.

105 Inoue, T. & Node, K. (2006) Vascular failure. *Journal of Hypertension*, 24, 2121–2130.

106 Arai, H. (2014) Oxidative modification of lipoproteins. *Subcellular Biochemistry*, 77, 103–114.

107 Leopold, J.A. & Loscalzo, J. (2008) Oxidative mechanisms and atherothrombotic cardiovascular disease. *Drug Discovery Today: Therapeutic Strategies*, 5, 5–13.

108 Shatrov, V.A. & Brüne, B. (2003) Induced expression of manganese superoxide dismutase by non-toxic concentrations of oxidized low-density lipoprotein (oxLDL) protects against oxLDL-mediated cytotoxicity. *Biochemical Journal*, 374, 505–511.

109 Kinscherf, R., Deigner, H.P., Usinger, C. *et al.* (1997) Induction of mitochondrial manganese superoxide dismutase in macrophages by oxidized LDL: its relevance in atherosclerosis of human and heritable hyperlipidemic rabbits. *FASEB Journal*, 11, 1317–1328.

110 Konior, A., Schramm, A., Czesnikiewicz-Guzik, M. *et al.* (2014) NADPH oxidase in vascular pathology. *Antioxidants and Redox Signaling*, 20, 2794–2814.

111 Kinkade, K., Streeter, J. & Miller, F.J. (2013) Inhibition of NADPH oxidase by apocynin attenuates progression of atherosclerosis. *International Journal of Molecular Sciences*, 14, 17017–17028.

112 Gray, S.P., Di Marco, E., Okabe, J. *et al.* (2013) NADPH oxidase 1 plays a key role in diabetes mellitus-accelerated atherosclerosis. *Circulation*, 127, 1888–1902.

113 Sheehan, A.L., Carrell, S., Johnson, B. *et al.* (2011) Role of Nox1 NADPH oxidase in atherosclerosis. *Atherosclerosis*, 216, 321–326.

114 Judkins, C.P., Diep, H., Broughton, B.R. *et al.* (2010) Direct evidence of a role for Nox2 in superoxide production, reduced nitric oxide bioavailability, and early atherosclerotic plaque formation in ApoE$^{-/-}$ mice. *American Journal of Physiology. Heart and Circulatory Physiology*, 298, H24–H32.

115 Barry-Lane, P.A., Patterson, C., van der Merwe, M. *et al.* (2001) p47phox is required for atherosclerotic lesion progression in ApoE$^{-/-}$ mice. *Journal of Clinical Investigation*, 108, 1513–1522.

116 Violi, F., Pignatelli, P., Pignata, C. *et al.* (2013) Reduced atherosclerotic burden in subjects with genetically determined low oxidative stress. *Arteriosclerosis, Thrombosis, and Vascular Biology*, 33, 406–412.

117 Violi, F., Sanguigni, V., Carnevale, R. *et al.* (2009) Hereditary deficiency of gp91phox is associated with enhanced arterial dilatation: results of a multicenter study. *Circulation*, 120, 1616–1622.

118 Sorescu, D., Weiss, D., Lassegue, B. *et al.* (2002) Superoxide production and expression of nox family proteins in human atherosclerosis. *Circulation*, 105, 1429–1435.

119 Xu, S., Chamseddine, A.H., Carrell, S. *et al.* (2014) Nox4 NADPH oxidase contributes to smooth muscle cell phenotypes associated with unstable atherosclerotic plaques. *Redox Biology*, 2, 642–650.

120 Guzik, T.J., Chen, W., Gongora, M.C. *et al.* (2008) Calcium-dependent Nox5 nicotinamide adenine dinucleotide phosphate oxidase contributes to vascular oxidative stress in human coronary artery disease. *Journal of the American College of Cardiology*, 52, 1803–1809.

121 Föstermann, U. & Sessa, W.C. (2012) Nitric oxide synthase: regulation and function. *European Heart Journal*, 33, 829–837.

122 Feron, O., Dessy, C., Moniotte, S. *et al.* (1999) Hypercholesterolemia decreases nitric oxide production by promoting the interaction of caveolin and endothelial nitric oxide synthase. *Journal of Clinical Investigation*, 103, 897–905.

123 Crabtree, M.J., Hale, A.B. & Channon, K.M. (2011) Dihydrofolate reductase protects endothelial nitric oxide synthase from uncoupling in tetrahydrobiopterin deficiency. *Free Radical Biology and Medicine*, 50, 1639–1646.

124 Daugherty, A., Rateri, D.L., Lu, H. *et al.* (2004) Hypercholesterolemia stimulates angiotensin peptide synthesis and contributes to atherosclerosis through the AT1A receptor. *Circulation*, 110, 3849–3857.

125 Chalupsky, K. & Cai, H. (2004) Endothelial n: critical for nitric oxide bioavailability and role in angiotensin-II uncoupling of endothelial nitric oxide synthase. *Proceedings of the National Academy of Sciences of the United States of America*, 102, 9056–9061.

126 Nickenig, N.W. (2003) AT1 receptors in atherosclerosis: biological effects including growth, angiogenesis, and apoptosis. *European Heart Journal Supplements*, 5, A9–A13.

127 Stocker, R. & Keaney, J.F. Jr., (2005) New insights on oxidative stress in the artery wall. *Journal of Thrombosis and Haemostasis*, 3, 1825–1834.

128 Wang, W., Hein, T.W., Zhang, C. *et al.* (2011) Oxidized low-density lipoprotein inhibits nitric oxide-mediated coronary arteriolar dilation by up-regulating endothelial arginase-I. *Microcirculation*, 18, 36–45.

129 Abdelsamie, S.A., Li, Y., Huang, Y. *et al.* (2011) Oxidized LDL immune complexes stimulate collagen IV production in mesangial cells via Fc gamma receptors I and III. *Clinical Immunology*, 139, 258–266.

130 Galis, Z.S., Surkhova, G.K., Lark, M.W. *et al.* (1994) Increased expression of matrix metalloproteinases and matrix degrading activity in vulnerable regions of human atherosclerotic plaques. *Journal of Clinical Investigation*, 94, 2493–2503.

131 Lin, J., Kakkar, V. & Lu, X. (2014) Impact of matrix metalloproteinases on atherosclerosis. *Current Drug Targets*, 15, 442–453.

132 Kamanna, V.S., Ganji, S.H. & Kashyap, M.L. (2013) Myeloperoxidase and atherosclerosis. *Current Cardiovascular Risk Reports*, 7, 102–107.

133 Shao, B., Tang, C., Sinha, A. *et al.* (2014) Human with atherosclerosis have impaired ACA1 cholesterol efflux and enhanced high-density lipoprotein oxidation by myeloperoxidase. *Circulation Research*, 114, 1733–1742.

134 Piper, H.M., Meuter, K. & Schäfer, C. (2003) Cellular mechanisms of ischemia-reperfusion injury. *Annals of Thoracic Surgery*, 75, S644–S648.

135 Frank, A., Bonney, M., Bonney, S. *et al.* (2012) Myocardial ischemia reperfusion injury – from basic science to clinical bedside. *Seminars in Cardiothoracic and Vascular Anesthesia*, 16, 123–132.

136 Ruiz-Meana, M. & Garcia-Dorado, D. (2009) Pathophysiology of ischemia-reperfusion injury: new therapeutic options for acute myocardial infarction. *Revista Española de Cardiología*, 62, 199–209.

137 Jennings, R.B., Sommers, H.M., Smyth, G.A. *et al.* (1960) Myocardial necrosis by temporary occlusion of a coronary artery in the dog. *Archives of Pathology*, 70, 68–78.

138 Bing, R.J., Danforth, W.H. & Ballard, F.B. (1960) Physiology of myocardium. *JAMA*, 172, 438–444.

139 Dhalla, N.S. & Duhamel, T.A. (2007) The paradox of reperfusion in the ischemic heart. *Heart and Metabolism*, 37, 31–34.

140 Zweier, J.L. & Talukder, M.A. (2006) The role of oxidants and free radicals in reperfusion injury. *Cardiovascular Research*, 70, 181–190.

141 Mozaffari, M.S., Liu, J.Y., Abebe, W. *et al.* (2013) Mechanisms of load dependency of myocardial ischemia reperfusion injury. *American Journal of Cardiovascular Disease*, 3, 180–196.

142 Stride, N., Larsen, S., Hey-Mogensen, M. *et al.* (2013) Impaired mitochondrial function in chronically ischemic human heart. *American Journal of Physiology. Heart and Circulatory Physiology*, 304, H1407–H1414.

143 Kalogeris, T., Baines, C.P., Krenz, M. *et al.* (2012) Cell biology of ischemia-reperfusion injury. *International Review of Cell and Molecular Biology*, 298, 229–317.

144 Kalra, B.S. & Roy, V. (2012) Efficacy of metabolic modulators in ischemic heart disease: an overview. *The Journal of Clinical Pharmacology*, 52, 292–305.

145 Van den Brom, C.E., Bulte, C.S.E., Loer, S.A. *et al.* (2013) Diabetes, perioperative ischemia and volatile anaesthetics: consequences of derangements of myocardial substrate metabolism. *Cardiovascular Diabetology*, 12, 42–54.

146 Clanachan, A.S. (2006) Contribution of protons to post-ischemic $Na^{(+)}$ and $Ca^{(2+)}$ overload and left ventricular mechanical dysfunction. *Journal of Cardiovascular Electrophysiology*, 17(Suppl 1), S141–S148.

147 Reimer, K.A. & Jennings, R.B. (1981) Energy metabolism in the reversible and irreversible phases of severe myocardial ischemia. *Acta Medica Scandinavica. Supplementum*, 651, 19–27.

148 Saini-Chohan, H.K., Goyal, R.K. & Dhalla, N.S. (2010) Involvement of sarcoplasmic reticulum in changing intracellular calcium due to Na^+/K^+-ATPase inhibition in cardiomyocyte. *Canadian Journal of Physiology and Pharmacology*, 88, 702–715.

149 Liu, T. & O'Rourke, B. (2009) Regulation of mitochondrial Ca^{2+} and its effects on energetics and redox balance in normal and failing heart. *Journal of Bioenergetics and Biomembranes*, 41, 127–132.

150 Halestrap, A.P. & Pasdois, P. (2009) The role of the mitochondrial permeability transition pore in heart disease. *Biochimica et Biophysica Acta*, 1787, 1402–1415.

151 Javadov, S., Karmazyan, M. & Escobales, N. (2009) Mitochondrial permeability transition pore opening as a promising therapeutic target in cardiac diseases. *Journal of Pharmacol and Experimental Therapeutics*, 330, 670–678.

152 Kandzari, D.E., Dery, J.P., Armstrong, P.W. *et al.* (2005) MC-1 (pyridoxal 5′-phosphate): novel therapeutic applications to reduce ischaemic injury. *Expert Opinion on Investigational Drugs*, 14, 1435–1442.

153 Kandzari, D.E., Labinaz, M., Cantor, W.J. *et al.* (2003) Reduction of myocardial ischemic injury following coronary intervention (the MC-1 to Eliminate Necrosis and Damage trial). *American Journal of Cardiology*, 92, 660–664.

154 Dhalla, N.S., Takeda, S. & Elimban, V. (2013) Mechanisms of the beneficial effects of vitamin B6 and pyrodoxal 5-phosphate on cardiac performance in ischemic heart disease. *Clinical Chemistry and Laboratory Medicine*, 51, 535–543.

155 Engerson, T.D., McKelvey, T.G., Rhyne, D.B. *et al.* (1987) Conversion of xanthine dehydrogenase to oxidase in ischemic rat tissue. *Journal of Clinical Investigation*, 79, 1564–1570.

156 Omar, H.A., Cherry, P.D., Mortelliti, M.P. *et al.* (1991) Inhibition of coronary artery superoxide dismutase attenuates endothelium-dependent and -independent nitrovasodilator relaxation. *Circulation Research*, 69, 601–608.

157 Raghuvanshi, R., Kaul, A., Bhakuni, P. *et al.* (2007) Xanthine oxidase as a marker of myocardial infarction. *Indian Journal of Clinical Biochemistry*, 22, 90–92.

158 Lee, B.E., Toledo, A.H., Anaya-Prado, R. *et al.* (2009) Allopurinol, xanthine oxidase, and cardiac ischemia. *Journal of Investigative Medicine*, 57, 902–909.

159 Maejima, J., Kuroda, J., Matsushima, S. *et al.* (2011) Regulation of myocardial growth and death by NADPH oxidase. *Journal of Molecular and Cellular Cardiology*, 50, 408–416.

160 Matsushima, S., Kuroda, J., Ago, T. *et al.* (2013) Broad suppression of NADPH oxidase activity exacerbates ischemia/reperfusion injury through inadvertent downregulation of hypoxia-inducible factor-1α and upregulation of peroxisome proliferator-activated receptor-α. *Circulation Research*, 112, 1135–1149.

161 Fukui, T., Yoshiyama, M., Hanatani, A. *et al.* (2001) Expression of p22phox and gp91phox, essential components of NADPH oxidase, increases after myocardial infarction. *Biochemical and Biophysical Research Communications*, 281, 1200–1206.

162 Krijnen, P.A., Meischl, C., Hack, C.E. *et al.* (2003) Increased Nox2 expression in human cardiomyocytes after acute myocardial infarction. *Journal of Clinical Pathology*, 56, 194–199.

163 Loukogeorgakis, S.P., Van den Berg, M.J., Sofat, R. *et al.* (2010) Role of NADPH oxidase in endothelial ischemia/reperfusion injury in humans. *Circulation*, 121, 2310–2316.

164 Zeng, M., Wei, X., Wu, Z. *et al.* (2013) NF-kB-mediated induction of autophagy in cardiac ischemia/reperfusion injury. *Biochemical and Biophysical Research Communications*, 436, 180–185.

165 Frangogiannis, N.G., Smith, C.W. & Entman, M.L. (2002) The inflammatory response in myocardia; infarction. *Cardiovascular Research*, 53, 31–47.

166 Sun, Y. (2009) Myocardial repair/remodeling following infarction: roles of local factors. *Cardiovascular Research*, 81, 482–490.

167 Pasnik, J. & Zeman, K. (2009) Role of the neutrophil in myocardial ischemia-reperfusion injury. *Journal of Organ Dysfunction*, 5, 193–207.

168 Harlan, J.M. & Winn, R.K. (2002) Leukocyte-endothelial interactions: clinical trials of anti-adhesion therapy. *Critical Care Medicine*, 30, S214–S219.

169 Matejikova, J., Kucharska, J., Pinterova, M. *et al.* (2009) Protection against ischemia-induced ventricular arrhythmias and myocardial dysfunction conferred by preconditioning in the rat heart: involvement of mitochondrial K(ATP) channels and reactive oxygen species. *Physiological Research*, 58, 9–19.

170 Salvemini, D., Wang, Z.Q., Zweier, J.L. *et al.* (1999) A nonpeptidyl mimic of superoxide dismutase with therapeutic activity in rats. *Science*, 286, 304–306.

171 Jeong, E.M., Liu, M., Sturdy, M. *et al.* (2012) Metabolic stress, reactive oxygen species, and arrhythmia. *Journal of Molecular and Cellular Cardiology*, 52, 454–463.

172 Estes, N.A. (2011) Predicting and preventing sudden cardiac death. *Circulation*, 124, 651–656.

173 Chugh, S.S., Havmoeller, R., Narayanan, K. *et al.* (2014) Worldwide epidemiology of atrial fibrillation: a Global Burden of Disease 2010 Study. *Circulation*, 129(8), 837–847.

174 Schnabel, R.B., Sullivan, L.M., Levy, D. *et al.* (2009) Development of a risk scores for atrial fibrillation (Framingham Heart Study): a community-based cohort study. *Lancet*, 373, 739–745.

175 Pang, H., Ronderos, R., Perez-Riera, A.R. *et al.* (2011) Reverse atrial electrical remodeling: a systematic review. *Cardiology Journal*, 18, 625–631.

176 Issac, T.T., Dokainish, H. & Lakkis, N.M. (2007) Role of inflammation in initiation and perpetuation of atrial fibrillation: a systematic review of the published data. *Journal of the American College of Cardiology*, 50, 2021–2028.

177 Miragoli, M., Gaudesius, G. & Rohr, S. (2006) Electronic modulation of cardiac impulse conduction by myofibroblasts. *Circulation Research*, 98, 801–810.

178 Cucoranu, I., Clempus, R., Dikalova, A. *et al.* (2005) NAD(P)H oxidase 4 mediates transforming growth factor-beta1-induced differentiation of cardiac fibroblasts into myofibroblasts. *Circulation Research*, 97, 900–907.

179 Ozaydin, M., Peker, O., Erdogan, D. *et al.* (2008) N-acetylcysteine for the prevention of postoperative atrial fibrillation: a prospective, randomized, placebo-controlled pilot study. *European Heart Journal*, 29, 625–631.

180 Negi, S., Sovari, A.A. & Dudley, S.C. Jr., (2010) Atrial fibrillation: the emerging role of inflammation and oxidative stress. *Cardiovascular & Hematological Disorders: Drug Targets*, 10, 262–268.

181 Gao, G. & Dudley, S.C. Jr., (2009) Redox regulation, NF-kB, and atrial fibrillation. *Antioxidants and Redox Signaling*, 11, 2265–2277.

182 Wolin, M.S. & Gupte, S.A. (2005) Novel roles for nox oxidases in cardiac arrhythmia and oxidized glutathione export in endothelial function. *Circulation Research*, 97, 612–614.

183 Youn, J.Y., Zhang, J., Zhang, Y. *et al.* (2013) Oxidative stress in atrial fibrillation: an emerging role of NADPH oxidase. *Journal of Molecular and Cellular Cardiology*, 62, 72–79.

184 Mihm, M.J., Yu, F., Carnes, C.A. *et al.* (2001) Impaired myofibrillar energetics and oxidative injury during human atrial fibrillation. *Circulation*, 104, 174–180.

185 Henry, L. & Ad, N. (2008) The maze procedure: a surgical intervention for ablation of atrial fibrillation. *Heart and Lung*, 37, 432–439.

186 Kim, Y.M., Kattach, H., Ratnatunga, C. *et al.* (2008) Association of atrial nicotinamide adenine dinucleotide phosphate oxidase activity with the development of atrial fibrillation after cardiac surgery. *Journal of the American College of Cardiology*, 51, 68–74.

187 Kim, Y.H., Lim, D.S., Lee, J.H. *et al.* (2003) Gene expression profiling of oxidative stress on atrial fibrillation in humans. *Experimental and Molecular Medicine*, 35, 336–349.

188 Neuman, R.B., Bloom, H.L., Shukrullah, I. *et al.* (2007) Oxidative stress markers are associated with persistent artial fibrillation. *Clinical Chemistry*, 53, 1652–1657.

189 Zhang, J., Young, J.Y., Kim, A.Y. *et al.* (2012) Nox4-dependent hydrogen peroxide overproduction in human atrial fibrillation and HL-1 atrial cells: relationship to hypertension. *Frontiers in Physiology*, 3, 140.

190 Reilly, S.N., Jayaram, R., Nahar, K. *et al.* (2011) Atrial sources of reactive oxygen species vary with the duration and substrate of atrial fibrillation: implications for the antiarrhythmic effect of statins. *Circulation*, 124, 1107–1117.

191 Yoo S, Aistrup GL, Browne S, *et al.* (2013). Targeted inhibition of NADPH oxidase in the poterior left atrium by Nox shRNA decreases formation of atrial fibrillation substrate in heart failure. In: American Heart Association 2013 Scientific sessions: Core 4. Heart Rhythm disorders and resuscitation science; 17–19 Nov 2013; Dallas, TX, USA; Circulation; A15234.

192 Zhang, Y., Shimizu, H., Siu, K.L. *et al.* (2014) NADPH oxidase 4 induces cardiac arrhythmic phenotype in zebrafish. *Journal of Biological Chemistry*, 289, 23200–23208.

193 Weiss, J.N., Nivala, M., Garfinkel, A. *et al.* (2011) Alterations and arrhythmias: from cell to heart. *Circulation Research*, 108, 98–112.

194 Carnes, C.A., Janssen, P.M., Ruehr, M.L. *et al.* (2007) Atrial glutathione content, calcium current and contractility. *Journal of Biological Chemistry*, 282, 28063–28073.

CHAPTER 18

Oxidative stress and antioxidant imbalance: respiratory disorders

Surinder K. Jindal[1,2]

[1] Formerly, Professor of Pulmonary Medicine, Postgraduate Institute of Medical Education & Research, Chandigarh, India
[2] Pulmonary Medicine, Jindal Clinics, Chandigarh, 160020 India

---

THEMATIC SUMMARY BOX

After going through this chapter, a student should know and understand about:

- The role of oxidative stress and reactive oxygen species (ROS) in the etio-pathogenesis of
    i. viral and bacterial respiratory infections, human immunodeficiency virus (HIV) infection, tuberculosis, cystic bronchiectasis and sepsis induced acute lung injury

    ii. airway diseases such as chronic obstructive pulmonary disease (COPD) with particular reference to tobacco smoking and air pollution

    iii. respiratory allergies and bronchial asthma

    iv. interstitial lung diseases (ILDs) including idiopathic pulmonary fibrosis (IPF), asbestosis and sarcoidosis

    v. lung cancer including smoking and asbestos exposure induced cancer

    vi. pulmonary arterial hypertension (PAH) and

    vii. respiratory muscle dysfunction

- The role of antioxidant drugs, herbs and mimics in the treatment and prevention of lung diseases

---

Summary

Oxidative stress is shown to play an important role in the pathogenesis of various respiratory ailments. Respiratory infections of all kinds including tuberculosis and human immunodeficiency virus (HIV) infection are shown to generate increased levels of reactive oxygen species (ROS) and decrease the endogenous production of antioxidant enzymes. Oxidative stress in airway disease, particularly in chronic obstructive pulmonary disease (COPD) is associated with accelerated lung ageing and

---

Oxidative Stress and Antioxidant Protection: The Science of Free Radical Biology and Disease, First Edition.
Edited by Donald Armstrong and Robert D. Stratton.
© 2016 John Wiley & Sons, Inc. Published 2016 by John Wiley & Sons, Inc.

cellular senescence. It is also responsible for systemic manifestations and comorbidities of COPD, and as well as for corticosteroid resistance by suppression of activated anti-inflammatory effects. A large number of studies have shown the significance of oxidative stress in other lung diseases, such as pulmonary fibrosis, asbestosis, lung cancer, pulmonary hypertension, respiratory muscle dysfunction, and others. The role of oxidative stress in the disease etiology and pathogenesis forms the basis for treatments with various exogenous antioxidant agents and mimics. Various dietary nutrients such as vitamins C and E, enzymatic drugs and herbal preparations have been tried in different diseases. Some benefits have been shown in experimental animal and small clinical studies. It seems that the antioxidant drugs and mimics have a potential supplementary role in disease-management and/or prevention of some of the respiratory ailments.

Introduction

The lungs remain at high risk of developing oxidative stress because they are directly exposed to ambient environment, and therefore to high oxygen concentrations. The alveolar macrophages constitute the first line of defense against environmental insults. The activated macrophages produce a large amount of ROS causing damage to the respiratory system. The normal antioxidant defenses which operate to counterbalance the oxidative stress are also impaired due to inhalational toxins. The oxidant:antioxidant imbalance is known to play an important role in the pathogenesis and/or clinical manifestations of different pulmonary disorders such as respiratory infections, asthma, COPD, lung cancer, idiopathic pulmonary fibrosis (IPF) and others.[1,2]

Respiratory infections

Different microorganisms including the bacteria, viruses and fungi are responsible for respiratory infections and lung injury. The important bacterial pathogens include the *Streptococcus pneumoniae, Haemophilus influenzae, Staphylococcus aureus, Pseudomonas aeruginosa, Acinetobacter* species, and several others. Most of these organisms cause lung injury by damaging the lipids, proteins and deoxyribonucleic acids (DNAs) of alveolar cells by producing reactive oxygen intermediates (ROI) such as superoxide anion, hydrogen peroxide and peroxynitrite, and so on. *H. influenza* is an important cause of exacerbations of COPD. The organism causes damage but itself has the capacity to withstand the effects of ROS produced by the inflammatory cells.[3] Endotoxins produced by different bacteria in particular cause extensive damage to the lungs sometimes producing a clinical picture of sepsis, acute respiratory distress syndrome (ARDS) and acute respiratory failure. There is significant evidence to suggest the involvement of various oxidants in the pathogenesis of sepsis induced acute lung injury and ARDS.[4]

Patients with HIV are shown to have significant oxidative stress.[5] The oxidative stress is also shown to enhance with introduction of anti-retroviral therapy thereby causing difficulties in treatment.[5] It is also responsible for impaired T-cell homeostasis due to decrease of Interleukin-7 (IL-7) responsiveness.[6] Respiratory syncytial virus is found to generate significant ROS *in vitro* to cause lower respiratory tract infection in children. Rhinoviruses infection is also associated with increased levels of superoxide formation and production of intercellular adhesion molecule (ICAM-I).

Tuberculosis is an important infection where oxidative stress has a significant role to play. Both the bacillary load and the disease severity are shown to significantly correlate with high oxidant and low antioxidant concentration.[7] It has been therefore suggested that exogenous antioxidant administration helps in fast recovery.[7] Oxidative stress is also implicated in the development of drug-resistance and multi-drug resistant tuberculosis.

Airway diseases

Chronic obstructive pulmonary disease, an important cause of global health burden is the third most common cause of death worldwide. There is high oxidative stress in airway diseases such as COPD and asthma.[8] There is an increase in the burden of superoxide anions and hydrogen peroxide (H_2O_2) radicals in COPD which continuously drives the airway inflammation.[9]

Tobacco smoking is the most important risk factor of COPD. A large number of ROS are present in the cigarette smoke. Smoking is shown to deplete antioxidants; the concentration of ascorbate and vitamin E are decreased in the smokers. The shift in the oxidant–antioxidant balance results in cellular injury, increased cell lysis and epithelial permeability. There is a strong relationship of tobacco use with oxidative stress and exacerbation of symptoms of COPD.[10]

Tobacco smoke is also rich in nitric oxide. The nitrogen species produce peroxynitrite, decrease antioxidant capacity and further augment oxidative stress. Both macrophages and neutrophils are increased in smokers, generate ROS and reactive nitrogen species (RNS). It results in increased oxidative and nitrative burden in the lungs as well as in the systemic circulation, contributing to different tobacco-induced diseases, particularly the COPD.

Air pollution is another important risk factor for COPD. The particulate matter present in the ambient and household air pollution, produce oxidants and oxidative stress. The organic compounds, metals and probably bacterial endotoxins present in air pollutants are associated with oxidant generation. Also, ozone, a secondary air pollutant produced through photochemical reaction, aggravates oxidative stress by depleting antioxidants and causing lipid peroxidation. Tobacco smoking and air pollution exposures are also responsible for continued deterioration of lung function and acute exacerbations of COPD. In the presence of an acute exacerbation, there is release of large amounts of ROS by the activated peripheral blood neutrophils.

The ROS generation through lipid peroxidation and other mechanisms may result in the formation of reactive carbonyls, (carbonyl stress) considered to be responsible for chronic disease and ageing. Carbonyl levels are shown to relate to COPD severity assessed by decline in forced expiratory volume in 1 second (FEV_1), that is, the volume of air forcefully exhaled in the first second, a measure of airway-obstruction.

Oxidative stress in COPD is responsible for increased airway inflammation. It activates transcription factor, nuclear factor-kappa B (NF-κB) as well as some signaling molecules. Further, the ROS also activate TGF-β signaling pathways which themselves induce oxidative stress. The carbonyl stress in COPD also impacts the different signaling pathways associated with increase in oxidative damage.

The oxidative stress in COPD is associated with accelerated lung ageing and cellular senescence.[11] These changes are possibly mediated through increased expression of the enzyme matrix metallo-proteinase-9 (MMP9) which causes elastin degradation and emphysematous changes in the lungs. Oxidative stress has been blamed for systemic manifestations and comorbidities of COPD, such as the cardiovascular disease and metabolic syndrome. It is also linked to skeletal muscle wasting and weakness.

Oxidative stress produces some degree of corticosteroid resistance. It suppresses the activated anti-inflammatory genes and therefore, the anti-inflammatory action of corticosteroids. The observation has important therapeutic implication in the use of antioxidant treatment to improve response to corticosteroids for acute exacerbation.

Bronchial asthma is the other important cause of airway obstruction. It is a disease which affects individuals of all age groups, from early childhood to old age. Air pollutants play an important role in pathogenesis of allergic rhinitis and asthma. Besides air pollutants, different allergens, infections and other stimuli can act as triggers of asthma, as well as result in increase in intracellular ROS through the NADPH enzyme activation.

Endogenous production of ROS is increased in allergic asthma.[12] The intracellular levels of ROS, particularly the H_2O_2, result in different alterations of intracellular signaling events depending upon the level of oxidative stress.[8] While the low levels of ROS act upon the antioxidant defense system through the Nrf2 activation, the higher levels cause inflammatory cellular response. Still higher levels of ROS may result in serious cytotoxic effects, apoptosis and cell necrosis.

Interstitial lung diseases

The interstitial lung disease (ILD) is a group of a large number of heterogeneous diseases with common clinical and radiological manifestations. The interstitial lung involvement can occur secondarily to a variety of systemic diseases such as connective tissue disorders, occupational lung disease, sarcoidosis and hypersensitivity pneumonias. Idiopathic interstitial pneumonia (IIP) is an important group of primary ILDs which include six different kinds of disorders of idiopathic origin. Usual interstitial pneumonia (UIP), also called as IPF is the most serious and progressive IIP. While IPF is one of the most important ILD, all ILDs cannot be considered as synonymous of IPF.

IPF is a progressive and disabling disease which results in marked respiratory distress leading to death within a short span of 3–4 years. There is no satisfactory treatment of IPF available at present. There is evidence to link the ROS levels with the pathogenesis of IPF in humans.[13] Marked oxidative stress has been also demonstrated in the experimental models of pulmonary fibrosis, such as fibrosis induced by administration of asbestos, silica and bleomycin. It is also shown that the bleomycin and asbestos-induced fibrosis can be prevented by antioxidants such as N-acetylcysteine (NAC), deferoxamine and other agents.[14]

Oxidative stress is also believed as one of the important mechanism which links ageing with IPF as well as COPD.[15] Matrix metallo-proteinases (MMPs) are matrix degradation enzymes (proteinases) which play an important role in the development of lung injury and pulmonary fibrosis. The MMPs are activated and up-regulated by ROS. The MMPs and their tissue inhibitor counterparts (TIMPs) regulate the degradation and turnover of extracellular matrix which is important for the stability and strength of lung structure and architecture. Excessive matrix deposition and impairment of extracellular matrix degradation is likely responsible for pulmonary fibrosis. Different studies show different results with reference to the role of MMPs and TIMPs pulmonary fibrosis. The broncho-alveolar lavage fluid shows higher MMP3, MMP7, MMP8, and MMP9 levels in IPF patients with earlier mortality.[16] It may be concluded that the MMPs–TIMPs balance is important in the pathogenesis as well as outcome of pulmonary fibrosis.

Sarcoidosis is another common and important ILD of unknown etiology. It is a multisystem disease which primarily affects the lungs and mediastinum with formation of noncaseating granulomas. There is newly emerging evidence that the markers of oxidative stress are increased. As an example, the malondialdehyde (MDA) which is an end product of lipid peroxidation, and oxidized low density lipoprotein were increased in sarcoidosis while the total antioxidant level was significantly lower.[17] It has been proposed that ROS promote inflammation and fibrosis in sarcoidosis by multiple mechanisms.

Asbestosis and lung cancer

Asbestos is commonly used for its fire-resistance properties for various industrial and manufacturing applications. It is a nondestructible fiber which when inhaled can remain in the lungs and initiate fibrogenic and neoplastic reactions. Asbestos fiber is responsible for diseases such as asbestosis (interstitial fibrotic lung disease), pleural plaques and malignancies such as mesothelioma and lung cancer.[18]

Formation of ROS, and to a lesser extent the RNS are important in the inflammation and fibrosis following asbestos inhalation.[19] Asbestos fiber induced oxidative stress stimulates IL-6, a key mediator in chronic inflammation and fibrosis. Increased IL-6 levels are shown in the bronchoalveolar lavage fluid of patients with long term asbestos exposure and fibrosis. The asbestos fibers also trigger H_2O_2 production in alveolar macrophages as well as other target cells such as the lung epithelial and mesothelial cells. The mitochondrial ROS production and oxidative stress cause DNA damage and cell death.

The various signaling pathways, DNA damage and disordered cell replication induced by asbestos lead to tumor development, lung cancers and mesothelioma.[20] Besides the asbestos induced cancers, oxidative stress is also important in tumor development and progression of other types of lung cancer.[21] Lipid peroxidation is the most common form of oxidative stress described in lung cancer. Tobacco smoking and environmental pollutants are known to promote oxidant burden and reduce the antioxidant levels. Oxidative stress is the most important factor implicated in smoking induced cancer.[22] ROS are also linked to tumor invasion, angiogenesis (i.e., the tumor associated) new vessel formation and metastases to other organs. These actions are mediated through dysregulation of metalloproteinases, angiogenic factors and adhesion molecules, released by abnormal signaling by ROS. The ROS also promote tumor cell survival and resistance to cancer chemotherapy and radiotherapy. There is newer evidence to suggest that chemotherapy may suppress the ROS in non small cell lung cancer.[23]

Pulmonary arterial hypertension

Pulmonary arterial hypertension (PAH) is a complex disease of multiple etiologies. While pulmonary hypertension can develop secondarily to cardio-pulmonary diseases, connective tissue disorders, HIV infection and other diseases, the etiology of primary pulmonary hypertension (PPH) is uncertain. There are limited options of treatment of PPH, and the prognosis is poor. The exact pathogenesis is not clear. There is increasing evidence to suggest the role of ROS and oxidative stress in its pathogenesis.[24] Several different markers of oxidative stress are shown to increase in plasma and urine of patients with both secondary and primary forms of PAH in different studies.

ROS are known to alter the levels of vasoactive mediators and enhance calcium signaling. This is likely to up-regulate the growth factors and induce pro-proliferative signaling responsible for enhanced proliferation, growth of fibroblasts and matrix deposition in the walls of pulmonary vessels. These changes may result in pulmonary vascular remodeling, increased vasoconstriction and aggravation of pulmonary hypertension. Further, drugs like bosentan used for treatment of pulmonary hypertension, antagonizes endothelin receptors, attenuates the oxidative and nitrosative stress, and prevents progression of pulmonary hypertension.[25]

Respiratory muscle dysfunction

The respiratory muscles, in particular the diaphragm and chest-wall muscles are important for an effective ventilation. Diaphragmatic atrophy and contractile dysfunction occurs in critically ill patients on prolonged mechanical ventilation. The degree of diaphragmatic atrophy is directly proportional to length of mechanical ventilation. There is increased muscle protein breakdown which affects muscle strength and endurance. Hypoventilation and respiratory failure results due to muscle weakness/atrophy.

The disuse of skeletal muscles of locomotion promotes oxidative injury. Muscle atrophy is caused due to the oxidative stress induced injury.[26] The oxidative stress in muscle disuse conditions has been attributed to the interaction of different pathways of oxidant production, such as the xanthine oxidase and nitric oxide synthase, NADPH oxidase and increased cellular levels of reactive iron.

Role of antioxidants in the management of lung diseases

The role of oxidative stress in various diseases forms the basis of potential treatment with antioxidant drugs. Therapeutically, two kinds of approaches have been tried: (i) Reduction of exposure to pro-oxidants, and (ii) Use of antioxidants to neutralize the oxidative burden. In case of lung diseases, reduction of exposure to pro-oxidants such as ozone and nitrites is shown to decrease asthma exacerbations. Similarly, tobacco cessation may reduce the occurrence of exacerbations of COPD.

Antioxidants have been tried for treatment and/or prevention of most of the lung diseases listed earlier in this chapter. There are several antioxidant drugs and food supplements which are now available. The list includes glutathione, vitamins A, C, and E, antioxidant enzymes and Nrf2 mediated antioxidants. In addition, many herbal preparations and other antioxidant mimics have been used in one or another form. Some of these specific indications for pulmonary conditions are:

Infections

The Nrf2 mediated antioxidant system is shown to prevent the oxidant injury and inflammation. Drugs like ketamine, propofol and ketofol, used for sedation are shown to reduce inflammation of endotoxaemia in rodents in experimental studies. The effect in humans has not been substantiated.

The dietary supplement sulforaphane is shown to increase the antioxidant defenses and lung health in HIV-1 infected rats. In humans, the dietary antioxidant for example, the gold kiwi fruit is shown to reduce symptoms of upper respiratory tract infection.[27]

A variety of Chinese and Japanese herbal medicines have been used for influenza virus infection for their anti-inflammatory and antioxidant effects. It has been shown that the levels of antioxidants increase whereas oxidants such as MDA and NO levels decrease significantly following anti-TB therapy. It has been suggested that antioxidants (vitamins C and E) supplementation might help in fast recovery of a tuberculosis patient. Herbs have been also tried for treatment of tuberculosis to enhance the cell mediated immunity and effects of anti-tubercular therapy.

Chronic obstructive pulmonary disease

N-acetyl cysteine, a mucolytic agent with antioxidant properties has been tried in COPD patients. It was shown to increase glutathione concentrations within the alveolar macrophages and inhibit ROS release from neutrophils. Several clinical trials indicate that NAC may reduce the occurrence of exacerbations and improve small airway function.[28] Small benefits were also seen with carbocisteine and erdosteine. The new Nrf2-activators seem to show encouraging results. For example, sulforaphane, a natural constituent of vegetables such as broccoli is under trial at present.

Interstitial lung diseases

Antioxidants such as NAC have shown some benefit in IPF – a progressive and fatal illness.[29] Use of NAC had shown improvements in oxygen saturation and CT images of IPF patients. No significant clinical benefits could, however, be attributed to its use. Antioxidant drugs such as carnosine and NAC have been used to supplement the immunosuppressive and antifibrotic therapy of IPF. Some of the antioxidant mimics, such as the metalloporphyrins have shown beneficial effects in radiation and bleomycin-induced fibrosis in experimental animals. One can hope that one of these drugs may find its use in IPF therapy in future.

Exogenous antioxidants such as NAC, pentoxifylline and quercetin have been used for sarcoidosis in small studies. Pentoxifylline, commercially available in the market is used for several medical indications because of its antioxidant properties. There is no consistency of results.

Other lung diseases

Antioxidants have been used as therapeutic agents in small studies to slow down the rate of dysfunction and atrophy due to muscle disuse. Addition of vitamins A, C, E, carotenoids or selenium did not show any reduction in lung cancer risk. The antioxidants for pulmonary hypertension hold promise for future use. At least four groups of antioxidant strategies (enzymatic ROS scavengers and regulators, small chemical ROS scavengers, ROS generation inhibitors and Nrf2 activator) are described.[30] Their exact role in treatment remains to be established.

A large number of Ayurvedic (traditional Indian medical system) preparations have been tried as antioxidant mimics.[31] *Ocimum sanctum* (holy basil or *"tulsi"*), a plant considered sacred by the Hindus is shown to possess excellent antioxidant activity. Its leaves are routinely used as home-remedies for cough, common-cold and fevers. Some animal studies have also shown it to have anti-cancer effects. *Curcuma longa* (turmeric), a commonly used spice in the Indian kitchens has anti-cancer, anti-inflammatory and anti-septic properties. Its active ingredient, curcumin is more potent that α-tocopherol in its antioxidant effects.

There are several other herbs and/or mineral based *Ayurvedic* drugs which have been used for different systemic and pulmonary ailments. Some such examples include: *Withania somnifera (winter cherry or "ashwagandh"),* gold-ash (*"Swarna bhasma"), Aloe vera (Aloaceae), Piper nigrum* (black pepper) and some poly-herbal preparations such as *"triphala." Ashwagandha* has been a particularly popular herb of interest which has been used for centuries for various respiratory illnesses, arthritis and other chronic diseases. The active compounds in the plant (*withanolides and sitoindosides*) are shown to accrue good antioxidant properties such as increasing the activity and/or levels of free radical scavenging enzymes and glutathione peroxidase.

There is significant evidence to show that the various antioxidant mimics are known to accrue potentially useful benefits in experimental animal studies as well as some small clinical benefits. They may have an additional supplementary role in various treatment protocols. For the present, the conclusive proof for their role in disease management and prevention remains enigmatic.

Multiple choice questions

1 Important bacterial pathogens include all but which bacteria?
   a. *Streptococcus pneumoniae*
   b. *Haemophilus influenzae*

   c. *Lactobacillus acidophilus*
   d. *Staphylococcus aureus*

2 Tobacco smoking and air pollution exposures are also responsible for continued deterioration of lung function and acute exacerbations of COPD.
   a. True
   b. False

3 The pathogenesis of primary pulmonary hypertension is:
   a. Heart disease
   b. Connective tissue disorder
   c. Infection
   d. Unknown

References

1 Galli, F., Conese, M., Maiuri, L. *et al.* (2014) Introduction to oxidative stress and anti-oxidant therapy in respiratory disorders. In: Ganguly, N.K., *et al.* (eds), *Studies on Respiratory Disorders.* Humana Press, New York.

2 Park, H.S., Kim, S.R. *et al.* (2009) Impact of oxidative stress on lung diseases. *Respirology*, 14(1), 27–38.

3 Harrison, A., Bakaletz, L.O. & Munson, R.S. Jr., (2012) Haemophilus influenzae and oxidative stress. *Frontiers in Cellular and Infection Microbiology*, 2(40), 28.

4 Sarma, J.V. & Ward, P.A. (2011) Oxidants and redox signaling in acute lung injury. *Comprehensive Physiology*, 1(3), 1365–81.

   The article describes the formation of oxidants in the lung cells in response to the causes of acute lung injury. These include the bacterial and viral infections, hypoxic and hyperoxic conditions and other environmental exposures.

5 Sharma, B. (2014) Oxidative stress in HIV patients receiving antiretroviral therapy. *Current HIV Research*, 12(1), 13–21.

   There is significant disturbance of antioxidant defenses and therefore redox imbalance which contributes to progression of HIV infection. It is also suggested that the introduction of antiretroviral therapy may enhance oxidant stress due to antioxidant dysfunction.

6 Kalinowska, M., Bazdar, D.A., Lederman, M.M. *et al.* (2013) Decreased IL-7 responsiveness is related to oxidative stress in HIV disease. *PloS One*, 8(3), e58764.

7 Mohod, K., Dhok, A. & Kumar, S. (2011) Status of oxidants and antioxidants in pulmonary tuberculosis with varying bacillary load. *Journal of Experimental Sciences*, 2(6), 35–37.

   There is significant relationship of bacillary load and disease severity with high oxidant and low antioxidant concentrations. Antioxidant supplementation may have beneficiary role in the fast recovery of a patient of tuberculosis.

8 Holguin, F. (2013) Oxidative stress in airway diseases. *Annals of the American Thoracic Society*, 10(Suppl), S150–S157.

   Airway oxidative stress in COPD and asthma occur from environmental pro-oxidants and inflammatory cell infiltration of the airways. It is associated with increased severity and poorer lung function. But oxidative stress is a continuously changing dynamic process rather than a single predictable threshold for occurrence of the airway disease.

9 Barnes, P.J. (2014) Cellular and molecular mechanisms of chronic obstructive pulmonary disease. *Clinics in Chest Medicine*, 35(1), 71–86.

COPD is characterized with inflammation and infiltration by alveolar macrophages, neutrophils and T lymphocytes. The inflammation is driven by increased oxidative stress and may also promote the occurrence of systemic manifestations, co-morbidities and lung cancer.

10 Zuo, L., He, F., Sergakis, G.G. *et al.* (2014) Interrelated role of cigarette smoking, oxidative stress and immune response in COPD and corresponding treatments. *American Journal of Physiology. Lung Cellular and Molecular Physiology*, 307(3), L205–L218.

Literature search of published studies searched from PubMed/Medline databases reveal that cigarette smoking impairs the immune system, increases oxidative stress and causes COPD exacerbations. Multiple approaches including pharmacotherapy targeting inflammation and oxidative stress are therefore recommended.

11 Ito, K. & Barnes, P.J. (2009) COPD as a disease of accelerated lung aging. *Chest*, 135, 173–180.

12 Jiang, L., Diaz, P.T., Best, T.M. *et al.* (2014) Molecular characterization of redox mechanisms in allergic asthma. *Annals of Allergy, Asthma, and Immunology*, 113(2), 137–142.

There is an increased production of endogenous ROS in allergic asthma. The redox mechanisms appear to play the key role in pathogenesis. The ROS over production is shown to cause abnormalities of DNA and protein function, as well as augment bronchial hyper-responsiveness – the key physiological abnormality in asthma.

13 Cheresh, P., Kim, S.J., Tulasiram, S. *et al.* (2013) Oxidative stress and pulmonary fibrosis. *Biochimica et Biophysica Acta*, 1832(7), 1028–1040.

Oxidative stress is an important mechanism of fibrosis of the lungs. There are multiple mechanisms for fibrous overlapping with molecular pathways in other organs. Environmental toxins and mitochondrial/NADPH oxidase of inflammatory and lung target cells are some of the important sources of oxidative stress. Apoptosis of lung epithelial cells by mitochondria and p53-regulated death pathways constitute some of the important mechanisms.

14 Teixeira, K.C. *et al.* (2008) Attenuation of bleomycin-induced lung injury and oxidative stress by N-acetylcysteine plus deferoxamine. *Pulmonary Pharmacology & Therapeutics*, 21(2), 309–316.

15 Faner, R., Rojas, M., Macnee, W. *et al.* (2012) Abnormal lung aging in chronic obstructive pulmonary disease and idiopathic pulmonary fibrosis. *American Journal of Respiratory and Critical Care Medicine*, 186(4), 306–313.

Both IPF and COPD are diseases of older age groups, therefore raising the possibility of abnormal ageing mechanisms in their pathogenesis. Oxidative stress is one of the important ageing mechanism which is abnormal in IPF and COPD.

16 McKeown, S. *et al.* (2009) MMP expression and abnormal lung permeability are important determinants of outcome in IPF. *European Respiratory Journal*, 33(1), 77–84.

17 Ivanisevic, J., Kotur-Stevuljevic, J., Stefanovic, A. *et al.* (2012) Dyslipidemia and oxidative stress in sarcoidosis patients. *Clinical Biochemistry*, 45(9), 677–682.

18 Banks, D.E. & Dedhia, H.V. (2011) Health effects of asbestos exposure. In: Jindal, S.K. (ed), (Chief ed.) *Text Book of Pulmonary and Critical Care Medicine*. Vol. 2 Chap 108. Jaypee Brothers Medical Publishers, New Delhi, India.

The chapter describes the various respiratory and other diseases due to occupational asbestos exposure; the pathogenesis of asbestos related interstitial lung disease and tumors such as lung cancer and mesothelioma.

19 Liu, G., Cheresh, P. & Kamp, D.W. (2013) Molecular basis of asbestos-induced lung disease. *Annual Review of Pathology*, 8, 161–87.

The mechanisms of asbestos induced interstitial lung disease (asbestosis) and malignancies depend upon the fiber type, lung clearance and genetic predisposition. Oxidative stress constitutes an important pathogenic pathway. There have been new insights into molecular basis of asbestos induced diseases which propose the development of novel therapeutic targets for these otherwise incurable problems.

20 Heintz, N.H., Janssen-Heininger, Y.M.W. & Mossman, B.T. (2010) Asbestos, lung cancers and mesotheliomas from molecular approaches to targeting tumor survival pathways. *American Journal of Respiratory Cell and Molecular Biology*, 42, 133–139.

21 Reuter, S., Gupta, S.C., Chaturvedi, M.M. *et al.* (2010) Oxidative stress, inflammation and cancer: how are they linked? *Free Radical Biology and Medicine*, 49(11), 1603–1616.

22 Smith, C.J., Perfetti, T.A. & King, J.A. (2006) Perspectives on pulmonary inflammation and lung cancer risk in cigarette smokers. *Inhalation Toxicology*, 18(9), 667–677.

23 Wakabayashi, T., Kawashima, T. & Matsuzawa, Y. (2014) Evaluation of reactive oxygen metabolites in patients with non-small cell lung cancer after chemotherapy. *Multidisciplinary Respiratory Medicine*, 9(1), 44.

   ROS levels are significantly increased in patients with nonsmall cell lung cancer which decrease following chemotherapy. It was also seen that the ROS suppression was present in responders but not in nonresponders to chemotherapy which implied the important role of ROS in prediction of clinical outcome to treatment.

24 Wong, C.M., Bansal, G., Pavlickova, L. *et al.* (2013) Reactive oxygen species and antioxidants in pulmonary hypertension. *Antioxidants and Redox Signaling*, 18(14), 1789–1796.

   There is evidence from human studies to support the presence of increased oxidative stress in patients with pulmonary hypertension. In animal models of pulmonary hypertension, a variety of antioxidant compounds have shown beneficial effects supporting the role of ROS to prevent disease progression.

25 Rafikova, O., Rafikov, R., Kumar, S. *et al.* (2013) Bosentan inhibits oxidative and nitrosative stress and rescues occlusive pulmonary hypertension. *Free Radical Biology and Medicine*, 56, 28–43.

   Pulmonary hypertension is a fatal disease. Bosentan is one of the few drugs of some marginal benefit in its treatment. The drug antagonizes the endothelin-1 receptors expressed at high levels in these patients. It is concluded that the inhibition of ET-1 signaling attenuates the oxidative and nitrosative stress associated with pulmonary hypertension.

26 Hafner, S., Radermacher, P., Frick, M. *et al.* (2014) Hyperglycemia, oxidative stress and the diaphragm: a link between chronic co-morbidity and acute stress? *Critical Care*, 18(3), 149.

   Respiratory muscles dysfunction which occurs in patients on prolonged, mechanical ventilation is attributable to aggravated oxidative and nitrosative stress. Possibly, the underlying co-morbidities (e.g., diabetes) further increase this acute stress situation (such as sepsis or mechanical ventilation).

27 Hunter, D.C., Skinner, M.A., Wolber, F.M. *et al.* (2012) Consumption of gold kiwi fruit reduces severity and duration of selected upper respiratory tract infection symptoms and increases plasma vitamin C concentration in healthy older adults. *British Journal of Nutrition*, 108(7), 1235–1245.

   Gold kiwi fruit contains important nutrient antioxidant such as vitamins C and E, folate, polyphenols and carotenoids. Its consumption reduced the severity of symptoms of upper respiratory tract infections in the elderly. No changes were detected in the body's innate immunity or inflammatory biomarkers.

28 Santus P, Corsico A, Solidoro P, et al. 2014. Oxidative stress and respiratory system: pharmacological and clinical reappraisal of N-acetyl cysteine. COPD, 11(6), 705–17.

   N-acetyl cysteine, an antioxidant drug has been used for its mucolytic properties in patients with COPD. Several clinical trials show that the drug may reduce the rate of acute exacerbations and improve the lung function.

29 Bast, A., Weseler, A.R., Haenen, G.R. *et al.* (2010) Oxidative stress and antioxidants in interstitial lung disease. *Current Opinion in Pulmonary Medicine*, 16(5), 516–520.

   The knowledge of the importance of ROS in the pathogenesis of IPF has raised the hope of treatment with antioxidants which have shown some benefits. The antioxidants such as NAC also protect against oxygen toxicity.

30 Suzuki, Y.J., Steinhorn, R.H. & Gladwin, M.T. (2013) Antioxidant therapy for the treatment of pulmonary hypertension. *Antioxidants and Redox Signaling*, 18(14), 1723–1726.

> *A number of antioxidant drugs have potential in the treatment armamentarium of various forms of pulmonary hypertension. It is important to undertake well designed clinical trials to evaluate the efficacy of antioxidant drugs. At least four groups of antioxidant strategies may find their use in treatment: enzymatic ROS scavengers and regulators; small chemical ROS scavengers; inhibitors of ROS generation; Nrf2 activators.*

31 Scartezzini, P. & Speroni, E. (2000) Review on some plants of Indian traditional medicine with antioxidant activity. *Journal of Ethnopharmacology*, 71, 23–43.

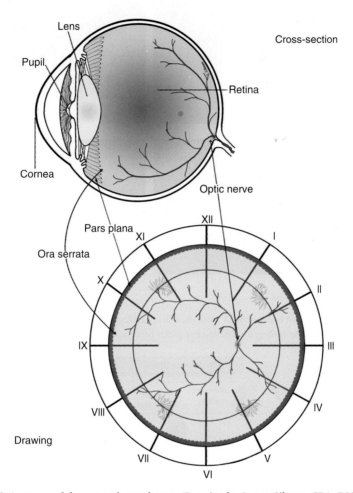

Figure 6.1 Anatomy of the eye and vasculature. (Drawing by James Gilman, CRA, FOPS.)

Oxidative Stress and Antioxidant Protection: The Science of Free Radical Biology and Disease, First Edition.
Edited by Donald Armstrong and Robert D. Stratton.
© 2016 John Wiley & Sons, Inc. Published 2016 by John Wiley & Sons, Inc.

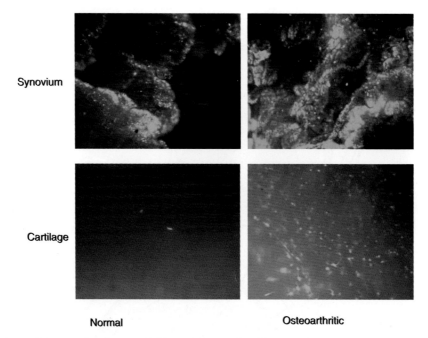

Figure 13.1 Normal and osteoarthritic synovium and cartilage transfected with scAAV packaged with genome coding for fluorescent green protein 10 days postinjection.

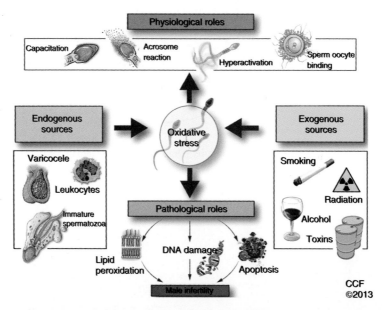

Figure 15.2 Potential generators of reactive oxygen species leading to oxidative stress in the male comprise endogenous and exogenous sources. Physiological levels of reactive oxygen species play a role in sperm capacitation, acrosome reaction, hyperactivation, and sperm–oocyte binding. However, at pathological levels, reactive oxygen species causes lipid peroxidation, DNA damage, and apoptosis, which lead to detrimental effects on male fertility.

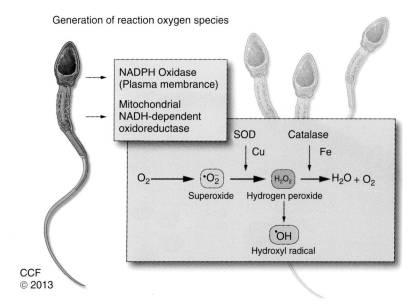

Figure 15.4 Reactive oxygen species (superoxide anion, hydrogen peroxide, and hydroxyl radical) are generated from oxidative processes in the plasma membrane and mitochondria of the male gamete. These reactions involve the SOD and catalase antioxidant enzymes along with copper and iron, respectively.

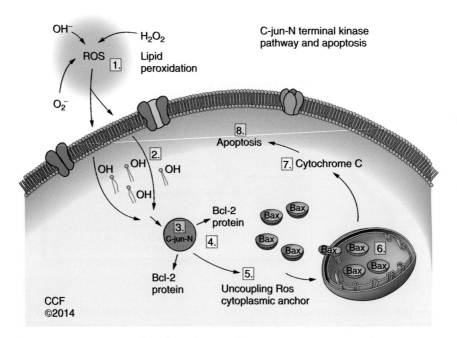

Figure 16.1 c-Jun N terminal kinase pathway and apoptosis. ROS act as secondary messengers that activate core apoptotic pathways via the activation of the c-Jun N-terminal kinase.

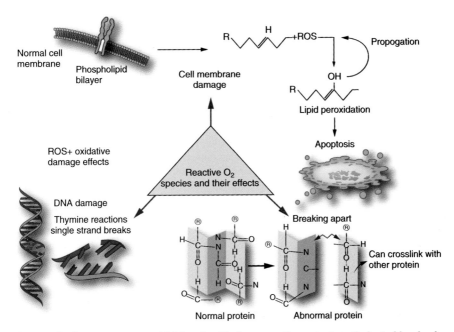

Figure 16.2 The consequences of ROS and oxidative stress. Exposure to pathological levels of ROS and oxidative stress causes cellular membrane and DNA damage along with protein manipulation.

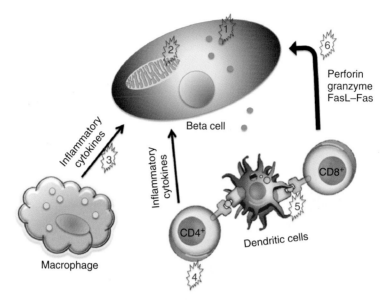

Figure 19.1 ROS participate in multiple stages during T1D development. (1) ROS directly induce β-cell dysfunction; (2) ROS facilitate programmed β-cell death; (3) ROS produced by macrophage directly induce β-cell destruction. (4) ROS promote CD4+ T-cell proliferation and secretion of inflammatory cytokines, which further induce β-cell damage. (5) During CD8+ T-cell activation, ROS participate in antigen cross-presentation from dendritic cells to CD8+ T cells. (6) CD8+ T cells destroy β cells through perforin, granzyme, and FasL–Fas pathways. ROS facilitate β-cell damage in all these pathways.

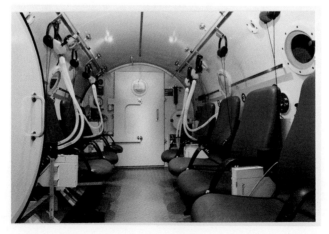

Figure 22.2 Multiplace chamber using compressed air and 100% oxygen hood.

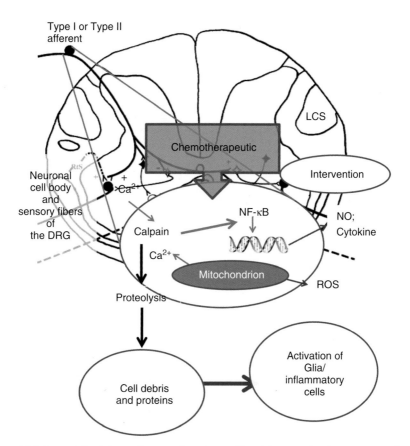

Figure 24.1 As a result of chemotherapeutic regimen: (1) calcium overload ensues, secondary to Na and Ca channel activation, resulting in calpain activation and proteolysis of the cytoskeletal (e.g., microtubules and neurofilaments). Calpain activity may also activate NF-κB; (2) downstream generation of inflammatory cytokines, NO, and exacerbation of ROS production from mitochondria, and increased Ca release; (3) this is aggravated by the attendant activation of glia (astrocytes and microglia) and immune effectors [see text for details]. Successful intervention to ameliorate NF-κB-induced pathology and oxidative stress would improve the efficacy and dosing regimens with chemotherapeutics. DRG = dorsal root ganglia.

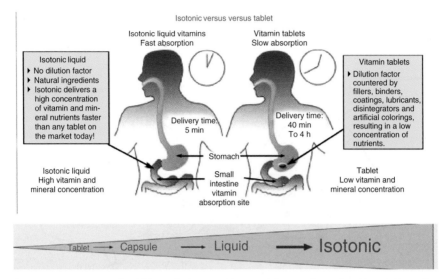

Figure 26.5 An isotonic liquid formulation provides much better absorption of OPC than other formulations. (Vijayalakslakshmi Nandakumar and Santosh K. Katiyar 2008. Reproduced with permission.)

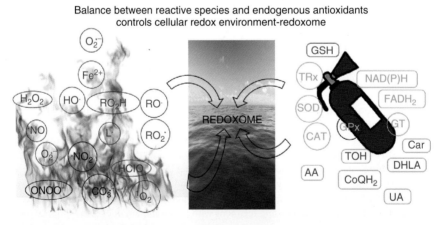

Figure 27.2 Redox-active molecules, reactive species, and antioxidants. Redox-active molecules/species (some of them listed here) and their involvement in redox-based pathways comprise the cellular redoxome. Redoxome is maintained by oxidation/reduction reactions, that is, electron shuttling; it is only natural that redox-active drugs may be best suited to restore it when perturbed in diseases.

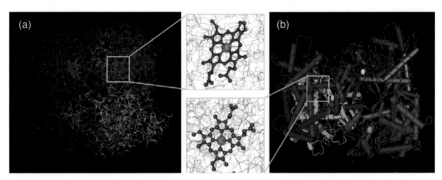

Figure 27.7 (a) Crystal structure of human erythrocyte catalase (PDB ID: 1QQW), and (b) crystal structure of human cytochrome P450 (PDB ID: 2F9Q) and their active sites. Pictures are created with Cn3D 4.3.1.[26-28] Porphyrin is a macrocyclic ring that encapsulates metal; in turn, it affords the highest stability to a metal complex, assuring no loss of metal where reactions of interest occur. It is only natural that such ligand has been used by nature for numerous proteins and enzymes, such as myoglobin, guanylyl cyclase, oxidases, oxygenases, prolyl hydroxylases, catalase, cytochrome P450 family of enzymes, and so on. For the same reason, we have chosen to modify a metalloporphyrin structure to be efficient catalyst for $O_2^{\bullet-}$ dismutation.

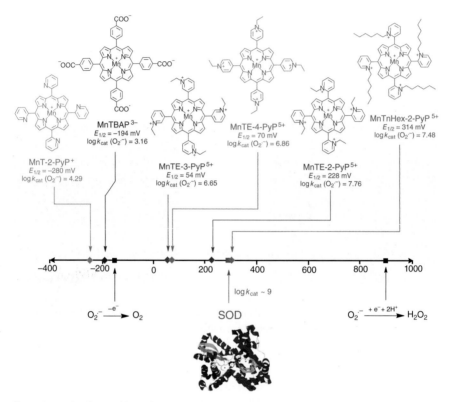

Figure 27.9 The design of porphyrin-based SOD mimics. Starting from unsubstituted MnT-4-PyP$^+$ (Figure 27.8), the pyridyl nitrogens were first alkylated giving rise to *para* analog MnTM-4-PyP^{5+} with fair SOD-like activity. To enhance electron-withdrawing effects, the nitrogens were then moved closer to the metal site from *para* into *ortho* positions. The MnTM(E)-2-PyP^{5+}, the first lead, was synthesized. It is still the most frequently studied compound.[2,23] Based on *ortho* pyridyl porphyrin, the imidazolyl analog (Figure 27.5) was subsequently synthesized and became the second lead – MnTDE-2-ImP^{5+}. In order to improve the bioavailability of highly charged compounds, the alkyl chains were then lengthened and the third lead, MnTnHex-2-PyP^{5+} was synthesized. Adapted from ref 126.

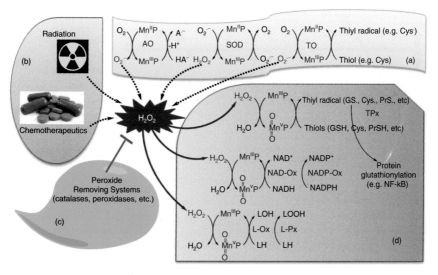

Figure 27.16 The role of H_2O_2 in MnP-related cellular pathways. The most potent SOD mimics are able to oxidize a number of biological molecules (those studied thus far listed here) in the presence of H_2O_2. AO, ascorbate oxidation; TO, thiol oxidation; TPx, thiol peroxidation; NAD-ox, NAD oxidation; NADP-ox, NADP oxidation; L-ox, lipid oxidation; L-Px, lipid peroxidation. Adapted from ref 216.

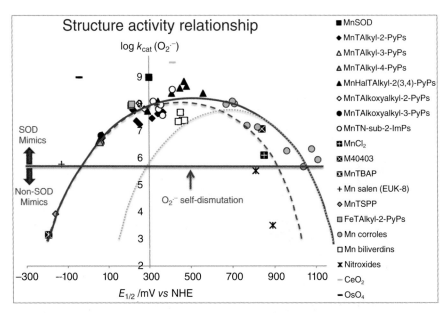

Figure 27.20 Structure–activity relationship between the log $k_{cat}(O_2^{\bullet-}$ and $E_{1/2}$ for redox couple involved (dashed red line for $Mn^{III}P/Mn^{II}P$ and dotted green line for $Mn^{IV}P/Mn^{III}P$). The relationship fits best for metal complexes and is not that good for nonmetal-based compounds such as nitroxides; its ability to affect superoxide dismutation is related to the fairly high rate constant for the reaction of nitroxide with protonated superoxide, very little of which is present at physiological pH. It seems that perhaps two relationships exist for two different redox couples and are slightly shifted based on different energetics of electron transfers involved with those couples. The maximum of the bell shape of the SAR describes the potential at which both steps of dismutation process occur at similar rates and where the $k_{cat}(O_2^{\bullet-})$ is in turn maximal. For those compounds that use $Mn^{III}P/Mn^{II}P$ redox couple, at more negative potentials, the metal +3 oxidation state is stabilized and cannot be reduced with $(O_2^{\bullet-}$ to start the dismutation process. At more positive potentials, Mn is stabilized in +2 oxidation state and cannot be oxidized with $(O_2^{\bullet-}$ in the first step of dismutation process. For those compounds that use $Mn^{IV}P/Mn^{III}P$ redox couple, such as corroles and biliverdins, the reverse is true. The first step would involve the oxidation of the metal site from Mn^{3+} to Mn^{4+} and the reduction of $(O_2^{\bullet-}$ followed by reduction of metal to Mn +3 oxidation resting state with concomitant oxidation of $(O_2^{\bullet-}$. To identify the compounds reader is directed to Table 27.1. Adapted from ref 2.

CHAPTER 19

Oxidative stress and type 1 diabetes

Chao Liu, Clayton E. Mathews, and Jing Chen

Department of Pathology, Immunology, and Laboratory Medicine, The University of Florida College of Medicine, Gainesville, FL, USA

THEMATIC SUMMARY BOX

At the end of this chapter, students should be able to:

- Tell the difference between type 1 and type 2 diabetes

- Define islets of Langerhans

- List cell types in the islets

- Define autoimmunity

- Describe why patients with type 1 diabetes are hyperglycemic

- List evidences for the role of ROS at β-cell level during the development of type 1 diabetes

- List major immune cells that mediate β-cell destruction and their mechanisms

- Describe role of ROS in the above-mentioned mechanisms

Introduction

Type 1 diabetes (T1D) has long been regarded as an autoimmune disease based on the presence of circulating autoantibodies and histological evidence of islet infiltrating immune cells. This disease occurs when self-reactive T cells infiltrate the pancreatic islets and destroy the insulin-producing β cells. When a threshold of β-cell mass has been destroyed, an insulin deficiency ensues and consequently leads to hyperglycemia. T1D was referred to as juvenile diabetes because when compared to type 2 diabetes (a disease usually caused by insulin resistance rather than insulin deficiency, and mainly affecting the adult population), the majority of T1D occurs in children and young adults. According to the "National Statistics Diabetes Report, 2014" published by the United States Centers for Disease Control and Prevention, during 2008–2009 an estimated 18,436 people younger than 20 years in the United States were diagnosed with T1D annually. Currently, there is no cure for T1D. Individuals with T1D require exogenous insulin by injection or infusion in order to maintain normal blood glucose levels. The administration of insulin does not prevent the

---

Oxidative Stress and Antioxidant Protection: The Science of Free Radical Biology and Disease, First Edition.
Edited by Donald Armstrong and Robert D. Stratton.

long-term complications of T1D. These associated pathologies include damage to the heart, blood vessels, eyes, and kidneys. Insulin injections can also bring the risk of the life-threatening side-effect hypoglycemia.

Free radicals are a double-edged sword in the development of T1D. On the one hand, free radicals are thought to be important messengers in signal transduction for cells to carry out normal function. On the other hand, free radicals participate in the destruction of β cells by the autoreactive immune system as well as in the damaging effects of hyperglycemia. This chapter reviews the role of free radicals during the pathogenesis of T1D, both in the immune system and on the target β cells.

Role of β-cell oxidative stress in T1D

The islets of Langerhans are cell clusters scattered throughout the pancreas. Islets contain four major cell types: α, β, δ, and PP cells. Among these, β cells secrete insulin upon stimulation by glucose or other secretagogues.[1] Insulin maintains normal blood glucose levels through regulating the balance of glucose uptake and gluconeogenesis. Insulin is first synthesized by β cells as a single-chain polypeptide called preproinsulin, Preproinsulin then experiences a series of processing reactions including cleavage of a signal sequence formation of three disulfide bonds to form proinsulin, the precursor of insulin, which is then converted to insulin and c-peptide by further cleavage.[2,3] Oxidative folding of the (pre)proinsulin chain allows proper formation of the three disulfide bonds that are critical for insulin function.[4] This process is catalyzed by oxidoreductases with hydrogen peroxide as a by-product.

Reactive oxygen species (ROS) contribute to β-cell dysfunction during the development of diabetes.[5] Although ROS-mediated β-cell dysfunction is majorly proposed as a mechanism for type 2 diabetes, during the development of T1D pancreatic islets are in an environment of enriched ROS produced by invading inflammatory cells that causes β-cell dysfunction and decreased insulin secretion.[6] ROS also participate in β-cell death during T1D development.

Autoimmune β-cell destruction is a fundamental feature in the pathogenesis of T1D. β cells are considered to be particularly vulnerable to oxidative damage compared to other tissues as a result of lower antioxidant defenses.[7,8] This exquisite sensitivity has been proposed to play a role in the pathogenesis of T1D.[8–14] The participation of ROS in T1D pathogenesis is also suggested by clinical studies, showing that T1D patients and antibody-positive, at-risk individuals bare an increase in oxidative stress compared to healthy controls or their first-degree relatives.[15,16] Production of ROS and downregulation of antioxidant defenses characterized by a decrease in reduced glutathione (GSH) and a progressive decline in the transcripts for catalase (CAS), superoxide dismutase (SOD), and thioredoxin (TRX) have been observed in apoptotic processes.[17–19] ROS is a mediator of β-cell death through apoptosis.[20,21]

Genetics of β cell sensitivity and resistance to ROS

A preponderance of evidence suggesting that ROS participate β-cell death during T1D development has come from studies using mouse models. The nonobese diabetic

(NOD) mouse strain is widely used in the study of T1D due to many similarities between spontaneous T1D in these mice and the human disease.[22] The NOD strain was first developed and inbred in Japan as a model of spontaneous T1D.[23] A closely related strain, alloxan-induced diabetes-resistant (ALR), was selected for resistance to alloxan-induced free radical-mediated diabetes.[24] Alloxan is a chemical that selectively destroys pancreatic β cells after uptake by the GLUT-2 glucose transporter.[25,26] Once inside of β cells, alloxan undergoes redox cycling and releases ROS to induce β-cell damage.[27,28] ALR mice are not only resistant to alloxan-induced diabetes but also resistant to autoimmune T1D. In addition, islets of these mice are resistant to death and destruction when exposed to inflammatory cytokines or diabetogenic T cells *in vitro*.[10] This resistance results in part from a constitutive upregulation of antioxidative capability in the ALR mice.[29] Genetic studies using NOD and ALR revealed loci associated with resistance/susceptibility to T1D in both nuclear and mitochondrial genomes.[30,31] A novel single-nucleotide polymorphism (SNP) in the mitochondrially encoded NADH dehydrogenase subunit 2 (*mt-Nd2*) distinguished the ALR mitochondrial genome from NOD and other related or commonly used strains.[32] Functional studies have revealed that the protective *mt-Nd2^a* allele is associated with lower mitochondrial ROS production under both basal and stimulated conditions.[33,34] Further investigation using mouse β-cell lines has confirmed that the *mt-Nd2^a* allele protects β cells against immune- and free radical-mediated attack.[35] *In vivo* evidence also supports the role of *mt-Nd2^a* in the protection against T1D at β-cell level through antioxidative defense mechanisms.[36]

A corresponding mitochondrial SNP exists in the human *mt-ND2* gene. This SNP, a nucleotide transversion from C to A, results in the same leucine to methionine substitution in the NADH dehydrogenase subunit-2 protein. The human *ND2^a* allotype has been associated with protection against a series of oxidative stress-related human diseases and physiological conditions, and it is associated with longevity,[37] lower plasma lipid levels,[38] reduced prevalence of myocardial infarction,[39] as well as protection against both type 1 and type 2 diabetes.[40,41] Our group has thoroughly studied the role of this human SNP within β cells under autoimmune attack. Using the cytoplasmic hybrid cell fusion technic (cybrid),[42] a human β-cell line BetaLox 5 was manipulated to harbor either *mt-ND2^a* or *mt-ND2^c* allele.[43] Cells with the protective *mt-ND2^a* allele resist inflammatory cytokines, death receptor stimulation, and diabetogenic T cells. This resistance is related with blunted mitochondrial ROS production and decreased autoimmune-induced apoptosis.[43]

ROS and autoimmunity in T1D

The immune system is the "defense department" of the body and serves to fight against invading foreign bodies, such as bacteria and viruses. Under certain abnormal conditions, the immune system can be engaged in misguided attacks on self-tissues. This phenomenon called "autoimmunity" results in clinically significant pathologies.[44–50] T1D is such an autoimmune disease wherein T cells mediate the destruction of insulin-producing β cells. Studies on pancreas from T1D patients[51] and NOD mice[52] suggest that the main player in T1D is the T cell. The finding that NOD mice deficient in CD8$^+$ T cells do not develop autoimmunity,[53–55] and that

CD8$^+$ T cells are a major component within the insulitic lesion in humans with T1D, suggesting an essential role of CD8$^+$ T cells in the pathogenesis of T1D. Yet CD8$^+$ T cells need the existence of CD4$^+$ T cells to enter pancreas, indicating that both major subsets of T cells are essential for the development of T1D. Participation of these T cells in T1D is also supported by the fact that both CD8$^+$ and CD4$^+$ autoreactive T cells were isolated from islets of diabetic NOD mice.[56–58] While CD8$^+$ T cells conduct killing of β cells through the release of perforin and granzyme B or via the FAS-L/FAS pathway[59], CD4$^+$ T cells, in addition to helping CD8$^+$ T cells to enter the pancreas, and also destroy β cells by releasing inflammatory cytokines and recruiting and activating macrophages to the islets.[60] Evidence from the NOD mouse model indicates that production of oxygen free radicals by macrophages can damage islet β cells, directly resulting in autoimmune type 1 diabetes in NOD mice.[9] Tse *et al.* reported that modulating redox balance led to decreased antigen-specific T-cell proliferation and IFN-γ synthesis by diminishing ROS production in the antigen-presenting cells. This further lead to reduced TNF-α levels produced by CD4$^+$ T cells and impaired effector function.[61] Ablation of ROS through the use of a SOD mimetic also inhibits CD8$^+$ T-cell proliferation, proinflammatory cytokine production, as well as functional suppression of both perforin and granzyme B production.[62] Direct evidence comes from a study showing that treatment of NOD mice with a SOD mimetic prevented the adoptive transfer of T1D by diabetogenic CD4$^+$ T-cell clone through blunting cytokine production and T-cell proliferation.[63] Thus, ROS could have a fundamental role in modulating immune systems in the T1D pathogenesis.

NADPH oxidase and T1D

Phagocyte NADPH oxidase (NOX2) is expressed in a wide variety of tissue/cell types and is well known as the free radical source of the "respiratory burst" by neutrophils and macrophages that is used to kill phagocytized pathogens. NOX2 is also critical for the production of ROS as biological messengers that participate in signal transduction within lymphocytes.[64,65] Although there are multiple reports supporting NOX2 as an immune suppressor in rheumatoid arthritis and experimental autoimmune encephalomyelitis (EAE), this enzyme supports the pathology of T1D.[66,67] To study the role of NOX2-derived ROS in T1D, we developed NOX2-deficient NOD mice, NOD-*Ncf1^{m1J}*.[67] These mice harbor a functionally inactive p47phox (neutrophil cytosolic factor 1) subunit. NOD-*Ncf1^{m1J}* mice have a reduced and delayed onset of T1D, and invasive leukocytic infiltrates are also rare in the islets of NOD-*Ncf1^{m1J}*. Such diabetes protection does not arise from better resistance of β cells to cytotoxic T cells (CTLs) nor from the combinations of proinflammatory cytokines, but instead from altered immune activation. Using adoptive transfer systems, pre-diabetic T cells from NOD-*Ncf1^{m1J}* or NOD were injected into immune compromised NOD-*Scid* mice. In the recipient mice, diabetes onset was significantly delayed in hosts receiving T cells with a deficiency in p47phox.[66] We originally speculated that altered CD4$^+$ T-helper cell differentiation contributed to this protection as naïve CD4$^+$ T cells from NOD-*Ncf1^{m1J}* exhibited significant ablation of IFN-γ production and increased

IL-17 secretion.[67] As Th1 responses play the dominant pathogenic role in T1D, such skewing away from Th1 in NOD-*Ncf1^{m1J}* CD4+ T cells may have resulted in an altered cytokine milieu in the draining lymph node and pancreatic islets.[68,69] However, in our follow-up study, we identified that deficient function of CTL protects the NOD-*Ncf1^{m1J}* mice from developing T1D.[66] We found that the lack in NOX2 activity results in a reduction in CTL pathogenicity. NOX2 deficiency or inhibition divested the CTL of lytic activity by suppressing the master transcriptional factor T-bet, which compromised the production of granzyme B, IFN-γ, and TNF-α. NOX2 promotes T-bet expression by enhancing the redox-sensitive mTORc1 pathway upon TCR activation.[70] Since the CTL performs as the major effector cells in T1D, ablation of CTL effector will have a significant impact upon T1D development.

In addition, ROS also promote T1D development in macrophages. Self-reactive BDC2.5 CD4+ T cell clones fail to transfer diabetes in NOD-*Ncf1^{m1J}* mice.[66] This failure of disease transfer highlights the indispensable role of NOX2 in host-derived effector cells, namely macrophages, during CD4+ T cell-mediated β-cell killing.[71] Similarly, bone marrow-derived macrophages from NOD-*Ncf1^{m1J}* mice are less capable of producing the proinflammatory cytokines IL-1β and TNF-α after stimulation with LPS.[67]

NOX2 has also been proposed to be critical to dendritic cell-mediated diabetogenic CD8+ T cell priming through cross-presentation. Cross-presentation is the process by which dendritic cells present exogenous antigens on MHC class I molecules for recognition by CD8+ T cells.[72] This process is required for naïve CD8+ CTL to recognize and be activated by non-self antigens. With the presence of specific T-helper cells, dendritic cells will cross-prime CTL to initiate efficient cytotoxic responses.[73–75]

Cross-priming is required for the spontaneous onset of T1D in NOD mice.[76] It has been proposed that NOX2 facilitates dendritic cell cross-presentation by adjusting the antigen degradation rate. In dendritic cell cross–presentation, exogenous protein is transported out of the phagosome and into cytosol. These full-length peptides are then further processed and loaded into MHC Class I molecules; thus, the relative integrity of antigen is required for efficient cross-membrane transportation.[77] During this process, NADPH oxidase is recruited to the phagosome and produces ROS to neutralize protons, which results in a sustained phagosomal pH and suppression of acidophilic proteolytic enzymes[78]. In NOX2-deficient mice or chronic granulomatous disease (CGD) patients, dendritic cells show much faster phagosomal acidification and significant deficiency in CD8+ T-cell activation. Thus, it would be a plausible hypothesis that the insufficient CTL priming in CGD mice, or patients, delays or prevents onset of CTL-mediated autoimmune disease such as T1D.

Conclusions

During the pathogenesis of T1D, autoimmune T cells mediate destruction of pancreatic insulin-producing β cells, resulting in insulin deficiency and hyperglycemia. Upon activation, CD8+ T cells conduct killing by releasing granzyme B and perforin, as well as through death receptor pathways. ROS are indispensable for initiation of this process by antigen cross-presentation for CD8+ T-cell priming. For CD4+ T cells, which perform β-cell killing by releasing cytokines, ROS facilitate cell proliferation and cytokine release. At the β-cell level, ROS participate in signaling of cell death induced

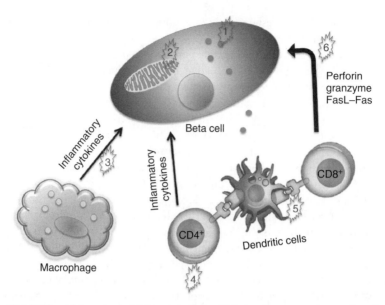

Figure 19.1 ROS participate in multiple stages during T1D development. (1) ROS directly induce β-cell dysfunction; (2) ROS facilitate programmed β-cell death; (3) ROS produced by macrophage directly induce β-cell destruction. (4) ROS promote CD4+ T-cell proliferation and secretion of inflammatory cytokines, which further induce β-cell damage. (5) During CD8+ T-cell activation, ROS participate in antigen cross-presentation from dendritic cells to CD8+ T cells. (6) CD8+ T cells destroy β cells through perforin, granzyme, and FasL–Fas pathways. ROS facilitate β-cell damage in all these pathways. *(See color plate section for the color representation of this figure.)*

by the above mechanisms, and also directly cause β-cell dysfunction and consequently reduce insulin secretion. Therefore, ROS play multiple roles during the development of T1D. Figure 19.1 concluded the role of ROS during the pathogenesis of T1D.

Multiple choice questions

1 Type 1 diabetes only occurs in children.
   a. True
   b. False

2 During the development of type 1 diabetes, ROS: (Choose all that apply)
   a. Induces β-cell dysfunction
   b. Contributes to β-cell destruction
   c. Promotes cytokine production from CD4+ T cells
   d. Signals CD8+ T-cell activation

3 Which of the following statement(s) is/are true? (Choose all that apply)
   a. Patients with type 1 diabetes need exogenous insulin replacement to maintain normal blood glucose
   b. Insulin treatment cures type 1 diabetes
   c. There is no genetic evidence showing ROS participate in type 1 diabetes pathogenesis

References

1 Bratanova-Tochkova, T.K., Cheng, H., Daniel, S. *et al.* (2002) Triggering and augmentation mechanisms, granule pools, and biphasic insulin secretion. *Diabetes*, 51(Suppl 1), S83–S90.

2 Orci, L., Ravazzola, M., Amherdt, M. *et al.* (1986) Conversion of proinsulin to insulin occurs coordinately with acidification of maturing secretory vesicles. *Journal of Cell Biology*, 103(6 Pt 1), 2273–2281.

3 Steiner, D.F., Cunningham, D., Spigelman, L. *et al.* (1967) Insulin biosynthesis: evidence for a precursor. *Science*, 157(3789), 697–700.

4 Zito, E., Chin, K.T., Blais, J. *et al.* (2010) ERO1-beta, a pancreas-specific disulfide oxidase, promotes insulin biogenesis and glucose homeostasis. *Journal of Cell Biology*, 188(6), 821–832.

5 Sakai, K., Matsumoto, K., Nishikawa, T. *et al.* (2003) Mitochondrial reactive oxygen species reduce insulin secretion by pancreatic beta-cells. *Biochemical and Biophysical Research Communications*, 300(1), 216–222.

6 Lightfoot, Y.L., Chen, J. & Mathews, C.E. (2012) Oxidative stress and beta cell dysfunction. *Methods in Molecular Biology*, 900, 347–362.

7 Lenzen, S., Drinkgern, J. & Tiedge, M. (1996) Low antioxidant enzyme gene expression in pancreatic islets compared with various other mouse tissues. *Free Radical Biology and Medicine*, 20(3), 463–466.

8 Tiedge, M., Lortz, S., Drinkgern, J. *et al.* (1997) Relation between antioxidant enzyme gene expression and antioxidative defense status of insulin-producing cells. *Diabetes*, 46(11), 1733–1742.

9 Horio, F., Fukuda, M., Katoh, H. *et al.* (1994) Reactive oxygen intermediates in autoimmune islet cell destruction of the NOD mouse induced by peritoneal exudate cells (rich in macrophages) but not T cells. *Diabetologia*, 37(1), 22–31.

10 Mathews, C.E., Graser, R.T., Savinov, A. *et al.* (2001) Unusual resistance of ALR/Lt mouse beta cells to autoimmune destruction: role for beta cell-expressed resistance determinants. *Proceedings of the National Academy of Sciences of the United States of America*, 98(1), 235–240.

11 Mathews, C.E., Suarez-Pinzon, W.L., Baust, J.J. *et al.* (2005) Mechanisms underlying resistance of pancreatic islets from ALR/Lt mice to cytokine-induced destruction. *Journal of Immunology*, 175(2), 1248–1256.

12 Nerup, J., Mandrup-Poulsen, T., Molvig, J. *et al.* (1988) Mechanisms of pancreatic beta-cell destruction in type I diabetes. *Diabetes Care*, 11(Suppl 1), 16–23.

13 Suarez-Pinzon, W.L., Mabley, J.G., Strynadka, K. *et al.* (2001) An inhibitor of inducible nitric oxide synthase and scavenger of peroxynitrite prevents diabetes development in NOD mice. *Journal of Autoimmunity*, 16(4), 449–455.

14 Tabatabaie, T., Waldon, A.M., Jacob, J.M. *et al.* (2000) COX-2 inhibition prevents insulin-dependent diabetes in low-dose streptozotocin-treated mice. *Biochemical and Biophysical Research Communications*, 273(2), 699–704.

15 Rocic, B., Vucic, M., Knezevic-Cuca, J. *et al.* (1997) Total plasma antioxidants in first-degree relatives of patients with insulin-dependent diabetes. *Experimental and Clinical Endocrinology and Diabetes*, 105(4), 213–217.

16 Varvarovska, J., Racek, J., Stetina, R. *et al.* (2004) Aspects of oxidative stress in children with type 1 diabetes mellitus. *Biomedicine and Pharmacotherapy*, 58(10), 539–545.

17 Briehl, M.M. & Baker, A.F. (1996) Modulation of the antioxidant defence as a factor in apoptosis. *Cell Death and Differentiation*, 3(1), 63–70.

18 Briehl, M.M., Cotgreave, I.A. & Powis, G. (1995) Downregulation of the antioxidant defence during glucocorticoid-mediated apoptosis. *Cell Death and Differentiation*, 2(1), 41–46.

19 Buttke, T.M. & Sandstrom, P.A. (1994) Oxidative stress as a mediator of apoptosis. *Immunology Today*, 15(1), 7–10.

20 Cnop, M., Welsh, N., Jonas, J.C. *et al.* (2005) Mechanisms of pancreatic beta-cell death in type 1 and type 2 diabetes: many differences, few similarities. *Diabetes*, 54(Suppl 2), S97–S107.

21 Pirot, P., Cardozo, A.K. & Eizirik, D.L. (2008) Mediators and mechanisms of pancreatic beta-cell death in type 1 diabetes. *Arquivos Brasileiros de Endocrinologia e Metabologia*, 52(2), 156–165.

22 Thayer, T.C., Wilson, S.B. & Mathews, C.E. (2010) Use of nonobese diabetic mice to understand human type 1 diabetes. *Endocrinology and Metabolism Clinics of North America*, 39(3), 541–561.

23 Makino, S., Kunimoto, K., Muraoka, Y. *et al.* (1980) Breeding of a non-obese, diabetic strain of mice. *Jikken Dobutsu*, 29(1), 1–13.

24 Ino, T., Kawamoto, Y., Sato, K. *et al.* (1991) Selection of mouse strains showing high and low incidences of alloxan-induced diabetes. *Jikken Dobutsu*, 40(1), 61–67.

25 Bloch, K.O., Zemel, R., Bloch, O.V. *et al.* (2000) Streptozotocin and alloxan-based selection improves toxin resistance of insulin-producing RINm cells. *International Journal of Experimental Diabetes Research*, 1(3), 211–219.

26 Elsner, M., Gurgul-Convey, E. & Lenzen, S. (2008) Relation between triketone structure, generation of reactive oxygen species, and selective toxicity of the diabetogenic agent alloxan. *Antioxidants and Redox Signaling*, 10(4), 691–699.

27 Lenzen, S. (2008) The mechanisms of alloxan- and streptozotocin-induced diabetes. *Diabetologia*, 51(2), 216–226.

28 Szkudelski, T. (2001) The mechanism of alloxan and streptozotocin action in B cells of the rat pancreas. *Physiological Research*, 50(6), 537–546.

29 Mathews, C.E. & Leiter, E.H. (1999) Constitutive differences in antioxidant defense status distinguish alloxan-resistant and alloxan-susceptible mice. *Free Radical Biology and Medicine*, 27(3–4), 449–455.

30 Chen, J., Lu, Y., Lee, C.H. *et al.* (2008) Commonalities of genetic resistance to spontaneous autoimmune and free radical-mediated diabetes. *Free Radical Biology and Medicine*, 45(9), 1263–1270.

31 Mathews, C.E., Graser, R.T., Bagley, R.J. *et al.* (2003) Genetic analysis of resistance to type-1 diabetes in ALR/Lt mice, a NOD-related strain with defenses against autoimmune-mediated diabetogenic stress. *Immunogenetics*, 55(7), 491–496.

32 Mathews, C.E., Leiter, E.H., Spirina, O. *et al.* (2005) mt-Nd2 Allele of the ALR/Lt mouse confers resistance against both chemically induced and autoimmune diabetes. *Diabetologia*, 48(2), 261–267.

33 Gusdon, A.M., Votyakova, T.V. & Mathews, C.E. (2008) mt-Nd2a suppresses reactive oxygen species production by mitochondrial complexes I and III. *Journal of Biological Chemistry* 283(16): 10690–10697.

34 Gusdon, A.M., Votyakova, T.V., Reynolds, I.J. *et al.* (2007) Nuclear and mitochondrial interaction involving mt-Nd2 leads to increased mitochondrial reactive oxygen species production. *Journal of Biological Chemistry*, 282(8), 5171–5179.

35 Chen, J., Gusdon, A.M. & Mathews, C.E. (2008) mt-Nd2a protects a beta cell line against immune- and free radical-mediated cell death. *Diabetes*, 57(S1), A180.

36 Chen, J., Gusdon, A.M., Piganelli, J. *et al.* (2010) mt-Nd2(a) Modifies resistance against autoimmune type 1 diabetes in NOD mice at the level of the pancreatic beta-cell. *Diabetes*, 60(1), 355–359.

37 Kokaze, A., Ishikawa, M., Matsunaga, N. *et al.* (2005) Longevity-associated mitochondrial DNA 5178 C/A polymorphism is associated with fasting plasma glucose levels and glucose tolerance in Japanese men. *Mitochondrion*, 5(6), 418–425.

38 Lal, S., Madhavan, M. & Heng, C.K. (2005) The association of mitochondrial DNA 5178 C > a polymorphism with plasma lipid levels among three ethnic groups. *Annals of Human Genetics*, 69(Pt 6), 639–644.

39 Takagi, K., Yamada, Y., Gong, J.S. *et al.* (2004) Association of a 5178C → A (Leu237Met) polymorphism in the mitochondrial DNA with a low prevalence of myocardial infarction in Japanese individuals. *Atherosclerosis*, 175(2), 281–286.

40 Liou, C.W., Chen, J.B., Tiao, M.M. *et al.* (2012) Mitochondrial DNA coding and control region variants as genetic risk factors for type 2 diabetes. *Diabetes*, 61(10), 2642–2651.

41 Uchigata, Y., Okada, T., Gong, J.S. *et al.* (2002) A mitochondrial genotype associated with the development of autoimmune-related type 1 diabetes. *Diabetes Care*, 25(11), 2106.

42 Chen, J., Hattori, Y., Nakajima, K. *et al.* (2006) Mitochondrial complex I activity is significantly decreased in a patient with maternally inherited type 2 diabetes mellitus and hypertrophic cardiomyopathy associated with mitochondrial DNA C3310T mutation: a cybrid study. *Diabetes Research and Clinical Practice*, 74(2), 148–153.

43 Lightfoot, Y.L., Chen, J. & Mathews, C.E. (2012) mt-Nd2a protects human and mouse beta cells against immune-mediated cell death. *Diabetes*, 61(S), A67.

44 Lee, P.Y., Kumagai, Y., Xu, Y. *et al.* (2011) IL-1alpha modulates neutrophil recruitment in chronic inflammation induced by hydrocarbon oil. *Journal of Immunology*, 186(3), 1747–1754.

45 Lee, P.Y., Li, Y., Kumagai, Y. *et al.* (2009) Type I interferon modulates monocyte recruitment and maturation in chronic inflammation. *American Journal of Pathology*, 175(5), 2023–2033.

46 Li, Y., Lee, P.Y., Kellner, E.S. *et al.* (2010) Monocyte surface expression of Fcgamma receptor RI (CD64), a biomarker reflecting type-I interferon levels in systemic lupus erythematosus. *Arthritis Research & Therapy*, 12(3), R90.

47 Weinstein, J.S., Delano, M.J., Xu, Y. *et al.* (2013) Maintenance of anti-Sm/RNP autoantibody production by plasma cells residing in ectopic lymphoid tissue and bone marrow memory B cells. *Journal of Immunology*, 190(8), 3916–3927.

48 Xu, Y., Lee, P.Y., Li, Y. *et al.* (2012) Pleiotropic IFN-dependent and -independent effects of IRF5 on the pathogenesis of experimental lupus. *Journal of Immunology*, 188(8), 4113–4121.

49 Xu, Y., Zeumer, L., Reeves, W.H. *et al.* (2014) Induced murine models of systemic lupus erythematosus. *Methods in Molecular Biology*, 1134, 103–130.

50 Zhuang, H., Han, S., Xu, Y. *et al.* (2014) Toll-like receptor 7-stimulated tumor necrosis factor alpha causes bone marrow damage in systemic lupus erythematosus. *Arthritis & Rheumatology*, 66(1), 140–151.

51 Willcox, A., Richardson, S.J., Bone, A.J. *et al.* (2009) Analysis of islet inflammation in human type 1 diabetes. *Clinical and Experimental Immunology*, 155(2), 173–181.

52 Leiter, E.H. (2001) The NOD mouse: a model for insulin-dependent diabetes mellitus. In: *Current Protocols in Immunology* 24:15.9:15.9.1–15.9.23.

53 Katz, J., Benoist, C. & Mathis, D. (1993) Major histocompatibility complex class I molecules are required for the development of insulitis in non-obese diabetic mice. *European Journal of Immunology*, 23(12), 3358–3360.

54 Serreze, D.V., Leiter, E.H., Christianson, G.J. *et al.* (1994) Major histocompatibility complex class I-deficient NOD-B2mnull mice are diabetes and insulitis resistant. *Diabetes*, 43(3), 505–509.

55 Sumida, T., Furukawa, M., Sakamoto, A. *et al.* (1994) Prevention of insulitis and diabetes in beta 2-microglobulin-deficient non-obese diabetic mice. *International Immunology*, 6(9), 1445–1449.

56 DiLorenzo, T.P., Graser, R.T., Ono, T. *et al.* (1998) Major histocompatibility complex class I-restricted T cells are required for all but the end stages of diabetes development in nonobese diabetic mice and use a prevalent T cell receptor alpha chain gene rearrangement. *Proceedings of the National Academy of Sciences of the United States of America*, 95(21), 12538–12543.

57 Haskins, K., Portas, M., Bergman, B. *et al.* (1989) Pancreatic islet-specific T-cell clones from nonobese diabetic mice. *Proceedings of the National Academy of Sciences of the United States of America*, 86(20), 8000–8004.

58 Wong, F.S., Visintin, I., Wen, L. *et al.* (1996) CD8 T cell clones from young nonobese diabetic (NOD) islets can transfer rapid onset of diabetes in NOD mice in the absence of CD4 cells. *Journal of Experimental Medicine*, 183(1), 67–76.

59 Thomas, H.E., Trapani, J.A. & Kay, T.W. (2010) The role of perforin and granzymes in diabetes. *Cell Death and Differentiation*, 17(4), 577–585.

60 Cantor, J. & Haskins, K. (2007) Recruitment and activation of macrophages by pathogenic CD4 T cells in type 1 diabetes: evidence for involvement of CCR8 and CCL1. *Journal of Immunology*, 179(9), 5760–5767.

61 Tse, H.M., Milton, M.J., Schreiner, S. *et al.* (2007) Disruption of innate-mediated proinflammatory cytokine and reactive oxygen species third signal leads to antigen-specific hyporesponsiveness. *Journal of Immunology*, 178(2), 908–917.

62 Sklavos, M.M., Tse, H.M. & Piganelli, J.D. (2008) Redox modulation inhibits CD8 T cell effector function. *Free Radical Biology and Medicine*, 45(10), 1477–1486.

63 Piganelli, J.D., Flores, S.C., Cruz, C. *et al.* (2002) A metalloporphyrin-based superoxide dismutase mimic inhibits adoptive transfer of autoimmune diabetes by a diabetogenic T-cell clone. *Diabetes*, 51(2), 347–355.

64 Bedard, K. & Krause, K.H. (2007) The NOX family of ROS-generating NADPH oxidases: physiology and pathophysiology. *Physiological Reviews*, 87(1), 245–313.

65 Thamsen, M. & Jakob, U. (2011) The redoxome: Proteomic analysis of cellular redox networks. *Current Opinion in Chemical Biology*, 15(1), 113–119.

66 Thayer, T.C., Delano, M., Liu, C. *et al.* (2011) Superoxide production by macrophages and T cells is critical for the induction of autoreactivity and type 1 diabetes. *Diabetes*, 60(8), 2144–2151.

67 Tse, H.M., Thayer, T.C., Steele, C. *et al.* (2010) NADPH oxidase deficiency regulates Th lineage commitment and modulates autoimmunity. *Journal of Immunology*, 185(9), 5247–5258.

68 Anderson, M.S. & Bluestone, J.A. (2005) The NOD mouse: a model of immune dysregulation. *Annual Review of Immunology*, 23, 447–485.

69 Bending, D., De La Pena, H., Veldhoen, M. *et al.* (2009) Highly purified Th17 cells from BDC2.5NOD mice convert into Th1-like cells in NOD/SCID recipient mice. *Journal of Clinical Investigation*, 119(3), 565–572.

70 Liu, C. & Mathews, C.E. (2013) Roles of ROS in the CD8+ T cell priming and cytotoxic functions. In: *73rd Scientific Sessions*. American Diabetes Association.

71 Calderon, B., Suri, A. & Unanue, E.R. (2006) In CD4+ T-cell-induced diabetes, macrophages are the final effector cells that mediate islet beta-cell killing: studies from an acute model. *American Journal of Pathology*, 169(6), 2137–2147.

72 Heath, W.R. & Carbone, F.R. (2001) Cross-presentation, dendritic cells, tolerance and immunity. *Annual Review of Immunology*, 19, 47–64.

73 Cui, Z.R. & Qiu, F. (2007) CD4(+) T helper cell response is required for memory in CD8(+) T lymphocytes induced by a poly(I : C)-adjuvanted MHC I-restricted peptide epitope. *Journal of Immunotherapy*, 30(2), 180–189.

74 Kurts, C., Robinso, B.W. & Knolle, P.A. (2010) Cross-priming in health and disease. *Nature Reviews Immunology*, 10(6), 403–414.

75 Smith, C.M., Wilson, N.S., Waithman, J. *et al.* (2004) Cognate CD4(+) T cell licensing of dendritic cells in CD8(+) T cell immunity. *Nature Immunology*, 5(11), 1143–1148.

76 de Jersey, J., Snelgrove, S.L., Palmer, S.E. *et al.* (2007) Beta cells cannot directly prime diabetogenic CD8 T cells in nonobese diabetic mice. *Proceedings of the National Academy of Sciences of the United States of America*, 104(4), 1295–1300.

77 Amigorena, S. & Savina, A. (2010) Intracellular mechanisms of antigen cross presentation in dendritic cells. *Current Opinion in Immunology*, 22(1), 109–117.

78 Savina, A., Jancic, C., Hugues, S. *et al.* (2006) NOX2 controls phagosomal pH to regulate antigen processing during cross-presentation by dendritic cells. *Cell*, 126(1), 205–218.

CHAPTER 20

Metabolic syndrome, inflammation, and reactive oxygen species in children and adults

William E. Winter[1] and Janet H. Silverstein[2]

[1] Department of Pathology, Immunology and Laboratory Medicine, University of Florida, Gainesville, FL, USA
[2] Department of Pediatrics, University of Florida College of Medicine, Gainesville, FL, USA

THEMATIC SUMMARY BOX

At the end of this chapter, students should be able to:

- Define and recognize the metabolic syndrome

- Relate insulin resistance to the pathologies of the metabolic syndrome

- Explain how hyperinsulinism causes component disorders of the metabolic syndrome

- Clarify how relative insulinopenia causes component disorders of the metabolic syndrome

- Inter-relate (1) decreased insulin action and hyperglycemia, and (2) the development of a prooxidant state

---

Introduction

The metabolic syndrome is a group of reproducible, recognizable characteristics that relate central obesity and insulin resistance to increased cardiovascular risk as well as to other problems. The National Institutes of Health states that (the metabolic syndrome is) "… the name for a group of risk factors linked to overweight and obesity that increase the patient's chance for heart disease and other health problems such as diabetes and stroke."[1] Other terms applied to the metabolic syndrome include the insulin resistance syndrome, syndrome X, the diabesity syndrome, morbesity syndrome, the cardiometabolic syndrome, and the dysmetabolic syndrome X (which is assigned the ICD-9 code 277.7). In addition to heart disease, adults with the

---

Oxidative Stress and Antioxidant Protection: The Science of Free Radical Biology and Disease, First Edition.
Edited by Donald Armstrong and Robert D. Stratton.
© 2016 John Wiley & Sons, Inc. Published 2016 by John Wiley & Sons, Inc.

metabolic syndrome are at risk for stroke and peripheral vascular disease. Obesity also predisposes to a variety of cancers.[2,3]

A wide variety of definitions for the metabolic syndrome exist among the World Health Organization (WHO), The European Group for the Study of Insulin Resistance (EGIR), the National Cholesterol Education Program (NCEP), the International Diabetes Federation (IDF), the National Heart, Lung and Blood Institute (NHLBI), and American Association for Clinical Endocrinologists (AACE). However, all of these organizations identify four common characteristics of the metabolic syndrome in their definitions: (i) elevated plasma glucose, (ii) hypertriglyceridemia, (iii) low levels of high-density lipoprotein (HDL), and (iv) hypertension.

While predominantly considered a disease of adults (with ~25% of adult Americans diagnosed with the metabolic syndrome), insulin resistance can begin in childhood and produce the full spectrum of the disease, with increased arterial stiffness, impaired endothelial function, and abnormal brachial artery reactivity.[4,5] All of these latter factors are risk factors for the development of cardiovascular disease in adults.

Although myocardial infarction due to the metabolic syndrome has not been reported in childhood, the pediatric origins of atherosclerotic cardiovascular disease (ASCVD) have been identified in several large longitudinal studies.[6–8]

It is estimated that ~6–39% of overweight or obese adolescents suffer from the metabolic syndrome.[9] In the authors' experience in a university-based pediatric endocrinology clinic, obese children (some even less than 10 years of age) can be affected with type 2 diabetes, hypertriglyceridemia, hypoalphalipoproteinemia (e.g., low HDL), hypertension, elevated uric acid, acanthosis nigricans, and, in adolescent girls, polycystic ovarian syndrome (PCOS). In short, the metabolic syndrome is just as much a pediatric disease as it is a disease of adults.

The biology of insulin resistance

To understand insulin resistance, we must understand insulin action. Insulin action is the product of the absolute insulin concentration in the interstitium (e.g., the insulin available for binding to the insulin receptors) multiplied by the responsiveness of the tissues to insulin (specifically the liver, adipose tissue, and skeletal muscle insulin sensitivity). Insulin sensitivity is a consequence of the insulin receptor binding of insulin and the generation of the postreceptor signals that involve multiple pathways that include, for example, insulin-receptor substrate 1 (IRS1) and insulin-receptor substrate 2 (IRS2). Plasma insulin concentrations serve as a proxy for interstitial insulin concentrations because interstitial insulin concentrations cannot ordinarily be measured.

Type 1 diabetes results from absolute insulinopenia, whereas type 2 diabetes results from insulin resistance and relative insulinopenia. Relative insulinopenia means that the absolute insulin concentration may be elevated yet the insulin concentrations are inadequate to compensate for the degree of tissue insulin resistance. Most cases of type 2 diabetes develop because of insulin resistance and an inability of the β cells to maintain adequate insulin action. The predecessor of many cases of type 2 diabetes is the metabolic syndrome.

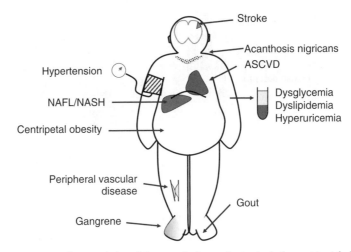

Figure 20.1 Common characteristics of the metabolic syndrome include centripetal obesity, dysglycemia, dyslipidemia, and hyperuricemia (detectable in the patient's plasma), hypertension, liver disease [non-alcoholic fatty liver (NAFL) or non-alcoholic steatohepatitis (NASH)], gout [manifested as pain in the great toe (podagra)], acanthosis nigricans and atherosclerotic cardiovascular disease (ASCVD), stroke, and peripheral vascular disease that can predispose to gangrene.

Central obesity is a common characteristic of the metabolic syndrome (Figure 20.1). Central obesity is diagnosed clinically by the measurement of increased abdominal circumference and an increased waist-to-hip ratio (WHR). A 2007 study in adults found a significant relationship between increasing waist circumference (WC) and WHR and the risk for cardiovascular disease.[10] While fat is increased both subcutaneously and intra-abdominally, the intra-abdominal fat is most closely correlated with the development of insulin resistance. This may be the result of the delivery of high concentrations of free fatty acids (FFAs) from the omentum to the liver via the portal vein. According to the Randle hypothesis, the body burns FFAs preferably over carbohydrate.[11] FFAs supply more energy per unit weight than carbohydrate. Recent studies demonstrate that diacyl glycerol (DAG) produced from the uptake of FFA's induces insulin resistance by altering the second messenger signaling of the insulin–insulin receptor complex (Figure 20.2).[12]

The consequences of insulin resistance result from (i) the direct consequences of hyperinsulinism (as an attempted physiologic compensation for insulin resistance) or (ii) inadequate insulinization (e.g., decreased insulin action) despite hyperinsulinism.

The consequences of hyperinsulinism

The direct adverse consequences of hyperinsulinism include atherogenesis, hypertension, increased plasminogen activator inhibitor-1 (PAI-1) concentrations (which increases coagulability and risk of clotting), hyperuricemia, hyperandrogenism,

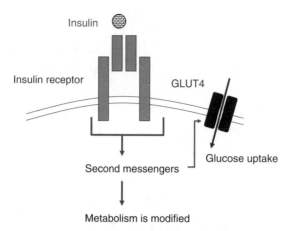

Figure 20.2 Insulin binds to the insulin receptor to initiate signaling. Second messengers cause alternations in metabolism (e.g., increased glycolysis and glycogen synthesis and suppressed gluconeogenesis) and the movement of the insulin-responsive glucose transporter (GLUT4) from the cytoplasmic pool to the plasma membrane facilitating glucose uptake into the skeletal muscle and adipose tissue.

and acanthosis nigricans. Furthermore, hypertension, increased PAI-1 concentrations, hyperuricemia, and hyperandrogenism contribute to the development of atherosclerosis.

There are many epidemiologic studies that demonstrate strong associations between elevated insulin levels and increased risks of atherosclerosis. Nevertheless, there is controversy as to whether insulin itself is atherogenic (e.g., insulin is a potent growth factor) or whether elevated insulin levels mark the insulin-resistant state without being intrinsically pathogenic.[13] There are at least four mechanisms whereby hyperinsulinism causes hypertension: hyperinsulinism (i) causes sodium retention (and water follows sodium to expand circulating blood volume), (ii) increases sympathetic tone causing vasoconstriction increasing peripheral vascular resistance, (iii) causes vascular hypertrophy (possibly by acting as a growth factor) that leads to vasoconstriction and increased peripheral vascular resistance, and (iv) increases the activity of the Na^+/K^+ ATPase pump, again causing vasoconstriction and increased peripheral vascular resistance.

Although it had been thought that adipose tissue was metabolically inactive, indeed adipose tissue produces a range of products from cytokines [including proinflammatory IL-1, IL-6, and tumor necrosis factor-α (TNF-α)], anti-inflammatory adiponectin (known specifically as a type of "adipokine"), and procoagulants.[14] Visceral obesity is associated with high concentrations of the inflammatory cytokines and low concentrations of adiponectin. Increased production of PAI-1 by adipose tissue in cases of the metabolic syndrome contributes to a prothrombotic state. Inflammation and rupture of the shoulder region of the atheromatous plaque (the region near to the vessel wall) with the release of the prothrombotic core of the plaque is the mechanism underlying acute coronary artery obstruction causing myocardial infarction and acute cerebral arterial obstruction (causing ~70–80% of strokes).[15]

Usually, tissue plasminogen activator (tPA) is released by injured tissue. tPA activates plasminogen to plasmin. In turn, plasmin degrades fibrin. However, high levels of PAI-1 inhibit plasmin favoring a persistence of fibrin and delayed clearance of thrombi. Because the metabolic syndrome is an inflammatory state, we will review how higher levels of fibrinogen and factor VIII also contribute to a propensity for thrombosis.

Hyperuricemia causes gout, which is a classic (and rather ancient) finding in the metabolic syndrome. As noted earlier, hyperinsulinism leads to sodium hyperabsorption by the kidney tubules.[16] Because uric acid and sodium reabsorption are linked, hyperinsulinism also causes increased uric acid reabsorption. Various studies suggest that the atherogenic effect of hyperuricemia is mediated by disturbed endothelial function.[17] Children with the metabolic syndrome have higher uric acid levels than children without the metabolic syndrome.[18]

Hyperandrogenism may be clinically evident during the physical examination of women with the metabolic syndrome. Hyperinsulinism suppresses follicular stimulating hormone (FSH) and stimulates luteinizing hormone (LH). Increased LH, which stimulates testosterone production by the theca cells of the ovary, and the direct effects of hyperinsulinism on the ovary lead to elevated ovarian androgen production. Hyperinsulinism reduces sex hormone-binding globulin (SHBG) levels. The combination of increased androgen levels and reduced SHBG levels produces an elevation in free androgen levels causing clinical hyperandrogenism. SHBG levels are reduced in children with the metabolic syndrome.[19]

Elevated androgens are atherogenic causing hypercoagulability, increased LDL-C levels, reduced HDL-C levels, and elevated blood pressure. Furthermore, women with the metabolic syndrome are frequently affected with the PCOS manifested by hirsutism, irregular menses, amenorrhea, and infertility.

Through the direct effect of hyperinsulinism on melanocytes, hyperpigmentation can occur in areas of the body where skin abrades on skin, such as the neck, axilla, groin, and breasts in women. In severe cases, the hyperpigmentation can be found on the face and cheeks in addition to the neck. The diagnostic term that describes such hyperpigmentation accompanied by a velvety change in the skin and skin tags is acanthosis nigricans.

The consequences of inadequate insulinization

The consequences of inadequate insulinization focus on the development of dysglycemia and dyslipidemia. Decreased insulin action disinhibits hormone-sensitive lipase (HSL) in adipose tissue permitting the breakdown of adipose tissue triglyceride into free fatty acids. These free fatty acids travel to the liver (bound to albumin) to be off-loaded to the hepatocytes (Figure 20.3). Within the hepatocytes, increased delivery of free fatty acids stimulates triglyceride synthesis. These hepatic triglycerides are incorporated into very low-density lipoprotein (VLDL). This produces hypertriglyceridemia in the plasma. Because of a subsequent disruption in very low-density lipoprotein–high-density lipoprotein (VLDL–HDL) biology, HDL levels decline. Furthermore, although LDL cholesterol levels may not rise, there are an increased number of apolipoprotein B (apo B) particles that raise LDL density and LDL molar

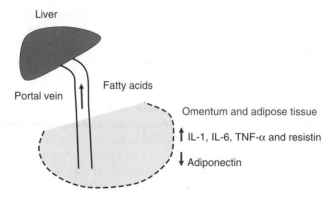

Figure 20.3 This figure illustrates the delivery of free fatty acids (FFAs) from the omentum to the liver via the portal circulation. The omentum and adipose tissue secrete IL-1, IL-6, TNF-α, and resistin, whereas adiponectin secretion is deficient.

concentrations in the plasma. Such "dense" LDL particles are highly atherogenic. Also atherogenic is the low HDL concentration, elevation in total apo B and hypertriglyceridemia.

Another mechanism of hypertriglyceridemia is the decreased activity of lipoprotein lipase (LPL), which is dependent on insulin for normal levels of activity. LPL is produced by adipose tissue and skeletal muscle. LPL then becomes attached to the luminal side the endothelium. When chylomicrons or VLDL bind to LPL via apo CII, LPL cleaves triglycerides in the chylomicrons and VLDL. The resulting free fatty acids and monoglycerides are then taken up by the adipose tissue and skeletal muscle. In the absence of normal levels of insulin action, clearance of chylomicrons and VLDL is impaired leading to hypertriglyceridemia.

Inflammation and the liver: prooxidant hepatocellular damage

The excess deposition of fatty acids in the liver during their transit through the liver causes hepatic steatosis. Such steatosis causes nonalcoholic fatty liver (NAFL). If there is continued steatosis, inflammation can result because of the prooxidant state produced by the high levels of hepatic fatty acids that trigger steatohepatitis. Cirrhosis can progress to frank liver failure and/or hepatocellular carcinoma. Steatosis is observed in obese children as well as in adults.[20]

Inadequate insulinization ultimately and commonly causes type 2 diabetes, which is generally a later development in the natural history of the untreated metabolic syndrome. For many reasons, obesity causes insulin resistance. The factors that contribute to the insulin resistance state include the following: (i) an abundance of plasma free fatty acids traveling particularly from the omentum to the liver via the portal system fostering insulin resistance (with decreased glycolysis and increased β-oxidation), (ii) deficient adipose production of insulin-sensitizing adipokines such as adiponectin and an excess of adipokines, producing insulin resistance such as resistin; (iii) relative excess of cortisol (possibly from increased expression of adipose tissue 11-β hydroxysteroid dehydrogenase type I that converts cortisone to cortisol); (iv) the ectopic deposition of fat in the liver, skeletal muscle, and the β cells, and (v) locally produced TNF-α from fatty infiltration of skeletal muscle that impairs skeletal muscle's response to insulin. The cellular infiltration of fat is accomplished by macrophages that actively produce inflammatory cytokines that contribute to necroinflammation (the combination of necrosis and inflammation).

Insulin resistance initially causes a slight initial rise in blood glucose. To maintain normoglycemia, this is immediately countered by a rise in the secretion of insulin by the β cells, therefore, returning the glucose to normal without the development of any detectable hyperglycemia. With continued and increasing degrees of insulin resistance, insulin secretion must continue to increase to maintain glycemic homeostasis. When the β cell is no longer able to produce the quantities of insulin needed to counter the insulin resistance, relative insulinopenia supervenes leading to dysglycemia. Such relative β-cell failure may result from exhaustion, β-cell infiltration by fat, and oxidant damage or a genetic limitation in insulin secretory capacity.

In adults, the first evidence of dysglycemia is impaired glucose tolerance [e.g., a 2 h glucose level on oral glucose tolerance testing (OGTT) of 140–199 mg/dl]. Impaired fasting glucose (e.g., a fasting plasma glucose of 100–125 mg/dl) can then follow in adults. In obese children, impaired fasting glucose is more frequent than impaired glucose tolerance. In adults, the progression of such prediabetic states to frank diabetes is first evident in a 2-h glucose on OGTT of ≥200 mg/dl followed temporally by a rise in fasting plasma glucose to 126 mg/dl or greater. The last glycemic index to become abnormal in adults is hemoglobin A1c (e.g., ≥6.5%). On the other hand, some children with type 2 diabetes have an abnormal oral glucose tolerance test and normal hemoglobin A1c, whereas other children with type 2 diabetes have an abnormal hemoglobin A1c and a normal oral glucose tolerance test.

Hyperglycemia (e.g., a 2-h OGTT glucose ≥200 mg/dl; fasting plasma glucose ≥126 mg/dl; or hemoglobin A1c ≥6.5%) in an asymptomatic patient must be confirmed on a second day to diagnose diabetes mellitus. Any combination of such abnormal tests can be used. In a symptomatic patient, a single recognition of hyperglycemia establishes the diagnosis of diabetes mellitus such as diabetic ketoacidosis (DKA), hyperglycemic, hyperosmolar coma (HHC), an elevated fasting plasma glucose, or 2-h OGTT plasma glucose.

Diabetes and reactive oxygen species

The consequence of decreased insulin action and hyperglycemia establishes a highly prooxidant state in the patient that can foster inflammatory disorders such as atherosclerosis and nonalcoholic steatohepatitis (NASH) as a consequence of free radical damage. Inflammation plays a major role in atherogenesis. As noted earlier, the region of the atherosclerotic plaque that is most likely to fracture and release the plaque's thrombogenic core is the shoulder region. The shoulder is near the junction of the arterial wall and the cap of the plaque. In the shoulder region are a large number of inflammatory cells such as macrophages and lymphocytes. With increased inflammation, activation of these cells, particularly the macrophages, leads to the secretion of matrix metalloproteinases (MMPs) that weaken the cap. This encourages fracture of the cap at the shoulder region. Fracture of the cap and exposure of plasma to the plaque's core trigger thrombosis. With a sufficient degree of acute ischemia, downstream infarction ensues. In the heart, this is manifested as myocardial infarction. In the central nervous system, this is manifested as ischemic stroke. In the lower extremities, ischemia contributes to the development of peripheral vascular disease whose most severe manifestation is gangrene.

A novel paradigm is the insulin–glucose–inflammation model: as glucose rises and/or insulin's action declines, inflammation is triggered. Increased glucose levels increase proinflammatory transcription factors [e.g., nuclear factor-kappa B (NF-κB), AP-1 (activator protein-1), I kappa B kinase-alpha and IKK B]; the expression of the NADPH oxidase subunit p47phox; interleukins IL-1 and IL-6; TNF-α; adhesion molecules ICAM-1 and E-selectins; MMP; and reactive oxidation species with the oxidation of nucleic acids, protein, carbohydrates, and lipids.[21–23] Furthermore increased O_2^- inactivates nitric oxide (NO), which is a potent and vital vasodilator that is critical for the maintenance of normal tissue perfusion. Consequently, vasoconstriction can ensue. Increased glucose also decreases cytoplasmic I kappa B, which normally regulates NF-κB.

On the other hand, decreased glucose and increased insulin action decreases NF-κB [with likely decreases in vascular endothelial growth factor (VGEF), TNF-α, and IL-6], AP-1, and early growth response-1 (Egr-1). AP-1 regulates MMPs that lyse collagen and other matrix proteins. Egr-1 regulates PAI-1 and tissue factor expression countering their usual procoagulant effects. Decreased glucose and increased insulin action also reduce the expression of the adhesion molecule ICAM-1, monocyte chemotactic protein-1 (MCP-1), C-reactive protein (CRP), reactive oxygen species (ROS), and the NADPH oxidase subunit p47phox. Finally, increased insulin action increases cytoplasmic I kappa B to downregulate the activity of NF-κB.

Of interest, elevated levels of laboratory markers of inflammation are identified in subjects with the metabolic syndrome. CRP is, on average, elevated in people with the metabolic syndrome.[24] In addition, ferritin, a potential marker of acute inflammation, is also commonly increased in the metabolic syndrome.[25]

Conclusion

The metabolic syndrome occurs in both children and adults. There is a significant inflammatory component to the metabolic syndrome that plays a major role in the pathogenesis of arteriosclerotic cardiovascular disease and steatohepatitis.[26] Decreased insulin action and hyperglycemia contribute to this proinflammatory state.

The prevention of obesity is the best public health approach to the prevention of the metabolic syndrome. Weight loss and physical fitness improve insulin sensitivity. For children and adults with obesity, achieving a reduced body weight and increasing physical activity are of benefit in preventing and, lacking prevention, treating the metabolic syndrome.

Multiple choice questions

1 A married woman of reproductive age with irregular menstrual periods has had unprotected sex with her husband for more than 1 year without becoming pregnant. Her body mass index is elevated as is her blood pressure and fasting plasma glucose. What therapy is most likely to benefit all of her medical problems?
   a. An anti-inflammatory medication
   b. Weight loss
   c. Insulin
   d. Vegetarian diet
   e. Atkins (high-fat) diet

2 What is the adverse effect of the metabolic syndrome on the liver?
   a. Steatosis
   b. Inflammation.
   c. Both a and b
   d. Neither a nor b.

3 An obese 55-year-old, hypertensive, diabetic man develops severe, acute pain in his right large toe. What is the most likely cause of the toe pain?
   a. Trauma
   b. Peripheral vascular disease
   c. Neuropathy
   d. Type 2 diabetes
   e. Gout

4 A woman brings her obese, hypertensive 10-year-old son to the physician complaining that "she can't get her son's neck clean" despite extensive washing. What is the most likely explanation for this complaint?
   a. Acanthosis nigricans
   b. Acne
   c. Staphylococcal skin infection
   d. Tattoos were applied
   e. Melanoma

References

1 U.S. Department of Health & Human Services, National Institutes of Health, National Heat, Lung and Blood Institute. What is Metabolic syndrome? http://www.nhlbi.nih.gov/health/dci/Diseases/ms/ms_whatis.html [accessed on 17 November 2015].
2 Matsuda, M. & Shimomura, I. (2013) Increased oxidative stress in obesity: implications for metabolic syndrome, diabetes, hypertension, dyslipidemia, atherosclerosis, and cancer. *Obesity Research & Clinical Practice*, 7(5), e330–e341.
3 Gallagher, E.J. & LeRoith, D. (2013) Epidemiology and molecular mechanisms tying obesity, diabetes, and the metabolic syndrome with cancer. *Diabetes Care*, 36 Suppl 2:S233–9.
4 Urbina, E.M., Kimball, T.R., Khoury, P.R. *et al.* (2010) Increased arterial stiffness is found in adolescents with obesity or obesity-related type 2 diabetes mellitus. *Journal of Hypertension*, 28, 1692–1698 2:S233–S239.
5 Meyer, A.A., Kundt, G., Steiner, M. *et al.* (2006) Impaired flow-mediated vasodilation, carotid artery intima-media thickening, and elevated endothelial plasma markers in obese children: the impact of cardiovascular risk factors. *Pediatrics*, 117, 1560–1567.

6 Berenson, G.S., Srinivasan, S.R., Bao, W. *et al.* (1998) Association between multiple cardio-vascular risk factors and atherosclerosis in children and young adults: The Bogalusa Heart Study. *New England Journal of Medicine*, 338, 1650–1656.

7 McGill, H.C. Jr., McMahan, C.A., Zieske, A.W. *et al.* (2000) Association of coronary heart disease risk factors with microscopic qualities of coronary atherosclerosis in youth. *Circulation*, 102, 374–379.

8 McGill, H.C. Jr., McMahan, C.A., Zieske, A.W. *et al.* (2000) Pathobiological Determinants of Atherosclerosis in Youth (PDAY) Research Group. Associations of coronary heart disease risk factors with the intermediate lesion of atherosclerosis in youth. *Arteriosclerosis, Thrombosis, and Vascular Biology*, 20, 1998–2004.

9 Reinehr, T., de Sousa, G., Toschke, A.M. *et al.* (2007) Comparison of metabolic syndrome prevalence using eight different definitions: a critical approach. *Archives of Disease in Childhood*, 92(12), 1067–1072.

10 de Koning, L., Merchant, A.T., Pogue, J. *et al.* (2007) Waist circumference and waist-to-hip ratio as predictors of cardiovascular events: meta-regression analysis of prospective studies. *European Heart Journal*, 28(7), 850–856.

11 Elks, M.L. (1990) Fat oxidation and diabetes of obesity: the Randle hypothesis revisited. *Medical Hypotheses*, 33(4), 257–260.

12 Shulman, G.I. (2014) Ectopic fat in insulin resistance, dyslipidemia, and cardiometabolic disease. *New England Journal of Medicine*, 371(12), 1131–1141.

13 Jandeleit-Dahm, K.A. & Gray, S.P. (2012) Insulin and cardiovascular disease: biomarker or association? *Diabetologia*, 55(12), 3145–3151.

14 Piya, M.K., McTernan, P.G. & Kumar, S. (2013) Adipokine inflammation and insulin resistance: the role of glucose, lipids and endotoxin. *Journal of Endocrinology*, 216(1), T1–T15.

15 Libby, P. (2013) Mechanisms of acute coronary syndromes and their implications for therapy. *New England Journal of Medicine*, 368(21), 2004–2013.

16 Puig, J.G. & Martínez, M.A. (2008) Hyperuricemia, gout and the metabolic syndrome. *Current Opinion in Rheumatology*, 20(2), 187–91.

17 Gustafsson, D. & Unwin, R. (2013) The pathophysiology of hyperuricaemia and its possible relationship to cardiovascular disease, morbidity and mortality. *BMC Nephrology*, 14, 164.

18 Valle, M., Martos, R., Cañete, M.D. *et al.* (2014) Association of serum uric acid levels to inflammation biomarkers and endothelial dysfunction in obese prepubertal children. *Pediatric Diabetes*, 16(6):441–7.

19 Agirbasli, M., Agaoglu, N.B., Orak, N. *et al.* (2009) Sex hormones and metabolic syndrome in children and adolescents. *Metabolism*, 58(9), 1256–1262.

20 Duarte, M.A. & Silva, G.A. (2011) Hepatic steatosis in obese children and adolescents. *Jornal de Pediatria*, 87(2), 150–156.

21 Bastard, J.P., Maachi, M., Lagathu, C. *et al.* (2006) Recent advances in the relationship between obesity, inflammation, and insulin resistance. *European Cytokine Network*, 17(1), 4–12.

22 Ruan, H. & Pownall, H.J. (2009) The adipocyte IKK/NFkappaB pathway: a therapeutic target for insulin resistance. *Current Opinion in Investigational Drugs*, 10(4), 346–352.

23 Shah, A., Mehta, N. & Reilly, M.P. (2008) Adipose inflammation, insulin resistance, and cardiovascular disease. *JPEN Journal of Parenteral and Enteral Nutrition*, 32(6), 638–44.

24 Lim, S., Lee, H.K., Kimm, K.C. *et al.* (2005) C-reactive protein level as an independent risk factor of metabolic syndrome in the Korean population. CRP as risk factor of metabolic syndrome. *Diabetes Research and Clinical Practice*, 70(2), 126–33.

25 Abril-Ulloa, V., Flores-Mateo, G., Solà-Alberich, R. *et al.* (2014) Ferritin levels and risk of metabolic syndrome: meta-analysis of observational studies. *BMC Public Health*, 14, 483.

26 Kraja, A.T., Chasman, D.I., North, K.E. *et al.* (2014) Pleiotropic genes for metabolic syndrome and inflammation. *Molecular Genetics and Metabolism*, 112(4), 317–338.

CHAPTER 21

Oxidative stress in chronic pancreatitis

Shweta Singh and Bechan Sharma
Department of Biochemistry, Faculty of Science, University of Allahabad, Allahabad, 211002 India

---

THEMATIC SUMMARY BOX

At the end of this chapter, students should be able to:

- Describe the main features of pancreatitis

- Outline the pathogenesis of acute and chronic pancreatitis

- Analyze the contribution of oxidative stress

- Discuss antioxidant therapy of chronic pancreatitis

Summary

Chronic pancreatitis (CP) is a multifactorial and multistep disease. Apart from male gender and early onset of disease, there are many other documented modifiable risk factors of CP. The pathobiological mechanisms underlying the phenotypic expression of CP are complex. Hereditary factors and mutations in some genes have been found to be associated with the initiation or severity of the disease in CP patients. However, the exact mechanisms of pathogenesis of pancreatitis are still not clear. CP has been shown to induce oxidative stress (OS) in the patients. In addition to the role of different gene variants contributing to this disease, CP induced production of free radicals in excess and generation of OS are also reported to play a critical role in onset of the disease, its severity, and progression. It appears that OS through altered xenobiotic metabolism may be directly or indirectly involved in the etiology of pancreatic disease. Consequently, the mutated alleles of genes-encoding enzymes metabolizing these substances and their metabolites may act as disease modifiers. This chapter presents a recent account of CP-induced free radical production, alterations in the levels of antioxidative enzymes, and possible remedies via application of antioxidants (both in single and in different combinations) for effective therapy of the disease.

Oxidative Stress and Antioxidant Protection: The Science of Free Radical Biology and Disease, First Edition.
Edited by Donald Armstrong and Robert D. Stratton.
© 2016 John Wiley & Sons, Inc. Published 2016 by John Wiley & Sons, Inc.

Introduction

Reactive species include reactive oxygen species (ROS), reactive nitrogen species (RNS), and other carbon-centered molecules that are unstable chemicals generated in biological systems under normal physiological as well as pathophysiological conditions. When a reactive molecule contains one or more unpaired electrons, the molecule is termed as free radical. These potentially harmful free radicals are scavenged by antioxidants. Many of these antioxidants are either essential nutrients or enzymes that maintain the critical balance between generation and destruction of free radicals.[1] Deficiencies of antioxidants as observed in patients with both temperate and tropical pancreatitis led to the hypothesis that pancreatitis was caused by OS, defined as a state of potential tissue injury due to imbalance between injurious free radicals and protective antioxidants.[2] OS may be important in the pathogenesis of ethanol-induced pancreatic injury, but radiation, exposure to cigarette smoke, medication, or trauma may also stimulate the generation of free radicals, which subsequently may result in damage to lipids, proteins, or nucleic acids. In this chapter, CP-induced alterations in the levels of key players of antioxidative defense system and associated factors are described. In addition, the role of antioxidants such as selenium, allopurinol, β-carotene, N-acetylcysteine (NAC), and pentoxifylline as well as some phytochemicals in amelioration of OS-mediated severity of pancreatic disease in CP patients is also discussed.

Pancreas, chronic pancreatitis (CP), and symptoms

The pancreas, being a glandular organ, plays a significant role in the digestive system and is responsible for secretion of pancreatic juice containing digestive enzymes, which are responsible for the digestion and absorption of nutrients such as carbohydrates, proteins, and lipids in the small intestine. It is also a key component of the endocrine system of vertebrates and produces several hormones that are primarily involved in maintenance of glucose homeostasis in the body. In humans, it is located in the abdominal cavity behind the stomach next to the small intestine.

The β-cells of pancreas release insulin (which causes decrease in blood glucose level), α-cells of pancreas secrete glucagon (responsible for increase in blood glucose level), Δ(delta)-cells secrete somatostatin (which regulate functions of α- and β-cells), and γ(gamma)-cells secrete pancreatic polypeptide into the blood stream to perform their desired functions in the body.

Pancreatitis is a complex disorder and is defined as an inflammation of the pancreas. Pancreatic inflammation occurs in two forms: acute pancreatitis (AP) and CP. AP is generally characterized by edema and inflammatory infiltration, and, in severe cases, by necrosis and hemorrhage.[3] AP is associated with abdominal pain that radiates into the back, swollen, and tender abdomen, fever, increased rate of heart beat, nausea and vomiting whereas most CP is associated with pancreatic-head enlargement, parenchymal calcification, cholecystitis, pancreatic stones, fibrosis, and pancreatic exocrine and endocrine dysfunction.

Long-lasting inflammation of the pancreas is called CP. The symptoms of CP are similar to that of AP. In addition, weight loss may occur due to poor absorption of food as enzymes from pancreas are not released properly to digest the food. CP patients may be predisposed to diabetes if insulin-producing β-cells of pancreas are damaged. The main causes of occurrence of CP include gallstones, heavy consumption of alcohol, smoking, infections, trauma, metabolic disorders, surgery, and even genetics. CP occurring due to injury may trigger excessive production of free radicals in the acinar cells of pancreas causing oxidation of polyunsaturated fatty acids (PUFAs) and proteins constituting the cellular membrane of pancreas.

CP-induced production of oxidative stress

CP is known to generate free radicals and to produce oxidative stress in patients. The OS is caused by an imbalance between the production of free radical species (FRS) and antioxidant defense system of an organism. This defense system is the ability of any biological system to readily detoxify or neutralize the reactive oxygen intermediates or easily repair the resulting damage. One source of reactive oxygen under normal conditions in humans is the leakage of activated oxygen (superoxide) from mitochondria during oxidative phosphorylation. However, *Escherichia coli* mutants that lack an active electron transport chain produce as much hydrogen peroxide (H_2O_2) as wild-type cells, indicating that other enzymes contribute to the bulk of oxidants in these organisms. One possibility is that multiple redox-active flavoproteins contribute a small portion to the overall production of oxidants under normal conditions. Other enzymes capable of producing superoxide anions are xanthine oxidase, nicotinamide adenine dinucleotide phosphate (NADPH) oxidases, and cytochromes P450. H_2O_2, one of the sources of ROS, is produced by a wide variety of enzymes including several oxidases. ROS play important roles in cell signaling, a process termed as redox-sensitive signaling. ROS are also involved in regulation of redox-sensitive transcription factors and other molecules. The defense cells of the body such as leukocytes also produce ROS and protect humans from many invading pathogenic microorganisms. In normal cases, the level of ROS is kept in balance by a potential antioxidant defense system present in the body. The best studied cellular antioxidants are the enzymes superoxide dismutase (SOD), catalase, and glutathione peroxidase (GPx). Some other well-studied antioxidants include the peroxiredoxins and the recently discovered sulfiredoxin. Other enzymes that have antioxidant properties (though this is not their primary role) include paraoxonase, glutathione *S*-transferases (GST), and aldehyde dehydrogenases (ADH).[4]

The OS has been recognized as a major component in the chain of pathogenic events that cause late complications in other diseases such as diabetes mellitus.[5] Several workers have opined that it may also be considered as a major contributor to vascular and neurological complications in patients with diabetes mellitus.[6–12] The involvement of OS in the cytotoxicity and genotoxicity of arsenic-induced acute pancreatitis has been reported.[5,6] It could be due to generation of NO causing DNA damage and activating poly(ADP-ribose)polymerase (PARP), a major cause of damage of islet cells in diabetics.[4,13]

Alcoholic pancreatitis and oxidative stress

As mentioned earlier, OS is caused by a combination of increased production of ROS and impaired antioxidant capacity. ROS consist of a group of highly reactive intermediary oxygen metabolites that are generated in the course of oxygen metabolism. ROS have many important biological functions, such as regulation of redox-sensitive transcription factors, redox-sensitive signal transduction pathways, and direct interaction with various molecules.[14–16] Alcoholic pancreatitis can be caused by various etiological factors. However, about 80% of all cases are related to either bile stones or excessive alcohol consumption. Alcohol-induced damage is, therefore, a relevant model to study the mechanism and pathophysiology of pancreatitis. The role of Nrf2 in preventing ethanol-induced oxidative stress and lipid accumulation has been demonstrated by Wu *et al.*[17] Alcohol's toxicity is mediated through the action of alcohol itself or through its oxidative and nonoxidative metabolism.[18]

It is reported that during CP, the level of ROS and RNS generation is enhanced.[4,6] ROS and RNS are highly unstable, and hence whenever they are in excess, they react with cellular biomolecules (primarily unsaturated fatty acids and proteins). The plasma membrane of any cell or the membrane of subcellular organelle such as mitochondria is mainly composed of PUFAs. These PUFAs are highly susceptible to attack by ROS/RNS.

Environmental factors induction of ROS/RNS production, pancreatic inflammation, and cellular injuries

The PUFAs have been shown to react with ROS, particularly with hydroxyl free radicals. The lipid peroxides thus formed cause significant disintegration of plasma membrane and necrosis of pancreatic cells.[19] The actions of ROS on mitochondrial membrane result in considerable disruption of the mitochondrial membrane potential, leading to cytochrome c release, and subsequent mitochondrial DNA fragmentation and mitochondrial damage.

Alcohol consumption is one of the major environmental factors that significantly influences the clinical condition of CP. Alcohol metabolism results in accumulation of acetaldehyde, acetate, NADH, and fatty acid ethyl ester (FAEE). The oxidative or nonoxidative metabolites of alcohol are responsible for alcohol-mediated toxicity. Changes in cellular and mitochondrial NAD^+ and NADH levels also occur. Excessive alcohol consumption results in pancreatic damage through a number of potential mechanisms.[20] Oxidative metabolism of alcohol catalyzed by alcohol dehydrogenase produces acetaldehyde due to the oxidation of ethanol, which is further oxidized into acetate. The ROS produced by oxidative metabolism of alcohol have been shown to be implicated in inhibition of secretion of pancreatic acinar cells, resulting in the onset of a cascade of cellular events leading to the development of inflammation and other adverse key changes in cellular functions. Another environmental factor, cigarette smoke, is a known inducer of ROS production and acts as one of the major risk elements for the onset and severity of CP.[18,21-23] The nonoxidative metabolism of ethanol, primarily catalyzed by fatty acid ethyl ester synthase (FAEES), is responsible for the production of FAEE.[18,23]

In addition to showing hazardous effect, the ROS/RNS have been shown to act as second messengers in intracellular signaling. They participate in redox-regulated signaling cascades, which involve mitogen-activated protein kinases (MAPKs) as well as nuclear factor-kappa B (NF-κB). The redox-sensitive kinases or transcription factors such as MAPKs, NF-κB, and activator protein-1 (AP-1) are reported to regulate the genes, which are known to be involved in migration and adhesion processes, such as chemotactic cytokines (chemokines) and intercellular adhesion molecules (ICAMs).[24-26] As reviewed by Robles *et al.*, these molecules may be associated with the damage of pancreatic exocrine cells via inflammatory reactions and ROS production.[18] They have found that ROS produced by nonpancreatic cells can also damage the pancreatic cells by mediating an inflammatory response through activation of NF-κB and AP-1.[26,27]

Application of antioxidants in amelioration of ROS/RNS-mediated pancreatic inflammation

Establishment of redox balance is highly complicated, requiring sophisticated regulation of scavenger bioavailability and of ROS/RNS generation. The major cellular ROS scavenger in the pancreas is GSH (a tripeptide consisting of glutamate, cysteine, and glycine). GSH reductase catalyzes the transfer of an electron from NADPH to the oxidized glutathione (GSSG) molecule, to recycle back to its reduced form, that is, GSH.[28,29] GSH is also involved in phase-II metabolism (via conjugation reactions) of some electrophilic xenobiotics, thereby helping their removal from cells.[27,29]

The application of selenium, D-α-tocopherol, vitamins C and E, β-carotene, and methionine have been shown to reduce the pain due to CP as well as increase the activities of antioxidant enzymes such as SOD, erythrocyte catalase, urinary amylase as well as number of CD4+ cells, ratio of CD4+ to CD8 cells, and levels of TNF-α, IL-6, and IL-8.[18,30] In a clinical trial study, the treatment of CP patients with melatonin has been reported to reduce production of OS and cytokine.[31] The application of NAC in CP patients has been reported to neutralize ROS and RNS and to protect key body organs (kidney, lungs, and liver) from OS.[32,33] The roles of oxidative stress and antioxidant therapy in chronic kidney disease and hypertension have been demonstrated by Vaziri.[33] Some antioxidants, mostly naturally occurring compounds, have undergone trials, with some agents conferring beneficial effects in CP patients. However, there is conflicting and insufficient clinical data available to support the routine use in humans.[18,30] For example, an analysis of the effect of the application of these antioxidants in various combinations (selenium, vitamin C, β-carotene, NAC, and α-tocopherol; vitamin C, NAC, and Antoxyl Forte; or vitamins A, E, and C) in CP patients has been reported to show no significant improvement in clinical health of CP patients.[18,34,35]

Phytochemicals as chemoprevention regimen against CP

Many epidemiological studies and evidence from population studies available for over last 10 years have demonstrated that phytochemicals present in fruits and vegetables provide protective effects against CP. The phytochemicals such as rottlerin, ellagic acid, embelin, lycopene, curcumin, isothiocyanates, capsaicin, green tea, and resveratrol have been shown to act to both attenuate the production of ROS in CP and afford prevention of the disease.[36]

Conclusion

OS plays a critical role in both acute and chronic pancreatic inflammation processes. CP-induced generation of ROS/RNS from numerous enzymatic systems, including xanthine oxidase, nitric oxide synthase, CYP2E1, and NADPH oxidase, oxidize a wide range of biomolecules and cause serious tissue damage. The application of different antioxidants in single or in various combinations has resulted mixed responses. Though it offers protection from OS-mediated damage of pancreatic cells, yet in some clinical trials, these antioxidants have not shown significant improvement in the status of clinical health of CP patients. The effect of a combination of different vitamins has also not displayed encouraging response toward recovery or protection in CP patients. Yet, the treatment of CP patients with some phytochemicals has shown an encouraging response in the recovery. A clear understanding of oxidative stress-related pathophysiology of CP may be useful to clinicians toward determination of the most appropriate therapy of the disease by targeting specific mediators during CP and monitoring OS biomarkers during treatment.[17–19]

Multiple choice questions

1 Pancreatitis is characterized by each of the following except:
   a. Abdominal pain
   b. Systemic inflammation
   c. Pancreatic endocrine and exocrine dysfunction
   d. Slow heart rate

2 The main causes of chronic pancreatitis include all except:
   a. Alcohol
   b. Pancreatic duct obstruction by gallstones
   c. Infection
   d. Deep vein thrombosis

3 Chronic alcohol abuse results in pancreatic damage through this mechanism:
   a. Oxidation products of ethanol including acetaldehyde and acetate
   b. Fatty acid ethyl esters catalyzed by fatty acid ethyl ester synthase
   c. Both a and b
   d. Neither a nor b

References

1 Malorni, W., Campesi, I., Straface, E. *et al.* (2007) Redox features of the cell: a gender perspective. *Antioxidant Redox Signal*, 9, 1779–1801.

2 Winterbourn, C.C., Bonham, M.J., Buss, H. *et al.* (2003) Elevated protein carbonyls as plasma markers of oxidative stress in acute pancreatitis. *Pancreatology*, 3, 375–382.

3 Cavestro, G.M., Comparato, G., Nouvenne, A. *et al.* (2005) Genetics of chronic pancreatitis. *Journal of Pancreas*, 13, 53–59.

4 Sharma, B., Singh, S. & Siddiqi, N.J. (2014) Biomedical implications of heavy metals induced imbalances in redox systems. *Biomedical Research International*, 2014, 640–754.

5 Rin, K., Kawaguchi, K., Yamanaka, K. *et al.* (1995) DNA-strand breaks induced by dimethylarsinic acid, a metabolite of inorganic arsenics, are strongly enhanced by superoxide anion radicals. *Biological and Pharmaceutical Bulletin*, 18(1), 45–48.

6 Yamanaka, K., Takabayashi, F., Mizoi, M. *et al.* (2001) Oral exposure of dimethylarsinic acid, a main metabolite of inorganic arsenics, in mice leads to an increase in 8-oxo-2′-deoxyguanosine level, specifically in the target organs for arsenic carcinogenesis. *Biochemical and Biophysical Research Communications*, 287(1), 66–70.

7 Kolb, H. & Kolb-Bachofen, V. (1992) Type I insulin dependent diabetes mellitus and nitric oxide. *Diabetologia*, 35(8), 796–797.

8 Baynes, J.W. (1996) Role of oxidative stress in development of complications in diabetes. *Diabetes*, 40(4), 405–412.

9 Giugliano, D., Ceriello, A. & Paolisso, G. (1996) Oxidative stress and diabetic vascular complications. *Diabetes Care*, 19(3), 257–267.

10 van Dam, P.S., van Asbeck, B.S., Erkelens, D.W. *et al.* (1995) The role of oxidative stress in neuropathy and other diabetic complications. *Diabetes/Metabolism Reviews*, 11(3), 181–192.

11 Parthiban, A., Vijayalingam, S., Shanmugasundaram, K.R. *et al.* (1995) Oxidative stress and the development of diabetic complications – antioxidants and lipid peroxidation in erythrocytes and cell membrane. *Cell Biology International*, 19(12), 987–993.

12 Williamson, J.R., Chang, K., Frangos, M. *et al.* (1993) Hyperglycemic pseudohypoxia and diabetic complications. *Diabetes*, 42(6), 801–813.

13 Nourooz-Zadeh, J., Rahimi, A., Tajaddini-Sarmadi, J. *et al.* (1997) Relationships between plasma measures of oxidative stress and metabolic control in NIDDM. *Diabetologia*, 40(6), 647–653.

14 Apte, M.V., Pirola, R.C. & Wilson, J.S. (2010) Mechanisms of alcoholic pancreatitis. *Journal Gastroenterology and Hepatology*, 25, 1816–1826.

15 Ji, C. (2012) Mechanisms of alcohol-induced endoplasmic reticulum stress and organ injuries. *Biochemistry Research International*, 2012, 216450.

16 Cote, G.A. (2011) Alcohol and smoking as risk factors in an epidemiology study of patients with chronic pancreatitis. Clinical gastroenterology and hepatology: the official clinical practice. *Journal of the American Gastroenterological Association*, 9, 266–273.

17 Wu, K.C., Liu, J. & Klaassen, C.D. (2012) Role of Nrf2 in preventing ethanol-induced oxidative stress and lipid accumulation. *Toxicology and Applied Pharmacology*, 262, 321–329.

18 Robles, L., Vaziri, N.D. & Ichii, H. (2013) Role of oxidative stress in the pathogenesis of pancreatitis: effect of antioxidant therapy. *Pancreatitic Disorders and Therapy*, 3(1), 112.

19 Robertson, R.P. & Harmon, J.S. (2006) Diabetes, glucose toxicity, and oxidative stress: a case of double jeopardy for the pancreatic islet beta cell. *Free Radicals Biology and Medicine*, 41, 177–184.

20 Gukovskaya, A.S., Mouria, M., Gukovsky, I. *et al.* (2002) Ethanol metabolism and transcription factor activation in pancreatic acinar cells in rats. *Gastroenterology*, 122, 106–118.

21 Pandol, S.J., Apte, M.V., Wilson, J.S. *et al.* (2012) The burning question: why is smoking a risk factor for pancreatic cancer? *Pancreatology*, 12(4), 344–349.

22 Grassi, D., Desideri, G., Ferri, L. *et al.* (2010) Oxidative stress and endothelial dysfunction: say NO to cigarette smoking. *Current Pharmaceutical Design*, 16(23), 2539–2550.

23 Law, R., Parsi, M., Lopez, R. *et al.* (2010) Cigarette smoking is independently associated with chronic pancreatitis. *Pancreatology*, 10, 54–59.

24 Robles, L., Vaziri, N.D. & Ichii, H. (2013) Role of oxidative stress in the pathogenesis of pancreatitis: effect of antioxidant therapy. *Pancreatic Disorders and Therapy*, 3(1), 112.

25 Conde de la Rosa, L., Schoemaker, M.H., Vrenken, T.E. *et al.* (2005) Superoxide anions and hydrogen peroxide induce hepatocyte death by different mechanisms: involvement of JNK and ERK MAP kinases. *Journal of Hepatology*, 12, 12.

26 Chan, Y.C. & Leung, P.S. (2007) Angiotensin II type 1 receptor-dependent nuclear factor-kappa B activation-mediated proinflammatory actions in a rat model of obstructive acute pancreatitis. *Journal of Pharmacology and Experimental Therapy*, 323, 10–18.

27 Surh, Y.J., Kundu, J.K., Na, H.K. *et al.* (2005) Redox-sensitive transcription factors as prime targets for chemoprevention with anti-inflammatory and antioxidative phytochemicals. *Journal of Nutritions*, 135(12 Suppl), 2993S–3001S.

28 Leung, P.S. & Chan, Y.C. (2009) Role of oxidative stress in pancreatic inflammation. *Antioxidants and Redox Signaling*, 11(1), 135–165.

29 Wang, W. & Ballatori, N. (1998) Endogenous glutathione conjugates: occurrence and biological functions. *Pharmacology Reviews*, 50(3), 335–352.

30 Du, W.D., Yuan, Z.R., Sun, J. *et al.* (2003) Therapeutic efficacy of high dose vitamin-C on acute Pancreatitis and its potential mechanisms. *World Journal of Gastroenterology*, 9, 2565–2569.

31 Rodriguez, C., Mayo, J.C., Sainz, R.M. *et al.* (2004) Regulation of antioxidant enzymes: a significant role for melatonin. *Journal of Pineal Research*, 36, 1–9.

32 Milewski, J., Rydzewska, G., Degewska, M. *et al.* (2006) N-acetylcysteine does not prevent post-endoscopic retrograde cholangiopancreatographyhyperamylasemia and acutepancreatitis. *World Journal of Gastroenterology*, 12, 3751–3755.

33 Vaziri, N.D. (2004) Roles of oxidative stress and antioxidant therapy in chronic kidney disease and hypertension. *Current Opinion on Nephrology and Hypertension*, 13, 93–99.

34 Sateesh, J. (2009) Effect of antioxidant therapy on hospital stay and complications in patients with early acute pancreatitis: a randomized controlled trial. *Tropical Gastroenterology*, 30, 201–206.

35 Bansal, D. (2011) Safety and efficacy of vitamin-based antioxidant therapy in patients with severe acute Pancreatitis: a randomized controlled trial. *Saudi Journal of Gastroenterology*, 17, 174–119.

36 Breddy, S.R. & Srivastava, S.K. (2013) Pancreatic cancer chemoprevention by phytochemicals. *Cancer Letters*, 334(1), 86–94.

CHAPTER 22

Wound healing and hyperbaric oxygen therapy physiology: oxidative damage and antioxidant imbalance

James R. Wilcox[1] and D. Scott Covington[2]

[1] *Clinical Research & Physician Education, Serena Group, Cambridge, MA 02140, USA*
[2] *Medical Affairs, Healogics, Inc, Jacksonville, FL 32256, USA*

THEMATIC SUMMARY BOX

At the end of this chapter, students should be able to:

• Describe how hypoxia impedes wound healing

• Differentiate between acute and chronic inflammation

• Describe pathways leading to apoptosis, necrosis, cell death and disease

• Explain how HBO therapy contributes to wound healing

• Describe how oxygen is transported during HBO therapy

• Explain how an imbalance in the micro-wound environment increases MMP production

---

Introduction

One of the basic pathways to nonhealing of wounds is the interplay between tissue hypoperfusion, resulting hypoxia, and infection. Chronic hypoxia both within the wound and periwound environment impedes wound healing by numerous mechanisms that act concurrently. One of the challenges of advanced wound care is identifying the extent to which local hypoxia contributes to the abnormal healing, and then correcting that hypoxia to the extent possible. During the past 40 years, a large body of research and clinical evidence has shown that intermittent oxygenation of hypoperfused tissue, which can only be achieved by exposure to hyperbaric oxygen (HBO), mitigates many of these impediments and sets in motion a cascade of responses that contributes to wound healing.[1]

Oxidative Stress and Antioxidant Protection: The Science of Free Radical Biology and Disease, First Edition.
Edited by Donald Armstrong and Robert D. Stratton.
© 2016 John Wiley & Sons, Inc. Published 2016 by John Wiley & Sons, Inc.

Adequate molecular oxygen is required for a wide range of biosynthetic processes essential to normal healing. Molecular oxygen is required for hydroxylation of proline during collagen synthesis and cross linking as well as for the production of reactive oxygen species (ROS) during the respiratory burst occurring within leukocytes that phagocytizes bacteria. While short-term hypoxia is one stimulus for angiogenesis in wound healing, adequate local oxygen levels are required to sustain an effective angiogenic response and for the reconstruction of the dermal matrix. Recent research has shown that oxygen also plays an important role in cell signaling events necessary for tissue repair, which further explains the fragile dynamic between oxygen availability and increased demands for oxygen during wound healing.[2]

Hyperbaric oxygen environment

Hyperbaric oxygenation of tissue is achieved when a patient breathes 100% oxygen in an environment of elevated atmospheric pressure typically in the range of 2.0–3.0 atmospheres absolute (ATA). This can occur in a monoplace (single patient) chamber typically compressed with 100% oxygen or less frequently in a multiplace (multiple patient) chamber typically compressed with air with the patient breathing 100% oxygen through a specially designed hood or mask. Both can increase PO2 values in excess of 1700 mm Hg (Figures 22.1 and 22.2).[3,4]

Oxygen is transported by the blood in two different ways: (i) chemically bound to hemoglobin in erythrocytes and (ii) physically dissolved in plasma according to Henry's law, namely, that the gas tension of oxygen in blood and tissue will increase as the partial pressure of oxygen increases in the alveoli. This will lead to increased oxygen availability through increased oxygen transport dissolved in the plasma and ultimately in the tissues.

Hemoglobin transport of oxygen is limited by chemical binding whereas the hemoglobin concentration of oxygen dissolved in plasma is only limited by the partial pressure of oxygen in the alveoli, which is determined by the partial pressure

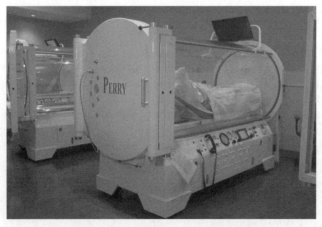

Figure 22.1 Monoplace chamber using compressed 100% oxygen.

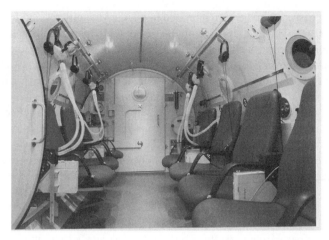

Figure 22.2 Multiplace chamber using compressed air and 100% oxygen hood. *(See color plate section for the color representation of this figure.)*

of oxygen in the inspired gas. At three ATA, the amount of oxygen dissolved in plasma can theoretically reach ≈2240 mm Hg, an amount large enough to maintain life in the absence of hemoglobin. This was demonstrated experimentally in 1960 by Boerema who, by exchange transfusion with Ringer's lactate, exsanguinated the blood from young pigs while inside a hyperbaric chamber. In the absence of hemoglobin, the pigs survived only on the oxygen dissolved in their plasma.[5]

When the body is at rest, it normally consumes about 6 ml of oxygen per 100 ml of blood, but of this amount only 0.3 ml is transported in the plasma. When the pressure is raised to 2 ATA of pure oxygen, the plasma oxygen level is raised to 4.4 ml. Thus, oxygen saturation of the tissue is considerably enhanced by the use of HBO therapy. At 3 ATA approximately 6.4 volumes percent of oxygen are physically dissolved in plasma, which is sufficient to sustain life even in the absence of hemoglobin.

When the amount of dissolved oxygen in plasma is increased so does the diffusion distance of oxygen from the capillaries. The oxygen diffusion distance increases approximately fourfold on the arterial end (64–247 μm) of the capillary and doubles that distance at the venous end (36–64 μm) with the increased PaO_2 from breathing 100% oxygen at 3 ATA compared to air at 1 ATA.[6]

Hyperbaric oxygen therapy (HBOT) provides additional oxygen to the hypoxic tissue and supports tissue healing through a variety of mechanisms. The immediate effects of HBOT occurring during treatments improve wound metabolism in the setting of acute and chronic hypoxia. These relatively short lived effects that contribute to wound healing include the following: (i) Improved local tissue oxygenation, leading to improved cellular energy metabolism. (ii) Increased collagen and other extracellular matrix (ECM) protein deposition and epithelization. (iii) Decreased local tissue edema due to vasoconstriction of vessels in nonischemic tissues. (iv) Improved leukocyte bacterial killing and suppressed exotoxin production. (v) Increased effectiveness of antibiotics that require oxygen for active transport across microbial cell membranes. These effects, while important, would not by themselves account for the degree of improvement in wound healing seen in most ulcers treated with HBOT.

Over the past 15 years, research has led to a somewhat different understanding of the role of HBOT, which has focused on the role of HBOT plays in altering the balance of ROS and reactive nitrogen species (RNS) within the wound, fundamentally altering the wound environment and its healing response. In this context, HBOT also must be thought of as providing oxygen as a cell-signaling agent.[7,8] Achieving these beneficial effects requires a minimum tissue oxygen tension of approximately 200 mm Hg which can only be achieved with use of HBO therapy. These effects include: (i) Enhanced growth factor and growth factor receptor site production, especially platelet derived growth factor (PDGF) and vascular endothelial growth factor (VEGF), which are helpful in wound matrix development and angiogenesis; (ii) Altered leukocyte β-integrin receptor sensitivity, helpful in mitigating ischemia reperfusion injuries which occurs in many chronic and acute wounds; (iii) Reducing inflammation and apoptosis which occur in many acute ischemic wound models; (iv) Activating stem cell metabolism and releasing them into circulation from bone marrow reservoirs. These physio-pharmacological effects have been observed both *in vitro* and *in vivo*.[9–12]

HBOT is complicated by the dual nature of oxygen in being essential to life, and at the same time, being toxic in excess. Hyperoxygenation enhances ROS and RNS mediated pathways, which can be expressed as both positive and negative for wound healing. While the body's defenses against excessive ROS are highly effective, the body can be overcome with prolonged exposures at high oxygen tensions, as occurs with HBOT. This leads to two of the greatest impediments to the application of HBOT, which are neurologic and pulmonary manifestations of oxygen toxicity. Oxygen tolerance limits are defined in order to avoid these manifestations. To negate these negative effects, the application of HBOT is limited to 3 ATA of 100% oxygen. Acute and chronic wounds are generally treated at 2.0–2.5 ATA, for at these levels daily exposures do not produce pulmonary symptoms.[3]

A normal by-product of cellular metabolism, ROS rate of production is increased in the presence of increased oxygen availability, which occurs during an HBOT exposure.[13] Not only does this primary generation of ROS occur, but also secondary generation of reactive oxygen intermediate molecules are produced, adding to oxidant injury, such as in neutrophil oxidant production.

Even so, the role of HBOT in free radical mediated tissue injury is not well defined. Researchers have demonstrated that HBOT enhances antioxidative defense mechanisms in some animal studies. Hyperoxia causes an increase in nitric oxide (NO) synthesis as part of a response to oxidative stress. Mechanisms for neuronal nitric oxide synthase (nNOS) activation include augmentation in association with a chaperone protein called heat shock protein 90 (Hsp90) and intracellular entry of calcium.[14]

Superoxide dismutase (SOD) is an important enzyme found in human cells that inactivates superoxides, the most common free radicals in the body, responsible for destruction of cells. HBOT stimulates SOD production thereby helping the body to rid itself of the byproducts of inflammation and damaging free radicals.

Wound environment

Acute wound healing is an orderly and efficient process which is characterized by four distinct but overlapping phases: hemostasis, inflammation, proliferation and remodeling. In nonhealing chronic wounds, this efficient and orderly process has been lost due to the absence of the hemostasis phase, causing these ulcers to become locked into a state of chronic inflammation that is self-sustaining and characterized by abundant neutrophil infiltration with associated ROS and destructive enzymes. Healing can only occur after this prolonged inflammatory phase is broken and the wound's micro-wound environment is once again in balance, thus allowing the wound to proceed through the remaining stages (Figure 22.3).[15,16]

All chronic wounds are similar in that each is characterized by one or more persistent inflammatory stimuli: repetitive trauma, ischemia, or low-grade bacterial contamination. Once the skin barrier is broken and bacterial colonization occurs, inflammatory molecules from bacteria, such as endotoxin, platelet products, such as transforming growth factor-β (TGF-β), or fragments of ECM molecules, such as fibronectin, stimulate proliferation of inflammatory cells (neutrophils and microphages) to enter the wound. These activated inflammatory cells secrete proinflammatory cytokines such as tumor necrosis factor-α (TNF-α) and interleukin-1β (IL-1β).[16]

These proinflammatory cytokines stimulate synthesize of matrix metalloproteinases (MMPs) and suppress tissue inhibitors of matrix metalloproteinase (TIMP).[16] The chronic wound environment has been proven to contain elevated protease (MMP) levels and decreased levels of protease inhibitors (TIMP).[16] MMPs play an important role in all phases of wound healing by promoting cell migration, breaking down extracellular matrix, and remodeling. An imbalance in the microwound environment with increasing numbers of MMP's and decrease TIMP's levels is associated with the degradation of collagen and reduction of growth factors and growth factor receptor sites as well as other vital components of the extracellular matrix. Once growth factors are degraded, communication between the various cells participating in the wound healing process stops and wound healing is delayed. As the inflammatory cycle is prolonged, it amplifies the proinflammatory cytokine cascade leading to a wound environment which is absent of DNA synthesis. This is related to the low mitotic activity, excessive levels of inflammatory cytokines, high levels of proteases, and excessive ROS found in chronic wound fluid which in turn results in senescent or mitotically incompetent cells.[17–19]

Once the inflammatory cycle is prolonged, it creates a "vicious cycle" which is characteristic of all chronic wounds.[16] The self-sustaining nature of this vicious cycle has far reaching effects. As the wound shifts from hypoxic to ischemic over time, the lack of available oxygen for metabolism leads to anaerobic metabolism which reduces concentrations of adenosine triphosphate (ATP) resulting in metabolic acidosis. The reduction of ATP facilitates increased lactate with decreasing pH which excites nociceptors and produces activation of pH-sensitive ion channels, resulting in pain. However, if there is sufficient oxygen for aerobic metabolism, then the by-product acid is metabolized. As a result, pain is alleviated or disappears. This accounts for higher levels of associated pain in chronic wounds when compared to acute wounds.[20–22]

Free radicals are known to cause cell damage and to function as inhibitory factors in the healing process.[20] The production of ROS within a chronic wound can originate from several potential sources. During healing, various inflammatory cells, such as neutrophils, macrophages, endothelial cells, fibroblasts, and

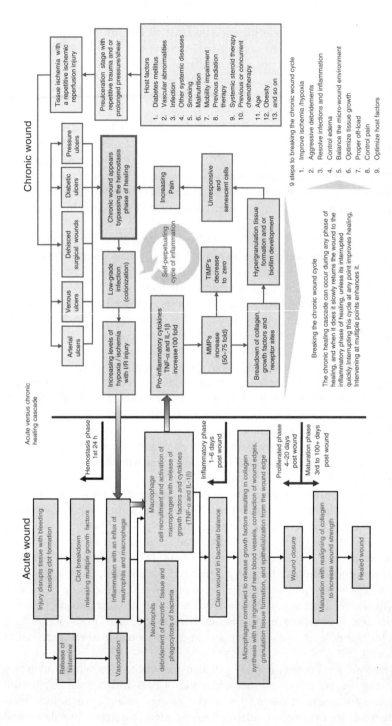

Figure 22.3 Acute wound healing is an orderly process. Interruption of this process leads to prolonged inflammation and a vicious cycle of further injury and continued inflammation. Breaking this chronic wound cycle requires diligent intervention at multiple points.

in particular senescent fibroblasts, which are prominent in chronic wounds, are capable of producing superoxide.[15] In chronic wounds, activated neutrophils and macrophages produce extremely large amounts of superoxide and its derivatives via the inducible phagocytic isoform of NADPH oxidases.[21] When polymorphonuclear neutrophils (PMNs) are recruited and activated at the wound site they consume an increased amount of oxygen, which is converted into ROS, in a process known as a "respiratory" or "oxidative" burst. This burst requires the consumption of large amounts of molecular oxygen, increasing oxygen consumption by at least 50% resulting in the generation of superoxide anions.[22] Most of the superoxide anions formed are converted into hydroxyl radicals via the iron-catalyzed Fenton reaction, creating a second source of ROS. Iron is released from hemoglobin by degraded erythrocytes, ferritin, and hemosiderin.[23] An example of this is venous ulcerations which have excessive iron deposition in the skin, which is used in a Fenton reaction to produce excessive amounts of ROS, in this case hydroxyl radicals. These highly toxic hydroxyl free radicals also enhance the synthesis and activation of even more matrix-degrading metalloproteinase.[24] The presence of excessive reactive oxygen metabolites are not only highly toxic to surrounding tissue, they also increase MMPs while decreasing TIMP levels, creating a highly aggressive chronic wound environment that inhibits healing.

Conclusion

Quality care of the wounded patient requires a detailed understanding of both the normal and pathologic healing processes. As noted, inflammation and oxidative stress are critical components of healing success, and when either under- or overexpressed, healing fails. Students of the healing sciences and providers alike must be mindful of these "two-edged swords" in their understanding of the patient's recalcitrant clinical response. Finally, insight into the importance of oxidative balance will more effectively guide the provider in the selection of appropriate advanced therapies in this era of cost-sensitive, personalized care planning.

Multiple choice questions

1 When polymorphonuclear neutrophils are activated at a wound site they increase oxygen consumption by at least what percentage?
   a. 25%
   b. 50%
   c. 75%
   d. 100%

2 What cells are capable of produce superoxides?
   a. Neutrophils
   b. Macrophages
   c. Fibroblasts
   d. All the above

3 At 3 ATA the amount of oxygen dissolved in plasma can support life in the absence of hemoglobin.
   a. True
   b. False

References

1 Ishii, Y., Miyanaga, Y., Shimojo, H. *et al.* (1999) Effects of hyperbaric oxygen on procollagen messenger RNA levels and collagen synthesis in the healing of rat tendon laceration. *Tissue Engineering*, 5(3), 279–286.

2 Sen, C.K. (2009) Wound healing essentials: let there be oxygen. *Wound Repair and Regeneration*, 7(1), 1–18.

3 Undersea and Hyperbaric Medical Society Hyperbaric Oxygen Committee. (2014) *Hyperbaric oxygen therapy indications: the Hyperbaric Oxygen Therapy Committee report* 13th edition, Weaver, L.K. ed., Best Publishing Company, North Palm Beach, FL, USA, pp 241–6.

4 Warriner, R. (2011) Physiology of hyperbaric oxygen treatment. In: Larson-Lohr, V., Josefsen, L. & Wilcox, J. (eds), *Hyperbaric Nursing and Wound Care*. Best Publishing Company, Palm Beach Gardens, FL, pp. 11–31.

5 Boerema, I., Meyne, N.G., Brummelkamp, W.H. *et al.* (1960) Life without blood. *Nederlands tijdschrift voor geneeskunde*, 104, 949–954.

6 Krogh, A. (1919) The number and distribution of capillaries in muscle with calculations of the oxygen pressure head necessary for supplying tissue. *Journal of Physiology*, 52(6), 409–415.

7 Thom, S.R. (2011) Hyperbaric oxygen: its mechanisms and efficacy. *Plastic and Reconstructive Surgery*, 127(Suppl 1), 131S–141S.

8 Thom, S.R. (2010) The impact of hyperbaric oxygen on cutaneous wound repair. In: Sen, C.K. (ed), *Advances in Wound Care*. Vol. 1. Mary Ann Liebert, Inc, Publishers, New Rochelle, NY, pp. 321–327.

9 Thom, S.R., Bhopale, V.M., Velazquez, O.C. *et al.* (2006) Stem cell mobilization by hyperbaric oxygen. *American Journal of Physiology – Heart and Circulatory Physiology*, 290(4), H1378–H1386.

10 Gallagher, K.A., Liu, Z.J., Xiao, M. *et al.* (2007) Diabetic impairments in NO-mediated endothelial progenitor cell mobilization and homing are reversed by hyperoxia and SDF-1 alpha. *Journal of Clinical Investigation*, 117(5), 1249–1259.

11 Liu, Z.J. & Velazquez, O.C. (2008) Hyperoxia, endothelial progenitor cell mobilization, and diabetic wound healing. *Antioxidant & Redox Signaling*, 10(11), 1869–1882.

12 Milovanova, T.N., Bhopale, V.M., Sorokina, E.M. *et al.* (2009) Hyperbaric oxygen stimulates vasculogenic stem cell growth and differentiation *in vivo*. *Journal of Applied Physiology*, 106(2), 711–728.

13 McCord, J.M. & Firdovich, I. (1978) The biology and pathology of oxygen radicals. *Annals of Internal Medicine*, 89(1), 122–127.

14 Thom, S.R., Bhopale, V., Fisher, D. *et al.* (2002) Stimulation of nitric oxide synthase in cerebral cortex due to elevated partial pressure of oxygen: an oxidative stress response. *Journal of Neurobiology*, 51(2), 85–100.

15 Eming, S.A., Krieg, T. & Davidson, J.M. (2007) Inflammation in wound repair: molecular and cellular mechanisms. *Journal for Investigative Dermatology*, 127(3), 514–525.

16 Mast, B.E. & Schultz, G.S. (1996) Interactions of cytokines, growth factors, and proteases in acute and chronic wound. *Wound Repair and Regeneration*, 4(4), 411–420.

17 Kim, T.J., Freml, L., Park, S.S. *et al.* (2007) Lactate concentrations in incisions indicate ischemic-like conditions may contribute to postoperative pain. *The Journal of Pain*, 8(1), 59–66.

18 Naves, L.A. & McCleskey, E.W. (2005) An acid-sensing ion channel that detects ischemic pain. *Brazilian Journal of Medical and Biological Research*, 38(11), 1561–1569.

19 Koban, M., Leis, S., Schultze-Mosgau, S. *et al.* (2003) Tissue hypoxia in complex regional pain syndrome. *Pain*, 104(1), 149–157.

20 Latha, B. & Babu, M. (2001) The involvement of free radicals in burn injury: a review. *Burns*, 27(4), 309–317.

21 Babior, B.M. (1978) Oxygen dependent microbial killing by phagocytes. *New England Journal of Medicine*, 298(13), 659–668.

22 Rabkin, J.M. & Hunt, T.K. (1988) Infections and oxygen. In: Davis, J.C. & Hunt, T.K. (eds), *Problem Wounds: The Role of Oxygen*. Elsevier.

23 Thomas, C.E., Morehouse, L.A. & Aust, S.D. (1985) Ferritin and superoxide-dependent lipid peroxidation. *Journal of Biological Chemistry*, 260(6), 3275–3280.

24 Saarialho-Kere, U.K. (1998) Patterns of matrix metalloproteinase and TIMP expression in chronic ulcers. *Archives of Dermatological Research*, 290(1), S47–S54.

CHAPTER 23

Radiobiology and radiotherapy

Justin Wray and Judith Lightsey

Department of Radiation Oncology, University of Florida College of Medicine, Gainesville, FL, USA

THEMATIC SUMMARY BOX

At the end of this chapter, students should be able to:

- Describe the formation of radiation-induced DNA damage

- Name the two types of radiation-induced DNA damage

- Name the 4 R's of radiobiology

- List the types of DNA damage that occur following radiation and the repair pathway that repairs the damage

- List the phases of the cell cycle and the cell cycle checkpoints

- List the strategies to overcome tumor radioresistance

- Describe the advantages and disadvantages of chemical radiosensitizers

---

A brief history of radiation therapy

To begin the discussion of the role of free radicals in radiation, it is important to have a framework of what radiation and radiation therapy is. Ionizing radiation was first described by Wilhelm Conrad Rontgen in December of 1895, and he performed the first public X-ray of a hand shortly thereafter. Following this discovery, there was a rush of interest and several famous researchers including Antoine Henry Becquerel, Leopold Freund, and Pierre and Marie Curie began using X-rays and gamma rays therapeutically for both benign and malignant tumors.[1] Unfortunately, there was much skin toxicity with low-energy X-rays and gamma-ray production from elements such as radium when trying to treat deep tumors, so their utility was reserved primarily for superficial tumors and diagnostic imaging. Clinically and dosimetrically, X-rays are measured in Grays (Gy). In 1975, the measurement was officially named after L.H. Gray, who died in 1965. One Gray is specifically the amount of X-rays to deposit one joule of energy per kilogram of water.

The modern era of radiation therapy began in the 1960s when the first high-energy X-ray machines began production. These linear accelerators, or Linacs,

---

Oxidative Stress and Antioxidant Protection: The Science of Free Radical Biology and Disease, First Edition.
Edited by Donald Armstrong and Robert D. Stratton.
© 2016 John Wiley & Sons, Inc. Published 2016 by John Wiley & Sons, Inc.

are capable of producing megavoltage X-rays, which allow the modern radiation oncologist to treat deep seated tumors easily without damaging the overlying skin to a prohibitive degree.[1] Compounding on these advances, computer planning systems and the advent of computed tomography scanning were introduced in the 1970s and 1980s. These advances using X-ray technology led to the ability to plan treatment based on 3D anatomy and monitor treatment using image guidance systems. In the current state of the art, X-ray therapy is used in some form or manner in the treatment of most cancers. In coordination with these technological advances, much time and effort has been invested into describing the biological effect of the delivery of X-ray therapy.

Mechanism of radiation

Radiation damages tissue in multiple ways, but the primary mechanism of cell death, carcinogenesis, and mutations is DNA damage. X-rays cause DNA damage by either direct or indirect mechanisms. Both of these mechanisms are based on the physics of radiation and molecular excitation, which leads to ionization when an orbital electron is ejected from an atom or molecule. If this occurs to the DNA itself, it is direct damage. In treatment with X-rays, this mechanism accounts for approximately one-third of the damage that takes place. Of note, in other forms of radiation including carbon ion, α particle, and neutron therapy, this is the dominant mechanism of DNA damage. The other two-thirds of the DNA damage after ionizing radiation that takes place is of more importance to this text.

Not only do X-rays damage DNA directly, they also interact with all of the other molecules within a cell and tissue. The prominent molecule in any functioning tissue is water (H_2O). It is well known that when an outer electron of H_2O is excited and ejected, it forms a primary ionized free radical, $H_2O^{\bullet+}$. This ion has a short half-life of approximately 10^{-10} s before it reacts with another nearby water molecule to form a hydroxyl radical ($OH^{\bullet}$) in the following equation:

$$H_2O^{\bullet+} + H_2O \rightarrow H_3O^+ + OH^{\bullet}$$

Once this occurs, the hydroxyl radical is highly reactive and has a longer half-life with the ability to diffuse over a short distance. If this occurs near a DNA strand it will induce damage, accounting for the other two-thirds of the total damage above. This DNA damage is the driving force when using radiation as a therapy in cancer treatment and as described earlier, the hydroxyl free radical is responsible for the majority of the damage imposed by X-rays.[2]

Biology in cancer

DNA damage after X-ray treatment is responsible for the majority of the effects including cell death, carcinogenesis, and mutations. It is a conundrum as to how then can radiation therapy be used to treat tumors and not cause the same effect in the adjacent normal tissue. The answer to this was discovered in the 1920s and 1930s in France.

At the time, radiation was being utilized to sterilize ram's testicles. If the treatment was given in a single dose, the ram had scrotal skin toxicity that led to increased complications. Although the treatment was given in smaller doses over a period of weeks, the side effects were minimal and tolerable. This was then taken into cancer therapy and shown to be effective. The division of a dose of radiation into many smaller doses given over a longer time is termed fractionation. This is the mainstay of radiation treatment in the modern era (Figure 23.1).

4 R's: repair, redistribution, reoxygenation, and repopulation

Much research has been performed to further understand this principle since the 1930s. Out of decades of research, an overarching theory has emerged to explain how fractionation is possible. This is commonly termed the 4 R's of radiobiology or fractionation. They are repair, redistribution, repopulation, and reoxygenation.

Repair is the first R. This term refers to the ability of normal or tumor tissue to repair the damage that the X-ray causes directly to the DNA molecule. As earlier, this damage is induced by both direct ionization of the DNA strand and that caused by the hydroxyl free radical generated by the ionization of adjacent water molecules. As a refresher, DNA is composed of two complimentary strands, which are opposite of each other, and made up of a chain of the nucleotides adenosine, guanine, cytosine, and thymine. When the DNA is damaged, many different types of damage

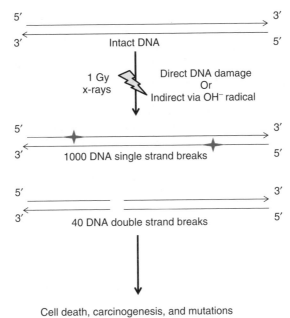

Figure 23.1 Radiation effect on DNA.

will occur, which can be classified into two broad categories: single-strand breaks and double-strand breaks.

Single-strand breaks occur when only nucleotides in one stand of the DNA molecule are damaged. The two main types of damage and the pathways which repair with them are as follows: (i) damage to the nucleotide base repaired by base excision repair; and (ii) the addition of bulky adduct molecules and damage to the nucleoside repaired by nucleotide excision repair. All of these damages may lead to dysfunction when the cell attempts to transcribe the DNA into RNA to make proteins or replicate the DNA before cell division. Furthermore, if unrepaired, these damages lead to permanent mutations in the DNA code.

Double-strand breaks can occur by several mechanisms that will not be described here, but all lead to the severing of the DNA strand. Although this is less common and occurs at a ratio of approximately 1:25 compared to single-strand breaks, double-strand breaks are the most lethal DNA damage to the cell and account for the majority of cell death. The two major pathways involved in repairing these lesions are homologous recombination and nonhomologous end joining.

In the cancer cell, it is evolutionarily advantageous to lose some of this DNA repair ability. Cancer cells divide at a high rate, and when they lose a certain level of repair capability, it allows for increased natural selection, which causes mutations that drive further growth and metastatic potential. The adjacent normal tissue that surrounds the cancer cells maintains the ability to accurately repair its DNA. This discrepancy accounts for the ability of X-rays to damage both normal tissue and cancer DNA, but only the cancer cell is killed during fractionation. Giving a smaller dose of X-rays over multiple fractions allows the normal tissue to have time to utilize repair pathways while the cancer cell has lost much of this response. Thus, causing cancer cell death while maintaining normal tissue integrity.

The second R is redistribution. Just as it is advantageous for cancer cells to lose the ability to accurately repair their DNA, it is also advantageous to gain the ability to divide quickly. During cell division, there are multiple phases including Gap 1 (G1), Synthesis (S), Gap 2 (G2), and Mitosis (M). Evolutionarily, the normal cell has multiple checkpoints that stop the cell from moving through and dividing if there is DNA damage. The primary checkpoints are at the G1/S border, in S, at the G2/M border, and during chromosome separation (decatenation). Each one of these checkpoints has specific pathways that will not be described here. In most cancer cells, there are mutations in one or more of these pathways that leads to dysfunction. This allows the cell to progress through the cell cycle at a higher rate (Figure 23.2).

Each of the cell cycles also have different levels of radiation sensitivity. S is the most resistant and G2 and M are the most sensitive. Since tumor cells have likely gained the ability to continue to divide after radiation therapy, the separation of the X-ray dose into fractions also allows the cancer cells to move into different cell cycles throughout treatment. Thus, the cells that are in radioresistant phases at the time of one treatment will have the opportunity to move into a more radiosensitive cycle during a subsequent treatment. This R is also sometimes referred to as reassortment.

The third R is reoxygenation. As tumors grow, there is a limit to their size based on blood flow. They must gain the ability to recruit blood vessels to continue to gain mass. During this process, the tumor continues to divide; thus, there is a ratio of cell death to tumor growth that is directly related to the amount of blood flow recruitment that

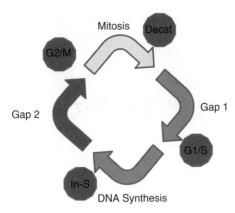

Figure 23.2 Cell cycle and cell cycle checkpoints.

has occurred. In tumors that are dividing rapidly, they will quickly outgrow blood supply there. Many of the cells that die in the center of the tumor succumb to a lack of energy that is driven by a lack of oxygen. The result of this is a tumor that has adequate blood flow around the exterior and minimal blood flow in the center. Thus, the inner core of the tumor often becomes necrotic as the majority of the cells succumbing to hypoxia reside in the center.

As described earlier, X-rays cause the majority of their DNA damage utilizing the indirect effect of hydroxyl radicals. This process is directly dependent on the oxygen concentration in the cell. Hypoxic environments are more radioresistant than well-vascularized environments. During radiation therapy over multiple fractions, the outer, well-oxygenated, region of the tumor is radiosensitive. As multiple fractions are given and the tumor regresses, the blood supply to the inner core increases with decreased demand to the dying outer shell. The inner core then becomes reoxygenated and has an increased level of radiosensitivity that it would not have had if the treatments were given in a single fraction.

This fourth R is repopulation. Repopulation is beneficial to the tumor. It refers to the ability of the tumor to divide between fractions. Under normal treatment circumstances, treatments at every 24 h, there is not enough time for this to play an important role. But, in patients to have a prolonged treatment, break and interfraction time is increased; it may become problematic. It is also an important concept because in certain cancers, such as head and neck squamous cell carcinoma, studies have indicated that these tumors may have accelerated repopulation. They may begin to increase their growth rate as treatment progresses because more radioresistant clones of the tumor are selected for. Clinically, this is the justification for decreased interfraction time or hyperfractionation where more than one X-ray treatment is given daily.

The impact of free radicals

Overall, free radicals play a major role in the production of DNA damage by radiation therapy via the hydroxyl free radical. Clinically, there has been much research and trial into finding both new methods and drugs to improve the therapeutic effect on cancer cells and ways to protect the normal tissue from damage using free radical modulation.

The role of oxygen in clinical radiation therapy

For years, scientists have tried to take advantage of the role oxygen plays in enhancing radiation cell kill. As stated previously, the indirect effect of radiation on living cells requires the presence of oxygen. Tumor cells that are hypoxic are more radioresistant. Several mechanisms have been used in an effort to overcome this radioresistance, including hyperbaric oxygen and the use of high linear energy transfer (LET) radiations, such as neutrons and heavy ions. Still another approach is the use of chemical radiosensitizers. Each of these approaches will be discussed in turn.

Hyperbaric oxygen

Several clinical trials were performed in the mid-1900s to evaluate the efficacy of hyperbaric oxygen in enhancing the therapeutic ratio of radiation therapy. Some of the studies did show an improvement in local control and also an increase in late normal tissue damage. Churchill-Davidson *et al.* demonstrated an advantage of hyperbaric oxygen in improving oxygenation of cells, but their study was nonrandomized.[3] A study by van den Brenk *et al.* showed an advantage for hyperbaric oxygen, but the study was criticized for using a suboptimal fractionation schedule for the conventional arm.[4] The Medical Research Council conducted a large multicenter trial that demonstrated a significant improvement in the control of carcinoma of the cervix.[5] Hyperbaric oxygen never became widely used partly because it is difficult to use and also because chemical radiosensitizers started to be developed allowing for a much simpler way to achieve radiosensitization.

High linear energy transfer radiation

Linear energy transfer is the energy transferred per unit length of a track. Some radiations are densely ionizing, that is, they produce a large number of ionizations over a specified distance. The advantage of this is that high LET radiation is more efficient at causing reproductive death. Thus, for high LET radiation, there is no shoulder on the cell survival curve. The other advantage of high LET radiation is that they are less dependent on the presence of oxygen. Thus, radioresistance due to hypoxia will be less marked for these treatments than for conventional radiations. Examples of high LET radiation include protons, neutrons, and α particles.

Neutrons in radiotherapy

Neutrons were first introduced for use in radiotherapy in the 1930s, but were subsequently abandoned due to severe late effects. They were later reintroduced in the 1950s. Most of the clinical studies did not show an advantage for neutrons over X-rays except in a few sites such as prostate cancer, salivary gland tumors, and soft tissue sarcomas. A study by Catterall and Bewley demonstrated increased local control in head and neck tumors irradiated with fast neutrons.[6] Neutrons have never been widely adopted by the radiotherapy community.

Protons in radiotherapy

In addition to a higher relative biologic effectiveness (RBE), protons have a superior dose distribution compared to X-rays. Initial studies showed advantages for protons in the treatment of choroidal melanoma of the eye, sarcoma of the base of skull, and chordomas. Over the past decade, the use of protons has exploded in the United States with new centers opening every year. Protons are now used to treat a wide variety of tumors. The most common tumors treated with protons are prostate cancer, head and neck cancer, and pediatric malignancies.

Chemical radiosensitizers

Radiosensitizers are agents that make cells more sensitive to radiation. Many compounds have been shown to enhance the effect of radiation. The problem is most of these compounds enhance both tumor cells and normal tissues to a similar degree and thus offer no advantage. Some radiosensitizers selectively sensitize hypoxic cells to radiation. This is a desirable effect since tumor cells are more likely to be hypoxic than normal cells. The best studied of this class of sensitizers are the nitroimidazoles such as misonidazole, etanidazole, and nimorazole. Early studies with misonidazole showed that the agent was selectively toxic to hypoxic cells. Hypoxic cells in the presence of 10 mM of misonidazole have a radiosensitivity approaching that of aerated cells. Unfortunately, results from several randomized clinical trials were not nearly as impressive as the laboratory data. These studies showed either no benefit or only minimal benefit although one trial showed a significant improvement in local–regional control for one subgroup in the study.[7,8] The inability of misonidazole to improve outcomes in clinical trials has been attributed inadequate doses due to dose-limiting toxicity. The drug can cause peripheral neuropathy that can progress to central nervous system toxicity.

Etanidazole (SR-2508) is more hydrophilic than misonidazole (dissolves in water better) and thus does not cross the blood–brain barrier. Therefore, neurotoxicity is not a dose-limiting factor. Controlled clinical trials by the RTOG in the United States and a multicenter trial in Europe showed no benefit for etanidazole when added to conventional therapy.

Another nitroimidazole, nimorazole, is much less toxic than misonidazole. The drug can be administered with each radiation treatment. The Danish Head and Neck Cancer Group (DAHANCA) conducted a phase III trial of nimorazole versus placebo for squamous cell carcinoma of the supraglottic larynx and pharynx. There was a statistically significant improvement in locoregional control. Nimorazole has not been widely used outside of Denmark.

Conclusion

Oxygen plays a crucial role in killing of tumor cells with radiation therapy. Radiation therapy is more effective at cell killing in the presence of oxygen. Hypoxia in tumor cells causes radioresistance. Strategies to overcome tumor hypoxia continue to be developed. Some promising approaches include using drugs that selectively kill hypoxic cells. Examples include mitomycin C and tirapazamine. Ongoing challenges include how to identify patients most likely to benefit from these interventions.

Multiple choice questions

1 The major, about two-thirds, damage to DNA of cells exposed to ionizing radiation comes from:
   a. Direct damage to the nuclear DNA
   b. Direct damage to the mitochondrial DNA
   c. Indirect damage by way of absorption of energy by water
   d. Indirect damage by way of absorption by polyunsaturated fatty acids

2 The four Rs of radiation biology include all but the following:
   a. Repair
   b. Reaction
   c. Redistribution
   d. Reoxygenation
   e. Repopulation

3 During which cell cycle phase are cells most resistant to radiation therapy?
   a. M, mitosis
   b. G1, gap 1
   c. S, synthesis
   d. G2, gap 2

References

1 del Regato, J.A. (1993) *Radiological Oncologists: The Unfolding of a Medical Specialty.* Radiological Centennial, Inc., Reston, VA.
2 Hall, E.J. & Garcia, A.J. (2012). In: Mitchell, C.W. (ed), *Radiobiology for the Radiologist.* Lippincott, Williams and Wilkins, Philadelphia, PA.
3 Churchill-Davidson, I., Sanger, C. & Thomlinson, R.H. (1957) Oxygenation in radiotherapy: II. Clinical application. *British Journal of Radiology*, 30, 406–421.

4 van den Brenk, H.A.S., Madigan, J.P. & Kerr, R.C. (1964) Experience with megavoltage irradiation of advanced malignant disease using high pressure oxygen. In: Boerema, I., Brunnelkamp, W.H. & Meijne, N.G. (eds), *Clinical Application of Hyperbaric Oxygen*. Elsevier, Amsterdam, pp. 144.

5 Watson, E.R., Halnan, K.E., Dische, S. *et al.* (1978) Hyperbaric oxygen and radiotherapy: A Medical Research Council trial in carcinoma of the cervix. *British Journal of Radiology*, 51, 879–887.

6 Catterall, M. & Bewley, D.K. (1979) *Fast Neutrons and the Treatment of Cancer*. Academic Press, London.

7 Dische, S. (1985) Chemical sensitizers for hypoxic cells: a decade of experience in clinical radiotherapy. *Radiotherapy and Oncology*, 3, 97–115.

8 Overgaard, J., Hansen, H.S., Anderson, A.P. *et al.* (1989) Misonidazole combined with split-course radiotherapy in the treatment of invasive carcinoma of the larynx and pharynx: report from the DAHANCA 2 study. *International Journal of Radiation Oncology, Biology, and Physics*, 16(4), 1065–1068.

CHAPTER 24

Chemotherapy-mediated pain and peripheral neuropathy: impact of oxidative stress and inflammation

Hassan A.N. El-Fawal[1,2], Robert Rembisz[1,2], Ryyan Alobaidi[1,3], and Shaker A. Mousa[2]

[1] Neurotoxicology Laboratory, Albany College of Pharmacy and Health Sciences, Albany, NY, USA
[2] The Pharmaceutical Research Institute, Albany College of Pharmacy and Health Sciences, Rensselaer, NY 12144, USA
[3] Pathology Department, King Saud University, Riyadh, KSA

THEMATIC SUMMARY BOX

At the end of this chapter, students should be able to:

- Describe chemotherapy-induced peripheral neuropathy (CIPN)

- Describe classes of chemotherapeutics inducing neuropathy

- Explain pathways and potential mechanisms involved in CIPN

- Explain studies dealing with pain and peripheral neuropathy

- Describe the current status in the management of CIPN

Introduction

Chemotherapy-induced peripheral neuropathy (CIPN) is one of the common side effects of chemotherapeutic agents from the taxane, vinca alkaloid, and platinum classes (Table 24.1).[1-13] It is considered a major dose-limiting effect for many of these agents, which affects the course of efficacious cancer therapy.[14] The incidence of CIPN varies according to the specific cancer and its stage, as well as according to the class and number of chemotherapeutic agents being used. It may reach 7% incidence with single agent and up to 38% incidence with combination regimens.[14,15] Factors such as patient age, dose, duration of therapy, and presence of pre-existing conditions such as diabetes mellitus, may also modify CIPN incidence. The manifestations of CIPN may be reversible or irreversible, and they differ depending on which neurons (autonomic, motor, or sensory) are affected.[16] Initially peripheral sensory fiber activation results in an increased burning sensation, hyperalgesia, and hypersensitivity to touch followed by decreased or absence of touch and pain sensation with

---

Oxidative Stress and Antioxidant Protection: The Science of Free Radical Biology and Disease, First Edition.
Edited by Donald Armstrong and Robert D. Stratton.
© 2016 John Wiley & Sons, Inc. Published 2016 by John Wiley & Sons, Inc.

Table 24.1 Commonly used chemotherapy agents associated with peripheral neuropathy.

Chemotherapy agent class	Sensory symptoms	Motor symptoms	References
Taxanes Paclitaxel (Taxol®) Docetaxel (Taxotere®) Abraxane®	Mild to moderate numbness, tingling, burning/stabbing pain of hands and feet.	Weakness of distal muscles with high cumulative doses of paclitaxel and docetaxel.	1–6
Vinca alkaloids Vincristine (Oncovin®) Vinorelbine (Navelbine®)	Mild to moderate numbness, tingling, burning/stabbing pain of hands and feet.	Weakness of distal muscles, decreased deep tendon reflexes, and foot drop.	1–4
Platinums Cisplatin (Platinol®) Carboplatin (Paraplatin®) Oxaliplatin (Eloxatin®)	Mild to moderate numbness and tingling of hands and feet can occur after prolonged use (4–6 months).	Weakness is rare but can occur with high doses of cisplatin and oxaliplatin.	9–12

the progression of neuropathy to involve the neuronal cell bodies in the dorsal root ganglia.[16] Additionally cancer pain by itself is recognized as a health care problem.

This chapter reviews the classes of chemotherapy associated with peripheral neuropathy, mechanisms involved, and potential strategies for pain management and prevention of peripheral neuropathy, without compromising anticancer efficacy of chemotherapy.

History of chemotherapy-induced peripheral neuropathy

The antineoplastic classes of chemotherapeutic agents, the taxanes and platinum compounds, are well established as effective chemotherapeutics. These agents are believed to be effective against testicular, ovarian, breast, colorectal, and lung cancers. As a result since January 2012 there are over 13 million documented cancer survivors in the USA alone.[17] The goal of these efficacious medications is to improve the survival rate and quality of life of the cancer patient, while remaining safe treatment options.

First identified in the 1960s, the taxanes – natural products isolated from the bark of the rare pacific yew, *Taxus brevifolia* – include paclitaxel, docetaxel, and cabazitaxel. Their mode of action primarily involves enhancing tubulin dimer activity, leading to increased microtubule assembly, and stabilizing existing microtubules by inhibiting their depolymerization.[18] This interferes with the late G2 mitotic phase of the cell cycle, essentially freezing the cell division process where distortion of mitotic spindles occurs, causing chromosome breakage and cell death.[18] It is in the context of targeting microtubules that taxane-induced peripheral neuropathy may be appreciated because neurons have more microtubules to support axonal trafficking via fast anterograde (FAT) and retrograde transport systems.

Also in the 1960s a group of researchers at Michigan State University discovered the biological action of platinum complexes, which eventually led to the FDA approval of the first platinum agent, cisplatin, in 1978.[19] The platinum group includes the drugs cisplatin, carboplatin, and oxaliplatin, which inhibit DNA synthesis by formation of DNA cross-links that denature the double helix, disrupting the function of DNA.[20] However, the site of action for platinum compounds is not cancer-cell specific and leads to toxicities.

The primary toxicities of greatest concern from taxanes and the platinum-based drugs are those associated with induction of hypersensitivity reactions and peripheral neuropathy.

Pathways involved in CIPN

The chemotherapy drug oxaliplatin provides a case in point. A third-generation platinum analog, it is considered a significant drug in the treatment of colorectal cancer. Through its binding and formation of cross-links between strands of DNA, it forms DNA adducts with resulting inhibition of DNA transcription and replication.[21]

Oxaliplatin-induced neuropathy (OIN) is a dose-related side effect, which occurs in almost 40% of patients treated with the compound. To date there are no established neuroprotective or treatment options to mitigate the development of CIPN, and there is a lack of sensitive assessment methods.[22,23]

Many studies have confirmed the role of the cell receptors in OIN pathogenesis. Some studies suggest that neuropathy is related to alterations in the sodium channel-dependent action potential because the oxalate (an oxaliplatin metabolite) alters the functional properties of voltage-gated sodium channels, resulting in a prolonged open state of the channels and hyper-excitability of sensory neurons.[24] It is this hyper-excitation that underlies the increased nociception and hyperalgesia. However as a consequence downstream signal transduction and activation of intracellular effectors mediate degenerative changes, the first of which may be increased oxidative stress.[15] Intimately involved in these signaling pathways, as has recently become evident, is nuclear factor-kappa B (NF-κB).

Nuclear factor kappa B is a group of structurally related proteins that are composed of various combinations of dimers that are members of the NF-κB/Rel family of proteins. This family is characterized by the presence of a highly conserved 300 amino acid Rel homology domain (RHD). This domain is responsible for dimerization, DNA binding, and association with inhibitor kappa B (IκB) proteins. Many variants of this domain exist with varying structures, the most predominant one being p65/p50.[25] In an unstimulated cell, NF-κB circulates in the cytosol bound to IκB, which inhibits the activity of NF-κB. The inhibitory actions of IκB involve several mechanisms: it sequesters the dimer in the cytosol, facilitates its dissociation from DNA, and exports NF-κB from the nucleus. Upon stimulation IκB is phosphorylated and degraded by IκB-kinase complex (IκK). When NF-κB is liberated it translocates into the nucleus, where it interacts with co-activator proteins or binds to specific sequences in the promoter region to regulate gene expression.[26]

Multidisciplinary research illustrated that NF-κB might be a key player molecule in the pathogenesis of neuropathic pain. NF-κB inhibits catechol-O-methyltransferase (COMT) expression in astrocytes, which are known to metabolize catecholamine and thereby act as a key modulator of dopaminergic and adrenergic/noradrenergic neurotransmission. So low COMT activity is associated with increased pain sensitivity and perception.[27] Also protein kinase C-associated kinase (PKK) is a critical regulator of NF-κB signaling, and overexpression of PKK activates NF-κB signaling and vice versa. This may explain the increases in the activity of protein kinase C in supra-spinal regions in CIPN cases.[28]

Intracellular stress conditions have been shown to result in the activation of NF-κB. Several studies have revealed the role of mitochondria in generation of reactive oxygen species (ROS) and the ensuing regulation of NF-κB signaling.[29] The converse has also been demonstrated because NF-κB activity increases nitric oxide production, and so increases intracellular oxidative stress.[30]

Convergence of mechanisms

Although no one mechanism may be responsible for the initiation of CIPN, there appears to be a convergence on a common pathway involving calcium overload and downstream activation of proteolysis and/or apoptosis and oxidative stress. As summarized in Figure 24.1, calpain is believed to participate in several neurodegenerative disorders and in glutamate-induced excitotoxicity and toxic neuropathies,[31–33] including paclitaxel-, vincristine-, or oxaliplatin-induced CIPN.[15,34] This activation of calpain may be a consequence of activation of calcium voltage-activated channels, as indicated above, or be secondary to increased sodium channel activation in these neurons. Calpain activation also results in exacerbation of neurotoxicity through NF-κB activation. For example, glutamate has been shown to activate NF-κB via calpain in neurons.[34,35] This may be mediated by IκB degradation.[36] Interestingly the chemotherapeutic agent bortezomib induces IκB degradation via calpain activation.[37] Calpain's participation in CIPN neurodegeneration induced by paclitaxel has been

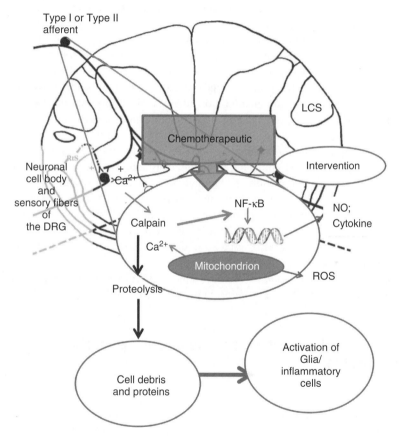

Figure 24.1 As a result of chemotherapeutic regimen: (1) calcium overload ensues, secondary to Na and Ca channel activation, resulting in calpain activation and proteolysis of the cytoskeletal (e.g., microtubules and neurofilaments). Calpain activity may also activate NF-κB; (2) downstream generation of inflammatory cytokines, NO, and exacerbation of ROS production from mitochondria, and increased Ca release; (3) this is aggravated by the attendant activation of glia (astrocytes and microglia) and immune effectors [see text for details]. Successful intervention to ameliorate NF-κB-induced pathology and oxidative stress would improve the efficacy and dosing regimens with chemotherapeutics. DRG = dorsal root ganglia. *(See color plate section for the color representation of this figure.)*

demonstrated by the use of a calpain inhibitor,[38] while oxaliplatin has been shown to induce apoptosis of HeLa cells via calpain activation.[39] These studies strongly implicate calpain as a mediator of neurotoxicity by direct proteolysis, secondary to oxidative stress, as well as mediating NF-κB activation. In addition, calpain leads to the dissociation of cardiolipin and cytochrome c, which leads to the activation of caspases as well as increasing mitochondrial permeability and calcium release. These activities are inhibited by calpain inhibitors and inhibitors of ROS and nitric oxide generation.[40]

The contribution of glia, most notably astrocytes and microglia, to CIPN and neuropathic pain has only been considered recently. In a study by Ji *et al.* it was demonstrated that inhibition of astrocytes, but not microglia, significantly reduced vincristine-induced CIPN in a rat model. This study also provided evidence that oxidative stress mediated the development of spinal astrocyte activation, and activated astrocytes dramatically increased interleukin (IL)-1β expression that induced NMDA receptor activation in spinal neurons to intensify pain transmission.[41] In a nerve injury rat model with tactile allodynia, Miyoshi *et al.* demonstrated the increased release of IL-18 from microglia and the upregulation of IL-18 receptors in astrocytes.[42] The functional inhibition of IL-18 signaling pathways suppressed injury-induced tactile allodynia and decreased the phosphorylation of NF-κB in spinal astrocytes and the induction of astroglial markers. In a similar model, astrocyte production of pro-inflammatory cytokines TNF-α, IL-1β, and NF-κB activation was reported.[43] These effects could be ameliorated by administration of recombinant erythropoietin. Consistent with the involvement of astrocytes and NF-κB in the production of neuropathy and pain is the study by Tchivileva *et al.*[27] of COMT whose activity is believed to modulate pain perception by the breakdown of catecholamines. The authors demonstrated that TNF-α inhibits COMT expression and activity in astrocytes by inducing NF-κB complex recruitment to the specific κB binding site, thereby enhancing pain signaling.

Taken together, this evidence provides a plausible convergence of mechanisms in the precipitation of CIPN and therefore suggests targets for the amelioration of CIPN.

Classes of chemotherapy associated peripheral neuropathy

Taxanes

Paclitaxel, docetaxel, and cabazitaxel are all formulated in vehicles that will cause hypersensitivity reactions in a person sensitive to the vehicle. Hypersensitivity reactions occur in up to 30% of patients receiving taxane chemotherapy without pre-medication. Therefore these three taxanes all require heavy pre-medication as prophylaxis for potential allergic reactions. This can be acutely manifested as chest pain, bronchospasm, dyspnea, hypotension, urticaria, skin reactions, and angioedema.[44] Even in the presence of pre-medication there is potential for hypersensitivity reactions of varying severity. Use of the pre-medication decreases the chances of having a reaction to 1–3%.[45] The incidence of acute anaphylactoid reactions decreases from paclitaxel to docetaxel and then cabazitaxel. The severity of the allergic reactions also decreases from paclitaxel to docetaxel to cabazitaxel.

Paclitaxel is a hydrophobic compound formulated with Cremophor EL for solubility and entrance into the body. Cremophor EL is known for causing the hypersensitivity reactions associated with administration of paclitaxel; it leaches plasticizers from the intravenous tubing used for administration. This releases histamine, causing hypersensitivity reactions. This harmful solvent requires special tubing for administration and pre-medication in order to try to avoid hypersensitivity reactions. Cremophor EL has also been shown to cause axonal degeneration and demyelination, contributing to the neurotoxicity of paclitaxel preparations in addition to the documented direct peripheral neuropathy induced by paclitaxel alone.[46]

In phase I trials, paclitaxel caused hypersensitivity reactions in 32 of 301 patients (11%). Of these 32 patients, 13 (41%) received the proper pre-medication to prevent hypersensitivity reactions.[45] In another study paclitaxel was found to cause severe hypersensitivity reactions in <4% of patients.[47] Severe hypersensitivity reactions associated with paclitaxel, such as anaphylaxis, may prove to be fatal. During a phase I trial one patient experienced asystole and a severe hypotensive state, resulting in death.[48] Great caution must be exercised in administering paclitaxel due to the reactions associated with its solvent.[48]

Abraxane® is a novel albumin-bound nanoparticle formulation of paclitaxel. This new formulation eliminates the need for the Cremophor solvent and allows paclitaxel to be given without the need for premedications. In a multicenter phase II clinical trial Abraxane was given to 63 patients with metastatic breast cancer. There was no pre-medication given or special drug administration required, and no hypersensitivity reactions were reported.[49] Gradishar *et al.* compared Abraxane to paclitaxel regarding hypersensitivity reactions.[50] Five hypersensitivity reactions in the paclitaxel group were noted (three allergic reaction, two chest pain), and none occurred in the Abraxane group. In another study Abraxane was infused over 2 h to 25 patients, and no patients received pre-medication; no patients experienced hypersensitivity reactions.[51] Although Abraxane nanoformulation eliminates hypersensitivity reactions associated with the solvent, it does not eliminate peripheral neuropathies associated with paclitaxel administration.

Docetaxel is a semi-synthetic analogue of paclitaxel and is formulated with polysorbate 80 as its vehicle. In phase II studies, while patients manifested hypersensitivity reactions in 41 out of 271 (15%) patients using paclitaxel, only 25 out of 496 patients (5%) using docetaxel manifested similar reactions.[52] These same studies revealed sensory neuropathy in 29 out of 271 patients (11%) on paclitaxel, but only 5 out of 496 patients (1%) using docetaxel developed neuropathy.[52] Cellax is a carboxymethylcellulose-based polymer conjugate nanoparticle formulation of docetaxel. Cellax has proven to have a better pharmacokinetic profile and increased efficacy compared to docetaxel.[53] This novel nanoformulation eliminates the need for polysorbate 80 as a vehicle used for docetaxel, but adverse effects to docetaxel remain.

Cabazitaxel is the newest of the taxane class, designed to protect against resistance seen in patients taking docetaxel or paclitaxel. It is required that pre-medication also be given with cabazitaxel (Table 24.2). Hypersensitivity reactions are seen with

Table 24.2 Taxanes and hypersensitivity reaction modulation.

Taxane compound	Vehicle used	CYP enzyme metabolism	Pre-medication for hypersensitivity reactions
Paclitaxel	Cremophor EL (polyoxyethylated castor oil + dehydrated alcohol)	2C8, 3A4	• Dexamethasone 20 mg PO/IV at 12 and 6 h before dose ↓dose to 10 mg in advanced-HIV • Diphenhydramine 50 mg IV given 30–60 min before chemo • H2-blocker given 30–60 min before chemo Cimetidine 300 mg IV Famotidine 20 mg IV Ranitidine 50 mg IV
Docetaxel	Polysorbate 80, ethanol	3A4	• Dexamethasone 8 mg PO day before, day of, and day after chemo
Cabazitaxel	Polysorbate 80	3A4, 3A5, 2C8 (minor)	• All given 30 min prior to chemo: Diphenhydramine 25 mg IV Dexamethasone 8 mg IV Ranitidine 50 mg IV
Abraxane	Nanoparticle albumin-bound	2C8, 3A4	No pre-medications required
Cellax	PEGylated carboxymethylcellulose nanoparticle	3A4	No pre-medications required

cabazitaxel administration. In multicenter phase II trials using cabazitaxel as treatment for taxane-resistant metastatic breast cancer, 4% of patients ($n = 67$) developed hypersensitivity reactions despite the use of pre-medications. One patient withdrew from the study due to a grade 4 allergic reaction.[54] In a phase I trial involving 21 patients treated with cabazitaxel, 2 mild hypersensitivity reactions were reported.[55]

Platinums

In contrast to the taxanes there is a paucity of evidence regarding hypersensitivity reactions with platinum agents. Reactions to cisplatin and carboplatin are uncommon, but there are a few case reports of hypersensitivity reactions to these agents. Two patients treated for lung cancer with cisplatin experienced symptoms of chest pressure, cough, flushing, urticarial, and tachycardia as part of a hypersensitivity reaction to the medication.[56] In a review of 180 patients given cisplatin as part of a regimen to treat colorectal cancer, 15% developed a hypersensitivity reaction to the medication while 2.2% developed a reaction of grade 3–4.[57] Over half of these patients experienced symptoms of the reaction within the first hour of infusion; rash, itching, flushing, chest discomfort, and dyspnea were among the most common symptoms documented.

Oxaliplatin is a third generation platinum agent that has shown a greater potential to induce hypersensitivity reactions, compared to cisplatin and carboplatin. However in a 10-year retrospective review of 1224 oxaliplatin-treated patients, 308 (25%) of these patients developed symptoms consistent with hypersensitivity reactions. Of the 308 patients, 63% experienced only mild itching and small area erythema, while 113 of the 308 patients had significant reactions that manifested as bronchospasm and facial swelling, starting within 10 min of infusion.[58]

Vinca alkaloid

Drugs from the class of vinca alkaloid including vincristine, vinblastine, and others play a major role in cancer management via their high affinity binding to tubulin, which causes a disruption of the mitotic spindles and arrest of cells in metaphase. Neuropathy is the dose-limiting side effect of all vinca alkaloids, being the most severe with vincristine and relatively less severe with vinblastine and vindesine.[59] Vincristine produces a mixed motor, sensory neuropathy, and autonomic neuropathy, with numbness and tingling.[60]

The cause of the neuropathy is most likely from impaired microtubule function involved in axonal transport where both myelinated and unmyelinated fibers are affected, which might result in impaired regeneration of nerves injured by other causes.[61]

In an animal model of vincristine neuropathy a distinct disorganization of the axonal microtubule cytoskeleton along with accumulation of neurofilaments in the soma of spinal ganglion cells was demonstrated, suggesting impaired anterograde axonal transport.[62]

Induction of peripheral neuropathy

Peripheral sensory neuropathy is inflammation or degeneration of sensory nerves in the periphery.[63] Clinically neuropathy is graded on a scale of 1–4: grade 1 is numbness, shooting pain, loss of deep tendon reflexes; grade 2 starts to see a limit in activities of day-to-day living such as preparing food, shopping, or using the telephone; grade 3 symptoms are more severe and activities such as feeding, bathing, dressing, and using the toilet are very difficult; and grade 4 is very painful, severe and life threatening, and immediate intervention is necessary.

The peripheral neuropathy with attendant pain that is caused by taxanes and platinum compounds is a main concern of these drug classes. Peripheral neuropathy, primarily sensory, is characterized by paresthesia of the extremities such as numbness, tingling, and shooting pain, and typically starts from the toes and fingers and works its way up to the feet and hands.[64] These symptoms adversely affect the health-related quality of life of patients.[65] The ability to walk can be impaired–patients often have trouble picking things up or holding objects, and they lose feeling in their extremities.[66] CIPN frequently leads to discontinuation of treatment or need for a lower dose. Patients cannot tolerate the undesirable symptoms of peripheral neuropathy, and this interferes with giving optimal cancer treatment to the patient.

Paclitaxel is known to cause peripheral neuropathy. In a phase III trial of 225 patients treated with paclitaxel 5 patients (2%) developed grade 3 sensory neuropathy; there was no reported grade 4 sensory neuropathy or motor neuropathy in any patients.[50] In a study of 67 patients treated with cabazitaxel, 17% developed sensory neuropathy.[54] In a phase I trial of 21 patients treated with cabazitaxel, 24% had transient neurosensory disorders. In the same trial no severe neurotoxicity was found in any of the patients.[55]

Like paclitaxel Abraxane administration precipitates peripheral neuropathy in patients. Paik *et al.* conducted a study with 25 patients receiving Abraxane over a 2-h infusion compared to another patient group that infused Abraxane over 30 min.[51] In the 2-h group treatment was discontinued in 5 patients (20%) due to toxic neuropathy compared to 14 patients (36%) in the 30-min group. Grade 2 neuropathy was present in 12% of patients in the 2-h arm and 33% of patients in the 30-min arm. Grade 3 neuropathy was seen in 16% of patients in the 2-h arm as compared to 23% of patients in the 30-min arm. Although Abraxane administration induces peripheral neuropathy, this study suggests extending the infusion time of the drug as a potential way to lower the incidence and grade of peripheral neuropathy. In a phase II clinical trial grade 2 and 3 peripheral neuropathy were reported in 19% and 11% of patients, respectively, and no grade 4 neuropathy was noted; however 8% of patients had to discontinue Abraxane treatment due to intolerable sensory neuropathy.[49] In a phase III trial Abraxane caused grade 3 peripheral neuropathy in 24 of 229 patients; no grade 4 peripheral neuropathy or motor neuropathy was seen.[50]

Van der Hoop *et al.* conducted a study comparing the incidence of neurotoxicity between different treatment groups with or without cisplatin.[67] In the group without cisplatin grade 2 neuropathy was seen in 3% of patients, and no grade 3 neuropathy was reported. In the group receiving cisplatin as part of the chemotherapy regimen grade 2 neuropathy was seen in 21% of patients, and grade 3 neuropathy was seen in 4% of patients. Total neuropathy was noted in 25% of patients in the group without cisplatin and 47% of patients in the groups with cisplatin. Severe neurotoxicity was seen in 4% of patients treated with cisplatin. This side effect was rare, but these patients exhibited a sensorimotor neuropathy with walking disability leading to a poor quality of life. From this study it is seen that adding cisplatin to a chemotherapy regimen increases not only the incidence of neurotoxicity but also the severity of it.

Since taxanes and platinums are used frequently and are believed to be effective, important concerns have been raised regarding the toxicities that they precipitate such as hypersensitivity reactions, myelo-suppression, and painful peripheral neuropathy. These harmful toxicities lead to dose reduction and treatment discontinuation in patients.[44] These patients are at a major disadvantage because their cancer continues to grow stronger in absence of treatment. Defining the mechanisms of toxicities and potential targets for their amelioration has become a concern in an effort to improve therapeutic regimens and efficacious treatment.

Classifications of CIPN based on the severity

CIPN is classified as an axonopathy, primarily of sensory fibers, hence its gradual reversibility on cessation of administration.[40] However recent studies suggest the involvement of the cell bodies of the dorsal root ganglia, as well as glia, astrocytes, and microglia. The pathogenesis of CIPN and the mechanism of associated pain are not clearly defined, but preclinical and *in vitro* studies provide some insight regarding how the development and progression of CIPN occurs. Ions such as potassium, sodium, and calcium play a key role in the transmission of pain in the peripheral nervous system.[15]

Kawakami *et al.* performed an experiment in rats to elucidate the effects of paclitaxel treatment on voltage gated calcium channels in the peripheral neurons.[68] Treatment with paclitaxel showed significant enhancement of calcium channels in dorsal root ganglia neurons. These neurons are mediators in the pain pathway. Treatment with paclitaxel also showed a much higher incidence of hyperalgesia and allodynia.

Increase in the number and activity of the calcium channels is thought to evoke the pain response associated with CIPN. The actions of $\alpha 2\delta 1$ subunit of the calcium channels are thought to play a part in the mechanism by which painful peripheral neuropathy occurs.[15] The subunit's role is to upregulate expression of calcium channels and increase calcium current density.[69] In one study the level of $\alpha 2\delta 1$ subunits of the calcium channels in dorsal root ganglia showed a significant increase in the paclitaxel-treated group.[69] In a separate study by Xiao *et al.* the levels of the $\alpha 2\delta 1$ subunit in dorsal root ganglia and dorsal spinal cord were examined; the levels did not show a significant increase, but in the dorsal spinal cord there were increased levels of this protein.[70] Gabapentin is an antagonist of the calcium channels located in the peripheral neurons and is found to bind to the $\alpha 2\delta 1$ subunit.[71] Gabapentin's pharmacologic profile makes it a candidate to reverse the increases in voltage-dependent calcium channels associated with paclitaxel treatment. Gabapentin has been found to reverse hyperalgesia and reduce calcium currents associated with peripheral neuropathy.[69]

The taxanes bind to β-tubulin to exhibit their effects in cancer cells, and β-tubulin is abundantly represented in neurons as part of the FAT and retrograde axonal transport (RAT) machinery.[40] Taxane administration causes dysfunction of axonal microtubules leading to disruption of the axoplasmic transport, eventually leading to interruption in axonal maintenance and retrograde trophic signaling.[15,40] Mitochondrial injury is also related to pain associated with CIPN. Oxaliplatin and paclitaxel-induced peripheral neuropathy are shown to cause decreased rates of oxygen consumption and decreased production of ATP in mitochondria.[72] It should be noted that mitochondria in support of axonal transport needs and neurotransmitter release are among the rare organelles found outside the cell body and transported by energy-dependent FAT. It has been demonstrated that paclitaxel opens up membrane pores of mitochondria, leading to calcium release. Excess calcium release from mitochondria in the peripheral nervous system leads to activation of both necrotic and apoptotic pathways and neuropathy.[15,40] In addition, increases in the number of the skin's Langerhans cells, effective antigen-presenting cells, are seen with administration of paclitaxel. This activation of these cells can lead to release of nitric oxide, inflammatory cytokines, and the activation of inflammatory effectors that contribute to painful neuropathy.

Prior or current treatment with platinums and taxanes, single dose amount, infusion duration, and cumulative dosage are all factors contributing to severity and incidence of CIPN.[73] For example sensory neuropathy induced by cisplatin can be delayed for weeks in its development after infusion of the drug,[74] while oxaliplatin causes painful neuropathy that manifests very quickly after infusion.[75,76]

Amelioration of pain: current status

Currently there is no available treatment that is clinically proven to help manage the pain associated with CIPN. Although many studies are being conducted on how to treat CIPN, the only current option is to lower the dose or discontinue treatment with the causative agents until CIPN is reversed, prior to resuming treatment. This highlights the compelling need to prevent or ameliorate CIPN because the etiological factors causing these side effects are effective and necessary in anticancer treatment. Below is a brief review of recent or current options being investigated.

Duloxetine

Duloxetine is a serotonin and norepinephrine reuptake inhibitor. Serotonin and norepinephrine are key neurotransmitters in suppressing the transmission of pain signals in the periphery.[77] Although serotonin through 5-HT2A receptors has been shown to sensitize peripheral pain receptors, it is thought to reduce spinal nociceptive transmission through this receptor.[15] A randomized phase III, double-blinded, placebo-controlled trial evaluated the effectiveness of duloxetine in treating CIPN. The patients treated with duloxetine reported an average decrease in pain of score 1.06 (95% CI, 0.72–1.40). The placebo group had a mean decrease in score of only 0.34 (95% CI, 0.01–0.66). In the duloxetine-treatment group, 59% of patients reported a decrease in pain and 38% reported no change in pain.[77] Based on the results of this study it seems that there is some benefit to duloxetine for helping ease the pain associated with CIPN. Due to safety, tolerability, and an increase in patients' quality of life, the authors reported that the "study results strongly suggest that duloxetine treatment is associated with a clinically meaningful improvement in chemotherapy-induced peripheral neuropathic pain."[77] Further studies are warranted to see if safe and effective pain relief can be achieved using duloxetine for painful CIPN.

Acetyl-L-carnitine (ALC)

Acetyl-L-carnitine (ALC) is a naturally occurring compound that increases acetyl-CoA in mitochondria to help eliminate toxic metabolic byproducts.[78] ALC was found to be beneficial in preventing mitochondrial injury due to paclitaxel and oxaliplatin treatment.[72] In a small phase II study 25 patients with grade 2 or 3 neuropathy due to paclitaxel or cisplatin treatment received 1 g of oral ALC three times daily.[79] Neuroprotective benefits were seen with sensory and motor improvement in most patients. A larger, placebo-controlled trial also assessed ALC's use in the improvement of pain associated with CIPN. In this study 208 patients received ALC, and 201 patients received placebo. Pain improvement scores were assessed after 12 and 24 weeks. After 12 weeks of ALC treatment, improvement in CIPN was negligible and comparable to placebo. After 24 weeks of treatment with ALC symptoms of CIPN were actually worse, and the functional status of patients was lower than baseline.[78]

Several agents have been shown to prevent experimental vincristine neuropathy. Similar to taxane and platinum-based compound-induced neuropathies, ALC has been shown in animal models to be protective against vincristine-induced neuropathies.[78]

BAK gel

A gel formed of 10 mg baclofen, 40 mg amitriptyline, and 20 mg ketamine was studied for its effect on reducing pain in CIPN. Baclofen is a GABAb receptor agonist that leads to the inhibition of signals sent through first order afferent neurons.[80] Baclofen has been shown to increase the pain threshold by inhibition of the neuronal signaling.[80] Amitriptyline is a tricyclic antidepressant with various mechanisms of action that are thought to help decrease peripheral nerve pain. It inhibits voltage gated sodium, potassium, and calcium channels and inhibits reuptake of norepinephrine and serotonin.[81] Amitriptyline's action is seen to elevate peripheral sensory thresholds in mechanical, heat, and cold stimuli.[81] Ketamine is a NMDA receptor antagonist, and phosphorylation of this receptor has been observed following neuropathic injury.[82] The combination of these three agents and their various pharmacologic effects provides a synergistic effect on treating pain. Patients who used this topical treatment showed a trend toward healing of sensory neuropathy; improvement in symptoms, such as shooting pain and tingling in fingers and hands and the ability to hold objects, was observed.[83] There are potential problems for the patient in being able to comply in using topical gel treatment for a more severe form of painful CIPN. It would be a challenge to open up the tube, squeeze out the gel, and apply the cream with so much pain in your hands. For mildly painful CIPN this treatment seems appropriate and useful. It may stop burning and tingling in hands and feet soon after application.

Vitamin E

Vitamin E depletion is seen in sensory neuropathy and following treatment by neurotoxic chemotherapeutic agents.[84] There have been several studies attempting to delineate the effectiveness of vitamin E for neuroprotection (Table 24.3). Argyriou *et al.* found vitamin E to have neuroprotective effects in CIPN.[85] This study used a high dose of vitamin E, which may have helped. Pace *et al.* described a decrease in paresthesia and deep tendon reflex alteration in patients taking vitamin E.[86] However the conclusions drawn by the authors may be overambitious considering the small number of patients involved. Afonseca *et al.* described no difference between placebo and vitamin E.[87] This was supported by a larger study by Kottschade *et al.* where no significant difference between vitamin E and placebo in providing neuroprotection was found.[88]

Other combinations of natural supplements

In cisplatin-induced peripheral neuropathy in rats, combination of monosodium glutamate with individual antioxidants including resveratrol, α-lipoic acid, and coenzyme Q_{10} demonstrated neuroprotective effects.[89]

Table 24.3 Vitamin E and chemotherapy-induced peripheral neuropathy (CIPN).

Study	Vitamin E dose (mg), frequency	Results
Argyriou *et al.*[85]	300, Twice daily	• Vitamin E group ($n = 20$): 1 mild, 3 moderate peripheral neuropathies • Placebo group ($n = 15$): 3 mild, 5 moderate, 3 severe peripheral neuropathies
Pace *et al.*[86]	400, Daily	• Vitamin E group ($n = 17$): decreased tendon reflexes and parasthesias in 6 patients (35.3%) • Placebo group ($n = 24$): decreased tendon reflexes and paresthesia in 13 patients (54.2%)
Afonseca *et al.*[87]	400, Daily	• Vitamin E group ($n = 18$): 83% developed peripheral neuropathy grade 1 or 2 • Placebo group ($n = 16$): 68% developed peripheral neuropathy grade 1 or 2
Kottschade *et al.*[88]	300, Daily	• Vitamin E group ($n = 96$): 34% developed sensory neuropathy grade ≥ 2 • Placebo group ($n = 93$): 29% developed sensory neuropathy grade ≥ 2

Oxaliplatin-induced painful peripheral neuropathy in mice appears to be mediated at least in part via oxidative stress, an effect that was prevented by flavonoids.[90]

The potential utility of various natural products including vitamin E, ALC, glutamine, glutathione, vitamin B6, omega-3 fatty acids, α-lipoic acid, and *n*-acetyl cysteine in CIPN were assessed in various clinical trials.[91]

Small molecule inhibitor

Recently we identified a dual ROS and NF-κB inhibitor small molecule(OT-404) that effectively inhibits cancer cell proliferation and tumor angiogenesis of either chemo-sensitive or chemo-resistant human cancer cells and also enhanced cancer cell sensitivity to different chemotherapy *in vitro* and in nude mice implanted with various human cancer cell types.[92] These findings provided evidence for the potential of dual ROS and NF-κB inhibition in cancer management from the anticancer efficacy standpoint as well as showed the potential in overcoming CIPN-associated adverse effects.[92]

Summary and conclusions

CIPN is one of the common side effects of chemotherapeutic agents from the taxane, vinca alkaloid, and platinum classes. It is considered a major dose-limiting effect for many of these agents, which affects the course of efficacious cancer therapy.[14] The incidence of CIPN varies according to the specific cancer and its stage, as well as according to the class and number of chemotherapeutic agents being used. Based on 24 different clinical trials no agent has shown clear positive evidence to be recommended at this stage for the management of CIPN. Consequently, the standard of care for CIPN includes dose reduction and/or discontinuation of chemotherapy treatment, which would compromise disease management.

The management of CIPN remains an important challenge, and clearly future studies are required before reaching definitive recommendations for the use of antioxidant or other supplements. However, based on the mechanisms involved in CIPN, potent ROS and NF-κB inhibitors might have potential in managing the efficacy of chemotherapy and adverse effects associated with it.

Multiple choice questions

1 What are the classes of chemotherapy associated with peripheral neuropathy?
  a. Taxanes
  b. Platinum
  c. Vinca alkaloid
  d. All of the above

2 Which statement is correct concerning chemotherapy-induced peripheral neuropathy (CIPN)?
  a. Inflammation does not contribute
  b. Oxidative stress does not contribute
  c. Chemotherapy dose reduction and/or discontinuation are required
  d. Chemotherapy dose increase and continuation are allowed

3 The major mechanism involved in CIPN is:
   a. Suppression of oxidative stress and inflammation
   b. Stimulation of oxidative stress and inflammation
   c. Elevation of angiogenesis growth factors
   d. Elevation of nerve growth factors

4 Common pathways involved in the development of CIPN may include:
   a. Calcium overload
   b. Calpain activation
   c. NF-κB activation
   d. Apoptosis
   e. All of the above

5 The reversibility of CIPN suggests that it is primarily:
   a. A neuronopathy
   b. An axonopathy
   c. A myelinopathy
   d. b and c
   e. "b" progressing to "a"

6 The most common manifestation of CIPN is:
   a. Sensory
   b. Motor
   c. Autonomic
   d. Sensory with secondary motor involvement

7 CIPN due to paclitaxel is compounded by:
   a. Its dilutent Cremophor EL induces neuropathy
   b. It has an extremely long half-life
   c. It has a high incidence of hypersensitivity reactions (HSR)
   d. a and c

8 Some of the limitations of paclitaxel have been overcome by nanoformulation. This is:
   a. Docetaxel
   b. Cabazitaxel
   c. Abraxane
   d. Bortezomib

9 Neurons are ideal targets of vinca alkaloid and taxane chemotherapeutics because these agents target:
   a. Neurofilaments
   b. Microglia
   c. Microtubule assemblies
   d. Glial filaments

10 Inflammation associated with CIPN may be exacerbated due to involvement of:
   a. Astrocytes
   b. Microglia
   c. Downstream cytokine elaboration
   d. All of the above

11 Evidence suggests that catecholamines may be associated with hyperalgesia transmission because of:
   a. A reduction in esterase activity
   b. An increased synthesis of norepinephrine
   c. A reduction in catechol-O-methyl transferase (COMT)
   d. A loss of autonomic fibers

12 The cross-linking of DNA by platinum compounds likely results in neuropathy because:
   a. Neurons will not divide
   b. It results in protein deficits, thereby affecting maintenance
   c. It results in mutations and promotes tumor growth
   d. None of the above

13 Hypersensitivity reactions (HSR) to chemotherapeutics are believed to be due to:
   a. Their being derived or partially derived from natural products
   b. The vehicle
   c. Contaminants
   d. Albumin

14 Acute manifestation of HSR may include:
   a. Chest pain
   b. Bronchospasm and dyspnea
   c. Hypotension
   d. Urticaria
   e. All of the above

15 Duloxetine is believed to reduce central pain transmission by:
   a. Increasing serotonin release
   b. Inhibiting serotonin reuptake
   c. Acting as a direct serotonin receptor agonist
   d. Inhibition of presynaptic monoamine oxidase

References

1 Chaudhry, V., Chaudhry, M., Crawford, T.O. *et al.* (2003) Toxic neuropathy in patients with pre-existing neuropathy. *Neurology*, 60(2), 337–340.
2 Bristol-Myers Squibb Company. (2011) TAXOL® (paclitaxel) INJECTION. http://www .accessdata.fda.gov/drugsatfda_docs/label/2011/020262s049lbl.pdf [accessed on 16 September 2015].
3 Sanofi-Aventis. (2013) TAXOTERE® Prescribing Information. http://products.sanofi.us/ Taxotere/taxotere.html [accessed on 16 September 2015].
4 Abraxis BioScience, LLC. (2013) Abraxane prescribing information. http://abraxane.com/ downloads/Abraxane_PrescribingInformation.pdf [accessed on 16 September 2015].
5 GlaxoSmithKline. (2002) NAVELBINE® prescribing information. http://www.fda.gov/ ohrms/dockets/ac/04/briefing/4021b1_10_vinorelbine%20label.pdf [accessed on 16 September 2015].
6 Drugs.com. (2015) Oncovin®. http://www.drugs.com/monograph/oncovin.html [accessed on October 28, 2015].

7 Sanofi-Aventis U.S. LLC. (2009) Eloxatin (oxaliplatin) prescribing information. http://www.accessdata.fda.gov/drugsatfda_docs/label/2009/021492s011,021759s009lbl.pdf [accessed on 16 September 2015].

8 Bristol-Myers Squibb Company. (2010) Cisplatin prescribing information. http://www.accessdata.fda.gov/drugsatfda_docs/label/2011/018057s080lbl.pdf [accessed on 16 September 2015].

9 Chaudhry, V., Rowinsky, E.K., Sartorius, S.E. *et al.* (1994) Peripheral neuropathy from taxol and cisplatin combination chemotherapy: clinical and electrophysiological studies. *Annals of Neurology*, 35(3), 304–311.

10 Wampler, M., Hamel K., Topp K. (2005) Impairments in postural control resulting from taxane-induced peripheral neuropathy in women with breast cancer. International Society of Posture and Gait Research Annual Meeting. Marseilles, France.

11 Dougherty, P.M., Cata, J.P., Cordella, J.V. *et al.* (2004) Taxol-induced sensory disturbance is characterized by preferential impairment of myelinated fiber function in cancer patients. *Pain*, 109(1–2), 132–142.

12 Butler, L., Bacon, M., Carey, M. *et al.* (2004) Determining the relationship between toxicity and quality of life in an ovarian cancer chemotherapy clinical trial. *Journal of Clinical Oncology*, 22(12), 2461–2468.

13 Pfizer New Zealand Ltd. (2010) Carboplatin injection data sheet. http://www.medsafe.govt.nz/profs/datasheet/c/Carboplatininj.pdf [accessed on 16 September 2015].

14 Wolf, S., Barton, D., Kottschade, L. *et al.* (2008) Chemotherapy-induced peripheral neuropathy: prevention and treatment strategies. *European Journal of Cancer*, 44(11), 1507–1515.

15 Jaggi, A.S. & Singh, N. (2012) Mechanisms in cancer-chemotherapeutic drugs-induced peripheral neuropathy. *Toxicology*, 291(1–3), 1–9.

16 Armstrong, T.S. & Grisdale, K.A. (2006) Peripheral neuropathy. In: Camp-Sorrell, D. & Hawkins, R.A. (eds), *Clinical Manual for the Oncology Advanced Practice Nurse*, 2nd edn. Oncology Nursing Society, Pittsburgh, PA, pp. 909–918.

17 de Moor, J.S., Mariotto, A.B., Parry, C. *et al.* (2013) Cancer survivors in the United States: prevalence across the survivorship trajectory and implications for care. *Cancer Epidemiology, Biomarkers & Prevention*, 22(4), 561–570.

18 Horwitz, S.B., Cohen, D., Rao, S. *et al.* (1993) Taxol: mechanisms of action and resistance. *Journal of the National Cancer Institute Monographs*, 15, 55–61.

19 Rosenberg, B., Van Camp, L., Grimley, E.B. *et al.* (1967) The inhibition of growth or cell division in *Escherichia coli* by different ionic species of platinum(IV) complexes. *The Journal of Biological Chemistry*, 242(6), 1347–1352.

20 Fuertes, M.A., Castilla, J., Alonso, C. *et al.* (2003) Cisplatin biochemical mechanism of action: from cytotoxicity to induction of cell death through interconnections between apoptotic and necrotic pathways. *Current Medicinal Chemistry*, 10(3), 257–266.

21 Saif, M.W. & Reardon, J. (2005) Management of oxaliplatin-induced peripheral neuropathy. *Therapeutics and Clinical Risk Management*, 1(4), 249–258.

22 Vincenzi, B., Frezza, A.M., Schiavon, G. *et al.* (2013) Identification of clinical predictive factors of oxaliplatin-induced chronic peripheral neuropathy in colorectal cancer patients treated with adjuvant Folfox IV. *Supportive Care in Cancer*, 21(5), 1313–1319.

23 Park, S.B., Lin, C.S. & Kiernan, M.C. (2012) Nerve excitability assessment in chemotherapy-induced neurotoxicity. *Journal of Visualized Experiments*, 62, 3439. doi:10.3791/3439

24 Grolleau, F., Gamelin, L., Boisdron-Celle, M. *et al.* (2001) A possible explanation for a neurotoxic effect of the anticancer agent oxaliplatin on neuronal voltage-gated sodium channels. *Journal of Neurophysiology*, 85(5), 2293–2297.

25 Liu, S.F. & Malik, A.B. (2006) NF-kappa B activation as a pathological mechanism of septic shock and inflammation. *American Journal of Physiology. Lung Cellular and Molecular Physiology*, 290(4), L622–L645.

26 Matsuda, N. & Hattori, Y. (2006) Systemic inflammatory response syndrome (SIRS): molecular pathophysiology and gene therapy. *Journal of Pharmacological Sciences*, 101(3), 189–198.

27 Tchivileva, I.E., Nackley, A.G., Qian, L. *et al.* (2009) Characterization of NF-kB-mediated inhibition of catechol-O-methyltransferase. *Molecular Pain*, 5, 13.

28 Kim, S.W., Oleksyn, D.W., Rossi, R.M. *et al.* (2008) Protein kinase C-associated kinase is required for NF-kappaB signaling and survival in diffuse large B-cell lymphoma cells. *Blood*, 111(3), 1644–1653.

29 Janes, K., Doyle, T., Bryant, L. *et al.* (2013) Bioenergetic deficits in peripheral nerve sensory axons during chemotherapy-induced neuropathic pain resulting from peroxynitrite-mediated post-translational nitration of mitochondrial superoxide dismutase. *Pain*, 154(11), 2432–2440.

30 Salminen, A. & Kaarniranta, K. (2010) Genetics vs. entropy: longevity factors suppress the NF-kappaB-driven entropic aging process. *Ageing Research Reviews*, 9(3), 298–314.

31 Goll, D.E., Thompson, V.F., Li, H. *et al.* (2003) The calpain system. *Physiological Reviews*, 83(3), 731–801.

32 Shields, D.C., Tyor, W.R., Deibler, G.E. *et al.* (1998) Increased calpain expression in activated glial and inflammatory cells in experimental allergic encephalomyelitis. *Proceedings of the National Academy of Sciences of the United States of America*, 95(10), 5768–5772.

33 Crocker, S.J., Smith, P.D., Jackson-Lewis, V. *et al.* (2003) Inhibition of calpains prevents neuronal and behavioral deficits in an MPTP mouse model of Parkinson's disease. *The Journal of Neuroscience*, 23(10), 4081–4091.

34 Benbow, J.H., Mann, T., Keeler, C. *et al.* (2012) Inhibition of paclitaxel-induced decreases in calcium signaling. *The Journal of Biological Chemistry*, 287(45), 37907–37916.

35 Scholzke, M.N., Potrovita, I., Subramaniam, S. *et al.* (2003) Glutamate activates NF-kappaB through calpain in neurons. *The European Journal of Neuroscience*, 18(12), 3305–3310.

36 Schaecher, K., Goust, J.M. & Banik, N.L. (2004) The effects of calpain inhibition on IkB alpha degradation after activation of PBMCs: identification of the calpain cleavage sites. *Neurochemical Research*, 29(7), 1443–1451.

37 Li, C., Chen, S., Yue, P. *et al.* (2010) Proteasome inhibitor PS-341 (bortezomib) induces calpain-dependent IkappaB(alpha) degradation. *The Journal of Biological Chemistry*, 285(21), 16096–16104.

38 Wang, M.S., Davis, A.A., Culver, D.G. *et al.* (2004) Calpain inhibition protects against Taxol-induced sensory neuropathy. *Brain*, 127(Pt 3), 671–679.

39 Anguissola, S., Kohler, B., O'Byrne, R. *et al.* (2009) Bid and calpains cooperate to trigger oxaliplatin-induced apoptosis of cervical carcinoma HeLa cells. *Molecular Pharmacology*, 76(5), 998–1010.

40 El-Fawal, H.A.N. (2011) Neurotoxicology. In: Nriagu, J.O. (ed), *Encyclopedia of Environmental Health*. Vol. 4. Elsevier, Burlington, pp. 87–106.

41 Ji, X.T., Qian, N.S., Zhang, T. *et al.* (2013) Spinal astrocytic activation contributes to mechanical allodynia in a rat chemotherapy-induced neuropathic pain model. *PloS One*, 8(4), e60733.

42 Miyoshi, K., Obata, K., Kondo, T. *et al.* (2008) Interleukin-18-mediated microglia/astrocyte interaction in the spinal cord enhances neuropathic pain processing after nerve injury. *The Journal of Neuroscience*, 28(48), 12775–12787.

43 Jia, H., Feng, X., Li, W. *et al.* (2009) Recombinant human erythropoietin attenuates spinal neuroimmune activation of neuropathic pain in rats. *Annals of Clinical and Laboratory Science*, 39(1), 84–91.

44 Verstappen, C.C., Heimans, J.J., Hoekman, K. *et al.* (2003) Neurotoxic complications of chemotherapy in patients with cancer: clinical signs and optimal management. *Drugs*, 63(15), 1549–1563.

45 Weiss, R.B., Donehower, R.C., Wiernik, P.H. *et al.* (1990) Hypersensitivity reactions from taxol. *Journal of Clinical Oncology*, 8(7), 1263–1268.

46 Gelderblom, H., Verweij, J., Nooter, K. *et al.* (2001) Cremophor EL: the drawbacks and advantages of vehicle selection for drug formulation. *European Journal of Cancer*, 37(13), 1590–1598,

47 Joerger, M. (2012) Prevention and handling of acute allergic and infusion reactions in oncology. *Annals of Oncology*, 23(Suppl 10), x313–319.

48 Kadoyama, K., Kuwahara, A., Yamamori, M. *et al.* (2011) Hypersensitivity reactions to anticancer agents: data mining of the public version of the FDA adverse event reporting system, AERS. *Journal of Experimental & Clinical Cancer Research*, 30, 93.

49 Ibrahim, N.K., Samuels, B., Page, R. *et al.* (2005) Multicenter phase II trial of ABI-007, an albumin-bound paclitaxel, in women with metastatic breast cancer. *Journal of Clinical Oncology*, 23(25), 6019–6026.

50 Gradishar, W.J., Tjulandin, S., Davidson, N. *et al.* (2005) Phase III trial of nanoparticle albumin-bound paclitaxel compared with polyethylated castor oil-based paclitaxel in women with breast cancer. *Journal of Clinical Oncology*, 23(31), 7794–7803.

51 Paik, P.K., James, L.P., Riely, G.J. *et al.* (2011) A phase 2 study of weekly albumin-bound paclitaxel (Abraxane(R)) given as a two-hour infusion. *Cancer Chemotherapy and Pharmacology*, 68(5), 1331–1337.

52 Verweij, J., Clavel, M. & Chevalier, B. (1994) Paclitaxel (Taxol) and docetaxel (Taxotere): not simply two of a kind. *Annals of Oncology*, 5(6), 495–505.

53 Ernsting, M.J., Foltz, W.D., Undzys, E. *et al.* (2012) Tumor-targeted drug delivery using MR-contrasted docetaxel – carboxymethylcellulose nanoparticles. *Biomaterials*, 33(15), 3931–3941.

54 Pivot, X., Koralewski, P., Hidalgo, J.L. *et al.* (2008) A multicenter phase II study of XRP6258 administered as a 1-h i.v. infusion every 3 weeks in taxane-resistant metastatic breast cancer patients. *Annals of Oncology*, 19(9), 1547–1552.

55 Dieras, V., Lortholary, A., Laurence, V. *et al.* (2013) Cabazitaxel in patients with advanced solid tumours: results of a Phase I and pharmacokinetic study. *European Journal of Cancer*, 49(1), 25–34.

56 Randall, J.M., Bharne, A.A. & Bazhenova, L.A. (2013) Hypersensitivity reactions to carboplatin and cisplatin in non-small cell lung cancer. *Journal of Thoracic Disease*, 5(2), E53–E57.

57 Siu, S.W., Chan, R.T. & Au, G.K. (2006) Hypersensitivity reactions to oxaliplatin: experience in a single institute. *Annals of Oncology*, 17(2), 259–261.

58 Polyzos, A., Tsavaris, N., Gogas, H. *et al.* (2009) Clinical features of hypersensitivity reactions to oxaliplatin: a 10-year experience. *Oncology*, 76(1), 36–41.

59 Pace, A., Bove, L., Nistico, C. *et al.* (1996) Vinorelbine neurotoxicity: clinical and neurophysiological findings in 23 patients. *Journal of Neurology, Neurosurgery and Psychiatry*, 61(4), 409–411.

60 Jackson, D.V., Wells, H.B., Atkins, J.N. *et al.* (1988) Amelioration of vincristine neurotoxicity by glutamic acid. *The American Journal of Medicine*, 84(6), 1016–1022.

61 Pan, Y.A., Misgeld, T., Lichtman, J.W. *et al.* (2003) Effects of neurotoxic and neuroprotective agents on peripheral nerve regeneration assayed by time-lapse imaging in vivo. *The Journal of Neuroscience*, 23(36), 11479–11488.

62 Topp, K.S., Tanner, K.D. & Levine, J.D. (2000) Damage to the cytoskeleton of large diameter sensory neurons and myelinated axons in vincristine-induced painful peripheral neuropathy in the rat. *The Journal of Comparative Neurology*, 424(4), 563–576.

63 Magrinelli, F., Zanette, G. & Tamburin, S. (2013) Neuropathic pain: diagnosis and treatment. *Practical Neurology*, 13(5), 292–307.

64 Mols, F., Beijers, T., Lemmens, V. *et al.* (2013) Chemotherapy-induced neuropathy and its association with quality of life among 2- to 11-year colorectal cancer survivors: results from the population-based PROFILES registry. *Journal of Clinical Oncology*, 31(21), 2699–2707.

65 Shimozuma, K., Ohashi, Y., Takeuchi, A. *et al.* (2012) Taxane-induced peripheral neuropathy and health-related quality of life in postoperative breast cancer patients undergoing adjuvant chemotherapy: N-SAS BC 02, a randomized clinical trial. *Supportive Care in Cancer*, 20(12), 3355–3364.

66 Speck, R.M., DeMichele, A., Farrar, J.T. *et al.* (2012) Scope of symptoms and self-management strategies for chemotherapy-induced peripheral neuropathy in breast cancer patients. *Supportive Care in Cancer*, 20(10), 2433–2439.

67 van der Hoop, R.G., van der Burg, M.E., ten Bokkel Huinink, W.W. *et al.* (1990) Incidence of neuropathy in 395 patients with ovarian cancer treated with or without cisplatin. *Cancer*, 66(8), 1697–1702.

68 Kawakami, K., Chiba, T., Katagiri, N. *et al.* (2012) Paclitaxel increases high voltage-dependent calcium channel current in dorsal root ganglion neurons of the rat. *Journal of Pharmacological Sciences*, 120(3), 187–195.

69 Gribkoff, V.K. (2006) The role of voltage-gated calcium channels in pain and nociception. *Seminars in Cell & Developmental Biology*, 17(5), 555–564.

70 Xiao, W., Boroujerdi, A., Bennett, G.J. *et al.* (2007) Chemotherapy-evoked painful peripheral neuropathy: analgesic effects of gabapentin and effects on expression of the alpha-2-delta type-1 calcium channel subunit. *Neuroscience*, 144(2), 714–720.

71 Marais, E., Klugbauer, N. & Hofmann, F. (2001) Calcium channel alpha(2)delta subunits-structure and Gabapentin binding. *Molecular Pharmacology*, 59(5), 1243–1248.

72 Zheng, H., Xiao, W.H. & Bennett, G.J. (2011) Functional deficits in peripheral nerve mitochondria in rats with paclitaxel- and oxaliplatin-evoked painful peripheral neuropathy. *Experimental Neurology*, 232(2), 154–161.

73 Argyriou, A.A., Koltzenburg, M., Polychronopoulos, P. *et al.* (2008) Peripheral nerve damage associated with administration of taxanes in patients with cancer. *Critical Reviews in Oncology/Hematology*, 66(3), 218–228.

74 Cersosimo, R.J. (1989) Cisplatin neurotoxicity. *Cancer Treatment Reviews*, 16(4), 195–211.

75 Joseph, E.K. & Levine, J.D. (2009) Comparison of oxaliplatin- and cisplatin-induced painful peripheral neuropathy in the rat. *The Journal of Pain*, 10(5), 534–541.

76 van den Bent, M.J., van Raaij-van den Aarssen, V.J., Verweij, J. *et al.* (1997) Progression of paclitaxel-induced neuropathy following discontinuation of treatment. *Muscle & Nerve*, 20(6), 750–752.

77 Smith, E.M., Pang, H., Cirrincione, C. *et al.* (2013) Effect of duloxetine on pain, function, and quality of life among patients with chemotherapy-induced painful peripheral neuropathy: a randomized clinical trial. *JAMA*, 309(13), 1359–1367.

78 Hershman, D.L., Unger, J.M., Crew, K.D. *et al.* (2013) Randomized double-blind placebo-controlled trial of acetyl-L-carnitine for the prevention of taxane-induced neuropathy in women undergoing adjuvant breast cancer therapy. *Journal of Clinical Oncology*, 31(20), 2627–2633.

79 Bianchi, G., Vitali, G., Caraceni, A. *et al.* (2005) Symptomatic and neurophysiological responses of paclitaxel- or cisplatin-induced neuropathy to oral acetyl-L-carnitine. *European Journal of Cancer*, 41(12), 1746–1750.

80 Franek, M., Vaculin, S. & Rokyta, R. (2004) GABA(B) receptor agonist baclofen has non-specific antinociceptive effect in the model of peripheral neuropathy in the rat. *Physiological Research*, 53(3), 351–355.

81 Duale, C., Daveau, J., Cardot, J.M. *et al.* (2008) Cutaneous amitriptyline in human volunteers: differential effects on the components of sensory information. *Anesthesiology*, 108(4), 714–721.

82 Ultenius, C., Linderoth, B., Meyerson, B.A. *et al.* (2006) Spinal NMDA receptor phosphorylation correlates with the presence of neuropathic signs following peripheral nerve injury in the rat. *Neuroscience Letters*, 399(1–2), 85–90.

83 Barton, D.L., Wos, E.J., Qin, R. *et al.* (2011) A double-blind, placebo-controlled trial of a topical treatment for chemotherapy-induced peripheral neuropathy: NCCTG trial N06CA. *Supportive Care in Cancer*, 19(6), 833–841.

84 Weijl, N.I., Hopman, G.D., Wipkink-Bakker, A. *et al.* (1998) Cisplatin combination chemotherapy induces a fall in plasma antioxidants of cancer patients. *Annals of Oncology*, 9(12), 1331–1337.

85 Argyriou, A.A., Chroni, E., Koutras, A. *et al.* (2005) Vitamin E for prophylaxis against chemotherapy-induced neuropathy: a randomized controlled trial. *Neurology*, 64(1), 26–31.

86 Pace, A., Giannarelli, D., Galie, E. *et al.* (2010) Vitamin E neuroprotection for cisplatin neuropathy: a randomized, placebo-controlled trial. *Neurology*, 74(9), 762–766.

87 Afonseca, S.O., Cruz, F.M., Cubero Dde, I. *et al.* (2013) Vitamin E for prevention of oxaliplatin-induced peripheral neuropathy: a pilot randomized clinical trial. *São Paulo Medical Journal*, 131(1), 35–38.

88 Kottschade, L.A., Sloan, J.A., Mazurczak, M.A. *et al.* (2011) The use of vitamin E for the prevention of chemotherapy-induced peripheral neuropathy: results of a randomized phase III clinical trial. *Supportive Care in Cancer*, 19(11), 1769–1777.

89 Bhadri, N., Sanji, T., Madakasira Guggilla, H. *et al.* (2013) Amelioration of behavioural, biochemical, and neurophysiological deficits by combination of monosodium glutamate with resveratrol/alpha-lipoic acid/coenzyme Q10 in rat model of cisplatin-induced peripheral neuropathy. *The Scientific World Journal*, 2013, 565813.

90 Azevedo, M.I., Pereira, A.F., Nogueira, R.B. *et al.* (2013) The antioxidant effects of the flavonoids rutin and quercetin inhibit oxaliplatin-induced chronic painful peripheral neuropathy. *Molecular Pain*, 9, 53.

91 Schloss, J.M., Colosimo, M., Airey, C. *et al.* (2013) Nutraceuticals and chemotherapy induced peripheral neuropathy (CIPN): a systematic review. *Clinical Nutrition*, 32(6), 888–893.

92 Rebbaa, A., Patil, G., Yalcin, M. *et al.* (2013) OT-404, multi-targeted anti-cancer agent affecting tumor proliferation, chemo-resistance, and angiogenesis. *Cancer Letters*, 332(1), 55–62.

CHAPTER 25

Grape polyphenol-rich products with antioxidant and anti-inflammatory properties

Eduarda Fernandes[1], Marisa Freitas[1], Renan C. Chisté[1], Elena Falqué[2], and Herminia Domínguez[3]

[1] *UCIBIO-REQUIMTE, Department of Chemical Sciences, Faculty of Pharmacy, University of Porto, 4050-313 Porto, Portugal*

[2] *Departamento de Química Analítica, Facultad de Ciencias, Universidade de Vigo (Campus Ourense), 32004 Ourense, Spain*

[3] *Departamento de Enxeñería Química, Facultad de Ciencias, Universidade de Vigo (Campus Ourense), 32004 Ourense, Spain*

THEMATIC SUMMARY BOX

At the end of this chapter, students should be able to:

- Describe the basic chemical composition of grapes, wines, and wine by-products

- Classify the phenolic compounds in grapes and wines

- List some phenolic molecules with health-benefiting properties

- Understand the meaning of dietary antioxidants

- Recognize the compounds present in grape extracts and its derivative products that may contribute to its antioxidant and anti-inflammatory effects

- List the immune cells that could be modulated by grape extracts and its derivative products

- Mention the steps in the cascade of activation of NF-κB, where grape extracts and its derivative products may interfere

- List the advantages of novel green extraction processes

- Describe the major features and variables affecting pressurized extraction processes

- List some technologies for the concentration and purification of phenolics

---

Oxidative Stress and Antioxidant Protection: The Science of Free Radical Biology and Disease, First Edition.
Edited by Donald Armstrong and Robert D. Stratton.
© 2016 John Wiley & Sons, Inc. Published 2016 by John Wiley & Sons, Inc.

Bioactive compounds in grape, wine, and wine by-products: phenolic compounds

Different groups of substances are present in grape, must, wine, and derived by-products, which includes water, sugars, organic acids, nitrogen compounds, minerals, vitamins, phenolic, and aromatic substances. Sugars, glucose and fructose, are the most abundant ones in grape and must, and practically disappear in wine. Organic acids possess a carboxylic group (–COOH) and sometimes an alcohol group (–OH). Tartaric, malic, and citric acids play an essential role on the maturation of berries and on the wine quality. When the grapes mature, the acid content decreases while the sugar concentration increases. The most relevant nitrogen compounds are amino acids, polypeptides, and proteins, which are the precursors of some volatile compounds. The cations K^+, Ca^{2+}, Mg^{2+}, and Na^+ and the anions PO_4^{3-}, SO_4^{2-}, and Cl^- are the main minerals in grapes and wines. Vitamin C and small quantities of vitamin B group are also present, essential for yeast to carry out the alcoholic fermentation. Phenolic compounds are responsible for color, taste, and aroma of grapes and wines. Some volatile compounds are present in grapes, whereas others are formed during alcoholic and malolactic fermentations.

Epidemiological studies carried out by Renaud and de Lorgeril showed that the consumption of 20–30 g alcohol/day could reduce the risk of coronary heart disease (CHD) in France in comparison with other countries.[1] This observation of low CHD death rates despite high intake of dietary cholesterol and saturated fat was labeled as "French paradox" (later also called "Mediterranean paradox" or "North–South paradox"). The specific mechanism of the "French paradox" has not yet been identified, but the beneficial effects of moderate consumption of wine on cardiovascular system are due to ethanol and phenolic content.[2]

Grapes (*Vitis vinifera* L.) are important sources of phenolic compounds. In grapes, most phenolics are associated with skins and seeds, their content being mainly dependent on grape variety, berry ripening degree, vine growing methods, and other viticulture practices. Most of all phenolics in wine are originated in the grape berry, and its concentration is influenced by the winemaking process or aging and storage conditions.[3] White wines present low phenolic content, since the skins and seeds are separated from the juice. On the contrary, in red wine vinification, the musts are fermented in the presence of solid parts, and finished red wines present higher phenolic content, reaching 1500–3500 mg/L.[4]

Phenolic compounds, as phytochemicals, play a major role in plants' and vegetables' defense systems, but particularly they are known to have several health-benefit properties.[4] According to the number of the basic units present in the molecule, phenolic compounds can be classified as "simple phenols" and "polyphenols" if one or several aromatic groups are present, respectively. Phenolic compounds in grapes and wines can also be divided into two groups (Table 25.1): flavonoid and nonflavonoid compounds. The antiradical and antioxidant properties are determined by the structure and by the position and degree of hydroxylation of the ring structure.

Flavonoids or proanthocyanidins are predominant in skins and seeds. The basic carbon structure contains 15 carbons and is endowed with two aromatic rings linked by a 3-carbon bridge. The most abundant classes of soluble polyphenols in grape

Table 25.1 Classification of phenolic compounds from grapes and wines.

Flavonoids		Nonflavonoids	
Subclass	Example	Subclass	Example
Catechins (flavan-3-ols)	Catechin, epicatechin, gallocatechin, procyanidins, etc	Hydroxybenzoic acids and derivatives	Vanillic acid, gallic acid, vanillin, syringaldehyde, etc
Flavonols	Quercetin, kaempferol, myricetin, etc	Hydroxycinnamic acids and derivatives	p-coumaric acid, ferulic acid, chlorogenic acid, caffeic acid, coniferaldehyde, sinapaldehyde, etc
Anthocyanins	Cyanidin, delphinidin, peonidin, malvidin etc	Hydrolyzable tannins	Gallotannins, ellagitannins
Condensed tannins, derivatives	Oligomers and polymers of flavan-3-diols (catechin or epicatechin derivatives)	Stilbenes	Resveratrol, piceid, viniferin, etc
		Other	Tyrosol, hydroxytyrosol, tryptophol, etc

berries are the condensed tannins or polymers of flavan-3-ol, followed by the anthocyanins, also located in the skin, allowing differentiating between red and white grapes. Malvidin-3-glucoside is the most important in *V. vinifera* grape. The third main group is flavonols, present as free forms in wines or bonded as glucosides, galactosides, rutinosides (ramnose + glucose, as found in rutin) and glucuronides in grapes.

Predominant phenolic acids include hydroxybenzoic and hydroxycinnamic acids, either occurring in the free or conjugated forms, usually as esters. Stilbenes are phenolic compounds displaying two aromatic rings linked by an ethane bridge, and the most abundant and well known is *trans*-resveratrol, a phytoalexin produced by plants and mainly contained in the skins of grapes, with a broad range of health benefits.[5] Red wine contains polyphenols that are present in the skin and seeds of grapes, and some of the various phenolic antioxidants present in red wine (Figure 25.1) are the most potent agent to protect health human.

The total phenolic composition was traditionally determined by specific spectrophotometric analysis; however, when the identification and quantification of each substance is needed, a high-performance liquid chromatographic technique (with refractive index, UV–vis absorption or fluorometric detection, or coupled to mass spectrometry) is employed.

Antioxidant and anti-inflammatory properties of grape extracts

Antioxidant properties

Dietary antioxidants have a broad scope and may be defined as substances in foods that significantly decrease the adverse effects of reactive oxygen species (ROS) and

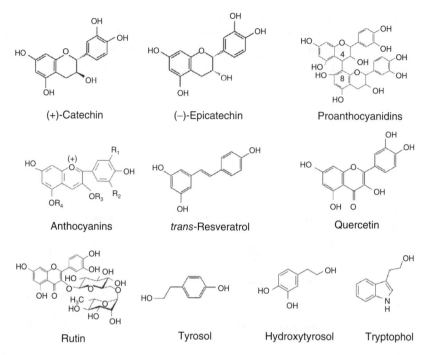

Figure 25.1 Chemical structures of main phenolic compounds present in grapes and wines.

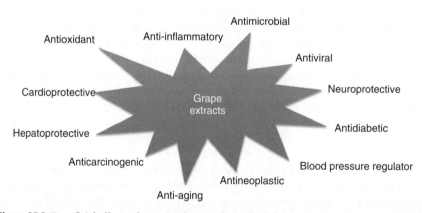

Figure 25.2 Beneficial effects of grape and grape-derived products.

reactive nitrogen species (RNS) on normal physiological functions in humans. The beneficial effects of grape and grape-derived products (Figure 25.2) have been associated with the presence of bioactive components, such as the phenolic compounds typically including resveratrol, anthocyanins, catechins, phenolic acids, and procyanidins. The *in vitro* and *in vivo* antioxidant properties of phenolic compounds have been extensively reported.[6–10]

According to the data gathered by Yang and Xiao, some *in vitro* and *in vivo* studies suggest that grape phenolics may decrease the risk of cardiovascular disease through the modulation of cellular redox state, prevention of low-density lipoprotein (LDL) oxidation, improvement of endothelial function, lowering blood pressure and inflammation, and inhibition of platelet aggregation.[10] Furthermore, the cardioprotective and vasoprotective properties mediated by grape phenolics may be associated with their angiogenic, antihypercholesterolemic, antiatherosclerotic, antiarrhythmic, and antidiabetic actions, with mechanisms involved in reduced ventricular remodeling and increased cardiac functions.[10] In general, these risk-preventing effects are claimed to be related to the inhibition or decrease in oxidation reactions via scavenging ROS and RNS.

The extracts obtained from grape, including pulp, skin, pomace, seed, and leaf extracts, possess high free radical scavenging capacities, as measured by the assays of oxygen radical absorbance capacity (ORAC), 2,2-diphenyl-1-picrylhydrazyl (DPPH), 2,2′-azino-bis(3-ethylbenzothiazoline-6-sulphonic acid) (ABTS), ferric reducing antioxidant power (FRAP), crocin bleaching assay (CBA), and thiobarbituric acid reactive substances (TBARS).[6,8,9] Moreover, some complex phenolics also show significant chelating activity on transition metal ions, which are known as lipid peroxidation inducers. Regarding the different parts of grape, the highest scavenging capacity was reported in grape seeds, followed by skin, and the flesh displayed the lowest scavenging capacity.[8]

The antioxidant potential of phenolic compounds from grape has also been demonstrated in various model systems, such as in rat liver microsomes, where the antioxidant activity (*ex vivo*) of resveratrol, from grape skin and seeds, was associated with the scavenging capacity against ROS, decrease in TBARS levels, and decrease in DNA damage.[6] Moreover, resveratrol was also reported to protect tissues from oxidative stress when catechol estrogen-induced damage is present.[6] Resveratrol is widely recognized as an excellent antioxidant, mainly due to the high correlation with its concentration and the antioxidant capacity of grapes.[10]

In the same sense, a positive correlation was found between anthocyanin contents and antioxidant activities of red grape extracts, grape juices, and red wines. Additionally, anthocyanins from grape skin exhibited a decrease in TBARS and carbonyl group levels in Wistar rats.[6]

Concerning other phenolic compounds, it was also reported that flavonoids from Concord grape juice (CGJ) presented antioxidant effects in healthy adults (*in vivo* experiment), where an increase in serum oxygen radical absorbance capacity was observed, while plasma protein carbonyls and LDL oxidation decreased after the intake of CGJ.[11] Grape seed proanthocyanidins inhibited polyunsaturated fatty acid peroxidation, presented higher scavenging capacity than vitamins C and E (*in vitro* assay), and were reported to protect DNA against oxidative damage.[7]

It is reported that dietary intake of grape antioxidants may prevent lipid peroxidation and inhibit ROS production.[9] As an example, a 4-week dietary supplementation of grape seed extract (600 mg/day) showed a decrease in oxidative stress and improved the reduced glutathione (GSH)/oxidized glutathione (GSSG) ratio, as well as the total antioxidant status in a double-blinded, randomized crossover human trial.[12] These authors suggest that grape seed extract may have a therapeutic role in decreasing cardiovascular risk.

Anti-inflammatory properties

Inflammation comprises a complex and highly variable set of processes that represent a response against cell injury, irritation, and pathogen invasions, as well as a mechanism for eliminating damaged and necrotic cells.[8] Under normal physiological conditions, a short period of acute inflammation can overcome negative effects on injured tissue. However, when inflammation becomes chronic or lasts too long, it can be harmful and may lead to several diseases. In addition, it has been estimated that approximately 15% of all cancers occur as a consequence of inflammatory processes.[13]

Grapes and its derivatives have shown significant anti-inflammatory effects on rats, mice, and humans, with the effect attributed to its phenolic content.[8] The contributive molecules may be resveratrol, anthocyanins, catechins, phenolic acids, and procyanidins.[8] It is important to note that the inflammatory process is orchestrated by many molecules, and the modulation of one or a few could not be enough. At the tissue level, the manifestations of inflammation, often called its cardinal signs, are characterized by redness, heat, pain, swelling, and loss of function, which result from local responses of immune, vascular, and parenchymal cells to infection or injury.

The acute inflammatory response rapidly delivers leukocytes and plasma proteins to sites of injury. Neutrophils are the first cells to arrive at the affected location. These cells clear the invaders through the production of an array of reactive species. It has been shown that grapes and its derivatives modulate the inflammatory process through the decrease in the reactive species production by inhibiting crucial enzymes, as nicotinamide adenine dinucleotide phosphate (NADPH) oxidase or, as discussed earlier, by the ability to scavenge ROS/RNS.[8,14] Moreover, resveratrol, one of the major phenolic compounds of grapes and red wine, modulates several human immune cell functions, namely, in peripheral blood mononuclear cells, T lymphocytes, and natural killer cells.[15] Within minutes of tissue injury, the immune cells activate transcription factors such as nuclear factor-kappa B (NF-κB) and activator protein-1 (AP-1), which are evolutionarily conserved eukaryotic transcription factors, acting independently or coordinately, that are responsible for the expression of genes involved in inflammatory response.[13]

Given that NF-κB is a known redox-sensitive transcription factor, functioning as proinflammatory signaling in the downstream of oxidative stress, finding a compound that inhibits its activation could be an excellent alternative anti-inflammatory agent. In this sense, the ability to inhibit the activity of NF-κB, *in vitro* and *in vivo*, of various phenolic compounds of grapes, namely resveratrol, has been described in the literature.[13,16] In resting cells, NF-κB, which is composed mainly of two proteins, p50 and p65, is present within the cytoplasm in an inactive state, bound to its inhibitory protein IκBα. However, a number of proinflammatory stimuli [cytokines such as tumor necrosis factor (TNF)-α or interleukin-1 (IL-1), oxidative stress, and infectious agents] can activate NF-κB in different cell types. These inflammatory stimulations can initiate an intracellular signaling cascade leading to IκBα phosphorylation, by IκB kinase (IKK) complex, with its subsequent dissociation from NF-κB and degradation by the proteasome. Once liberated from its inhibitory protein, NF-κB translocates to the nucleus (Figure 25.3). Nevertheless, several studies suggest that nuclear translocation of NF-κB may take place even in the absence of IκBα degradation.[16] Grapes and its phenolic components inhibit NF-κB activation by blocking some of the multiple steps of the above-mentioned cascade of the activation pathway of NF-κB. An interesting work of Schubert *et al.* showed that red wine inhibited p65 nuclear translocation, and that inhibition was accompanied by inhibition of p65 binding activity in bovine aortic endothelial cell cultures.[17] It seems that resveratrol is able to block phosphorylation and degradation of IκBα, and subsequent nuclear translocation and DNA binding of NF-κB subunits. In addition to the suppression of NF-κB, grapes and its derivatives inhibited the activation of AP-1. The transcription factor AP-1 can be produced by about 18 different dimeric combinations of proteins from the Jun and Fos family, Jun dimerization partners (JDP1 and JDP2) and the closely related activating transcription factor (ATF2, LRF1/ATF3, and B-ATF) subfamilies, which are bZIP proteins. The AP-1 activity controls both basal and inducible transcription of several genes containing AP-1 sites (consensus sequence 5′-TGAG/CTCA-3′), also known as TPA-responsive elements (TREs). The activation of AP-1 is mediated predominantly via the mitogen-activated protein kinase (MAPK) cascades (Figure 25.3).[13] Once again, the anti-inflammatory activity of grapes could be attributed to the inhibition of the activation of AP-1 or may be partially ascribed to the inhibition of the associated upstream kinases. As example, pretreatment of human adipocytes with grape powder extract for 1 h decreased TNF-α-mediated phosphorylation of c-Jun, a component of AP-1 and a downstream target of Jun-NH$_2$-terminal kinase (JNK).[18] Some of the major phenolic compounds of grapes were also able to inactivate directly or indirectly AP-1, such as resveratrol, quercetin, catechin among others.[19]

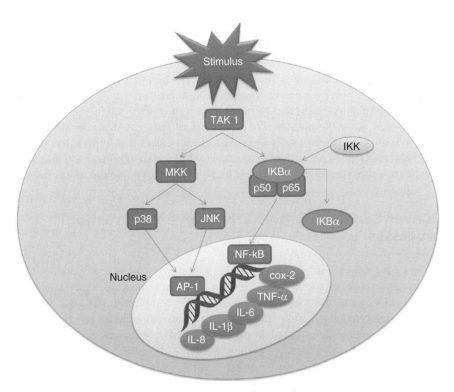

Figure 25.3 General signaling mechanisms associated with inflammatory processes.

Several studies have shown that activation of NF-κB and AP-1 is responsible for increasing the expression of inflammatory cytokines such as TNF-α, IL-6, IL-1β, IL-8, and monocyte chemoattractant protein (MCP)-1 and inflammatory proteins such as toll-like receptor (TLR)-2.[18] It has been observed that several phenolic compounds of grapes were able to decrease the expression of these proinflammatory cytokines/chemokines, in different cell types such as RAW macrophages, Jurkat T-cells, and peripheral blood mononuclear cells. The molecular mechanisms involved in the modulation of cytokine activity could be through the inhibition of transcription factors NF-κB and AP-1 and reduction of MAPK activity.[19] Once again, the grape compound that had more impact on literature about this issue is resveratrol. NF-κB and AP-1 were also responsible for the expression of other proteins apart from cytokines, which initiate the inflammatory response such as cyclooxygenase (COX)-2. COX-2 isoforms are associated with the production of prostaglandins and thromboxanes (TXs), collectively termed prostanoids, formed after cleavage of arachidonic acid from the cell membrane.[20] The modulation of inflammatory process by grape derivatives could also occur through the inhibition of COX-2 activity, which was found to be reduced in the 2,4,6-trinitrobenzenesulfonic acid-induced colitis in Wistar rats exposed to 1% grape juice.[21] Inclusively, the grape derivatives, namely grape seed proanthocyanidins, are considered promising agents against a broad range of cancer cells due to its inhibitory effect on overexpression of COX-2 and prostaglandin E_2 receptors.[9] Novel strategies of anti-inflammatory therapy are the modulation of the cell growth and apoptosis processes. In addition, the inhibition of cell cycle progression is a possible target for chemopreventive agents.

Apoptosis, a programmed cell death, is an important cellular process that participates in the altered homeostasis of various pathophysiological conditions. Grape extracts are able to induce cell death. Despite the importance of this issue to the anti-inflammatory activity of grapes, the studies in literature about apoptosis and cell antiproliferative effects are mainly in cancer cell lines. In fact, it is clearly demonstrated that grape phenolics have some interesting effects as chemopreventives.[10] This effect of grape phenolic compounds could be interesting in the modulation of chronic inflammation, since the induction of inflammatory cells' apoptosis could result on the resolution of inflammation. Nevertheless, there is a lack of information about the effect of grapes and its derivatives in antiproliferative and apoptosis of inflammatory cells. Once again, resveratrol is the most studied phenolic compound in this field.

Conventional and alternative solvent extraction processes

The selective separation, recovery, and concentration of active components from grape, grape products, and by-products are a key step to provide highly active and increasingly demanded ingredients for functional foods, nutraceuticals, and pharmaceutical products. Solid–liquid extraction is a unit operation useful for the selective separation of components from plant materials, based on the mass transfer of solutes from a solid matrix to a fluid.

Traditional extraction processes using acetone, methanol, ethanol, and water are widely employed for extracting phenolic components at atmospheric pressure.

Toxicity concerns favor the use of ethanol–water mixtures and hot water although yields and selectivity are not optimal. An influential variable is temperature, which cannot be increased indefinitely, because instability of phenolic compounds may occur by oxidation, hydrolysis, and isomerization. Other variables influencing the yield, rate, and selectivity are the particle size, moisture content, and extraction time.

Conventional extraction with organic solvents may require large solvent volume and prolonged times. Novel extraction techniques have been developed to overcome these limitations and have been promoted by increasingly restrictive environmental, toxicological, and health regulations. Improved efficient and selective novel extraction methods are based on the use of safer solvents and are performed under pressurized conditions, which can also be enhanced by intensification strategies, that is, ultrasound and microwave. Table 25.2 summarizes these technologies reported to extract phenolic compounds from grape products and by-products.

Table 25.2 Techniques proposed for the extraction of the phenolic fraction from grapes, grapes products, and by-products and the major variables affecting the process.

Material/solutes	Extraction technique	Operational conditions
Grapes and grape pomace/PhC	Conventional solvent extraction (CSE)	Pressure: atmospheric Temperature: 25–50 °C Time: 30–90 min Solvent: ethanol or methanol water solutions
Grape pomace/PhC	Ultrasound-assisted extraction (UAE)	Pressure: atmospheric Temperature: 40 °C Time: 30 min Solvent: 50% aqueous ethanol Frequency: 25 kHz
Grape pomace/PC	Accelerated solvent extraction (ASE)	Pressure: 6–10 MPa Temperature: 40–140 °C Time: 20 min Solvents: acetone, ethanol, water
Grape pomace/PhC	Enzyme-assisted extraction (EAE)	Pressure: atmospheric Temperature: 50 °C Enzyme: cellulase, xylanase, pectinase, protease Time: 8 h
Grape seeds/SA, VA, GA, HBA, PA, pCA, Q, C, EC, EGC, EGCG, R	Supercritical fluid extraction (SFE)	Pressure: 20–35 MPa Temperature: 40–60 °C Time: 240 min Cosolvent: 5–20% (w/w) ethanol
Grape pomace/ C, EC, K, M, R, AC, CT	Pressurized hot water extraction (PHWE)	Pressure: up to 1.5 MPa Temperature: 50–200 °C Time: 5–30 min

C, catechin; EC, epicatechin; EGC, epigallocatechin; EGCG, epigallocatechingallate; SA, syringic acid; VA, vanillic acid; GA, gallic acid; HBA, *p*-hydroxybenzoic acid; PA, protocatechuic acid; pCA, *p*-coumaric acid; PC, procyanidins; PhC, phenolic compounds; Q, quercetin; R, resveratrol AC: anthocyanins; CT: condensed tannins; K: kaempferol; M: myricetin.

Pressurized liquid extraction

Pressurized liquid extraction or accelerated solvent extraction is performed at high temperature and pressure to maintain the solvent in liquid state, offering improved mass transfer and lower solvent requirements than traditional extraction. The structure, activity, and properties of the extracts are usually unaffected in an oxygen and light-free environment, and microorganisms and enzymes may be inactivated.

The adequate selection of pressure, temperature, solvent, solid/solvent ratio, and time is required. Ethanol–water mixtures are less toxic and can be more efficient than mixtures containing methanol or acetone. Thermal degradation of flavonoids can occur at temperatures higher than 100 °C and for prolonged time. The development of brown, polar, odoriferous, and antioxidant compounds at high extraction temperatures suggested the participation of Maillard reactions in ethanolic and in water extracts.

When the solvent used is water, subcritical water extraction, superheated water extraction, high-temperature water extraction, and pressurized hot water extraction (PHWE) are alternative names. Water exists in the subcritical state at 100–374 °C. Most studies have been performed in batch processes, but a semicontinuous process has also been developed to maximize the recovery of anthocyanins and procyanidins from red grape pomace.[22]

High temperature decreases the dielectric constant of water, also enhances diffusivities, facilitating the transport of solutes from the solid matrix, and may aid in the breakdown of cellular constituents. However, a compromise with thermal degradation must be reached. The release of hydroxycinnamates from cell walls is favored by temperature, but anthocyanins can undergo degradation reactions. Stepwise pressure/temperature increases can be useful to recover molecules with different degrees of polymerization and structure, that is, galloylated compounds (50–100 °C), flavanol dimers and trimers, and gallic acid (150 °C) and oligomeric fractions (>150 °C).[23]

Supercritical fluid extraction (SFE) using CO_2

Carbon dioxide under supercritical conditions, SC-CO_2, shows favorable physico-chemical and transport properties in comparison with conventional solvents. The characteristics of the final product are improved by high selectivity, low processing temperatures, and absence of solvent in extracts. SC-CO_2 extracts are regarded as natural and do not require additional sterilization. The major disadvantages of SC-CO_2 are the high pressure required and the poor solvent power because of its low polarity.

The economic feasibility of large-scale supercritical fluid extraction (SFE) operations for the recovery of phenolic compounds from grape seeds and bagasse was confirmed.[24,25]

The careful selection of the most important variables of the SC-CO_2 extraction is recommended. Samples require proper drying without affecting the structure of the solid or the thermal stability of the solutes. Mass transfer is facilitated by higher grinding degree, but the performance of fixed beds during leaching could be limited by too fine particles. Pressure and temperature determine the solubility equilibrium and solvent density increases with pressure, but beyond a certain threshold, the increased solvent viscosity could reduce the diffusion coefficients. The presence of a modifier increases the solvent density and its interaction with solutes. The commonly used modifiers are alcohols (ethanol, methanol, isopropanol).

The extracts can be fractionated into multiple-stage separators. In addition a stepwise increase in the pressure enables the selective extraction of compounds, that is, the oil from grape seeds with pure SC-CO2; monomeric polyphenols with methanol-modified subcritical CO_2; dimers, trimers, and procyanidins with pure methanol.[26]

Concentration and purification

Several technologies with different degrees of sophistication (including solvents, resins, and membranes) have been used to obtain purified compounds or fractions with biological activity from crude solvent extracts. A simple purification process based on the sequential extraction of the raw materials or extracts with solvents of different polarity has been applied to the extraction of antioxidant compounds from plant foods.

Membrane processes offer advantages derived from the low-energy requirement, no additives, mild operating conditions, separation efficiency, and ease of scaling up. Ultrafiltration and nanofiltration have been used to concentrate aqueous extracts from grape pomace. Ultrafiltration was proposed for tailoring grape anthocyanins according to their molecular weight.[27] The polymeric forms are recovered in the retentates, whereas permeates contained low-MW compounds.[28]

Extraction followed by chromatographic separation was used for the purification of oligomeric procyanidins.[29] Liquors obtained by pressing distilled grape pomace were refined by solvent extraction of concentrates and adsorption–desorption of permeates onto commercial polymeric resins.[27]

Conclusions

Undoubtedly, grapes and its derivative products are a reliable and rich source of phytochemicals, whose individual and summated biological activities suggest potentially beneficial effects for human health and as nutritional supplements and easily accessible sources of natural antioxidants and anti-inflammatory compounds. Nevertheless, further research is required using the whole extracts of grape or grape derivatives in order to understand the additive and synergistic molecular mechanisms of its bioactive compounds *in vitro* and *in vivo*, instead of just its activity *per se*. It is also important to clarify the mode and dosage of application that maximize the anti-inflammatory benefits.

Novel extraction procedures using green nontoxic solvents are technically available and their implementation is recommended based on environmental, toxicological, and functional perspectives.

Multiple choice questions

1 The major phenolic compounds in grapes and wines are present in the lowest concentration in:
   a. The skin
   b. The pulp
   c. The seeds

2 Grapes and its derivatives modulate the inflammatory process by:
   a. Inhibiting NF-κB activity
   b. Increasing COX-2 activity
   c. The production of reactive species

3 In order to choose a selective solvent for extracting an apolar bioactive from grape by-products to be incorporated into a nutraceutical product, which of these solvent extraction process could you recommend?
   a. Conventional dichloromethane extraction
   b. Supercritical CO_2 extraction
   c. Enzyme-assisted aqueous extraction

References

1 Renaud S, de Lorgeril M. (1992). Wine, alcohol, platelets, and the French paradox for coronary heart disease. *Lancet* 339(8808):1523–1526. http://www.ncbi.nlm.nih.gov/pubmed/1351198.

2 Estruch, R. & Lamuela-Raventós, R.M. (2014) Wine, alcohol, polyphenols and cardiovascular disease. *Nutrition and Aging*, 2, 101–109.

3 Lachman, J., Sulc, M., Faitová, K. *et al.* (2009) Major factors influencing antioxidant contents and antioxidant activity in grapes and wines. *International Journal of Wine Research*, 1, 101–121.

4 Ebeler, S.E. (2013) Wine and cancer. In: Carkeet, C., Grann, K., Randolph, R.K., *et al.* (eds), *Phytochemicals: Health Promotion and Therapeutic Potential*. CRC Press Taylor & Francis, Boca Raton, pp. 21–38 ISBN: 978-1-4665-5162-6.

5 Pollack, R.M. & Crandall, J.P. (2013) Resveratrol: therapeutic potential for improving cardiometabolic health. *American Journal of Hypertension*, 26(11), 1260–1268.

6 Yadav, M., Jain, S., Bhardwaj, A. *et al.* (2009) Biological and medicinal properties of grapes and their bioactive constituents: an update. *Journal of Medicinal Food*, 12(3), 473–484.

7 Nassiri-Asl, M. & Hosseinzadeh, H. (2009) Review of the pharmacological effects of *Vitis vinifera* (grape) and its bioactive compounds. *Phytotherapy Research*, 23, 1197–1204.

8 Xia, E.Q., Deng, G.F., Guo, Y.J. *et al.* (2010) Biological activities of polyphenols from grapes. *International Journal of Molecular Sciences*, 11, 622–646.

9 Zhou, K. & Raffoul, J.J. (2012) Potential anticancer properties of grape antioxidants. *Journal of Oncology*, 2012, 803294 8 pages.

10 Yang, J. & Xiao, Y. (2013) Grape phytochemicals and associated health benefits. *Critical Reviews in Food Science and Nutrition*, 53(11), 1202–1225.

11 O'Byrne, D.J., Devaraj, S., Grundy, S.M. *et al.* (2002) Comparison of the antioxidant effects of Concord grape juice flavonoids and α-tocopherol on markers of oxidative stress in healthy adults. *American Journal of Clinical Nutrition*, 76(6), 1367–1374.

12 Kar, P., Laight, D., Rooprait, H.K. *et al.* (2009) Effects of grape seed extract in type 2 diabetic subjects at high cardiovascular risk: a double blind randomized placebo controlled trial

examining metabolic markers, vascular tone, inflammation, oxidative stress and insulin sensitivity. *Diabetic Medicine*, 26(5), 526–531.

13 Kundu, J.K., Shin, Y.K. & Surh, Y.J. (2006) Resveratrol modulates phorbol ester-induced pro-inflammatory signal transduction pathways in mouse skin in vivo: NF-kappaB and AP-1 as prime targets. *Biochemical Pharmacology*, 72(11), 1506–1515.

14 Castilla, P., Dávalos, A., Teruel, J.L. *et al.* (2008) Comparative effects of dietary supplementation with red grape juice and vitamin E on production of superoxide by circulating neutrophil NADPH oxidase in hemodialysis patients. *The American Journal of Clinical Nutrition*, 87(4), 1053–1061.

15 Falchetti, R., Fuggetta, M.P., Lanzilli, G. *et al.* (2001) Effects of resveratrol on human immune cell function. *Life Sciences*, 70(1), 81–96.

16 Kundu, J.K. & Surh, Y.J. (2004) Molecular basis of chemoprevention by resveratrol: NF-kappaB and AP-1 as potential targets. *Mutation Research*, 555(1–2), 65–80.

17 Schubert, S.Y., Neeman, I. & Resnick, N. (2002) A novel mechanism for the inhibition of NFκB activation in vascular endothelial cells by natural antioxidants. *FASEB Journal*, 16(14), 1931–1933.

18 Chuang, C.C. & McIntosh, M.K. (2011) Potential mechanisms by which polyphenol-rich grapes prevent obesity-mediated inflammation and metabolic diseases. *Annual Review of Nutrition*, 31, 155–176.

19 Serafini, M., Peluso, I. & Raguzzini, A. (2010) Flavonoids as anti-inflammatory agents. *Proceedings of the Nutrition Society*, 69(3), 273–278.

20 Ramon, S., Woeller, C.F. & Phipps, R.P. (2013) The influence of Cox-2 and bioactive lipids on hematological cancers. *Current Angiogenesis*, 2(2), 135–142.

21 Marchi, P., Paiotti, A.P.R., Neto, R.A. *et al.* (2014) Concentrated grape juice (G8000™) reduces immunoexpression of iNOS, TNF-alpha, COX-2 and DNA damage on 2,4,6-trinitrobenzene sulfonic acid-induced-colitis. *Environmental Toxicology and Pharmacology*, 37(2), 819–827.

22 Monrad, J.K., Srinivas, K., Howard, L.R. *et al.* (2012) Design and optimization of a semicontinuous hot-cold extraction of polyphenols from grape pomace. *Journal of Agricultural and Food Chemistry*, 60(22), 5571–5582.

23 Vergara-Salinas, J.R., Bulnes, P., Zuniga, M.C. *et al.* (2013) Effect of pressurized hot water extraction on antioxidants from grape pomace before and after enological fermentation. *Journal of Agricultural and Food Chemistry*, 61(28), 6929–6936.

24 Prado, J.M., Dalmolin, I., Carareto, N.D.D. *et al.* (2012) Supercritical fluid extraction of grape seed: process scale-up, extract chemical composition and economic evaluation. *Journal of Food Engineering*, 109(2), 249–257.

25 Farias-Campomanes, A.M., Rostagno, M.A. & Meireles, M.A.A. (2013) Production of polyphenol extracts from grape bagasse using supercritical fluids: yield, extract composition and economic evaluation. *Journal of Supercritical Fluids*, 77, 70–78.

26 Murga, R., Ruiz, R., Beltrán, S. *et al.* (2000) Extraction of natural complex phenols and tannins from grape seeds by using supercritical mixtures of carbon dioxide and alcohol. *Journal of Agricultural and Food Chemistry*, 48(8), 3408–3412.

27 Díaz-Reinoso, B., González-López, N., Moure, A. *et al.* (2010) Recovery of antioxidants from industrial waste liquors using membranes and polymeric resins. *Journal of Food Engineering*, 96(1), 127–133.

28 Kalbasi, A. & Cisneros-Zevallos, L. (2007) Fractionation of monomeric and polymeric anthocyaninins from concord grape (*Vitis labrusca* L.) juice by membrane ultrafiltration. *Journal of Agricultural and Food Chemistry*, 55(17), 7036–7042.

29 Huh, Y.S., Hong, T.H. & Hong, W.H. (2004) Effective extraction of oligomeric proanthocyanidin (OPC) from wild grape seeds. *Biotechnology and Bioprocess Engineering*, 9(6), 471–475.

CHAPTER 26

Isotonic oligomeric proanthocyanidins

Hanbo Hu[1,2] and Donald Armstrong[3,4]

[1] Pulmonary Division, Department of Medicine, University of Florida, Gainesville, FL, USA
[2] Malcom Randall Veteran Affairs Medical Center, Gainesville, FL, USA
[3] Department of Biotechnical and Clinical Laboratory Sciences, State University of New York at Buffalo, Buffalo, NY, USA
[4] Department of Ophthalmology, University of Florida School of Medicine, Gainesville, FL, USA

THEMATIC SUMMARY BOX

At the end of this chapter, students should be able to:

- Define oligomeric proanthocyanidins (OPCs)

- Show how OPCs scavenge free radicals and participate in vitamins C and E recycling

- Understand the health properties or beneficial actions of OPCs

- Describe the major mechanisms of OPCs on the disease prevention

- List the effects of OPCs on the cardiovascular system

- Know the major mechanisms of OPCs on the cancer prevention and therapy

- Describe the major pathways of OPCs as anti-allergy and anti-inflammatory agents

- Show the difference of the isotonic delivery system from tablets

---

Introduction

A free radical (FR) is an especially reactive atom or group of atoms that has one or more unpaired electrons. After production, it exists for a brief period of time before reacting to produce a more stable molecule. This high reactivity of FR empowers it to target other molecules easily in the biological system such as human cells. It has been estimated that each cell in the body suffers from about 10,000 FR "hits" every day. These FR attacks may create a series of chain reactions, consequently inducing a lot of damage until they can be terminated by antioxidants. The degree of cell or organ damage often depends on how well the cells are protected by antioxidants (Chapter 1). The imbalance of FRs and antioxidants leads to oxidative stress (OS), which has been found to contribute to over 100 diseases or medical conditions as we discussed in previous chapters, and a brief summary is shown in Figure 26.1.

Oxidative Stress and Antioxidant Protection: The Science of Free Radical Biology and Disease, First Edition.
Edited by Donald Armstrong and Robert D. Stratton.
© 2016 John Wiley & Sons, Inc. Published 2016 by John Wiley & Sons, Inc.

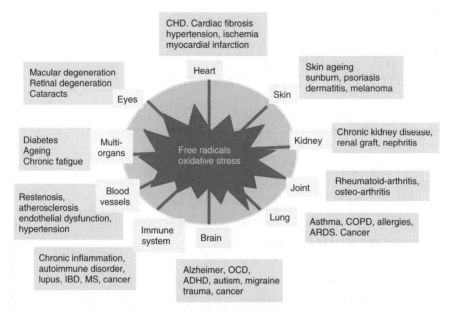

Figure 26.1 Oxidative stress contributes to many diseases.

Although the body or cells produce some antioxidants, we are largely dependent on the diet to supply many antioxidants to scavenge or remove FRs produced in the body. There are many dietary supplements of antioxidants including vitamins, minerals, amino acids, and coenzymes. Vitamins C and E, selenium, and carotenoids such as β-carotene, lycopene, lutein, zeaxanthin, and astaxanthin are commonly used. These antioxidant dietary supplements are discussed in a great detail in other chapters of this book. The major focus of this chapter will be on oligomeric proanthocyanidins (OPCs). It is our hope that the data or information summarized here will provide an integrative picture about the characteristics of OPCs and their beneficial actions on human health.

OPCs are plant-based flavonol compounds that are abundant in a variety of flowers, hops, leaves, fruits, berries, and nuts of many plant species. They are also found in beans and in red wine. OPCs are some of the best flavonoids and have been used in clinics for many years. Given the depletion of soil in modern agriculture and the damage of many antioxidants during food processing, nutritional supplementation of antioxidants such as OPCs is becoming essential for reducing the oxidative injury and improving health. Many clinicians are now recommending and offering efficacious antioxidant supplementation including OPCs more frequently since recognizing the deficient eating habits of their patients and the growing scientific evidence behind oxidative stress and antioxidants. Among different OPC food supplements, isotonic OPCs are isotonic-capable food supplements that are made from a combination of bilberry, grape seed, red wine and pine bark extracts, and citrus extract, consisting of OPCs, anthocyanins, resveratrol, and other phenolic acids. It is now well accepted that isotonic OPCs provide exceptional nutritional benefits to human health mainly due to antioxidant actions.

The discovery of OPCs

In the winter of 1534–1535, the French explorer Jacques Cartier and his crew were trapped for months by freezing weather during an expedition up the St. Lawrence River. They had to spend the whole winter there, a fatal situation for them. Their gums receded to the extent that they lost their teeth. Their skin turned blotchy; they could not breathe well; and their legs became swollen, lost strength, and turned black. Many of them died with much red fluid around their heart. Fortunately, they finally got help from a native Indian. With his instruction, they made tea from the skin and needles of a tree named "Aneda" (arborvitae). The men drank this liquid and recovered within 1 week! They were recommended to put a poultice of the decoction on the afflicted body parts. In this way, the remaining crew managed to survive. They all suffered from scurvy, a common affliction in the explorers at that time which had no known prevention or cure.

In the mid-1900s, Dr. Jacques Masquelier of the University of Bordeaux read the chronicle of Cartier's and began a scientific study to discover the active components in the pine bark that gave the lifesaving power. Dr. Masquelier found that pine bark contains some components, which offered their own independent health-promoting properties and enhanced the activity of vitamin C. He isolated OPCs and found that OPCs are unique in their ability to work in both aqueous and lipid phase to provide excellent antioxidant support.

The characteristics of OPCs

OPCs are phenolic compounds that exist in the form of flavan-3-ol units. Considering monomers, catechin or epicatechin, as one bioflavonoid unit, compounds containing two and three monomers are called "dimers" and "trimers," respectively. OPCs are organic clusters of dimers, trimers, tetramers, and pentamers. It is known that the larger proanthocyanidin clusters are not more effective. For example, tetramers, but not hexamers of OPCs, inhibited lung angiotensin-I-converting enzyme.[1] Polymeric proanthocyanidins are hard to pass the intestinal tract into the blood and so do not have the same nutritional value. When we compare the chemical structure of OPCs with vitamin C, we can see that the unique formula of OPCs can provide much more free electrons to quench the FR reaction in terminating chain reactions. Studies have shown that OPCs work synergistically to be up to 20 times more powerful than vitamin C and 50 times than vitamin E in neutralizing free radicals.[2] In addition, OPCs from different plants work together well and synergistically improve other antioxidant effects by donation of electrons to regenerate functional vitamins C and E in the cells as shown as Figure 26.2. Some grape seed extract, red wine extract, or other plant's extract contain OPCs or proanthocyanidins. It has been shown that OPCs from maritime pine bark extract (MPBE) can double the concentrations of superoxide dismutase, catalase, and glutathione inside the cells.[3,4]

There is an isotonic form of MPBE available. MPBE is a natural plant extract from the bark of the French maritime pine tree and the most clinically researched and potent bioflavonoid. Most of the clinical results discussed in this chapter are from

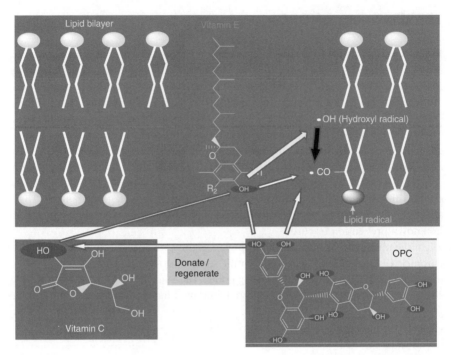

Figure 26.2 OPC is an electron donor that regenerates vitamins C and E and has direct action on hydroxyl and lipid radicals.

MPBE. Isotonic OPCs can provide up to 90–98% absorption by oral drink, much higher than tablets (10–40%), and exert maximum action to neutralize FRs.

The health-promoting properties of OPCs

With the knowledge of oxidative damage and understanding the chemical characters of OPCs, we can understand how OPCs protect us from head to toe, inside out. These beneficial effects of OPCs are as follows:

Maintain cellular health, protecting every cell from FR attacks

As shown in Figure 26.3, ROS react with vital cell components to induce cellular membrane injury, protein cross-links, fragmentation, and DNA damage, which can be prevented by antioxidants: vitamins A, C, and E; glutathione; ferritin; and others. OPCs provide more electrons to FRs, which directly protects cells from injury. OPCs also help to regenerate vitamins C, E, and A, indirectly neutralizing FRs to terminate the corresponding chain reactions and the consequent vicious cycle. In this manner, OPCs have the potential to more effectively prevent many chronic diseases associated with FRs-induced injury compared to other antioxidants (Figure 26.3).

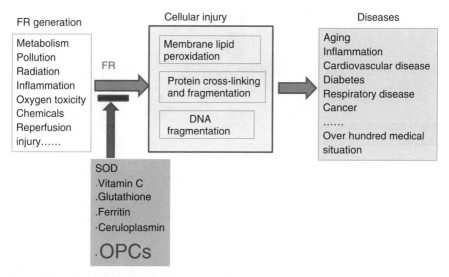

Figure 26.3 Antioxidants provide electrons which prevent ROS reactions with vital cell components. OPCs also help regenerate vitamin C, vitamin E, vitamin A, indirectly neutralizing FRs to terminate corresponding chain reactions and the consequent vicious cycle.

Strengthen and repair connective tissue

Groups of cells are connected to one another with connective tissue, and collagens are the most important component of this type of tissue. Improving collagen's longevity in an enzymatic environment is critical to the longevity of a good cellular and tissue connection such as dentin bonding, vascular strength, and organ health, which in turn maintains overall body health. OPCs have strong affinity to bind collagen and elastin, inhibit collagenase, help collagen repair, and rebuild their molecular connection correctly. One of the earliest discoveries about OPCs was their ability to strengthen capillaries, which reduces capillary permeability and thereby prevents the leakage and microbleeding from a number of injury factors. OPCs have been proven to improve symptoms of chronic venous insufficiency by 80–90% and used for this vascular disease in Europe for many years. OPCs may also effectively cross-link collagen and improve its biological stability in time periods as short as 10 s. The use of OPCs as a priming agent is a clinically feasible and promising approach to improving the durability of current dentin bonding systems. In addition, OPCs are taken as an oral cosmetic to help prevent and even reverse the wrinkles, which may be due to their actions on connective tissue.

Improve blood circulation and reduce the risk of cardiovascular diseases

Several mechanisms as discussed below are responsible for the protecting action of OPCs from cardiovascular diseases (Figure 26.4):

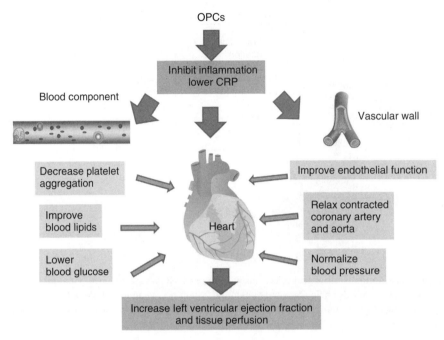

OPCs

Inhibit inflammation
lower CRP

Blood component

Vascular wall

Decrease platelet
aggregation

Improve endothelial function

Improve
blood lipids

Heart

Relax contracted
coronary artery
and aorta

Lower
blood glucose

Normalize
blood pressure

Increase left ventricular ejection fraction
and tissue perfusion

Figure 26.4 OPC protects the cardiovascular system through reduction of inflammation.

Protection of endothelial function

Endothelial function is considered one of the best indicators of vascular health. OPCs significantly ameliorate endothelial function in metabolic syndrome patients and healthy fatty diet volunteers, as shown by the significant increase in carotid arterial blood flow and much better tissue perfusion. This was shown to be related to the protection of endothelium from oxidant injury because 52% reduction of C-reaction protein and enhancement of eNOS-mediated NO release were observed. This endothelium-associated functional improvement might lead to vasodilation including coronary artery, ultimately increasing blood flow and tissue perfusion.

Reduction of low-density lipoprotein (LDL)

It has been reported that oxidation of low-density lipoprotein (LDL) plays the critical role in the pathogenesis of atherosclerosis. OPCs lower the LDL level and also prevent LDL oxidation, which blocks cholesterol deposits in macrophages or vascular cells and consequently stops the formation of "foam" cells and atherosclerotic plaques. In addition, OPCs can combine with omega-3 polyunsaturated fatty acids (PUFA) supplementation to reduce the narrowing of coronary artery from 80% to 46% in 3 months, as shown in one of our patients suffering from coronary atherosclerosis.

Inhibition of platelets aggregation

The platelet accumulation at the sclerotic plaque area is one of the major risk factors for heart attack and stroke. OPCs inhibit adhesion and aggregation of platelets by inhibition of the activity of 5-lipoxygenase and the formation of thromboxane A2 and other clotting compounds. This action of OPCs prevents thrombus formation and reduces the risk of embolism. OPCs are more efficient than aspirin for the inhibition of smoking-induced platelet aggregation without the common risk of increased bleeding as with aspirin.[5]

Antidiabetic actions

OPCs in MPBE have been shown to significantly lower the fasting plasma glucose level and improve postprandial hyperglycemia without increase in the insulin level in type II diabetes patients. MPBE also exhibited the most potent inhibitory action on alpha-glucosidase compared to green tea extract (~4 times) and to the diabetes control drug acarbose (~190 times). It also significantly improves the LDL in type II diabetes patients and enhances endothelial function. Furthermore, lower urinary albumin level as a marker of end-stage renal disease was observed at 8 weeks in OPCs-treated type II diabetes patients, indicating improvement of diabetic complications. Another clinical trial with 1289 type II patients has shown that MPBE improves capillary resistance and reduces leakage into the retina, contributing to the beneficial action on this common diabetic complication. In addition, OPCs in MPBE also improve ulcer healing in diabetic microangiopathy. Apart from these actions on type II diabetic patients, both isotonic OPCs and OPCs in MPBE show the beneficial action on type I diabetes patients by clinical observations and in animal models.[6]

Lowering blood pressure

OPCs have been demonstrated to lower systemic blood pressure in hypertensive animal models and patients. The mechanisms by which OPCs reduce blood pressure are related to NO release mentioned earlier, relaxation of systemic arteries (~78% relaxation of contracted arteries), and inhibition of angiotensin-converting enzyme. Proanthocyanidins also decrease thromboxane A2 level by the inhibition of thromboxane A2 biosynthesis, which reduces endogenous vasoconstriction during pathological conditions such as hypertension. In addition, OPC-mediated inhibition of platelet activation can help maintain normal blood viscosity and vascular resistance, thereby lowering blood pressure.

Prevention of heart attack

Numerous reports have shown that OPCs and resveratrol improve the survival of ischemia–reperfusion injury during heart attack. OPCs supplementation enhances the SOD activity increasing superoxide dismutation. In a recent human study, 2-month treatment of an isotonic OPC food supplement was found to significantly increase the left ventricular ejection fraction in patients with metabolic syndrome.[7]

Taken together, OPCs have strong beneficial action on the cardiovascular systems to prevent common diseases such as hypertension, atherosclerosis, and diabetic vasculopathy. A healthy heart and circulatory system are essential to the maintenance of cellular or organ functions, given the role of providing oxygen and other nutrients to every cell in the body. The combination of isotonic OPCs with multivitamins and omega-3 PUFAs is considered as a very efficient supplementation to help maintain cardiovascular health.

Immune and inflammation regulation

It is well known that FRs play a critical role in the inflammatory propagation and immune regulation. OPCs, as some of the most potent FR scavengers, can interact with different molecules involved in the inflammatory response or immune reaction. For example, allergy is a common medical problem in the United States, and allergic rhinitis has affected the life quality of more than 50 million Americans. Asthma is one of major diseases requiring visits to the emergency department, which may result in missed days of school for children and of work for adults. Isotonic OPCs with resveratrol and multivitamins together can significantly reduce the symptoms of allergic diseases and largely improve the life quality as adjunct treatment or even replacement of medicine. This beneficial effect of OPCs might be produced by at least four mechanisms. First, OPCs reduce the production of histamine by the inhibition of histidine decarboxylase, which catalyzes the synthesis of histamine from the amino acid histidine. Second, OPCs block the histamine release from mast cells (by 72%), which is mediated through the FR scavenging properties of OPCs. Third, OPCs are able to increase the uptake and reuptake of histamine into storage granules, where it is ineffective at causing allergic symptoms. All these show that OPCs prevent histamine's release from the mast cells, which is more efficient than strategies that target the receptors of histamine on the effector cells. Fourth, OPCs decrease the activation of inflammatory cells and thereby reduce the inflammatory cytokine release such as IL-4, IL-5, IL-13, and TNF-α. They also inhibit cyclooxygenase-2 (Cox-2), blocking the inflammatory response to different stimuli or danger factors. In our own studies, isotonic OPCs combined with other isotonic antioxidants, resveratrol, and multivitamins are effective against allergies in many patients without side effects such as drowsiness, dry membranes, and secondary infection induced by often used conventional medications including steroid. This combination also improved lung function and helped reduce the dosage of oral steroid in some asthma patients and even discontinued the use of rescue inhalers in some patients.

Furthermore, a recent human study has shown that oral isotonic OPCs for 2 months significantly decrease the blood level of the inflammation biomarker, C-reaction protein, by 52% in patients with metabolic syndrome. We have also observed the significant benefits of isotonic OPCs for patients suffering from allergic rhinitis, sinusitis, asthma, and arthritis including rheumatoid arthritis. In addition, OPCs in MPBE significantly prevent the cramps and muscular pain in normal subjects, venous-deficiency patients, athletes, claudicants, and diabetic macroangiopathy patients.

In animal experiments, OPCs dose-dependently reduced histamine release from rat peritoneal mast cells (RPMC) triggered by anti-DNP IgE and also inhibited the protein expression and secretion of TNF-α and interleukin-6 in anti-DNP IgE-stimulated RPMC. Moreover, OPCs decreased anti-DNP IgE-induced calcium uptake into RPMC and suppressed nuclear factor-kappa B activation, which consequently reduced the inflammatory response or immune reaction.

Prevention of cancer and as an adjunct to cancer therapy

It has been reported that any type of cancer is associated with the modifications of 300–500 human genes, and that the deregulation of cell signaling pathways at

multiple steps leads to cancer phenotype. Thus, a proper management of tumorigenesis requires the development of multitargeted therapies. Several adverse effects
associated with current cancer therapies and the thirsts for multitargeted safe anti-
cancer drug instigate the use of natural polyphenols. It is well accepted that OPCs
may have a blend of anticarcinogenic, proapoptotic, antiangiogenic, antimetastatic,
immunomodulatory, and antioxidant activities. The actions and related mechanisms
of OPCs used for tumor preventions include the following:

(a) OPCs help protect against the very early causes of cancer by (i) Neutralizing
    cancer-causing FRs and in this way protecting against the DNA damage by
    hydroxyl radical, decreasing FR-induced NF-κB activation, and preventing
    activation of oncogenes. (ii) Inactivating carcinogens by an increase in some
    enzymes that detoxify carcinogen. For example, OPCs inhibit the formation
    of reactive metabolites that are produced from NNK, a nitrosamine present in
    tobacco and very mutagenic and carcinogenic. (iii) Boosting the immune system
    so that any mutated cells can be destroyed before becoming cancerous. This
    action may be produced in several ways such as enhancing production of NK
    cells, protecting the existing immune components from injury, and restoring the
    decreased levels of certain protective cytokines.

(b) OPCs inhibit tumor growth and induce apoptosis of different human cancer cell
    lines *in vitro* and *in vivo*. The cancer cell proliferation was reduced and even com-
    pletely inhibited by OPCs in a dose-dependent manner. The beauty is that OPCs
    inhibit the growth of breast, stomach, and lung cancer cells, but maintain the
    growth and viability of normal cells. Induction of apoptosis or cell cycle arrest
    can be an excellent approach to inhibit the promotion and progression of car-
    cinogenesis. OPCs from grape seeds even induce apoptosis of up to 75% tumor
    cells with high metastasis character *in vivo* and *in vitro* in only 48 h.

(c) OPCs attenuate the metastasis of human oral, prostate, breast, and colon cancer
    cell lines. This may be due to reduction of the tendency of cancer cells to stick
    together and adhere to other sites. In addition, by downregulation of both vas-
    cular endothelial growth factor and angiopoietin, OPCs inhibit angiogenesis in
    growing tumor, which is important in supporting tumor cell proliferation and
    metastasis.

(d) In addition, OPCs can effectively reduce the damage to human liver cells induced
    by chemotherapy. Therefore, OPCs are very helpful as adjuncts to cancer
    chemotherapy. We observed that pretreatment with isotonic OPCs combined
    with isotonic resveratrol significantly protected cancer patients from hair loss
    and a decrease in white blood cells during chemotherapy. In combination with
    other therapies, OPCs have helped to save many patients' lives from several
    end-stage cancers and also significantly improved their life quality.

Antiaging and aging-related diseases

Living better and longer is a dream of most people. It has been reported that
FRs play a critical role in aging and aging-related degenerative diseases. In some
studies, OPCs were found to delay the aging process and to extend the life span of
senescent-accelerated mice. OPCs and resveratrol also produce beneficial action on

aging-related diseases such as Alzheimer's disease, Parkinson's disease, Huntington's disease, epilepsy, atherosclerosis, hypertension, and end-stage renal diseases.

Dosage and safety

The effective dosage of OPCs varies with the goal and the individuals. A dose of 100–300 mg (2–4 mg/kg) a day is mostly often recommended for the prevention of chronic inflammation and cardiovascular diseases as well as in adjunct therapy. A dose of 2–4 g/day is the dosage that produces significant benefit for cancer patients. Animal studies did not find any toxicity effects at 50–200 times these routine human dosages for 3 months or for 1 year.

Why isotonic?

The absorption of flavonoids in humans is a complex process. For example, the monomeric flavonoid constituents such as anthocyanins, catechin, quercetin, and hesperidin appear in the blood stream, and their metabolites subsequently appear in the urine within the first 2 h.[8,9] OPCs in MPBE can be completely absorbed into the blood stream, but it requires up to 8 h.[9] Isotonic OPCs have the same osmotic pressure as body fluid. Such isotonic nutritional supplements can be digested more quickly and efficiently, and they are distributed throughout the body faster with higher bioavailability than other formulations (Figure 26.5). Also, the buffer of isotonic OPCs protects nutrients from hydrochloric acid-mediated degradation in

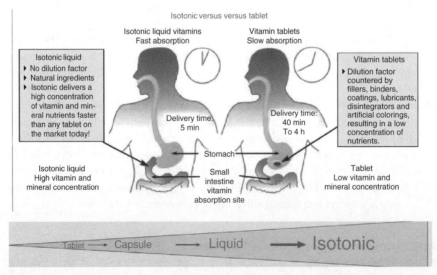

Figure 26.5 An isotonic liquid formulation provides much better absorption of OPC than other formulations.[10] (Vijayalakslakshmi Nandakumar and Santosh K. Katiyar 2008. Reproduced with permission.) *(See color plate section for the color representation of this figure.)*

the stomach. In addition, an isotonic solution works as already digested food that trigger osmoreceptors located at the distal end of the stomach, which opens the duodenal sphincter, releasing the isotonic OPCs into the duodenum in minutes. The antioxidant benefits of isotonic OPCs have been observed in 10 min in human and maintain a higher level for 4 h, which is much faster and more efficient to produce antioxidant action compared to tablets.[7]

Multiple choice questions

1 Oligomeric polymer means
  a. single monomer
  b. a few monomers
  c. Many monomers

2 In restoring the antioxidant properties of vitamins C and E, oligomeric proanthocyanidins act as electron:
  a. Donors
  b. Acceptors

3 Isotonic means
  a. Equal blood pressure
  b. Equal osmotic pressure
  c. Equal temperature

References

1 Ottaviani, J.L. *et al* (2006) Procyanidin structure defines the extent and specificity of angiotensin I converting enzyme inhibition. *Biochimie*, 88(3–4), 359–365.
2 Shi, J. *et al.* (2003) Polyphenolics in grape seeds-biochemistry and functionality. *Journal of Medicinal Food*, 6(4), 291–299.
3 Berryman, A.M. *et al.* (2004) Influence of treatment of diabetic rats with combinations of pycnogenol, beta-carotene, and alpha-lipoic acid on parameters of oxidative stress. *Journal of Biochemical and Molecular Toxicology*, 18(6), 345–352.
4 Seifried, H.E. *et al.* (2007) A review of the interaction among dietary antioxidants and reactive oxygen species. *Journal of Nutrition and Biochemistry*, 18(9), 567–579.
5 Pütter, M. *et al.* (1999) Inhibition of smoking-induced platelet aggregation by aspirin and Pycnogenol. *Thrombosis Research*, 95(4), 155–161.
6 Parveen, K. *et al* (2013) Modulatory effects of Pycnogenol in a rat model of insulin-dependent diabetes mellitus: biochemical, histological, and immunohistochemical evidences. *Protoplasma*, 250(1), 347–360.
7 Cesarone, M.R. *et al.* (2009) Accelerated antioxidant bioavailability of OPC-3 bioflavonoids administered as isotonic solution. *Phytotherapy Research*, 23(6), 775–777.
8 Dávalos, A. *et al.* (2006) Quercetin is bioavailable from a single ingestion of grape juice. *International Journal of Food Sciences and Nutrition*, 57(5–6), 391–398.
9 Grimm, T. *et al.* (2006) Single and multiple dose pharmacokinetics of maritime pine bark extract (Pycnogenol) after oral administration to healthy volunteers. *BMC Clinical Pharmacology*, 6, 4.
10 Nandakumar, V. *et al.* (2008) Multi-targeted prevention and therapy of cancer by proanthocyanidins. *Cancer Lett.* 269(2), 378–387.

CHAPTER 27

Superoxide dismutase mimics and other redox-active therapeutics

Ines Batinic-Haberle and Artak Tovmasyan

Department of Radiation Oncology, Duke University School of Medicine, Durham, NC, USA

THEMATIC SUMMARY BOX

At the end of this chapter, students should be able to:

- Understand the redox character of biological processes comprising cellular redoxome

- Understand the role and mechanism of action of superoxide dismutases

- Explain what is an SOD mimic and describe how to design SOD mimics and explain its therapeutic relevance

- Explain the complexity of the reactivities of SOD mimics versus specificity of SOD enzymes toward superoxide

- Understand the relevance of SOD mimicking *in vivo* in normal versus cancer cell

- Describe different classes of SOD mimics and other redox-active drugs that are not SOD mimics but SOD mimicking constitutes the basis for their therapeutic effects

---

List of abbreviations

SOD   Superoxide dismutase

Peroxynitrite $ONOO^- + ONOOH$, but given its $pK_a = 6.6$ at pH 7.8, peroxynitrite exists predominantly as $ONOO^-$

$O_2^{\bullet-}$ Superoxide

$^{\bullet}NO$ Nitric oxide

$CO_3^{\bullet-}$ Carbonate radical

MnP   Mn porphyrin

$(H_2O)_2Mn^{III}P$ Mn(III) diaqua porphyrin

$(H_2O)Mn^{II}P$ Mn(II) monoaqua porphyrin

Note: When cationic porphyrins bear 5+ charge, they have Mn in +3 oxidation state ($Mn^{III}P^{5+}$) and with 4+ total charge Mn is in +2 ($Mn^{II}P^{4+}$) or +4 oxidation states ($O=Mn^{IV}P^{4+}$). **Also note that** in the formulas below 2, 3, and 4 relate to *ortho, meta,* and *para* isomers, respectively

MnT-2-PyP$^+$ Mn(III) *meso*-tetrakis(2-pyridyl)porphyrin

Oxidative Stress and Antioxidant Protection: The Science of Free Radical Biology and Disease, First Edition.
Edited by Donald Armstrong and Robert D. Stratton.
© 2016 John Wiley & Sons, Inc. Published 2016 by John Wiley & Sons, Inc.

MnTAlkyl-2(3, 4)-PyPs Mn(III) *meso*-tetrakis(*N*-alkylpyridinium-2(3, 4)-yl)porphyrins (Note that alkyl=methyl for AEOL10112; alkyl=ethyl for AEOL10113 aka BMX-010 aka FBC-007)

MnTnBuOE-2-PyP^{5+} Mn(III) *meso*-tetrakis(*N*-(*n*-butoxyethyl)pyridinium-2-yl) porphyrin (BMX-001)

MnTDE-2-ImP^{5+} Mn(III) tetrakis(*N,N'*-diethylimidazolium-2-yl)porphyrin (AEOL 10150)

MnHalTAlkyl-2(3,4)-PyP^{5+}s Mn(III) β-halogenated *meso*-tetrakis(*N*-alkylpyridinium -2(3,4)-yl)porphyrins

MnTAlkoxyalkyl-2-PyP^{5+}s Mn(III) *meso*-tetrakis(*N*-alkoxyalkylpyridinium-2-yl) porphyrins

MnTAlkoxyalkyl-3-PyP^{5+}s Mn(III) *meso*-tetrakis(*N*-alkoxyalkylpyridinium-3-yl) porphyrins

MnTN-sub-2-ImP^{5+}s Mn(III) *meso*-tetrakis(*N*-substituted imidazolium-2-yl) porphyrins

M40403 Cyclic polyamine

MnTBAP^{3-} Mn(III) *meso*-tetrakis(4-carboxylatophenyl)porphyrin

Mn^{III}Br$_8$TCPP^{3-} Mn(III) β-octabromo-*meso*-tetrakis(4-carboxylatophenyl)porphyrin (also Mn^{III}Br$_8$TBAP^{3-})

MnTSPP^{3-} Mn(III) *meso*-tetrakis(4-sulfonatophenyl)porphyrin

Mn^{III}Br$_8$TSPP^{3-} Mn(III) β-octabromo-*meso*-tetrakis(4-sulfonatophenyl)porphyrin

Mn^{II}Br$_8$TM-4-PyP^{4+} Mn(II) β-octabromo-*meso*-tetrakis(*N*-methylpyridinium-4-yl) porphyrin

MnBr$_8$TM-3-PyP^{4+} Mn(II) β-octabromo-*meso*-tetrakis(*N*-methylpyridinium-3-yl) porphyrin

[MnBV^{2-}]$_2$ Mn(III) biliverdin

Mn^{III}C Mn(III) corrole

Mn^{IV}C Mn(IV) corrole

MnDiM-4-PyMAn-corrole^{3+} Mn(III) *meso-trans*-di(*N*-methylpyridinium-4-yl)-mono(anisyl)corrole

MnTrF$_5$Ph-β-(SO$_3$)$_2$-corrole^{2-} Mn(III) *meso*-tris(pentafluorophenyl)-β-bis(sulfonato) corrole

Salen *N,N'*-bis-(salicylideneamino)ethane

EUK-8 Mn(III) salen

FeP Fe porphyrin

(OH)(H$_2$O)Fe^{III}P Fe(III) monohydroxo monoaqua porphyrin

(OH)Fe^{II}P Fe(II) monohydroxo porphyrin

FeTE-2(or 3)-PyP^{5+} Fe(III) *meso*-tetrakis(*N*-ethylpyridinium-2(or 3)-yl)porphyrin

FeTM-4-PyP^{5+} Fe(III) *meso*-tetrakis(*N*-methylpyridinium-4-yl)porphyrin

FeTnBuOE-2-PyP^{5+} Fe(III) *meso*-tetrakis(*N*-(*n*-butoxyethyl)pyridinium-2-yl)por phyrin

FeTSPP^{3-} (Fe(III) *meso*-tetrakis(4-sulfonatophenyl)porphyrin

FeTMSP^{7-} Fe(III) *meso*-tetrakis(2,4,6-trimethyl-3,5-disulfonatophenyl)porphyrin

Tempol 4-OH-2,2,6,6,-tetramethylpiperidine-1-oxyl

Mito-CP Mito-carboxypropyl

MitoQ Mitochondrially targeted redox cycling quinone

$E_{1/2}$ Half-wave reduction potential

NHE   Normal hydrogen electrode
HIF-1α   Hypoxia inducible factor-1
NF-κB   Nuclear factor-κB
AP-1   Activator protein-1
TF   Transcription factor
Nrf-2   Nuclear factor (erythroid-derived 2)-like 2
VEGF   Vascular endothelial growth factor
PTEN   Phosphoinositide 3-phosphatase
Trx   Thioredoxin
HO-1   Hemeoxygenase-1
CAT   Catalase
IL   Interleukins
GPx   Glutathione peroxidase
GST   Glutathione *S*-transferase
GR   Glutathione reductase
MCP-1   Monocyte chemoattractant protein-1
TD50   Median toxic dose of a drug at which toxicity occurs in 50% of cases

Introduction – redoxome

Major metabolic pathways such as mitochondrial respiration, Krebs cycle, glycolysis, fatty acid oxidation, and actions of endogenous antioxidants are redox based and involve electron shuttling among the biomolecules (Figure 27.1).

The physiological redox environment is maintained by an interplay/balance between the endogenous antioxidative defenses of low-molecular weights (e.g., glutathione (GSH), cysteine, ascorbic acid, tetrahydrobiopterin, tocopherol, NADPH) and high-molecular weights (e.g., SOD enzymes, catalases, GPx, GR, GST, peroxiredoxins) and reactive species (Figure 27.2). It is critical to note here that not all reactive species are radicals, that is, have unpaired electrons. Highly reactive nonradicals are peroxynitrite (both protonated and unprotonated forms present *in vivo*, $ONOOH + ONOO^-$, $pK_a = 6.14$); adduct between peroxynitrite and CO_2, $ONOOCO_2^-$; hydrogen peroxide ($pK_a = 11.65$, deprotonated species, HO_2^- is much more reactive yet present at very low levels *in vivo*); hypochlorite ($pK_a = 7.5$, protonated and deprotonated forms are present *in vivo*, $HClO + ClO^-$); thiols (protonated and deprotonated species $RSH + RS^-$ are present *in vivo* (GSH has $pK_a = 9.2$, protein thiols have $pK_a \sim 8.5$, but many proteins have domains that lower the pK_a so that under physiological conditions they are deprotonated and thus more reactive); and ascorbate (with $pK_a = 4.2$ it is present *in vivo* predominantly as monodeprotonated HA^-). Among radicals there are superoxide radical, $O_2^{\bullet-}$ ($pK_a = 4.8$); nitric oxide, $^{\bullet}NO$; carbonate radical, $CO_3^{\bullet-}$; nitrogen dioxide radical, $^{\bullet}NO_2$; hydroxy radical $^{\bullet}OH$; thiyl radical $RS^{\bullet}$; ascorbyl radical, $HA^{\bullet}$; metal ions, such as Cu^{2+}, $Mn^{2+/3+}$, $Fe^{2+/3+}$, and so on. Rather than using term "free radicals," the reader is advised that it is safer to use the term "reactive species." It is then again more correct to use "reactive species" than "reactive oxygen species" and "reactive nitrogen species," as often particular species, such as $^{\bullet}NO$ and $ONOO^-$, belong to both categories. The interplay between

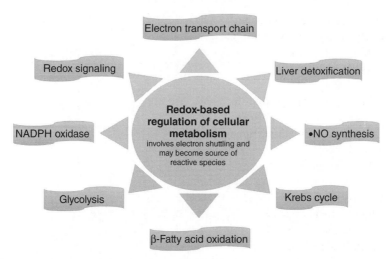

Figure 27.1 Cellular metabolism is redox controlled. Listed are some of the major metabolic pathways that involve electron shuttling among biomolecules. Some of the electron shuttling would eventually, intentionally (supporting signaling pathways) or not (such as mitochondrial respiration at complexes I and III where electron from ubiquinol would hit the surrounding oxygen and reduce it one-electronically to $O_2^{\bullet-}$), give rise to reactive oxygen, nitrogen, sulfur, selenium, and chlorine species (RS). Endogenous antioxidative defenses, for example, catalase, families of superoxide dismutases (SOD), glutathione peroxidases (GPx), and peroxyredoxins, are in charge of maintaining low nanomolar levels of reactive species (RS), that is, physiological redox environment (Figure 27.2). *If levels of RS increase, as a consequence of cellular injury, a cascade of signaling events are upregulated with the goal to restore normal redox environment. Redox-active pathways along with reactive species and endogenous low- and high-molecular antioxidants are now recognized as* **redoxome and define cellular redox environment.** *Redoxome is as critical for cell metabolism as are proteome and genome.*

antioxidants and reactive species unambiguously involves electron shuttling. Any perturbation in the balance between RS and antioxidants, resulting in diseased cell, leads to an increase in the levels of reactive species above the nanomolar levels – the condition best described and widely recognized as oxidative stress. In order to restore the normal redox environment (i.e., heal the cell), the appropriate redox-active signaling pathways get activated. Often the signaling cascades would involve the amplification of initial oxidative insult (inducing secondary oxidative stress), which may either lead to cell healing, if of modest magnitude, or cause the cell death if excessive. *It is only natural that the re-establishment of physiological redoxome may be best achieved with those drugs that are redox active and whose redox properties are biologically compatible, that is, that can easily exchange electrons with components of redoxome.* The redox-active compounds are frequently those bearing metal sites where reactions of interest occur. Indeed, major endogenous antioxidative defenses (e.g., families of SODs, family of catalases, cytochrome P450 family of enzymes, and heme oxygenase), as well as those systems producing reactive species critical for resolving injuries (e.g., nitric oxide synthases and NADPH oxidases) have metals at their active sites. The one-electronically reduced oxygen, the superoxide $O_2^{\bullet-}$, is produced under physiological levels at several

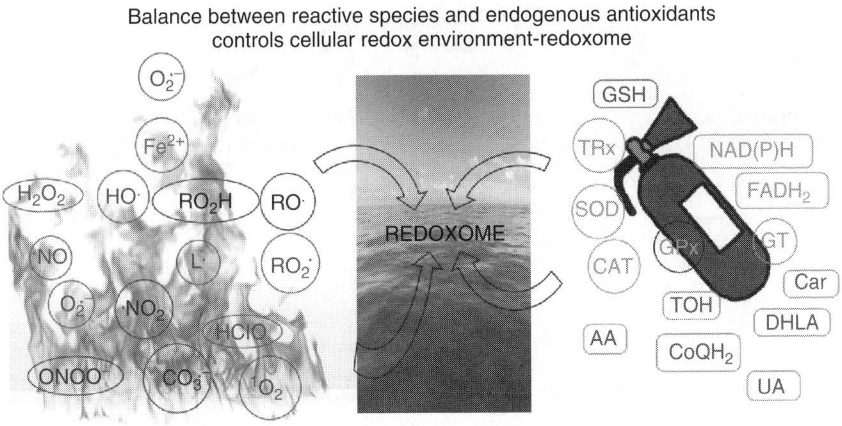

Figure 27.2 Redox-active molecules, reactive species, and antioxidants. Redox-active molecules/species (some of them listed here) and their involvement in redox-based pathways comprise the cellular redoxome. Redoxome is maintained by oxidation/reduction reactions, that is, electron shuttling; it is only natural that redox-active drugs may be best suited to restore it when perturbed in diseases. *(See color plate section for the color representation of this figure.)*

sites, one of them being mitochondrial respiration, that is, electron transport chain shown in Figure 27.1. Some of the other sources including the family of NADPH oxidases are listed in Figure 27.3.

Superoxide dismutases

The most crucial enzymes of the first-line defense against oxidative stress in all living organisms are members of the family of superoxide dismutases. Three isoforms exist in mammalian organisms: (i) extracellular CuZnSOD; (ii) cytosolic and intermembrane space CuZnSOD; and (iii) mitochondrial matrix, MnSOD. Some other organisms have other SODs with Fe or Ni at active sites.[1] Life is not possible without MnSOD, knockout mice live only few hours after birth, while heterozygous mice, with 50% less of MnSOD, have age-dependent increase in oxidative DNA damage. The CuZnSOD knockout mice exhibit accelerated aging and higher incidence of tumors. All three isoforms catalyze the dismutation of superoxide, $O_2^{\bullet-}$. The dismutation describes the disproportionation of $O_2^{\bullet-}$ into its one-electronically oxidized, oxygen, O_2 in the first step (Eq. (27.1)) and one-electronically reduced form, hydrogen peroxide, H_2O_2 in the second step (Eq. (27.2)). Both the oxidation and reduction steps of superoxide dismutation, along with concomitant reduction and re-oxidation of enzyme (whereby enzyme/catalyst is recovered) happen at the same diffusion-limited rates of $\sim 10^9 M^{-1} s^{-1}$. Both rates are identical because the electron transfer happens at the metal-centered reduction potential ($E_{1/2}$), which is at the midway between the potential for $O_2^{\bullet-}$ oxidation at -180 mV versus NHE (normal hydrogen electrode) and potential for $O_2^{\bullet-}$ reduction at $+890$ mV versus NHE; in turn both reactions, described by equations (27.1) and (27.2), have the

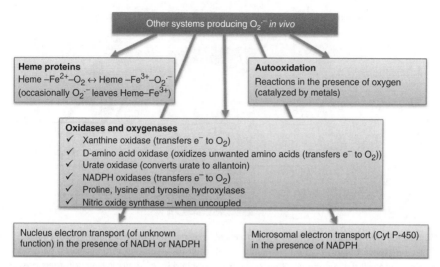

Figure 27.3 In addition to the electron transport chain, several other metabolic pathways (some of which are listed here) produce superoxide and subsequently its progeny and contribute to oxidative stress. The $O_2^{\bullet-}$ production via such pathways is enhanced if redox environment is perturbed. Some enzymes would produce superoxide under pathological conditions such as family of nitric oxide synthases (NOS), and some would produce under both pathological and physiological conditions such NADPH oxidases. For example, family of nitric oxide synthases produces $^{\bullet}NO$ under physiological conditions; yet in the absence of reducing equivalents (tetrahydrobiopterin) such as in case of oxidative stress, they would produce $O_2^{\bullet-}$. Under oxidative stress and with excessive $^{\bullet}NO$ production, the action of cytochrome oxidase complex IV, the terminal enzyme of ETC, may be blocked due to the nitrosylation of the Fe protoporphyrin active site.

same magnitude of thermodynamic facilitation.[2] Expressed as log $k_{cat}(O_2^{\bullet-})$, the rate constant was determined by several groups to be in the range of 8.84–9.30 and represents ~3 orders of magnitude enhancement over $O_2^{\bullet-}$ nonenzymatic self-dismutation (Table 27.1).[2–7] The dismutation of superoxide catalyzed by enzymes maintains the superoxide anion at physiological nanomolar levels. In the absence of enzyme, $O_2^{\bullet-}$ would self-dismute with a fairly high rate constant of $\sim 5 \times 10^5 \, M^{-1} \, s^{-1}$. The oxidation of $O_2^{\bullet-}$ with concomitant reduction of the enzyme (Eq. (27.1)) takes place at the metal site, manganese or copper, Mn or Cu. Mn gets reduced from Mn^{3+} to Mn^{2+}. In the case of CuZnSOD, Cu^{2+} gets reduced to Cu^+, while redox-inactive Zn supports the dismutation process:

$$Mn^{III}SOD^+ + O_2^{\bullet-} + e^- \leftrightarrow Mn^{II}SOD + O_2 \; k_{ox}(O_2^{\bullet-}) \tag{27.1}$$

$$Mn^{II}SOD + 2H^+ + O_2^{\bullet-} \leftrightarrow Mn^{III}SOD^+ + H_2O_2 + e^- \; k_{red}(O_2^{\bullet-}) \tag{27.2}$$

It is important to note that all is well in our body, as long as H_2O_2, formed during $O_2^{\bullet-}$ dismutation, is continuously removed whereby being maintained at nanomolar levels. Under physiological conditions, numerous systems such as catalase, families of peroxidases, and peroxiredoxins (supported by cellular reducing equivalents in the form of thiols, such as glutathione,

Table 27.1 Metal-centered reduction potential $E_{1/2}$ versus NHE (for M^{III}/M^{II} redox couple, M being metal, except when indicated otherwise), $\log k_{cat}(O_2^{\bullet-})$ for the catalysis of $O_2^{\bullet-}$ dismutation , $\log k_{red}(ONOO^-)$ for the one-electron reduction of $ONOO^-$ to $^{\bullet}NO_2$) and the lipophilicity of redox-active drugs expressed in terms of log value of the partition between n-octanol and water, $\log P_{OW}$.

Redox-active compounds	$E_{1/2}$/mV versus NHE[a]	$\log k(O_2^{\bullet-})$[b]	$\log k_{red}$ (ONOO$^-$)	$\log P_{OW}$[c]
Mn porphyrins				
MnTM-2-PyP^{5+}	+220	7.79	7.28	−8.16[c]
MnTE-2-PyP^{5+}	+228	7.76 (cyt c), 7.73 (p.r.)[d]	7.53	−7.79[c]
MnTnBu-2-PyP^{5+}	+254	7.25	7.11	−6.19[e]
MnTnHex-2-PyP^{5+}	+314	7.48	7.11	−3.84[e]
MnTnHep-2-PyP^{5+}	+342	7.65		−3.18[e]
MnTnOct-2-PyP^{5+}	+367	7.71	7.15	−2.32[e]
MnTMOE-2-PyP^{5+}	+251	8.04 (p.r.)	7.36	−7.52[c]
MnTMOHex-3-PyP^{5+}	+68	6.78		−5.45[e]
MnTnBuOE-2-PyP^{5+}	+277	7.83		−4.10[e]
MnBr$_8$TM-3-PyP^{4+}	+468	>8.85		
MnTCl$_5$TE-2-PyP^{4+}	+560	8.41		
MnTDE-2-ImP^{5+}	+346	7.83 (p.r.)	7.43	
MnTTEG-2-ImP^{5+}	+412	8.55		
MnMImPh$_3$P^{2+}		6.92		4.78
[MnBV^{2-}]$_2$	+460[f]	7.4		
[MnBVDME]$_2$	+450[f]	7.7		
[MnMBVDME]$_2$	+440[f]	7.36		
[MnBVDT^{2-}]$_2$	+470[f]	7.4		
MnTBAP^{3-}	−194	3.16	5.02	
MnTSPP^{3-}	−160	3.93		
MnTCHP$^+$	−200 to −400		5–6[109]	
AEOL11207		5.23		
Fe porphyrins				
FeTM-2-PyP^{5+}	+212	7.95		
FeTE-2-PyP^{5+}	+215	8.00		
FP15			6.80[67]	
cis-FeTM-4-Py$_2$Ph$_2$P^{2+}		7[110]		
INO-4885(WW-85)		>8[g]	~7[111]	
Mn salens				
EUK-8	−130	5.78 (cyt c)[29]		
EUK-134	~−130	5.78		
EUK-189	~−130	5.78		−0.90[112]
EUK-207	~−130	IC50 = 0.48 (NBT assay)		−1.41[113]
EUK-418		IC50 = 1.73 (NBT assay)		0.518[112]
Cyclic polyamine				
M40403	+525 (ACN) +840 (methanol)	7.08		
Nitroxides				
Tempol	+810[f]	<3 (pH 7.8)		
Tempone	+918[f]	<3 (pH 7.8)		
4-Carboxy-Tempo	+805[f]	7.54 (pH 5.4)		

Table 27.1   (*Continued*)

Redox-active compounds	$E_{1/2}$/mV versus NHE[a]	log $k(O_2^{\bullet-})$[b]	log k_{red} (ONOO$^-$)	log P_{OW}[c]
Mn corroles				
MnDiM-4-PyMAn-corrole^{3+}	~+700[h]	8.11		
MnTrF$_5$Ph-β(SO$_3$)$_2$-corrole^{2-}	+1040[h]	5.68	4.93	
FeTrF$_5$Ph-β(SO$_3$)$_2$-corrole^{2-}	+1050[h]	6.48	6.48	
Metals, metal ions, and oxides				
OsO$_4$		9.14 (pH 5.1–8.7)[36]		
CeO$_2$ (3–5 nm particles)		9.55[114]		
Nano-Pt		48.9 μM (IC50, WST-1 assay)[115]		
Mn^{2+}	+850[i]	6.11 (cyt c), 6.28 (p.r.)		
Metallotexaphyrin				
Gd(III) texaphyrin (XCYTRIN™)	−41[i] [31]			
Mn complex, SRI110		None	6.20	
Natural compounds (polyphenols)				
Curcumin		115 μM (IC50, NBT assay)[116]		
Honokiol		5.5[117]		
MitoQ	−105 (MitoQ/UQH$^\bullet$) water	8.30 $k_{ox}(O_2^{\bullet-})$		3.44 (37 °C) n-Octanol/PBS
SOD enzymes	~+300	8.84–9.30	3.97	
Self-dismutation		5.7[k]		

For comparison, the values for some other compounds listed in Figures 27.5, 27.6, 27.21, and 27.22 are also given. In the absence of SOD enzyme, $O_2^{\bullet-}$ self-dismutes at pH 7.0 with fairly high rate constant of $k(O_2^{\bullet-}$self-dismutation$) \sim 5 \times 10^5$ M^{-1} s^{-1}. Therefore, the compounds cannot be described as SOD mimics, if they disproportionate $O_2^{\bullet-}$ with a rate constant equal to or lower than 5×10^5 M^{-1} s^{-1} (log $k(O_2^{\bullet-}) \leq 5.7$). Data are taken from Refs.[2, 25, 35].

[a]$E_{1/2}$ is determined in 0.05 M phosphate buffer (pH 7.8, 0.1 M NaCl).

[b]k_{cat} was determined by cytochrome c assay in 0.05 M potassium phosphate buffer (pH 7.8, at 25 ± 1 °C); for detailed comparison of validity of cytochrome c versus NBT assay for k_{cat} determination, see Ref.[119].

[c]Data obtained from the relationship R_f versus log P_{OW} (log $P_{OW} = 12.207 \times R_f - 8.521$) and direct determinations of log P_{OW} for Mn(III) N-alkoxyalkylpyridylporphyrins.[120]

[d]p.r., Pulse radiolysis.

[e]Determined experimentally using n-butanol and water biphasic system and converted to log P_{OW} according to the equation log $P_{OW} = 1.55 \times P_{BW} - 0.54$; P_{BW} is the partition between n-butanol and water.

[f]The one-electron reduction potential refers to Mn^{IV}/Mn^{III} couple with Mn biliverdin analogs and RNO$^+$/RNO$^\bullet$ redox couple with nitroxides.

[g]Estimation based on determined k_{cat} ($O_2^{\bullet-}$) for MnTTEG-2-PyP^{5+} and the relationship between the $k_{cat}(O_2^{\bullet-})$ for FePs and MnPs.

[h]$E_{1/2}$ data associated with the Mn^{IV}/Mn^{III} reduction potential. Based on the data obtained in acetonitrile versus Ag/AgCl for M^{IV}/M^{III} redox couple with 0.3 M tetrabutylammonium perchlorate (for MnDiM-4-PyMAn-Corrole^{3+}) or 0.1 M TBAP/TBAPF$_6$ (for FeTrF$_5$Ph-β(SO$_3$)$_2$-Corrole^{2-}) as electrolyte and the similarity of $E_{1/2}$ in aqueous medium and acetonitrile, the $E_{1/2}$ versus NHE in aqueous medium was estimated.

[i]Oxidation potential only, Mn^{III}/Mn^{II} redox couple is irreversible.

[j]Reduction potential only in N,N-dimethylformamide.

[k]At physiological pH.

and supporting enzymes such as glutathione reductase) maintain levels of H_2O_2 under physiological nanomolar conditions. Yet, within the last decade, it became clear that in cancer such systems are frequently downregulated and often accompanied by an increase in MnSOD levels; in turn H_2O_2 levels may be increased.[8–11]

As soon as the importance of SOD enzymes was recognized, the search for their low-molecular weight mimics has started. The enzymes themselves are not suitable as therapeutics due to the nature of protein, which imposes the antigenicity and prevents them to cross-cellular membranes. The search has accelerated as it becomes obvious that the superoxide is involved in the production of numerous other reactive species. Some of those such as $ONOO^-$, $^{\bullet}OH$, and lipid radicals are even more hazardous than superoxide. Some of the reactive species arising from $O_2^{\bullet-}$ are listed in Figure 27.4.

Reactive oxygen species such as hydroxyl radical, $^{\bullet}OH$, can induce lipid peroxidation by hydrogen abstraction from fatty acids. Lipid peroxidation will give rise to highly oxidizing lipid hydroperoxides, lipid alkoxyl, and peroxyl radicals.[13] Reactive oxygen species are further involved in the production of sulfur-reactive species such as (glutathione or cysteine) thiyl radicals, which again can induce lipid peroxidation.[13] It is crucial to note here that reactive species are not only free radicals. Many highly reactive species are nonradicals, most important among them are hydrogen peroxide, H_2O_2 (proton dissociation constant $pK_a = 11.65$, still deprotonated species, anionic HO_2^- has the most oxidizing power) and peroxynitrite.[12] Under physiological conditions, both protonated and deprotonated species with different stability, reactivity, and transport across cellular membranes, will exist.

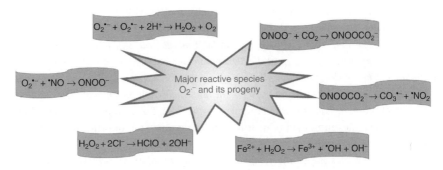

Figure 27.4 The involvement of superoxide in the production of some of the major reactive species contributing to oxidative stress. Dismutation of $O_2^{\bullet-}$ leads to the formation of peroxide, a major signaling and damaging species, maintained under physiological conditions at nanomolar levels. With any free low-molecular weight Fe^{2+} species around (e.g., aqua or carboxylato complexes), H_2O_2 will produce the most oxidizing, yet shortly-lived hydroxyl radical $^{\bullet}OH$. When $^{\bullet}OH$ is formed in the vicinity of nucleic acids (RNA and DNA), the major oxidative damage will occur. By the action of myeloperoxidase, H_2O_2 will produce another strongly oxidizing hypochlorous acid, which is under physiological conditions in equilibrium with deprotonated and reactive form, ClO^-. $O_2^{\bullet-}$ would react with $^{\bullet}NO$ at diffusion-limited rates of $>10^9\ M^{-1}\ s^{-1}$ to form highly damaging peroxynitrite, predominantly in $ONOO^-$ form.[12] $ONOO^-$ would *in vivo* make an adduct with CO_2, which would decompose to form two highly oxidizing radicals, $CO_3^{\bullet-}$ and $^{\bullet}NO_2$.

With $pK_a = 6.8$, at pH 7.4, ~80% of peroxynitrite will be in the anionic form; conversely, at pH 6.2 (*e.g.*, inside a macrophage phagocytic vacuole), up to 80% will be in the protonated HNOO form; HNOO will decompose to give rise to radicals, •OH and •NO$_2$. Finally, a strongly oxidizing hypochlorite is not a radical either. With $pK_a = 7.53$, ~50% of hypochlorous acid, HClO, is deprotonated under physiological condition and exists as ClO$^-$ (Figure 27.4). For details on the chemistry, biology, and medicine of reactive species, see Ref.[13].

What is an SOD mimic?

Several different classes of SOD mimics were synthesized and explored during decades since the discovery of superoxide dismutase by McCord and Fridovich at the end of 1960s.[14] It is important to note that a compound could only be called SOD mimic if it catalyzes the O$_2^{•-}$ dismutation (Eqs (27.1) and (27.2)) with a rate constant that is higher than the k for O$_2^{•-}$ dismutation of $\sim 5 \times 10^5$ M^{-1} s^{-1} [log k_{cat}(O$_2^{•-}$) = 5.7]. The frequently studied Mn porphyrin-based SOD mimics are presented in Figure 27.5. Other classes of SOD mimics are presented in Figure 27.6. Some of the compounds, initially developed to be SOD mimics, are nonmetal based, such as nitroxides (see the section "Redox-active compounds other than SOD mimics" and Figure 27.21). While nitroxides exhibited therapeutic potential, the very low k_{cat}(O$_2^{•-}$) ~ 3 at physiological pH values precludes them to be categorized as SOD mimics.

Many compounds could undergo only one step of O$_2^{•-}$ dismutation process, that is, oxidize or reduce O$_2^{•-}$ while closing the catalytic cycle with other reactive species, such as ONOO$^-$, or cellular reductants, such as ascorbate or glutathione or cysteine. *Importantly, such compounds, while not being SOD mimics, can still restore the physiological redox environment of a cell. Even the SOD mimics may not in vivo act as catalysts of O$_2^{•-}$ dismutation. They could in turn couple with other reactive species or cellular reductants while eliminating superoxide. The reactive species in the immediate environment and their concentrations would control the type of reaction of a redox-active drug.* For example, the rubredoxin oxidoreductase (desulfoferrodoxin) undergoes only one step of dismutation process and thus acts as O$_2^{•-}$ reductase.[18]

The structures of some of the compounds bearing metal sites could be appropriately tuned to dismute O$_2^{•-}$ under enzymatic (thermodynamic and kinetic) conditions; thus, they are considered SOD mimics (Figures 27.5 and 27.6). The nature has developed different SOD enzymes in such a way that, regardless of the type of metal site, they oxidize and reduce O$_2^{•-}$ at identical diffusion-limited rates of ~2×10^9 M^{-1} s^{-1} (see above).[4,6,7] Four different metal sites (Mn, Cu, Fe, or Ni) are found in SOD isoforms: MnSOD, CuZnSOD, FeSOD, and NiSOD. The diffusion-limited process has been facilitated by the metal-centered reduction potential being identical for all types of metal centers. It is placed at the midway potential ($E_{1/2}$ ~+300 mV vs NHE) between the potential for O$_2^{•-}$ oxidation, at −180 mV versus NHE, and O$_2^{•-}$ reduction at +890 mV versus NHE (see the section "Design of an SOD mimic").[2]

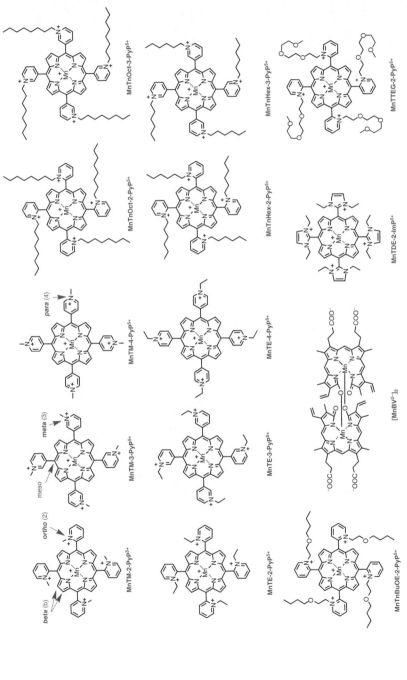

Figure 27.5 MnP-based "true" SOD mimics, that is, compounds that catalyze $O_2^{\bullet-}$ dismutation with k_{cat} ($O_2^{\bullet-}$) higher than k for $O_2^{\bullet-}$ self-dismutation of $\sim 5 \times 10^5$ M^{-1} s^{-1} at pH 7.[13]

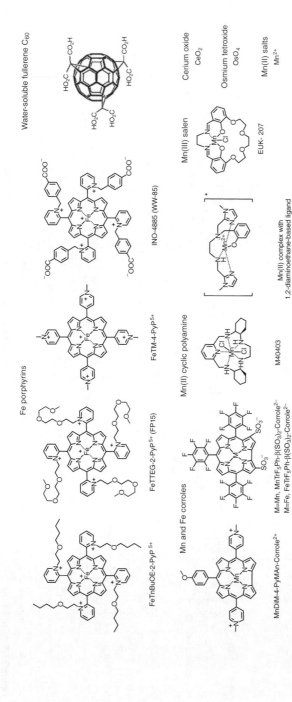

Figure 27.6 SOD mimics other than Mn porphyrins. Only "true" metal-bearing SOD mimics are listed referring to compounds that catalyze $O_2^{\bullet-}$ dismutation with k_{cat} ($O_2^{\bullet-}$) higher than k for $O_2^{\bullet-}$ self-dismutation of $\sim 5 \times 10^5$ M^{-1} s^{-1} at pH 7.[13] The structures of Fe porphyrins, Mn and Fe corroles, Mn cyclic polyamine, M40403 (GC4403), water-soluble fullerene, and Mn salen EUK-207 (of cyclic structure that enhances its stability toward loss of Mn) are shown. Metal salts, for example, those of Mn, Ce, and Os, are also potent SOD mimics. Cerium dioxide comes in a form of ceria nanoparticles. While very potent SOD mimic in aqueous setting (k_{cat} as high as that of SOD enzyme), the OsO_4 is too toxic for therapeutic purposes. The Mn^{2+} ion ligated with different low-molecular weight ligands is a fair SOD mimic; yet its clinical development might be precluded due to the neurotoxicity described as manganism.[15–17]

Mn, Cu, Fe or Ni have very different aqueous chemistry. Yet, the specific conformation of the amino acids and aqua or hydroxo ligands around the metal sites of each of those SOD enzymes is different, in order to provide identical thermodynamics and kinetics for the catalysis of dismutation process.[1] In a very elegant work, Frances-Muller's group has shown that the replacement of Fe with Mn, and vice versa, in SOD active sites results in the loss of enzyme activity.[19] In addition to the thermodynamic property that indicates whether the reaction is possible or not, there is a kinetic component that describes the rate at which the $O_2^{\bullet-}$ dismutation occurs. The X-ray crystallography of SOD enzyme shows that there is a tunnel made of positively charged (cationic) amino acid residues, which provides electrostatic guidance for the approach of negatively charged anionic $O_2^{\bullet-}$ toward the active metal site and contributes to the diffusion-limited rate of $O_2^{\bullet-}$ dismutation.[3,5] Of note, cambialistic SOD enzymes exist in some anaerobic bacteria (e.g., *Propionibacterium shermanii*) who wisely employ Mn under aerobic and Fe under anaerobic conditions.[20]

Design of an SOD mimic – introduction

Our group has established a rational approach to the design of SOD mimics and reported the very first structure–activity relationship, which relates kinetics to the thermodynamics of the $O_2^{\bullet-}$ dismutation process. The thermodynamics is described by $E_{1/2}$ of $Mn^{III}P/Mn^{II}P$, where III and II indicate that Mn is in +3 or +2 oxidation state, respectively. The kinetics is described by the rate constant for the catalysis of $O_2^{\bullet-}$ dismutation, $k_{cat}(O_2^{\bullet-})$.[2,21–25] *We have recently provided evidence that such SAR (structure–activity relationship) is applicable to any class of SOD mimics (see below). Therefore, such approach in the design of porphyrin-based SOD mimics justifies a more detailed explanation.*

We have early on realized that the following criteria need to be met for a powerful SOD mimic, that is, compound with high $k_{cat}(O_2^{\bullet-})$: *(1) it must be a stable complex between metal and ligand that would assure the integrity of the metal site where reactions of interest occur, that is, no loss of metal should occur in vivo; and (2) the thermodynamics and kinetics of $O_2^{\bullet-}$ dismutation should be as close as possible to those of SOD enzymes affording in turn (i) the thermodynamic property, $E_{1/2}$ to be as close as possible to +300 mV and (ii) kinetic facilitation for the reaction that is, appropriate electrostatics and sterics for the approach of anionic superoxide to metal site to be achieved.* Then, the other important aspect of an SOD mimic must be addressed: its bioavailability. Even if the SOD mimic has high $k_{cat}(O_2^{\bullet-})$, the therapeutic efficacy may be missing if it does not reach the target location within diseased organism: tissue and/or cell. The most stable metal complexes are those where metal is surrounded with a macrocyclic ring. The porphyrin is such a macrocyclic ring that encapsulates metal and prevents its loss. Therefore, nature has chosen it as active site in numerous enzymes and proteins, which are critical for our survival including catalase and cytochrome P450 (Figure 27.7).

Porphyrin is not only a strong ligand, preventing the loss of metal under nearly all conditions, but it also has several positions (eight *beta* positions on pyrrolic rings and

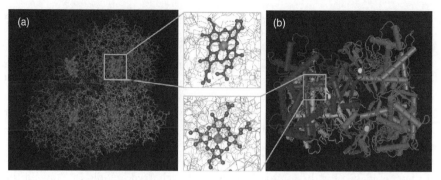

Figure 27.7 (a) Crystal structure of human erythrocyte catalase (PDB ID: 1QQW), and (b) crystal structure of human cytochrome P450 (PDB ID: 2F9Q) and their active sites. Pictures are created with Cn3D 4.3.1.[26–28] Porphyrin is a macrocyclic ring that encapsulates metal; in turn, it affords the highest stability to a metal complex, assuring no loss of metal where reactions of interest occur. It is only natural that such ligand has been used by nature for numerous proteins and enzymes, such as myoglobin, guanylyl cyclase, oxidases, oxygenases, prolyl hydroxylases, catalase, cytochrome P450 family of enzymes, and so on. For the same reason, we have chosen to modify a metalloporphyrin structure to be efficient catalyst for $O_2^{\bullet-}$ dismutation. (*See color plate section for the color representation of this figure.*)

four *meso* positions on four methane bridges) where modifications could be made to tune its properties (indicated in Figure 27.5). Within design strategies, modified porphyrin structures have been developed such as biliverdin analogs, texaphyrins with extended macrocyclic rings, and corroles with a shrunken porphyrin ring (one less carbon in the ring).[29–35] Among modified porphyrins corroles are the most promising compounds because of their extreme metal/ligand stability and high $k_{cat}(O_2^{\bullet-})$ (Figure 27.6).[33–35] Also, metal oxides have been successfully explored as SOD mimics, such as CeO_2 in the form of nanoparticles.[36–39]

Phase I. Mimicking the thermodynamics and kinetics of enzymatic $O_2^{\bullet-}$ dismutation establishes the first lead and efficacious SOD mimic – MnTE-2-PyP^{5+}

Phase I relates to the efforts to modify the porphyrin structure to achieve highest $k_{cat}(O_2^{\bullet-})$ as depicted in Figure 27.9. *Though similar rational approach has not widely been used with other redox-active drugs, except with corroles to a limited extent, it is a most appropriate strategy (if not the only one) for designing powerful SOD mimics.*[35] The nonsubstituted porphyrins, MnTPP$^+$ and MnT-2(or 4)-PyP$^+$ (Figure 27.8), have $E_{1/2}$ in a range of −200 to −280 mV versus NHE. Such $E_{1/2}$ stabilizes Mn in its +3 oxidation state and would prevent those Mn porphyrins to be reduced from +3 to +2 Mn oxidation state in a first step in order to start the dismutation process.

In order to facilitate the reduction, that is, to increase the willingness of Mn site to accept electrons from $O_2^{\bullet-}$, one needs to remove electron density from Mn by attaching electron-withdrawing groups to porphyrin. The very first Mn porphyrin (MnP)

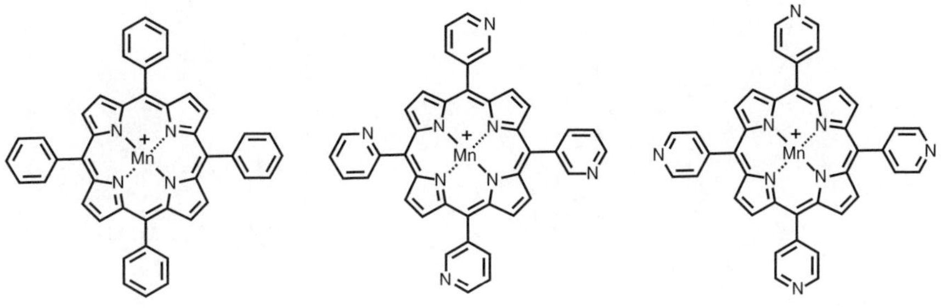

MnTPP⁺   MnT-3-PyP⁺   MnT-4-PyP⁺

Figure 27.8 Phase I of the design of porphyrin-based SOD mimic started from nonsubstituted Mn phenyl- and pyridylporphyrins, Mn(III) *meso-tetrakis*-phenylporphyrin, MnTPP⁺, and Mn(III) *meso-tetrakis*(pyridinium-2 (3 or 4)-yl)porphyrins, MnT-2(3 or 4)-PyP⁺. In these complexes, Mn is in its +3 oxidation state and is bound to four pyrrolic nitrogens.[24] Two of these form coordinated bonds with Mn – sharing the electrons with Mn. The other two nitrogens are deprotonated and are thus negatively charged and in turn provide one electron each to neutralize Mn 3+ charge. Consequently, one charge is left on Mn³⁺ center in a resting state. The appropriate thermodynamics and kinetics for the catalysis of $O_2^{\bullet-}$ dismutation has been adjusted by alkylation of the pyridyl nitrogens with alkyl carbocations. In turn, the nitrogens end up carrying cationic charges. Those charges pull the electron density from the Mn site, making it electron deficient and in turn ready to accept electrons from anionic $O_2^{\bullet-}$ in the first step of dismutation process. Moreover, the charges impose favorable electrostatics attracting anionic superoxide. Electrostatics accounts for ~2 orders of magnitude in the value of $k_{cat}(O_2^{\bullet-})$.[2,23]

that had reasonably high log $k_{cat}(O_2^{\bullet-})$ of 6.58 was produced when methyl groups were placed on all *para* (position 4, see Figure 27.5) pyridyl nitrogens of MnT-4-PyP⁺; such modification introduced four cationic charges on pyridyl nitrogens that withdrew the electron density from Mn site to such a magnitude that the $E_{1/2}$ was shifted by 340 mV from −280 to +60 mV versus NHE (Figure 27.9). This resulted in a fair log $k_{cat}(O_2^{\bullet-})$ of 6.58. While still not a very good SOD mimic, the compound potentially bears another disadvantage. It is planar and imposes toxicity because of its intercalation into nucleic acids.[21]

The intercalation of such planar Mn porphyrin into nucleic acids, resulting in a loss of its SOD-like activity, has been demonstrated. The study was performed on *Escherichia coli* cell extract; only upon removal of nucleic acids from the cell extract was the SOD-like activity fully restored.[21] While intercalation may be beneficial if the compound is used as an anticancer agent, the potential for systemic toxicity needs further investigation. Due to such observations, the methyl and ethyl groups were subsequently shifted from *para* to *ortho* pyridyl nitrogens. As *ortho* pyridyl nitrogens are located closer to Mn than are the *para* ones, the $E_{1/2}$ was shifted for another 160 mV from +60 mV to +220 mV versus NHE (or +228 for ethyl compound), very close to the $E_{1/2}$ of SOD enzyme. The $E_{1/2}$ correlates well with other thermodynamic properties that describe the electron density at metal site: the third proton dissociation constant of pyrrolic nitrogens, pK_{a3}, and of the first axially coordinated waters, pK_{a1}.[24,40] The closer location of charges in the vicinity of Mn site in *ortho* relative to *para* position

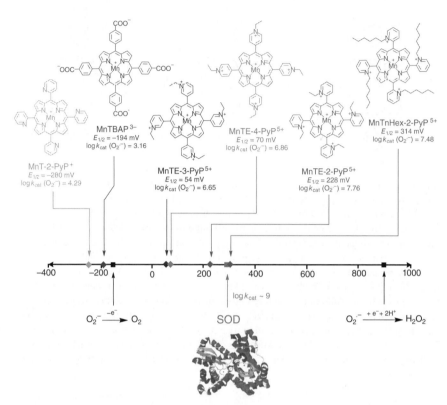

Figure 27.9 The design of porphyrin-based SOD mimics. Starting from unsubstituted MnT-4-PyP⁺ (Figure 27.8), the pyridyl nitrogens were first alkylated giving rise to *para* analog MnTM-4-PyP⁵⁺ with fair SOD-like activity. To enhance electron-withdrawing effects, the nitrogens were then moved closer to the metal site from *para* into *ortho* positions. The MnTM(E)-2-PyP⁵⁺, the first lead, was synthesized. It is still the most frequently studied compound.[2,23] Based on *ortho* pyridyl porphyrin, the imidazolyl analog (Figure 27.5) was subsequently synthesized and became the second lead – MnTDE-2-ImP⁵⁺. In order to improve the bioavailability of highly charged compounds, the alkyl chains were then lengthened and the third lead, MnTnHex-2-PyP⁵⁺ was synthesized. Adapted from ref 216. (*See color plate section for the color representation of this figure.*)

enhances the electrostatic facilitation of the dismutation process. Consequently, the log $k_{cat}(O_2^{\bullet-})$ was enhanced for another log unit from 6.58 (MnTM-4-PyP⁵⁺) to 7.79 (MnTM-2-PyP⁵⁺). Due to the vicinity of alkyl groups to *beta* pyrrolic hydrogens, the free rotation of pyridyl rings was precluded. In turn, the four atropisomers exist in all "*ortho*"-substituted N-pyridylporphyrins.[41] The alkyl pyridyl chains form a cavity around the metal site contributing to the fairly bulky structure of "*ortho*" Mn porphyrins; their bulkiness is dependent on the length of alkyl chains. Importantly, such conformation prevents their intercalation into nucleic acids; in turn their *in vivo* loss of SOD-like activity is prevented. *The discovery of an "ortho" effect presents the breakthrough in the development of Mn porphyrin-based SOD mimics.* Several other Mn and Fe analogs have been developed by us and others, and either they are in clinical trials or the feasibility of clinical trials is being pursued, which is discussed in detail later.

As anticipated, based on the properties similar to those of SOD enzymes, our first lead compound and thus far the most frequently studied porphyrin-based SOD mimic MnTE-2-PyP^{5+} was found to reduce and oxidize $O_2^{\bullet-}$ at similar rates of $\sim 5.75 \times 10^7\,M^{-1}\,s^{-1}$.[42] *We have experienced the fortunate coincidence that electron-withdrawing groups have introduced both appropriate thermodynamic and kinetic facilitation for the reactions of cationic Mn porphyrin with anionic reactive species, $O_2^{\bullet-}$. Such substituents (i) pulled the electron density from Mn site allowing it to be easily reduced with $O_2^{\bullet-}$ in the first step and as easily oxidized with $O_2^{\bullet-}$ in a second step of dismutation; and (ii) facilitate the $O_2^{\bullet-}$ dismutation on electrostatic grounds where species of opposing charges are favoring reacting with each other. Another fortunate coincidence (at least from therapeutic point of view), realized years later, is that not only $O_2^{\bullet-}$ is anionic, but it turns out that many other reactive species which can interact with Mn porphyrins and are biologically relevant in oxidative stress are anionic also: $ONOO^-$, ClO^-, HO_2^-, deprotonated glutathione (GS^-), deprotonated protein thiols (RS^-), and deprotonated ascorbate (HA^-).*

Within our developmental strategies, different types of substitutions with electron-withdrawing groups have been introduced at both *meso* and *beta* positions. A successful modification of the porphyrin core was performed with electron-withdrawing halogens (chlorines and bromines) at *beta* pyrrolic positions.[43,44] Insertion of eight bromines at all *beta* pyrrolic positions of MnTM-3-PyP^{5+} (and MnTM-4-PyP^{5+}) gave rise to octabrominated *meta* MnBr$_8$TM-3-PyP^{4+} and *para* porphyrin, Mn MnBr$_8$TM-4-PyP^{4+}.[43,45] The electron-withdrawing effect was of such magnitude that it facilitated the reduction of Mn^{3+} to Mn^{2+} to such an extent that the complex is finally stabilized in resting state as Mn^{II}P. In turn, Mn has enough electron density, does not care for obtaining additional electrons, and thus binds ligands weakly including its own porphyrin ligand. In turn, Mn^{II}P loses Mn even at physiological low hydrogen conditions of pH ~ 7.8. Regardless of the lack of practical use of such compounds, the ability to reach the log $k_{cat}(O_2^{\bullet-})$ as high as that of SOD enzymes was demonstrated. Such data have proven that even with low-molecular weight compounds, the proper thermodynamic and kinetic properties of the metal site may be achieved to afford the $k_{cat}(O_2^{\bullet-})$ of enzymatic magnitude. The same has been demonstrated with another class of low-molecular weight Mn complexes, Mn(II) cyclic polyamines.[46–49]

The "*ortho*" effect facilitated the development of SOD mimics – second lead, MnTDE-2-ImP^{5+} was identified

The discovery of a remarkable impact of *ortho* effect on SOD-like activity of metalloporphyrin initiated the design of *di-ortho N,N'*-dialkylimidazolyl series of compounds. One of the members of the series MnTDE-2-ImP^{5+} (AEOL10150) has been identified as the second lead Mn porphyrin and is now in preclinical studies as a radioprotector.[42,50–55] It was also in a phase I clinical safety/toxicity studies on ALS patients and was well tolerated.[56] It has the same high log $k_{cat}(O_2^{\bullet-})$ as MnTE-2-PyP^{5+} (Table 27.1). It is bulkier as it has two ethyl groups on each imidazolyl ring, which are placed on both sides of porphyrin plane; consequently, the compound has no atropisomers (Figure 27.6). The cationic charge is delocalized within ring. Its bulkiness renders it some advantages with respect to less interactions with biomolecules and, in turn, lower toxicity than MnTE-2-PyP^{5+}.[57] However, a higher dosing is required to compensate for the lower intracellular accumulation.[42,57,58]

The "*ortho*" effect further led to the development of different other *ortho* porphyrins with us and others having either Mn or Fe as active sites. The examples are Mn(III) PEG-ylated pyridyl and imidazolyl compounds and the analogous Fe(III) PEG-ylated pyridyl porphyrin, FP-15; all are cationic bearing 5+ charges.[59,64,65,67] The Fe(III) analog INO-4885, known also as WW-85, has *ortho* pyridyl nitrogens substituted with benzoates and is thus anionic with four negative charges on periphery.[23,60–62] The FP-15 has been frequently used in different *in vivo* models.[23,63–68] INO-4885, the Fe(III) *ortho* carboxylatophenylpyridyl analog, is in clinical development for the prevention of contrast-induced nephropathy by *Inotek Pharmaceuticals Corporation* (*FierceBiotech*, 2014).

Structure–activity relationships for Mn porphyrins

A number of substituted porphyrinic compounds of different charges, stericity, size, and shape were analyzed over years (Figure 27.10). A first linear structure–activity relationship (SAR) between $E_{1/2}$ and $\log k_{cat}(O_2^{\bullet-})$ was established in 1999 in the range of $E_{1/2}$ studied.[24] More compounds were subsequently synthesized or obtained from commercial sources and analyzed on their redox properties and SOD-like activities ($\log k_{cat}(O_2^{\bullet-})$). The availability of a variety of compounds allowed us to distinguish between three SARs (Figure 27.11) that relate the thermodynamics (ability to catalyze $O_2^{\bullet-}$ dismutation) and kinetics (speed of dismutation process) of the catalysis driven by Mn porphyrins, which bear either (i) cationic charges, (ii) anionic charges, or (iii) no charges at the periphery.[69]

Furthermore, the availability of three compounds with essentially identical values of $E_{1/2}$, but of either cationic (MnTE-2-PyP^{5+}) or anionic (MnBr$_8$TSPP^{3-}) charge or lacking charge at the periphery (MnBr$_8$T-2-PyP$^+$), allowed us to quantify the remarkable contribution of electrostatics – of >2 orders of magnitude – in the catalysis of $O_2^{\bullet-}$ dismutation (Figures 27.11 and 27.12).

Finally, the effect of charge distribution was also studied. The remarkable difference of 220-fold was found between two compounds that have the same total charge and heterocyclic substituents of same 5-membered ring size and same number of carbon and nitrogen atoms. Yet the charges are differently distributed within *meso* imidazolyl relative to *meso* pyrazolyl rings (Figure 27.13).[70] In turn, $E_{1/2}=-4$ mV, resulting in significant reduction in $k_{cat}(O_2^{\bullet-})$, but the pyrazolium compound precluded protection of superoxide dismutase-deficient *E. coli* when growing aerobically. Its SOD-like activity is just slightly beyond the k for $O_2^{\bullet-}$ self-dismutation [$\log k_{self\text{-}dismutation}(O_2^{\bullet-}) = 5.70$].

Phase II. Improving the Mn porphyrin bioavailability – third lead compound MnTnHex-2-PyP^{5+} was identified

Throughout our developmental strategies, the medical audience questioned the *in vivo* location of Mn porphyrins. Can such highly positively charged compound cross

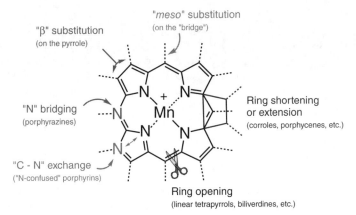

Figure 27.10 The substitutions of the porphyrin ring aimed to develop potent SOD mimics. Different substitutions were done on different *meso* and *beta* positions of porphyrin core. Also the carbons were replaced with nitrogens at *beta* and *meso* positions. The porphyrins of shrunken core, that is, corroles, and those of extended core, that is, porphycenes were synthesized by us and others also.

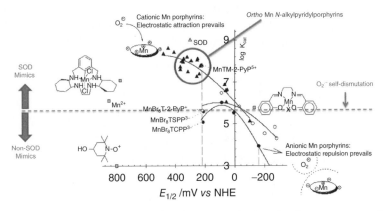

Figure 27.11 Structure–activity relationships for Mn porphyrins. The three structure–activity relationships were established between the $E_{1/2}$ for Mn^{III}P/Mn^{II}P redox couple and log $k_{cat}(O_2^{\bullet-})$ for Mn porphyrins that have either cationic charges on periphery (<u>triangles</u>), anionic charges on periphery (<u>circles</u>), or no charges on periphery (<u>open squares</u>). All complexes have +1 charge on metal site in resting (stable) state, that is, Mn +3 oxidation state. Few compounds have Mn in +2 oxidation state in resting state, that is MnBr$_8$TM-3(or 4)PyP^{4+}. Adapted from ref 216.

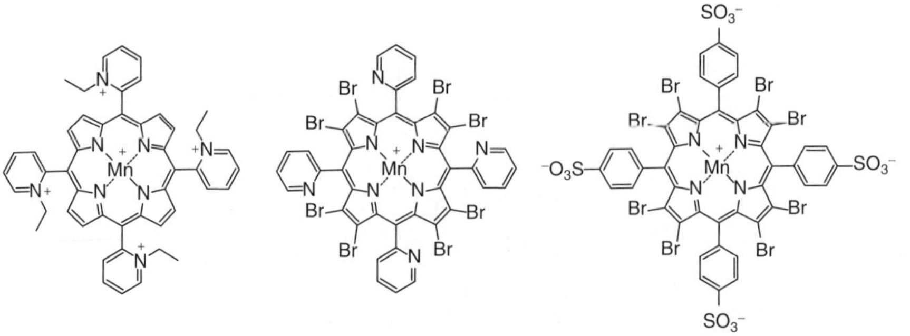

MnTE-2-PyP^{5+}
$E_{1/2}$ = 228 mV
logk_{cat}(O2$^-$) = 7.76

MnBr$_8$-2-PyP$^+$
$E_{1/2}$ = 219 mV
logk_{cat}(O2$^-$) = 5.63

MnBr$_8$TSPP^{3-}
$E_{1/2}$ = 209 mV
logk_{cat}(O2$^-$) = 5.15

Figure 27.12 The impact of electrostatics on $O_2^{\bullet-}$ dismutation. The electrostatic effects account for differences of more than two orders of magnitude in the catalysis of $O_2^{\bullet-}$ dismutation. The difference is higher between the porphyrins that are cationic (MnTE-2-PyP^{5+}) and anionic (MnBr$_8$TSPP^{3-}) (400-fold) than is between the Mn porphyrins that are cationic (MnTE-2-PyP^{5+}) and neutral (MnBr$_8$T-2-PyP$^+$) on the periphery (130-fold).[43]

the plasma and mitochondrial membranes? Do they mimic cytosolic or mitochondrial SOD, or perhaps localize outside of the cell and mimic extracellular CuZnSOD (EC-SOD). Therefore, in a Phase II of drug development (Figure 27.14), we concentrated on the ways to enhance the tissue and intracellular accumulation of Mn porphyrins. The data from a very first study, where Mn porphyrins had been tested on their ability to protect SOD-deficient *E. coli* when growing aerobically, indicated that in addition to high log $k_{cat}(O_2^{\bullet-})$, drug bioavailability is a second major factor that affects the magnitude of the therapeutic effects of Mn porphyrins. Mn porphyrins with highest $k_{cat}(O_2^{\bullet-})$ are pentacationic; owing to such high charge, they are very

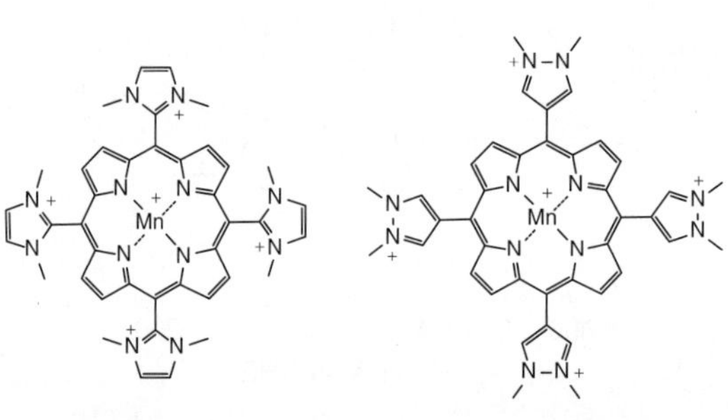

MnTDM-2-ImP^{5+}
$E_{1/2}$ ~ 320 mV
logk_{cat}(O2$^-$) ~ 8.19

MnTDM-2-PzP^{5+}
$E_{1/2}$ ~ -4 mV
logk_{cat}(O2$^-$) ~ 5.83

Figure 27.13 The impact of charge distribution on the SOD-like activity of MnP-based SOD mimics. The charge distribution contributed to 220-fold difference in $k_{cat}(O_2^{\bullet-})$ between compounds that appear similar on the first sight (with same number and types of atoms in their structures), with five-membered rings attached at *meso* positions. While both compounds bear two nitrogen atoms and three carbon atoms in each of their five-membered rings, those rings are differently organized. Different organization in turn results in different proximity of five positive charges to Mn site and in markedly different SOD-like activities.[70]

hydrophilic. This property indeed limits their accumulation *in vivo*, mostly from limited transport across the lipid membranes of mitochondria and blood-brain barrier. In a second phase of drug design (Figure 27.14), the modifications were implemented on *ortho* pyridyl core structure to increase their lipophilicity. Thus, the alkyl chains were lengthened from methyl to *n*-octyl.[71]

Such modifications preserve the cationic charges in close vicinity of Mn site and, in turn, provide the appropriate $E_{1/2}$ and log $k_{cat}(O_2^{\bullet-})$ to maintain high SOD-like activity. The one order of magnitude increase in lipophilicity, achieved by lengthening the alkyl chains by each CH_2 group, was demonstrated. The ~4 orders of magnitude increase in lipophilicity by lengthening the alkyl chains from ethyl (MnTE-2-PyP^{5+}) to *n*-hexyl (MnTnHex-2-PyP^{5+})contributed to up to 120-fold increase in therapeutic efficacy.[23] MnTnHex-2-PyP^{5+} became our third lead Mn porphyrin; it was forwarded to numerous *in vitro* and *in vivo* studies.[2,23,25,72–74] Its disadvantage was attributed to its toxicity, which is due to the interplay of enhanced cellular accumulation, micellar property, and (relative to MnTE-2-PyP^{5+}) 86 mV higher $E_{1/2}$. Consequently, it localizes in membranes where it disrupts their integrity. Still, its remarkable efficacy compensated in part for its toxicity, resulting in a better therapeutic window when compared to MnTE-2-PyP^{5+}.[73] Its clinical development is slowed down due to the lack of appropriate patenting and licensing. The HPLC/fluorescence and LCMS/MS methods were subsequently developed to provide the direct proof of Mn porphyrins' accumulation within cell and cellular fragments such as membranes, nucleus, cytosol, and mitochondria.[75–79] The ability of lipophilic analog, MnTnHex-2-PyP^{5+}, to cross blood–brain barrier and accumulate in brain parts (hippocampus, cortex, cerebellum, thalamus, brain stem, olfactory tube, and caudate-putamen) has been demonstrated.[75–79] As the Zn analogs are fluorescent, the imaging techniques were used to follow their accumulation within subcellular fragments.[80,81] Such data help us to understand the accumulation of Mn porphyrins in different fragments including cytosol, nucleus, lysosomes, and endoplasmic reticulum.[80,81] More hydrophilic Zn porphyrins, such as ZnTM-2-PyP^{4+}, would localize in lysosomes, while lipophilic ZnTnHex-2-PyP^{4+} distributes into mitochondria (reportedly next to cytochrome *c* oxidase at inner mitochondrial membrane), endoplasmic reticulum, and membranes.[80,81] The comparison can be done as the only structural difference among Zn and Mn porphyrins is the lack of the single positive charge on Zn site. With longer alkyl analogs (e.g., MnTnHex-2-PyP^{5+} and MnTnBuOE-2-PyP^{5+}), this difference is minimized as the alkyl chains hinder the single charge on metal site.[82,83] The correctness of such reasoning has been further substantiated with very similar lipophilicity of longer alkyl chain analogs of pentacationic Mn^{III}P and their reduced tetracationic Mn^{II}P analogs.[82] The data obtained with HPLC/fluorescence and LCMS/MS agree well with imaging techniques and with UV/vis measurements of the distribution of Mn porphyrins in the cytosol and membranes of *E. coli*.[84]

The LCMS/MS method was subsequently used for comprehensive pharmacokinetic studies, which were conducted with different lead Mn porphyrins via different routes, intravenous, subcutaneous, intraperitoneal, and oral. MnTE-2-PyP^{5+}, MnTnHex-2-PyP^{5+}, and MnTnBuOE-2-PyP^{5+} were studied in plasma and different organs.[74–76] Their accumulation owing to their pentacationic charge is fairly low in brain but is much higher with lipophilic analogs than with less lipophilic MnTE-2-PyP^{5+}. While originally reported as orally available, the oral availability

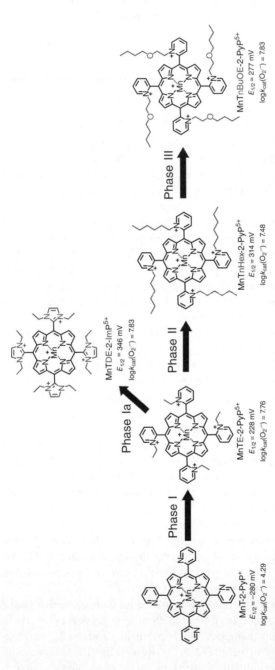

Figure 27.14 Phases of the design of Mn porphyrin-based SOD mimics. In phase I, the critical impact of *ortho* cationic charges on the log $k_{cat}(O_2^{\bullet-})$ was recognized. This feature was then preserved in all subsequent analogs while the nature of pyridyl substituents was modified to optimize bioavailability and toxicity. In phase II, the lipophilicity of MnTE-2-PyP⁵⁺ was enhanced, the MnTnHex-2-PyP⁵⁺ was synthesized. Its enhanced efficacy, due in major part to its orders of magnitude higher accumulation within cell and mitochondria, overcomes its increased toxicity. The higher toxicity is, at least in part, due to its higher cellular accumulation. In phase III, the insertion of oxygen atoms into alkyl chains suppressed the toxicity in MnTnBuOE-2-PyP⁵⁺ relative to MnTnHex-2-PyP⁵⁺ without reducing the lipophilicity of the molecule. Adapted from ref 216.

has been reassessed and is low but is almost sevenfold higher with lipophilic compounds: 0.6% (MnTE-2-PyP^{5+}) vs 4% (MnTnBuOE-2-PyP^{5+}) [Tovmasyan *et al.*, in preparation]. A high bioavailability was achieved via subcutaneous route that will be employed in anticipated 2016 clinical trials. Both heart and brain mitochondrial accumulation of different Mn analogs was assessed.[75,76] Again the more lipophilic MnTnHex-2-PyP^{5+} accumulates 3.6-fold more into heart mitochondria relative to cytosol while the ratio is 1.6 with MnTE-2-PyP^{5+}.[75–79] A similar ratio was found in brain mitochondria for lipophilic Mn porphyrins. Due to the low levels in brain mitochondria, the ratio could not have been assessed for MnTE-2-PyP^{5+}.

Phase III. Suppressing the toxicity of amphiphilic MnTnHex-2-PyP^{5+} – fourth lead compound MnTnBuOE-2-PyP^{5+} was identified

In Phase III of development, the goal was to reduce the MnTnHex-2-PyP^{5+}toxicity. The micellar properties of MnTnHex-2-PyP^{5+} were suppressed by the introduction of oxygen atoms into all *N*-alkylpyridyl chains (Figure 27.15).[42,92,93] The alkyl chains form the cavity around the Mn site, which by itself suppresses the solvation of Mn porphyrin. Introduction of oxygen atoms with two electron pairs at the end of each *n*-hexyl chain suppressed the molecule lipophilicity significantly.[93] Yet, if the oxygens are introduced deeper into the chains (where the oxygen solvation was minimized), the high lipophilicity of the molecule was preserved.[92] Thus, the fourth lead compound MnTnBuOE-2-PyP^{5+} was established. Relative to MnTnHex-2-PyP^{5+}, it is four-to-fivefold less toxic to mice.[92] MnTnBuOE-2-PyP^{5+} also accumulates less in liver that likely contributes to its lower toxicity. This may, at least in part, be due to the lesser ability of oxygen-bearing chains to align with phospholipids in cellular membranes and interrupt their integrity. The MnTnBuOE-2-PyP^{5+} has been used in several animal models and is being aggressively developed toward clinical trials as radioprotector of normal tissue.[94,95] The exciting data have been obtained in a glioblastoma multiforme and head and neck cancer subcutaneous mouse xenograft model where this compound acted as a radiosensitizer of tumor.[94,95] It also offered remarkable radioprotection of salivary glands and oral mucosa in a noncancer-bearing mouse.[94,95] Thus, the phase I/II clinical trials are anticipated to commence in January 2016 on radioprotection of normal brain and salivary glands and oral mucosa in glioma and head and neck cancer patients.

The activities of SOD mimics other than catalysis of O$_2^{\bullet-}$ dismutation

Three important issues need to be considered with regard to the in vivo actions of SOD mimics and relevant mechanistic conclusions: Firstly, since SOD enzymes and SOD mimics have similar thermodynamics and kinetics for O$_2^{\bullet-}$ dismutation,   they ought to have similar reactivities toward other species also.   Yet,   the

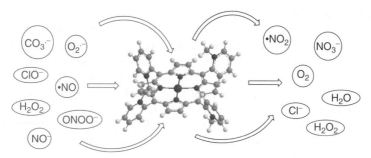

Figure 27.15 The reactivity of Mn porphyrin-based SOD mimics toward different reactive species. Thus far, the reactivity toward $O_2^{\bullet-}$, $ONOO^-$, $CO_3^{\bullet-}$, ClO^-, $^\bullet NO$, HNO, and H_2O_2 was assessed for many Mn porphyrins. The data were published or reported at meetings.[2,23] The data on $O_2^{\bullet-}$ and $ONOO^-$ are given in Table 27.1. The most potent SOD mimics, such as MnTE-2-PyP^{5+}, have rate constant with $O_2^{\bullet-}$ somewhat higher than with $ONOO^-$. The reactions with ClO^- occur with similar rate constants to those with $ONOO^-$.[2,25,85] While not listed here, the MnP-based SOD mimics are also reactive toward lipid-reactive species; no quantification is available. The reactivities toward thiols, simple and protein thiols, and toward ascorbate HA^- has been quantified in part.[2,40,83,86–91,213,214]

protein-structured SOD enzyme, but not small molecules such as Mn porphyrin, precludes the approach of the species other than $O_2^{\bullet-}$ to its metal site at any significant rate and is thus highly specific to $O_2^{\bullet-}$. Still even SOD enzymes can react with other species such as $ONOO^-$ or cysteines or H_2O_2 but at orders of magnitude lower rates than with $O_2^{\bullet-}$.[96–99] With no steric constrains of an enzyme, the high reactivity of Mn porphyrins toward many reactive species has been demonstrated.[2] While the $k_{cat}(O_2^{\bullet-})$ of potent SOD mimics is still higher than the rate constant for any other species determined thus far, the reactivities toward $ONOO^-$ and ClO^- are only slightly lower than $k_{cat}(O_2^{\bullet-})$.[85] While not quantified, the reactivities toward lipid radicals are probably very high also.[100,101] Thus, any conclusions on exclusive involvement of $O_2^{\bullet-}$ in the mechanistic considerations of low-molecular weight SOD mimics must be supported with genetic experiments on MnSOD and CuZnSOD knockout, transgenic, or overexpressor animals. *Secondly,* $O_2^{\bullet-}$ is a mild reductant and oxidant; therefore, it can be reduced and oxidized with SOD mimic at a fairly low biologically relevant potential. This in turn means that SOD enzymes and their efficacious mimics operate at mild potentials without imposing danger to the surrounding biological molecules. Critical to this fact – and not realized at the point the design of SOD mimics was launched – that such potentials appear to be ideal for coupling with numerous other species also, including cellular reductants and thiols, involved in fine modulation of cellular redox environment. *The type of action that would predominate in vivo would depend on the levels of oxidative stress, concentration, localization of an SOD mimic, and the species it encounters. Such action is still speculated and may not be SOD mimicking; yet the data demonstrate that potent SOD mimics would undergo readily all other reactions also when compared to less redox-active drugs. Thirdly,* even when a redox-active compound is not an SOD mimic, it may still be able to restore physiological redox environment, such as, for example, nitroxides and MitoQ (see also section "Redox-active compounds other than SOD mimics"). Nitroxides are able to undergo oxidation to oxo-ammonium cation with carbonate anion radical,

$CO_3^{\bullet-}$ (decomposition product of $ONOO^-$ adduct with CO_2). The oxo-ammonium cation can cycle back to nitroxide with $O_2^{\bullet-}$.[2,23,102] Similar is the case with quinone MitoQ, which can be reduced with $O_2^{\bullet-}$ to semiquinone radical MitoQH$^{\bullet}$. The MitoQH$^{\bullet}$ can then undergo disproportionation to MitoQ and quinol MitoQH$_2$. MitoQH$_2$ can reduce $ONOO^-$ while oxidizing itself to MitoQH$^{\bullet}$.[2,23,102] Finally, the anionic Mn(III) *meso-tetrakis*(4-carboxylatophenyl)porphyrin, MnTBAP^{3-}, and Mn complexes developed by Salvemini, Neumann, Patel and their colleagues, cannot be reduced with $O_2^{\bullet-}$ in the first step (and are thus not SOD mimics). Yet they can be oxidized with highly oxidizing species such as $ONOO^-$ or ClO^-; in the reverse reaction, they could possibly remove $O_2^{\bullet-}$.[102] In turn, they could favorably affect redox environment also.

The interpretation of data where synthetic redox-active compounds are used aiming at definite mechanistic conclusions, regardless of the class of compound tested, must be accompanied with genetic studies, employing transgenic, knockout, and overexpressor animals.

The data critical for the therapeutic efficacy of SOD mimics such as $\log k_{cat}(O_2^{\bullet-})$, $E_{1/2}$ for Mn^{III}P/Mn^{II}P redox couple, and lipophilicity (in terms of distribution between *n*-octanol and water, $\log P_{OW}$) are summarized in Table 27.1. As our research, along with the insight into the biology of oxidative stress, advances it became clear that those Mn porphyrins, whose Mn site is electron deficient and have biologically compatible $E_{1/2}$, react not only with $O_2^{\bullet-}$ but with other electron-donating reactive species ($ONOO^-$, $CO_3^{\bullet-}$, ClO^-, HO_2^-), cellular reductants (ascorbate, glutathione, cysteine, tetrahydrobiopterin), and redox-active protein sites such as thiols (Figures 27.15–27.17). In most cases, the reaction involves binding of ligand, except with $O_2^{\bullet-}$ where electron hopping happens via predominantly outer-sphere mechanism.[42,103] Mn porphyrins can also bind $^{\bullet}$NO as well as its protonated one-electron reduced species HNO.[104,105] The biological relevance of HNO gained lately a considerable interest.[106] With either $^{\bullet}$NO or HNO, Mn porphyrins would give rise to (NO)Mn^{II}P complexes.

Reactions with peroxynitrite and hypochlorite. Upon binding of highly oxidizing reactive species such as $ONOO^-$, MnPs would undergo one-electron transfer to form $O=Mn^{IV}P$ species with Mn in +4 oxidation state (Eq. (27.3)). In turn, $ONOO^-$ will be reduced to $^{\bullet}$NO$_2$ radical. The highly oxidizing $O=Mn^{IV}P$ species would be quickly reduced back to Mn^{III}P at the expense of cellular reductants (ascorbate, cysteine, glutathione) (Eq. (27.8)).[107,108] Due to biologically favorable $E_{1/2}$ for Mn^{III}P/Mn^{II}P redox couple and high intracellular levels of ascorbate and thiols, in addition to $O_2^{\bullet-}$ (Eq. (27.5)), the Mn^{III}P would be easily and more likely reduced to Mn^{II}P with those cellular reductants (Eq. (27.6)). The Mn^{II}P formed would subsequently undergo two-electron reaction with $ONOO^-$ giving rise to benign nitrite, NO_2^- (Eq. (27.4)). It has been also demonstrated that reduced Mn^{II}TE-2-PyP^{4+} reacts at similar rates with $ONOO^-$ (Eq. (27.4)) as does the Mn^{III}P in +3 oxidation state (Eq. (27.3)).[108] Therefore, the O_2^- removal may be coupled with $ONOO^-$ reduction and with ascorbate oxidation as demonstrated in equations (27.3)–(27.7). The rate constants for the reactions of Mn^{III}P with $ONOO^-$ were determined for a range of Mn porphyrins (Table 27.1).

$$Mn^{III}P^{5+} + ONOO^- \leftrightarrow O = Mn^{IV}P^{4+} + {}^\bullet NO_2 k_{red1}(ONOO^-) \tag{27.3}$$

$$Mn^{II}P^{4+} + ONOO^- \leftrightarrow O = Mn^{IV}P^{4+} + NO_2{}^- k_{red2}(ONOO^-) \tag{27.4}$$

$$Mn^{III}P^{5+} + O_2^{\bullet-} \leftrightarrow Mn^{II}P^{4+} + O_2 k_{ox}(O_2^{\bullet-}) \tag{27.5}$$

$$Mn^{III}P^{5+} + HA^- \leftrightarrow Mn^{II}P^{4+} + HA^\bullet (HA^\bullet \rightarrow A^{\bullet-} + H^+) k_{ox}(HA^-) \tag{27.6}$$

$$Mn^{II}P^{4+} + 2H^+ + O_2^{\bullet-} \leftrightarrow Mn^{III}P^{5+} + H_2O_2 \; k_{red}(O_2^{\bullet-}) \tag{27.7}$$

$$O = Mn^{IV}P^{4+} + HA^- + H^+ \leftrightarrow Mn^{III}P^{5+} + A^{\bullet-} + H_2O \; k_{red}(O = Mn^{IV}P^{4+}) \tag{27.8}$$

The rate constants determined for one-electron reaction of MnP with ClO^- are of similar magnitude as those with $ONOO^-$ (Table 27.1).[85] The compounds that are powerful SOD mimics are also powerful reductants of $ONOO^-$ and ClO^-. This is so because of the crucial role of the binding of those reactive species to Mn site in a first step of reduction, which in turn parallels $E_{1/2}$ of $Mn^{III}P/Mn^{II}P$ redox couple involved in $O_2^{\bullet-}$ dismutation as follows. The $E_{1/2}$ values of $O=Mn^{IV}P/Mn^{III}P$ redox couple (involved in the electron transfer from Mn +3 oxidation state to $ONOO^-$ bound to it), however, are essentially identical for all Mn complexes as found by us and others (Table 27.1).[40] Thus, the reactivities of Mn porphyrins towards $ONOO^-$ and ClO^- depend solely upon their binding. The binding of either $ONOO^-$ or ClO^- is proportional to the electron density of the metal site and thus relates to the first proton dissociation constant of axial waters, the pK_{a1}. Mn and Fe porphyrins have axially bound either aqua or hydroxo ligands, depending on the pH and the value of pK_{a1}. Those Mn porphyrins that are more electron deficient tend to bind the oxygen of a water molecule stronger. Subsequently, the hydrogen atom is loosely bound to oxygen and water is easily deprotonated at lower pH values, the pK_{a1} is in turn lower. In addition to axial waters, the pyrrolic nitrogens of a porphyrin ligand have their own proton equilibria. Their pK_a values parallel the pK_a values of axial waters of metalloporphyrins as both depend on the electronic properties of a porphyrin ring. We have previously reported that pK_a values of pyrrolic nitrogens are linearly related to $E_{1/2}$ of $Mn^{III}P/Mn^{II}P$ redox couple.[24] We have recently reported that the pK_{a1} values of the first axial water are also linearly related to $E_{1/2}$ of $Mn^{III}P/Mn^{II}P$ redox couple.[40] *Therefore, the reductions of ClO^- and $ONOO^-$ by MnP are controlled by same parameters as the dismutation of $O_2^{\bullet-}$.*

Reactions with ${}^\bullet NO$ and HNO. Cationic Mn(III) N-substituted pyridylporphyrins react with ${}^\bullet NO$ and HNO.[104,105] In either case, they give rise to the same product, nitrosylated compound, $(NO)Mn^{II}P$. Under aerobic conditions, nitrogen oxides (such as N_2O_3) get released very slowly and $Mn^{III}P$ is re-generated.[104] $MnTBAP^{3-}$ reacts with HNO also, but with ${}^\bullet NO$ only under reducing conditions.[121]

Catalase-like activity. The catalase-like activities of Mn porphyrins (dismutation of H_2O_2 to oxygen O_2 and H_2O) are very low and practically insignificant (Eqs (27.9) and (27.10)) accounting for at most 0.06% of activity of catalase enzyme [$\log k_{cat}(H_2O_2)$].[89] The two-electron reduction and oxidation during H_2O_2 dismutation process can occur not only through equations involving $O=Mn^VP=O$ species (Eqs (27.9) and (27.10)) but also through redox cycling between $Mn^{II}P$ (formed *in*

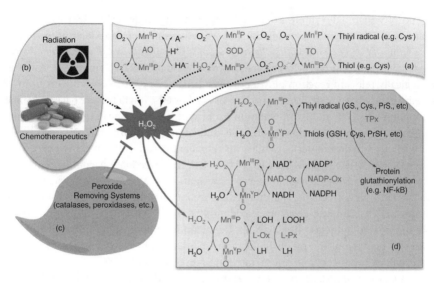

Figure 27.16 The role of H_2O_2 in MnP-related cellular pathways. The most potent SOD mimics are able to oxidize a number of biological molecules (those studied thus far are listed here) in the presence of H_2O_2. AO, ascorbate oxidation; TO, thiol oxidation; TPx, thiol peroxidation; NAD-ox, NAD oxidation; NADP-ox, NADP oxidation; L-ox, lipid oxidation; L-Px, lipid peroxidation. Adapted from ref 216. *(See color plate section for the color representation of this figure.)*

vivo for example in reaction with ascorbate, Eq. (27.6)), and $O=Mn^{IV}P$.[216]

$$Mn^{III}P^{5+} + H_2O_2 \leftrightarrow O = Mn^V P = O^{3+} + 2H^+ \qquad (27.9)$$

$$O = Mn^V P = O^{3+} + H_2O_2 + 2H^+ \leftrightarrow Mn^{III}P^{5+} + O_2 + 2H_2O \qquad (27.10)$$

The catalase-like activity of Fe porphyrins is somewhat higher, but occurs with lower turnover number, due to faster oxidative degradation of Fe versus Mn porphyrins. Higher turnover number could be achieved with more stable corrole complexes [see the section "Other SOD mimics"].[89]

Reactivity toward biological targets (NADH, NADPH, lipids, protein thiols, etc.) that involves H_2O_2. While catalase-like activity is insignificant, it appears that H_2O_2 is involved in the mechanism of action of SOD mimics via GPx- and cysteine oxidase-like actions.[2,122,123] It has been initially suggested by Piganelli's group that Mn porphyrins can oxidize cysteine of the p50 subunit of NF-κB.[88,124,125] It has been recently substantiated by Tome's group that Mn porphyrins can S-glutathionylate protein thiols of p50 and p65 subunits of NF-κB and of complexes I, III, and IV.[2,122,123] Data demonstrated the key involvement of H_2O_2 and GSH.[87-89,123-125] We have subsequently explored H_2O_2-related aqueous chemistry and have summarized the role of peroxide in Mn porphyrins action in Figure 27.16.[123] In the presence of peroxide, MnP can oxidize several other targets, such as NADH and NADPH, as well as lipids such as shown in Figure 27.16.

Yet, MnP can produce H_2O_2 also (during cycling with cellular reductants and $O_2^{\bullet-}$) and subsequently reuse it to oxidize biotargets (see the following paragraph and Figures 27.16 and 27.17). This agrees well with the results from Tome's group: no change in H_2O_2 levels were demonstrated upon the treatment of lymphoma cells with

Mn porphyrin and dexamethasone – a H_2O_2 producing system.[88,124,125] This indicates that any H_2O_2 formed is consumed in subsequent oxidation/peroxidation processes.

Cycling of Mn porphyrin with cellular reductants – ascorbate and thiols resulting in the production of H_2O_2. Such cycling is demonstrated in Figures 27.16 and 27.17. Another biologically relevant reaction of Mn porphyrins is the one with oxygen. Once $Mn^{III}P$ is reduced to $Mn^{II}P$ with either ascorbate or thiol or $O_2^{\bullet-}$, it will recycle back to $Mn^{III}P$. Reduction to $Mn^{II}P$ and its reoxidation occur with those species that are in vicinity of MnP and at high concentration. The reduction can occur with $O_2^{\bullet-}$, ascorbate, thiol, or tetrahydrobiopterin, while the reoxidation can occur with oxygen, $O_2^{\bullet-}$, $ONOO^-$, ClO^-, HO_2^-, lipid species, and so on. The reoxidation could be one- or two-electron reaction. The one-electron reoxidation of $Mn^{II}P$ with oxygen will give rise to $O_2^{\bullet-}$, which can then either in enzymatic manner or via self-dismutation produce H_2O_2. The one-electron reoxidation of $Mn^{II}P$ with $O_2^{\bullet-}$ will produce H_2O_2 in a superoxide reductase-like manner. Reoxidation with oxygen is more likely than with $O_2^{\bullet-}$ because of the much higher intracellular levels of oxygen. In either case, H_2O_2 will be produced, as demonstrated in several studies.[87,88,90]

The impact of MnP/H_2O_2 on cancer versus normal cell. H_2O_2 produced *in vivo* by MnP or by other means can be used by MnP to catalyze the oxidation of biological targets, such as thiols of p50 and p65 subunits of NF-κB with its subsequent inactivation (Figure 27.17).[88,90,123] If the magnitude of their oxidation is modest, the NF-κB-driven secondary oxidative stress will be suppressed and, in turn, cells will be rescued from excessive inflammation.[129,130] If, however, the levels of H_2O_2 are high, such as in cancer cell and further enhanced during cancer radio- or chemotherapy, the NF-κB (tumor) suppression may be of such magnitude that it will result in cell death. The cell death could be enhanced via inactivation of complexes I and III of electron transport chain by MnP.[88] Mild transient (normal tissue) versus continuous and extensive inhibition of NF-κB (tumor) may be compared to the use of aspirin/ibuprofen versus steroids (prednisone, dexamethasone). All three therapies suppress activation of NF-κB but result to various degrees of resolution of inflammation.[131] Therapy with aspirin and ibuprofen is efficacious only under mild inflammatory conditions. When inflammation is excessive, only steroid (prednisone/dexamethasone) would work, such is the case with pinched nerves or brain tumors. The steroid therapy would cause major suppression of NF-κB-driven inflammation and would thus be very efficacious on a short run. Yet on a long run, the extensive suppression of major transcription factor, NF-κB, would cause extensive metabolic damage such as induction of diabetes.[131]

The difference between the cancer and normal cell is depicted in Figure 27.17. The cancer cell, relative to the normal cell, is under constant oxidative stress, that is, inflamed, as it has the perturbed balance between the endogenous antioxidants and reactive species.[2,102] Frequently, the enzymes in charge of maintaining H_2O_2 at nanomolar levels are downregulated, while MnSOD is upregulated. Consequently, the peroxide levels get increased, which in turn results in high sensitivity of cancer cell to any further increase in oxidative stress. Such sensitivity of cancer cell is frequently exploited in cancer radio- and/or chemotherapy. These therapeutic modalities kill cancer cell via excessive production of reactive species. Recently, we and others are exploring the use of a joint MnP/ascorbate system in cancer therapy, alone or in combination with radiation, to enhance H_2O_2 production and cancer cell death.[123,214,215]

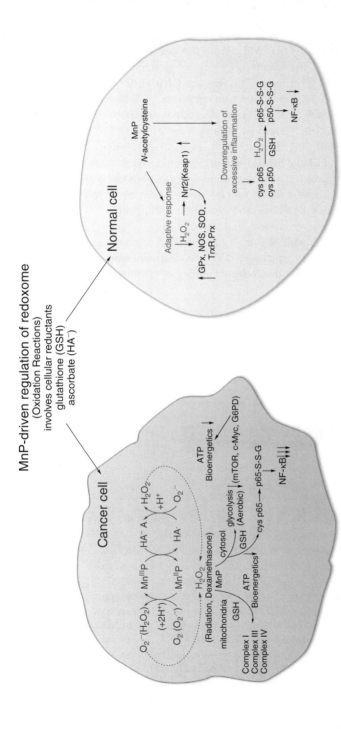

Figure 27.17 Differential role of peroxide in the mechanism(s) of action(s) of Mn porphyrins in cancer versus normal cell. Cancer cell is already under oxidative stress and is vulnerable to any additional increase in it. This is frequently due to the perturbed balance between SOD enzymes and H_2O_2-removing enzymes, which results in high H_2O_2 levels. The sensitivity to additional increase in oxidative stress has been employed in reactive species-producing cancer treatments such as radio- and chemotherapy. Normal cell has a variety of endogenous antioxidants to fight oxidative stress unless overwhelming. Most recent data indicate higher MnP levels in tumor than in normal tissue.[214,215] Thus, given higher H_2O_2 and MnP levels, the yield of the reactions of the MnP in cancer and normal cells is differential and results in cancer cell death vs normal cell healing. The p50 and p65 are NF-κB subunits; p65-S-S-G and p50-S-S-G are glutathionylated subunits; cys p65 and cys p50 relate to cysteines of p50 and p65 subunits; complexes I, III, and IV are complexes of mitochondrial respiration; GSH, glutathione; NOS, nitric oxide synthase; GPx, glutathione peroxidase; TrxR, thioredoxin reductase; Prx, peroxiredoxin; HA⁻, monodeprotonated physiologically relevant form of ascorbic acid. The Nrf2/Keap1 pathway, which controls levels of endogenous antioxidative defenses, seems to be involved. For details on pathways listed, please refer to Forum Issue of *Antioxid Redox Signal* on "SOD therapeutics", 2014.[88,127,128,216]

Interaction with signaling proteins and other cellular proteins

Several studies provided evidence that Mn porphyrins are able to suppress cellular transcription (Figure 27.18). Impact on signaling proteins may be one of the major mechanisms of MnP actions. This was seen with HIF-1α, AP-1, AIF, SP-1, and NF-κB.[74,87,88,138–142] The impact on other proteins with their subsequent inactivation, such as complexes I and III of mitochondrial respiratory chain, was also demonstrated.[124] MnTE-2-PyP^{5+} and its analogs, such as MnTnBuOE-2-PyP^{5+} (Jaramillo *et al.*, unpublished), directly react with the cysteine residues of NF-κB subunits and complexes I and III and oxidize and/or *S*-glutathionylate them either in cytosol (p65 and p50) or in nucleus (p50), preventing their DNA binding.[88,124] Indeed aqueous solution chemistry showed that cationic Mn(III) *N*-alkylpyridylporphyrins can oxidize thiols (glutathione, cysteine, and *N*-acetylcysteine) directly and in a pH-dependent manner in aqueous system and in a cell-free medium.[86,123] Such one-electron oxidation of thiols with Mn^{III}P is only possible with those compounds that have $E_{1/2} > 0$ mV versus NHE, such as *ortho, meta,* and *para* isomeric Mn(III) *N*-substituted pyridyl- or *diortho N,N'*-disubstituted imidazolylporphyrins. Those that

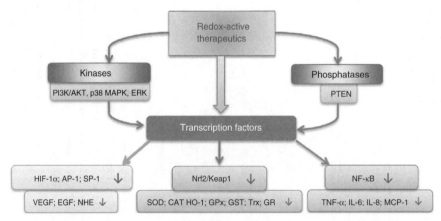

Figure 27.18 The interaction of SOD mimics with key cellular proteins. Only some of them are listed. The knowledge on new interactions emerges constantly. The interactions with HIF-1α, AP-1, SP-1, and NF-κB were demonstrated with several classes of SOD mimics (see manuscripts published in 2014, Forum Issue on "SOD Therapeutics").[2,122,127,132–135] The action upon Nrf2/Keap1 (with subsequent upregulation of numerous endogenous antioxidative defense systems, some of which are indicated here) has been reported with Mn(II) cyclic polyamine, nitroxide, and natural product curcumin. While the Nrf2/Keap1 was not directly assessed as a result of treatment with MnTnHex-2-PyP^{5+}, the upregulation of most of the listed enzymes was seen in kidney ischemia/reperfusion rat model; the impact was enhanced when MnP was given along with *N*-acetylcysteine – H$_2$O$_2$ producing system.[136,137] The action upon protein thiols is direct, while the actions on other pathways may be indirect and waits further exploration.[86,88] The inhibition of Na$^+$/H$^+$ exchanger (NHE in this figure only, otherwise normal hydrogen electrode) was shown in rat streptozotocin diabetes nephropathy model.[130] Adapted from ref 2.

have negative values of $E_{1/2} < 0$ mV versus NHE (e.g., MnTBAP^{3-} or MnTSPP^{3-}) can perhaps still oxidize thiols but through another redox couple, O=Mn^{IV}P/Mn^{III}P. Also Mn(III) corroles and Mn(II) cyclic polyamines (e.g., M40403) could not directly oxidize thiols. Yet, *in vivo* in an oxidizing environment MnTBAP^{3-}, MnTSPP^{3-} and Mn and Fe corroles, as well as cationic Mn porphyrins, could be oxidized to O=Mn^{IV}P with a variety of reactive species including ONOO$^-$, H$_2$O$_2$, and lipid-reactive species. Such highly oxidizing O=Mn^{IV}P species can in turn oxidize or *S*-glutathionylate thiols. This could perhaps explain why *in vivo* MnTBAP^{3-} suppresses NF-κB expression,[121] but not in a cell-free medium where there is no species that could first oxidize it. Possibly, MnTBAP^{3-} could inactivate NF-κB via H$_2$S/NO/HNO pathways studied by Miljkovic/Filipovic *et al.*[121,143] Under conditions of increased oxidative stress, such as in cancer or in cancer treated with radio- and chemotherapy or with MnP/ascorbate H$_2$O$_2$-producing system, MnP can utilize H$_2$O$_2$ to catalyze protein oxidation or peroxidation employing O=Mn^{IV}P/Mn^{III}P redox couple also.[123,213,214]

Mn porphyrins were not yet explored on their ability to activate Nrf2 via oxidation of Keap1 cysteines. Such a possibility has been addressed with another type of SOD mimic, Mn(II) cyclic polyamine, and nitroxide.[143] Nrf2 controls the regulation of endogenous antioxidative defenses; thus, its activation may be the most appropriate therapeutic effect of a drug. The data on MnP have been provided by Dorai's group in a rat kidney I/R model. The rats were treated with combined multicomponent drug system (MnTnHex-2-PyP^{5+}/growth factors/amino acids of Krebs cycle). The adaptive response was demonstrated as a significant upregulation of CuZnSOD, MnSOD, EC-SOD, lactoperoxidase, catalase, several peroxyredoxins, thioredoxin reductase, and so on, was observed. The effect was more pronounced when *N*-acetylcysteine was added that might have cycled with MnTnHex-2-PyP^{5+}. Consequently, H$_2$O$_2$ might have been produced resulting in an adaptive response such that Keap1 thiol would be oxidized and Nrf2 would be activated.[136] Considering the impact of curcumin and Mn(II) cyclic polyamine on Nrf2/Keap1 pathway (for further discussion see Ref.[2]), the data on kidney ischemia/reperfusion suggest that such a pathway might have also been involved in MnP action in rat kidney I/R injury.

Apart from the direct action of SOD mimics on signaling protein thiols, it has been proposed that Mn porphyrins can scavenge reactive species and eliminate the signal for activation of transcription factors that are postulated in Refs[138,139]. Whether the magnitude of such action(s) is high enough to be biologically relevant or not is still to be explored.

Which type of reaction(s) will an SOD mimic undergo *in vivo*?

This is the most difficult and poorly understood issue with SOD mimics and many other redox-active drugs. It can be safely said that *the type and the magnitude of the reactions of SOD mimics and other redox-active drugs with biological targets will be controlled by the location of an SOD mimic, by its concentration, and the concentration and type of reactive species (low-molecular weight or protein bound) it will encounter.* Fairly accurate conclusions may be drawn on the mechanisms of action only if the studies using pharmacological

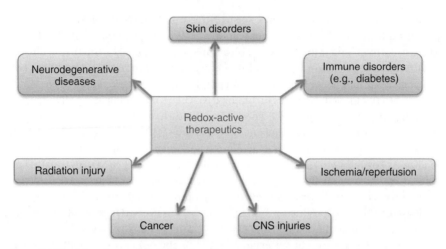

Figure 27.19 Major classes of diseases where Mn porphyrin-based SOD mimics were successfully tested. Studies are done on cells, rodents, and nonhuman primates. Pulmonary radioprotection was tested on nonhuman primates.[146] More detailed list of diseases was provided in Refs.[2,22, 25]. Some of the therapeutic effects of other drugs are listed in several review manuscripts.[2,22,25,127,133,147]

approaches (with SOD mimic or other redox-active drugs) are combined with those using MnSOD or Cu,ZnSOD genetically modified animals.[139,144,145]

Therapeutic effects of Mn porphyrins

Therapeutic effects of Mn porphyrins are summarized in Figure 27.19. Mn porphyrin-based SOD mimics have exhibited therapeutic efficacy in numerous cellular and animal studies (including nonhuman primates) in different models of diseases where the redox environment has been perturbed and which conditions, therefore, have oxidative stress in common. Some of the examples are injuries of central nervous system, cancer, radiation injury, immune disorders such as diabetes and neurodegenerative disorders. All therapeutic effects are summarized in Refs[2,22,25,127,133]. The therapeutic effects are explained in detail in the manuscripts of the Forum Issue of *Antioxidants & Redox Signaling* 2014 dedicated to SOD therapeutics.[2,122,127,132–135]

Other SOD mimics

Most recently, we have shown that the SAR established for Mn porphyrins is valid for all classes of SOD mimics. *Such evidence provided in Figure 27.20 supports the notion that the approach to the design of porphyrin-based SOD mimics is of much wider relevance.* In other words, superoxide does not care with whom it exchanges electrons as long as it is energetically favored, that is, if it happens at $E_{1/2}$ somewhere in between the $E_{1/2}$ for $O_2^{\bullet-}$ reduction and oxidation. The more the $E_{1/2}$ of SOD mimic is close to

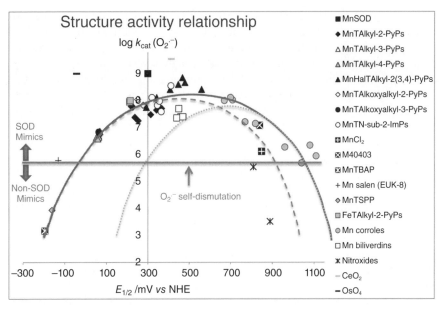

Figure 27.20 Structure–activity relationship between the log $k_{cat}(O_2^{\bullet-})$ and $E_{1/2}$ for redox couple involved (dashed line for $Mn^{III}P/Mn^{II}P$ and dotted line for $Mn^{IV}P/Mn^{III}P$). The relationship fits best for metal complexes and is not that good for nonmetal-based compounds such as nitroxides; its ability to affect superoxide dismutation is related to the fairly high rate constant for the reaction of nitroxide with protonated superoxide, very little of which is present at physiological pH. It seems that perhaps two relationships exist for two different redox couples and are slightly shifted based on different energetics of electron transfers involved with those couples. The maximum of the bell shape of the SAR describes the potential at which both steps of dismutation process occur at similar rates and where the $k_{cat}(O_2^{\bullet-})$ is in turn maximal. For those compounds that use $Mn^{III}P/Mn^{II}P$ redox couple, at more negative potentials, the metal +3 oxidation state is stabilized and cannot be reduced with $O_2^{\bullet-}$ to start the dismutation process. At more positive potentials, Mn is stabilized in +2 oxidation state and cannot be oxidized with $O_2^{\bullet}$ in the first step of dismutation process. For those compounds that use $Mn^{IV}P/Mn^{III}P$ redox couple, such as corroles and biliverdins, the reverse is true. The first step would involve the oxidation of the metal site from Mn^{3+} to Mn^{4+} and the reduction of $O_2^{\bullet-}$ followed by reduction of metal to Mn +3 oxidation resting state with concomitant oxidation of $O_2^{\bullet-}$. To identify the compounds reader is directed to Table 27.1. Adapted from ref 2. (*See color plate section for the color representation of this figure.*)

the midway potential of ~+300 mV versus NHE, the more potent it is, as it is able to equally well reduce and oxidize superoxide. So no half-reaction of dismutation process will be precluded owing to stabilization of Mn in either +3 (if $E_{1/2} < -180$ mV and thus cannot be reduced with $O_2^{\bullet-}$) or in +2 oxidation state (if $E_{1/2} > +800$ mV, and thus cannot be oxidized with $O_2^{\bullet-}$). The larger is the distance from the midway potential, the slower is the dismutation because of the stabilization of Mn in one of the two oxidation states. While most of Mn and Fe porphyrins employ $Mn^{III}P/Mn^{II}P$ redox couple for $O_2^{\bullet-}$ dismutation, Mn biliverdin and its analogs and Mn corroles employ Mn^{IV}/Mn^{III} redox couple with a remarkable success.[2,29]

Structures of SOD mimics other than Mn porphyrins are listed in Figure 27.6.

Fe(III) porphyrins. The chemistry of Mn and Fe aqua ions and their porphyrin complexes is very different. Thus, under physiological conditions, the cationic Fe porphyrins exist as monohydroxo species, $(OH)FeTE-2-PyP^{4+}$. In such a state, the Fe site has one water molecule and one OH^- axially ligated. Mn analog, however, under identical conditions has two axially coordinated waters. The presence of OH^- ligand in FeP modifies its potential to be identical to the one of MnP; in turn under physiological pH both Fe and Mn porphyrins have identical log $k_{cat}(O_2^{\bullet-})$ values listed in Table 27.1.[24,83] The Fe porphyrins are listed in Figure 27.6. FePs are more prone to undergo oxidation with H_2O_2 than Mn analogs, whereby they lose metal; consequently released Fe may impose toxicity as proven in our studies.[24,83,89] As demonstrated in *E. coli* study, when used at very low $\leq 1\,\mu M$ levels (up to 100-fold lower concentrations than those of analogous MnPs), the release of tiny amounts of Fe was protective as it replaced the Fe lost during $O_2^{\bullet-}$-driven oxidation and inactivation of Fe–S cluster-based proteins (dehydrates such as aconitase, Krebs cycle enzyme) and other Fe proteins as well.[149] It is intriguing, however, that many reports showed the protective effects of Fe porphyrins ($FeTM-4-PyP^{5+}$, $FeTSPP^{5+}$, FP-15, INO-4885, or WW-85) in animal models of diseases where they were administered at fairly high levels comparable to those of MnPs.[23,60–68] Still the *ortho* Fe(III) *N*-benzylpyridylporphyrin INO-4885(WW-85) showed protection on μg/kg scale. It is presently in clinical trials for contrast-induced nephropathy.

Mn(III) biliverdin and its analogs. The Mn(III) biliverdin is listed in Figure 27.5. The beneficial impact of heme oxygenase suggested that Mn(III) biliverdin and Mn(III) bilirubin may act as SOD mimics. Indeed, the high log $k_{cat}(O_2^{\bullet-})$ were reported for Mn biliverdin dimethyl ester and several other analogs.[29] The most unexpected observation was that Mn(III) complex with biliverdin is a dimer where presence of oxygen stabilized Mn in +4 oxidation state. Thus, instead of using $Mn^{III}P/Mn^{II}P$ redox couple, Mn biliverdin and its analogs use $(Mn^{IV}BV^-)_2/(Mn^{III}BV^{2-})_2$ redox couple for the catalysis of $O_2^{\bullet-}$ dismutation. The use of this couple where Mn cycles between +3 and +4 oxidation states, first established with Mn biliverdins, was later demonstrated also with Mn(III) corroles.[2]

Mn(III) and Fe(III) corroles. Because corrole ligand is tri-anionic ligand (relative to two-anionic porphyrin ligand), it forms three covalent bonds with Mn that neutralize the charge on the Mn site. In turn, it forms more stable complexes relative to porphyrin ligand. The metallocorrole is further stabilized during the catalysis of $O_2^{\bullet-}$ dismutation as it cycles from lower Mn +3 to higher Mn +4 oxidation state. Such redox cycling is similar to the one observed with Mn(III) biliverdin. The most potent corrole is di-anionic compound with two sulfonato groups on *beta* pyrrolic positions.[35,150] The complex is anionic that seems to suppress its transport across the lipid membranes into central nervous system and into mitochondria.[150] Based on Murphy's and Skulachev's research, to reach mitochondria, the drug must be (i) positively charged to be driven there by mitochondrial negative potential; and (ii) lipophilic to cross two mitochondrial membranes.[151–153] Metallocorrole with log $k_{cat}(O_2^{\bullet-})$ as high as 8.11 was synthesized (Figure 27.6 and Table 27.1). In comparison with MnPs, the elimination of $ONOO^-$ with metallocorroles is catalytic. In the first step, Mn(III) corrole gets oxidized with $ONOO^-$ two-electronically to O=Mn(V) compound. This highly valent Mn corrole is an oxidant that is strong enough to oxidize $ONOO^-$ while being reduced back to Mn(III) corrole.[35] HNO_2 was made in both steps.

Similar to Mn porphyrins, the metallocorroles are not $O_2^{\bullet-}$ specific and reportedly reduce $ONOO^-$.[154–157] Also, the Fe analog has fair catalase-like activity, which is several fold higher than that of Fe porphyrin.[158,159] Additionally, Fe corrole is more stable toward oxidative degradation with H_2O_2 and thus has higher turnover number than Fe porphyrin.[158,159] The *in vivo* H_2O_2 levels are so low – in between nanomolar and submicromolar – that the impact of H_2O_2 on metalloporphyrin degradation may be irrelevant. If levels of H_2O_2 relevant to porphyrin degradation were ever reached, the cell would undergo death. Thus, *in vivo* relevance of catalase-like activity of redox-active drugs such as Mn porphyrins and Fe corroles needs further investigation. A few recent studies indicate that in both normal (but sick) and cancer cells, MnPs employ H_2O_2 to modify protein thiols whereby affecting the cellular transcription activity and restoring physiological redox environment. Therefore, if MnP-based therapy requires H_2O_2, the logical question arises: Is there any advantage of designing an MnP with catalase-like activity?

Mn salen derivatives. Doctrow's group has developed Mn salen derivatives for therapeutic purposes. Such compounds have modest SOD-like activities (Table 27.1).[160] While of limited metal/ligand stability, Mn salens have showed therapeutic efficacy in many models of diseases. The cyclic crown ether structure was added to Mn salen core in EUK-207 (Figure 27.6) to improve metal/ligand stability and in turn prevent Mn loss. In an elegant study on Cryptococcus *neoformans,* the Mn salen EUK-8 and ascorbate were the only compounds capable of rescuing the MnSOD-deficient mutant when exposed to elevated temperatures.[161] None of the Mn porphyrins studied, cationic (*N*-alkylpyridylporphyrins) and anionic ($MnTBAP^{3-}$) were successful. The data suggest that perhaps Mn salen delivers Mn to the mitochondria. Similar to other SOD mimics, Mn salens can also react with species other than $O_2^{\bullet-}$, such as $ONOO^-$, ClO^-, and $^\bullet NO$.[23]

Mn(II) cyclic polyamines. It is important to note that this class of very potent SOD mimics (some of which approaching the activity of SOD enzyme) has Mn in +2 oxidation state, which renders it low metal ligand stability.[2,23,46–49] Apart from $O_2^{\bullet-}$, it reportedly can react with $^\bullet NO$ dismuting it to NO^+ and NO^- (HNO).[162] The most studied M40403 is listed in Figure 27.6.

Mn(II) complex with 1,2-diaminoethane-based ligand. The Mn(II) complex listed in Figure 27.6 has appropriate thermodynamics for $O_2^{\bullet-}$ dismutation, with $E_{1/2} = +440\,mV$ versus NHE to operate as SOD mimic as indicated in studies of Policar's group on activated macrophages.[163] The plus charge on Mn site may facilitate the approach of anionic $O_2^{\bullet-}$.

Mn(II) low-molecular weight complexes such as Mn(II) phosphate and Mn(II) lactate. Some living organisms, such *Lactobacillus plantarum*, do not have SOD enzymes but accumulate millimolar levels of manganese. Mn(II) lactate reportedly has the SOD-like activity, only 65-fold lower than that of SOD enzymes.[164] The fair SOD-like activity of low-molecular Mn(II) complexes, log $k_{cat}(O_2^{\bullet-})$ of 6–7, was reported by several groups.[29,165,166] Some of the published data suggest that some metal complexes are therapeutically efficacious as they serve as Mn transporters.[2,23,102,167] Thus, with *C. neoformans,* Mn salen but not $MnCl_2$ was efficacious in protecting MnSOD-deficient strain from oxidative stress caused by temperature elevation.[161] The protection of SOD-deficient *E. coli* and *Saccharomyces cerevisiae* was reported also at levels of $\geq 0.5\,mM$ Mn.[168]

Metal oxides. Metal oxides are listed in Figure 27.6. Osmium tetroxide aqueous solution has a very high log $k_{cat}(O_2^{\bullet-}) \sim 9$; yet OsO_4 is very toxic. Nanoparticles of cerium dioxide have log $k_{cat}(O_2^{\bullet-})$ of 9.55 due to their unique structure similar to nitrone spin trap. Nanoceria has been tested in numerous models of diseases. The activity toward other species such as $^{\bullet}NO$, $ONOO^-$, and H_2O_2 has been reported (see Ref.[169] for review). While such nanoparticles carry pharmacological potential, caution must be exercised due to their cationic charge, potential to aggregate, and potential to release extremely toxic Ce^{4+}. The recently reported results suggested that prolonged oral exposure to nanoceria has the potential to cause genetic damage, biochemical alterations, and histological changes after retention in vital organs of rats.[169,170] Research on biocompatibility of pharmaceutical preparations is needed because conclusions on nanoceria toxicity remain uncertain and controversial.

Is therapeutic efficacy proportional to SOD-like activity of SOD mimics?

Are there data demonstrating that the higher SOD-like activity results in higher therapeutic effects – a frequently asked question? The studies on SOD-deficient *E. coli* and *S. cerevisiae* prove that the more potent SOD mimic is, the higher the protection of those organisms is offered.[40,134] Most recently, two studies also substantiate that fact. Only potent SOD mimics are able to suppress spontaneous lipid peroxidation of brain homogenates.[40] The magnitude of suppression parallels the metal-centered $E_{1/2}$, which in turn parallels log $k_{cat}(O_2^{\bullet-})$, that is, the SOD-like activity (see structure–activity relationship in Figures 27.11 and 27.20). Thus, at 20 μM, MnTE-2-PyP^{5+} fully suppressed lipid peroxidation, while MnTBAP^{3-} was inefficacious up to 200 μM. When breast cancer cell line MCF-7 was exposed to MnP/ascorbate – a peroxide producing system – again those Mn porphyrins that are more potent SOD mimics are more efficacious in killing cancer cells.[171] The most recent data showed that MnTE-2-PyP^{5+}/radiation /ascorbate and MnTnBuOE-2-PyP^{5+}/radiation/ascorbate, but not MnTBAP^{3-}/radiation /ascorbate, profoundly suppressed 4T1 tumor growth in a mouse flank model.[215] A different mechanism seems to be involved: in lipid peroxidation, the antioxidative actions in elimination of the lipid-reactive species are likely operative, while prooxidative H_2O_2-driven peroxidation reactions catalyzed by MnPs account for cancer cell killing and tumor growth delay (see earlier sections). As already noted, and because it operates at a fairly mild reduction potential (which allows it to oxidize and reduce superoxide equally well), within the redox environment of a cell, the efficacious SOD mimic carries the potential to be an equally able (pro)oxidant and antioxidant (reductant). *Data on therapeutic efficacy suggest that development of compounds to be powerful SOD mimics may still be the most appropriate strategy in designing powerful drugs to suppress oxidative stress injuries and restore the physiological redox environment.*

Redox-active synthetic compounds other than SOD mimics

Some of the redox-active synthetic drugs, which are not SOD mimics, frequently used in animal models and progressed into clinical trials are presented in Figure 27.21.

Mangafodipir and its mixed Mn/Ca analog, Ca$_4$Mn(DPDP)$_5$. Mangafodipir, a manganese(II) complex with dipyridoxyl diphosphate, has been developed as a contrast agent, but is protective in cardiac infarction also (Figure 27.21) due reportedly to its MnSOD-like, catalase-like and glutathione reductase-like activities.[172–175] The ability to scavenge $O_2^{\bullet-}$ (but not ability to catalyze $O_2^{\bullet-}$ dismutation) was determined originally by EPR in the reaction with DMPO.[172] Later, its SOD-like activity in a

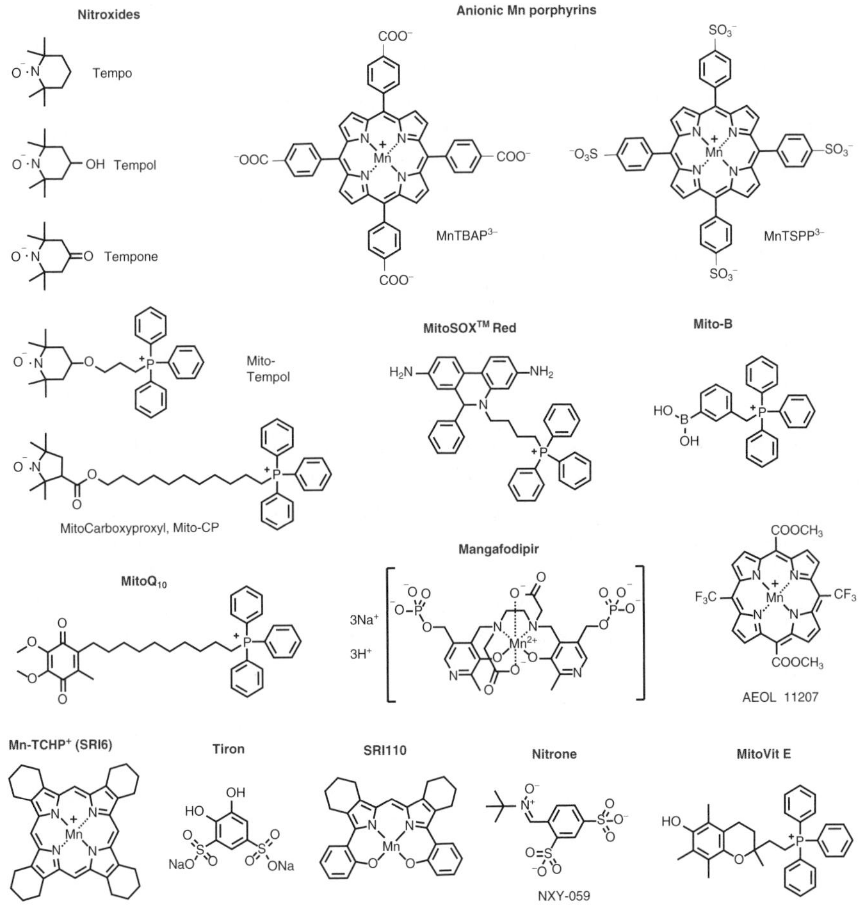

Figure 27.21 Other redox-active therapeutics. Listed are compounds that are not able to both reduce and oxidize $O_2^{\bullet-}$. Yet, during redox cycling, they can still affect *in vivo* levels of $O_2^{\bullet-}$ by removing it via oxidizing or reducing it. Due to its high oxidizing power, some compounds can be oxidized with $ONOO^-$ and cycle back with cellular reductants or perhaps other species (e.g., MnTBAP^{3-} or AEOL11207).

nitrobluetetrazolium (NBT) assay was expressed as 680 U/mg and compared with the "SOD-like activity" of MnTBAP^{3-}, which was expressed as 1700 U/mg.[119,176] As MnTBAP^{3-} has no SOD-like activity, the SOD-like activity of mangafodipir and its mixed Mn/Ca calmangafodipir analog is questionable.[167] No data are available to support its exclusive mitochondrial matrix location. The replacement of 80% of Mn with Ca in Ca$_4$Mn(DPDP)$_5$ reportedly improved the therapeutic efficacy of a compound. The therapeutic effects against myelosuppressive actions of oxaliplatin and in myocardial infarction may be due to other mechanism(s) of action(s). The compound is in phase II clinical trial as a chemotherapy adjunct in patients with colorectal cancer.

MnTBAP^{2-}. Mn(III) *meso-tetrakis*(4-carboxylatophenyl)porphyrin, MnTBAP^{3-} (Figure 27.21) was incorrectly identified as SOD mimic decades ago and subsequently and frequently used *in vivo* to show the involvement of O$_2^{\bullet-}$ in different animal models of diseases.[177–179] The aqueous chemistry studies on log k_{cat}(O$_2^{\bullet-}$) and the data on O$_2^{\bullet-}$ – specific aerobic growth of *E. coli* and *S. cerevisiae* (where MnTBAP^{3-} has no efficacy in rescuing those organisms when growing aerobically) unambiguously identified the lack of SOD-like activity of MnTBAP^{3-}; log k_{cat}(O$_2^{\bullet-}$) = 3.16 is below the value for O$_2^{\bullet-}$ self-dismutation.[134,167] The catalase-like activity has also been assigned to it, while there is none.[89] Based on early misassignment, MnTBAP^{3-} has been frequently used by numerous researchers to show the involvement of combined impact of O$_2^{\bullet-}$ and H$_2$O$_2$ on metabolic pathways.[121,177,179] Regardless of the incorrect assignment of the mechanism, such studies have demonstrated its therapeutic potential and thus justify efforts to figure out what types of actions are involved in the efficacy of MnTBAP^{3-}.[126,148,180–202] To complicate things further, often impure MnBAP^{3-} from different commercial sources was used in studies. Such preparations contain Mn aqua/hydroxo/acetato species, which in their own right have the SOD-like activity.[167,203] The use of impure samples obstructs the efforts to understand the therapeutic efficacy of MnTBAP^{3-}. Yet, three other studies demonstrated its efficacy where a very pure sample of MnTBAP^{3-} was compared to efficacious cationic SOD mimics (MnTE-2-PyP^{5+} and MnTnHex-2-PyP^{5+}).[121,178] The impact of MnTBAP^{3-} and MnTnHex-2-PyP^{5+} on NF-κB pathways, but not on SOD-like activity, was demonstrated in the suppression of spinal cord ischemia/reperfusion injury under identical dosing regimens.[121] Studies are in progress to examine the magnitude of the effect and gain further insight into the mechanistic issues involved. The role of ONOO$^-$ is also possible.[121] MnTBAP^{3-} is able to reduce it with k_{red}(ONOO$^-$) of 1.05×10^5 M^{-1} s^{-1}, which is still more than 2 orders of magnitude slower than of MnTnAlkyl-2-PyP^{5+} (Table 27.1).

MnTSPP^{2-}, Mn(III) *meso-tetrakis*(4-sulfonatophenyl)porphyrin (Figure 27.21). With $E_{1/2} = -160$ mV versus NHE, this MnP can hardly catalyze O$_2^{\bullet-}$ dismutation. Additional unfavorable electrostatics, where anionic charges on periphery repulse anionic O$_2^{\bullet-}$, reduces its ability to catalyze O$_2^{\bullet-}$ dismutation to a very low log k_{cat}(O$_2^{\bullet-}$) = 3.93, below the one for O$_2^{\bullet-}$ self-dismutation.[23]

Mn(III) 5,15-bis(methoxycarbonyl)-10,20-bis-trifluoromethylporphyrin (AEOL11207) (Figure 27.21). Although two CF$_3$ electron-withdrawing groups were attached directly to the porphyrin *meso* positions, the SOD-like activity of this compound is still not significant due, at least in part, to the lack of electrostatic facilitation for the reaction with O$_2^{\bullet-}$. It is reportedly ~350-fold less potent SOD mimic than

MnTDE-2-ImP^{5+} (with log $k_{cat}(O_2^{\bullet-})$ ~ 5.23).[50] Due to its lipophilicity, it is orally available and reportedly helpful in animal models of oxidative stress.[51] AEOL11207 likely has the modest ability to reduce ONOO$^-$. Interestingly, in an epilepsy model, it partially rescues complex II of mitochondrial respiration with concomitant increase in ATP levels in MnSOD$^{-/-}$ knockout mice (B6D2F2, able to live 8–10 days), without impacting complex I. The inability to inhibit complexes I and III (observed with MnTE-2-PyP^{5+}) supports the notion that such Mn porphyrin relative to cationic MnTAlkyl-2-PyP^{5+} is less redox-able to act on protein thiols and perhaps acts through different pathways.[51]

Manganese(III) complex of bis(hydroxyphenyl)dipyrromethene, SRI110 (Figure 27.21). While having no SOD-like activity, the compound does have fair ability to scavenge ONOO$^-$ with $k_{red}(ONOO^-) = 1.6 \times 10^6\ M^{-1}\ s^{-1}$ (Table 27.1).[204]

Nitroxides. Some representatives of nitroxides are shown in Figure 27.21. While not SOD mimics under physiological pH, nitroxides could be oxidized one-electronically to oxo-ammonium cation with decompositional products of ONOOCO$_2^-$: CO$_3^{\bullet-}$ and $^{\bullet}$NO$_2$. Yet, they could affect the levels of O$_2^{\bullet-}$ because the oxo-ammonium cation reacts rapidly with O$_2^{\bullet-}$ closing the catalytic cycle.[23] Nitroxide could be reduced one-electronically by ascorbate and thiols to hydroxylamine. The oxo-ammonium cation could also be reduced to hydroxylamine two-electronically with alcohols, thiols, and so on. Redox properties of nitroxides are also used in imaging of oxidative stress *in vivo*.[205,206] Recent data indicate that nitroxides could affect the Nrf2/Keap1 pathway.[207] This pathway controls the levels of endogenous antioxidative defenses such as catalases, SOD enzymes, peroxidases, and so on. The oxidation of Keap1 cysteines could occur via reduction of oxo-ammonium cation to nitroxide. The redox cycling of nitroxide to hydroxylamine is less likely, but not excluded.

Nitrones. The di-sulfonated PBN, NXY-059 (Figure 27.21), failed to produce an effect in stroke clinical trials. Presently, the NXY-059 is in development as anticancer agent.[208] Substantial evidence indicated its therapeutic efficacy in blast-induced brain injury.[209] Modified versions are synthesized to target mitochondria.[210] Nitrones could be oxidized to nitroxides one-electronically with carbon-centered radicals. They are now in development as anticancer agents.

MitoQ. MitoQ (Figure 27.21) was designed to mimic redox properties of ubiquinone, a site of O$_2^{\bullet-}$ production in electron transport chain. The molecule contains a redox-active unit, quinone, lipophilic alkyl chain of 10-carbon atoms, and cationic triphenylphosphonium ion. Studies by Skulachev's and Murphy's groups taught us that a compound must bear the lipophilic and cationic components to reach mitochondria.[151–153] MitoQ has been an excellent tool in demonstrating the role of mitochondria in oxidative stress. While not an SOD mimic, MitoQ can still modulate cellular redox status. It can oxidize O$_2^{\bullet-}$ with log $k_{ox}(O_2^{\bullet-})$ ~8. The semiquinone radical MitoQH$^{\bullet}$ can disproportionate into MitoQ and Mito quinol MitoQH$_2$.[211] The MitoQH$_2$ can rapidly reduce ONOO$^-$.

Redox-active natural compounds

Extensive research is ongoing on natural compounds. Many of those exhibit significant therapeutic effects. Some of those such as curcumin likely work through activating Nrf2/Keap1 pathway and thus upregulating endogenous antioxidative defenses. Indeed those compounds that have polyphenol moieties (many of them present in tea) may act as prooxidants inducing adaptive response. Such compounds can undergo one-electron reaction to form radicals that can be oxidized and in turn reduce oxygen to superoxide. Some of those are listed in Figure 27.22.

Purity of drugs

Over years, we have shown that many commercial sources do not supply MnTBAP^{3-} and MnTE-2-PyP^{5+}of adequate purities.[167,203,212] Caution needs to be exercised in drug development. An essential requirement is to have a pure drug of known identity to assign all the therapeutic effects to it and not to its impurity. The most recent identity issue occurring on TIC$_{10}$ anticancer drug, already in clinical trials, further supports the need to exercise such cautions.

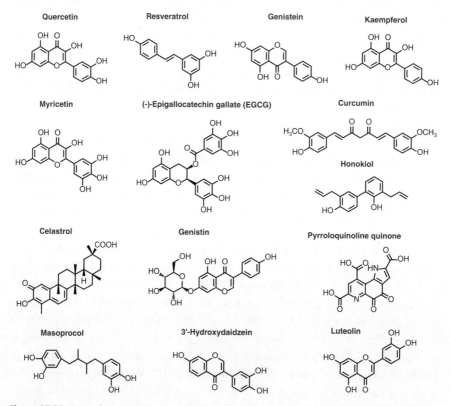

Figure 27.22 Natural compounds that exhibit therapeutic effects. Some of those are attributed incorrectly to the SOD-like activity.[118]

Pharmacokinetics and bioavailability of redox-active drugs and SOD mimics

A recent example with curcumin (Figure 27.22) points to the need, before progressing with clinical trials, to thoroughly investigate the drug pharmacokinetic (PK) behavior. Curcumin failed in Alzheimer's clinical trial as it does not even cross the blood barrier to enter the targeted central nervous system. Rarely thorough PK studies are reported. The comprehensive PK and bioavailability studies on Mn porphyrins via different routes of administration and different dosing regimens have been performed on plasma, tissues, mitochondria, nucleus, cytosol, and so on.[74–79] As lipophilic Mn(III) N-substituted pyridylporphyrins possess five positive charges and four lipophilic chains, they distribute within mitochondria as demonstrated with heart and brain mitochondria where they seem to mimic MnSOD.[76–79] Due to the cationic charges, they are attracted to phosphates, which allows them to accumulate in phosphate-rich tissues (brain, spinal cord) and cross the blood–brain barrier. The recent imaging studies on fluorescent Zn analogs help us understand the accumulation and localization of Mn porphyrins (see also phase II in the design of SOD mimics).[80,81]

Summary

- Cellular metabolism is largely redox based, that is, involves shutting of electrons among biomolecules.
- Superoxide dismutases (SOD) are key endogenous antioxidative defenses essential for all aerobic life. Along with systems that remove the product of $O_2^{\bullet-}$ dismutation, H_2O_2, they are gatekeepers of our redoxome, that is, they maintain physiological redox environment of a cell.
- $O_2^{\bullet-}$ itself is not a very damaging species, but it gives rise to different species some of which are highly damaging, such as hydroxyl radical, lipid radicals, and $ONOO^-$.
- Once the critical importance of $O_2^{\bullet-}$ and in turn SOD enzymes to our health became obvious, different classes of mimics of SOD enzymes have been sought. Some of those are as active as SOD enzymes as judged by the rate constant for the catalysis of $O_2^{\bullet-}$ dismutation.
- When designing SOD mimics, the thermodynamic (the ability to accept and give electron to $O_2^{\bullet-}$) and kinetic properties (the rate of electron transfer) of SOD enzyme were mimicked. In addition, the high stability of such molecules with regard to the loss of redox-active metal center was sought.
- Apart from SOD mimics, different other redox-active compounds have been developed and exhibit therapeutic effects in cellular and animal models of diseases.
- Due to key role that mitochondria have in diseases, the mitochondrially targeted redox-active drugs, bearing cationic charge(s) and lipophilic components in their structures, have been of special interest.
- SOD enzymes and SOD mimics of the same redox properties and electrostatics should theoretically undergo the same reactions. Yet those reactions are sterically precluded with large SOD protein structures. An SOD mimic, however, is not $O_2^{\bullet-}$ –

specific and undergoes a variety of other reactions with high rate constants. Some of those may even predominate in its mechanism(s) of action(s).

- In addition to its redox properties and stability, the biodistribution of a redox-active drug to a targeted location is a second major requirement for a potential therapeutic drug.
- While the type of action of an SOD mimic that predominates *in vivo* (depending on the levels of reactive species) is still speculated and may not be SOD mimicking, it seems that potent SOD mimics would more readily undergo reactions with a beneficial therapeutic outcome than less potent SOD mimics. Thus, aiming to synthesize redox-active drugs of high SOD-like activity may still be the best strategy to develop potent therapeutics.
- While the ability of SOD mimics to undergo a variety of reactions may be favored for therapeutic purposes, when one needs to conclude whether $O_2^{\bullet-}$ is involved in the mechanism of a certain disease, the use of genetically modified organisms, in addition to pharmacological approaches, is essential.

Acknowledgment

Authors are grateful to numerous researchers whose expert and invaluable work with Mn porphyrins and other redox-active drugs on cellular and animal models is driving the field of redoxome and drug development. The research on SOD mimics has been supported by numerous agencies most so by NIH/NCI (1R03-NS082704-01, NIH U19AI067798), DCI Core Grant P30-CA14236-29, Duke University CTSA 1UL1RR024128-01, William H Coulter Foundation, The Preston Robert Tisch Brain Tumor Center at Duke, BioMimetix JVLLC, NC Biotechnology Agency and IBH General Research funds.

Multiple choice questions

1 Cellular metabolism is largely based on redox reactions which always:
   a. Involve reactions with oxygen
   b. Involve shuttling of electrons between molecules
   c. Involve metal complexes
   d. Involve organic molecules

2 Important properties for SOD mimic design include all except:
   a. Thermodynamic property
   b. Kinetic property
   c. Stability
   d. Bioavailability
   e. Paramagnetism property

3 Drugs which target mitochondria have:
   a. Cationic charges and lipophilic components
   b. Anionic charges and lipophilic components
   c. Cationic charges and nucleophilic components
   d. Anionic chargers and nucleophilic components

4 Which equation involves a two-electron reaction?
  a. $Mn^{III}P^{5+} + ONOO^- \leftrightarrow O{=}Mn^{IV}P^{4+} + {}^\bullet NO_2$
  b. $Mn^{II}P^{4+} + ONOO^- \leftrightarrow O{=}Mn^{IV}P^{4+} + NO_2{}^-$
  c. $Mn^{III}P^{5+} + O_2^{\bullet-} \leftrightarrow Mn^{II}P^{4+} + O_2$
  d. $Mn^{II}P^{4+} + 2H^+ + O_2^{\bullet-} \leftrightarrow Mn^{III}P^{5+} + H_2O_2$

5 Which reactions will an SOD mimic perform *in vivo*?
  a. Catalysis of $O_2^{\bullet-}$ dismutation
  b. Catalysis of H_2O_2 dismutation
  c. Reduction of $ONOO^-$
  d. Reduction of HClO
  e. Oxidation of thiols
  f. Oxidation of ascorbate

6 Which factor(s) will affect the type of reactions occurring *in vivo*:
  a. Redox environment of the cell: normal versus cancer cell
  b. Reactivity of SOD mimic described by rate constants
  c. Co-localization of an SOD mimic with reactive species
  d. Concentrations of reactive species and SOD mimic

7 Can redox-active compounds with no SOD activity produce beneficial therapeutic effects?
  a. Yes
  b. No

8 Can compounds other than SOD mimic affect levels of superoxide *in vivo*?
  a. Yes
  b. No

9 Which of these species are free radicals with unpaired electron:
  a. Superoxide, $O_2^{\bullet-}$
  b. Peroxynitrite, $ONOO^-$
  c. Hydrogen peroxide H_2O_2
  d. Hypochlorite, HClO
  e. Nitric oxide, ${}^\bullet NO$
  f. Nitrogen dioxide, ${}^\bullet NO_2$
  g. Glutathione, GSH
  h. Metal ions, Mn^{3+}, Fe^{3+}

References

1 Abreu, I.A. & Cabelli, D.E. (2010) Superoxide dismutases – a review of the metal-associated mechanistic variations. *Biochimica et Biophysica Acta*, 1804, 263–274.
2 Batinic-Haberle, I., Tovmasyan, A., Roberts, E.R. *et al.* (2014) SOD therapeutics: latest insights into their structure-activity relationships and impact on the cellular redox-based signaling pathways. *Antioxidants and Redox Signaling*, 20, 2372–2415.
3 Desideri, A., Falconi, M., Parisi, V. *et al.* (1988) Is the activity-linked electrostatic gradient of bovine Cu, Zn superoxide dismutases conserved in homologous enzymes irrespective of the number and distribution of charges? *Free Radical Biology and Medicine*, 5, 313–317.
4 Ellerby, R.M., Cabelli, D.E., Graden, J.A. *et al.* (1996) Copper-zinc superoxide dismutase: why not pH-dependent? *Journal of the American Chemical Society*, 118, 6556–6561.

5 Getzoff, E.D., Tainer, J.A., Weiner, P.K. *et al.* (1983) Electrostatic recognition between super-oxide and copper, zinc superoxide dismutase. *Nature*, 306, 287–290.

6 Klug-Roth, D., Fridovich, I. & Rabani, J. (1973) Pulse radiolytic investigations of superoxide catalyzed disproportionation. Mechanism for bovine superoxide dismutase. *Journal of the American Chemical Society*, 95, 2786–2790.

7 Vance, C.K. & Miller, A.F. (1998) A simple proposal that can explain the inactivity of metal-substituted superoxide dismutases. *Journal of the American Chemical Society*, 120, 461–467.

8 Buettner, G.R. (2011) Superoxide dismutase in redox biology: the roles of superoxide and hydrogen peroxide. *Anti-Cancer Agents in Medicinal Chemistry*, 11, 341–346.

9 Buettner, G.R., Ng, C.F., Wang, M. *et al.* (2006) A new paradigm: manganese superoxide dismutase influences the production of H2O2 in cells and thereby their biological state. *Free Radical Biology and Medicine*, 41, 1338–1350.

10 Buettner, G.R., Wagner, B.A. & Rodgers, V.G. (2013) Quantitative redox biology: an approach to understand the role of reactive species in defining the cellular redox environment. *Cell Biochemistry and Biophysics*, 67, 477–483.

11 Ng, C.F., Schafer, F.Q., Buettner, G.R. *et al.* (2007) The rate of cellular hydrogen peroxide removal shows dependency on GSH: mathematical insight into in vivo H2O2 and GPx concentrations. *Free Radical Research*, 41, 1201–1211.

12 Radi, R. (2013) Peroxynitrite, a stealthy biological oxidant. *Journal of Biological Chemistry*, 288, 26464–26472.

13 Halliwell, B. & Gutteridge, J.M.C. (2007) *Free Radicals in Biology and Medicine*. Oxford University Press, New York.

14 McCord, J.M. & Fridovich, I. (1969) Superoxide dismutase. An enzymic function for ery-throcuprein (hemocuprein). *Journal of Biological Chemistry*, 244, 6049–6055.

15 Bagga, P. & Patel, A.B. (2012) Regional cerebral metabolism in mouse under chronic manganese exposure: implications for manganism. *Neurochemistry International*, 60, 177–185.

16 Hudnell, H.K. (1999) Effects from environmental Mn exposures: a review of the evidence from non-occupational exposure studies. *Neurotoxicology*, 20, 379–397.

17 Roth, J.A. (2006) Homeostatic and toxic mechanisms regulating manganese uptake, retention, and elimination. *Biological Research*, 39, 45–57.

18 Coulter, E.D., Emerson, J.P., Kurtz, D.M. *et al.* (2000) Superoxide reactivity of rubredoxin oxidoreductase (desulfoferrodoxin) from *Desulfovibrio vulgaris*: A pulse radiolysis study. *Journal of the American Chemical Society*, 122, 11555–11556.

19 Vance, C.K. & Miller, A.F. (2001) Novel insights into the basis for Escherichia coli superoxide dismutase's metal ion specificity from Mn-substituted FeSOD and its very high E(m). *Biochemistry*, 40, 13079–13087.

20 Gabbianelli, R., Battistoni, A., Polizio, F. *et al.* (1995) Metal uptake of recombinant cambialistic superoxide dismutase from Propionibacterium shermanii is affected by growth conditions of host *Escherichia coli* cells. *Biochemical and Biophysical Research Communications*, 216, 841–847.

21 Batinic-Haberle, I., Benov, L., Spasojevic, I. *et al.* (1998) The ortho effect makes manganese(III) meso-tetrakis(N-methylpyridinium-2-yl)porphyrin a powerful and potentially useful superoxide dismutase mimic. *Journal of Biological Chemistry*, 273, 24521–24528.

22 Batinic-Haberle, I., Rajic, Z., Tovmasyan, A. *et al.* (2011) Diverse functions of cationic Mn(III) N-substituted pyridylporphyrins, recognized as SOD mimics. *Free Radical Biology and Medicine*, 51, 1035–1053.

23 Batinic-Haberle, I., Reboucas, J.S. & Spasojevic, I. (2010) Superoxide dismutase mimics: chemistry, pharmacology, and therapeutic potential. *Antioxidants and Redox Signaling*, 13, 877–918.

24 Batinić-Haberle, I., Spasojević, I., Hambright, P. *et al.* (1999) Relationship among redox potentials, proton dissociation constants of pyrrolic nitrogens, and in vivo and in vitro superoxide dismutating activities of manganese(III) and iron(III) water-soluble porphyrins. *Inorganic Chemistry*, 38, 4011–4022.

25 Tovmasyan, A., Sheng, H., Weitner, T. *et al.* (2013) Design, mechanism of action, bioavailability and therapeutic effects of mn porphyrin-based redox modulators. *Medical Principles and Practice*, 22, 103–130.

26 Ko, T.P., Safo, M.K., Musayev, F.N. *et al.* (2000) Structure of human erythrocyte catalase. *Acta Crystallographica Section D: Biological Crystallography*, 56, 241–245.

27 Rowland, P., Blaney, F.E., Smyth, M.G. *et al.* (2006) Crystal structure of human cytochrome P450 2D6. *Journal of Biological Chemistry*, 281, 7614–7622.

28 Wang, Y., Geer, L.Y., Chappey, C. *et al.* (2000) Cn3D: sequence and structure views for Entrez. *Trends in Biochemical Sciences*, 25, 300–302.

29 Spasojevic, I., Batinic-Haberle, I., Stevens, R.D. *et al.* (2001) Manganese(III) biliverdin IX dimethyl ester: a powerful catalytic scavenger of superoxide employing the Mn(III)/Mn(IV) redox couple. *Inorganic Chemistry*, 40, 726–739.

30 Arambula, J.F., Preihs, C., Borthwick, D. *et al.* (2011) Texaphyrins: tumor localizing redox active expanded porphyrins. *Anti-Cancer Agents in Medicinal Chemistry*, 11, 222–232.

31 Sessler, J.L. & Miller, R.A. (2000) Texaphyrins: new drugs with diverse clinical applications in radiation and photodynamic therapy. *Biochemical Pharmacology*, 59, 733–739.

32 Young, S.W., Qing, F., Harriman, A. *et al.* (1996) Gadolinium(III) texaphyrin: a tumor selective radiation sensitizer that is detectable by MRI. *Proceedings of the National Academy of Sciences of the United States of America*, 93, 6610–6615.

33 Aviv, I. & Gross, Z. (2008) Iron(III) corroles and porphyrins as superior catalysts for the reactions of diazoacetates with nitrogen- or sulfur-containing nucleophilic substrates: synthetic uses and mechanistic insights. *Chemistry*, 14, 3995–4005.

34 Eckshtain, M., Zilbermann, I., Mahammed, A. *et al.* (2009) Superoxide dismutase activity of corrole metal complexes. *Dalton Transactions: An International Journal of Inorganic Chemistry*, 38, 7879–7882.

35 Okun, Z. & Gross, Z. (2012) Fine tuning the reactivity of corrole-based catalytic antioxidants. *Inorganic Chemistry*, 51, 8083–8090.

36 Goldstein, S., Czapski, G. & Heller, A. (2005) Osmium tetroxide, used in the treatment of arthritic joints, is a fast mimic of superoxide dismutase. *Free Radical Biology and Medicine*, 38, 839–845.

37 Heckert, E.G., Karakoti, A.S., Seal, S. *et al.* (2008) The role of cerium redox state in the SOD mimetic activity of nanoceria. *Biomaterials*, 29, 2705–2709.

38 Korsvik, C., Patil, S., Seal, S. *et al.* (2007) Superoxide dismutase mimetic properties exhibited by vacancy engineered ceria nanoparticles. *Chemical Communications (Cambridge, England)*, 10, 1056–1058.

39 Tantra, R., Cackett, A., Peck, R. *et al.* (2012) Measurement of redox potential in nanoecotoxicological investigations. *Jounal of Toxicology*, 2012, 270651.

40 Tovmasyan, A., Carballal, S., Ghazaryan, R. *et al.* (2014) Rational design of superoxide dismutase (SOD) mimics: the evaluation of the therapeutic potential of new cationic Mn porphyrins with linear and cyclic substituents. *Inorganic Chemistry*, 53, 11467–11483.

41 Spasojevic, I., Menzeleev, R., White, P.S. *et al.* (2002) Rotational isomers of N-alkylpyridyl porphyrins and their metal complexes. HPLC separation, H-1 NMR and X-ray structural characterization, electrochemistry, and catalysis of O-2(•-) disproportionation. *Inorganic Chemistry*, 41, 5874–5881.

42 Batinic-Haberle, I., Spasojevic, I., Stevens, R.D. *et al.* (2004) New class of potent catalysts of O2.-dismutation. Mn(III) ortho-methoxyethylpyridyl- and di-ortho-methoxyethylimidazo

lylporphyrins. *Dalton Transactions: An International Journal of Inorganic Chemistry*, 11, 1696–1702.

43 DeFreitas-Silva, G., Reboucas, J.S., Spasojevic, I. *et al.* (2008) SOD-like activity of Mn(II) beta-octabromo-meso-tetrakis(N-methylpyridinium-3-yl)porphyrin equals that of the enzyme itself. *Archives of Biochemistry and Biophysics*, 477, 105–112.

44 Kachadourian, R., Batinić-Haberle, I. & Fridovich, I. (1999) Syntheses and superoxide dismuting activities of partially (1-4) β-chlorinated derivatives of manganese(III) meso-tetrakis(N-ethylpyridinium-2-yl)porphyrin. *Inorganic Chemistry*, 38, 391–396.

45 Batinic-Haberle, I., Liochev, S.I., Spasojevic, I. *et al.* (1997) A potent superoxide dismutase mimic: manganese beta-octabromo-meso-tetrakis-(N-methylpyridinium-4-yl) porphyrin. *Archives of Biochemistry and Biophysics*, 343, 225–233.

46 Aston, K., Rath, N., Naik, A. *et al.* (2001) Computer-aided design (CAD) of Mn(II) complexes: superoxide dismutase mimetics with catalytic activity exceeding the native enzyme. *Inorganic Chemistry*, 40, 1779–1789.

47 Maroz, A., Kelso, G.F., Smith, R.A. *et al.* (2008) Pulse radiolysis investigation on the mechanism of the catalytic action of Mn(II)-pentaazamacrocycle compounds as superoxide dismutase mimetics. *Journal of Physical Chemistry A*, 112, 4929–4935.

48 Riley, D.P., Lennon, P.J., Neumann, W.L. *et al.* (1997) Toward the rational design of superoxide dismutase mimics: Mechanistic studies for the elucidation of substituent effects on the catalytic activity of macrocyclic manganese(II) complexes. *Journal of the American Chemical Society*, 119, 6522–6528.

49 Salvemini, D., Wang, Z.Q., Zweier, J.L. *et al.* (1999) A nonpeptidyl mimic of superoxide dismutase with therapeutic activity in rats. *Science*, 286, 304–306.

50 Liang, L.P., Huang, J., Fulton, R. *et al.* (2007) An orally active catalytic metalloporphyrin protects against 1-methyl-4-phenyl-1,2,3,6-tetrahydropyridine neurotoxicity in vivo. *Journal of Neuroscience*, 27, 4326–4333.

51 Liang, L.P., Waldbaum, S., Rowley, S. *et al.* (2012) Mitochondrial oxidative stress and epilepsy in SOD2 deficient mice: Attenuation by a lipophilic metalloporphyrin. *Neurobiology of Disease*, 45, 1068–1076.

52 McGovern, T., Day, B.J., White, C.W. *et al.* (2011) AEOL10150: a novel therapeutic for rescue treatment after toxic gas lung injury. *Free Radical Biology and Medicine*, 50, 602–608.

53 O'Neill, H.C., Orlicky, D.J., Hendry-Hofer, T.B. *et al.* (2011) Role of reactive oxygen and nitrogen species in olfactory epithelial injury by the sulfur mustard analogue 2-chloroethyl ethyl sulfide. *American Journal of Respiratory Cell and Molecular Biology*, 45, 323–331.

54 O'Neill, H.C., White, C.W., Veress, L.A. *et al.* (2010) Treatment with the catalytic metalloporphyrin AEOL 10150 reduces inflammation and oxidative stress due to inhalation of the sulfur mustard analog 2-chloroethyl ethyl sulfide. *Free Radical Biology and Medicine*, 48, 1188–1196.

55 Rabbani, Z.N., Batinic-Haberle, I., Anscher, M.S. *et al.* (2007) Long-term administration of a small molecular weight catalytic metalloporphyrin antioxidant, AEOL 10150, protects lungs from radiation-induced injury. *International Journal of Radiation Oncology, Biology, and Physics*, 67, 573–580.

56 Orrell, R.W. (2006) AEOL-10150 (Aeolus). *Current Opinion in Investigational Drugs*, 7, 70–80.

57 Sheng, H., Enghild, J.J., Bowler, R. *et al.* (2002) Effects of metalloporphyrin catalytic antioxidants in experimental brain ischemia. *Free Radical Biology and Medicine*, 33, 947–961.

58 Okado-Matsumoto, A., Batinic-Haberle, I. & Fridovich, I. (2004) Complementation of SOD-deficient *Escherichia coli* by manganese porphyrin mimics of superoxide dismutase activity. *Free Radical Biology and Medicine*, 37, 401–410.

59 Batinic-Haberle, I., Spasojevic, I., Stevens, R.D. *et al.* (2006) New PEG-ylated Mn(III) porphyrins approaching catalytic activity of SOD enzyme. *Dalton Transactions: An International Journal of Inorganic Chemistry*, 4, 617–624.

60 Genovese, T., Mazzon, E., Esposito, E. *et al.* (2009) Effects of a metalloporphyrinic peroxynitrite decomposition catalyst, ww-85, in a mouse model of spinal cord injury. *Free Radical Research*, 43, 631–645.

61 Maybauer, D.M., Maybauer, M.O., Szabo, C. *et al.* (2011) The peroxynitrite catalyst WW-85 improves pulmonary function in ovine septic shock. *Shock*, 35, 148–155.

62 Maybauer, D.M., Maybauer, M.O., Szabo, C. *et al.* (2011) The peroxynitrite catalyst WW-85 improves microcirculation in ovine smoke inhalation injury and septic shock. *Burns*, 37, 842–850.

63 Obrosova, I.G., Mabley, J.G., Zsengeller, Z. *et al.* (2005) Role for nitrosative stress in diabetic neuropathy: evidence from studies with a peroxynitrite decomposition catalyst. *FASEB Journal*, 19, 401–403.

64 Radovits, T., Beller, C.J., Groves, J.T. *et al.* (2012) Effects of FP15, a peroxynitrite decomposition catalyst on cardiac and pulmonary function after cardiopulmonary bypass. *European Journal of Cardio-Thoracic Surgery*, 41, 391–396.

65 Radovits T, Seres L, Gero D, et al. (2007) The peroxynitrite decomposition catalyst FP15 improves ageing-associated cardiac and vascular dysfunction. *Mechanisms of Ageing and Development* 128: 173-181.

66 Soriano, F.G., Lorigados, C.B., Pacher, P. *et al.* (2011) Effects of a potent peroxynitrite decomposition catalyst in murine models of endotoxemia and sepsis. *Shock*, 35, 560–566.

67 Szabo, C., Mabley, J.G., Moeller, S.M. *et al.* (2002) Part I: pathogenetic role of peroxynitrite in the development of diabetes and diabetic vascular complications: studies with FP15, a novel potent peroxynitrite decomposition catalyst. *Molecular Medicine*, 8, 571–580.

68 Szabo, G., Loganathan, S., Merkely, B. *et al.* (2012) *Catalytic peroxynitrite decomposition improves reperfusion injury after heart transplantation.* Vol. 143. *J Thorac Cardiovasc Surg*, pp. 1443–1449.

69 Reboucas, J.S., DeFreitas-Silva, G., Spasojevic, I. *et al.* (2008) Impact of electrostatics in redox modulation of oxidative stress by Mn porphyrins: protection of SOD-deficient *Escherichia coli* via alternative mechanism where Mn porphyrin acts as a Mn carrier. *Free Radical Biology and Medicine*, 45, 201–210.

70 Reboucas, J.S., Spasojevic, I., Tjahjono, D.H. *et al.* (2008) Redox modulation of oxidative stress by Mn porphyrin-based therapeutics: the effect of charge distribution. *Dalton Transactions: An International Journal of Inorganic Chemistry*, 9, 1233–1242.

71 Batinić-Haberle, I., Spasojević, I., Stevens, R.D. *et al.* (2002) Manganese(III) meso-tetrakis (ortho-N-alkylpyridyl)porphyrins. Synthesis, characterization, and catalysis of $O_2^{\bullet-}$ dismutation. *Journal of the Chemical Society, Dalton Transactions*, 2002, 2689–2696.

72 Gauter-Fleckenstein, B., Reboucas, J.S., Fleckenstein, K. *et al.* (2014) Robust rat pulmonary radioprotection by a lipophilic Mn N-alkylpyridylporphyrin, MnTnHex-2-PyP5+. *Redox Biology*, 2, 400–410.

73 Pollard, J.M., Reboucas, J.S., Durazo, A. *et al.* (2009) Radioprotective effects of manganese-containing superoxide dismutase mimics on ataxia-telangiectasia cells. *Free Radical Biology and Medicine*, 47, 250–260.

74 Weitner, T., Kos, I., Sheng, H. *et al.* (2013) Comprehensive pharmacokinetic studies and oral bioavailability of two Mn porphyrin-based SOD mimics, MnTE-2-PyP(5+) and MnTnHex-2-PyP(5+). *Free Radical Biology and Medicine*, 58, 73–80.

75 Spasojevic, I., Chen, Y., Noel, T.J. *et al.* (2008) Pharmacokinetics of the potent redox-modulating manganese porphyrin, MnTE-2-PyP(5+), in plasma and major organs of B6C3F1 mice. *Free Radical Biology and Medicine*, 45, 943–949.

76 Spasojevic, I., Chen, Y., Noel, T.J. *et al.* (2007) Mn porphyrin-based superoxide dismutase (SOD) mimic, MnIIITE-2-PyP5+, targets mouse heart mitochondria. *Free Radical Biology and Medicine*, 42, 1193–1200.

77 Spasojevic, I., Li, A., Tovmasyan, A. *et al.* (2010) Accumulation of Porphyrin-based SOD Mimics in Mitochondria is Proportional to Their Lipophilicity: S-cerevisiae Study of ortho Mn(III) N-alkylpyridylporphyrins. *Free Radical Biology and Medicine*, 49, S199–S199.

78 Spasojevic, I., Miryala, S., Tovmasyan, A. *et al.* (2011) Lipophilicity of Mn(III) *N*-alkylpyridylporphyrins dominates their accumulation within mitochondria and therefore *in vivo* efficacy. A mouse study. *Free Radical Biology and Medicine*, 51, S98.

79 Spasojevic, I., Weitner, T., Tovmasyan, A. *et al.* (2013) Pharmacokinetics, brain hippocampus and cortex, and mitochondrial accumulation of a new generation of lipophilic redox-active therapeutic, Mn(III) *meso* tetrakis(*N*-n-butoxyethylpyridinium-2-yl) porphyrin, MnTnBuOE-2-PyP^{5+}, in comparison with its Ethyl and N-hexyl Analogs, MnTE-2-PyP^{5+} and MnTnHex-2-PyP^{5+}. *Free Radical Biology and Medicine*, 65, S132.

80 Ezzeddine, R., Al-Banaw, A., Tovmasyan, A. *et al.* (2013) Effect of molecular characteristics on cellular uptake, subcellular localization, and phototoxicity of Zn(II) N-alkylpyridylporphyrins. *Journal of Biological Chemistry*, 288, 36579–36588.

81 Odeh, A.M., Craik, J.D., Ezzeddine, R. *et al.* (2014) Targeting Mitochondria by Zn(II)N-Alkylpyridylporphyrins: The Impact of Compound Sub-Mitochondrial Partition on Cell Respiration and Overall Photodynamic Efficacy. *PLoS One*, 9, e108238.

82 Spasojevic, I., Kos, I., Benov, L.T. *et al.* (2011) Bioavailability of metalloporphyrin-based SOD mimics is greatly influenced by a single charge residing on a Mn site. *Free Radical Research*, 45, 188–200.

83 Tovmasyan, A., Weitner, T., Sheng, H. *et al.* (2013) Differential coordination demands in Fe versus Mn water-soluble cationic metalloporphyrins translate into remarkably different aqueous redox chemistry and biology. *Inorganic Chemistry*, 52, 5677–5691.

84 Kos, I., Benov, L., Spasojevic, I. *et al.* (2009) High lipophilicity of meta Mn(III) N-alkylpyridylporphyrin-based superoxide dismutase mimics compensates for their lower antioxidant potency and makes them as effective as ortho analogues in protecting superoxide dismutase-deficient Escherichia coli. *Journal of Medicinal Chemistry*, 52, 7868–7872.

85 Carballal, S., Valez, V., Batinic-Haberle, I. *et al.* (2013) Reactivity and cytoprotective capacity of the synthetic catalytic antioxidants Mnporphyrins towards peroxynitrite and hypochlorite. *Free Radical Biology and Medicine*, 65(Supplement 2), S121–S122.

86 Batinic-Haberle, I., Spasojevic, I., Tse, H.M. *et al.* (2012) Design of Mn porphyrins for treating oxidative stress injuries and their redox-based regulation of cellular transcriptional activities. *Amino Acids*, 42, 95–113.

87 Evans, M.K., Tovmasyan, A., Batinic-Haberle, I. *et al.* (2014) Mn porphyrin in combination with ascorbate acts as a pro-oxidant and mediates caspase-independent cancer cell death. *Free Radical Biology and Medicine*, 68, 302–314.

88 Jaramillo, M.C., Briehl, M.M., Crapo, J.D. *et al.* (2012) Manganese porphyrin, MnTE-2-PyP5+, Acts as a pro-oxidant to potentiate glucocorticoid-induced apoptosis in lymphoma cells. *Free Radical Biology and Medicine*, 52, 1272–1284.

89 Tovmasyan, A., Maia, C.G., Weitner, T. et al. (2015) A comprehensive evaluation of catalase-like activity of different classes of redox-active therapeutics. *Free Radical Biology and Medicine*, 86, 308–321.

90 Ye, X., Fels, D., Tovmasyan, A. *et al.* (2011) Cytotoxic effects of Mn(III) N-alkylpyridylporphyrins in the presence of cellular reductant, ascorbate. *Free Radical Research*, 45, 1289–1306.

91 Batinic-Haberle, I., Rajic, Z. & Benov, L. (2011) A combination of two antioxidants (an SOD mimic and ascorbate) produces a pro-oxidative effect forcing *Escherichia coli* to adapt via induction of oxyR regulon. *Anti-Cancer Agents in Medicinal Chemistry*, 11, 329–340.

92 Rajic, Z., Tovmasyan, A., Spasojevic, I. *et al.* (2012) A new SOD mimic, Mn(III) ortho N-butoxyethylpyridylporphyrin, combines superb potency and lipophilicity with low toxicity. *Free Radical Biology and Medicine*, 52, 1828–1834.

93 Tovmasyan, A.G., Rajic, Z., Spasojevic, I. *et al.* (2011) Methoxy-derivatization of alkyl chains increases the in vivo efficacy of cationic Mn porphyrins. Synthesis, characterization, SOD-like activity, and SOD-deficient *E. coli* study of meta Mn(III) N-methoxyalkylpyridylporphyrins. *Dalton Transactions: An International Journal of Inorganic Chemistry*, 40, 4111–4121.

94 Weitzel, D.H., Tovmasyan, A., Ashcraft, K.A. *et al.* (2014) Radioprotection of the brain white matter by Mn(III) N-butoxyethylpyridylporphyrin-based superoxide dismutase mimic, MnTnBuOE-2-PyP^{5+}. *Molecular Cancer Therapeutics*, 14(1), 70–79.

95 Ashcraft, K.A, Boss, M.-K., Tovmasyan, A. *et al.* (2015) A novel manganese-porphyrin superoxide dismutase-mimetic widens the therapeutic margin in a pre-clinical head and neck cancer model. *International Journal of Radiation Oncology*Biology*Physics*, 93, 892–900.

96 Alvarez, B., Demicheli, V., Duran, R. *et al.* (2004) Inactivation of human Cu,Zn superoxide dismutase by peroxynitrite and formation of histidinyl radical. *Free Radical Biology and Medicine*, 37, 813–822.

97 Ansenberger-Fricano, K., Ganini, D., Mao, M. *et al.* (2013) The peroxidase activity of mitochondrial superoxide dismutase. *Free Radical Biology and Medicine*, 54, 116–124.

98 Araujo-Chaves, J.C., Yokomizo, C.H., Kawai, C. *et al.* (2011) Towards the mechanisms involved in the antioxidant action of MnIII [meso-tetrakis(4-N-methyl pyridinium) porphyrin] in mitochondria. *Journal of Bioenergetics and Biomembranes*, 43, 663–671.

99 Winterbourn, C.C., Peskin, A.V. & Parsons-Mair, H.N. (2002) Thiol oxidase activity of copper, zinc superoxide dismutase. *Journal of Biological Chemistry*, 277, 1906–1911.

100 Bloodsworth, A., O'Donnell, V.B., Batinic-Haberle, I. *et al.* (2000) Manganese-porphyrin reactions with lipids and lipoproteins. *Free Radical Biology and Medicine*, 28, 1017–1029.

101 Rubbo, H. & Radi, R. (2008) Protein and lipid nitration: role in redox signaling and injury. *Biochimica et Biophysica Acta*, 1780, 1318–1324.

102 Miriyala, S., Spasojevic, I., Tovmasyan, A. *et al.* (2012) Manganese superoxide dismutase, MnSOD and its mimics. *Biochimica et Biophysica Acta*, 1822, 794–814.

103 Spasojevic, I., Batinic-Haberle, I., Reboucas, J.S. *et al.* (2003) Electrostatic contribution in the catalysis of O2*- dismutation by superoxide dismutase mimics. MnIIITE-2-PyP5+ versus MnIIIBr8T-2-PyP+. *Journal of Biological Chemistry*, 278, 6831–6837.

104 Spasojevic, I., Batinic-Haberle, I. & Fridovich, I. (2000) Nitrosylation of manganese(II) tetrakis(N-ethylpyridinium-2-yl)porphyrin: a simple and sensitive spectrophotometric assay for nitric oxide. *Nitric Oxide*, 4, 526–533.

105 Alvarez, L., Suarez, S.A., Bikiel, D.E. *et al.* (2014) Redox potential determines the reaction mechanism of HNO donors with Mn and Fe porphyrins: defining the better traps. *Inorganic Chemistry*, 53, 7351–7360.

106 Eberhardt, M., Dux, M., Namer, B. *et al.* (2014) H2S and NO cooperatively regulate vascular tone by activating a neuroendocrine HNO-TRPA1-CGRP signalling pathway. *Nature Communications*, 5, 4381.

107 Ferrer-Sueta, G., Hannibal, L., Batinic-Haberle, I. *et al.* (2006) Reduction of manganese porphyrins by flavoenzymes and submitochondrial particles: a catalytic cycle for the reduction of peroxynitrite. *Free Radical Biology and Medicine*, 41, 503–512.

108 Ferrer-Sueta, G., Quijano, C., Alvarez, B. *et al.* (2002) Reactions of manganese porphyrins and manganese-superoxide dismutase with peroxynitrite. *Methods in Enzymology*, 349, 23–37.

109 Rausaria, S., Ghaffari, M.M., Kamadulski, A. *et al.* (2011) Retooling manganese(III) porphyrin-based peroxynitrite decomposition catalysts for selectivity and oral activity:

a potential new strategy for treating chronic pain. *Journal of Medicinal Chemistry*, 54, 8658–8669.

110 Kasugai, N., Murase, T., Ohse, T. *et al.* (2002) Selective cell death by water-soluble Fe-porphyrins with superoxide dismutase (SOD) activity. *Journal of Inorganic Biochemistry*, 91, 349–355.

111 Jiao, X.Y., Gao, E., Yuan, Y. *et al.* (2009) INO-4885 [5,10,15,20-tetra[N-(benzyl-4'-carboxy late)-2-pyridinium]-21H,23H-porphine iron(III) chloride], a peroxynitrite decomposition catalyst, protects the heart against reperfusion injury in mice. *Journal of Pharmacology and Experimental Therapeutics*, 328, 777–784.

112 Rosenthal, R.A., Huffman, K.D., Fisette, L.W. *et al.* (2009) Orally available Mn porphyrins with superoxide dismutase and catalase activities. *Journal of Biological Inorganic Chemistry*, 14, 979–991.

113 Melov, S., Doctrow, S.R., Schneider, J.A. *et al.* (2001) Lifespan extension and rescue of spongiform encephalopathy in superoxide dismutase 2 nullizygous mice treated with superoxide dismutase-catalase mimetics. *Journal of Neuroscience*, 21, 8348–8353.

114 Colon, J., Herrera, L., Smith, J. *et al.* (2009) Protection from radiation-induced pneumonitis using cerium oxide nanoparticles. *Nanomedicine*, 5, 225–231.

115 Kim, J., Takahashi, M., Shimizu, T. *et al.* (2008) Effects of a potent antioxidant, platinum nanoparticle, on the lifespan of *Caenorhabditis elegans*. *Mechanisms of Ageing and Development*, 129, 322–331.

116 Kunchandy, E. & Rao, M.N.A. (1990) Oxygen radical scavenging activity of curcumin. *International Journal of Pharmaceutics*, 58, 237–240.

117 Dikalov, S., Losik, T. & Arbiser, J.L. (2008) Honokiol is a potent scavenger of superoxide and peroxyl radicals. *Biochemical Pharmacology*, 76, 589–596.

118 Kancheva, V.D. & Kasaikina, O.T. (2013) Bio-antioxidants - a chemical base of their antioxidant activity and beneficial effect on human health. *Current Medicinal Chemistry*, 20, 4784–4805.

119 Batinic-Haberle I, Reboucas JS, Spasojevic I. 2011. Response to Rosenthal et al.. *Antioxidants and Redox Signaling* 14: 1174-1176.

120 Kos, I., Reboucas, J.S., DeFreitas-Silva, G. *et al.* (2009) Lipophilicity of potent porphyrin-based antioxidants: comparison of ortho and meta isomers of Mn(III) N-alkylpyridylporphyrins. *Free Radical Biology and Medicine*, 47, 72–78.

121 Celic, T., Spanjol, J., Bobinac, M. *et al.* (2014) Mn porphyrin-based SOD mimic, MnTnHex-2-PyP and non-SOD mimic, MnTBAP suppressed rat spinal cord ischemia/reperfusion injury via NF-kappaB pathways. *Free Radical Research*, 48(12), 1–35.

122 Batinic-Haberle, I. & Spasojevic, I. (2014) Complex chemistry and biology of redox-active compounds, commonly known as SOD mimics, affect their therapeutic effects. *Antioxidants and Redox Signaling*, 20, 2323–2325.

123 Tovmasyan, A., Weitner, T., Jaramillo, M. *et al.* (2013) We have come a long way with Mn porphyrins: from superoxide dismutation to H_2O_2-driven pathways. *Free Radical Biology and Medicine*, 65, S133.

124 Jaramillo, M.C., Briehl, M.M., Batinic Haberle, I. *et al.* (2013) Inhibition of the electron transport chain via the pro-oxidative activity of manganese porphyrin-based SOD mimetics modulates bioenergetics and enhances the response to chemotherapy. *Free Radical Biology and Medicine*, 65, S25.

125 Jaramillo, M.C., Briehl, M.M. & Tome, M.E. (2010) Manganese porphyrin glutathionylates the p65 subunit of NF-κB to potentiate glucocorticoid-induced apoptosis in lymphoma. *Free Radical Biology and Medicine*, 49, S63.

126 Niwa, K., Porter, V.A., Kazama, K. *et al.* (2001) A beta-peptides enhance vasoconstriction in cerebral circulation. *American Journal of Physiology. Heart and Circulatory Physiology*, 281, H2417–H2424.

127 Delmastro-Greenwood, M.M., Tse, H.M. & Piganelli, J.D. (2014) Effects of metalloporphyrins on reducing inflammation and autoimmunity. *Antioxidants and Redox Signaling*, 20, 2465–2477.

128 Delmastro-Greenwood, M.M., Votyakova, T., Goetzman, E. *et al.* (2013) Mn porphyrin regulation of aerobic glycolysis: implications on the activation of diabetogenic immune cells. *Antioxidants and Redox Signaling*, 19, 1902–1915.

129 Archambeau, J.O., Tovmasyan, A., Pearlstein, R.D. *et al.* (2013) Superoxide dismutase mimic, MnTE-2-PyP(5+) ameliorates acute and chronic proctitis following focal proton irradiation of the rat rectum. *Redox Biology*, 1, 599–607.

130 Khan, I., Batinic-Haberle, I. & Benov, L.T. (2009) Effect of potent redox-modulating manganese porphyrin, MnTM-2-PyP, on the Na(+)/H(+) exchangers NHE-1 and NHE-3 in the diabetic rat. *Redox Report*, 14, 236–242.

131 Takada, Y., Bhardwaj, A., Potdar, P. *et al.* (2004) Nonsteroidal anti-inflammatory agents differ in their ability to suppress NF-kappaB activation, inhibition of expression of cyclooxygenase-2 and cyclin D1, and abrogation of tumor cell proliferation. *Oncogene*, 23, 9247–9258.

132 Jumbo-Lucioni, P.P., Ryan, E.L., Hopson, M.L. *et al.* (2014) Manganese-based superoxide dismutase mimics modify both acute and long-term outcome severity in a drosophila melanogaster model of classic galactosemia. *Antioxidants and Redox Signaling*, 20, 2361–2371.

133 Sheng, H., Chaparro, R.E., Sasaki, T. *et al.* (2014) Metalloporphyrins as therapeutic catalytic oxidoreductants in central nervous system disorders. *Antioxidants and Redox Signaling*, 20, 2437–2464.

134 Tovmasyan, A., Reboucas, J.S. & Benov, L. (2014) Simple biological systems for assessing the activity of superoxide dismutase mimics. *Antioxidants and Redox Signaling*, 20, 2416–2436.

135 Holley, A.K., Xu, Y., Noel, T. *et al.* (2014) Manganese superoxide dismutase-mediated inside-out signaling in HaCaT human keratinocytes and SKH-1 mouse skin. *Mary Ann Liebert, Inc. Antioxidants and Redox Signaling*, 20(15), 2347–2360.

136 Dorai, T., Fishman, A.I., Ding, C. *et al.* (2011) Amelioration of renal ischemia-reperfusion injury with a novel protective cocktail. *Journal of Urology*, 186, 2448–2454.

137 Cohen, J., Dorai, T., Ding, C. *et al.* (2013) The administration of renoprotective agents extends warm ischemia in a rat model. *Journal of Endourology*, 27, 343–348.

138 Moeller, B.J., Batinic-Haberle, I., Spasojevic, I. *et al.* (2005) A manganese porphyrin superoxide dismutase mimetic enhances tumor radioresponsiveness. *International Journal of Radiation Oncology, Biology, and Physics*, 63, 545–552.

139 Zhao, Y., Chaiswing, L., Oberley, T.D. *et al.* (2005) A mechanism-based antioxidant approach for the reduction of skin carcinogenesis. *Cancer Research*, 65, 1401–1405.

140 Miriyala, S., Thipakkorn, C., Xu, Y. *et al.* (2011) 4-Hydroxy-2-nonenal mediates Aifm2 release from mitochondria: an insight into the mechanism of oxidative stress mediated retrograde signaling. *Free Radical Biology and Medicine*, 51, S30.

141 Tse, H.M., Milton, M.J. & Piganelli, J.D. (2004) Mechanistic analysis of the immunomodulatory effects of a catalytic antioxidant on antigen-presenting cells: implication for their use in targeting oxidation-reduction reactions in innate immunity. *Free Radical Biology and Medicine*, 36, 233–247.

142 Bottino, R., Balamurugan, A.N., Tse, H. *et al.* (2004) Response of human islets to isolation stress and the effect of antioxidant treatment. *Diabetes*, 53, 2559–2568.

143 Miljkovic, J., Filipovic, M. (2016) *HNO/thiol biology as a therapeutic target*. In "Redox-active therapeutics". Springer-Verlag. In press.

144 Dhar, S.K., Tangpong, J., Chaiswing, L. *et al.* (2011) Manganese superoxide dismutase is a p53-regulated gene that switches cancers between early and advanced stages. *Cancer Research*, 71, 6684–6695.

145 Zhao, Y., Xue, Y., Oberley, T.D. *et al.* (2001) Overexpression of manganese superoxide dismutase suppresses tumor formation by modulation of activator protein-1 signaling in a multistage skin carcinogenesis model. *Cancer Research*, 61, 6082–6088.

146 Garofalo, M.C., Ward, A.A., Farese, A.M. *et al.* (2014) A pilot study in rhesus macaques to assess the treatment efficacy of a small molecular weight catalytic metalloporphyrin antioxidant (AEOL 10150) in mitigating radiation-induced lung damage. *Health Physics*, 106, 73–83.

147 Batinic Haberle, I., Tovmasyan, A. & Spasojevic, I. (2013) The complex mechanistic aspects of redox-active compounds, commonly regarded as SOD mimics. *BioInorg React Mech*, 9(1-4), 35–58.

148 Pan H, Shen K, Wang X, et al. 2014. Protective effect of metalloporphyrins against cisplatin-induced kidney injury in mice. *PLoS One* 9(1): e86057.

149 Gu, M. & Imlay, J.A. (2013) Superoxide poisons mononuclear iron enzymes by causing mismetallation. *Molecular Microbiology*, 89, 123–134.

150 Okun, Z., Kuperschmidt, L., Youdim, M.B. *et al.* (2011) Cellular uptake and organ accumulation of amphipolar metallocorroles with cytoprotective and cytotoxic properties. *Anti-Cancer Agents in Medicinal Chemistry*, 11, 380–384.

151 Liberman, E.A., Topaly, V.P., Tsofina, L.M. *et al.* (1969) Mechanism of coupling of oxidative phosphorylation and the membrane potential of mitochondria. *Nature*, 222, 1076–1078.

152 Murphy, M.P. (2008) Targeting lipophilic cations to mitochondria. *Biochimica et Biophysica Acta*, 1777, 1028–1031.

153 Murphy, M.P. & Smith, R.A. (2007) Targeting antioxidants to mitochondria by conjugation to lipophilic cations. *Annual Review of Pharmacology and Toxicology*, 47, 629–656.

154 Haber, A., Mahammed, A., Fuhrman, B. *et al.* (2008) Amphiphilic/bipolar metallocorroles that catalyze the decomposition of reactive oxygen and nitrogen species, rescue lipoproteins from oxidative damage, and attenuate atherosclerosis in mice. *Angewandte Chemie-International Edition*, 47, 7896–7900.

155 Kupershmidt, L., Okun, Z., Amit, T. *et al.* (2010) Metallocorroles as cytoprotective agents against oxidative and nitrative stress in cellular models of neurodegeneration. *Journal of Neurochemistry*, 113, 363–373.

156 Mahammed, A. & Gross, Z. (2006) Iron and manganese corroles are potent catalysts for the decomposition of peroxynitrite. *Angewandte Chemie International Edition in English*, 45, 6544–6547.

157 Okun, Z., Kupershmidt, L., Amit, T. *et al.* (2009) Manganese corroles prevent intracellular nitration and subsequent death of insulin-producing cells. *ACS Chemical Biology*, 4, 910–914.

158 Mahammed, A. & Gross, Z. (2010) Highly efficient catalase activity of metallocorroles. *Chemical Communications (Cambridge, England)*, 46, 7040–7042.

159 Mahammed, A. & Gross, Z. (2011) The importance of developing metal complexes with pronounced catalase-like activity. *Catalysis Science & Technology*, 1, 535–540.

160 Rosenthal, R.A., Fish, B., Hill, R.P. *et al.* (2011) Salen Mn complexes mitigate radiation injury in normal tissues. *Anti-Cancer Agents in Medicinal Chemistry*, 11, 359–372.

161 Giles, S.S., Batinic-Haberle, I., Perfect, J.R. *et al.* (2005) Cryptococcus neoformans mitochondrial superoxide dismutase: an essential link between antioxidant function and high-temperature growth. *Eukaryotic Cell*, 4, 46–54.

162 Filipovic, M.R., Duerr, K., Mojovic, M. *et al.* (2008) NO dismutase activity of seven-coordinate manganese(II) pentaazamacrocyclic complexes. *Angewandte Chemie-International Edition*, 47, 8735–8739.

163 Bernard, A.S., Giroud, C., Ching, H.Y. *et al.* (2012) Evaluation of the anti-oxidant properties of a SOD-mimic Mn-complex in activated macrophages. *Dalton Transactions: An International Journal of Inorganic Chemistry*, 41, 6399–6403.

164 Archibald, F.S. & Fridovich, I. (1982) The scavenging of superoxide radical by manganous complexes: in vitro. *Archives of Biochemistry and Biophysics*, 214, 452–463.

165 Barnese, K., Gralla, E., Valentine, J. *et al.* (2012) Biologically relevant mechanism for catalytic superoxide removal by simple manganese compounds. *Proceedings of the National Academy of Sciences of the United States of America*, 109, 6892–6897.

166 Barnese, K., Gralla, E.B., Cabelli, D.E. *et al.* (2008) Manganous phosphate acts as a super-oxide dismutase. *Journal of the American Chemical Society*, 130, 4604–4606.

167 Reboucas, J.S., Spasojevic, I. & Batinic-Haberle, I. (2008) Pure manganese(III) 5,10,15,20-tetrakis(4-benzoic acid)porphyrin (MnTBAP) is not a superoxide dismu-tase mimic in aqueous systems: a case of structure-activity relationship as a watchdog mechanism in experimental therapeutics and biology. *Journal of Biological Inorganic Chemistry*, 13, 289–302.

168 Al-Maghrebi, M., Fridovich, I. & Benov, L. (2002) Manganese supplementation relieves the phenotypic deficits seen in superoxide-dismutase-null *Escherichia coli*. *Archives of Bio-chemistry and Biophysics*, 402, 104–109.

169 Caputo, F., De Nicola, M. & Ghibelli, L. (2014) Pharmacological potential of bioactive engi-neered nanomaterials. *Biochemical Pharmacology*, 92, 112–130.

170 Kumari, M., Kumari, S.I. & Grover, P. (2014) Genotoxicity analysis of cerium oxide micro and nanoparticles in Wistar rats after 28 days of repeated oral administration. *Mutagenesis*, 29, 467–479.

171 Tovmasyan, A., Roberts, E.R.H., Yuliana, Y. *et al.* (2014) The role of ascorbate in therapeu-tic actions of cationic Mn porphyrin-based SOD mimics. *Free Radical Biology and Medicine*, 76(Supplement 1), S94–S99.

172 Brurok, H., Ardenkjaer-Larsen, J.H., Hansson, G. *et al.* (1999) Manganese dipyridoxyl diphosphate: MRI contrast agent with antioxidative and cardioprotective properties? *Biochemical and Biophysical Research Communications*, 254, 768–772.

173 Karlsson, J.O., Brurok, H., Eriksen, M. *et al.* (2001) Cardioprotective effects of the MR con-trast agent MnDPDP and its metabolite MnPLED upon reperfusion of the ischemic porcine myocardium. *Acta Radiologica*, 42, 540–547.

174 Karlsson, J.O., Kurz, T., Flechsig, S. *et al.* (2012) Superior therapeutic index of calman-gafodipir in comparison to mangafodipir as a chemotherapy adjunct. *Translational Oncology*, 5, 492–502.

175 Karlsson, J.O., Mortensen, E., Pedersen, H.K. *et al.* (1997) Cardiovascular effects of MnD-PDP and MnCl2 in dogs with acute ischaemic heart failure. *Acta Radiologica*, 38, 750–758.

176 Bedda, S., Laurent, A., Conti, F. *et al.* (2003) Mangafodipir prevents liver injury induced by acetaminophen in the mouse. *Journal of Hepatology*, 39, 765–772.

177 Faulkner, K.M., Liochev, S.I. & Fridovich, I. (1994) Stable Mn(III) porphyrins mimic super-oxide dismutase in vitro and substitute for it in vivo. *Journal of Biological Chemistry*, 269, 23471–23476.

178 Batinic-Haberle, I., Cuzzocrea, S., Reboucas, J.S. *et al.* (2009) Pure MnTBAP selectively scavenges peroxynitrite over superoxide: comparison of pure and commercial MnTBAP samples to MnTE-2-PyP in two models of oxidative stress injury, an SOD-specific *Escherichia coli* model and carrageenan-induced pleurisy. *Free Radical Biology and Medicine*, 46, 192–201.

179 Day, B.J., Fridovich, I. & Crapo, J.D. (1997) Manganic porphyrins possess catalase activity and protect endothelial cells against hydrogen peroxide-mediated injury. *Archives of Bio-chemistry and Biophysics*, 347, 256–262.

180 Cui YY, Qian JM, Yao AH, *et al.* 2012. SOD mimetic improves the function, growth, and survival of small-size liver grafts after transplantation in rats. *Transplantation* 94: 687-694.

181 Aladag, M.A., Turkoz, Y., Sahna, E. *et al.* (2003) The attenuation of vasospasm by using a SOD mimetic after experimental subarachnoidal haemorrhage in rats. *Acta Neurochirurgica*, 145, 673–677.

182 Bao, F., DeWitt, D.S., Prough, D.S. *et al.* (2003) Peroxynitrite generated in the rat spinal cord induces oxidation and nitration of proteins: reduction by Mn (III) tetrakis (4-benzoic acid) porphyrin. *Journal of Neuroscience Research*, 71, 220–227.

183 Bao, F. & Liu, D. (2004) Hydroxyl radicals generated in the rat spinal cord at the level produced by impact injury induce cell death by necrosis and apoptosis: protection by a metalloporphyrin. *Neuroscience*, 126, 285–295.

184 Cuzzocrea, S., Costantino, G., Mazzon, E. *et al.* (1999) Beneficial effects of Mn(III)tetrakis (4-benzoic acid) porphyrin (MnTBAP), a superoxide dismutase mimetic, in zymosan-induced shock. *British Journal of Pharmacology*, 128, 1241–1251.

185 Hachmeister, J.E., Valluru, L., Bao, F. *et al.* (2006) Mn (III) tetrakis (4-benzoic acid) porphyrin administered into the intrathecal space reduces oxidative damage and neuron death after spinal cord injury: a comparison with methylprednisolone. *Journal of Neurotrauma*, 23, 1766–1778.

186 Lee, B.I., Chan, P.H. & Kim, G.W. (2005) Metalloporphyrin-based superoxide dismutase mimic attenuates the nuclear translocation of apoptosis-inducing factor and the subsequent DNA fragmentation after permanent focal cerebral ischemia in mice. *Stroke*, 36, 2712–2717.

187 Leski, M.L., Bao, F., Wu, L. *et al.* (2001) Protein and DNA oxidation in spinal injury: neurofilaments--an oxidation target. *Free Radical Biology and Medicine*, 30, 613–624.

188 Levrand, S., Vannay-Bouchiche, C., Pesse, B. *et al.* (2006) Peroxynitrite is a major trigger of cardiomyocyte apoptosis in vitro and in vivo. *Free Radical Biology and Medicine*, 41, 886–895.

189 Li, Y., Wende, A.R., Nunthakungwan, O. *et al.* (2012) Cytosolic, but not mitochondrial, oxidative stress is a likely contributor to cardiac hypertrophy resulting from cardiac specific GLUT4 deletion in mice. *FEBS Journal*, 279, 599–611.

190 Ling, X. & Liu, D. (2007) Temporal and spatial profiles of cell loss after spinal cord injury: reduction by a metalloporphyrin. *Journal of Neuroscience Research*, 85, 2175–2185.

191 Liu, D., Bao, F., Prough, D.S. *et al.* (2005) Peroxynitrite generated at the level produced by spinal cord injury induces peroxidation of membrane phospholipids in normal rat cord: reduction by a metalloporphyrin. *Journal of Neurotrauma*, 22, 1123–1133.

192 Liu, D., Ling, X., Wen, J. *et al.* (2000) The role of reactive nitrogen species in secondary spinal cord injury: formation of nitric oxide, peroxynitrite, and nitrated protein. *Journal of Neurochemistry*, 75, 2144–2154.

193 Liu, D., Shan, Y., Valluru, L. *et al.* (2013) Mn (III) tetrakis (4-benzoic acid) porphyrin scavenges reactive species, reduces oxidative stress, and improves functional recovery after experimental spinal cord injury in rats: comparison with methylprednisolone. *BMC Neuroscience*, 14, 23.

194 Loukili, N., Rosenblatt-Velin, N., Li, J. *et al.* (2011) Peroxynitrite induces HMGB1 release by cardiac cells in vitro and HMGB1 upregulation in the infarcted myocardium in vivo. *Cardiovascular Research*, 89, 586–594.

195 Luo, J., Li, N., Paul Robinson, J. *et al.* (2002) Detection of reactive oxygen species by flow cytometry after spinal cord injury. *Journal of Neuroscience Methods*, 120, 105–112.

196 Patel, M., Day, B.J., Crapo, J.D. *et al.* (1996) Requirement for superoxide in excitotoxic cell death. *Neuron*, 16, 345–355.

197 Pires, K.M., Ilkun, O., Valente, M. *et al.* (2014) Treatment with a SOD mimetic reduces visceral adiposity, adipocyte death, and adipose tissue inflammation in high fat-fed mice. *Obesity (Silver Spring)*, 22, 178–187.

198 Suresh, M.V., Yu, B., Lakshminrusimha, S. *et al.* (2013) The protective role of MnTBAP in oxidant-mediated injury and inflammation in a rat model of lung contusion. *Surgery*, 154, 980–990.

199 Valluru, L., Diao, Y., Hachmeister, J.E. *et al.* (2012) Mn (III) tetrakis (4-benzoic acid) porphyrin protects against neuronal and glial oxidative stress and death after spinal cord injury. *CNS & Neurological Disorders Drug Targets*, 11, 774–790.

200 Wu, L., Shan, Y. & Liu, D. (2012) Stability, disposition, and penetration of catalytic antioxidants Mn-porphyrin and Mn-salen and of methylprednisolone in spinal cord injury. *Central Nervous System Agents in Medicinal Chemistry*, 12, 122–130.

201 Yu, D., Neeley, W.L., Pritchard, C.D. *et al.* (2009) Blockade of peroxynitrite-induced neural stem cell death in the acutely injured spinal cord by drug-releasing polymer. *Stem Cells*, 27, 1212–1222.

202 Cao, Y., Fujii, M., Ishihara, K. *et al.* (2013) Effect of a peroxynitrite scavenger, a manganese-porphyrin compound on airway remodeling in a murine asthma. *Biological and Pharmaceutical Bulletin*, 36, 850–855.

203 Reboucas, J.S., Spasojevic, I. & Batinic-Haberle, I. (2008) Quality of potent Mn porphyrin-based SOD mimics and peroxynitrite scavengers for pre-clinical mechanistic/therapeutic purposes. *Journal of Pharmaceutical and Biomedical Analysis*, 48, 1046–1049.

204 Rausaria, S., Kamadulski, A., Rath, N.P. *et al.* (2011) Manganese(III) complexes of bis(hydroxyphenyl)dipyrromethenes are potent orally active peroxynitrite scavengers. *Journal of the American Chemical Society*, 133, 4200–4203.

205 Davis, R.M., Mitchell, J.B. & Krishna, M.C. (2011) Nitroxides as cancer imaging agents. *Anti-Cancer Agents in Medicinal Chemistry*, 11, 347–358.

206 Fujii, H., Sato-Akaba, H., Kawanishi, K. *et al.* (2011) Mapping of redox status in a brain-disease mouse model by three-dimensional EPR imaging. *Magnetic Resonance in Medicine*, 65, 295–303.

207 Ben Yehuda Greenwald, M., Anzi, S., Ben Sasson, S. *et al.* (2014) Can nitroxides evoke the Keap1-Nrf2-ARE pathway in skin? *Free Radical Biology and Medicine*, 77, 258–269.

208 Floyd, R.A., Chandru, H.K., He, T. *et al.* (2011) Anti-cancer activity of nitrones and observations on mechanism of action. *Anti-Cancer Agents in Medicinal Chemistry*, 11, 373–379.

209 Du X, Ewert DL, Cheng W, et al. 2013. Effects of antioxidant treatment on blast-induced brain injury. *PLoS One* 8(11): e80138.

210 Hardy, M., Poulhes, F., Rizzato, E. *et al.* (2014) Mitochondria-targeted spin traps: synthesis, superoxide spin trapping, and mitochondrial uptake. *Chemical Research in Toxicology*, 27, 1155–1165.

211 Maroz, A., Anderson, R.F., Smith, R.A. *et al.* (2009) Reactivity of ubiquinone and ubiquinol with superoxide and the hydroperoxyl radical: implications for in vivo antioxidant activity. *Free Radical Biology and Medicine*, 46, 105–109.

212 Reboucas, J.S., Kos, I., Vujaskovic, Z. *et al.* (2009) Determination of residual manganese in Mn porphyrin-based superoxide dismutase (SOD) and peroxynitrite reductase mimics. *Journal of Pharmaceutical and Biomedical Analysis*, 50, 1088–1091.

213 Bueno-Janice, J., Tovmasyan, A., and Batinic-Haberle I. (2015) Comprehensive study of GPx activity of different classes of redox-active therapeutics – implications for their therapeutic actions. Free Radical Biology and Medicine, 87, S86–S87.

214 Tovmasyan, A., Sampaio, R.S., Boss, M-K. *et al.* (2015) Anticancer therapeutic potential of Mn porphyrin/ascorbate system. Free Radical Biology and Medicine, 89, 1231–1247.

215 Tovmasyan, A., Bueno-Janice, J., Boss, M-K. et al. (2015) Mn porphyrin-based SOD mimic and vitamin C enhance radiation-induced tumor growth inhibition. Free Radical Biology and Medicine, 87, S97.

216 Batinic-Haberle, I., Tovmasyan, A., Spasojevic, I. (2015) An educational overview of the chemistry, biochemistry and therapeutic aspects of Mn porphyrins – From superoxide dismutation to H_2O_2-driven pathways. Redox Biology, 5, 43–65.

CHAPTER 28

Herbal medicine: past, present, and future with emphasis on the use of some common species

Aneela Afzal[1] and Mohammad Afzal[2]

[1] *Advanced Imaging Research Center, Oregon Health and Sciences University, Portland, OR, USA*
[2] *Department of Biological Sciences, Faculty of Science, Kuwait University, Safat, Kuwait*

THEMATIC SUMMARY BOX

At the end of this chapter, students should be able to:

- Describe the main herbal medicine traditions

- Describe the main herbal medications

- Describe some of the main biochemical components of herbal medications

---

Herbal medicines or alternate medicines, that are used as dietary supplements in the United States, are the only choice to most people of the world. Herbal medicines are not only cheaper and more easily accessible to most of the world community than regular medications, but are also effective since the active components are present in their natural environment and may have synergistic effects with the cometabolites present in herbs. The herbs are also a source of new drug discovery and can give lead compounds that can be modified to useful medicines. In this review, we discuss the past, present, and future of herbal medicine.

Early Islamic herbal medicine

Herbal medicine was promoted as early as 864 CE by Rhazes (864–930 CE), the discoverer of alcohol and sulfuric acid, who contributed more than 200 works on science and his medical encyclopedia *Kitab al-Hawi* that comprised 23 volumes. For hundreds of years, his work on surgery and therapy was the basis of medical curriculum in many Western universities. He was the first to offer an idea to establish hospitals.

Subsequently Avicenna (980 CE) further strengthened the foundation of medicine when he wrote more than 450 books, 40 of them dealing with medicine. Avicenna's standard textbook of medicine is called Cannon of Medicine (*Kitab*

Oxidative Stress and Antioxidant Protection: The Science of Free Radical Biology and Disease, First Edition.
Edited by Donald Armstrong and Robert D. Stratton.
© 2016 John Wiley & Sons, Inc. Published 2016 by John Wiley & Sons, Inc.

al-Qanun fi al-Tibb) that comprised 14 volumes and was accepted as a standard text of medicine in Europe and Islamic world, till late 17th century. Avicenna was the first individual to explain diabetes and facial paralysis in detail. It was also the great Avicenna who extolled the advantages of exercise as the life-blood of human health and promoted whole grain diet, together with many recipes for good health practices that are solicited and practiced even today.

In addition to the early herbal medicine practitioners, Ali Ibn al-Abbas al-Majusi, a great Islamic scholar, was born (994 CE). He composed comprehensive and well-organized compendia in early medical literature. He wrote two books, one of which was called "The Complete Book of Medical Art" (*Kitab Kamil al-Sina'ah al-Tibbiyah*) that described numerous diseases and their treatment in a most organized and systematic manner. At about the same time of al-Majusi, another Islamic scholar, Abu al-Qasim Khalaf ibn al-Abbas al-Zahrawi wrote more than 30 books on medical practices and composed treatments for 325 different diseases. His most popular book, named "Managing Medical Knowledge" (*Kitab al-Tasrif li-man*) was frequently consulted in medical practice.

Averroes who was another master of the science of medicine, wrote *Kitab al-Kulyat fi al-Tibb*. His work was spread over 20 volumes with 20,000 pages on a variety of subjects. He also wrote a commentary on the Cannon of Medicine, contributed by Avicenna. Thus, Avicenna together with others such as Rhazes, Ali Ibn al-Abbas, and Averroes established the basis for modern Islamic medicine. Today Islamic medicine, also called traditional or alternative medicine, is practiced in most Islamic countries, and according to the World Health Organization (WHO), more than 80% of the population in all Islamic countries practice Islamic medicine.

Ayurvedic, siddha, and traditional Chinese medicine

Ayurveda means knowledge of life. Ayurvedic medicine originated from Hinduism and Buddhism and is divided into three doshas (bodily humors making up one's constitution) and five subdoshas.[1] Ayurvedic medicine which dates back to 2000–3000 B.C. believes that an imbalance between these doshas (Vata, Pitta, and Kapha) creates disease. Ayurvedic medicine is considered one of the oldest systems of medicine. Tibetan, Chinese, and Greek medicine all evolved from Ayurvedic medicine. All these systems of medicine mostly use plant-derived products, but some animal products such as bones, milk, and gallstones are also used. People have benefited from this holistic medicine for centuries. Unlike the Islamic medicine, however, there is no systematic compilation on the Ayurvedic system of medicine. With the new technologies available and the knowledge of human genome, today, the old axioms and treatments are being eroded extensively and must be given a chance to survive and be explored.

Present eminence of herbal medicine

In one form or another, 80–90% of the world population consumes herbal medicine. According to a WHO report, three-quarters of the world's population depend on herbal medicine or alternate medicine for their primary health care. In fact, in developing countries, alternate medicine and homeopathic medicine are the only medications available to its people. The picture is not much different in the Western world, where an increasing number of people are turning to herbal medicine only because it is a natural system of medicine. Therefore, in the United States alone, the herbal medicine business, mostly sold under the name of "dietary supplement," has crossed a hefty 13 billion USD, annually. More than 30% of the US population uses herbal medicine in one way or the other, and this figure is increasing day by day.

Presently, there is a complete renaissance in the use of natural products as medicament since people are frequently turning to natural remedies, traditionally called alternate medicine, nutraceuticals, Islamic and/or Ayurvedic medicine, and so on. Lately, there seems to be an upsurge in the use and an appreciation of the benefits of natural products for medicinal purpose. As an example, even the spices, used for several centuries as food complements, are recognized as useful drugs, and researchers have started to look into the biology of spices. In the following, the pharmacological (medicinal) relevance of some spices is described.

Curcuma longa, Zingiberacease

Curcuma longa belongs to the genus *Curcuma* (Zingiberaceae), with more than 70 varieties. Because of its color, it has also been named as "Asian gold," from the golden pigment in the spice.[2] Due to the pharmacological activities of curcumin, it has attracted worldwide attention. Yet, far and near, in its original form, it is still used as an exotic spice and for centuries, its use has not changed from the grinding stone to the cooking vessel.

First isolated in 1851, curcumin is the main constituent of *Curcuma*. So far, more than 100 curcuminoids have been identified from different *Curcuma* species. In addition, more than 250 natural terpenoids have been identified from *Curcuma* species and a large number of synthetic curcuminoids are being rapidly added to the list.[3] The curcuminoids have found use in acute ischemic injuries, where the mechanism of action of curcumin has been postulated through leptin release.[4] Curcumin in combination with aspirin and sulforaphane is implicated in the chemoprevention of pancreatic cancer and its mechanism of action has been reported by Thakkar *et al.*[5] Many workers have studied structure–activity relationships of synthetic curcuminoids and their derivatives.[6]

Being the major component of *C. longa*, curcumin has received major attention due to its numerous biological activities. It shows anti-inflammatory, anticancer, antioxidant, antileukemic, antiallergic, antiarthritic, antimicrobial, cytotoxic, hepatoprotective, neuroprotective, immune protective, cell signaling, anti-atherosclerotic, and anti-autoimmune effects.[7–15] Other effects include diabetic wound healing and the preventive (alleviation) of anorexia, sinusitis, allergies, aging, obesity, psoriasis, Alzheimer's disease, Parkinson's disease, cardiac diseases, and many more.[16] Thus, it is one of the most researched molecules, straight from kitchen to research laboratory with a huge potential as a life-saving drug.

A polymeric nanocurcumin has been used for the treatment of cancer.[17] Bisdemethoxycurcumin nanoparticles have been synthesized that show anticancer activity *in vitro*.[18] Curcumin is also known to potentiate other anticancer drugs such as doxorubicin while at the same time it alleviates doxorubicin's side effects such as the onset of cardiomyopathy and heart failure in cancer patients.[19] Curcumin downgrades cisplatin, 5-fluorouracil, and paclitaxel toxicity and thus succors cancer patients taking cisplatin medication.[20,21]

Presently, more than 4000 phytoterpenoids have been identified and more than 500 terpenoids have been recognized in Zingiberaceae alone, and several are associated with *C. longa*. In *C. longa*, more than 250 terpenoids, monoterpenoids and sequiterpenoids (bisabolene, curcumene) with interesting biological activities, have been acknowledged. *C. longa* rhizome is known to be 60 times more effective in anti-inflammatory activities than that of phenylbutazone, a drug for controlling inflammation. Aromatic turmerone from *C. longa* has been shown to induce apoptosis in human hepatocellular carcinoma.[22] Some of the terpenoids from *C. longa* have a close structural similarity with taxol (paclitaxel), an anticancer, antineoplastic agent, a triterpenoid isolated from *Taxus brevifolia* (Pacific yew). The macrocyclic ring of taxol and procurcumenol show structural similarity (Figure 28.1).

The anticancer and anti-inflammatory activity of *C. longa* has been recognized due to curcumin, but the co-occurring antitumor bisabolene-type sesquiterpenes are equally important and exhibit therapeutic activities. A synergic role of curcumanoids, phospholipids, and terpenoids in *C. longa* has been suggested.[23] Curcumin itself, a potent inhibitor of COP9 signalosome (CSN), is being tested for activities against colon, breast, lung, and prostate cancer. Synergistic action of curcumin with xanthorrhizol in the treatment of breast cancer has been reported. At the same time, a synergistic action of curcumin with taxol, docosahexaenoic acid, genistein sulfinosine, celecoxib, and phenylisothiocyanate has been examined.[21] All these pieces of information suggest compelling evidence for curcumin as a wonder drug.

Nigella sativa

Nigella sativa (NS) is extensively used in the Middle East and South Asian countries for all kind of ailments. It has been used to control diabetes, hypertension, cancer (leukemia, liver, lung, kidney, prostate, breast, cervix, and skin), inflammation, hepatic disorders, arthritis, kidney disorders, cardiovascular complications, neurode-generative disorders (Parkinsons's disease), dermatological conditions, and many others.[24] Thymoquinone is the major component of this plant that has attracted worldwide attention. A gas chromatography/mass spectroscopy analysis of the NS seed extract has shown it to be a mixture of 32 volatile terpenes. Thymoquinone (TQ), dithymoquinone (DTQ), *trans*-anethol, *p*-cymene, limonine, and carvone have been identified as major terpenes in the NS seeds.[25] Both TQ and DTQ are cytotoxic for various types of tumors.[26] In addition, diterpenes, triterpene, and terpene alkaloids have been identified in NS seeds (Figure 28.2).

In animal studies, NS shows dose-dependent suppression of nociceptive pain response and anti-cestodes (tapeworm) activity. These activities occur through TQ that works through indirect activation of the µ1- and κ-opioid receptor subtypes.[27] The antihypertensive effect of TQ and other constituents of NS are also protective agents against the chromosomal aberrations induced by schistosomiasis.[28] These compounds are used in the control of systemic blood pressure, as an anticholinergic, an antihistaminic, a tracheal relaxant, an antiasthmatic, and as a treatment of other allergic conditions.[29–31]

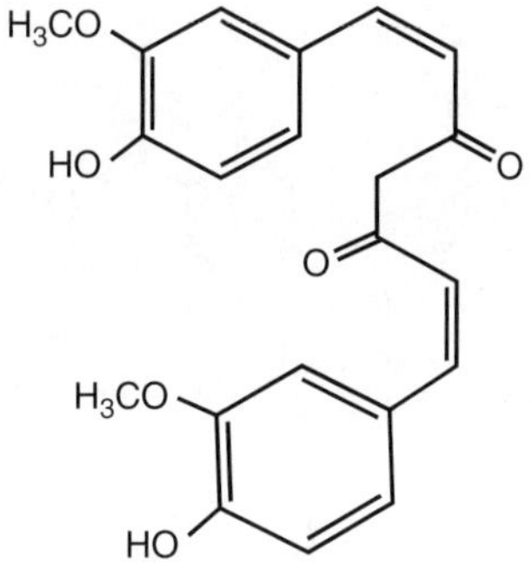

Curcumin

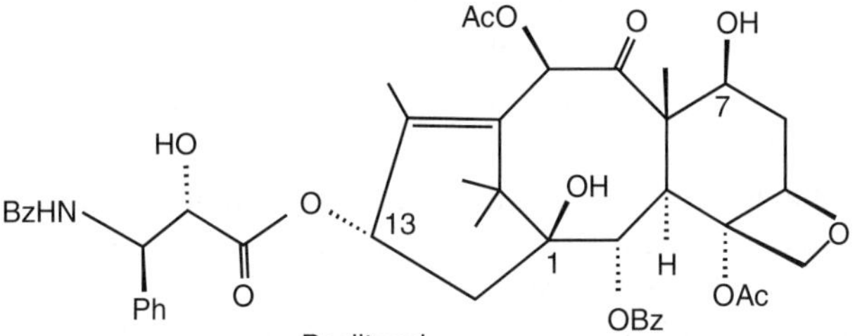

Paclitaxel

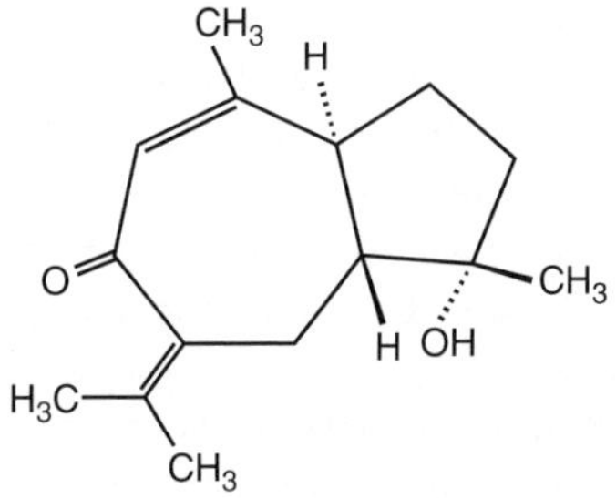

Procurcumenol

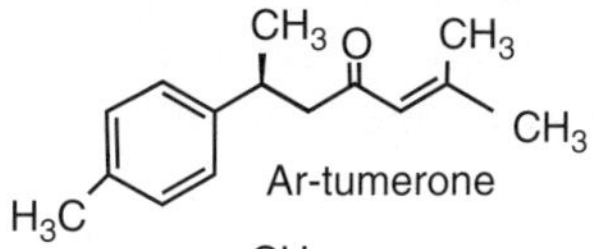

Ar-tumerone

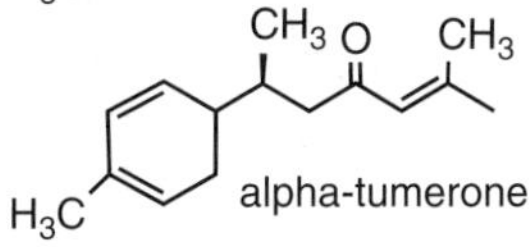

alpha-tumerone

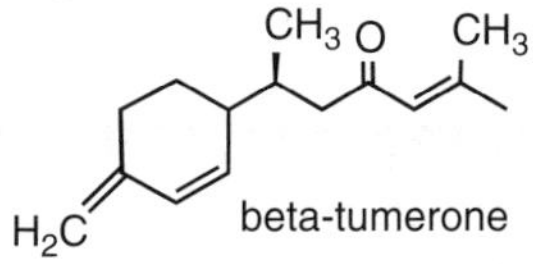

beta-tumerone

Figure 28.1 Molecular structures of some biologically active components in *Cucuma longa except paclitaxel is in Taxus brevifolia.*

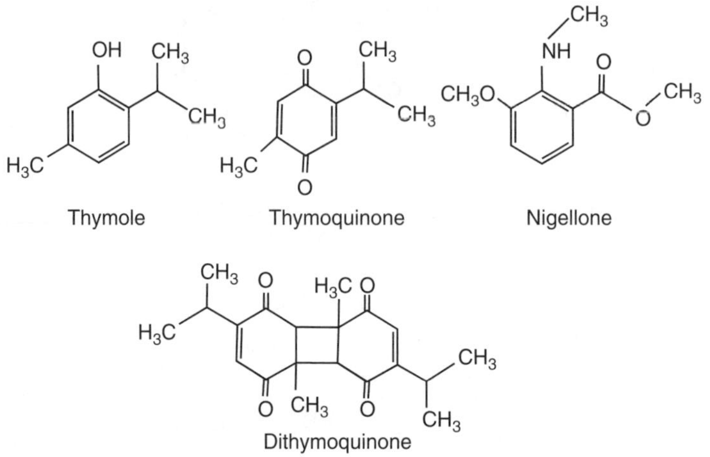

Figure 28.2 Molecular structures of some components in *Nigella sativa*.

A number of antitumor compounds have been identified from NS. These compounds are TQ; alpha-hederin (a triterpene); isopropylmethylphenols; dolabellane-type diterpene alkaloid; and nigellamine A3, A4, A5, and C.[32–34] Numerous types of cancers such as Ehrlich ascites carcinoma (EAC), Dalton's lymphonia ascites (DLA) and Sarcoma-180 (S-180) cells, colon carcinoma, pancreatic carcinoma, and hepatic carcinoma have been treated with NS extracts.[35] TQ protects the benzo-a-pyrene-induced clastogenic activity in rats. 20-Methylcycloanthrene-induced fibrocarcinoma is inhibited by NS extracts.[36] Benefit in the treatment of methylnitrosourea-induced inflammation, carcinogenesis, and oxidative stress has been reported for NS and honey supplementation.[37] The antioxidant and prooxidant properties of TQ have been substantiated by augmented TQ-mediated scavenging of superoxide anion.[38]

TQ has antitumor activity against pancreatic carcinoma (PC). However, for PC, the dose of TQ has to be high. Therefore, many attempts have been made to study structure–activity relationship by synthesizing TQ analogs, and some of these compounds have shown potent antitumor activity against PC.[39] Gemcitabine- or oxaliplatin-induced activation of nuclear factor-kappa B is revoked by TQ, resulting in the chemosensitization of pancreatic tumors to conventional therapeutics.[40]

The molecular mechanism of the action of TQ in colon cancer has been suggested. Colon cancer is inhibited in G1 cell cycle phase, and apoptosis is mediated by TQ.[41] NS also has the ability to detoxify 1,2-dimethylhydrazine (DMH), a colon cancer inducer.[42] The pre-neoplastic lesions for colon cancer have been investigated, and it was found that colon cancer in postinitiation stage can be prevented by volatile components of NS seeds.[43]

Zinzibar officinale (Zingiberaceae, ginger)

Ginger is an exotic spice with an age-old reputation for its pungent flavor and many therapeutic properties. Ginger is effectively used in treating many ailments

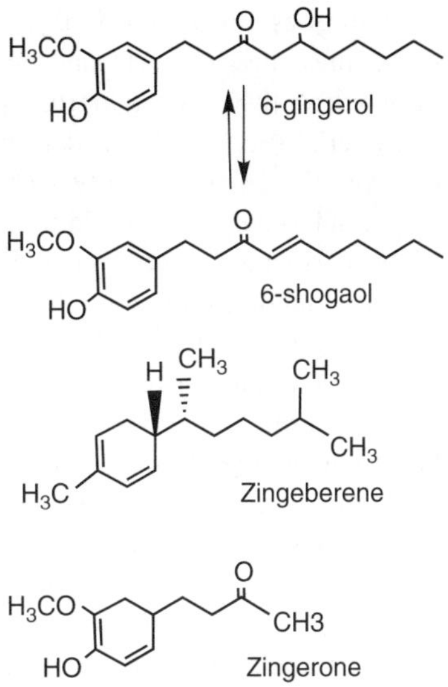

Figure 28.3 Molecular structures of some of the components of ginger.

from respiratory infections and digestive ailments to morning sickness, general nausea, and flatulence. Since its gingerol content works as an anticoagulant and as a blood-thinning agent, ginger is believed to have the potential of decreasing the likelihood of coronary thrombosis or stroke by reducing LDL cholesterol.[44]

The main components in ginger are three types of gingerols: 6-, 8-, and 10-gingerols. 6-Gingerol potentiates prostaglandin F_2-α induced muscle contraction and inhibits breast and colon cancer.[45,46] 6-Gingerol is the most common component in ginger; when it is heated, it gives 6-shogaol. A cysteine-conjugated metabolite of 6-shogaol is a novel compound with chemoprotective anticancer properties.[47] Thus, cysteine-conjugated 6-shogaol inhibits colon and breast, and cervical cancer.[48,49] 6-Shogaol has several other pharmacological and physiological activities such as alleviation of dementia, anti-inflammatory, analgesic, gastroprotective, cardiotonic, hepatoprotective, and antipyretic effects.[50,51] Many analogs of gingerol and shogaol have been synthesized and some components are shown in Figure 28.3.[52]

Zingiberene from ginger is a monocyclic sesquiterpene that has anticancer properties with a pronounced therapeutic potential in treating oxidative damage caused by neurodegenerative diseases, while zingerone is protective against acute lung injury.[53,54] Due to its powerful antioxidant properties, zingerone has been indicated in the prevention of irritable bowel syndrome, ROS-mediated DNA damage, and colon cancer.[55,56]

Future potential of herbal medicine

What was not known yesterday is known today, and the future looks even brighter. Herbal medicine has been practiced for centuries, but in the past it was believed that a certain plant concoction works for a certain condition and people used to take it without questions asked. With the advent of chromatographic and other separation techniques, these extracts have been separated into new drugs. More than 50% of the drugs currently in use are natural products or their modifications. Each plant, fungus, or bacterial species has its own metabolic pathways synthesizing innumerable new compounds. Currently, only 10% of the world's higher plants have been analyzed or partially analyzed. In addition to terrestrial plants, the phytoplankton, sponges, and sea flora have yet to be explored for new drugs. Within what has been analyzed, only the major components have been identified, with the minor components still remaining to be identified. Therefore, what is known today is only the tip of the iceberg. A wealth of new drugs and chemical compounds and new treatments for currently untreatable diseases awaits discovery in the coming few years. The current knowledge of metabolomics will be of great help in the exploration of new drugs.

In the past, there was a feeble knowledge of separation and identification of natural products. Today, we have sophisticated techniques available and these are getting better day by day. Extraction techniques need to be further developed so that unstable components that can become lead compounds for drug discovery can be isolated and identified. Supercritical extraction has been developed for unstable molecular extraction, but the technique is unavailable in most parts of the world.

Modern identification techniques are playing a vital role in adding new compounds to our knowledge. The high-powered 2D proton (^{1}H) and ^{13}C NMR spectroscopy coupled with the MS/MS techniques have completely revolutionized identification of new compounds, even in extremely small amounts. The latest GC/MS and LC/MS techniques are being used in finger printing of complex mixtures. It is especially true when new and striking derivatizing agents have been discovered that have made the use of GC/MS–MS and LC/MS–MS indispensable in the drug industry and quantitation of natural products. The latest TLC/MS technique has been added to the list. This has eliminated the isolation by conventional column chromatography, and the compound can be identified straight from the TLC separated spot using GC/MS or LC/MS.

With further refinement of the extracts, chromatographic and instrumental techniques hold a very bright future for new drug discovery from the untapped treasures under the oceans and the terrestrial plants. We remain convinced that the future for natural therapy is exceptionally bright and this field of research is ever expanding. The world resources should be put into use to tap the unexplored natural wealth and put them to human use.

Multiple choice questions

1 What proportion of the world's population depends on herbal medicine for primary health care?
   a. One quarter
   b. One half
   c. Three quarters

2 Curcumin has been shown to have:
   a. Anti-inflammatory activity
   b. Antioxidant activity

    c. Anticarcinogenic activity
    d. Neuroprotective and hepatoprotective activity
    e. All the above

3 Thymoquinone (TQ) and dithymoquinone (DTQ) are terpenes found in:
    a. *Curcuma longa*
    b. *Nigella sativa*
    c. *Zinzibar officinale*

References

1 Hankey, A. (2001) Ayurvedic physiology and etiology: Ayurvedo Amritanaam. The doshas and their functioning in terms of contemporary biology and physical chemistry. *Journal of Alternative and Complementary Medicine*, 7, 567–574.

2 Prasad, S., Tyagi, A.K. & Aggarwal, B.B. (2014) Recent developments in delivery, bioavailability, absorption and metabolism of curcumin: the golden pigment from golden spice. *Cancer Research and Treatment : Official Journal of Korean Cancer Association*, 46, 2–18.

3 Asti, M., Ferrari, E., Croci, S. *et al.* (2014) Synthesis and characterization of (68)Ga-labeled curcumin and curcuminoid complexes as potential radiotracers for imaging of cancer and Alzheimer's disease. *Inorganic Chemistry*, 53, 4922–4933.

4 Deng, Z.H., Liao, J., Zhang, J.Y. *et al.* (2013) Localized leptin release may be an important mechanism of curcumin action after acute ischemic injuries. *The Journal of Trauma and Acute Care Surgery*, 74, 1044–1051.

5 Thakkar, A., Sutaria, D., Grandhi, B.K. *et al.* (2013) The molecular mechanism of action of aspirin, curcumin and sulforaphane combinations in the chemoprevention of pancreatic cancer. *Oncology Reports*, 29, 1671–1677.

6 Zhang, Y., Zhao, L., Wu, J. *et al.* (2014) Synthesis and evaluation of a series of novel asymmetrical curcumin analogs for the treatment of inflammation. *Molecules*, 19, 7287–7307.

7 Mehta, H.J., Patel, V. & Sadikot, R.T. (2014) Curcumin and lung cancer-a review. *Targeted Oncology*, 9(4), 295–310.

8 Rajan, I., Rabindran, R., Jayasree, P.R. *et al.* (2014) Antioxidant potential and oxidative DNA damage preventive activity of unexplored endemic species of *Curcuma*. *Indian Journal of Experimental Biology*, 52, 133–138.

9 Chaudhuri, J., Chowdhury, A.A., Biswas, N. *et al.* (2014) Superoxide activates mTOR-eIF4E-Bax route to induce enhanced apoptosis in leukemic cells. *Apoptosis : An International Journal on Programmed Cell Death*, 19, 135–148.

10 Chang, R., Sun, L. & Webster, T.J. (2014) Short communication: selective cytotoxicity of curcumin on osteosarcoma cells compared to healthy osteoblasts. *International Journal of Nanomedicine*, 9, 461–465.

11 Palipoch, S., Punsawad, C., Koomhin, P. *et al.* (2014) Hepatoprotective effect of curcumin and alpha-tocopherol against cisplatin-induced oxidative stress. *BMC Complementary and Alternative Medicine*, 14, 111.

12 Kim, K.T., Kim, M.J., Cho, D.C. *et al.* (2014) The neuroprotective effect of treatment with curcumin in acute spinal cord injury: laboratory investigation. *Neurologia Medico-Chirurgica*, 54, 387–394.

13 Anderson, G. & Maes, M. (2014) Oxidative/nitrosative stress and immuno-inflammatory pathways in depression: treatment implications. *Current Pharmaceutical Design*, 20, 3812–3847.

14 Chen, J., Wang, F.L. & Chen, W.D. (2014) Modulation of apoptosis-related cell signalling pathways by curcumin as a strategy to inhibit tumor progression. *Molecular Biology Reports*, 41, 4583–4594.

15 Shin, H.S., Han, J.M., Kim, H.G. *et al.* (2014) Anti-atherosclerosis and hyperlipidemia effects of herbal mixture, Artemisia iwayomogi Kitamura and *Curcuma longa* Linne, in apolipoprotein E-deficient mice. *Journal of Ethnopharmacology*, 153, 142–150.

16 Shishodia, S. (2013) Molecular mechanisms of curcumin action: gene expression. *BioFactors*, 39, 37–55.

17 Naksuriya, O., Okonogi, S., Schiffelers, R.M. *et al.* (2014) Curcumin nanoformulations: a review of pharmaceutical properties and preclinical studies and clinical data related to cancer treatment. *Biomaterials*, 35, 3365–3383.

18 Francis, A.P., Murthy, P.B. & Devas, T. (2014) Bis-demethoxy curcumin analog nanoparticles: synthesis, characterization, and anticancer activity in vitro. *Journal of Nanoscience and Nanotechnology*, 14, 4865–4873.

19 Hosseinzadeh, L., Behravan, J., Mosaffa, F. *et al.* (2011) Curcumin potentiates doxorubicin-induced apoptosis in H9c2 cardiac muscle cells through generation of reactive oxygen species. *Food and Chemical Toxicology : An International Journal Published for the British Industrial Biological Research Association*, 49, 1102–1109.

20 Ferguson, J.E. & Orlando, R.A. (2014) Curcumin Reduces Cytotoxicity of 5-Fluorouracil Treatment in Human Breast Cancer Cells. *Journal of Medicinal Food*, 18(4), 497–502.

21 Bava, S.V., Sreekanth, C.N., Thulasidasan, A.K. *et al.* (2011) Akt is upstream and MAPKs are downstream of NF-kappaB in paclitaxel-induced survival signaling events, which are down-regulated by curcumin contributing to their synergism. *The International Journal of Biochemistry & Cell Biology*, 43, 331–341.

22 Cheng, S.B., Wu, L.C., Hsieh, Y.C. *et al.* (2012) Supercritical carbon dioxide extraction of aromatic turmerone from *Curcuma longa* Linn. induces apoptosis through reactive oxygen species-triggered intrinsic and extrinsic pathways in human hepatocellular carcinoma HepG2 cells. *Journal of Agricultural and Food Chemistry*, 60, 9620–9630.

23 Mercanti, G., Ragazzi, E., Toffano, G. *et al.* (2014) Phosphatidylserine and curcumin act synergistically to down-regulate release of interleukin-1beta from lipopolysaccharide-stimulated cortical primary microglial cells. *CNS & Neurological Disorders Drug Targets*, 13(5), 792–800.

24 Khan, M.A., Chen, H.C., Tania, M. *et al.* (2011) Anticancer activities of *Nigella sativa* (black cumin). *African Journal of Traditional, Complementary, and Alternative Medicines : AJTCAM/African Networks on Ethnomedicines*, 8, 226–232.

25 Nickavar, B., Mojab, F., Javidnia, K. *et al.* (2003) Chemical composition of the fixed and volatile oils of *Nigella sativa* L. from Iran. *Zeitschrift fur Naturforschung, Section C: Biosciences*, 58, 629–631.

26 Worthen, D.R., Ghosheh, O.A. & Crooks, P.A. (1998) The in vitro anti-tumor activity of some crude and purified components of blackseed, Nigella sativa L. *Anticancer Research*, 18, 1527–1532.

27 Akhtar, M.S. & Riffat, S. (1991) Field trial of Saussurea lappa roots against nematodes and Nigella sativa seeds against cestodes in children. *JPMA. The Journal of the Pakistan Medical Association*, 41, 185–187.

28 Aboul-Ela, E.I. (2002) Cytogenetic studies on Nigella sativa seeds extract and thymoquinone on mouse cells infected with schistosomiasis using karyotyping. *Mutation Research*, 516, 11–17.

29 Ahmed, O.G. & El-Mottaleb, N.A. (2013) Renal function and arterial blood pressure alterations after exposure to acetaminophen with a potential role of *Nigella sativa* oil in adult male rats. *Journal of Physiology and Biochemistry*, 69, 1–13.

30 Al-Majed, A.A., Daba, M.H., Asiri, Y.A. *et al.* (2001) Thymoquinone-induced relaxation of guinea-pig isolated trachea. *Research Communications in Molecular Pathology and Pharmacology,* 110, 333–345.

31 Kalus, U., Pruss, A., Bystron, J. *et al.* (2003) Effect of Nigella sativa (black seed) on subjective feeling in patients with allergic diseases. *Phytotherapy Research : PTR,* 17, 1209–1214.

32 Kumara, S.S. & Huat, B.T. (2001) Extraction, isolation and characterisation of antitumor principle, alpha-hederin, from the seeds of Nigella sativa. *Planta Medica,* 67, 29–32.

33 Michelitsch, A., Rittmannsberger, A., Hufner, A. *et al.* (2004) Determination of isopropylmethylphenols in black seed oil by differential pulse voltammetry. *Phytochemical Analysis: PCA,* 15, 320–324.

34 Morikawa, T., Xu, F., Ninomiya, K. *et al.* (2004) Nigellamines A3, A4, A5, and C, new dolabellane-type diterpene alkaloids, with lipid metabolism-promoting activities from the Egyptian medicinal food black cumin. *Chemical & Pharmaceutical Bulletin,* 52, 494–497.

35 Salomi, N.J., Nair, S.C., Jayawardhanan, K.K. *et al.* (1992) Antitumour principles from *Nigella sativa* seeds. *Cancer Letters,* 63, 41–46.

36 Badary, O.A., Abd-Ellah, M.F., El-Mahdy, M.A. *et al.* (2007) Anticlastogenic activity of thymoquinone against benzo(a)pyrene in mice. *Food and Chemical Toxicology : An International Journal Published for the British Industrial Biological Research Association,* 45, 88–92.

37 Mabrouk, G.M., Moselhy, S.S., Zohny, S.F. *et al.* (2002) Inhibition of methylnitrosourea (MNU) induced oxidative stress and carcinogenesis by orally administered bee honey and Nigella grains in Sprague Dawely rats. *Journal of Experimental & Clinical Cancer Research: CR,* 21, 341–346.

38 Badary, O.A., Taha, R.A., Gamal el-Din, A.M. *et al.* (2003) Thymoquinone is a potent superoxide anion scavenger. *Drug and Chemical Toxicology,* 26, 87–98.

39 Banerjee, S., Azmi, A.S., Padhye, S. *et al.* (2010) Structure-activity studies on therapeutic potential of Thymoquinone analogs in pancreatic cancer. *Pharmaceutical Research,* 27, 1146–1158.

40 Banerjee, S., Kaseb, A.O., Wang, Z. *et al.* (2009) Antitumor activity of gemcitabine and oxaliplatin is augmented by thymoquinone in pancreatic cancer. *Cancer Research,* 69, 5575–5583.

41 Gali-Muhtasib, H., Diab-Assaf, M., Boltze, C. *et al.* (2004) Thymoquinone extracted from black seed triggers apoptotic cell death in human colorectal cancer cells via a p53-dependent mechanism. *International Journal of Oncology,* 25, 857–866.

42 Harzallah, H.J., Grayaa, R., Kharoubi, W. *et al.* (2012) Thymoquinone, the *Nigella sativa* bioactive compound, prevents circulatory oxidative stress caused by 1,2-dimethylhydrazine in erythrocyte during colon postinitiation carcinogenesis. *Oxidative Medicine and Cellular Longevity,* 2012, 854065.

43 Salim, E.I. & Fukushima, S. (2003) Chemopreventive potential of volatile oil from black cumin (*Nigella sativa* L.) seeds against rat colon carcinogenesis. *Nutrition and Cancer,* 45, 195–202.

44 Pancho, L.R., Kimura, I., Unno, R. *et al.* (1989) Reversed effects between crude and processed ginger extracts on PGF2 alpha-induced contraction in mouse mesenteric veins. *Japanese Journal of Pharmacology,* 50, 243–246.

45 Kimura, I., Pancho, L.R., Koizumi, T. *et al.* (1989) Chemical structural requirement in gingerol derivatives for potentiation of prostaglandin F2 alpha-induced contraction in isolated mesenteric veins of mice. *Journal of Pharmacobio-Dynamics,* 12, 220–227.

46 Kimura, I., Kimura, M. & Pancho, L.R. (1989) Modulation of eicosanoid-induced contraction of mouse and rat blood vessels by gingerols. *Japanese Journal of Pharmacology,* 50, 253–261.

47 Warin, R.F., Chen, H., Soroka, D.N. *et al.* (2014) Induction of lung cancer cell apoptosis through a p53 pathway by [6]-shogaol and its cysteine-conjugated metabolite M2. *Journal of Agricultural and Food Chemistry,* 62, 1352–1362.

48 Chen, H., Soroka, D.N., Zhu, Y. *et al.* (2013) Cysteine-conjugated metabolite of ginger component [6]-shogaol serves as a carrier of [6]-shogaol in cancer cells and in mice. *Chemical Research in Toxicology*, 26, 976–985.

49 Liu, Q., Peng, Y.B., Qi, L.W. *et al.* (2012) The cytotoxicity mechanism of 6-shogaol-treated HeLa human cervical cancer cells revealed by label-tree shotgun proteomics and bioinformatics analysis. *Evidence-Based Complementary and Alternative Medicine : eCAM*, 2012, 278652.

50 Kim, S.M., Kim, C., Bae, H. *et al.* (2015) 6-Shogaol exerts anti-proliferative and pro-apoptotic effects through the modulation of STAT3 and MAPKs signaling pathways. *Molecular Carcinogenesis*, 54(10), 1132–1146.

51 Moon, M., Kim, H.G., Choi, J.G. *et al.* (2014) 6-Shogaol, an active constituent of ginger, attenuates neuroinflammation and cognitive deficits in animal models of dementia. *Biochemical and Biophysical Research Communications*, 449, 8–13.

52 Shih, H.C., Chern, C.Y., Kuo, P.C. *et al.* (2014) Synthesis of analogues of gingerol and shogaol, the active pungent principles from the rhizomes of *Zingiber officinale* and evaluation of their anti-platelet aggregation effects. *International Journal of Molecular Sciences*, 15, 3926–3951.

53 Togar, B., Turkez, H., Stefano, A. *et al.* (2015) Zingiberene attenuates hydrogen peroxide-induced toxicity in neuronal cells. *Human and Experimental Toxicology*, 34(2), 135–144.

54 Xie, X., Sun, S., Zhong, W. *et al.* (2014) Zingerone attenuates lipopolysaccharide-induced acute lung injury in mice. *International Immunopharmacology*, 19, 103–109.

55 Rajan, I., Narayanan, N., Rabindran, R. *et al.* (2013) Zingerone protects against stannous chloride-induced and hydrogen peroxide-induced oxidative DNA damage in vitro. *Biological Trace Element Research*, 155, 455–459.

56 Vinothkumar, R., Vinothkumar, R., Sudha, M. *et al.* (2014) Chemopreventive effect of zingerone against colon carcinogenesis induced by 1,2-dimethylhydrazine in rats. *European Journal of Cancer Prevention*, 23(5), 361–71.

CHAPTER 29

Ayurvedic perspective on oxidative stress management*

Priyanka M. Jadhav

Department of Pathology, Immunology and Laboratory Medicine, UF Diabetes Institute, College of Medicine, University of Florida, Gainesville, FL, USA

THEMATIC SUMMARY BOX

At the end of this chapter, students should be able to:

- Understand basics of Ayurveda related to the oxidative stress

- Outline the role of oxidative stress in various diseases in view of Ayurveda

- Infer pathophysiologic correlations of disease and its interpretation with Ayurvedic perspective

- Orient to different treatment methods for reducing oxidative stress in Ayurvedic management of diseases

- Illustrate some common chronic diseases and their interlink between oxidative stress and Ayurvedic perspective

Ayurveda[1]

Before discussing the topic of chapter, it is imperative to give[1] readers small introduction to the basis of this chapter, as some may not be aware of the traditional basis of this science called Ayurveda. There are four basic Vedas (ancient scriptures) described in Indian/Hindu literature, namely **Rigveda, Yajurveda, Samaveda**, and **Atharvaveda**. These scriptures are in Sanskrit language containing hymns, philosophy, and ways of living life, rituals, along with knowledge of diseases and their

---

*The aim of this chapter is to understand the basis of oxidative stress from Ayurvedic perspective and its management as per Ayurveda. This by no means replaces the need to consult a trained Ayurvedic practitioner for health advice and medication. The author specifically disclaims any liability, loss, or risk, personal or otherwise, that is incurred as a consequence, directly or indirectly, of the use and application of any of the contents of this chapter.

[1]"The further back you look, further forward you see." Winston Churchill

---

Oxidative Stress and Antioxidant Protection: The Science of Free Radical Biology and Disease, First Edition.
Edited by Donald Armstrong and Robert D. Stratton.
© 2016 John Wiley & Sons, Inc. Published 2016 by John Wiley & Sons, Inc.

treatment. The timeline for Veda is varied but generally regarded as 1500–2000 BC, some also date them back to 4000 BC. Irrespective of their time of origin, they are old and astonishing by the fact that they are the source of vast information wittily conveyed in a code to be deciphered. The code is necessarily use of Sanskrit language in very concise form as these hymns or slokas are to be recited and passed to pupils of next generation. Ayurveda is considered as a part of Atharvaveda that deals with health science. As very famously known, Ayurveda is regarded as "Science of Life." It is made up two words "Ayu + Veda"; "Ayu" is defined as

शरीरेयेन्द्रियसत्त्वात्मसंयोगोधारि जीवितम् |

नित्यगश्चनुबन्धच पर्ययैरायुरुच्यते (Charaka Samhita/Sutrasthana v1/Sloka 42)

which states that Ayu which is the living being or life is the dynamic composition of body, sense organs, mind, and soul, and then synonyms of Ayu are Dhaari (the one sustaining the body), Jivita (alive/animated), Nityaga (continuous in life), and Anubandha (continuous flow). "Veda" originates from "Vid," which means to understand, thus Vedas are the scriptures or the medium to understand and acquire knowledge. Thus, Ayurveda can be defined as the science that deals with well-being and treatment of diseases of not only body and sense organs (both sensory and locomotory) but also of mind and soul.

Basic concepts of Ayurveda

Ayurveda is majorly described in three main treatises, namely Charaka Samhita, Sushruta Samhita, and Vagbhata or Astanga Samgraha. These are called Bruhat Trayi, that is, the three main fundamental texts. These are also the texts that are studied by the students of Ayurveda all over the world along with many other scriptures written along the centuries.

The principles of Ayurveda are based on three Doshas, seven Dhatus, five Mahabhutas, which are responsible for health and disease.

Three Doshas described in Ayurveda are as follows:

Vata (V): All the movement in body and of the body is by the virtue of this dosha. It is responsible for the elimination of waste products at both cellular and excretory levels; it controls senses and makes the channels in the body for gross and subtle functions; it influences the digestive ability as per its status (whether vitiated or normal); and it leads the mind and controls thoughts, enthusiasm, and many other functions.

Pitta (P): Pitta is responsible for all conversions in body for digestion, skin color, eyesight, perspiration, bile pigmentation, and mental abilities.

Kapha (K): This dosha is responsible for strength, enthusiasm, overall health, and knowledge. All the moistness in body can be attributed to Kapha; the mucus lining is intact because of Kapha, nutrition in the brain, and so on.

रोगस्तु दोषवैषंयं दोषसाम्यमरोगता||(Vagbhat/Sutrasthana v1/Sloka 20)

The above sloka describes the importance of dosha in body, as it says that all the diseases are due to vitiation or imbalance in these three doshas, while the health is a result of the balance of the same.

The basis of the human body is made up of seven elements called as Dhatus, and they are as follows:

Rasa: This is the fluid in the body which is circulating apart from blood, that is, blood plasma. The major function is to nourish the body and carry waste products toward excretion.

Rakta: This is the blood in the body which is responsible for life and its loss can be fatal.

Maunsa: This is the musculature that holds the body and organs together. It also includes the smooth muscles of the body.

Meda: This is the fat or oily portion of the body responsible for nutrition of joints, bone, and organs.

Asthi: This is the bone structure and the skeleton for locomotion.

Majja: Majja is the bone marrow that is responsible for many vital functions such as bone strength, nutrition to vital organs, and intellectual qualities.

Shukra: This is the last dhatu and the essence of all. It is responsible for reproduction and all kinds of regeneration in body.

विकारो धातुवैषंयं साम्यं प्रकृतिरुच्यते|(Charaka Samhita/Sutrasthana v9/Sloka 4)[2]

Charaka explains the importance of the dhatus by stating that the diseases are resultant of the dysfunction and imbalance of dhatus and that normal state of health is achieved by balanced functional dhatus.

The five elements by the virtue of which the whole Universe is sustained and originated are as follows:

Prithvi: It is the element that gives form or matter to any object. This is related to the sense of smell.

Jal: It is the one that can bind the matter and bring it together and is related to taste.

Agni: It is responsible for the conversion or change or metabolism in current view. It has the prowess to change the matter from one form to another. It is related to vision or sight.

Vayu: This is the element responsible for all kinds of motion such as air and flow of matter and is related to the sense of touch.

Akash: This is related to the vacuum or space needed for any action to take place and is related to sound.

---

[2]Maintaining order rather than correcting disorder is the ultimate principle of wisdom. To cure disease after it has appeared is like digging a well when one feels thirsty or forging weapons after the war has already begun. – *Huangdi Neijing*, 2nd century BC; from book *Grain Brain* by David Perlmutter.

Role of oxidative stress in disease development: Ayurvedic perspective

Oxidative stress

Oxidative stress: It is associated frequently with tissue damage, and its role in many diseases is of crucial importance that is discussed in the various chapters.[1] The imbalance and disturbance between antioxidant defense mechanism and reactive oxygen species (ROS) production is the major key player in oxidative stress conditions. Oxygen is a diradical and thermodynamically wants to take on additional electron to produce two water molecules with lower free energy. But the unconventional distribution of electrons of oxygen, nonspin paired electrons (triplet state), makes it difficult for oxygen to accept a spin-matched pair of electron. Under normal conditions, the kinetic barrier to allow the spin reversal maintains homeostasis, unless an enzyme comes in play. The enzyme increases the period of contact between oxygen and substrate long enough to overcome the barrier, although this process occurs under normal conditions by nonenzymatic autoxidation. So at this point, oxygen accepts one electron at a time and this breaking up of electron pair results in free radical formation. Incomplete reduction of oxygen leads to the production of number of reactive intermediates such as superoxide radical, hydroxyl radical, and hydrogen peroxide. Since hydrogen peroxide is not a radical, free radicals in a misnomer, **Reactive oxygen metabolites (ROM) or active oxygen (AO)**, are used generally to address this family.[2] The damage conferred by the ROM can range from tissue damage to DNA damage and mutations.

Evolutionarily, organisms have developed multiple mechanisms to curtail the damage by ROM or by the by-products of oxidative metabolism. One of these is the ability to synthesize and assimilate molecules that would react and obliterate the active oxygen species. Free radicals are normally neutralized by efficient systems in the body that include the antioxidant enzymes (superoxide dismutase (SOD), catalase, and glutathione peroxidase) and the nutrient-derived antioxidant small molecules (vitamins C and E, carotenes, flavonoids, glutathione, uric acid, and taurine)[3]. These antioxidant compounds are shown to be present in plants too, for example, ascorbic acid[2], berberine[4], and epigallocatechin gallate[3]. Thus, they form a rich source of antioxidants.

Ayurveda

Ayurveda: It explains that the basis of the disease is the production of Ama.

उष्मणोऽल्पबलत्वेन् धातुमाद्यमपाचितम् ।

दुष्टमामाशयगतं रसामांम प्रचक्षते ॥ (Vagbhat/Sutrasthana v13/Sloka 25)

Ama is the undigested portion that is not limited to the gastrointestinal tract and if present chronically can spread throughout the body to cause various diseases. Ama is formed due to Agni-mandhya, which can be defined as the loss of appetite or the potency to metabolize at both gastrointestinal and cellular level. This correlates the concept of Ama, Agni-mandhya, metabolic dysfunction, and oxidative stress together.

Mitochondria in the cell are responsible for metabolizing 98% of the oxygen present in the body by an enzyme, cytochrome oxidase. This organelle represents the pitta or the agni in the body, which is responsible to bring about change or metabolize. It is shown that although highly insulated to keep the electrons flowing in proper channel, at two sites in the electron transport chain (Complex I and ubisemiquinone), the electrons may come in contact with the molecules of oxygen to form ROMs.[2] This also affects the micronutrition of the organelles, resulting in chronic starvation of the cells. This corresponds to Agni-mandhya state where the nutrition to each layer of dhatus, as explained earlier, fails and it results in vitiation of the doshas causing diseases by altering the normal status of the dhatus.

Pathophysiological basis of oxidative stress in Ayurveda

Fever: For all conditions where fever is a symptom, this would apply to both infectious and noninfectious conditions.

As a result of Agni-mandhya, which can be because of certain infection or chronic conditions, Ama is formed and it enters various small places or pathways in bodies called as Strotas to block them and cause fever as a symptom of other underlying conditions. Oxidative stress is induced in febrile conditions, leading to the release of ROMs and reactive nitrogen species (RNS), and thus leading to cellular injury via the peroxidation of membrane lipids and oxidative damage of proteins and DNA. The inflammation response associated with hypoxia causes further toxicity.[5,6]

Diabetes: In Ayurveda, Diabetes or Prameha is described to be of three major categories, corresponding to each dosha type i.e. vata, pitta and kapha. These major categories are then subdivided in various types as per the symptoms. Irrespective of the category or type of diabetes, the etiology is contribute to kleda, a form of Ama which is responsible for vitiating the doshas. Erythrocyte oxidative stress is amplified and reduction in plasma antioxidant status is observed in diabetic patients with increased levels of plasma protein glycation products.[7]

Cardiac diseases and hypertension: Although this is described as a consequence of chronic presence of Ama dosha in body, it has been shown that oxidative stress and renal tubulointerstitial inflammation play a major role in the pathogenesis of hypertension. Oxidative stress in kidney and vascular tissue is the primary mediator in the pathogenesis of angiotensin-induced hypertension.[8] Global metabolic disturbance is observed in all kinds of abnormalities in the tissues.[9]

Ayurvedic management of oxidative stress

The approach to treat oxidative stress in Ayurveda is at multiple levels and also being a holistic science, it is very personalized. For the purposes of summarization, it has been classified as follows.

Drug treatment

The current databases have abundant information on the antioxidant potential of numerous plant extracts in various diseases and preexisting conditions used in Ayurveda.

Some of these details are summarized here:

Sr. no.	Botanical name	Proposed role in oxidative stress
1	*Withania somnifera* (Linn.) Dunal (Solanaceae)	*W. somnifera* root powder administered to Sprague-Dawley rat model of pulmonary hypertension reduced the right ventricular pressure and showed improvement in inflammation, oxidative stress, endothelial dysfunction, and attenuation of proliferative marker and apoptotic resistance in lungs. [10] Withania coagulan treated streptozotocin-induced diabetic rat's kidneys, and serum showed decreased inflammatory cytokines and maintained antioxidant status, along with reduction and development of renal injury in diabetes.[11]
2	*Tinospora cordifolia* (Willd.) Miers. (Menispermaceae)	*T. cordifolia* extract against 6-hydroxydopamine (6-OHDA) lesion rat model of Parkinson's disease (PD) exhibited neuroprotective activity and reduced oxidative stress.[12]
3	*Terminalia chebula* Retz. (Combretaceae)	*T. chebula* protects the pheochromacytoma cells from damage by its strong antioxidant and anti-inflammatory activity.[13]
4	*Phyllanthus emblica* L. or *Emblica officinalis* Gaertn. (Phyllanthaceae)	*P. emblica* supplementation increases plasma antioxidant power and decreases oxidative stress in uremic patients when treated for 4 months.[14]
5	*Punica granatum* L. (Lythraceae)	The addition of *P. granatum extract* to simvastatin therapy in hypercholesterolemic patients improved oxidative stress and lipid status in the patient's serum. These antiatherogenic effects could reduce the risk for atherosclerosis development.[15]
6	*Bacopa monnieri* (L.) Pennell (Plantaginaceae)	*B. monnieri* extract prophylaxis renders the brain resistant to paraquat-mediated oxidative perturbations and thus protect against oxidative-mediated neuronal dysfunctions.[16]

Cleansing of body

The cleansing or the detoxification is the process by which the Ama from the various strotas or pathways in the body is either annihilated at the same location (if minor) or brought to the nearest exit of the body and purged out. This is all done on an individual basis depending on various factors such as the physical state of the patient for enduring the process; diseases state; and other underlying conditions, season, and so on.

The annihilation process (Shamana) is done majorly by medication and fasting. The fasting mentioned in Ayurveda is sequential and does not merely imply giving up food intake. It is a process where the drugs mentioned above may be given to accentuate the metabolism and overcome the oxidative stress. This process helps body to produce antioxidants to combat the stress. It is been shown that autophagy activation as a response to cellular starvation, glucose, and amino acid deprivation plays an important role in redox homeostasis. Autophagy contributes to clearing the cells of all irreversibly oxidized biomolecules (lipids, DNA, and protein), and thus have a role in antioxidant and DNA damage repair systems.[17]

The removal methods are called Shodhana. This is performed by medicated emetic therapy (Vamana), purgation therapy (Virechana), medicated enema (Basti), and bloodletting (Rakta mokshan). As explained earlier, in this method by means of oil application/oleation (Snehana) and heat application (Swedana), both medicated, the Ama from all over the body is brought to the nearest desired exit, that is, either stomach or large intestine and then expelled out. Owing to the benefits of this detoxification, Ayurveda advocates cleansing as per the season change.[18]

Some examples: For treatment of hyperlipidemia, a specific form of enema to reduce the Meda dhatu from body was administered and serum cholesterol, low-density lipoprotein (LDL), and apolipoprotein B were reduced after 21 days of treatment.[19] A pilot study on diabetic patients was conducted to see effects of Vamana and Virechana on blood sugar level (BSL). The sample size was too small to lead to concrete conclusions, yet the BSL was markedly reduced.[20] Virechana and Basti was performed on the patients of hypertension, and the regulation of blood pressure was observed.[21] Also the physiological changes induced by Vamana include improvement in appetite, regulation of bowel habits, and improvement in sleep pattern. It also reduced LDL and serum cholesterol positively affecting the blood pressure.[22,23]

Methods enabling the body to heal and be healthy

Apart from administering medication, there are multiple ways of enabling the antioxidant activity in the body to counter the oxidative stress. In other words, as per Ayurvedic perspective, these are the methods to bring the vitiated dosha to normalcy and maintain homeostasis. One of those treatments is the following:

> *Rasayan chikitsa*: Ayurveda has described single and combinations of herbs for increasing vitality; warding off diseases; and maintaining good health, skin, and body. Some of the famous ones are *Tinospora cordifolia* (Thunb.) Miers, *Withania somnifera* (L.) Dunal., *Phyllanthus embellica* L., *Terminalia chebula* Retz., and many others. Specific methods and administration instructions are mentioned specific to person, season, and age variations.

A study has shown increase in SOD after administration of rasayan therapy indicating its role toward altering the oxidative stress.[24] SOD belongs to family of antioxidant enzymes evolved for catalysis of toxic substances including superoxide radical. This therapy also enhances strotas' potency and detoxification.

Yoga: This has been studied in various diseases for its role in bringing equilibrium to the diseased state of body and also positively affecting the quality of life. Many such studies and meta-analysis have been conducted to show the widespread effect of yoga on the well-being of patients.[25–27]

Current advances

Unlike traditional medicine that utilizes whole plant powders for the treatment, current research focuses on the single compound found in these extracts. Some of the few purified compounds studied extensively for their role in attenuating oxidative stress in various diseases such as cancer, diabetes, hypertension, neurological disorders, and so on include epigallocatechin gallate, keampferol, resveratrol, ellagic acid, punicalagin, berberine, curcumin, and many others. Despite of all the efforts till date, more scientifically sound and strong studies are warranted to understand the complexities of Ayurvedic medication in current time. Efforts are ongoing in both India and internationally to address the issues of subjective bias and personalized therapy for Ayurvedic branch of medicine. Translating the knowledge of Ayurveda for public health benefit is a matter of team efforts and interdisciplinary approach, which warrants efficient infrastructure and funding support.

It is beyond the scope of this chapter to explain the complete therapeutics of oxidative stress utilized in Ayurveda, yet the chapter must acquaint the reader with basic principles and understanding of the same in the context of Ayurveda.

Multiple choice questions

1 Which of the following is false:
   a. Ayurveda is a holistic science that treats body, mind, and soul to attain health
   b. Oxidative stress can be attenuated by many plants and their extracts mentioned in Ayurveda
   c. There is no further need of RCTs and evidence for Ayurvedic therapy for various diseases
   d. Yoga can be instrumental in reducing oxidative stress and managing chronic diseases

2 Choose all that apply Ayurveda:
   a. It is an ancient science described in old scriptures called Vedas
   b. It describes that vitiation of three doshas are the cause of all diseases
   c. It mentions that oxidative stress explicitly and methods to assess it
   d. The loss of potency to metabolize or digest is the root cause of all diseases

References

1 Betteridge, D.J. (2000) What is oxidative stress? *Metabolism*, 49(2 Suppl 1), 3–8.
2 McCord, J.M. (2000) The evolution of free radicals and oxidative stress. *American Journal of Medicine*, 108(8), 652–659.
3 Dickinson, D., DeRossi, S., Yu, H. *et al.* (2014) Epigallocatechin-3-gallate modulates anti-oxidant defense enzyme expression in murine submandibular and pancreatic exocrine gland cells and human Hsg cells. *Autoimmunity*, 47(3), 177–184.

4 Li, Z., Geng, Y.N., Jiang, J.D. & Kong, W.J. (2014) Antioxidant and anti-inflammatory activities of berberine in the treatment of diabetes mellitus. *Evidence-based Complementary and Alternative Medicine*, 2014, 289264.

5 Hou, C.C., Lin, H., Chang, C.P., Huang, W.T. & Lin, M.T. (2011) Oxidative stress and pyrogenic fever pathogenesis. *European Journal of Pharmacology*, 667(1–3), 6–12.

6 Martynenko, A.Y., Timoshin, S.S., Fleishman, M.Y. & Lebed'ko, O.A. (2014) Proliferation and local manifestations of oxidative stress in mucous membrane of the ileum in patients at early stages of hemorrhagic fever with renal syndrome. *Bulletin of Experimental Biology and Medicine*, 156(3), 320–322.

7 Tupe, R.S., Diwan, A.G., Mittal, V.D., Narayanam, P.S. & Mahajan, K.B. (2014) Association of plasma proteins at multiple stages of glycation and antioxidant status with erythrocyte oxidative stress in patients with type 2 diabetes. *British Journal of Biomedical Science*, 71(3), 93–99.; quiz 138

8 Vaziri, N.D. & Rodríguez-Iturbe, B. (2006) Mechanisms of disease: oxidative stress and inflammation in the pathogenesis of hypertension. *Nature Clinical Practice Nephrology*, 2(10), 582–593.

9 Paulin, R. & Michelakis, E.D. (2014) The metabolic theory of pulmonary arterial hypertension. *Circulation Research*, 115(1), 148–164.

10 Kaur, G., Singh, N., Samuel, S.S. *et al.* (2014) Withania somnifera shows a protective effect in monocrotaline-induced pulmonary hypertension. *Pharmaceutical Biology*, 1, 147–57.

11 Ojha, S., Alkaabi, J., Amir, N. *et al.* (2014) Withania coagulans fruit extract reduces oxidative stress and inflammation in kidneys of streptozotocin-induced diabetic rats. *Oxidative Medicine and Cellular Longevity*, 2014, 201436.

12 Kosaraju, J., Chinni, S., Roy, P.D., Kannan, E., Antony, A.S. & Kumar, M.N. (2014) Neuroprotective effect of tinospora cordifolia ethanol extract on 6-hydroxy dopamine induced parkinsonism. *Indian Journal of Pharmacology*, 46(2), 176–180.

13 Gaire, B.P., Jamarkattel-Pandit, N., Lee, D. *et al.* (2013) Terminalia chebula extract protects Ogd-R induced Pc12 cell death and inhibits Lps induced microglia activation. *Molecules*, 18(3), 3529–3542.

14 Chen, T.S., Liou, S.Y. & Chang, Y.L. (2009) Supplementation of *Emblica officinalis* (amla) extract reduces oxidative stress in uremic patients. *American Journal of Chinese Medicine*, 37(1), 19–25.

15 Hamoud, S., Hayek, T., Volkova, N. *et al.* (2014) Pomegranate extract (Pomx) decreases the atherogenicity of serum and of human monocyte-derived macrophages (Hmdm) in simvastatin-treated hypercholesterolemic patients: a double-blinded, placebo-controlled, randomized, prospective pilot study. *Atherosclerosis*, 232(1), 204–210.

16 Hosamani, R., G. Krishna, and Muralidhara. Aug 2014 "Standardized Bacopa Monnieri Extract Ameliorates Acute Paraquat-Induced Oxidative Stress, and Neurotoxicity in Prepubertal Mice Brain." [In ENG]. *Nutr Neurosci* ().

17 Filomeni, G., De Zio, D. & Cecconi, F. (2014) Oxidative stress and autophagy: the clash between damage and metabolic needs. *Cell Death and Differentiation*, 22(3), 377–388.

18 Bhatted, S., Shukla, V.D., Thakar, A. & Bhatt, N.N. (2011) A study on Vasantika Vamana (therapeutic emesis in spring season) – a preventive measure for diseases of kapha origin. *Ayu*, 32(2), 181–186.

19 Auti, S.S., Thakar, A.B., Shukla, V.J. & Ravishankar, B. (2013) Assessment of Lekhana Basti in the management of hyperlipidemia. *Ayu*, 34(4), 339–345.

20 Jindal, N. & Joshi, N.P. (2013) Comparative study of Vamana and Virechanakarma in controlling blood sugar levels in diabetes mellitus. *Ayu*, 34(3), 263–269.

21 Shukla, G., Bhatted, S.K., Dave, A.R. & Shukla, V.D. (2013) Efficacy of Virechana and Basti Karma with Shamana therapy in the management of essential hypertension: a comparative study. *Ayu*, 34(1), 70–76.

22 Gupta, B., Mahapatra, S.C., Makhija, R. *et al.* (2012) Physiological and biochemical changes with Vamana procedure. *Ayu*, 33(3), 348–355.

23 Gupta, B., Mahapatra, S.C., Makhija, R. *et al.* (2011) Observations on Vamana procedure in healthy volunteers. *Ayu*, 32(1), 40–45.

24 Kuchewar, V.V., Borkar, M.A. & Nisargandha, M.A. (2014) Evaluation of antioxidant potential of Rasayana drugs in healthy human volunteers. *Ayu.*, 35(1), 46–49.

25 Ram, A., Banerjee, B., Hosakote, V.S., Rao, R.M. & Nagarathna, R. (2013) Comparison of lymphocyte apoptotic index and qualitative dna damage in yoga practitioners and breast cancer patients: a pilot study. *International Journal of Yoga*, 6(1), 20–25.

26 Littman, A.J., Bertram, L.C., Ceballos, R. *et al.* (2012) Randomized controlled pilot trial of yoga in overweight and obese breast cancer survivors: effects on quality of life and anthropometric measures. *Support Care Cancer*, 20(2), 267–277.

27 Balasubramaniam, M., Telles, S. & Doraiswamy, P.M. (2012) Yoga on our minds: a systematic review of yoga for neuropsychiatric disorders. *Front Psychiatry*, 3, 117.

CHAPTER 30

Clinical trials and antioxidant outcomes

Carlos Palacio and Arshag D. Mooradian

Department of Medicine, University of Florida College of Medicine, Jacksonville, FL, USA

THEMATIC SUMMARY BOX

At the end of this chapter, students should be able to:

- Describe the existing pathogenesis mechanisms of oxidative stress and its impact

- Differentiate how the free radical generating system is counterbalanced by a host of enzymatic and nonenzymatic antioxidant defense system

- Describe pathways leading to apoptosis, necrosis , cell death, and disease

- Define pathogenesis of oxidative stress

- Understand the different findings between observational studies and clinical trials assessing clinical outcomes associated with the consumption of antioxidant vitamins

- List the limitations of currently available trials on antioxidant use

- Describe the potential hazards of indiscriminate consumption of antioxidants

Introduction

One of the key metabolic processes is the mitochondrial respiration with its attendant increase in reactive oxygen species (ROS). This oxidative burst is counterbalanced with a myriad of endogenous and exogenous antioxidants that would allow optimal oxidative state to facilitate growth factor signaling, cellular growth, and functioning of mitochondria and endoplasmic reticulum (ER). Metabolic substrates such as dextrose and fatty acids are the principal drivers of mitochondrial respiration that can increase the two major known cellular stress pathways, namely the oxidative stress and ER stress.[1,2] Both of these stresses are implicated in a host of pathological states including atherosclerosis, cancer, and pancreatic β-cell failure. Antioxidants theoretically should ameliorate the oxidative stress but generally seem incapable of reducing ER stress.[3,4] In addition, the degree of amelioration of oxidative stress may be tissue specific such that they may have beneficial effect in one cell type while having

Oxidative Stress and Antioxidant Protection: The Science of Free Radical Biology and Disease, First Edition.
Edited by Donald Armstrong and Robert D. Stratton.

harmful effects in others. An example is that glutathione peroxidase (GPx-1) overexpression is protective in pancreatic β cells that are naturally deficient of GPx-1, while in other cells, such as skeletal muscle and liver, excess GPx-1 causes insulin resistance through suppressing essential ROS necessary for signal transduction.[5] Despite the biological plausibility of the effectiveness of the natural antioxidants in ameliorating oxidative stress, so far most of the interventional trials have failed to show favorable effects on clinical end points.[6,7] In this chapter, the large trials with antioxidants will be reviewed, the limitations of these studies discussed, and future directions in identifying a clinically effective antioxidant will be suggested.

Pathophysiology of oxidative pathways

Free radicals have pathophysiological role in a host of cardiovascular, neurodegenerative, and other chronic inflammatory diseases and have been implicated in a variety of adverse effects of therapeutic agents.[8]

Free radicals are obligatory by-products of cellular respiration that are essential for cellular function but can also compromise cellular viability. A free radical is an atom or molecule with a single unpaired electron.[8] Examples include nitric oxide ($^{\bullet}NO$), superoxide ($O_2^{\bullet-}$), hydroxyl radical ($^{\bullet}OH$), and lipid peroxy radical ($LOO^{\bullet}$). Of these, hydroxyl radical is the most reactive and nonselective species that can react with proteins, lipids, and DNA and has a very short lifetime (10^{-9} s). Through hydroxyl radical-initiated events, cardiotoxicity of the chemotherapeutic agent such as doxorubicin occurs.[8–10]

Superoxide is much less reactive and much more selective. Superoxide rapidly reacts with another molecule of superoxide to form hydrogen peroxide (H_2O_2) that can be rapidly destroyed by antioxidant enzymes (catalase, glutathione peroxidase), otherwise it is relatively stable *in vitro*. Superoxide can also interact with nitric oxide to form a very potent and reactive peroxynitrite.[8,11]

There are several enzymes that mediate the generation of superoxide and hydrogen either as a by-product of the reactions they facilitate or as the primary biological product. Example of the latter is the family of NADPH oxidases while the former category includes cytochrome P450, lipoxygenase, cyclooxygenase, and ATP-generating system of mitochondria.[8]

Superoxide has been implicated in both health-promoting and disease-promoting biochemical reactions. This free radical is the mechanism by which phagocytic cells kill pathogens. When the $O_2^{\bullet-}$ generation is impaired such as in chronic granulomatous disease (CGD) then the individual experiences repeated infections.[12,13] Superoxide generation is also implicated in reperfusion injury when an ischemic tissue such as the myocardium, cerebrum, or kidney is reperfused after a brief period of ischemia.[14,15] Potential interventions to ameliorate reperfusion injury include use of enzymatic scavengers of superoxide and hydrogen peroxide (superoxide dismutase and catalase), iron chelators (deferoxamine), and inhibitors of xanthine oxidase (e.g., allopurinol).[8]

Other ROS include the singlet oxygen and hypochlorous acid (HClO) that is formed from myeloperoxidase-mediated H_2O_2-dependent oxidation of chloride anion. Singlet oxygen is a mediator of the photodynamic therapy-related sensitization of cancerous cells to the therapeutic effects of irradiation and chemotherapy. Singlet oxygen is also implicated in the phototoxicity associated with certain drugs and diseases such as cutaneous porphyrias.[8]

Nitric oxide (•NO) is an important mediator of vasodilation and smooth muscle cell relaxation through its effects on soluble guanylyl cyclase (sGC) that converts guanosine triphosphate (GTP) to cyclic GMP.[16] The discovery of this pathway has led to the development of agents such as sildenafil to treat pulmonary hypertension and erectile dysfunction. However, (•NO) can also interact with $O_2^{•-}$ to form peroxynitrite that can nitrate tyrosyl groups in proteins, thereby altering (increasing or decreasing) protein function.[11] Peroxynitrite also oxidizes unsaturated fatty acids and DNA. The latter reaction can be monitored by measuring the 8-hydroxy-2-deoxyguanosine and 8-nitroguanine in biological fluids and are considered biomarkers of oxidative damage.[17] It is noteworthy that the peroxynitrite formation not only produces a potent cytotoxic species but also adversely affects vasodilation by removing the •NO.

To maintain optimal redox state within the cell, the free radical generating system is counterbalanced by a host of enzymatic and nonenzymatic antioxidant defense system, such as superoxide dismutases, catalases, and peroxidases; vitamins C and E; and adaptive genetic response to maximize the antioxidant capacity.[8] The thioredoxin (Trx) system, which is composed of NADPH, thioredoxin reductase (TrxR), and thioredoxin, is a key antioxidant system involved in DNA and protein repair.[18] In mammalian cells, the cytosolic and mitochondrial Trx systems, together with the glutathione (GSH)–glutaredoxin (Grx) system (NADPH, glutathione reductase, GSH, and Grx) control the cellular redox environment. The absence of a GSH–Grx system in some pathogenic bacteria makes the bacterial Trx system essential for survival and therefore a potential target for antibacterial drugs.[18]

Imbalance between the prooxidant and antioxidant levels could result in either oxidative stress or reductive stress. Thus, excess antioxidants may have deleterious effects. Identification of a clinically effective antioxidant may be able to change the natural history of these diseases.

Clinical trials

Tables 30.1 and 30.2 summarize the available major trials. They are categorized according to observational studies (Table 30.1) and interventional trials (Table 30.2). Each table summarizes the authors, population studied, median follow-up, agents studied, and summary results.

Clinical implications of research findings

The overall outcome of interventional trials does not support the indiscriminate use of antioxidant vitamins. In a meta-analysis of seven randomized trials of vitamin E treatment (81,788 patients) and of eight trials of β-carotene treatment (138,113 patients), there was no cardiovascular or mortality benefit of supplementation with

Table 30.1 Observational studies assessing clinical outcomes associated with the consumption of antioxidant vitamins.

References	Population	Follow-up	Antioxidants	Result
Kushi et al.[19]	34,486 postmenopausal women without history of cardiovascular disease aged 55–69 years	7 years	Vitamins A, C, and E	High-dietary vitamin E intake conferred protection. No association between vitamin A and C consumption and CAD death
Stampfer et al.[20]	87,245 female nurses aged 34–59 years without CAD and cancer	8 years	Wide range of nutrients, including vitamin E	Beneficial effect of vitamin E especially in the form of supplements and greater than 2 years of intake
Yochum et al.[21]	34,492 postmenopausal women without history of cardiovascular disease aged 55–69 years	7 years	Vitamins A, C, and E	Protective effect of vitamin E from foods on death from stroke. No protective role for supplemental vitamin E or other antioxidant
Klipstein-Grobusch et al.[22]	4,802 participants aged 55–95 years without history of myocardial infarction	4 years	β-Carotene, vitamins C and E	High dietary carotene intake may protect against myocardial infarction. Effect was more distinct in current and former smokers. No association with vitamin C or E
Losonczy et al.[23]	11,178 persons aged 67–105 years	8–9 years	Vitamins C and E supplementation	Beneficial effect of vitamin E; beneficial effect of vitamin C only when used with vitamin E on all-cause mortality and cardiac mortality

Study	Population	Duration	Supplement	Result
Knekt et al.[24]	5,133 Finnish men and women aged 30–69 years and initially without heart disease	12–16 years	β-Carotene, vitamins C and E	Beneficial effect of vitamins A and C only in women. Beneficial effect of vitamin E in women and men
Rimm et al.[25]	39,910 US male health professionals aged 40–75 years, without history of CAD, diabetes, and hypercholesterolemia	4 years	Vitamin C, β-carotene, and vitamin E	Beneficial effect of vitamin E
Todd et al.[26]	4,036 Scottish men and 3,833 women aged 40–59 years without history of heart disease	7.7 years	Vitamin C, β-carotene, and vitamin E	Beneficial effect of vitamin C and β-carotene only in men in reducing only coronary artery events. No beneficial effect of vitamin E
Enstrom et al.[27]	11,348 US adults aged 25–74 years	10 years	Vitamin C	Beneficial effect of vitamin C on all-cause mortality and cardiovascular mortality in males
Neuhouser et al.[28]	161,808 postmenopausal women	8.0 years in the clinical trial and 7.9 years observational study cohorts	Multivitamins	No evidence that multivitamin use influences the risk of common cancers, CVD, or total mortality

(Adapted from Ref. [4] with permission from the publisher.)

Table 30.2 Large interventional trials of antioxidant vitamins.

References	Population	Follow-up and design	Antioxidants	Result
Collaborative Group of the Primary Prevention Project. 2001[29]	4,495 Italians at risk of CVD	3.6 years. Randomized placebo controlled	Vitamin E 300 mg/day	No benefit
Yusuf et al.[30]	2,545 women 6,996 men, 55 years or older at high risk of CVD, 38% diabetic	4.5 years. Double-blinded, randomized trial. 2× 2-factorial design	Vitamin E 400 IU/day	No significant difference in death, myocardial infarction, or stroke
Yusuf et al.[30], Lonn et al.[31]	3,994 continued to take the study intervention, and 738 agreed to passive follow-up	7.0 years	Vitamin E 400 IU/day	No effect on cancer or major cardiovascular events. May increase the risk of heart failure
Salonen et al.[32]	520 men and postmenopausal women aged 45–69	3 years. Randomized double-blind placebo controlled	91 mg of d-α-tocopherol, or 250 mg of slow-release vitamin C, or combination or placebo	Progression of atherosclerosis was reduced in male participants who were smokers and were assigned to the combined treatment of vitamins E and C
Lee et al.[33,34]	39,876 women aged ≥ 45 years	2.1 years (β-carotene arm terminated early after a median duration of and followed for two additional years). The vitamin E arm was followed for an average of 10.1 years. Randomized, double-blind, placebo controlled	β-Carotene 50 mg, vitamin E 600 IU, and aspirin 100 mg given on alternate days either alone or in various combinations of the three	No harm or benefit demonstrated in the subjects receiving the β-carotene No benefit from alternate-day vitamin E in the primary prevention of cardiovascular disease or cancer
GISSI-Prevenzione Investigators[35]	11,324 Italian patients with myocardial infarction ≤3 months	3.5-years. Randomized placebo controlled	n-3 poly-unsaturated fatty acid 1 g daily, vitamin E 300 mg daily, both, or none	Vitamin E conferred no significant benefit

Study	Population	Design	Intervention	Results
Hennekens et al.[36]	22,071 male physicians, 40–84 years of age	12 years. Randomized, double-blind, placebo controlled	β-Carotene 50 mg tablets given on alternate days	No adverse or beneficial effect
Omenn et al.[37]	18,314 men and women at high risk for lung cancer	Randomized, double-blinded, placebo-controlled chemopreventive trial	30 mg of β-carotene and 25,000 IU of retinal palmitate taken daily or placebo	No cardioprotective effect of combination of β-carotene and vitamin A. β-Carotene (30 mg) + retinyl palmitate (35,000) per day increased lung cancer
Stephens et al.[38]	2,002 patients with angiographically proven coronary atherosclerosis	1.4-years. Randomized double-blind, placebo controlled	400 or 800 IU of vitamin E or placebo	Reduced rate of nonfatal myocardial infarction with 400 IU but not 800 IU of vitamin E. No difference in death. Number of events was small. Randomization was flawed.
The ATBC Cancer Prevention Study Group[39]	29,133 male smokers, 50–69 years of age	5–8 years. Randomized, double-blind, placebo controlled	α-Tocopherol, 50 mg/day and β-carotene 20 mg/day given alone or in a combination of both	No effect of vitamin E on CVD or risk of death from coronary heart disease. Risk of nonfatal myocardial infarction decreased in vitamin E only, not in vitamin E + carotene versus placebo. There was increased incidence of lung cancer and cardiovascular events observed in patients taking carotene. There was a slightly higher incidence of hemorrhagic strokes in patients on vitamin E

(continued)

Table 30.2 (*Continued*)

References	Population	Follow-up and design	Antioxidants	Result
Blot et al.[40]	29,584 Chinese participants aged 40–69 years	5.4-years. Double-blinded, randomized, placebo controlled	Daily retinol 5000 IU + zinc 22.5 mg or riboflavin 3.2 mg + niacin 40 mg, or ascorbic acid 120 mg + molybdenum 30 mg, or β-carotene 15 mg + selenium 50 mg + α-tocopherol 30 mg	Reduction in total mortality in cancer death and incidence especially for stomach cancer in those on the combination regimen of β-carotene, selenium, and α-tocopherol
Women's Antioxidant and CV Study Group[41–43]	8,171 US female health professionals aged ≥40 years with a history of CVD or three or more CVD risk factors	9.4 years mean duration of follow-up for the ascorbic acid, vitamin E, or β-carotene arm. 7.3 years follow-up for the folate/B_6/B_{12} arm. Double-blinded randomized placebo controlled	600 mg vitamin E every other day, 500 mg vitamin C daily, 50 mg of β-carotene every other day, or placebo. Of these women, 5,442 were also subsequently randomized to folic acid/vitamin B_6/vitamin B_{12} or placebo	No overall effects of ascorbic acid, vitamin E, or β-carotene on cardiovascular events. Combination of folic acid, vitamin B_6, and B_{12} did not reduce combined end point of total cardiovascular events despite significant lowering of homocysteine

Study	Participants	Duration and design	Intervention	Findings
Sesso et al.[44]	14,641 US male physicians who were initially aged ≥ 50 years	8 years. Double-blinded randomized placebo controlled	400 IU of vitamin E every other day and 500 mg of vitamin C daily compared with placebo	No effect on the incidence of major cardiovascular events. Vitamin E was associated with an increased risk of hemorrhagic stroke.
Belch et al.[45]	1276 adults aged ≥ 40 years with type 1 or type 2 diabetes and an ankle brachial pressure index of 0.99 or less but no symptomatic cardiovascular disease	6.7 years	Daily, 100 mg aspirin tablet plus antioxidant capsule; aspirin tablet plus placebo capsule; placebo tablet plus antioxidant capsule; or placebo tablet plus placebo capsule (antioxidant capsule contained α-tocopherol 200 mg, ascorbic acid 100 mg, pyridoxine 25 mg, zinc sulfate 10 mg, nicotinamide 10 mg, lecithin 9.4 mg, and sodium selenite 0.8 mg)	No evidence to support the use of aspirin or antioxidants in the primary prevention of cardiovascular events and mortality in people with diabetes. Statistically significant increase in the number of deaths from any cause in the antioxidant group compared with the no antioxidant group

(Adapted from Ref. [4] with permission from the publisher).

these antioxidants.[46] The dose range for vitamin E used was 50–800 IU, and for β-carotene it was 15–50 mg. Supplementation with β-carotene led to a small but significant increase in all-cause mortality (7.4% vs 7.0%, $p=0.003$) and a slight increase in cardiovascular deaths (3.4% vs 3.1%, $p=0.003$). Thus, some antioxidant vitamins, notably β-carotene, may be harmful.[37,39] Similarly, although ROS-mediated cardiac injury is implicated in the development of congestive heart failure (CHF), antioxidant vitamin trials in CHF have been disappointing and a meta-analysis concluded that vitamin E, given as monotherapy in certain patient categories, may contribute to the development or aggravation of heart failure.[47] Moreover, in one study, daily ingestion of vitamins C and E abrogated exercise-related enhancement of insulin sensitivity[48], and in the HATS trial, the HDL raising effect of nicotinic acid was blunted with high-dose vitamin E.[49]

There are also some adverse drug reactions associated with antioxidant vitamins. High intake of vitamin C can cause abdominal bloating and diarrhea, and in people with glucose-6-phosphate dehydrogenase deficiency it can cause hemolysis.[50] Use of vitamin C may also aggravate iron overload in those who carry hemochromatosis genes. Excessive use of vitamin A may promote osteoporosis and teratogenicity, and vitamin E may worsen retinitis pigmentosa and may increase the risk of hemorrhagic stroke.[50]

Given these observations, the decision to prescribe antioxidants should be based on an assessment of the patient's nutritional status. Supplementation may be more advisable in the malnourished people, older people in nursing homes, elderly, strict vegetarians, or people on calorie-restricted diets. Supplement intake should not exceed the recommended daily allowances.

Limitations of current clinical trials

Currently available trials have many limitations. Most trials were done in people with established pathology, and theoretically one can argue that preventing the disease would be a more realistic expectation than reversing the course once it is manifested in clinical symptoms.

A major drawback of the available interventional trials is the lack of measures of oxidation; thus, it is not clear if sufficient amounts of the prescribed antioxidants were absorbed or were biologically active. Lack of clinical biomarkers and inability to stratify subjects according to their oxidative load and inability to monitor oxidative load for effectiveness of the prescribed antioxidant are major limitations. Since oxidation can occur through multiple pathways, different antioxidants with different targets can be combined to study the effects on clinical outcomes. It is furthermore possible that no benefit or even a detrimental effect is associated with certain doses. For example, several studies of vitamin E suggested worse outcomes.[39,44] Certain studies saw benefit from dietary sources rather than supplements.[19,21,22] The latter observations could have been due to a healthy volunteer effect.

The inability to stratify and monitor patients according to oxidative load may have inadvertent adverse consequences due to overtreatment. Although there is biological plausibility of oxidative load as a causal mechanism in clinical disease, the existing treatment trials have failed to consistently show successful interventions targeting this mechanism. Further research should address optimal dosing of antioxidants, stratification of patients according to oxidative load, whether dietary sources of antioxidants have benefit over supplements, and clarification of mechanisms of adverse events. While observational studies showed promise, they are insufficient to establish cause and effect because of inherent internal validity limitations. Randomized, controlled clinical trials, which are needed to establish cause and effect, have not bridged the gap between antioxidants' scientific plausibility and the therapeutic value of treatment interventions.

Summary points

1 Metabolic substrates such as dextrose and fatty acids are the principal drivers of mitochondrial respiration that can increase the two major known cellular stress pathways, namely the oxidative stress and ER stress.
2 These stresses are implicated in a host of pathological states including atherosclerosis, cancer, and pancreatic β-cell failure.
3 To maintain optimal redox state within the cell, the free radical generating system is counterbalanced by a host of enzymatic and nonenzymatic antioxidant defense system, such as superoxide dismutases, catalases, and peroxidases; vitamins C and E; and adaptive genetic response to maximize the antioxidant capacity.
4 The outcomes of interventional trials do not support the indiscriminate use of antioxidant vitamins. In some cases, supplements are associated with worse outcomes.
5 The decision to prescribe antioxidants should be based on an assessment of the patient's nutritional status. Supplementation may be more advisable in the malnourished people, older people in nursing homes, elderly, strict vegetarians, or people on calorie-restricted diets. In general, supplement intake should not exceed the recommended daily allowances.

Multiple choice questions

1 Which of the following statements is true?
   i. Free radicals are obligatory by-products of cellular respiration that are essential for cellular function but can also compromise cellular viability
   ii. Antioxidants have been shown to have beneficial effects in observational trials
   iii. There is strong evidence supporting the use of vitamin C for the prevention of dementia
   a. i and ii
   b. ii and iii
   c. All of the above are true statements

2 Observational trials, in contrast to interventional trials, are useful in which of the following?
   a. Hypothesis generation
   b. Establishing cause-and-effect relationships
   c. Both a and b
   d. None of the above

3 Which of the following antioxidants has been associated with an increased risk of hemorrhagic stroke events?

a. Vitamin C

b. β-Carotene

c. Pyridoxine

d. Vitamin E

References

1 Horani, M.H., Haas, M.J. & Mooradian, A.D. (2006) Saturated, unsaturated, and trans-fatty acids modulate oxidative burst induced by high dextrose in human umbilical vein endothelial cells. *Nutrition*, 22(2), 123–127.

2 Sheikh-Ali, M. *et al.* (2010) Effects of antioxidants on glucose-induced oxidative stress and endoplasmic reticulum stress in endothelial cells. *Diabetes Research and Clinical Practice*, 87(2), 161–166.

3 Mooradian, A.D. & Haas, M.J. (2011) Glucose-induced endoplasmic reticulum stress is independent of oxidative stress: A mechanistic explanation for the failure of antioxidant therapy in diabetes. *Free Radical Biology and Medicine*, 50(9), 1140–1143.

4 Sheikh-Ali, M., Chehade, J.M. & Mooradian, A.D. (2011) The antioxidant paradox in diabetes mellitus. *American Journal of Therapeutics*, 18(3), 266–278.

5 Lubos, E., Loscalzo, J. & Handy, D.E. (2011) Glutathione peroxidase-1 in health and disease: from molecular mechanisms to therapeutic opportunities. *Antioxidants and Redox Signaling*, 15(7), 1957–1997.

6 Hasanain, B. & Mooradian, A.D. (2002) Antioxidant vitamins and their influence in diabetes mellitus. *Current Diabetes Reports*, 2(5), 448–456.

7 Mooradian, A.D. (2006) Antioxidants and diabetes. *Nestlé Nutrition Workshop Series. Clinical & Performance Programme*, 11, 107–122.; discussion 122-5

8 Kalyanaraman, B. (2013) Teaching the basics of redox biology to medical and graduate students: Oxidants, antioxidants and disease mechanisms. *Redox Biology*, 1(1), 244–257.

9 Lipshultz, S.E. *et al.* (2004) The effect of dexrazoxane on myocardial injury in doxorubicin-treated children with acute lymphoblastic leukemia. *New England Journal of Medicine*, 351(2), 145–153.

10 Heard, K.J. (2008) Acetylcysteine for acetaminophen poisoning. *New England Journal of Medicine*, 359(3), 285–292.

11 Beckman, J.S. *et al.* (1990) Apparent hydroxyl radical production by peroxynitrite: implications for endothelial injury from nitric oxide and superoxide. *Proceedings of the National Academy of Sciences of the United States of America*, 87(4), 1620–1624.

12 Grant, S.S. *et al.* (2012) Eradication of bacterial persisters with antibiotic-generated hydroxyl radicals. *Proceedings of the National Academy of Sciences of the United States of America*, 109(30), 12147–12152.

13 Kuhns, D.B. *et al.* (2010) Residual NADPH oxidase and survival in chronic granulomatous disease. *New England Journal of Medicine*, 363(27), 2600–2610.

14 McCord, J.M. (1985) Oxygen-derived free radicals in postischemic tissue injury. *New England Journal of Medicine*, 312(3), 159–163.

15 Odeh, M. (1991) The role of reperfusion-induced injury in the pathogenesis of the crush syndrome. *New England Journal of Medicine*, 324(20), 1417–1422.

16 Murad, F. (2006) Shattuck Lecture. Nitric oxide and cyclic GMP in cell signaling and drug development. *New England Journal of Medicine*, 355(19), 2003–2011.

17 Ohshima, H., Sawa, T. & Akaike, T. (2006) 8-nitroguanine, a product of nitrative DNA damage caused by reactive nitrogen species: formation, occurrence, and implications in inflammation and carcinogenesis. *Antioxidants and Redox Signaling*, 8(5–6), 1033–1045.

18 Lu, J. & Holmgren, A. (2014) The thioredoxin antioxidant system. *Free Radical Biology and Medicine*, 66, 75–87.

19 Kushi, L.H. *et al.* (1996) Dietary antioxidant vitamins and death from coronary heart disease in postmenopausal women. *New England Journal of Medicine*, 334(18), 1156–1162.

20 Stampfer, M.J. *et al.* (1993) Vitamin E consumption and the risk of coronary disease in women. *New England Journal of Medicine*, 328(20), 1444–1449.

21 Yochum, L.A., Folsom, A.R. & Kushi, L.H. (2000) Intake of antioxidant vitamins and risk of death from stroke in postmenopausal women. *American Journal of Clinical Nutrition*, 72(2), 476–483.

22 Klipstein-Grobusch, K. *et al.* (1999) Dietary antioxidants and risk of myocardial infarction in the elderly: the Rotterdam Study. *American Journal of Clinical Nutrition*, 69(2), 261–266.

23 Losonczy, K.G., Harris, T.B. & Havlik, R.J. (1996) Vitamin E and vitamin C supplement use and risk of all-cause and coronary heart disease mortality in older persons: the Established Populations for Epidemiologic Studies of the Elderly. *American Journal of Clinical Nutrition*, 64(2), 190–196.

24 Knekt, P. *et al.* (1994) Antioxidant vitamin intake and coronary mortality in a longitudinal population study. *American Journal of Epidemiology*, 139(12), 1180–1189.

25 Rimm, E.B. *et al.* (1993) Vitamin E consumption and the risk of coronary heart disease in men. *New England Journal of Medicine*, 328(20), 1450–1456.

26 Todd, S. *et al.* (1999) Dietary antioxidant vitamins and fiber in the etiology of cardiovascular disease and all-causes mortality: results from the Scottish Heart Health Study. *American Journal of Epidemiology*, 150(10), 1073–1080.

27 Enstrom, J.E., Kanim, L.E. & Klein, M.A. (1992) Vitamin C intake and mortality among a sample of the United States population. *Epidemiology*, 3(3), 194–202.

28 Neuhouser, M.L. *et al.* (2009) Multivitamin use and risk of cancer and cardiovascular disease in the Women's Health Initiative cohorts. *Archives of Internal Medicine*, 169(3), 294–304.

29 de Gaetano, G. (2001) Collaborative Group of the Primary Prevention, Low-dose aspirin and vitamin E in people at cardiovascular risk: a randomised trial in general practice. Collaborative Group of the Primary Prevention Project. *Lancet*, 357(9250), 89–95.

30 Yusuf, S. *et al.* (2000) Vitamin E supplementation and cardiovascular events in high-risk patients. The Heart Outcomes Prevention Evaluation Study Investigators. *New England Journal of Medicine*, 342(3), 154–160.

31 Lonn, E. *et al.* (2005) Effects of long-term vitamin E supplementation on cardiovascular events and cancer: a randomized controlled trial. *JAMA*, 293(11), 1338–1347.

32 Salonen, J.T. *et al.* (2000) Antioxidant Supplementation in Atherosclerosis Prevention (ASAP) study: a randomized trial of the effect of vitamins E and C on 3-year progression of carotid atherosclerosis. *Journal of Internal Medicine*, 248(5), 377–386.

33 Lee, I.M. *et al.* (1999) Beta-carotene supplementation and incidence of cancer and cardiovascular disease: the Women's Health Study. *Journal of the National Cancer Institute*, 91(24), 2102–2106.

34 Lee, I.M. *et al.* (2005) Vitamin E in the primary prevention of cardiovascular disease and cancer: the Women's Health Study: a randomized controlled trial. *JAMA*, 294(1), 56–65.

35 GISSI-I Prevenzione Investigators (1999) Dietary supplementation with n-3 polyunsaturated fatty acids and vitamin E after myocardial infarction: results of the GISSI-Prevenzione trial. Gruppo Italiano per lo Studio della Sopravvivenza nell'Infarto miocardico. *Lancet*, 354(9177), 447–455.

36 Hennekens, C.H. *et al.* (1996) Lack of effect of long-term supplementation with beta carotene on the incidence of malignant neoplasms and cardiovascular disease. *New England Journal of Medicine*, 334(18), 1145–1149.

37 Omenn, G.S. *et al.* (1996) Risk factors for lung cancer and for intervention effects in CARET, the Beta-carotene and retinol efficacy trial. *Journal of the National Cancer Institute*, 88(21), 1550–1559.

38 Stephens, N.G. *et al.* (1996) Randomised controlled trial of vitamin E in patients with coronary disease: Cambridge Heart Antioxidant Study (CHAOS). *Lancet*, 347(9004), 781–786.

39 ATBC Study Group (1994) The effect of vitamin E and beta carotene on the incidence of lung cancer and other cancers in male smokers. The Alpha-Tocopherol. Beta Carotene Cancer Prevention Study Group. *New England Journal of Medicine*, 330(15), 1029–1035.

40 Blot, W.J. *et al.* (1993) Nutrition intervention trials in Linxian, China: supplementation with specific vitamin/mineral combinations, cancer incidence, and disease-specific mortality in the general population. *Journal of the National Cancer Institute*, 85(18), 1483–1492.

41 Cook, N.R. *et al.* (2007) A randomized factorial trial of vitamins C and E and beta carotene in the secondary prevention of cardiovascular events in women: results from the Women's Antioxidant Cardiovascular Study. *Archives of Internal Medicine*, 167(15), 1610–1618.

42 Albert, C.M. *et al.* (2008) Effect of folic acid and B vitamins on risk of cardiovascular events and total mortality among women at high risk for cardiovascular disease: a randomized trial. *JAMA*, 299(17), 2027–2036.

43 Lin, J. *et al.* (2009) Vitamins C and E and beta carotene supplementation and cancer risk: a randomized controlled trial. *Journal of the National Cancer Institute*, 101(1), 14–23.

44 Sesso, H.D. *et al.* (2008) Vitamins E and C in the prevention of cardiovascular disease in men: the Physicians' Health Study II randomized controlled trial. *JAMA*, 300(18), 2123–2133.

45 Belch, J. *et al.* (2008) The prevention of progression of arterial disease and diabetes (POPADAD) trial: factorial randomised placebo controlled trial of aspirin and antioxidants in patients with diabetes and asymptomatic peripheral arterial disease. *BMJ*, 337, a1840.

46 Vivekananthan, D.P. *et al.* (2003) Use of antioxidant vitamins for the prevention of cardiovascular disease: meta-analysis of randomised trials. *Lancet*, 361(9374), 2017–2023.

47 Thomson, M.J., Frenneaux, M.P. & Kaski, J.C. (2009) Antioxidant treatment for heart failure: friend or foe? *QJM*, 102(5), 305–310.

48 Ristow, M. *et al.* (2009) Antioxidants prevent health-promoting effects of physical exercise in humans. *Proceedings of the National Academy of Sciences of the United States of America*, 106(21), 8665–8670.

49 Cheung, M.C. *et al.* (2001) Antioxidant supplements block the response of HDL to simvastatin-niacin therapy in patients with coronary artery disease and low HDL. *Arteriosclerosis, Thrombosis, and Vascular Biology*, 21(8), 1320–1326.

50 Hasnain, B.I. & Mooradian, A.D. (2004) Recent trials of antioxidant therapy: what should we be telling our patients? *Cleveland Clinic Journal of Medicine*, 71(4), 327–334.

CHAPTER 31

Statistical approaches to make decisions in clinical experiments

Albert Vexler and Xiwei Chen

Department of Biostatistics, School of Public Health and Health Professions, State University of New York at Buffalo, Buffalo, NY, USA

THEMATIC SUMMARY BOX

At the end of this chapter, students should be able to:

- Correctly formulate statistical hypotheses with respect to the aims of epidemiological and/or biomedical studies

- Construct and provide statistical decision-making test rules corresponding to practical experiments

- Use parametric and nonparametric likelihood testing techniques in applied researches

- Understand basic properties of likelihood ratio type tests in parametric and nonparametric manners

- Use basic test procedures and their components in practical statistical decision-making mechanisms

- Employ statistical software at a beginning level

Introduction, preliminaries, and basic components of statistical decision-making mechanisms

Often, experiments in biomedicine and other health-related sciences involve mathematically formalized tests, employing appropriate and efficient statistical procedures to analyze data. Mathematical strategies to make decisions via formal rules play important roles in medical and epidemiological discovery, in policy formulation, and in clinical practice. In this context, in order to make conclusions about populations on the basis of samples from those populations, clinical trials commonly require the application of the mathematical statistical discipline.

The aim of the scientific methods in decision theory is to simultaneously maximize quantified gains and minimize losses in reaching a conclusion. For example, statements of clinical experiments can request to maximize factors (gains) such as

Oxidative Stress and Antioxidant Protection: The Science of Free Radical Biology and Disease, First Edition.
Edited by Donald Armstrong and Robert D. Stratton.
© 2016 John Wiley & Sons, Inc. Published 2016 by John Wiley & Sons, Inc.

accuracy of diagnosis of medical conditions, faster healing, and greater patient satis-
faction, while minimizing factors (losses) such as efforts, durations of screening for
disease, more side effects, and costs of the experiments.

There are many constraints and formalisms to deal with while constructing statistical tests. An essential part of the test-constructing process is that statistical hypotheses should be clearly formulated with respect to the objectives of clinical studies.

Statistical hypotheses

Commonly, statistical hypotheses and the corresponding clinical hypotheses are associated but stated in different forms and orders. In most clinical experiments, we are
interested in tests regarding characteristics or distributions of one or more populations. In such cases, the statistical hypotheses must be very carefully formulated,
and formally and clearly stated, displaying, for example, the nature of associations
between characteristics or distributions of populations. For example, suppose that
the clinical hypothesis is that the population mean time to heal with an antibiotic is
different from the mean time to heal without the antibiotic. In this case, the statistical hypothesis to be tested should be that the population mean time to heal with an
antibiotic is equivalent to the mean time to heal without the antibiotic. Here, we will
test for the equivalence of parameters of populations. Note that one can ask to test
for distribution difference with/without the antibiotic.

The term *Null Hypothesis*, symbolized H_0, is commonly used to show our primary
statistical hypothesis. For example, when the clinical hypothesis is that a biomarker of
oxidative stress has different circulating levels with respect to patients with and without atherosclerosis, a null hypothesis can be proposed corresponding to the assumption that levels of the biomarker in individuals with and without atherosclerosis
are distributed equally. Note that the clinical hypothesis points out that we want to
indicate the discriminating power of the biomarker, whereas H_0 says there are not significant associations between the disease and biomarker's levels. The reason lies in the
ability to formulate H_0 clearly and unambiguously, as well as quantify and calculate
expected errors in decision-making procedures. If the null hypothesis were formed
in a similar manner to the clinical hypothesis, we probably could not unambiguously
determine which links between the disease and biomarker's levels we should test.

Common errors related to the statistical testing mechanisms

The null hypothesis is usually a statement to be tested. Commonly, the statistical testing procedure results in a decision to reject or not reject the null hypothesis. In the
context of testing statistical hypotheses, in order to provide a formal test procedure,
as well as compare mathematical strategies for making decisions (e.g., with respect
to statistical powers of tests), algorithms for monitoring test characteristics associated with the probability to reject a correct hypothesis should be considered. Here,
we define the statistical power of a test as the probability that H_0 is correctly rejected
when H_0 is false. In general, while developing and applying test procedures, the practical statistician faces the task of controlling the probability of the event that a test's
outcome requests to reject H_0 when in fact H_0 is correct, a Type I error. For example,
assume that L is the test statistic based on the observed data, C is a threshold, and the

decision rule is to reject H_0 for large values of L, that is, when $L > C$; then, the threshold should be defined such that $\Pr(L > C|H_0) = \alpha$, where α is a presumed significance level, i.e., the probability of committing a Type I error. Note that when we compare two statistical tests, we mean to compare powers of the tests, given that the rate of Type I error is fixed.

It is clear that in order to construct statistical tests, we must review the corresponding clinical study, formalizing objectives of the experiments and making assumptions in hypothesis testing. A violation of the assumptions can result in incorrect conclusions based on outputs of the test, as well as a vital malfunction of the Type I error control system. Moreover, should the user verify that the assumptions are satisfied, errors in the verifications can affect the Type I error control.

The practitioner may also be interested to consider another related type of error in statistical testing procedures. If H_0 is false but fails to be rejected, the incorrect decision of not rejecting H_0 is called a Type II error. The Type II error rate can be defined as $\Pr(L < C|H_1)$, when we assume that L is the test statistic based on the observed data, C is a threshold, and the decision rule is to reject H_0 when $L > C$. Type II errors may occur when the effect size, biases in testing procedures, and random variability combine to lead to results insufficiently inconsistent with H_0 to reject it.[1,2] Essentially, it is the dichotomization of the study results into the categories "significant" or "not significant" that leads to Type I and Type II errors. Although errors resulting from an incorrect classification of the study outputs would seem to be unnecessary and avoidable, the Neyman–Pearson (dichotomous) hypothesis testing, when the Type I error is under control, is ingrained in scientific research due to the apparent objectivity and definitiveness of the pronouncement of significance.[1,2]

p-Values

The traditional testing procedure assumes to define a test threshold and reject or not reject H_0 based on comparisons between values of test statistics and the threshold. An alternative approach to hypothesis testing is to obtain the *p*-value.

As a continuous data based measure of the compatibility between a hypothesis and data, a *p*-value is defined as the probability of obtaining a test statistic (a corresponding quantity computed from the data, such as, e. g., a *t*-statistic) at least as extreme or close to the one that was actually observed, assuming that H_0 is true.[3] *p*-Values can be divided into two major types: one-sided (upper and lower) and two-sided. Assuming there are no biases in the data collection or the data analysis procedure, an upper one-sided *p*-value is the conditional on the data probability under the test hypothesis that the test statistic will be no less than the observed value. Similarly, a lower one-sided *p*-value is the probability under the test hypothesis that the test statistic will be no greater than the observed value. The two-sided *p*-value is defined as twice the smaller of the upper and lower *p*-values.[2,4]

If the *p*-value is small, it can be interpreted that the sample produced a very rare result under H_0, that is, the sample result is inconsistent with the null hypothesis statement. On the other hand, a large *p*-value indicates the consistency of the sample result with the null hypothesis. At the pre-specified α significance level, the decision is to reject H_0 when the *p*-value is less than or equal to α; otherwise, the decision is to not reject H_0. Therefore, the *p*-value is the smallest level of significance at which

H_0 would be rejected. In addition to providing a decision-making mechanism, the p-value also sheds some data based light on the strength of the evidence against H_0.[5]

Misinterpretations of p-values are common in clinical trials and epidemiology. In one of the most common misinterpretations, p-values are erroneously defined as the probabilities of test hypotheses. In many situations, the probability of the test hypothesis can be computed, but it will almost always be far from the two-sided p-value.[3] Note that the p-values can be viewed as a random variable, uniformly distributed between 0 and 1 if the null hypothesis is true. For example, suppose that the test statistic L has a cumulative distribution function (CDF) F under H_0. Then, the p-value is the random variable $1 - F(L)$, which is uniformly distributed under H_0.[6]

Sorts of information applicable to construct test procedures

The interests of clinical investigators usually lead to the problem of mathematically expressing procedures, using statistical decision rules based on sample data to test statistical hypotheses. In this case, when the users construct the decision rules, two additional information resources can be incorporated. The first is a defined function that consists of the explicit, quantified gains and losses in reaching a conclusion and their relative weights. Frequently, this function determines the loss that can be expected corresponding to each possible decision. This type of information can incorporate a loss function into the statistical decision-making process.

The second source reflects prior information. Commonly, in order to derive prior information, researchers should consider past experiences in similar situations. The Bayesian methodology formally provides clear technique manuals on how to construct efficient statistical decision rules for various complex problems related to clinical experiments, employing prior information.[7,8]

Parametric approach

A clinical statistician may use a sort of technical statements related to the observed data, while constructing the corresponding decision rules. The above-mentioned types of information used for test constructing can induce the technical statements, which oftentimes are called assumptions regarding the distribution of data. The assumptions often define a fit of the data distribution to a functional form that is completely known, or known up to parameters, since a complete knowledge of the distribution of data can provide all the information investigators need for efficient applications of statistical techniques. However, in many scenarios, the assumptions are presumed and very difficult to prove, or to test for being proper. The simple, but widely used, assumptions in biostatistics are that data derived via a clinical study follow one of the commonly used distribution functions: the normal, lognormal, t, χ^2, gamma, F, binominal, uniform, Wishart, and Poisson. The data distribution function can be defined up to parameters.[9] For example, the normal distribution $N(\mu, \sigma^2)$ is the famous bell curve, where the parameters μ and σ^2 represent the mean and variance of the population from which the data were sampled. The values of the parameters μ and σ^2 may be assumed to be unknown. Mostly, in such cases, assumed functional forms of the data distributions are involved in making statistical decision rules via the use of statistics, which we name Parametric Statistics. If certain key assumptions are met, parametric methods can yield very simple, efficient, and powerful inferences.

Nonparametric approach

The statistical literature has widely addressed the issue that parametric methods are often very sensitive to moderate violations of parametric assumptions, and hence nonrobust.[10] The parametric assumptions can be tested in order to reduce the risk of applying a misleading parametric approach. Note that in order to test for parametric assumptions, a goodness-of-fit test, outlined in a later section of this chapter, can be applied. In this case, statisticians can try to verify the assumptions, while making decisions with respect to main objectives of the clinical study. This leads to very complicated topics, dealt with in multiple testing. For example, it turns out that a computation of the expected risk of making a wrong decision strongly depends on the errors that can be made by not rejecting the parametric assumptions. The complexity of this problem can increase when researchers examine various functional forms to fit the data distribution in order to apply parametric methods. A substantial body of theoretical and experimental literature has discussed the pitfalls of multiple testing, placing blame squarely on the shoulders of the many clinical investigators who examine their data before deciding how to analyze it, or neglect to report the statistical tests that may not have supported their objectives.[11] In this context, one can present different examples, both hypothetical and actual, to get to the heart of issues that especially arise in the health-related sciences. Note, also, that in many situations, due to the wide variety and complex nature of problematic real data (e.g., incomplete data subject to instrumental limitations of studies), statistical parametric assumptions are hardly satisfied, and their relevant formal tests are complicated or not readily available.[12]

Unfortunately, even clinical investigators trained in statistical methods do not always verify the corresponding parametric assumptions, nor attend to probabilistic errors of the corresponding verification, when they use well-known elementary parametric statistical methods, for example, the t-tests.

Thus, it is known that when the key assumptions are not met, the parametric approach may be extremely biased and inefficient when compared to its robust nonparametric counterparts. Statistical inference under the nonparametric regime offers decision-making procedures, avoiding or minimizing the use of the assumptions regarding functional forms of the data distributions.

In general, the balance between parametric and nonparametric approaches can boil down to expected efficiency versus robustness to assumptions. One very important issue is preserving the efficiency of statistical techniques through the use of robust nonparametric likelihood methods, minimizing required assumptions about data distributions.[5,10,13]

Remarks

A wealth of additional applied and theoretical materials related to statistical decision-making procedures may be found in a variety of scientific publications.[1,2,4,5,9,10,13–15]

This chapter is organized as follows: In Section *R: statistical software*, the statistical software R is outlined at a beginning level. The likelihood methodology is described in Section *Likelihood*. In Section *Tests on means of continuous data*, we show different tests for means of continuous data. Section *The exact likelihood ratio test for equality of two normal populations* reviews the exact likelihood ratio test for equality of two normal populations. Section *Empirical likelihood* introduces the empirical likelihood methodology. In Section *Receiver operating characteristic curve analysis*, we introduce common methods based on the receiver operating characteristic (ROC) curves used in biomedical and epidemiological researches to make statistical decisions.[16,17] Goodness-of-fit tests are reviewed in Section *Goodness-of-fit tests*. In Section *Wilcoxon rank-sum tests*, we review the Wilcoxon two-sample test, a nonparametric analogy to the two-sample *t*-test. Different tests for independence are introduced in Section *Tests for independence*. Section *Numerical methods for calculating critical values and powers of statistical tests* presents the method for Type I error control using Monte Carlo techniques. In Section *Concluding remarks*, we conclude this chapter with remarks.

R: statistical software

In this section we outline the use of R, a powerful and flexible statistical software language.[18–20] Examples of statistical techniques implemented using R codes are employed in this chapter material.

R is a free case-sensitive, command line-driven software for statistical computing and graphics. Once the R program is installed via *www.r-project.org* and starts up, the main input window and a short introductory message (which appears a little differently on each operating system) are presented.[19] For example, Figure 31.1 shows the main input window in the operating system Windows with a few menus available at the top. Below the header a blank line is presented, with a screen prompt symbol > in the left-hand margin, showing the place where commands should be typed.

For example, we consider a simple way to input data using the `c()` function, which creates a vector, a variable with one or more values of the same type. We input a data containing 3, 5, 10, and 7, as shown below.

```
> x<-c(3,5,10,7)
```

To see the value of *x*, we type in the name of the vector *x* after the prompt symbol and press the Return key.

```
> x
```

As a result, R provides the following output:

```
[1]   3   5 10   7
```

R can perform simple statistical calculations as well as very complex computations. Table 31.1 shows some simple commands that produce descriptive statistics of a vector *x* created based on a sample of measurements $X_1, \ldots, X_n$.

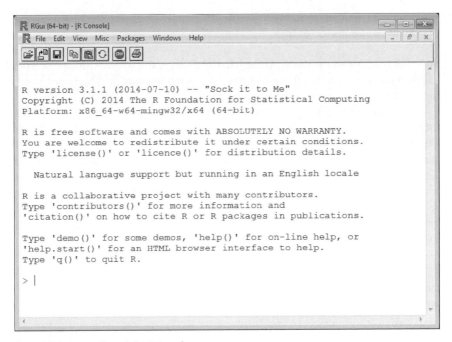

Figure 31.1 Screenshot of the R interface.

Table 31.1 The R commands that produce introductory descriptive statistics based on a numerical vector x.

Command	Explanations
mean(x, na.rm = FALSE)	Calculates the arithmetic mean of x.
sd(x, na.rm = FALSE)	Calculates the sample estimator of the standard deviation of x.
var(x, na.rm = FALSE)	Calculates the sample estimator of the variance of x.
sum(x, na.rm = FALSE)	Calculates the sum of the elements of x.

Instead of using the built-in functions such as those shown in Table 31.1, if a custom function need to be created to carry out some specific tasks, the function() command can be used. The following example shows a simple function mymean that determines the running mean of the first i, $i = 1, \ldots, n$ elements of a vector x, where n is the number of elements in x. Results are shown for x, specified above by applying the customized function.

```
> mymean <- function(x) {
+    tmp <- c()
+    for(i in 1:length(x)) tmp[i] <- mean(x[1:i])
+    return(tmp)
+ }
```

```
> mymean(x)
[1] 3.00 4.00 6.00 6.25
```

Note that the symbol + is shown at the left-hand side of the screen instead of > when it is working, meaning that the last command typed is incomplete. The function() can also be used to create complicated functions.

There are many packages in R that can be downloaded and installed from CRAN-like repositories or local files using the command install.packages ("packagename"), where packagename is the name of the package to be installed and must be in quotes; single or double quotes are both fine as long as they are not mixed. Once the package is installed, it can be loaded by issuing the command library(packagename), and commands available in the package can be accessed. Through an extensive help system built into R, a help entry for a specified command can be brought up via the help(commandname) command. As a simple example, we introduce the command EL.means in the EL library.

```
> install.packages("EL")
Installing pack-
age into 'C:/Users/xiwei/Documents/R/win-library/3.1'
(as 'lib' is unspecified)
trying URL 'http://cran.rstudio.com/bin/windows/contrib/3.1/
EL_1.0.zip'
Content type 'application/zip' length 53774 bytes (52 Kb)
opened URL
downloaded 52 Kb

package 'EL' successfully unpacked and MD5 sums checked

The downloaded binary packages are in
 C:\Users\xiwei\AppData\Local\Temp\Rtmp4uRCPS\downloaded
_packages
> library(EL)
> help(EL.means)
```

The use of the function EL.means provides possibilities to implement the empirical likelihood tests that are introduced in detail in Section *Empirical likelihood*.

As another concrete example, we show the mvrnorm command in the MASS library.

```
> install.packages("MASS")
Installing package into 'C:/Users/xiwei/Documents/R/win-
library/3.1'
(as 'lib' is unspecified)
trying URL 'http://cran.rstudio.com/bin/windows/contrib/3.1/
MASS_7.3-34.zip'
Content type 'application/zip' length 1083003 bytes (1.0 Mb)
opened URL
```

```
downloaded 1.0 Mb

package 'MASS' successfully unpacked and MD5 sums checked

The downloaded binary packages are in
 C:\Users\xiwei\AppData\Local\Temp\Rtmp4uRCPS\downloaded
_packages
> library(MASS)
> help(mvrnorm)
```

The mvrnorm command is very useful for simulating the data from a multivariate normal distribution. To illustrate, we simulate bivariate normal data with mean $(0, 0)^T$ and an identity covariance matrix with a sample size of 5.

```
> n <- 5  # define the sample size
> mu <- c(0,0)  # define the mean vector
> Sigma <- matrix(c(1,0,0,1), byrow=TRUE, ncol=2) # define
# covariance matrix
> set.seed(123)  # define the seed to fix the sample
> X <- mvrnorm (n, mu=mu, Sigma=Sigma) # generate data
> X
            [,1]           [,2]
[1,] -1.7150650 -0.56047565
[2,] -0.4609162 -0.23017749
[3,]  1.2650612  1.55870831
[4,]  0.6868529  0.07050839
[5,]  0.4456620  0.12928774
```

Likelihood

One of the traditional instruments used in medical experiments and drug development is the testing of statistical hypotheses based on the *t*-test or its different modifications. Despite the fact that these tests are straightforward with respect to their applications in clinical trials, it should be noted that there has been a huge literature on the criticism of *t*-test-type statistical tools. One major issue that has been widely recognized is the significant loss of efficiency of these procedures under different distributional assumptions. The legitimacy of *t*-test-type procedures also comes into question in the context of inflated Type I errors seen when data distributions differ from normal and the number of observations is fixed. This can pose serious problems when data based on biomarker measurements are available for statistical testing. The recent biostatistical literature has well addressed the arguments that show the values of biomarker measurements that tend to follow skewed distributions, for example, a lognormal distribution.[21] Hence, the use of *t*-test-type techniques in this setting is suboptimal and is accompanied by significant difficulties in controlling the corresponding Type I error.

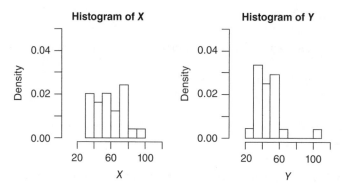

Figure 31.2 R data analysis output for measurements of HDL cholesterol levels (mg/dl) in healthy individuals.

Consider the following example based on the data from a study evaluating biomarkers related to atherosclerotic coronary heart disease:[22]

A cross-sectional population-based sample of randomly selected residents (aged 35–79) of Erie and Niagara counties of the state of New York, USA, was the focus of this experiment. The New York State Department of Motor Vehicles drivers' license rolls were employed as the sampling frame for adults between the ages of 35 and 65, whereas the elderly sample (aged 65–79) was randomly selected from the Health Care Financing Administration database. Participants provided a 12-h fasting blood specimen for biochemical analysis at baseline, and a number of characteristics were evaluated from fresh blood samples. The samples X and Y presented 50 measurements (mg/dl) of the biomarker "high-density lipoprotein (HDL) cholesterol" obtained from healthy patients. These measurements were divided into the two groups: X and Y. The following R code shows the input of the data and the construction of histograms of the data, as seen in Figure 31.2

```
X<-c(37.4,70.4,52.8,46.2,74.8,96.8,41.8,55.0,83.6,63.8,63.8,
52.8,46.2,37.4,50.6,74.8,46.2,39.6,70.4,30.8,74.8,61.6,30.8,
74.8,52.8)
Y<-c(44.0,35.2,110.0,63.8,44,26.4,52.8,30.8,39.6,44,48.4,39.6,
55,52.8,50.6,39.6,35.2,55,57.2,37.4,30.8,46.2,50.6,44,44)
> a<-min(c(X,Y))-20
> b<-max(c(X,Y))+20
> par(pty = "s",mfrow = c(1,2))
> hist(X,xlim = c(a,b),ylim = c(0,0.05),freq = FALSE)
> hist(Y,xlim = c(a,b),ylim = c(0,0.05),freq = FALSE)
```

Although one can reasonably expect the samples are from the same population, the t-test result shows a significant difference of their distributions, as demonstrated below via the use of the function t.test in R.

```
> t.test(X,Y)

  Welch Two Sample t-test

data:  X and Y
t = 2.1526, df = 47.704, p-value = 0.03644
alternative hypothesis: true difference in means is not equal
to 0
95 percent confidence interval:
  0.6657898 19.5742102
sample estimates:
mean of x mean of y
    57.20     47.08
```

Perhaps, in order to investigate reasons for this incorrect output of the t-test, the following issues may be taken into account:

The histograms displayed in Figure 31.2 indicate that the distributions of the X and Y are probably skewed. In a nonasymptotic context, when the sample sizes are relatively small, one can show that the t-test statistic is a product of likelihood ratio-type considerations, based on normally distributed observations.[9,14] That is, the t-test is a parametric test, and the parametric assumption seems to be violated in this example.

Thus, in many settings, it may be reasonable to propose an approach for developing statistical tests, attending to data distributions, in order to provide procedures that are as efficient as the t-test based on normally distributed observations. Toward this end, the likelihood methodology can be employed.

Likelihood ratio and its optimality

Now we turn to outlining the likelihood principle. When the forms of data distributions are assumed to be known, the likelihood principle is a central tenet for developing powerful statistical inference tools for use in clinical experiments. The *likelihood method*, or simply the *likelihood*, is arguably the most important concept for inference in parametric modeling when the underlying data are subject to different problems and limitations related to medical and epidemiological studies, for example, in the context of the analysis of survival data. Likelihood-based testing, as we know, was mainly founded and formulated in a series of fundamental papers published in the period of 1928–1938 by Jerzy Neyman and Egon Pearson.[23–26] In 1928, the authors introduced the generalized likelihood ratio test and its association with chi-squared statistics. Five years later, the Neyman–Pearson Lemma was introduced, showing the optimality of the likelihood ratio test.[24] These seminal works provided us with the familiar notions of simple and composite hypotheses and errors of the first and second kinds, thus defining formal decision-making rules for testing. Without loss of generality, the principle idea of the proof of the Neyman–Pearson Lemma can be shown by using the trivial inequality

$$(A - B)(I\{A \geq B\} - \delta) \geq 0, \tag{31.1}$$

for all A, B, where $\delta \in [0, 1]$ and $I\{\cdot\}$ denotes the indicator function. For example, suppose we would like to classify independent identically distributed (i.i.d.) biomarker measurements $\{X_i, i = 1, \ldots, n\}$ corresponding to hypotheses of the following form: H_0: X_1 is from a density function f_0, versus H_1: X_1 is from a density function f_1. In this context, to construct the likelihood ratio test statistic, we should consider the ratio between the joint density function of $\{X_1, \ldots, X_n\}$ obtained under H_1 and the joint density function of $\{X_1, \ldots, X_n\}$ obtained under H_0, and then define $\prod_{i=1}^{n} f_1(X_i) / \prod_{i=1}^{n} f_0(X_i)$ as the likelihood ratio. In this case, the likelihood ratio test is uniformly most powerful. This proposition directly follows from the expected value under H_0 of the inequality (31.1), where we define $A = \prod_{i=1}^{n} f_1(X_i) / f_0(X_i)$, B to be a test threshold (i.e., the likelihood ratio test rejects H_0 if and only if $L \geq B$), and δ is assumed to represent any decision rule based on $\{X_i, i = 1, \ldots, n\}$. The Appendix contains details of the proof. This simple proof technique was used to show optimal aspects of different statistical decision-making policies based on the likelihood ratio concept applied in clinical experiments.[27,28]

The likelihood ratio based on the likelihood ratio test statistic is the likelihood ratio test statistic

The Neyman–Pearson test concept, fixing the probability of a Type I error, comes under some criticism by epidemiologists. One of the critical points is about the importance of paying attention to Type II errors. For example, Freiman *et al.* pointed out results of 71 clinical trials that reported no "significant" differences between the compared treatments.[2] The authors found that in the great majority of these trials, the strong effects of new treatment were reasonable. The investigators in such trials inappropriately accepted the null hypothesis as correct, which probably resulted in Type II errors. In the context of likelihood ratio-based tests, we present the following result that demonstrates an association between the probabilities of Type I and II errors.

Suppose we would like to test for H_0 versus H_1, employing the likelihood ratio $L = f_{H_1}(D) / f_{H_0}(D)$ based on data D, where f_H defines a density function that corresponds to the data distribution under the hypothesis H. Say, for simplicity, we reject H_0 if $L > C$, where C is a presumed threshold. In this case, we can then show that

$$f_{H_1}^{L}(u) = u f_{H_0}^{L}(u), \tag{31.2}$$

where $f_H^L(u)$ is the density function of the test statistic L under the hypothesis H and $u > 0$. Details of the proof of this fact are shown in the Appendix. Thus, we can obtain the probability of a Type II error in the form of

$$\Pr\{\text{the test does not reject } H_0 | H_1 \text{ is true}\} = \Pr\{L \leq C | H_1 \text{ is true}\} = \int_0^C f_{H_1}^L(u)du$$

$$= \int_0^C u f_{H_0}^L(u)du.$$

Now, if, in order to control the Type I error, the density function $f_{H_0}^L(u)$ is assumed to be known, then the probability of the Type II error can be easily computed.

The likelihood ratio property $f_{H_1}^L(u)/f_{H_0}^L(u) = u$ can be applied to solve different issues related to performing the likelihood ratio test. For example, in terms of the bias of the test, one can request to find a value of the threshold C that maximizes

$$\Pr\{\text{the test rejects } H_0 | H_1 \text{ is true}\} - \Pr\{\text{the test rejects } H_0 | H_0 \text{ is true}\},$$

where the probability $\Pr\{\text{the test rejects } H_0 | H_1 \text{ is true}\}$ depicts the power of the test. This equation can be expressed as

$$\Pr\{L > C | H_1 \text{ is true}\} - \Pr\{L > C | H_0 \text{ is true}\} = \left(1 - \int_0^C f_{H_1}^L(u)\, du\right)$$
$$- \left(1 - \int_0^C f_{H_0}^L(u)\, du\right).$$

Let the derivative of this notation equal zero and solve the equation:

$$\frac{d}{dC}\left[\left(1 - \int_0^C f_{H_1}^L(u)\, du\right) - \left(1 - \int_0^C f_{H_0}^L(u)\, du\right)\right] = -f_{H_1}^L(C) + f_{H_0}^L(C) = 0.$$

By virtue of the property (31.2), this implies $-Cf_{H_0}^L(C) + f_{H_0}^L(C) = 0$ and then $C = 1$, which provides the maximum discrimination between the power and the probability of the Type I error in the likelihood ratio test.

In other words, the interesting fact is that the likelihood ratio $f_{H_1}^L / f_{H_0}^L$ based on the likelihood ratio $L = f_{H_1}/f_{H_0}$ comes to be the likelihood ratio, that is, $f_{H_1}^L(L)/f_{H_0}^L(L) = L$. Interpretations of this statement in terms of information, we leave to the reader's imagination.

Exercise 31.1

Given a sample of i.i.d. measurements $X_1, \dots, X_n$ from exponential distribution with the rate parameter λ, that is, $X_1, \dots, X_n \sim f(x) = \lambda \exp(-\lambda x)$, derive the likelihood ratio test statistics for the simple hypothesis $H_0 : \lambda = 1$ versus $H_1 : \lambda = 2$.

Maximum likelihood; is it the likelihood?

Various real-world data problems require considerations of statistical hypotheses with structures that depend on unknown parameters. In this case, the maximum likelihood method proposes to approximate the most powerful likelihood ratio, employing a proportion of the maximum likelihoods, where the maximizations are over values of the unknown parameters belonging to distributions of observations under the corresponding hypotheses. We shall assume the existence of essential maximum likelihood estimators. The influential Wilks' theorem provides the basic rationale as to why the maximum likelihood ratio approach has had tremendous success in statistical applications.[30] Wilks showed that under regularity conditions, asymptotic null distributions of maximum likelihood ratio test statistics are independent of nuisance parameters. That is, a Type I error in the maximum likelihood ratio tests can be controlled asymptotically, and approximations of the corresponding p-values can be computed.

Thus, if certain key assumptions are met, one can show that parametric likelihood methods are very powerful and efficient statistical tools. We should emphasize that the role of the discovery of the likelihood ratio methodology in statistical developments can be compared to the development of the assembly line technique of mass production. The likelihood ratio principle gives clear instructions and technique manuals on how to construct efficient statistical decision rules in various complex problems related to clinical experiments. For example, Vexler *et al.* developed a likelihood ratio test for comparing populations based on incomplete longitudinal data subject to instrumental limitations.[31]

Although many statistical publications continue to contribute to the likelihood paradigm and are very important in the statistical discipline (an excellent account can be found in Lehmann and Romano[14]), several significant questions naturally arise about the maximum likelihood approach's general applicability. Conceptually, there is an issue specific to classifying maximum likelihoods in terms of likelihoods that are given by joint density (or probability) functions based on data. Integrated likelihood functions, with respect to their arguments related to data points, are equal to 1, whereas accordingly integrated maximum likelihood functions often have values that are indefinite. Thus, while likelihoods present full information regarding the data, the maximum likelihoods might lose information conditional on the observed data. Consider this simple example:

Suppose we observe X_1, which is assumed to be from a normal distribution $N(\mu, 1)$ with mean parameter μ. In this case, the likelihood has the form $(2\pi)^{-0.5} \exp(-(X_1 - \mu)^2/2)$ and, correspondingly, $\int (2\pi)^{-0.5} \exp(-(X_1 - \mu)^2/2) dX_1 = 1$, whereas the maximum likelihood, that is, the likelihood evaluated at the estimated μ, $\hat{\mu} = X_1$ is $(2\pi)^{-0.5}$, which clearly does not represent the data and is not a proper density. This demonstrates that since the Neyman–Pearson lemma is fundamentally founded on the use of the density-based constitutions of likelihood ratios, maximum likelihood ratios cannot be optimal in general. That is, the likelihood ratio principle is generally not robust when the hypothesis tests have corresponding nuisance parameters to consider, for example, testing a hypothesized mean given an unknown variance.

An additional inherent difficulty of the likelihood ratio test occurs when a clinical experiment is associated with an infinite-dimensional problem and the number of unknown parameters is relatively large. In this case, Wilks' theorem should be re-evaluated, and nonparametric approaches should be considered in the contexts of reasonable alternatives to the parametric likelihood methodology.[32]

The ideas of likelihood and maximum likelihood ratio testing may not be fiducial and applicable in general nonparametric function estimation/testing settings. It is also well known that when key assumptions are not met, parametric approaches may be suboptimal or biased as compared to their robust counterparts across the many features of statistical inferences. For example, in a biomedical application, Gosh proved that the maximum likelihood estimators for the Rasch model are inconsistent, as the number of nuisance parameters increases to infinity (Rasch models are often employed in clinical trials that deal with psychological measurements, e.g., abilities, attitudes, and personality traits).[33] Due to the structure of likelihood functions based on products of densities, or conditional density functions, relatively insignificant errors in classifications of data distributions can lead to vital problems related to the applications of likelihood ratio type tests.[34] Moreover, one can note that, given the wide variety and complex nature of biomedical data (e.g., incomplete data subject to instrumental limitations or complex correlation structures), parametric assumptions are rarely satisfied. The respective formal tests are complicated, or oftentimes not readily available.

Exercise 31.2

Given a sample of i.i.d. measurements $X_1, \dots, X_n$, following an exponential distribution with the rate parameter λ, that is, $X_1, \dots, X_n \sim f(x) = \lambda \exp(-\lambda x)$, derive the maximum likelihood ratio test statistic for the composite hypothesis $H_0 : \lambda = 1$ versus $H_1 : \lambda \neq 1$.

Tests on means of continuous data

Does a sample mean equal a pre-specified population mean, or, alternatively, do two or more samples have the same population mean? These questions can be answered by the hypothesis testing of equal means.

Likelihood ratio test for the mean of normally distributed data

Given a random sample of i.i.d. observations $X_1, \dots, X_n$ from a normal population with the mean μ and the variance σ^2, we would like to test for the simple hypothesis

$$H_0 : \mu = \mu_0 \text{ versus } H_1 : \mu = \mu_1.$$

In this case, the likelihood function is

$$L = (\sigma \sqrt{2\pi})^{-n} \exp \left\{ \sum_{i=1}^{n} (X_i - \mu)^2 / (2\sigma^2) \right\}.$$

Therefore, the likelihood ratio has the form

$$\Lambda = \frac{(\sigma \sqrt{2\pi})^{-n} \exp \left\{ -\sum_{i=1}^{n} (X_i - \mu_0)^2 / (2\sigma^2) \right\}}{(\sigma \sqrt{2\pi})^{-n} \exp \left\{ -\sum_{i=1}^{n} (X_i - \mu_1)^2 / (2\sigma^2) \right\}}$$

$$= \exp \left\{ \left(2 (\mu_0 - \mu_1) \sum_{i=1}^{n} X_i - n(\mu_0^2 - \mu_1^2) \right) / (2\sigma^2) \right\}.$$

We reject H_0 if $\Lambda > C_\alpha$, where the constant C_α is selected for a specified value for the significance level α, the Type I error rate. To use this most powerful likelihood ratio test, we must know the values of σ^2, μ_0, and μ_1.

t-Type tests

As an example of the likelihood methodology, we present t-test-type decision rules, which are widely used in practice. Assuming the following constraints: (1) the data is a simple random sample from the population and each observation is independent of each other, and (2) the sample observations were drawn from a normal distribution, t-tests can be conducted to test means of continuous data. The one-sample t-test can be applied to test for the equality of the sample mean to a presumed value when the population variance is unknown or the sample size is small (no greater than 30 as a rule of thumb). To determine if two independent sets of data are significantly different from each other, the two-sample t-test can be applied. In the case of multivariate hypothesis testing, the multivariate t (Hotelling's t-squared) test, as a generalization of the Student's t-statistic, can be used.

One-sample t-tests

In testing the null hypothesis that the population mean is equal to a specified value μ_0, i.e., $H_0 : \mu = \mu_0$ versus $H_1 : \mu \neq \mu_0$, based on the observed data $X_1, \dots, X_n$, one uses the statistic

$$t = \frac{\overline{X} - \mu_0}{s/\sqrt{n}},$$

where n is the sample size, $\overline{X} = n^{-1} \sum_{i=1}^{n} X_i$ is the sample mean, and $s^2 = (n-1)^{-1} \sum_{i=1}^{n} (X_i - \overline{X})^2$ is the sample standard deviation. At the α significance level, the null hypothesis can be rejected if $|t| \geq t_{\alpha/2,n-1}$, where $t_{\alpha/2,n-1}$ is the $(1 - \alpha/2)$th quantile of t distribution with $n - 1$ degrees of freedom. Here, we define the pth quantile for a random variable as the value x, such that the probability that the random variable will be less than x is at most p and the probability that the random variable will be more than x is at least $1 - p$. Note that the population does not need to be normally distributed for a large sample size (>30, as a rule of thumb). By the central limit theorem, the distribution of the population of sample means, $\overline{X}$, will be approximately normal for a sufficiently large sample size.[35]

Two-sample t-tests

The two-sample t-test is used to determine if two independent population means are equal, that is, $H_0 : \mu_1 = \mu_2$ versus $H_1 : \mu_1 \neq \mu_2$. Given two samples of i.i.d. observations $X_{i1}, \dots, X_{in_i}$, $i = 1, 2$, we denote the sample size, the sample mean, and the unbiased estimator of the variance of the two samples as n_i, $\overline{X}_i = n_i^{-1} \sum_{j=1}^{n_i} X_{ij}$, and $s_i^2 = (n_i - 1)^{-1} \sum_{j=1}^{n_i} (X_{ij} - \overline{X}_i)^2$, $i = 1, 2$, respectively. The t-statistic to test whether the means are equal can be calculated as follows:

$$t = \frac{\overline{X}_1 - \overline{X}_2}{s_d}.$$

Based on the equivalence of the population variance in two groups, the equal variances case and the unequal variances case are considered separately, and the estimate of s_d can be calculated accordingly.

Equal variances: When the two distributions are assumed to have the same variance, the estimator is $s_d = s_p \sqrt{n_1^{-1} + n_2^{-1}}$, where the pooled standard deviation $s_p =$

$\sqrt{((n_1 - 1)s_1^2 + (n_2 - 1)s_2^2)/(n_1 + n_2 - 2)}$ is an estimator of the common standard deviation of the two samples. At the α significance level, the null hypothesis of equal means can be rejected if $|t| \geq t_{\alpha/2, n_1 + n_2 - 2}$, where $t_{\alpha/2, n_1 + n_2 - 2}$ is the $(1 - \alpha/2)$th quantile of t distribution with $n_1 + n_2 - 2$ degrees of freedom. Note that s_p^2 is an unbiased estimator of the common variance whether the population means are the same or not.

Unequal variances (Welch's t-test): When the two population variances are not assumed to be equal, the estimator is $s_d = \sqrt{s_1^2/n_1 + s_2^2/n_2}$. Note that in this case, s_d^2 is not a pooled variance. At the α significance level, the null hypothesis of equal means can be rejected if $|t| \geq t_{\alpha/2, df}$, where $t_{\alpha/2, df}$ is the $(1 - \alpha/2)$th quantile of t distribution with

$$df = \frac{(s_1^2/n_1 + s_2^2/n_2)^2}{(s_1^2/n_1)^2/(n_1 - 1) + (s_2^2/n_2)^2/(n_2 - 1)}$$

degrees of freedom.

Paired t-tests

In clinical trials, the generalized treatment effect can be used to compare treatments or interventions based on the difference in mean outcomes between pre- and post-treatment measurements. In the case of one paired sample, paired t-tests can be conducted to test for a paired difference. Given a paired sample $X_{k1}, \ldots, X_{kn}$ of pre-treatment $(k = 1)$ and post-treatment $(k = 2)$ measurements, to test whether the difference μ_D in means between post- and pre-treatment measurements is μ_0, the t statistic is

$$t = \frac{\overline{X}_D - \mu_0}{s_D/\sqrt{n}},$$

where n is the number of pairs, $X_{Di} = X_{2i} - X_{1i}$, $\overline{X}_D = n^{-1} \sum_{i=1}^{n} X_{Di}$, and $s_D^2 = (n - 1)^{-1} \sum_{i=1}^{n} (X_{Di} - \overline{X}_D)^2$ is the sample mean and sample variance of differences between all pairs, respectively. At the α significance level, the null hypothesis can be rejected if $|t| \geq t_{\alpha/2, n-1}$, where $t_{\alpha/2, n-1}$ is the $(1 - \alpha/2)$th quantile of t-distribution with $n - 1$ degrees of freedom.

We exemplify the use of the paired t-test with a real-life example of the effect of asthma education on pediatric patients' acute care visits.[15]

Example 31.1

The study sample consists of 32 patients who satisfy inclusion criteria and present over a period of time. The number of acute care visits during a year is recorded. After a standardized course of asthma training, the number of acute care visits for the following year is recorded again. The change per patient, that is, the before-and-after difference in the number of visits, was 1, 1, 2, 4, 0, 5, −3, 0, 4, 2, 8, 1, 1, 0, −1, 3, 6, 3, 1, 2, 0, −1, 0,3 ,2, 1, 3, −1, −1, 1, 1, and 5. It is of interest to test if the training affects the number of visits.[15]

The following R code can be used to carry out the two-tailed test $H_0 : \mu_D = 0$ against $H_1 : \mu_D \neq 0$:

```
> # input the data: difference (before-after)
> D <-c(1,1,2,4,0,5,-3,0,4,2,8,1,1,0,-1,3,6,3,1,2,0,-1,0,3,2,
  1,3,-1,-1,1,1,5)
> alpha <- 0.05 # pre-specified significance level
> # check the normality by the histogram
> hist(D,xlab="Difference",main="Histogram of before and after
  difference")
>
> # calculate the test statistic
> n <- length(D)  # the sample size, i.e., the number of pairs
> t.stat <- (mean(D)-0)/(sd(D)/sqrt(n))
> t.stat
[1] 4.034031
>
> # obtain the critical value and the p-value
> crit <- qt(1-alpha/2,df=n-1)
> crit
[1] 2.039513
> pval <- 2*(1-pt(t.stat,df=n-1))  # a two-sided test
> pval
[1] 0.0003323025
```

Alternatively, one may use the built-in function t.test setting paired=TRUE and alternative="two.sided" in R to conduct the two-sided paired *t*test. It yields the following output:

```
> t.test(D,rep(0,n),paired=TRUE,alternative="two.sided")

 Paired t-test

data:  D and rep(0, n)
t = 4.034, df = 31, p-value = 0.0003323
alternative hypothesis: true differ-
ence in means is not equal to 0
95 percent confidence interval:
 0.818888 2.493612
sample estimates:
mean of the differences
              1.65625
```

The form of the histogram of the differences shown in Figure 31.3 suggests an approximately normal shape, satisfying the normal distribution assumption. The test statistic is $t = 4.034$, which is greater than the critical value $t_{\alpha/2, df=31} = 2.04$ at the $\alpha = 0.05$ significance level. We can state that we are 95% sure that the asthma training was efficacious.

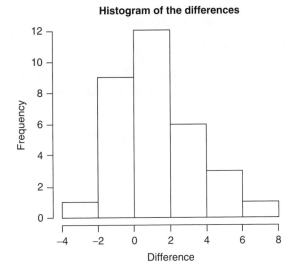

Figure 31.3 Histogram of the differences in the number of acute care visits pre- and post-asthma training.

Multivariate *t*-tests

Hypothesis testing of the equality of means can be constructed in the multivariate case, say, p-variate. Assuming that the samples are independently drawn from two multivariate normal distributions with the same covariance, i.e., for the ith sample, $\mathbf{X}_{ij} \sim N_p(\mu_i, \Sigma)$, $i = 1, 2$, $j = 1, \ldots, n_i$, the hypothesis is $H_0 : \mu_1 = \mu_2$ versus $H_a : \mu_1 \neq \mu_2$. Define $\overline{\mathbf{X}}_i = n_i^{-1} \sum_{j=1}^{n_i} \mathbf{X}_{ij}$, $i = 1, 2$, as the sample means and $\mathbf{W} = (n_1 + n_2 - 2)^{-1} \sum_{i=1}^{2} \sum_{j=1}^{n_i} (\mathbf{X}_{ij} - \overline{\mathbf{X}}_i)(\mathbf{X}_{ij} - \overline{\mathbf{X}}_i)^T$ as the unbiased pooled covariance matrix estimate, then the Hotelling's two-sample T^2 statistic is

$$t^2 = \frac{n_1 n_2}{n_1 + n_2} (\overline{\mathbf{X}}_1 - \overline{\mathbf{X}}_2)^T \mathbf{W}^{-1} (\overline{\mathbf{X}}_1 - \overline{\mathbf{X}}_2).$$

Note that the Hotelling's two-sample T^2 statistic follows the Hotelling's T^2 distribution with parameters p and $n_1 + n_2 - 2$; that is, $t^2 \sim T^2(p, n_1 + n_2 - 2)$. The null hypothesis is rejected if $t^2 > T^2(\alpha; p, n_1 + n_2 - 2)$ at the α significance level. (Here the matrix operation ^{T} means $\begin{bmatrix} a \\ b \end{bmatrix}^T = [a\ b]$ and $[a\ b]^T = \begin{bmatrix} a \\ b \end{bmatrix}$.)

In the following example, we will show how to simulate data in R using the function `mvrnorm` introduced in Section *R: statistical software* and conduct the Hotelling's two-sample T^2 test.

Example 31.2

We consider an example in which age and measurements of weight (in kg) are recorded for each patient in two independent groups. Assume that for the first group, $n_1 = 50$ patients' measurements are randomly sampled from the population $N_2\left(\begin{pmatrix} 25 \\ 65 \end{pmatrix}, \begin{pmatrix} 5 & 1 \\ 1 & 9 \end{pmatrix} \right)$, while for the second group, $n_2 = 70$ patients' measurements are randomly sampled from the population $N_2 \left(\begin{pmatrix} 25 \\ 70 \end{pmatrix}, \begin{pmatrix} 5 & 1 \\ 1 & 9 \end{pmatrix} \right)$. We simulate the data and test if the means are equal.

For each group, we assume bivariate normal data with a common population covariance matrix. The following R code can simulate bivariate normal data and conduct the Hotelling's two-sample T^2 test for the equality of means:

```
> # Check if packages are aleady installed.
> check.pkg <- c("ICSNP", "MASS") %in% rownames(installed
.packages())
> if(any(!check.pkg)) install.packages(c("ICSNP", "MASS")
[!check.pkg])
> # load packages
> library(ICSNP)
> library(MASS)
> # simulate data
> set.seed(123)
> n1 <- 50
> n2 <- 70
> Sigma <- matrix(c(5,1, 1,9), byrow=TRUE, ncol=2) # common
# covariance matrix
> X1    <- mvrnorm (n1, mu=c(25, 65), Sigma=Sigma)
> X2  <- mvrnorm (n2, mu=c(25, 70), Sigma=Sigma)
> X <- rbind(X1, X2)
> Group <- factor(rep(1:2, c(n1,n2)))
> HotellingsT2(X ~ Group)
```

It yields the following output:

```
Hotelling's two sample T2-test

data:  X by Group
T.2 = 35.3461, df1 = 2, df2 = 117, p-value = 9.823e-13
alternative hypothesis: true location difference is not equal
to c(0,0)
```

With the obtained p-value far less than 0.001, we reject the null hypothesis at the 0.05 significance level. We can conclude that we are 95% sure there is significant difference in the means of two bivariate data.

Exercise 31.3

Generate data from bivariate normal distributions based on $n_1 = 130$, $n_2 = 100$, $\mu_1 = (50\ 89)^T$, $\mu_2 = (45\ 92)^T$ and the common covariance matrix $\Sigma = \begin{pmatrix} 15 & -2 \\ -2 & 9 \end{pmatrix}$. Conduct the Hotelling's two-sample T^2 test at the $\alpha = 0.05$ significance level.

The exact likelihood ratio test for equality of two normal populations

Testing the equality of two independent normal populations is of practical importance. Let $\{X_1, \dots, X_n\}$ and $\{Y_1, \dots, Y_n\}$ be two independent random samples from normal populations $N(\mu_1, \sigma_1^2)$ and $N(\mu_2, \sigma_2^2)$, respectively. It is of interest to test

$$H_0 : \mu_1 = \mu_2 \text{ and } \sigma_1^2 = \sigma_2^2 \text{ versus } H_1 : \mu_1 \neq \mu_2 \text{ or } \sigma_1^2 \neq \sigma_2^2.$$

Pearson and Neyman[36] considered the likelihood ratio test

$$\lambda_{n,m} = \frac{\left[\sum_{i=1}^n \left(X_i - \overline{X}\right)^2 / n\right]^{n/2} \left[\sum_{j=1}^m \left(Y_j - \overline{Y}\right)^2 / m\right]^{m/2}}{\left\{\left[\sum_{i=1}^n \left(X_i - u\right)^2 + \sum_{j=1}^m \left(Y_j - u\right)^2\right] / (n+m)\right\}^{(n+m)/2}},$$

where $\overline{X}$, $\overline{Y}$, and u are the sample means of the X sample, the Y sample, and the combined sample, respectively. For $\lambda \in (0, 1)$, Zhang et al.[37] derived the exact distribution of $\lambda_{n,m}$ as

$$\Pr(\lambda_{n,m} \leq \lambda) = 1 - C \iint_D w_1^{(n-1)/2-1} w_2^{(m-1)/2-1} / \sqrt{1 - w_1 - w_2} \, dw_1 \, dw_2$$

$$= 1 - C \int_{r_1}^{r_2} w_1^{(n-3)/2} \int_{z/w_1^{n/m}}^{1-w_1} w_2^{(m-3)/2} / \sqrt{1 - w_1 - w_2} \, dw_2 \, dw_1,$$

where $z = \dfrac{\lambda^{2/m} n^{n/m} m}{(n+m)^{(n+m)/m}}$, $C = \dfrac{\Gamma((n+m-1)/2)}{\Gamma((n-1)/2)\Gamma((m-1)/2)\Gamma(1/2)}$,

$$D = \left\{(w_1, w_2) : w_1 > 0, w_2 > 0, w_1 + w_2 < 1, w_1^{n/2} w_2^{m/2} (n+m)^{(n+m)/2} / (n^{n/2} m^{m/2}) > \lambda\right\},$$

and $r_1 < r_2$ are the two roots (for the variable w_1) of

$$1 - w_1 - z/w_1^{n/m} = 0.$$

Note that the double integral can be computed using Gaussian quadrature, implemented with the R function `plrt`.

Empirical likelihood

One very important issue is to preserve efficiency of the statistical inference through the use of robust likelihood-type methods, while concurrently minimizing assumptions about the underlying distribution. Toward this end, the recent biostatistical literature has shifted focus toward robust and efficient nonparametric and semiparametric developments of various "artificial" or "approximate" likelihood techniques. These methods have a wide variety of applications related to clinical experiments. Many nonparametric and semiparametric approximations to powerful parametric likelihood procedures have been used routinely in both statistical theory and practice. Well-known examples include the quasi-likelihood method, approximations of

parametric likelihoods via orthogonal functions, techniques based on quadratic artificial likelihood functions, and the local maximum likelihood methodology.[38-41] Various studies have shown that artificial or approximate likelihood-based techniques
efficiently incorporate information expressed through the data, and have many of
the same asymptotic properties as those derived from the corresponding parametric
likelihoods. The empirical likelihood (EL) method is one of a growing array of artificial or approximate likelihood-based methods currently in use in statistical practice.[42]
Interest in and the resulting impact of EL methods continue to grow rapidly. Perhaps
more importantly, EL methods now have various vital applications in a large and
expanding number of areas of clinical studies.

A question of major interest to this section turns on the performance of EL constructs relative to ordinary parametric likelihood ratio-based procedures in the context of clinical experiments. Our desire to incorporate several recent developments
and applications in these areas in an easy-to-use manner provides one of the main
impetuses for this section. The EL method for testing has been dealt with extensively
in the literature within a variety of settings.[42-47]

Classical empirical likelihood

As background for the development of EL-type techniques, we first outline the
classical EL approach. The simple classical EL takes the form $\prod_{i=1}^{n}(F(X_i) - F(X_i-))$,
which is a functional of the cumulative distribution function F and i.i.d. observations X_i, $i = 1, \ldots, n$. This EL technique is *"distribution function-based."* [42] In the
distribution-free setting, an empirical estimator of this likelihood may take the
form of $L_p = \prod_{i=1}^{n} p_i$, where the components p_i, $i = 1, \ldots, n$, the estimators of the
probability weights, should maximize the likelihood L_p, provided that $\sum_{i=1}^{n} p_i = 1$
and empirical constraints based on $X_1, \ldots, X_n$ hold. For example, suppose we would
like to test the hypothesis

$$H_0 : E(g(X_1, \theta)) = 0 \text{ versus } H_1 : E(g(X_1, \theta)) \neq 0,$$

where $g(.,.)$ is a given function and θ is a parameter. Then, in a nonparametric man-
ner, we define the EL function of the form $EL(\theta) = L(X_1, \ldots, X_n|\theta) = \prod_{i=1}^{n} p_i$, where
$\sum_{i=1}^{n} p_i = 1$. Under the null hypothesis, the maximum likelihood approach requires
one to find the values of p_i $i = 1, \ldots, n$, that maximize the EL given the empirical
constraints $\sum_{i=1}^{n} p_i = 1$ and $\sum_{i=1}^{n} p_i g(X_i, \theta) = 0$ that present an empirical version of the
condition under H_0 that $E(g(X_1, \theta)) = 0$ (the null hypothesis is assumed to be rejected
when there are no $0 < p_1, \ldots, p_n < 1$ to satisfy the empirical constraints). In this case,
using Lagrange multipliers, one can show that

$$EL(\theta) = \sup_{0<p_1,p_2,\ldots,p_n<1, \sum p_i=1, \sum p_i g(X_i,\theta)=0} \prod_{i=1}^{n} p_i = \prod_{i=1}^{n} (n + \lambda g(X_i, \theta))^{-1}, \tag{31.3}$$

where λ is a root of $\sum g(X_i, \theta)(n + \lambda g(X_i, \theta))^{-1} = 0$. Since under H_1, the only constraint
under consideration is $\sum p_i = 1$, we have

$$EL = \sup_{0<p_1,p_2,\ldots,p_n<1, \sum p_i=1} \prod_{i=1}^{n} p_i = \prod_{i=1}^{n} n^{-1} = (n)^{-n}. \tag{31.4}$$

Combining equations (31.3) and (31.4), we obtain the EL ratio (ELR) test statistic $\text{ELR}(\theta) = \text{EL}/\text{EL}(\theta)$ for the hypothesis test of H_0 versus H_1. For example, when the function $g(u, \theta) = u - \theta$, the null hypothesis corresponds to the expectation.

Owen showed that the nonparametric test statistic $2 \log \text{ELR}(\theta)$ has an asymptotic chi-square distribution under the null hypothesis.[42] This result illustrates that Wilks' theorem-type results continue to hold in the context of this infinite-dimensional problem. Consequently, there are techniques for correcting forms of ELRs to improve the convergence rate of the null distributions of ELR test statistics to chi-square distributions. These techniques are similar to those applied in the field of parametric maximum likelihood ratio procedures.[45] The statement of the hypothesis testing above can easily be inverted with respect to providing nonparametric confidence interval estimators.

In terms of the accessibility of this method, it should be noted that the number of EL software packages continues to expand, particularly the R software packages. For example, `library(emplik)` and `library(EL)` of R packages that include the R function `el.test()` and `EL.test()`. These simple R functions can be very useful for the EL analysis of data from clinical studies.

For illustrative example, we revisit the HDL cholesterol data shown in Figure 31.2. Now, we use the empirical likelihood ratio test for means. The following R output shows the result of the empirical likelihood comparison between the means of the groups: X and Y.

```
> library(EL)
> EL.means(X,Y)

 Empirical likelihood mean difference test

data:  X  and  Y
-2 * LogLikelihood = 3.547, p-value = 0.05965
95 percent confidence interval:
 -0.4900842 19.0138090
sample estimates:
Mean difference
        10.17393
```

Perhaps, in this example, the ELR test outperforms the t-test that claims to reject the hypothesis $E(X) = E(Y)$, when X and Y are the measurements related to the same group of patients.

The classical EL methodology has been shown to have properties that make it attractive for testing hypotheses regarding parameters (e.g., moments) of distributions.[43,48] However, practicing statisticians working on clinical experiments, for example, case–control studies, commonly face a variety of distribution-free comparisons and/or evaluations over all distribution functions of complete and incomplete data subject to different types of measurement errors. In this framework, the *density-based* empirical likelihood methodology figures prominently.

Exercise 31.4

Generate a sample of i.i.d. measurements $X_1, \ldots, X_{25}$ from the following distributions: (1) normal distribution $N(0,1)$, and (2) $F(x) = (1 - \lambda \exp(-\lambda(x+1)))I\{x+1 > 0\}$, where λ is the rate parameter. Test $H_0 : E(X) = 0$ versus $H_1 : E(X) \neq 0$ at the $\alpha = 0.05$ significance level. In the case (2), is the result of the t-test application valid? Explain the reason.

Density-based empirical likelihood

According to the Neyman–Pearson lemma, density-based likelihood ratios can provide uniformly most powerful tests. Using this as a starting point, Vexler *et al.* proposed an alternative to the *"distribution function-based"* EL methodology.[49–53] The authors employed the approximate density-based likelihood, which has the following form:

$$L_f = \prod_{i=1}^{n} f(X_i) = \prod_{i=1}^{n} f_i, \ f_i = f(X_{(i)}),$$

where $X_{(1)} \leq X_{(2)} \leq \cdots \leq X_{(n)}$ are the order statistics based on $X_1, \ldots, X_n$, and $f_1, \ldots, f_n$ take on the values that maximize L_f given the empirical constraint corresponding to $\int f(u)du = 1$. This density-based EL approach was used successfully in order to construct efficient entropy-based goodness-of-fit test procedures.[29,50] The density-based EL methodology has been satisfactorily applied to develop a test for symmetry based on paired data. This test significantly outperforms classical procedures.[49] Gurevich and Vexler extended the density-based EL approach to a two-sample nonparametric likelihood ratio test.[51] Vexler and Yu used the density-based EL concept to present two group comparison principles based on bivariate data with a missing pattern as a consequence of data collection procedures.[52] Furthermore, the density-based EL methods were used to efficiently address nonparametric problems of complex composite hypothesis testing in children, in social/behavioral studies based on randomized prospective experiments.[53] In many practical settings, the density-based ELRs can provide simple and exact tests. Some distinctive characteristics of the density-based EL method test statistic as compared to the typical EL approach are summarized in Table 31.2.[48,54]

We note that Table 31.2 cannot correspond to all relevant EL constructions. For example, Hall and Owen developed large-sample methods for constructing *"distribution function-based"* EL confidence bands in problems of nonparametric density estimation.[55] Einmahl and McKeague proposed to localize the *"distribution function-based"* EL approach using one or more "time" variables implicit in the given null hypothesis.[56] Integrating the log-likelihood ratio over those variables, the authors constructed exact-test procedures for detecting a change in distribution, testing for symmetry about zero, testing for exponentiality, and testing for independence.

It is a common practice to conduct medical trials in order to compare a new therapy with a standard of care based on paired data consisting of pre- and post-treatment measurements. In such cases, there is often great interest in identifying treatment effects within each therapy group, as well as detecting a between-group difference. Nonparametric comparisons between distributions of new therapy and control groups, as well as detecting treatment effects within each group, may be based on

Table 31.2   Comparison of the classical EL and density-based EL approaches.

Characteristics	Owen[48]; Yu et al.[54]	The density-based EL method
Construction of the likelihood function	Distribution based	Density based
Usage of Lagrange multipliers method	Yes	Yes
Usage of constraints for maximization	Yes	Yes
Common focus of the test	Parameters (e.g., moments)	Overall distributions
Critical value	Asymptotic	Exact
The form of the test statistic	Numeric approach is required to calculate values of Lagrange multipliers	No numeric approach

multiple-hypothesis tests. To this end, one can create relevant tests combining, for example, the Kolmogorov–Smirnov test and the Wilcoxon signed-rank test. The use of the classical procedures commonly requires complex considerations about combining the known nonparametric tests and preserving the Type I error control and reasonable power of the resulting test. Alternatively, the density-based ELR technique provides a direct distribution-free approach for efficiently analyzing a variety of tasks occurring in clinical trials. The density-based EL method can easily be applied to test nonparametrically for different composite hypotheses. In this case, the density-based EL approach implies a standard scheme to develop highly efficient procedures, approximating nonparametrically the most powerful Neyman–Pearson test rules, given the aims of clinical studies. For example, Vexler *et al.* developed a density-based ELR methodology that was efficiently used to compare two therapy strategies for treating children's attention-deficit/hyperactivity disorder and severe mood dysregulation.[53,57] It was demonstrated that various composite hypotheses in a paired data setting (e.g., before vs. after treatment) can be tested with the density-based ELR tests, which give more emphasis to the overall distributional difference rather than to certain location parameter differences.

The R software can be employed in order to implement a computer program that realizes a density-based EL strategy. For example, programs of this type are presented in the *Statistics in Medicine* journal's Web domain *http://onlinelibrary.wiley.com/doi/10.1002/sim.4467/suppinfo*. Miecznikowski, Vexler, and Shepherd developed the R package "dbEmpLikeGOF" for nonparametric density-based likelihood ratio tests for goodness of fit and two-sample comparisons.[58] See also *http://cran.r-project.org/web/packages/dbEmpLikeNorm/* for the R package "dbEmpLikeNorm: Test for joint assessment of normality," developed by Drs. Shepherd, Tsai, Vexler, and Miecznikowski. The group of coauthors Tanajian, Vexler, and Hutson presented a package entitled "Novel and efficient density-based empirical likelihood procedures for symmetry and K-sample comparisons" in STATA, a general-purpose statistical software language.[59] It is available over the web at *http://sphhp.buffalo.edu/biostatistics/research-and-facilities/software/stata.html*.

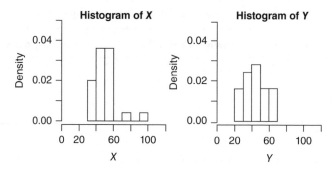

Figure 31.4 R data analysis output for measurements of HDL cholesterol levels (mg/dl) X and Y in the disease and healthy individuals, respectively.

In order to exemplify the density-based empirical likelihood method, we employ data from the clinical study that is mentioned in Section *Likelihood*. This study was designed as a case–control study of biomarkers for coronary heart disease. In accordance with the biomedical literature, the HDL biomarker has been suggested as having strong discriminatory ability for myocardial infarction (MI). To define cases, we consider the sample Y that consists of 25 measurements of the HDL biomarker on individuals who recently survived an MI. In order to represent controls, 25 HDL biomarker measurements on healthy subjects are denoted as $X_1, \dots, X_{25}$. The following R code inputs the data and constructs the histograms of the data, as shown in Figure 31.4:

```
> X<-c(96.8,57.2,37.4,44.0,55.0,41.8,46.2,41.8,41.8,59.4,44.0,
52.8,33.0,52.8,41.8,44.0,52.8,59.4,37.4,77.0,39.6,57.2,57.2,
41.8,39.6)
> Y<-c(26.4,33.0,30.8,35.2,44.0,48.4,61.6,41.8,26.4,28.6,55.0,
61.6,63.8,24.2,37.4,48.4,52.8,46.2,57.2,68.2,46.2,37.4,46.2,
52.8,35.2)
> a<-min(c(X,Y))-20
> b<-max(c(X,Y))+20
> par(pty="s",mfrow=c(1,2),oma=c(0,0,0,0),mar=c(0,4,0,0))
> hist(X,xlim=c(a,b),ylim=c(0,0.05),freq=FALSE)
> hist(Y,xlim=c(a,b),ylim=c(0,0.05),freq=FALSE)
```

The classical empirical likelihood ratio test can be conducted via the R function EL.means. With the p-value of 0.101 as shown below, we fail to reject that $E(X) = E(Y)$ at the 0.05 significance level.

```
> EL.means(X,Y)

Empirical likelihood mean difference test

data:  X  and  Y
-2 * LogLikelihood = 2.6898, p-value = 0.101
```

```
95 percent confidence interval:
 -1.066513 13.850197
sample estimates:
Mean difference
       5.720065
```

Thus, in this example, the ELR test cannot be used to demonstrate the discriminatory ability of the HDL biomarker with respect to the MI disease. In this case, the two-sample density-based empirical likelihood ratio test[51] shows the p-value <0.043, supporting rejection of the hypothesis regarding equivalence of distributions of X and Y. For the sake of completeness, the Appendix presents an example of R procedures for executing the two-sample density-based ELR test. In addition, Vexler *et al.* proposed a simple, but very efficient, density-based empirical likelihood ratio test for independence and provided the R code to run the procedure.[60]

Combinations of likelihoods to assemble composite tests and archive full information regarding data

Strictly speaking, "distribution-function/density-based" EL techniques and parametric likelihood methods are closely related concepts. This provides the impetus for an impressive expansion in the number of EL developments, based on combinations of likelihoods of different types.[61]

Consider a simple example, where we assume to observe independent couples given as (X, Y). In this case, the likelihood function can be denoted as $L(X, Y)$. Suppose values of X's are observed completely, whereas a proportion of the observed data for the Y's is incomplete. Assume a model of Y given X, $Y|X$, is well defined, for example, $Y_i = \beta X_i + \varepsilon_i$, where β denotes the model parameter and ε_i is a normally distributed error term, for $i = 1, \dots, n$. Then, we refer to Bayes' theorem to represent $L(X, Y) = L(Y|X)L(X)$, where $L(X)$ can be substituted by the EL to avoid parametric assumptions regarding distributions of X's.

In this context, Qin shows an inference on incomplete bivariate data using a method that combines the parametric model and ELs.[62] This method also incorporates auxiliary information from variables in the form of constraints, which can be obtained from reliable resources such as census reports. This approach makes it possible to use all available bivariate data, whether completely or incompletely observed. In the context of a group comparison, constraints can be formed based on null and alternative hypotheses, and these constraints are incorporated into the EL. This result was extended and applied to the following practical issues:

Malaria remains a major epidemiological problem in many developing countries. In endemic areas, an individual may have symptoms attributable either to malaria or to other causes. From a clinical viewpoint, it is important to attend to the next tasks: (i) to correctly diagnose an individual who has developed symptoms, so that the appropriate treatments can be given; (ii) to determine the proportion of malaria-affected cases in individuals who have symptoms, so that policies on intervention program can be developed. Once symptoms have developed in an individual, the diagnosis of malaria can be based on the analysis of the parasite levels in blood samples. However, even a blood test is not conclusive, as in endemic areas many healthy individuals can have parasites in their blood slides. Therefore, data from this type of study can be viewed as coming from a mixture distribution, with the components corresponding to malaria and nonmalaria cases. Qin and Leung constructed new EL procedures to estimate the proportion of clinical malaria using parasite-level data from a group of individuals with symptoms attributable to malaria.[63] Yu *et al.* and Vexler *et al.* proposed two-sample EL techniques based on incomplete data to analyze a Pneumonia Risk Study in an ICU Setting.[46,47] In the context of this study, the initial detection of ventilator-associated pneumonia (VAP) for inpatients at an intensive care unit requires composite symptom evaluation, using clinical criteria such as the clinical pulmonary infection score (CPIS). When CPIS is above a threshold value, bronchoalveolar lavage (BAL) is performed to confirm the diagnosis by counting actual bacterial pathogens. Thus, CPIS and BAL results are closely related, and both are important indicators of pneumonia, whereas BAL data are incomplete. Yu *et al.* and Vexler *et al.* derived EL methods to compare the pneumonia risks among treatment groups for such incomplete data.[46,47] In semi- and nonparametric contexts, including EL settings, Qin and Zhang showed that the full likelihood can be decomposed into the product of a conditional likelihood and a marginal likelihood, in a similar manner to the parametric likelihood considerations.[61] These techniques augment the study's power by enabling researchers to use any observed data and relevant information.

Receiver operating characteristic curve analysis

The ROC curves are useful visualization tools for illustrating the discriminant ability of biomarkers to distinguish between two populations: diseased and nondiseased. The ROC curve methodology was originally developed for radar signal detection theory and was extensively employed in psychological and, most importantly, medical research and epidemiology.[16,64]

Assume, without loss of generality, that $X_1, \ldots, X_n$ and $Y_1, \ldots, Y_m$ are measurements from the diseased and nondiseased populations, respectively. The observations $X_1, \ldots, X_n$ are i.i.d. and independent of i.i.d. measurements $Y_1, \ldots, Y_m$. Let F and G denote the CDFs of X and Y, respectively. The ROC curve $R(t)$ can be defined as $R(t) = 1 - F(G^{-1}(1 - t))$, where $t \in [0, 1]$.[65] It plots sensitivity (the true positive rate, $1 - F(t)$) against 1 minus specificity (the true negative rate, $1 - G(t)$) for various values of the threshold t. As an example, we consider three biomarkers with their corresponding ROC curves presented in Figure 31.5, where underlying distributions are $F_1 \sim N(0, 1)$, $G_1 \sim N(0, 1)$ for biomarker A (the diagonal line), $F_2 \sim N(0, 1)$, $G_2 \sim N(1, 1)$ for biomarker B (in a dashed line), and $F_3 \sim N(0, 1)$, $G_3 \sim N(10, 1)$ for biomarker C (in a dotted line), respectively.

The following R code plots the ROC curve, as shown Figure 31.5.

```
> t<-seq(0,1,0.001)
> R1<-1-pnorm(qnorm(1-t,0,1),0,1)    # biomarker 1
> R2<-1-pnorm(qnorm(1-t,1,1),0,1)    # biomarker 2
> R3<-1-pnorm(qnorm(1-t,10,1),0,1)    # biomarker 3
```

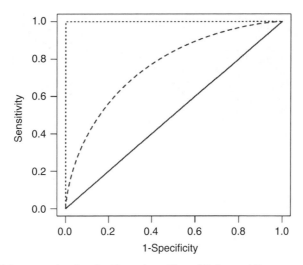

Figure 31.5 ROC curves related to the biomarkers. The solid diagonal line corresponds to the ROC curve of biomarker A, where $F_1 \sim N(0, 1)$ and $G_1 \sim N(0, 1)$. The dashed line displays the ROC curve of biomarker B, where $F_2 \sim N(0, 1)$ and $G_2 \sim N(1, 1)$. The dotted line close to the upper left corner plots the ROC curve for biomarker C, where $F_3 \sim N(0, 1)$ and $G_3 \sim N(10, 1)$.

```
> plot(R1,t,type="l",lwd=1.5,lty=1,cex.lab=1.1,ylab=
"Sensitivity",xlab="1-Specificity")
> lines(R2,t,lwd=1.5,lty=2)
> lines(R3,t,lwd=1.5,lty=3)
```

It can be seen that the farther apart the two distributions F and G fall, the more the ROC curve curves up to the top left corner. A perfect biomarker would have the ROC curve come close to the top left corner, and a biomarker without discriminability would result in a diagonal line in the ROC curve. We also observe that there exists a trade-off between specificity and sensitivity.

There exists extensive research on estimating the ROC curves from the parametric and nonparametric perspectives.[65–67] Assuming both the diseased and nondiseased populations are normally distributed, i.e., $F \sim N(\mu_1, \sigma_1^2)$ and $G \sim N(\mu_2, \sigma_2^2)$, the corresponding ROC curve can be expressed as

$$\text{ROC}(t) = \Phi[a + b\Phi^{-1}(t)],$$

where $a = (\mu_1 - \mu_2)/\sigma_1$, $b = \sigma_2/\sigma_1$, and Φ is the standard normal CDF. In this case the estimated ROC curve is obtained by substituting the maximum likelihood estimators (MLEs) of the normal parameters μ_1, μ_2, σ_1, and σ_2 into the formula. The nonparametric estimate of the ROC curve used the empirical distribution functions.[66,67] Define the empirical distribution function of F as

$$\hat{F}_n(t) = \frac{1}{n}\sum_{i=1}^{n} I\{X_i \leq t\},$$

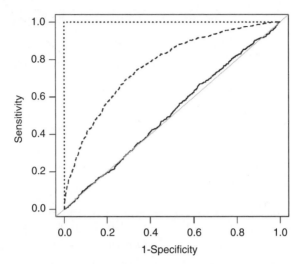

Figure 31.6 The nonparametric estimators of ROC curves of three different biomarkers based on samples of sizes 1000. The solid diagonal line corresponds to the nonparametric estimator of the ROC curve of biomarker A, where $F_1 \sim N(0,1)$ and $G_1 \sim N(0,1)$. The dashed line displays the nonparametric estimator of the ROC curve of biomarker B, where $F_2 \sim N(0,1)$ and $G_2 \sim N(1,1)$. The dotted line close to the upper left corner plots the nonparametric estimator of the ROC curve for biomarker C, where $F_3 \sim N(0,1)$ and $G_3 \sim N(10,1)$.

where $I\{\cdot\}$ denotes the indicator function and the empirical distribution function $\widehat{G}_m$ of G can be defined similarly. Substituting F and G by corresponding empirical estimates $\widehat{F}_n$ and $\widehat{G}_m$, respectively, the empirical estimator of the ROC curve is given by

$$\widehat{R}(t) = 1 - \widehat{F}_n(\widehat{G}_m^{-1}(1-t)),$$

which converges to $R(t)$ for a large sample.[66]

Figure 31.6 presents the nonparametric estimators of the ROC curves with sample sizes $n = m = 1000$ for three biomarkers described in Figure 31.5, that is, $F_1 \sim N(0,1)$, $G_1 \sim N(0,1)$ for biomarker A (the diagonal line), $F_2 \sim N(0,1)$, $G_2 \sim N(1,1)$ for biomarker B (in a dashed line), and $F_3 \sim N(0,1)$, $G_3 \sim N(10,1)$ for biomarker C (in a dotted line), respectively, using the following R code:

```
> if(!("pROC" %in% rownames(installed.packages())))  install
.packages("pROC")
> library(pROC)
> n<-1000
> set.seed(123)   # set the seed
> # Simulate data from the normal distribution
> X1<-rnorm(n,0,1)
> Y1<-rnorm(n,1,1)
> group<-cbind(rep(1,n),rep(0,n))
> measures<-c(X1,Y1)
> roc1<-roc(group, measures)
```

```
> plot(1-roc1$specificities,roc1$sensitivities,type="l",
ylab="Sensitivity",xlab="1-Specificity")
> abline(a=0,b=1,col="grey")   # add the diagonal line for
# reference
```

It should be noted the ROC curves are well approximated by the nonparametric estimators.

In health-related studies, the ROC curve methodology is commonly related to case–control type evaluations. As a type of observational study, case–control studies differentiate and compare two existing groups differing in outcome on the basis of some supposed causal attribute. For example, based on factors that may contribute to a medical condition, subjects can be grouped as the cases, for patients with condition/disease, and the controls, for patients without the condition/disease. For independent populations, for example, cases and controls, various parametric and nonparametric approaches have been proposed to evaluate the performance of biomarkers.[65–71]

Area under the ROC curve

A rough idea of the performance of the biomarkers can be obtained with the ROC curve. However, judgments solely based on the ROC curves are far from enough to precisely describe the diagnostic accuracy of biomarkers. The area under the ROC curve (AUC) is a common index of the diagnostic performance of a continuous biomarker. It measures the ability to discriminate between the control and the disease groups.[68,69] Bamber noted that the area under this curve is equal to $Pr(X > Y)$.[70] We prove this result in the Appendix. Values of AUCs can range from 0.5, in the case of no difference between distributions, to 1, where the two distributions are perfectly discriminated. For more details, see Kotz *et al.* for wide discussions regarding evaluations of the AUC-type objectives.[71]

Parametric approach for AUC testing

Under the normal assumptions, a closed form of the AUC is presented as

$$A = \Phi\left(\frac{\mu_1 - \mu_2}{\sqrt{\sigma_1^2 + \sigma_2^2}} \right),$$

where for the diseased population $X \sim N(\mu_1, \sigma_1^2)$ and for the nondiseased population $Y \sim N(\mu_2, \sigma_2^2)$, $\mu_1 \geq \mu_2$.[72] We provide the proof in the Appendix. By substituting maximum likelihood estimators for μ_i and σ_i^2, $i = 1, 2$ into the above formula, the maximum likelihood estimator of the AUC can be obtained correspondingly. Given the estimator of the AUC under the normal distribution assumption, one can easily construct confidence interval-based tests for the AUC using the delta method; see Kotz *et al.* for details.[71] In several cases, to achieve a fit between data distributions and the normal assumptions, a transformation of observations, for example, the Box–Cox transformation, can be recommended, before the above parametric approach is applied.[73] In general, when data distributions are different from the normal distribution function, the AUC can be expressed as $\Pr(X > Y) = \int G(x)dF(x)$, and evaluated in a similar manner to the technique above.[72]

Exercise 31.5

Biomarker levels were measured from diseased and healthy populations, providing i.i.d. observations $X_1 = 0.39$, $X_2 = 1.97$, $X_3 = 1.03$, $X_4 = 0.16$, which are assumed to be from a normal distribution, as well as i.i.d. observations $Y_1 = 0.42$, $Y_2 = 0.29$, $Y_3 = 0.56$, $Y_4 = -0.68$, $Y_5 = -0.54$, which are also assumed to be from a normal distribution, respectively. Please define the ROC curve. Obtain a formal notation of the AUC and estimate the AUC. What can be concluded about the discriminating ability of the biomarker with respect to the disease?

Hint: Values that may help you to approximate the estimated AUC: $\Pr(\xi < x) \approx 0.56$, when $x = 1$, $\xi = N(0.7, 4)$; $\Pr(\xi < x) \approx 0.18$, when $x = 1$, $\xi = N(0.9, 1)$; $\Pr(\xi < x) = 0.21$, when $x = 0$, $\xi = N(0.8, 1)$; $\Pr(\xi < x) \approx 0.18$, when $x = 0$, $\xi = N(0.9, 1)$.

Nonparametric approach for AUC testing

Conversely, a nonparametric estimator for the AUC based on continual biomarker values can be obtained as

$$\hat{A} = \frac{1}{mn}\sum_{i=1}^{n}\sum_{j=1}^{m} I(X_i > Y_j),$$

where X_i, $i = 1, \ldots, n$ and Y_j, $j = 1, \ldots, m$ are the observations for diseased and nondiseased populations, respectively.[74] It is equivalent to the well-known Mann–Whitney statistic, and the variance of this empirical estimator can be obtained using U-statistic theory.[75] The empirical likelihood method to construct the confidence interval estimation of the AUC was introduced by Qin and Zhou.[76] Replacing the indicator function by a kernel function, one can obtain a smoothed ROC curve.[77]

Exercise 31.6

Biomarker levels were measured from diseased and healthy populations, providing i.i.d. observations $X_1 = 0.39$, $X_2 = 1.97$, $X_3 = 1.03$, $X_4 = 0.16$, which are assumed to be from a continuous distribution, as well as i.i.d. observations $Y_1 = 0.42$, $Y_2 = 0.29$, $Y_3 = 0.56$, $Y_4 = -0.68$, $Y_5 = -0.54$, which are also assumed to be from a continuous distribution, respectively. Please define the ROC curve. Estimate the AUC nonparametrically. What can be concluded regarding the discriminating ability of the biomarker with respect to the disease?

Nonparametric comparison between two ROC curves

It is of great importance for the researchers to compare two biomarkers. If we use both diagnostic markers on the same m controls and n cases, we can represent the bivariate outcomes as (X_{1j}, X_{2j}) ($j = 1, \ldots, m$) and (Y_{1k}, Y_{2k}) ($k = 1, \ldots, n$), respectively. We denote the respective bivariate distributions by $F(x_1, x_2)$ and $G(y_1, y_2)$, and the marginals by $F_i(x_i)$ and $G_i(y_i)$, $i = 1, 2$, and we assume that the $m + n$ bivariate vectors are mutually independent. Denote the sensitivity at specificity p by $S_i(p)$, $i = 1, 2$, and define

$$\Delta = \int (S_1 p) - S_2(p))dW(p),$$

where W is a probability measure on the open unit interval. The parameter Δ allows one to compare sensitivities on a predefined range of specificities of clinical interest by adjusting the weight function W accordingly. When $W(p)$ is the uniform distribution on $(0, 1)$, the parameter Δ equals the difference of AUCs between two biomarkers.

Wieand *et al.* considered a nonparametric estimate of Δ in the form of

$$\hat{\Delta} = \int (\hat{S}_1 p) - \hat{S}_2(p))dW(p),$$

where $\hat{S}_i(p) = 1 - \hat{G}_i(\hat{\xi}_{ip})$, $\hat{G}_i$ is the empirical distribution of G_i and the sample quantile $\hat{\xi}_{ip}$ is the $[mp]$th order statistic among the m values of X_i, where $[mp]$ is the smallest integer that equals or exceeds mp.[67] Assume that W is a probability measure in $(0, 1)$ and that there exists $\varepsilon > 0$ such that W has a bounded derivative in $(0, \varepsilon)$ and $(1 - \varepsilon, 1)$. Suppose further that $G_i(\xi_{ip})$, for $i = 1, 2$, has continuous derivatives in $(0, 1)$, which are monotone in $(0, \varepsilon)$ and $(1 - \varepsilon, 1)$. Define $s_i(p) = S_i'(p) = -G_i'(\xi_{ip})/F_i'(\xi_{ip})$. Then, as $N = n + m$ tends to ∞ with $m/N \to \lambda$, for $0 < \lambda < 1$, $N^{1/2}(\hat{\Delta} - \Delta)$ tends to a normal distribution with variance $\sigma^2 = \sigma^{11} - 2\sigma^{12} + \sigma^{22}$, where

$$\sigma_{ii} = \int_0^1 \int_0^1 \{(1 - \lambda)^{-1} S_i(\max(p, q))(1 - S_i(\min(p, q)))$$
$$+ \lambda^{-1} s_i(p)s_i(q)(\min(p, q) - pq)\}dW(p)dW(q),$$

and

$$\sigma_{12} = (1 - \lambda)^{-1} \int\int (G(\xi_{1p}, \xi_{2q})) - (1 - S_1(p))(1 - S_2(q))dW(p)dW(q)$$
$$+ \lambda^{-1} \int\int (G(\xi_{1p}, \xi_{2q}) - pq)s_1(p)s_2(q)dW(p)dW(q).$$

Based on this asymptotic distribution of $\hat{\Delta}$, one can conduct a nonparametric proce-dure for testing $H_0 : \Delta = 0$ versus $H_1 : \Delta > 0$.[67] Note that the test proposed by Wieand *et al.* requires the estimation of densities, and selection of a satisfactory smoothing parameter may be problematic.

Best linear combinations based on values of multiple biomarkers

In practice, different markers are usually related to the disease, showing treatment effects in various magnitudes and different directions. For example, low levels of high-density lipoprotein (HDL)-cholesterol and high levels of thiobarbuturic acid reacting substances (TBARS), biomarkers of oxidative stress and antioxidant status, are indicators of coronary heart disease.[22] When multiple biomarkers are available, it is of great interest to seek a simple best linear combination (BLC) of biomarkers, such that the combined score achieves the maximum AUC or the maximum treatment effect over all possible linear combinations. Consider a study with d continuous-scale biomarkers yielding measurements $\mathbf{X}_i = (X_{1i}, \dots, X_{di})^T$, $i = 1, \dots, n$, on n diseased patients, and measurements $\mathbf{Y}_j = (Y_{1j}, \ldots, Y_{dj})^T$, $j = 1, \ldots, m$, on m nondiseased patients, respectively. It is of interest to construct effective one-dimensional combined scores of biomarkers' measurements, that is, $X(\mathbf{a}) = \mathbf{a}^T\mathbf{X}$ and $Y(\mathbf{a}) = \mathbf{a}^T\mathbf{Y}$, such that the AUC based on these scores is maximized over all possible linear combinations of biomarkers. Define $A(\mathbf{a}) = \Pr(X(\mathbf{a}) > Y(\mathbf{a}))$; the statistical problem is to estimate the maximum AUC defined as $A = A(\mathbf{a}_0)$, where $\mathbf{a}_0$ are the BLC coefficients satisfying $\mathbf{a}_0 = \arg\max_{\mathbf{a}} A(\mathbf{a})$. For simplicity, we assume that the first component of the vector $\mathbf{a}$ equals one.[78] For example, in the case of two biomarkers, i.e. $d = 2$, the AUC can be defined as $A(a) = \Pr(X_1 + aX_2 > Y_1 + aY_2)$.

Parametric method

Assuming $\mathbf{X}_i \sim N(\boldsymbol{\mu}_{\mathbf{X}}, \boldsymbol{\Sigma}_{\mathbf{X}})$, $i = 1, \ldots, n$ and $\mathbf{Y}_j \sim N(\boldsymbol{\mu}_{\mathbf{Y}}, \boldsymbol{\Sigma}_{\mathbf{Y}})$, $j = 1, \ldots, m$, Su and Liu derived the BLC coefficients $\mathbf{a}_0 \propto \boldsymbol{\Sigma}_C^{-1}\boldsymbol{\mu}$ and the corresponding optimal AUC as $\Phi(\omega^{1/2})$, where $\boldsymbol{\mu} = \boldsymbol{\mu}_{\mathbf{X}} - \boldsymbol{\mu}_{\mathbf{Y}}$, $\boldsymbol{\Sigma}_C = \boldsymbol{\Sigma}_{\mathbf{X}} + \boldsymbol{\Sigma}_{\mathbf{Y}}$, $\omega = \boldsymbol{\mu}^T\boldsymbol{\Sigma}_C^{-1}\boldsymbol{\mu}$, and Φ is the standard normal CDF.[79]

Based on Su and Liu's point estimator, we can derive the confidence interval estimation for the BLC-based AUC under multivariate normality assumptions.[80]

Exercise 31.7

Consider the simple bivariate normal case where $\mathbf{X}_i \sim N\left(\begin{pmatrix} \mu_{X_1} \\ \mu_{X_2} \end{pmatrix}, \begin{pmatrix} 1 & \rho_X \\ \rho_X & 1 \end{pmatrix}\right)$, $i = 1, \ldots, n$ and $\mathbf{Y}_j \sim N\left(\begin{pmatrix} \mu_{Y_1} \\ \mu_{Y_2} \end{pmatrix}, \begin{pmatrix} 1 & \rho_Y \\ \rho_Y & 1 \end{pmatrix}\right)$, $j = 1, \ldots, m$. Derive the best linear combination and the corresponding maximum AUC.

Nonparametric method

Chen *et al.* proposed to use kernal functions to develop the EL-based confidence interval estimation for the BLC-based AUC via construction of the empirical likelihood ratio (ELR) test statistic for testing the hypothesis $H_0 : A = A_0$ versus $H_1 : A \neq A_0$.[81] Let k be a symmetric kernel function and define $K_h(x) = \int_{-\infty}^{x/h} k(u)du$, $v_i(\mathbf{a}) = m^{-1}\sum_{j=1}^{m} K_h(\mathbf{a}^T\mathbf{X}_i - \mathbf{a}^T\mathbf{Y}_j)$, $i = 1, \ldots, n$, where h is the bandwidth parameter. Regarding to the kernel estimation, we refer the reader to the textbook of Silverman.[82] Let $\mathbf{p} = (p_1, p_2, \ldots, p_n)^T$ be a probability weight vector, $\sum_{i=1}^{n} p_i = 1$ and $p_i \geq 0$ for all $i = 1, \ldots, n$. The EL for the BLC-based AUC evaluated at the true value A_0 of AUC can be defined as

$$L(A_0) = \sup\left\{\prod_{i=1}^{n} p_i : \sum_{i=1}^{n} p_i = 1, \sum_{i=1}^{n} p_i v_i(\hat{\mathbf{a}}_0) = A_0\right\},$$

where $\hat{\mathbf{a}}_0$ satisfies $\sum_{i=1}^{n} p_i \partial v_i(\mathbf{a})/\partial \mathbf{a}|_{\mathbf{a}=\hat{\mathbf{a}}_0} = 0$. One can show (using Lagrange multipliers) that

$$p_i = \frac{1}{n}\frac{1}{1 + \lambda(v_i(\hat{\mathbf{a}}_0) - A_0)}, \quad i = 1, \ldots, n,$$

where the Lagrange multiplier λ is the root of

$$\frac{1}{n}\sum_{i=1}^{n} \frac{v_i(\hat{\mathbf{a}}_0)}{1 + \lambda(v_i(\hat{\mathbf{a}}_0) - A_0)} = A_0.$$

Under the alternative hypothesis, we have just the constraint $\sum_{i=1}^{n} p_i = 1$, and hence $L(A_0) = (1/n)^n$ at $p_i = 1/n$. Therefore, the empirical log-likelihood ratio test statistic is

$$l(A_0) = -2 \log \mathrm{ELR}(A_0) = 2 \sum_{i=1}^{n} \log(1 + \lambda(v_i(\hat{\mathbf{a}}_0) - A_0)).$$

We define $\hat{\mathbf{a}}_K = \arg\max_{\mathbf{a}} A_{m,n}^{K}(\mathbf{a})$, where $A_{m,n}^{K}(\mathbf{a}) = \sum_{i=1}^{n} v_i(\mathbf{a})/n$. Under some general conditions (see Chen *et al.*[81] for details), the asymptotic distribution of $l(A_0)$ under $H_0 : A = A_0$ is a scaled chi-square distribution with one degree of freedom, that is,

$$\gamma(A_0)l(A_0) \xrightarrow{d} \chi_1^2, \quad \text{as } n, m \to \infty,$$

where

$$\gamma(A_0) = \frac{m\hat{\sigma}^2}{(m+n)s^2}, \quad \hat{\sigma}^2 = n^{-1} \sum_{i=1}^{n} \left(v_i\left(\hat{\mathbf{a}}_K\right) - n^{-1} \sum_{j=1}^{n} v_j(\hat{\mathbf{a}}_K) \right)^2, \quad s^2 = \frac{m\hat{\sigma}_{10}^2 + n\hat{\sigma}_{01}^2}{m+n},$$

$$\sigma_{10}^2 = \mathrm{Cov}(K_h(\mathbf{a}_0^T\mathbf{X}_1 - \mathbf{a}_0^T\mathbf{Y}_1), \ K_h(\mathbf{a}_0^T\mathbf{X}_1 - \mathbf{a}_0^T\mathbf{Y}_2)),$$

$$\sigma_{01}^2 = \mathrm{Cov}(K_h(\mathbf{a}_0^T\mathbf{X}_1 - \mathbf{a}_0^T\mathbf{Y}_1), \ K_h(\mathbf{a}_0^T\mathbf{X}_2 - \mathbf{a}_0^T\mathbf{Y}_1)),$$

and $\hat{\sigma}_{10}^2$ and $\hat{\sigma}_{01}^2$ are the corresponding estimates.

Based on the asymptotic distribution of the statistic $l(A_0)$, the $100(1 - \alpha)\%$ empirical likelihood-based confidence interval for the maximum AUC can be constructed as

$$R_\alpha = \left\{ A_0 : \gamma\left(A_0\right) l(A_0) \le \chi_1^2(1 - \alpha) \right\},$$

where $\chi_1^2(1 - \alpha)$ is the $(1 - \alpha)$th quantile of the chi-square distribution with one degree of freedom. It gives a confidence interval with asymptotically correct coverage probability $1 - \alpha$, that is, $\Pr(A_0 \in R_\alpha) = 1 - \alpha + o(1)$. R code related to the problem described above can be found at following the Web domain: *http://www.sciencedirect.com/science/article/pii/S0167947314002710*.

Goodness-of-fit tests

Many statistical procedures are, strictly speaking, only appropriate when parametric assumptions about data distribution are made. If the distributions under the null hypothesis are completely known, then testing for goodness of fit is equivalent to testing for uniformity. In general, testing distribution assumptions for normality and uniformity is suggested and has been one of the major areas of continuing statistical research both theoretically and practically. Normal distributions are commonly assumed in applications of statistical procedures. For instance, in most situations, parametric linear regression analysis can be done where the errors are assumed to be normally distributed. Thus, tests for goodness of fit, especially tests for normality, have a very important role in clinical experiments. Testing composite hypotheses of normality (or other specified), i.e., H_0: the population is normally distributed versus H_1: the population is not normally distributed, is well addressed in statistical literature.[29] Included in the coverage are the Shapiro–Wilk test, the Kolmogorov–Smirnov test, and the Anderson–Darling test, among which the Shapiro–Wilk test is highly efficient and has the best power for a given significance.[83,84]

The Shapiro–Wilk test employs the null hypothesis principle to check whether a sample of i.i.d. observations $X_1, \ldots, X_n$ came from a normally distributed population. The test statistic is

$$W = \left(\sum_{i=1}^{n} a_i X_{(i)} \right)^2 \bigg/ \sum_{i=1}^{n} (X_i - \overline{X})^2 ,$$

where $X_{(i)}$ is the ith order statistic, i.e., the ith smallest number in the sample, and $\overline{X}$ is the sample mean; the constants a_i are given by

$$(a_1, \ldots, a_n) = \frac{m^T V^{-1}}{(m^T V^{-1} V^{-1} m)^{1/2}},$$

where $m = (m_1, \ldots, m_n)^T$ and $m_1, \ldots, m_n$ are the expected values of the order statistics of independent and identically distributed random variables sampled from the standard normal distribution and V is the covariance matrix of those order statistics. For example, when the sample size is 10, it can be obtained that $a_1 = 0.5739$, $a_2 = 0.3291$, $a_3 = 0.2141$, $a_4 = 0.1224$, and $a_5 = 0.0399$. The null hypothesis is rejected if W is below a predetermined threshold. Note that the Shapiro–Wilk test does not depend on the distribution of data under the null hypothesis and the corresponding critical value can be calculated numerically using Monte Carlo techniques, see Section *Numerical methods for calculating critical values and powers of statistical tests* for details.

It should be noted that the tests for normality can be subject to low power, especially when the sample size is small. Thus, a Q–Q plot is recommended for verification in addition to the test.[85]

The function `shapiro.test` in R makes it possible to implement the Shapiro–Wilk test for normality.

Example 31.3

We generate data from $N(0, 1)$, Uniform$(0, 1)$, and exp(1) with the sample size $n = 100$ using R and check the data for normality.

The following program shows how to simulate the normal distribution, the uniform distribution, and the exponential distribution in R, respectively. The function `shapiro.test` in R conducts the Shapiro–Wilk test for normality, yielding the test statistic and the corresponding p-value, as seen below:

```
> n<-100
>
> # N(0,1)
> x<-rnorm(n,mean=0,sd=1)
> shapiro.test(x)

        Shapiro-Wilk normality test

data:   x
```

```
W = 0.9876, p-value = 0.4825


> # Uniform(0,1)
> y<-runif(n,min=0,max=1)
> shapiro.test(y)

  Shapiro-Wilk normality test

data:  y
W = 0.9545, p-value = 0.001674

> # exp(1)
> z<-rexp(n,rate=1)
> shapiro.test(z)

  Shapiro-Wilk normality test

data:  z
W = 0.8113, p-value = 5.574e-10
```

For the data simulated from $N(0, 1)$, we fail to reject the null hypothesis at the 0.05 significance level with p-value = 0.4825, stating that there is no sufficient evidence to conclude the data are not from the normal distribution. For the data simulated from either Uniform(0, 1) or exp(1), we reject the null hypothesis at the 0.05 significance level with p-values of 0.0017 and <0.0001, respectively, stating that there are sufficient evidences to conclude the data are not from the normal distribution. For more details regarding goodness-of-fit tests, we refer the reader to Vexler *et al.* and Claeskens *et al.*[29,39,50]

Exercise 31.8

Simulate data from $N(1,2)$, Gamma(1/2,1), $t_{df=5}$ with the sample size $n = 15$, 25, and 50, respectively, and test for normality.

Wilcoxon rank-sum tests

As a nonparametric analog to the two-sample t-test, Wilcoxon rank-sum test (also called the Mann–Whitney U test or the Mann–Whitney–Wilcoxon test) can be used primarily when investigators do not want to, or cannot, assume that data distributions are normal.

Suppose that we have two samples of observations, containing i.i.d. measurements $X_1, \ldots, X_m$ and i.i.d. measurements $Y_1, \ldots, Y_n$, respectively. In practical applications, we often want to test the hypothesis that two populations are the same in the context of no location shift. To formulate, assuming that $X_1, \ldots, X_m \sim F(x - u)$,

$Y_1, \ldots, Y_n \sim F(y - v)$, that all $m + n$ observations are independent, and that $F(\cdot)$ is symmetric about zero, it is of interest to test $H_0 : \Delta = v - u = 0$ against $H_1 : \Delta = v - u > 0$. Let R_i be the rank of Y_i among all $m + n$ observations, where the rank refers to the ordinal number of the corresponding observation among a pre-ordered data set in ascending order. The Mann–Whitney statistic $W = \sum_{i=1}^{n} R_i - n(n + 1)/2 = \sum_{i=1}^{n} \sum_{i=1}^{n} I\{X_i < Y_j\}$ rejects H_0 for large values of W. It follows from one-sample U-statistics theory that, under the null hypothesis, W is asymptotically normal,[86] that is,

$$\frac{W - mn/2}{\sqrt{mn(m + n + 1)/12}} \xrightarrow{d} N(0, 1).$$

Therefore, a cutoff value for the α-level test can be found as $K_\alpha = mn/2 + 1/2 + z_\alpha \sqrt{mn(m + n + 1)/12}$, where the additional $1/2$ is added for a continuity correction.

The Wilcoxon rank-sum test has greater efficiency than the t-test on non-normal distributions, and it is nearly as efficient as the t-test on normal distributions.

Example 31.4

We consider the HDL data described in Figure 31.2 and conduct a Wilcoxon rank-sum test.

The function `wilcox.test` in R conducts the two-sample Wilcoxon test for equality on means. Note that the alternative can be revised to "less" or "greater" in terms of a one-sided test.

```
> X<-c(37.4,70.4,52.8,46.2,74.8,96.8,41.8,55.0,83.6,63.8,63.8,
52.8,46.2,37.4,50.6,74.8,46.2,39.6,70.4,30.8,74.8,61.6,30.8,
74.8,52.8)
> Y<-c(44.0,35.2,110.0,63.8,44,26.4,52.8,30.8,39.6,44,48.4,
39.6,55,52.8,50.6,39.6,35.2,55,57.2,37.4,30.8,46.2,50.6,44,44)
> wilcox.test(X,Y)

	Wilcoxon rank sum test with continuity correction

data:  X and Y
W = 432, p-value = 0.02065
alternative hypothesis: true location shift is not equal to 0
```

For the HDL data, the p-value of the Wilcoxon rank-sum test is 0.02065. Thus, we reject the null hypothesis at the 0.05 significance level and conclude a significant difference in the mean between two groups.

Tests for independence

Evaluations of relationships between pairs of variables, including testing for independence, are increasingly important. In this section, we introduce nonparametric tests for independence, including the Pearson correlation coefficient ρ, the Spearman's rank correlation coefficient ρ_S, the Kendall's rank correlation coefficient, data-driven rank techniques, the empirical likelihood-based method, and the density-based empirical likelihood ratio test.[56,60,87] Note that the Pearson correlation coefficient, the Spearman's rank correlation coefficient, and the Kendall's rank correlation coefficient focus on specific interdependence, for example, linear and/or monotone. In reality, the dependence structure may be more complex. For example, in the models $Y = 1/X$, $Y = \varepsilon/X$, or $Y = \varepsilon/X^2$, where ε is a random variable, X and Y are dependent in an inverse manner, and in the second and third cases, $E(XY)$ can be nonexistent. In this lies the difficulty of interpreting the correlation as a measure of dependence in general.

Assume we obtain a random sample of n pairs of observations $(X_1, Y_1), \ldots, (X_n, Y_n)$ from a continuous bivariate population. Let $F_X(x)$ and $F_Y(y)$ denote the marginal distribution functions of X and Y, respectively, and let $F_{XY}(x, y) = \Pr(X \leq x, Y \leq y)$ be the joint distribution function of the (X, Y) pairs. For future use, let R_i and S_i denote the rank of X_i and Y_i, $i = 1, \ldots, n$, respectively, and F_{Xn}, F_{Yn}, and F_n be the empirical distribution functions of F_X, F_Y, and F_{XY}, respectively. The null hypothesis of bivariate independence between X and Y can be formally stated as $H_0 : F(x, y) = F(x)F(y)$ for all $(x, y) \in R^2$ versus $H_1 : F(x, y) \neq F(x)F(y)$ for some $(x, y) \in R^2$.

Pearson correlation coefficient

Pearson's correlation coefficient ρ is the most familiar dependence concept that measures the linear dependence between a pair of variables (X, Y). It is defined in terms of moments as

$$\rho = \rho(X, Y) = \frac{\text{cov}(X, Y)}{\sigma_X \sigma_Y} = \frac{E((X - \mu_X)(Y - \mu_Y))}{\sqrt{E((X - \mu_X)^2)E((Y - \mu_Y)^2)}},$$

where $\text{cov}(X, Y)$ represents the covariance of X and Y and $\sigma_X, \sigma_Y > 0$ denotes the standard deviations of X and Y, respectively. The Pearson correlation is defined only if both of the standard deviations are finite and nonzero. Applying the Cauchy–Schwarz inequality to the definition of covariance, it can be easily shown that $-1 \leq \rho \leq 1$, where $\rho = 1$ indicates a perfect increasing linear relationship and $\rho = -1$ shows a perfect decreasing linear relationship. By substituting corresponding moment estimators, the sample Pearson correlation coefficient can be obtained as

$$r = \frac{\sum_{i=1}^{n}(X_i - \overline{X})(Y_i - \overline{Y})}{\sqrt{\sum_{i=1}^{n}(X_i - \overline{X})^2 \sum_{i=1}^{n}(Y_i - \overline{Y})^2}}, \quad -1 \leq r \leq 1.$$

If two variables are from bivariate normal distribution or the sample size is not very small, $t = \sqrt{(n - 2)/(1 - r^2)}\,r$, which has an asymptotic t-distribution with $n - 2$ degrees of freedom under H_0. Accordingly, we reject H_0 if $|t| \geq t_{\alpha/2, n-2}$, where $t_{\alpha/2, n-2}$ is the $(1 - \alpha/2)$th quantile of t-distribution with $n - 2$ degrees of freedom.

Spearman's rank correlation coefficient

Spearman's rank correlation coefficient ρ_S is the Pearson correlation between ranks of X and Y, that is, $\rho_S = \rho_S(X, Y) = \rho(F_X(X), F_Y(Y))$, $-1 \leq \rho_S \leq 1$. It accesses a monotonic relationship between two variables. Testing for independence is equivalent to the test $H_0 : \rho_S = 0$ versus $H_1 : \rho_S \neq 0$. The sample Spearman's rank correlation coefficient is

$$r_s = 1 - 6 \sum_{i=1}^{n} (R_i - S_i)^2 / (n(n^2 - 1)).$$

Note that if there are tied X values and/or tied Y values, each observation in the tied group is assigned with the average of the ranks associated with the tied group.

At the significance level α, we reject H_0 if $|r_s| \geq r_{s,\alpha/2}$, where $r_{s,\alpha/2}$ can be found by *qSpearman* in R.[89] For large sample sizes, we can also conduct the test based on the asymptotic t-distribution of the Pearson correlation coefficient between the ranked variables.

Kendall's rank correlation coefficient

The Kendall's rank correlation coefficient is a distribution-free measure of independence based on signs of products of differences, where

$$\tau = \Pr((X_1 - X_2)(Y_1 - Y_2) > 0) - \Pr((X_1 - X_2)(Y_1 - Y_2) < 0), \quad -1 \leq \rho_S \leq 1.$$

It is a measure of the relative difference between Pr{concordance} and Pr{discordance}. Testing for independence is equivalent to the test $H_0 : \tau = 0$ versus $H_1 : \tau \neq 0$. The Kendall statistic can be defined as

$$K = \sum_{i=1}^{n} \sum_{j=i+1}^{n} \text{sgn}\{(Y_j - Y_i)(X_j - X_i)\},$$

where $\text{sgn}\{x\} = 1$, if $x > 0$; 0, if $x = 0$; and -1, if $x < 0$.

Accordingly, at the significance level α, an exact test can be conducted, and we reject H_0 if $\overline{K} \geq k_{\alpha/2}$, where $\overline{K} = K/(n(n - 1)/2)$ and $k_{\alpha/2}$ can be found by *qKendall* in R.[89] Alternatively, the test can be conducted based on the asymptotic standard normal distribution of the standardized $K^* = (n(n - 1)(2n + 5)/18)^{-1/2}K$ under H_0. The null hypothesis is rejected if $|K^*| \geq z_{\alpha/2}$, where $z_{\alpha/2}$ is the $100(1 - \alpha/2)$th quantile of the standard normal distribution.

Example 31.5

We consider the HDL data described in Figure 31.2 and test for independence between X and Y.

The function `cor.test` conducts the test for independence between two samples. Note that the alternative can be revised to "less" or "greater" in terms of a one-sided test. The method option can be "Pearson," "Kendall," or "Spearman," based on the method chosen. The following shows the Pearson correlation test for the HDL data:

```
> cor.test(X,Y,alternative = "two.sided",method="pearson")

  Pearson's product-moment correlation

data:  X and Y
t = -1.5704, df = 23, p-value = 0.13
alternative hypothesis: true correlation is not equal to 0
95 percent confidence interval:
 -0.62897919  0.09571223
sample estimates:
      cor
-0.3111874
```

The p-value of the Pearson correlation test for the HDL data is 0.13. Thus, we fail to reject the null hypothesis at the 0.05 significance level and state that there is not sufficient evidence to conclude X and Y are independent.

Data-driven rank tests

Kallenberg *et al.* expressed the dependence between X and Y via Fourier coefficients of the grade representation against a very wide class of alternatives.[87] To this end, the distribution of $(F_X(X), F_Y(Y))$ is considered as exponential families with respect to the Lebesgue measure $[0, 1] \times [0, 1]$. It is assumed that the observed samples are distributed according to the joint density function given as

$$h(F_X(x), F_Y(y)) = c(\theta)exp \left\{ \sum_{j=1}^{k} \theta_j b_j(x^*) b_j(y^*) \right\},$$

where b_j denotes the jth orthonormal Legendre polynomial, $\boldsymbol{\theta} = (\theta_1, \ldots, \theta_k)^T$, and $c(\theta)$ is a normalizing constant.

Within exponential families, the null hypothesis corresponds to $\boldsymbol{\theta} = 0$, and the score test for testing $\boldsymbol{\theta} = 0$ against $\boldsymbol{\theta} \neq 0$ is given by rejecting for large values of $\left\{ n^{-1/2} \sum_{i=1}^{n} b_r \left(F_X(X_i) \right) b_s(F_Y(Y_i)) \right\}^2$.

A smooth test statistic

$$T_k = \sum_{j=1}^{k} V(j,j),$$

where

$$V(r,s) = \left\{ n^{-\frac{1}{2}} \sum_{i=1}^{n} b_r \left(\frac{R_i - 1/2}{n} \right) b_s \left(\frac{S_i - 1/2}{n} \right) \right\}^2,$$

can be obtained by replacing unknown distribution functions F_X and F_Y by corresponding empirical distribution functions and applying a correction for continuity. Accordingly, two different test statistics TS2 and V were proposed based on different selection of the order k.

A "diagonal" test statistic $TS2 = T_{S2}$ is useful in the case of the "diagonal" model, which contains only products of Legendre polynomials with the same order in both variables. Let $d(n)$ be a sequence of numbers tending to infinity as $n \to \infty$. In a similar manner to the modified Schwarz's rule,[90] the order is chosen as $S2 = \text{argmin}_k\{T_k - k\log(n) \geq T_j - j\log(n), 1 \leq j, k \leq d(n)\}$. The score test for testing $H_0 : \boldsymbol{\theta} = 0$ against $H_a : \boldsymbol{\theta} \neq 0$ in the exponential family is given by rejecting for large values of TS2.

Otherwise, the "mixed" products are involved and a "mixed" statistic can be used. Let $|\Lambda|$ denote the cardinality of Λ and $T_\Lambda = \sum_{(r,s)\in\Lambda} V(r,s)$, and search for a model $\Lambda^* = \text{arg max}_\Lambda\{T_\Lambda - |\Lambda|\log(n)\}$. If Λ^* is not unique, the first among those Λ^*'s that have smallest cardinality is chosen. Then, the "mixed" statistic of H_0 is $V = T_{\Lambda^*}$.

Empirical likelihood-based method

Einmahl *et al.* constructed a test statistic by localizing the empirical likelihood.[56] Let $L(\widetilde{F}_{XY}) = \prod_{i=1}^{n} \widetilde{P}(\{(X_i, Y_i)\})$, where $\widetilde{P}$ is the probability measure corresponding to F_{XY}. For $(x, y) \in R^2$, the local likelihood ratio test statistic is

$$R(x, y) = \frac{\sup\{L(\widetilde{F}_{XY}) : \widetilde{F}_{XY}(x, y) = \widetilde{F}_X(x)\widetilde{F}_Y(y)\}}{\sup\{L(\widetilde{F}_{XY})\}}.$$

Then,

$$\log R(x, y) = nP_n(A_{11})\log\frac{F_{Xn}(x)F_{Yn}(y)}{P_n(A_{11})} + nP_n(A_{12})\log\frac{F_{Xn}(x)(1 - F_{Yn}(y))}{P_n(A_{12})}$$

$$+ nP_n(A_{21})\log\frac{(1 - F_{Xn}(x))F_{Yn}(y)}{P_n(A_{21})} + nP_n(A_{22})\log\frac{(1 - F_{Xn}(x))(1 - F_{Yn}(y))}{P_n(A_{22})},$$

where P_n is the empirical probability measure, F_{Xn} and F_{Yn} are the corresponding marginal distribution functions, and $A_{11} = (-\infty, x] \times (-\infty, y]$, $A_{12} = (-\infty, x] \times (y, \infty)$, $A_{21} = (x, \infty) \times (-\infty, y]$, $A_{22} = (x, \infty) \times (y, \infty)$, and $0\log(\cdot/0) = 0$. Then, the distribution-free test statistic T_n of testing for independence, that is,

$$T_n = -2\int_{-\infty}^{\infty}\int_{-\infty}^{\infty} \log R(x, y)dF_{Xn}(x)dF_{Yn}(y),$$

can be obtained by forming integrals of the log-likelihood ratio statistic.

Density-based empirical likelihood ratio test

Vexler *et al.* considered a density-based empirical likelihood approach for creating nonparametric test statistics, which approximate a parametric Neyman–Pearson statistic to test the null hypothesis of bivariate independence against a wide class of alternatives.[60] The test statistic is defined as

$$VT_n = \prod_{i=1}^{n} n^{1-\beta_2}\widetilde{\Delta}_i([0.5n^{\beta_2}, 0.5n^{\beta_2}]),$$

where the function $[x]$ denotes the nearest integer to x, $0 < \beta_1 < 0.5, 0.75 < \beta_2 < 0.9$, and

$$\widetilde{\Delta}_i(m, r) \equiv (F_{Xn}(X_{(M_i+r)}) - F_{Xn}(X_{(M_i-r)}))^{-1}(F_n(X_{(M_i+r)}, Y_{(i+m)}) - F_n(X_{(M_i-r)}, Y_{(i+m)})$$

$$- F_n(X_{(M_i+r)}, Y_{(i-m)}) + F_n(X_{(M_i-r)}, Y_{(i-m)}) + n^{-\beta_1}).$$

with M_i that is an integer number such that $X_{(M_i)} = X_{(t_i)}$ ($X_{(t_i)}$ is the concomitant of the i-th order statistic $Y_{(i)}$, see [60] for details); $X_{(M_i+r)} = X_{(n)}$, if $M_i + r > n$; $X_{(M_i-r)} = X_{(1)}$, if $M_i - r > 1$ as well as $Y_{(i+m)} = Y_{(n)}$, if $i + m > n$; $Y_{(i-m)} = Y_{(1)}$, if $i - m < 1$. The proposed test is exact. The null hypothesis is rejected if $\log(VT_n) > C_\alpha$, where C_α is an α-level test threshold. It follows that $\Pr_{H_0}(\log(VT_n) > C_\alpha) = \Pr_{\{X_i\}_{i=1}^n,\{Y_i\}_{i=1}^n \sim \text{Uniform}[0,1]}(\log(VT_n) > C_\alpha H_0)$. The critical values for the proposed test can be accurately approximated using Monte Carlo techniques.[59,91] See references 59 and 91, and the next section for details. The power of the density-based EL test does not depend significantly on the choice of $\beta 1$ and $\beta 2$.

Numerical methods for calculating critical values and powers of statistical tests

Many statistical tests, including Shapiro–Wilk tests, data-driven rank techniques, the empirical likelihood-based method, the density-based empirical likelihood ratio test, t-test-type tests, and likelihood ratio tests are exact. Exact tests are well known to be simple, efficient, and reliable and to have finite sample Type I error control. Under the null hypothesis, distributions of test statistics for exact tests are independent of the underlying data distributions. For example, we consider the two-sample t-test with equal variances introduced in Section *Multivariate t-tests*. When the assumptions of normality and homogeneity of the variances are satisfied, the test is exact, i.e. the sampling distribution of $t = (\overline{X}_1 - \overline{X}_2)/(s_p\sqrt{n_1^{-1} + n_2^{-1}})$ under a true null hypothesis can be given exactly by the t-distribution with degrees of freedom $n_1 + n_2 - 2$ (we refer the notations to Section *Multivariate t-tests*.). Another concrete example of exact tests is the Wilcoxon rank-sum test, in which the test statistic is based on indicator functions $I\{\cdot\}$. Noticing the fact that $I\{Z_1 < Z_2\} = I\{F_Z(Z_1) < F_Z(Z_2)\}$ and $I\{Z_1 < -Z_2\} = I\{F_Z(Z_1) < F_Z(-Z_2)\} = I\{F_Z(Z_1) < 1 - F_Z(Z_2)\}$ under the symmetry of H_0-distribustion, where the random variables $F_Z(Z_1)$ and $F_Z(Z_2)$ have a uniform distribution under H_0, the distributions of the test statistics for the Wilcoxon rank-sum type test are independent of the distributions of observations. Due to the independence of the null distribution of test statistics on the data distribution, the critical values of exact tests can be computed exactly, without using asymptotic approximations.

Methods that can be used to calculate the critical values of exact tests include a classical technique based on Monte Carlo (MC) evaluations, an interpolation technique based on tabulated critical values, and a hybrid of the MC and the interpolation methods. The classical Monte Carlo strategy is a well-known approach for obtaining accurate approximations to the critical values of exact tests. The critical values can be calculated by simulating data for a relatively large number of MC repetitions; say, for example, from a standard normal distribution for one-sample tests and a Uniform$(0,1)$ distribution for two-sample and three-sample tests. The generated values of the test statistic L of the exact test of interest can be used to determine the critical value C_α at the desired significance level α. Assuming that the decision rule is to reject H_0 for large values of L, that is, when $L > C_\alpha$, then the critical value C_α can be obtained via calculating the $1 - \alpha$ quantile of the MC null distribution of L. However, the use of the MC technique can be computationally intensive in some testing situations. For example, a relatively large number of MC repetitions, which we define as M, are needed to evaluate critical values that correspond to the 1% significance level, since in this case the common 95% confidence interval of such evaluation can be calculated as $[0.01 \pm 1.96\sqrt{0.01(1-0.01)/M}]$. Another standard method applied in various statistical software routines is the interpolation technique based on tabulated critical values. Interpolation differs from MC method in that tables of critical values are calculated beforehand for an exact test of interest and for various sample sizes and significance levels. Therefore, the execution speed of the testing algorithm improves when the tables are provided for use within the testing algorithm. However, the interpolation method becomes less reliable when real data characteristics (e.g., sample sizes) differ from those used to tabulate the critical values. As an outgrowth of the methods described above, a hybrid method combines both interpolation and MC by means of the nonparametric Bayes concept. The hybrid method can be applied in a broad setting and is shown to be very efficient in the context of computations of exact-tests' critical values and powers.[57,91]

Concluding remarks

The necessity and danger of testing the statistical hypothesis

The ubiquitous use of statistical decision-making procedures' findings in the current medical literature displays the vital role that statistical hypothesis testing plays in clinical trials in different branches of biomedical sciences. The benefits and fruits of statistical tests based on mathematical probabilistic techniques in epidemiology or other health-related disciplines strongly depend on successful formal presentations of statements of problems and a description of their nature. Oftentimes, certain assumptions about the observations used for the tests provide the probability statements that are required for the statistical tests. These assumptions do not come for free, and ignoring their appropriateness can cause serious bias or inconsistency of statistical inferences, even when the test procedures themselves are carried out without mistakes. The sensitivity of the probabilistic properties of a test to the assumptions is referred to as the lack of robustness of the test.[88,92]

Various statistical techniques require parametric assumptions to define forms of data distributions to be known up to parameters' values. For example, in the t-test, the assumptions are that the observations of different individuals are realizations of independent, normally distributed random variables, with the same expected value and variance for all individuals within the investigated group. Such assumptions are not automatically satisfied, and for some assumptions, it may be doubted whether they are ever satisfied exactly. The null hypothesis H_0 and alternative hypothesis H_1 are statements that, strictly speaking, imply these assumptions, and which therefore are not each other's complement. There is a possibility that the assumptions are invalid, and neither H_0 nor H_1 is true. Thus, we can reject a statement related to clinical trials' interests just because the assumptions are not met. This issue is an impetus to depart from parametric families of data distributions and employ nonparametric test-strategies. Wilk and Gnanadesikan described and discussed graphical techniques based on the primitive empirical CDF and on quantile (Q–Q) plots, percent (P–P) plots, and hybrids of these, which are useful in assessing one-dimensional samples.[85] Statistical techniques such as the likelihood ratio test, the maximum likelihood ratio test, the ROC curve methodology, t-tests, and so on, can really assist in solving challenging epidemiology and biomedical problems. Several components regarding tests based on incomplete data or data subject to instrument limitation can be found in Vexler *et al.*[12] Reiser developed a corrected confidence interval for the AUC, adjusted for measurement errors.[93] In this chapter, we considered retrospective statistical problems, i.e. issues based on already collected data. There are also statistical mechanisms based on sequentially observed measurement.[94]

Appendix

The most powerful test: As discussed in Section *Likelihood*, the most powerful statistical decision rule is to reject H_0 if and only if $\prod_{i=1}^{n} f_1(X_i)/f_0(X_i) \geq B$. The term "most powerful" induces us to formally define how to compare statistical tests. Without loss of generality, since the ability to control the Type I error (TIE) rate of statistical tests has an essential role in statistical decision-making, we compare tests with equivalent probabilities of the TIE, $\Pr_{H_0}\{$test rejects $H_0\} = \alpha$, where the subscript H_0 indicates that we consider the probability given that the hull hypothesis is correct. The level of significance α is the probability of making a TIE. In practice, the researcher should choose a value of α, for example, $\alpha = 0.05$, before performing the test. Thus, we should compare the likelihood ratio test with δ, any decision rule based on $\{X_i, i = 1, \dots, n\}$, setting up $\Pr_{H_0}\{\delta$ rejects $H_0\} = \alpha$ and $\Pr_{H_0}\{\prod_{i=1}^{n} f_1(X_i)/f_0(X_i) \geq B\} = \alpha$. This comparison is with respect to the power $\Pr_{H_1}\{$test rejects $H_0\}$. Notice that to derive the mathematical expectation, in the context of a problem related to testing statistical hypotheses, one must define whether the expectation should be conducted under H_0- or H_1-regime. For example,

$$E_{H_1}\varphi(X_1, X_2, \dots X_n) = \int \varphi(x_1, x_2, \dots x_n) f_1(x_1, x_2, \dots x_n) dx_1 dx_2 \dots dx_n$$

$$= \int \varphi(x_1, x_2, \dots x_n) \prod_{i=1}^{n} f_1(x_i) \prod_{i=1}^{n} dx_i,$$

where the expectation is considered under the alternative hypothesis. The indicator $I\{C\}$ of the event C can be considered as a random variable with values 0 and 1. By virtue of the definition, the expected value of $I\{C\}$ is $EI\{C\} = 0 \times \Pr\{I\{C\} = 0\} + 1 \times \Pr\{I\{C\} = 1\} = \Pr\{I\{C\} = 1\} = \Pr\{C\}$.

Taking into account the comments mentioned above, we derive the expectation under H_0 of the inequality (31.1), where $A = \prod_{i=1}^{n} f_1(X_i)/f_0(X_i)$, B is a test threshold, and δ represents any decision rule based on $\{X_i, i = 1, \dots, n\}$. One can assume that $\delta = 0, 1$, and when $\delta = 1$ we reject H_0. Thus, we obtain

$$E_{H_0}\left(\left(\prod_{i=1}^{n} \frac{f_1(X_i)}{f_0(X_i)} - B\right) I\left\{\frac{f_1(X_i)}{f_0(X_i)} \geq B\right\}\right) \geq E_{H_0}\left(\left(\prod_{i=1}^{n} \frac{f_1(X_i)}{f_0(X_i)} - B\right)\delta\right).$$

And hence,

$$E_{H_0}\left(\prod_{i=1}^n \frac{f_1(X_i)}{f_0(X_i)} I\left\{\frac{f_1(X_i)}{f_0(X_i)} \geq B\right\}\right) - BE_{H_0}\left(I\left\{\frac{f_1(X_i)}{f_0(X_i)} \geq B\right\}\right)$$

$$\geq E_{H_0}\left(\prod_{i=1}^n \frac{f_1(X_i)}{f_0(X_i)}\delta\right) - BE_{H_0}(\delta),$$

where $E_{H_0}(\delta) = E_{H_0}(I\{\delta = 1\}) = \Pr_{H_0}\{\delta = 1\} = \Pr_{H_0}\{\delta \text{ rejects } H_0\}$. Since we compare the tests with the fixed level of significance

$$E_{H_0}\left(I\left\{\frac{f_1(X_i)}{f_0(X_i)} \geq B\right\}\right) = \Pr_{H_0}\left\{\frac{f_1(X_i)}{f_0(X_i)} \geq B\right\} = \Pr_{H_0}\{\delta \text{ rejects } H_0\} = \alpha,$$

we have

$$E_{H_0}\left(\prod_{i=1}^n \frac{f_1(X_i)}{f_0(X_i)} I\left\{\frac{f_1(X_i)}{f_0(X_i)} \geq B\right\}\right) \geq E_{H_0}\left(\prod_{i=1}^n \frac{f_1(X_i)}{f_0(X_i)}\delta\right). \tag{A.1}$$

Consider

$$E_{H_0}\left(\prod_{i=1}^n \frac{f_1(X_i)}{f_0(X_i)}\delta\right) = E_{H_0}\left(\prod_{i=1}^n \frac{f_1(X_i)}{f_0(X_i)}\delta(X_1, \ldots, X_n)\right)$$

$$= \int \prod_{i=1}^n \frac{f_1(x_i)}{f_0(x_i)}\delta(x_1, \ldots, x_n)f_0(x_1, \ldots, x_n)dx_1 \ldots dx_n$$

$$= \int \frac{\prod_{i=1}^n f_1(x_i)}{\prod_{i=1}^n f_0(x_i)}\delta(x_1, \ldots, x_n)\prod_{i=1}^n f_0(x_i)dx_1 \ldots dx_n = \int \delta(x_1, \ldots, x_n)\prod_{i=1}^n f_1(x_i)dx_1 \ldots dx_n$$

$$= E_{H_1}\delta = \Pr_{H_1}\{\delta \text{ rejects } H_0\}. \tag{A.2}$$

Since δ represents any decision rule based on $\{X_i, i = 1, \ldots, n\}$, including the likelihood ratio based test, equation (A.2) implies

$$E_{H_0}\left(\prod_{i=1}^n \frac{f_1(X_i)}{f_0(X_i)} I\left\{\prod_{i=1}^n \frac{f_1(X_i)}{f_0(X_i)} \geq B\right\}\right) = \Pr_{H_1}\left\{\prod_{i=1}^n \frac{f_1(X_i)}{f_0(X_i)} \geq B\right\}.$$

Applying this equation and (A.2) to (A.1), we complete to prove that the likelihood ratio test is a most powerful statistical decision rule.

The likelihood ratio property $f_{H_1}^L(u) = f_{H_0}^L(u)u$: in order to obtain this property, we consider

$$\Pr_{H_1}\{u - s \leq L \leq u\} = E_{H_1}I\{u - s \leq L \leq u\} = \int I\{u - s \leq L \leq u\}f_{H_1}$$

$$= \int I\{u - s \leq L \leq u\}\frac{f_{H_1}}{f_{H_0}}f_{H_0} = \int I\{u - s \leq L \leq u\}Lf_{H_0}.$$

This implies the inequalities

$$\Pr_{H_1}\{u - s \leq L \leq u\} \leq \int I\{u - s \leq L \leq u\}uf_{H_0} = u\Pr_{H_0}\{u - s \leq L \leq u\}$$

and

$$\Pr\nolimits_{H_1}\{u-s \le L \le u\} \ge \int I\{u-s \le L \le u\}(u-s)f_{H_0} = (u-s)\Pr\nolimits_{H_0}\{u-s \le L \le u\}.$$

Dividing these inequalities by s and employing $s \to 0$, we get $f_{H_1}^L(u) = f_{H_0}^L(u)u$, where $f_{H_0}^L(u)$ and $f_{H_1}^L(u)$ are the density functions of the statistic $L = f_{H_1}/f_{H_0}$ under H_0 and H_1, respectively.

The general form of the AUC: By the definition of the AUC (the area under the ROC curve) and the fact that F and G are CDFs of X and Y, respectively, the AUC can be expressed as

$$\int_0^1 \mathrm{ROC}(t)dt = \int_0^1 (1 - F(G^{-1}(1-t)))dt = \int_{-\infty}^\infty (1 - F(w))dG(w)$$

$$= 1 - \int_{-\infty}^\infty F(w)dG(w) = 1 - \Pr(X \le Y) = \Pr(X > Y).$$

The form of the AUC under the normal data distribution assumption: Assume $X \sim N(\mu_1, \sigma_1^2)$ and, for the nondiseased population, $Y \sim N(\mu_2, \sigma_2^2)$. Note that X and Y are independent. Consequently, we can obtain that

$$A = \Pr(X > Y) = \Pr(X - Y > 0) = 1 - \Pr\left\{ \frac{(X-Y)-(\mu_1-\mu_2)}{\sqrt{\sigma_1^2+\sigma_2^2}} \le -\frac{\mu_1-\mu_2}{\sqrt{\sigma_1^2+\sigma_2^2}} \right\}$$

$$= 1 - \Phi\left(-\frac{\mu_1-\mu_2}{\sqrt{\sigma_1^2+\sigma_2^2}} \right) = \Phi\left(\frac{\mu_1-\mu_2}{\sqrt{\sigma_1^2+\sigma_2^2}} \right).$$

R code: the two-sample (X and Y) density-based ELR test (for the details, see Ref. 51):

```
###########sample data with the sample sizes
n1=n2=25############
n1=25
n2=25
x<-sample(control,n1)
y<-sample(case,n2)
delta<-0.1
z<-c(x,y)
sx<-sort(x)
sy<-sort(y)
sz<-sort(z)
#############################################
#######obtaining the ELR based on the sample X###
#############################################
m<-c(round(n1^(delta+0.5)):min(c(round((n1)^(1-delta)),
round(n1/2))))   ###generate a vector       of "m"
a<-replicate(n1,m)            ###store repeated values of the
#vector "m"
```

```
rm<-as.vector(t(a))          ###transpose the previous
length(m)*n1 matrix and make it to be a vector
#rm<-rep(m, each = n1)        ###repeat the vector of "m"
#n1 times
L<-c(1:n1)- rm               ###order from (1-m) to (n1-m)
LL<-replace(L, L <= 0, 1 )   ###replace values that are
#<=0 with 1 when (1-m) <=0
U<-c(1:n1)+ rm               ###order from (1+m) to (n1+m)
UU<-replace(U, U > n1, n1)   ###replace values that are n1
#with n1 when (n1+m)>n1
xL<-sx[LL]   ###obtain x(i-m)
xU<-sx[UU]   ###obtain x(i+m)
F<-ecdf(z)(xU)-ecdf(z)(xL)   ### the empirical distribution
#function
F[F==0]<-1/(n1+n2)
I<-2*rm/(n1*F)       ### a (n1*length(m)) vector of (2*m)/
#(n1*empirical distribution function)
ux<-array(I, c(n1,length(m)))     ### make the previous vector
#as a n1*length(m) matrix
tstat1<-log(min(apply(ux,2,prod)))               ###get the
#part of the test statistic based on the sample X
###################################################
######obtaining the ELR based on the sample Y#######
###################################################
m<-c(round(n2^(delta+0.5)):min(c(round((n2)^(1-delta)),
round(n2/2))))    ###generate a vector of "m"
a<-replicate(n2,m)           ###store repeated values of the
#vector "m"
rm<-as.vector(t(a))          ###transpose the previous
#length(m)*n2 matrix and make it to be a vector
#rm<-rep(m, each = n2)            ###repeat the vector of "m"
#n2 times
L<-c(1:n2)-rm                ###order from (1-m) to (n2-m)
LL<-replace(L, L <= 0, 1 )      ###replace values that are
#<=0 with 1 when (1-m) <=0
U<-c(1:n2)+ rm               ###order from (1+m) to (n2+m)
#UU<-replace(U, U > n2, n2)      ###replace values that are
>n2 with n2 when (n2+m)>n2
yL<-sy[LL]   ###obtain y(i-m)
yU<-sy[UU]   ###obtain y(i+m)
F<-ecdf(z)(yU)-ecdf(z)(yL) ###the empirical distribution
#function
F[F==0]<-1/(n1+n2)
I<-2*rm/(n2*F)       ### the (n2*length(m)) vector of (2*m)/
#(n2*empirical distribution fuction)
uy<-array(I, c(n2,length(m)))                ### make the
#previous vector as a n2*length(m) matrix
```

```
tstat2<-log(min(apply(uy,2, prod)))        ###get ELR_Y
finalts<-tstat1+tstat2                     ### the final test
statistic log(V)
```

Multiple choice questions

1 Investigators are interested in evaluating a treatment A. They target to show that A can significantly improve health conditions of patients. How can the investigators formulate the null statistical hypothesis?
   a. The corresponding distribution function based on measurements from a group of patients without treatment A is different from that based on measurements from a group of patients with treatment A
   b. The corresponding distribution function based on measurements from a group of patients without treatment A equals to that based on measurements from a group of patients with treatment A
   c. The corresponding distribution function based on measurements from a group of patients without treatment A is smaller than that based on measurements from a group of patients with treatment A

2 Assume we observe i.i.d. measurements from a known distribution that depends on a simple parameter θ. We would like to test that $\theta = 0$ versus $\theta \neq 0$. What kind of testing strategies could you propose:
   a. Likelihood ratio
   b. t-Test
   c. Maximum likelihood ratio.
   d. Empirical likelihood ratio
   e. Density-based empirical likelihood ratio

3 Assume we observe independent measurements from normal distributions with the variances $\sigma_X^2 = \sigma_Y^2 = 1$. The diseased population is presented by a sample $X_1, \ldots, X_n$, the nondiseased population is presented by a sample $Y_1, \ldots, Y_m$. To evaluate the area under the ROC curve, it is better to use the following formal notation:

   a. $\Phi\left(\dfrac{\overline{X} - \overline{Y}}{\sqrt{2}} \right)$

   b. $\dfrac{1}{mn} \displaystyle\sum_{i=1}^{n} \sum_{j=1}^{m} I\{X_i > Y_j\}$

   c. $\dfrac{(\overline{X} - \overline{Y})^2}{\sum_{i=1}^{n}(X_i - \overline{X})^2 + \sum_{j=1}^{m}(Y_j - \overline{Y})^2}$

References

1 Rothman, K.J., Greenland, S. & Lash, T.L. (2008) *Modern Epidemiology*. Lippincott Williams & Wilkins.
2 Freiman, J.A., Chalmers, T.C., Smith, H. Jr., *et al.* (1978) The importance of beta, the type II error and sample size in the design and interpretation of the randomized control trial. Survey of 71 "negative" trials. *The New England Journal of Medicine*, 299(13), 690–694.

3 Goodman, S.N. (1999) Toward evidence-based medical statistics. 1: The P value fallacy. *Annals of Internal Medicine*, 130(12), 995–1004.

4 Berger, J.O. & Mohan, D. (1987) Testing precise hypotheses. *Statistical Science*, 2(3), 317–335.

5 Gibbons, J.D. & Chakraborti, S. (2011) *Nonparametric Statistical Inference*. Springer.

6 Sackrowitz, H. & Samuel-Cahn, E. (1999) P values as random variables – expected P values. *The American Statistician*, 53(4), 326–331.

7 Berger, J.O. (1985) *Statistical Decision Theory and Bayesian Analysis*. Springer.

8 Vexler, A., Tao, G. & Hutson, A.D. (2014) Posterior expectation based on empirical likelihoods. *Biometrika*, 101(3), 711–718.

9 Lindsey, J. (1996) *Parametric Statistical Inference*. Oxford Science Publications.

10 Freedman, D. (2009) *Statistical Models: Theory and Practice*. Cambridge University Press.

11 Austin, P.C., Mamdani, M.M., Juurlink, D.N. *et al.* (2006) Testing multiple statistical hypotheses resulted in spurious associations: a study of astrological signs and health. *Journal of Clinical Epidemiology*, 59(9), 964–969.

12 Vexler, A., Tao, G. & Chen, X. (2014) A toolkit for clinical statisticians to fix problems based on biomarker measurements subject to instrumental limitations: From repeated measurement techniques to a hybrid pooled-unpooled design. In: *Advanced Protocols in Oxidative Stress III*. Humana Press.

13 Wilcox, R.R. (2012) *Introduction to Robust Estimation and Hypothesis Testing*. Academic Press.

14 Lehmann, E.L. & Romano, J.P. (2006) *Testing Statistical Hypotheses*. Springer.

15 Riffenburgh, R.H. (2012) *Statistics in Medicine*, 3rd edn. Academic Press.

16 Pepe, M.S. (2003) *The Statistical Evaluation of Medical Tests for Classification and Prediction*. Oxford University Press.

17 Vexler, A., Schisterman, E.F. & Liu, A. (2008) Estimation of ROC curves based on stably distributed biomarkers subject to measurement error and pooling mixtures. *Statistics in Medicine*, 27(2), 280–296.

18 R Development Core Team. *R: A Language and Environment for Statistical Computing*. 2013. R Foundation for Statistical Computing, Vienna, Austria. URL http://www.R-project.org/. ISBN 3-900051-07-0.

19 Gardener, M. (2012) *Beginning R: The Statistical Programming Language*. John Wiley & Sons.

20 Crawley, M.J. (2012) *The R Book*. John Wiley & Sons.

21 Limpert, E., Stahel, W.A. *et al.* (2001) Log-normal distributions across the sciences: Keys and clues on the charms of statistics, and how mechanical models resembling gambling machines offer a link to a handy way to characterize log-normal distributions, which can provide deeper insight into variability and probability-normal or log-normal: That is the question. *BioScience*, 51(5), 341–352.

22 Schisterman, E.F., Faraggi, D., Browne, R. *et al.* (2001) Tbars and cardiovascular disease in a population-based sample. *Journal of Cardiovascular Risk*, 8(4), 219–225.

23 Neyman, J. & Pearson, E.S. (1928) On the use and interpretation of certain test criteria for purposes of statistical inference: Part II. *Biometrika*, 20, 263–294.

24 Neyman, J., Pearson, E. S. (1933) The testing of statistical hypotheses in relation to probabilities a priori. In *Mathematical Proceedings of the Cambridge Philosophical Society*, 29(4), 492–510. Cambridge University Press.

25 Neyman, J. & Pearson, E.S. (1936) *Contributions to the Theory of Testing Statistical Hypotheses I*. University Press, pp. 1–37.

26 Neyman, J. & Pearson, E.S. (1938) *Contributions to the Theory of Testing Statistical Hypotheses II*. University Press.

27 Vexler, A., Wu, C. & Yu, K.F. (2010) Optimal hypothesis testing: from semi to fully bayes factors. *Metrika*, 71(2), 125–138.

28 Vexler, A. & Wu, C. (2009) An optimal retrospective change point detection policy. *Scandinavian Journal of Statistics*, 36(3), 542–558.

29 Vexler, A. & Gurevich, G. (2010) Empirical likelihood ratios applied to goodness-of-fit tests based on sample entropy. *Computational Statistics & Data Analysis*, 54(2), 531–545.

30 Wilks, S.S. (1938) The large-sample distribution of the likelihood ratio for testing composite hypotheses. *The Annals of Mathematical Statistics*, 9(1), 60–62.

31 Vexler, A., Yu, J. & Hutson, A.D. (2011) Likelihood testing populations modeled by autoregressive process subject to the limit of detection in applications to longitudinal biomedical data. *Journal of Applied Statistics*, 38(7), 1333–1346.

32 Fan, J., Zhang, C. & Zhang, J. (2001) Generalized likelihood ratio statistics and Wilks phenomenon. *Annals of Statistics*, 29(1), 153–193.

33 Ghosh, M. (1995) Inconsistent maximum likelihood estimators for the rasch model. *Statistics & Probability Letters*, 23(2), 165–170.

34 Gurevich, G. & Vexler, A. (2010) Retrospective change point detection: from parametric to distribution free policies. *Communications in Statistics-Simulation and Computation*, 39(5), 899–920.

35 Box, G.E.P., Hunter, J.S. & Hunter, W.G. (1978) *Statistics for Experimenters*. Wiley, pp. 144.

36 Pearson, E.S. & Neyman, J. (1930) *On the Problem of Two Samples*. Imprimerie de l'university.

37 Zhang, L., Xu, X. & Chen, G. (2012) The exact likelihood ratio test for equality of two normal populations. *The American Statistician*, 66(3), 180–184.

38 Wedderburn, R.W.M. (1974) Quasi-likelihood functions, generalized linear models, and the Gauss–Newton method. *Biometrika*, 61(3), 439–447.

39 Claeskens, G. & Hjort, N.L. (2004) Goodness of fit via non-parametric likelihood ratios. *Scandinavian Journal of Statistics*, 31(4), 487–513.

40 Wang, J. (2006) Quadratic artificial likelihood functions using estimating functions. *Scandinavian Journal of Statistics*, 33(2), 379–390.

41 Fan, J., Farmen, M. & Gijbels, I. (1998) Local maximum likelihood estimation and inference. *Journal of the Royal Statistical Society: Series B (Statistical Methodology)*, 60(3), 591–608.

42 Owen, A. (1990) Empirical likelihood ratio confidence regions. *The Annals of Statistics*, 18(1), 90–120.

43 Qin, J. & Lawless, J. (1994) Empirical likelihood and general estimating equations. *The Annals of Statistics*, 22(1), 300–325.

44 Lazar, N. & Mykland, P.A. (1998) An evaluation of the power and conditionality properties of empirical likelihood. *Biometrika*, 85(3), 523–534.

45 Vexler, A., Liu, S., Kang, L. *et al.* (2009) Modifications of the empirical likelihood interval estimation with improved coverage probabilities. *Communications in Statistics-Simulation and Computation*, 38(10), 2171–2183.

46 Yu, J., Vexler, A. & Tian, L. (2010) Analyzing incomplete data subject to a threshold using empirical likelihood methods: an application to a pneumonia risk study in an ICU setting. *Biometrics*, 66(1), 123–130.

47 Vexler, A., Yu, J., Tian, L. *et al.* (2010) Two-sample nonparametric likelihood inference based on incomplete data with an application to a pneumonia study. *Biometrical Journal*, 52(3), 348–361.

48 Owen, A.B. (1988) Empirical likelihood ratio confidence intervals for a single functional. *Biometrika*, 75(2), 237–249.

49 Vexler, A., Gurevich, G. & Hutson, A.D. (2013) An exact density- based empirical likelihood ratio test for paired data. *Journal of Statistical Planning and Inference*, 143(2), 334–345.

50 Vexler, A., Shan, G., Kim, S. *et al.* (2011) An empirical likelihood ratio based goodness-of-fit test for inverse Gaussian distributions. *Journal of Statistical Planning and Inference*, 141(6), 2128–2140.

51 Gurevich, G. & Vexler, A. (2011) A two-sample empirical likelihood ratio test based on samples entropy. *Statistics and Computing*, 21(4), 657–670.

52 Vexler, A. & Yu, J. (2011) Two-sample density-based empirical likelihood tests for incomplete data in application to a pneumonia study. *Biometrical Journal*, 53(4), 628–651.

53 Vexler, A., Tsai, W.M., Gurevich, G. *et al.* (2012) Two-sample density-based empirical likelihood ratio tests based on paired data, with application to a treatment study of attention-deficit/hyperactivity disorder and severe mood dysregulation. *Statistics in Medicine*, 31(17), 1821–1837.

54 Yu, J., Vexler, A., Kim, S.E. *et al.* (2011) Two-sample empirical likelihood ratio tests for medians in application to biomarker evaluations. *Canadian Journal of Statistics*, 39(4), 671–689.

55 Hall, P. & Owen, A.B. (1993) Empirical likelihood confidence bands in density estimation. *Journal of Computational and Graphical Statistics*, 2(3), 273–289.

56 Einmahl, J.H.J. & McKeague, I.W. (2003) Empirical likelihood based hypothesis testing. *Bernoulli*, 9(2), 267–290.

57 Vexler, A., Tsai, W.M. & Malinovsky, Y. (2012) Estimation and testing based on data subject to measurement errors: from parametric to non-parametric likelihood methods. *Statistics in Medicine*, 31(22), 2498–2512.

58 Miecznikowski, J.C., Vexler, A. & Shepherd, L. (2013) dbEmpLikeGOF: An R package for nonparametric likelihood ratio tests for goodness-of-fit and two sample comparisons based on sample entropy. *Journal of Statistical Software*, 54(3), 1–19.

59 Vexler, A., Tanajian, H. & Hutson, A.D. (2014) Density-based empirical likelihood procedures for testing symmetry of data distributions and K-sample comparisons. *The Stata Journal*, 14(2), 304–328.

60 Vexler, A., Tsai, W.M. & Hutson, A.D. (2014) A simple density-based empirical likelihood ratio test for independence. *The American Statistician*, 68(3), 158–169.

61 Qin, J. & Zhang, B. (2005) Marginal likelihood, conditional likelihood and empirical likelihood: connections and applications. *Biometrika*, 92(2), 251–270.

62 Qin, J. (2000) Miscellanea. Combining parametric and empirical likelihoods. *Biometrika*, 87(2), 484–490.

63 Qin, J. & Leung, D.H.Y. (2005) A semiparametric two-component compound mixture model and its application to estimating malaria attributable fractions. *Biometrics*, 61(2), 456–464.

64 Green, D.M. & Swets, J.A. (1966) *Signal Detection Theory and Psychophysics*. Vol. 1. Wiley, New York.

65 Pepe, M.S. (1997) A regression modelling framework for receiver operating characteristic curves in medical diagnostic testing. *Biometrika*, 84(3), 595–608.

66 Hsieh, F., Turnbull, B.W. *et al.* (1996) Nonparametric and semiparametric estimation of the receiver operating characteristic curve. *The Annals of Statistics*, 24(1), 25–40.

67 Wieand, S., Gail, M.H., James, B.R. *et al.* (1989) A family of nonparametric statistics for comparing diagnostic markers with paired or unpaired data. *Biometrika*, 76(3), 585–592.

68 Pepe, M.S. & Thompson, M.L. (2000) Combining diagnostic test results to increase accuracy. *Biostatistics*, 1(2), 123–140.

69 McIntosh, M.W. & Pepe, M.S. (2002) Combining several screening tests: optimality of the risk score. *Biometrics*, 58(3), 657–664.

70 Bamber, D. (1975) The area above the ordinal dominance graph and the area below the receiver operating characteristic graph. *Journal of Mathematical Psychology*, 12(4), 387–415.

71 Kotz, S., Lumelskii, Y. & Pensky, M. (2003) The stress-strength model and its generalizations. In: *Theory and Applications*. World Scientific, Singapore.

72 Metz, C.E., Herman, B.A. & Shen, J.H. (1998) Maximum likelihood estimation of receiver operating characteristic (ROC) curves from continuousl-distributed data. *Statistics in Medicine*, 17(9), 1033–1053.

73 Box, G.E.P. & Cox, D.R. (1964) An analysis of transformations. *Journal of the Royal Statistical Society, Series B (Methodological)*, 26(2), 211–252.

74 Zhou, X.H., Obuchowski, N.A. & McClish, D.K. (2011) *Statistical Methods in Diagnostic Medicine.* Vol. 712. John Wiley & Sons.

75 Sering, R.J. (2009) *Approximation Theorems of Mathematical Statistics.* Vol. 162. John Wiley & Sons.

76 Qin, G. & Zhou, X.H. (2006) Empirical likelihood inference for the area under the ROC curve. *Biometrics,* 62(2), 613–622.

77 Zou, K.H., Hall, W.J. & Shapiro, D.E. (1997) Smooth non-parametric receiver operating characteristic (ROC) curves for continuous diagnostic tests. *Statistics in Medicine,* 16(19), 2143–2156.

78 Pepe, M.S., Cai, T. & Longton, G. (2006) Combining predictors for classification using the area under the receiver operating characteristic curve. *Biometrics,* 62(1), 221–229.

79 Su, J.Q. & Liu, J.S. (1993) Linear combinations of multiple diagnostic markers. *Journal of the American Statistical Association,* 88(424), 1350–1355.

80 Reiser, B. & Faraggi, D. (1997) Confidence intervals for the generalized ROC criterion. *Biometrics,* 53, 644–652.

81 Chen, X., Vexler, A. & Markatou, M. (2015) Empirical likelihood ratio confidence interval estimation of best linear combinations of biomarkers. *Computational Statistics & Data Analysis,* 82, 186–198.

82 Silverman, B.W. (1986) *Density Estimation for Statistics and Data Analysis.* Vol. 26. CRC Press.

83 Shapiro, S.S. & Wilk, M.B. (1965) An analysis of variance test for normality (complete samples). *Biometrika,* 52(3/4), 591–611.

84 Razali, N.M. & Wah, Y.B. (2011) Power comparisons of Shapiro-Wilk, Kolmogorov-Smirnov, Lilliefors and Anderson-Darling tests. *Journal of Statistical Modeling and Analytics,* 2(1), 21–33.

85 Wilk, M.B. & Gnanadesikan, R. (1968) Probability plotting methods for the analysis of data. *Biometrika,* 55(1), 1–17.

86 Hettmansperger, T.P. & McKean, J.W. (1978) Statistical inference based on ranks. *Psychometrika,* 43(1), 69–79.

87 Kallenberg, W.C.M. & Ledwina, T. (1999) Data-driven rank tests for independence. *Journal of the American Statistical Association,* 94(445), 285–301.

88 Cox, D.R. & Hinkley, D.V. (1979) *Theoretical Statistics.* CRC Press.

89 Wheeler, B. (2013) *SuppDists: Supplementary distributions. R Package Version* 1.1-9.1. http://CRAN.R-project.org/package=SuppDists

90 Inglot, T., Kallenberg, W.C.M. & Ledwina, T. (1997) Data driven smooth tests for composite hypotheses. *The Annals of Statistics,* 25(3), 1222–1250.

91 Vexler, A., Kim, Y.M., Yu, J. *et al.* (2014) Computing critical values of exact tests by incorporating Monte Carlo simulations combined with statistical tables. *Scandinavian Journal of Statistics,* 41(4), 1013–1030.

92 Wilcox, R.R. (1998) The goals and strategies of robust methods. *British Journal of Mathematical and Statistical Psychology,* 51(1), 1–39.

93 Reiser, B. (2000) Measuring the effectiveness of diagnostic markers in the presence of measurement error through the use of ROC curves. *Statistic in Medicine,* 19(16), 2115–2129.

94 Jennison, C. & Turnbull, B.W. (2010) *Group Sequential Methods with Applications to Clinical Trials.* CRC Press.

Webliographies

http://onlinelibrary.wiley.com/doi/10.1002/sim.4467/suppinfo – R programs of realization of the two-sample density-based empirical likelihood ratio tests based on paired data.

http://cran.r-project.org/web/packages/dbEmpLikeNorm/ – The R package "dbEmpLikeNorm": Test for joint assessment of normality.

http://cran.r-project.org/web/packages/dbEmpLikeGOF/index.html – The R package "dbEmpLikeGOF" for nonparametric density-based likelihood ratio tests for goodness of fit and two-sample comparisons.

http://sphhp.buffalo.edu/biostatistics/research-and-facilities/software/stata.html – The STATA package entitled "Novel and efficient density-based empirical likelihood procedures for symmetry and k-sample comparisons."

http://www.sciencedirect.com/science/article/pii/S0167947314002710 – The R function for obtaining the empirical likelihood ratio confidence interval estimation of best linear combinations of biomarkers.

Index

Oxidative Stress and Antioxidant Protection: The Science of Free Radical Biology and Disease, First Edition.
Edited by Donald Armstrong and Robert D. Stratton.
© 2016 John Wiley & Sons, Inc. Published 2016 by John Wiley & Sons, Inc.